HANDBOOK
OF
SUSTAINABLE POLYMERS
FOR
ADDITIVE MANUFACTURING

HANDBOOK OF SUSTAINABLE POLYMERS FOR ADDITIVE MANUFACTURING

Antonio Paesano

CRC Press
Taylor & Francis Group
Boca Raton London New York

CRC Press is an imprint of the
Taylor & Francis Group, an **informa** business

First edition published 2022
by CRC Press
6000 Broken Sound Parkway NW, Suite 300, Boca Raton, FL 33487-2742

and by CRC Press
2 Park Square, Milton Park, Abingdon, Oxon, OX14 4RN

Library of Congress Cataloging-in-Publication Data
A catalog record for this title has been requested

ISBN: 978-1-138-47888-6 (hbk)
ISBN: 978-1-032-11719-5 (pbk)
ISBN: 978-1-003-22121-0 (ebk)

DOI: 10.1201/9781003221210

Typeset in Times
by MPS Limited, Dehradun

Dedication

My wife, Kathleen
My children, Agostino, Francis, and Mario
My parents, Agostino and Maria
My sister, Rossanna, and brother, Mario

"Ye were not made to live like brutes, but to pursue virtue and knowledge."
Dante, The Divine Comedy, Inferno, Canto XXVI

Contents

Preface

This book is an overview of commercial and experimental sustainable polymers used as feedstocks for additive manufacturing, namely their physical and mechanical properties, suppliers, applications, current and future market, advantages and limitations.

Sustainable materials and additive manufacturing (AM), or 3D printing, have been two areas attracting strong and growing interest worldwide from an economic, scientific, technological, and societal standpoint.

This book for the first time provides students, academics, professionals, companies, investors, entrepreneurs, consultants, material formulators, product developers, materials scientists, and materials engineers with a comprehensive and current overview and analysis of the state of the art of commercial and experimental sustainable polymeric feedstocks for AM processes, with focus on their physical, mechanical, thermal, and electrical properties, processing characteristics, characterization techniques, applications, advantages, limitations, suppliers, current and future market data, and near-future outlook. Additionally, the interplay among material formulation, materials properties, process conditions, and performance of 3D printed items is elucidated.

The book's content was distilled from the most recent technical information published worldwide in books, scientific journals, conferences, data sheets, reports, patents, dissertations, market reports, studies, and trade magazines. More than 1,700 references are cited, enabling the readers to delve in every topic in the book, should they wish to.

The polymers discussed are poly(lactic acid), polyamide, polyhydroxyalkanoates, wood, cellulose, bamboo, lignin, natural fibers, carbohydrates, hydrogels, polybutylene succinate, food, and acrylates.

This volume contains 198 figures and 207 tables of material property data and prices gathered from selected sources. These data are valuable for design, stress analysis, material and product development, and cost analysis.

The content of this book is not tailored for a specific and specialized audience, and does not require a college-level knowledge of polymer science, chemistry, engineering, materials science, or manufacturing. All the topics are illustrated minimizing the technical terms, and a glossary of the mentioned technical terms is included. Also added at the end are the following:

- A list of companies and their websites involved in AM that comprises printer manufacturers, suppliers of materials, software, and services, research organizations, and universities.
- A list of ASTM and ISO test standards relevant to this book.

The R&D conducted in sustainable polymers for AM is remarkable in amount and creativity, covers a wide range of materials and processes, and has the potential to reduce the use of petrochemical resources, and increase the amount of recycling, biodegradation and composting. However, only the combinations of materials and processes with a competitive or adequate performance/cost ratio will be commercially successful, being sustainable polymers still more expensive than fossil-based polymer, broadly speaking. One advantage of AM from an R&D perspective is that it allows the development of new materials and fabrication processes at a lower cost than would be possible with conventional fabrication processes, and hence it ates accelerates the pace of innovation in polymers.

The intersection of AM and sustainable polymers enables to produce items with novel shapes, structures, and performance, and is fascinating from a technological perspective, because it combines the latest advances in various areas: computer-assisted design, engineering, physics, chemistry, artificial intelligence, robotics, industrial organization, materials science, manufacturing, computer-aided structural analysis, and so on. It is also notable that sustainable polymers for AM range from very old materials, such as flax, utilized from 30,000 BC, to recent materials, such as nanocellulose, patented in 1980s.

AM processes keep spreading across markets and applications, spanning from medicine to space to food to architecture, just to mention a few examples. AM feedstocks include polymers, metals, ceramics, and any combination of the above. Sustainable polymers are among them and currently important, because they are environmentally friendly, being derived from renewable sources such as plants, and bio degradable.

This book starts with a general introduction of polymers, sustainable polymers, and AM processes. It shifts to used and studied sustainable polymers and feedstocks for AM derived from them, by discussing the interplay among material formulation and properties, AM processing conditions, and performance of 3D printed articles. Suppliers, present and next applications, prices, and current and near-future markets are included for each material family described.

This book is for the present and the near future, because it is not only a snapshot of the current status of sustainable polymers for AM and their applications and relative AM processes, but also contains predictions and trends for their near future.

Acknowledgments

My wife Kathleen, an accomplished, smart, lovely, and multitalented professional in engineering and business, who enabled me to accomplish this book by patiently and taking on all the family responsibilities for quite a while. She has been my companion and best friend for 28 years, has gifted me and our children with unselfish love, attentive caring, brilliant intelligence, and tireless work in everything touching our family. She even found time to meticulously edit my book's reference lists.

My children, Agostino, Francis, and Mario, who helped me type tables, draw figures, and edit references and lists. They have started their life journey on the right path standing on their own legs and guided by righteous principles, and have made their parents proud of their behavior and achievements.

My father, Agostino, and mother, Maria, who have reached extraordinary success in their professions. They have dedicated their life to their children, and have been exemplary for moral and civic values, work ethic, integrity, tenacity, sacrifices, wisdom, and kindness. They have always encouraged, guided and believed in me.

My sister, Rossanna, and brother, Mario: two successful professionals, and role models for me. We have accomplished together beautiful and great things, and built a special bond among each other by sharing many happy times and facing together some difficult moments, always in armony and helping and caring for each other.

I am sincerely very grateful to the publishers, authors, academics, companies, professionals, and organizations who granted me permission to reproduce and include their own valuable information in this book: their names are too many to be listed here but not forgotten.

Last but not least, a special thanks to everybody at Taylor and Francis who worked with me and on my book, and, particularly, Allison Shatkin, senior publisher and my editor, for proposing me to write this book and supporting my vision for it, Gabrielle Vernachio, editorial assistant, Mayank Sharma, project manager, Carly Cassano, production editor, and the copy editing team.

Author Biography

Antonio Paesano is Additive Manufacturing (AM) Lead at The Boeing Company. He earned a M.S. in mechanical engineering and Ph.D. in materials engineering from the University of Naples Federico II (Italy), and M.S. in polymer science from the University of Ferrara (Italy). He is a certified Six Sigma Green Belt, and a trained Black Belt.

His work experience encompasses AM, engineering polymers, composite and advanced engineering materials, design, advanced manufacturing, testing and analysis, correlation material properties-processing-product performance, product development, metrology, applied statistics, and quality improvement. He holds six patents and several trade secrets, and authored and co-authored more than 50 technical papers and two book chapters in his areas of expertise. He was the scientific chair of international symposia on AM, sustainable polymers, and innovation in aerospace.

He has been an invited speaker in AM at international conferences and workshops. His paper on sustainable polymers for AM received the 2017 SAMPE/CAMX Award for "outstanding technical paper". He also received the 2018 Knowledge Management Award by the Boeing Technical Journal for his co-authored paper on fatigue testing and analysis, and the 2020 Boeing Innovation Award for improved processing of composite materials.

Throughout his career, he has identified, accelerated, and integrated innovation in products for transportation, construction, marine, sports, military, and aerospace. Prior to joining Boeing, he held positions in engineering, R&D, and management at Magnaghi Aeronautica SpA, Italian National Research Council, FIAT Research Center, University of Delaware, and The Dow Chemical Company. He has also worked as mentor, advisor, consultant, instructor to companies, academia, and government, and as adjunct faculty in Italy and the United States.

List of Abbreviations

2PVP	two-photon vat photopolymerization
3DP™	three-dimensional printing™
4DP	4D printing
ABS	acrylonitrile butadiene styrene
AESO	acrylated epoxidized soybean oil
AI	artificial intelligence
AM	additive manufacturing or 3D printing
APC	aliphatic polycarbonates
ASA	acrylonitrile styrene acrylate
ASTM	American Society for Testing and Materials
Avg	average
AWCO	acrylated waste cooking oil
BDO	butanediol
BE	bioextrusion
BJ	binder jetting
BMC	bulk molding compound
BP	bioprinting
C	carbon
CA	cellulose acetate
CAB	cellulose acetate butyrate
CAD	computer-aided design
CAM	computer-aided manufacturing
CAN	cellulose acetate nitrate
CAP	cellulose acetate propionate
CBAM	composite-based additive manufacturing technology
CC	contour crafting
CF	carbon fibers
CFF	continuous filament fabrication
CFRP	carbon fiber-reinforced polymer
CLF	cellulose/lignocellulose fibers
CLIP™	continuous liquid interface production
CLTE	coefficient of linear thermal expansion
CMC	carboxyl methyl cellulose
CNC	cellulose nanocrystal
CNF	cellulose nanofiber
CNT	carbon nanotube
CTE	coefficient of thermal expansion
CV	coefficient of variation
CVD	chemical vapor deposition
DCP	dicumyl peroxide
DDDA	dodecanedioic acid
DED	direct energy deposition
DGEBA	diglycidyl ether of bisphenol A
DI	deionized water
DIW	direct ink writing
DLP	digital laser processing
DMA	dynamic mechanical analysis

DMSO	dimethyl sulfoxide
DOE	design of experiments
DP	degree of polymerization
DPP	daylight polymer printing
DSC	differential scanning calorimetry
DTUL	distortion temperature under load
DW	direct writing
EAP	electroactive polymer
EB	ethyl benzene
EBB	extrusion-based bioprinting
EBP	extrusion-based bioprinting
EC	ethyl cellulose
ECH	epichlorohydrin
ECM	extracellular matrix
EMA	ethylene methyl acrylate
EMI	electromagnetic interference
EP	epoxy
EPDM	Ethylene propylene diene monomer rubber
EPR	ethylene propylene rubber
EPS	expanded polyestyrene
ETFE	ethylene tetrafluoroethylene
EVA	ethylene vinyl acetate
EVOH	ethylene vinyl alcohol
FDA	Food and Drug Administration
FDCA	furandicarboxylic acid
FDM™	fused deposition modeling
FEA	finite element analysis
FEAM	fiber encapsulation additive manufacturing
FEP	fluoroethylene propylene copolymer
FFF	fused filament fabrication
FGF	fused granular fabrication
FGM	functionally graded material
FR	fiber-reinforced; flame-retardant
FRPC	fiber reinforced polymer composite
FRP	fiber-reinforced plastic
FTIR	Fourier transform infrared spectrometer or spectroscopy
GelMA	methacrylated gelatin or gelatin methacrylate
GF	glass fibers
GMA	glycidyl methacrylate
H	hydrogen
HDPE	high-density polyethylene
HDT	heat deflection temperature
HEC	hydroxyethyl cellulose
HG	hydrogel
HIPS	high-impact polystyrene
HPC	hydroxypropyl cellulose
HPMC	hydroxypropyl methylcellulose
IJBP	inkjet-based bioprinting
ILs	ionic liquids
IM	injection molding
IoT	Internet of Things

IPN	interpenetrating polymer network
IR	infrared
ISO	International Standard Organization
LBP	laser-assisted bioprinting
LCP	liquid-crystalline polymer
LDM	liquid deposition modeling
LDPE	low-density polyethylene
LED	light-emitted diode
LIM	liquid injection molding
LLDPE	linear low-density polyethylene
LS	laser sintering
LTDW	laser transfer direct writing
L/D	length/diameter ratio in extruders
MA	maleic anhydride
MAPP	maleic anhydride polypropylene
MC	methyl cellulose
MCC	microcrystalline cellulose
MDPE	medium-density polyethylene
ME	material extrusion
MEK	methyl ethyl ketone
MEMS	microelectromechanical system
MF	melamine formaldehyde
MFC	microfibrillated cellulose
MFI	melt flow index
MFR	melt flow rate
MJ	material jetting
MJF	multi jet fusion
MW	weight average molecular weight
N	nitrogen
NFC	nanofibrillated cellulose
NMMO	N-methylmorpholine-N-oxide
NR	natural rubber
NC	nanocellulose
NMR	nuclear magnetic resonance spectroscopy
NVP	n-vinylpyrrolidone
O	oxygen
OSHA	Occupational Safety and Health Act
P3HB	poly(3-hydroxybutyrate)
PA	polyamide (nylon)
PAA	polyamic acid
PAEK	polyaryletherketone
PAI	polyamide imide
PAN	polyacrylonitride
PAS	polyarylsulfone
PB	polybutylene
PBAF	poly(butylene adipate-co-butylene 2,5-furandicarboxylate)
PBAS	polybutylene adipate-co-succinate
PBAT	polybutylene adipate-co-terephthalate
PBF	powder bed fusion
PBI	polybenzimidazol
PBS	polybutylene succinate

PBS	phosphate-buffered saline
PBSA	poly(butylene succinate-co-butylene adipate)
PBT	polybutylene terephthalate
PC	polycarbonate
PCL	polycaprolactone
PDLA	poly(D-lactide)
PDLLA	poly(D,L-lactide)
PDMA	pyromellitic dianhydride
PDMS	polydimethylsiloxane
PDO	propanediol
PE	polyethylene
PEEK	polyetheretherketone
PEET	polyetherester terephthalate
PEF	polyethylene furanoate
PEG	polyethylene glycol
PEGDA	poly(ethylene glycol) diacrylate
PEI	polyetherimide
PEK	poly(ether ketone)
PEKK	polyetherketoneketone
PEO	poly(ethylene oxide)
PES	polyethersulfone
PETG	polyethylene terephthalate glycol-modified
PET, PETE	polyethylene terephthalate
PF	phenol formaldehyde
PFA	perfluoroalkoxy resin
PGA	polyglycolic acid or polyglycolide
PHA	polyxydroxyalkanoate or poly(3-hydroxyalkanoate)
PHB	polyhydroxybutyrate, or poly(3-hydroxybutyrate)
PHBH	poly(3-hydroxybutyrate-co-3-hydroxyhexanoate)
PHBV	poly(hydroxybutyrate–co-hydroxyvalerate)
PHB-PHV	polyhydroxybutyrate-polyhydroxyvalerate
PHR	parts per hundred resin
PHV	polyhydroxyvalerate
PI	polyimide
PLA	poly(lactic acid), or polylactide
PLGA	poly(lactic-co-glycolic acid)
PLLA	poly(L-lactide)
PMC	polymer matrix composite
PMDA	pyromellitic dianhydride
PMMA	polymethyl methacrylate
PMP	polymethyl pentene
PNC	polymer-based nanocomposite
POC	point-of-care
POM	polyoxymethylene
PP	polypropylene
PPA	polyphthalamide
PPE	polyphenylene ether
PPO	polyphenylene oxide
PPS	polyphenylene sulphide
PPSF	polyphenylsulfone
PPSS	polyphenylene sulfide sulfone

PPSU	polyphenylsulfone
PS	polystyrene
PSU	polylsulfone
PTFE	polytetrafluoroethylene
PTMAT	polytetramethylene adipate terephthalate
PTT	polytrimethlene terephthalate
PU, PUR	polyurethane
PVA	polyvinyl alcohol
PVAc	polyvinyl acetate
PVC	polyvinyl chloride
PVDC	polyvinylidene chloride
PVDF	polyvinylidene fluoride
PVF	polyvinyl fluoride
PVOH	poly(vinyl alcohol)
PWPC	particulate wood polymer composite
R	radical
RIJ	reactive inkjet printing
R_a	arithmetical mean deviation of surface roughness
RTV	room-temperature vulcanizing
SAN	styrene acrylonitrile
SBR	synthetic butyl rubber
SBS	styrene butadiene styrene
SD	standard deviation
SEM	scanning electron microscopy
ShL	sheet lamination
SL	stereolithography (in this book)
SLA	stereolithography (outside this book)
SLCOM	selective lamination composite object manufacturing
SLS	selective laser sintering
SMA	styrene maleic anhydride
SMC	sheet molding compound
SMPs	shape-memory polymers
SP	sustainable polymer
TDI	toluene diisocyanate
TE	tissue engineering
TEM	transmission electron microscopy
TFE	polytetrafluoroethylene
TGA	thermogravimetric analysis
TMA	thermomechanical analysis
TP	thermoplastic
TPC-ET	thermoplastic copolymer elastomer
TPE	thermoplastic elastomer
TPS	thermoplastic starch
TPU	thermoplastic polyurethane
TS	thermoset, thermosetting
T_c	crystallization temperature
T_g	glass transition temperature
T_m	melting temperature
UF	urea formaldehyde
UHMWPE	ultrahigh-molecular-weight polyethylene
UPs	unsaturated polyesters

UTS	ultimate tensile strength
UV	ultraviolet
VA	vinyl acetate
VP	vat photopolymerization
VPBP	vat photopolymerization bioprinting
μ-PAD	microfluidic paper-based analytical devices

Glossary of Terms and Definitions

For abbreviations used in this glossary, see List of Abbreviations

acrylates: The esters, salts, and conjugate bases of acrylic acid and compounds derived from it, and characterized by the *acrylate ion* $CH_2=CHCOO^-$. They are the building blocks of acrylic polymers that are widely spread in numerous consumer products, and are also feedstocks for two families of AM processes: VP and MJ.

acrylic: Any compound (typically polymers) including acrylic acid $CH_2CHCOOH$ or acrylates. Acrylics are feedstocks for two families of AM processes: VP and MJ.

adaptive: Referred to a material or component able to change certain properties in a predictable manner due to the forces or stimuli acting on it (passive) or by means of built-in actuators (active).

addition (or chain) polymerization: Chemical reaction in which two or more monomers are added to each other to form, in a single reaction, a polymer that contains all atoms of all monomers by breaking the double-bonds between C atoms, and allowing them to link to the adjacent C atoms and form long chains. No by-products are formed in this reaction. Examples of addition polymers are polyethylene and polypropylene, employed in form of filaments as feedstocks for AM.

additive: Any substance added in minor quantity to a polymerin order to modify its properties in a specific way.

agar: Gelatinous substance obtained from various kinds of red seaweed, and used in biological culture media and as a thickener in foods.

aging: The process of and the results of exposing a material to natural or simulated environmental conditions over a more or less extensive period of time.

air gap or **raster to raster gap:** Distance between adjacent laid beads of extruded material.

alcohol: Organic compound featuring the general structure ROH, in which OH is the hydroxyl group and R is, f. e., C_nH_{2n+1} in the case of the simplest alcohols, such as methanol CH_3OH. Alcohols are valuable starting compounds for manufacturing resins, rubber, and plasticizers.

aldehyde: Organic compound containing a carbonyl group CO linked to a terminal C atom, and having the general formula R-CHO. Aldehydes are volatile liquids with sharp, penetrating odors that are widely used industrially as chemical building blocks in the synthesis of organic compounds. The most known aldehyde is formaldehyde.

alginate: Natural anionic copolymer polysaccharide extracted from brown seaweed and bacteria. Present in biobased inks for AM.

aliphatic: Compounds composed of C and H featuring molecules linked in an open chain that can be straight or branched but without rings. Examples are methane CH_4 and ethylene C_2H_4.

alkoxy group: Functional group containing an alkyl group bonded to an O atom through a single bond, and denoted as R–O. The simplest alkoxy group is the methoxy group CH_3O.

alkyl group: Functional group of an organic compound that contains only C and H atoms arranged in a chain. It has general formula C_nH_{2n+1}. One example is ethyl C_2H_5-.

alkylphenols: Family of organic compounds produced by the alkylation of phenols, being alkylation the transfer of an alkyl group from one molecule to another.

amine: Compound derived from ammonia (NH_3) by replacing one or more H atoms with organic groups.

amino acids: Organic compounds that contain amine ($-NH_2$) and carboxyl ($-COOH$) functional groups bonded to the same C atom.

amino: Substances containing $-NH$ or $-NH_2$ groups.

amorphous plastic, **amorphous polymer:** Plastic and polymer that feature only amorphous regions and no crystalline regions.

amorphous: Region of polymers in which the molecular chains are randomly arranged in coils and entanglements.

anion: Ion that has an electrical negative charge.

anisotropic: Typically referred to a material whose properties assume a different value when measured in different directions. Often, printed objects have physical and mechanical properties higher in the printing plane directions than in the direction perpendicular to it.

anisotropy: The property of a material to have anisotropic properties.

annealing: Heat treatment consisting in heating and slow cooling a part in order to remove internal stresses and toughen it, and ultimately improve its performance and extend its service life.

antigen: Substance, such as part of a virus or bacterium, that triggers the immune system to produce cells (*antibodies*) that specifically attack and try to destroy the antigen.

antigenicity: Ability to be specifically recognized by the antibodies generated as a result of the immune response to a given substance.

antioxidant: Substance added in the formulation of plastics to prevent or slow down the degradation of them induced by thermo-mechanical or thermo-oxidative conditions at room and elevated temperature, and hence to extend plastics' performance. Amines and phenolics are examples of antioxidants for plastics. Lignin is a natural polymer that can serve as an effective antioxidant in feedstocks for extrusion-based AM.

aromatic compound: Unsaturated organic compounds characterized by one or more planar rings of atoms joined by covalent bonds. A typical example is benzene, featuring a hexagonal six-carbon-atom structure with three double C bonds. Aromatic compounds usually yield plastics that are more thermally stable than aliphatic structures.

base: Substance that in aqueous solution increases the hydrogen ion concentration.

biobased: Substance composed or derived in whole or in part from biological products generated from the biomass. Biobased can be synonym of *sustainable* and *renewable* if the exploitation rate of the biobased substance does not surpass its replenishment rate by natural processes. A biobased *polymer* is not necessarily environmentally friendly nor biocompatible nor biodegradable (Vert et al. 2012). F. e. biobased polyethylene derives from sugar cane but is resistant to biodegradation.

biocompatible: Compatible with living tissues, i.e. that does not produce an immune or toxic response when in contact with the body or bodily fluids.

biodegradable plastic: "Degradable plastic in which the degradation results from the action of naturally occurring microorganisms such as bacteria, fungi, and algae." (ASTM D6400).

biodegradation: Degradation caused by enzymatic process resulting from the action of cells (Vert et al. 2012), and degradation by "the action of naturally occurring microorganisms such as bacteria, fungi, and algae." (ASTM D6400).

biofabrication: Production of complex living and non-living biological products from raw materials such as living cells, molecules, extracellular macromolecules, and biomaterials.

bioglass: Family of silica-based glasses featuring a three-dimensional SiO_2 network that is modified by incorporating CaO, Na_2O, and P_2O_5. Bioglass differ from traditional soda-lime–silica glass in these compositional features: (a) SiO_2 inferior to 60%, (b) higher Na_2O and CaO content, and (c) higher $CaO{:}P_2O_5$ ratio (Huang 2017).

bioink: Material suited for biofabricating biomaterials "that are either (i) directly processed with the cells as suspensions/dispersions, (ii) deposited simultaneously in a separate printing process, or (iii) used as a transient support material" (Groll et al. 2019). Bioinks are different from biomaterial inks.

biomass: Substances produced by living organism including animals, land and water plants, and microorganisms. It excludes material fossilized or located in geological formations.

biomaterial: Natural or synthetic material (including ceramics, metals, and polymers) that can be introduced into and stay in contact with a living tissue without damaging the tissue. Examples are materials for artificial joints.

biomineralization: Process by which cells of living organisms convert ions that are in solution into solid minerals, often to harden or stiffen existing tissues. These minerals form f. e. sea shells, and bones and teeth in mammals.

bio-origami: It refers to either 3D folded structures composed of biological materials such as proteins, DNA, and cells, or self-folding of polymeric hydrogels usable to engineer cell-laden hydrogels for long-term 3D cell culture.

bioplotting: Bioprinting consisting in extruding living cells that are in a hydrogen solution into a plotting medium.

biopolymers: "Macromolecules (including proteins, nucleic acids and polysaccharides) formed by living organisms." IUPAC 2014.

bioprinting or **3D bioprinting:** Version of biofabrication, and a group of AM processes consisting in printing structures and organs using viable biological molecules, biomaterials, and cells.

blend: See **polymer blend.**

bone grafting: Transplanting of bone tissue, a surgical procedure for fixing problems with bones or joints, and leading to bone regeneration.

brittle: Polymer that breaks under stress at room temperature and shows very small strain at failure.

carbene: Molecule containing a C atom with a valence of two and two unshared valence electrons.

carbohydrate: Biomolecule consisting of C, H, and O atoms, typically with H/O atom ratio of 2:1.

carbon nanotubes (CNTs): Rolled-up one-atom thick sheets (called *graphene*) of carbon that feature aspect ratio greater than 1,000 (that is nanometer-size diameter and micrometer-size length), and possess outstanding mechanical, electrical, and thermal properties.

carbonyl: CO group of atoms.

carboxyl: Group of atoms with formula COOH.

carboxylate: Conjugate base of a carboxylic acid.

carboxylic acid: Compound containing the carboxyl group COOH.

carboxymethylation: Addition of one or more carboxymethyl groups ($-CH_2-COOH$) to a compound.

catalyst: Substance added to reactants in order to cause, accelerate, inhibit, or delay a chemical reaction among them.

cation: Ion that has a positive electrical charge.

cell viability: Number of healthy cells in a sample, expressed as the ratio of the number of living cells over the total population of cells.

cellulose: Carbohydrate and high MW homopolymer with chemical formula $(C_6H_{12}O_5)_n$, and consisting of glucose (D-glucopyranose $C_6H_{12}O_6$) ring-shaped units joined together by β-1,4 glucosidic bonds.

chain polymerization: See **addition polymerization**.

chiral: Having the property of chirality.

chirality: The property of a molecule to be different from (non-superposable to) its mirror image.

chitosan: Linear, semi-crystalline polysaccharide, derived by removing the acetyl group (CH_3CO) from the molecule of chitin. Its formula is $(C_6H_{11}NO_4)_n$.

chondrocyte: Cell which has secreted the matrix of cartilage and become embedded in it.

coacervate: When a colloidal system separates into two liquid phases, the phase richer in colloid component is its coacervate.

coalescence: The disappearance of the boundary between two particles in contact that merge into one. It occurs between adjacent deposited filaments during extrusion-based AM, and powder particles during laser powder bed-based AM.

cold crystallization: Exothermic crystallization process consisting in self-nucleation of the crystalline phase. It is observed on heating a polymer that has been previously cooled too quickly to have time to crystallize, and occurs above its glass transition temperature and below its melting temperature, when the polymer chains become sufficiently mobile.

collagen: Chief protein of connective tissue in animals, and the most abundant protein in mammals. It features molecular formula $C_{65}H_{102}N_{18}O_{21}$.

colloid: Mixture of one substance microscopically dispersed and suspended throughout another substance.

colloidal: Relative to colloid.

compatibility: Ability of two or more compounds to mix together without separating.

compatibilizer: Polymer made of two parts to achieve compatibility between polymer A and B: one part is compatible with A, and the other part is compatible with B. They are employed to improve the mechanical properties of polymer blends.

compostable plastic: "Plastic that undergoes degradation by biological processes during composting to yield CO_2, water, inorganic compounds, and biomass at a rate consistent with other known compostable materials and leaves no visually distinguishable or toxic residue" (ASTM D6400). Examples of compostable plastics are two polymers for AM: PLA and PBS.

composting: Biological process in which organic wastes are degraded by microorganisms (bio-oxidation) into CO_2, water, minerals, and stabilized organic matter (compost or humus). A more detailed definition is reported in ASTM D6400.

compressive modulus of elasticity: The ratio of nominal compressive stress to the corresponding strain below the proportional limit of a material. Measured in MPa.

compressive strength at failure: The (nominal) compressive stress sustained by the test coupon when it fails if shattering occurs. Measured in MPa.

compressive strength: The maximum (nominal) compressive stress that a rigid plastic is capable of sustaining. It may not coincide with the compressive strength at failure. Measured in MPA.

condensation (or **step**) **polymerization:** Chemical reaction consisting in mixing two monomers with end groups that react with each other, and form long molecular chains, and release small molecules as by-products such as water or methanol. Examples of condensation polymers are polyamides and epoxy resins, both processed in AM.

conjugate base: Substance formed from an acid after its proton has been removed.

contact angle: It measures the wettability of a surface by a liquid: the lower the contact angle between a liquid and a surface, the more wetted that surface is.

copolymer: Polymer derived from two or more species of monomer. Examples are acrylates used in VP, and MJ.

coupling agent: Coupling agents are used in composite materials comprising a polymer matrix and inclusions (fibers or particles) embedded in the matrix: they are bifunctional molecules, in which one functionality reacts with the polymer and the other with the inclusion's surface, and bond the two together. They perform other functions as well. F.e. coupling agents can lower the amount of moisture that wood polymer composites absorb by decreasing gaps at the wood–matrix interface and by reducing the number of hydroxyl groups available for hydrogen bonding with water.

covalent bond: Type of chemical bond between atoms consisting in sharing electron pairs, and the most basic force keeping together the backbone of polymer molecules. An example is the C–C bond.

creep: Permanent strain resulting from the continuous application of stress over time.

critical strain energy release rate G_c: Quantifies the energy needed to increase a crack by a unit length in a sample of unit width, and depends on the amount and type of stress applied. Measured in J/m^2.

critical stress intensity factor K_c: Predicts failure due to crack propagation, and depends on the crack length, and the amount and type of stress applied to the material around the crack. The greater K_c, the more fracture resistant is the material. Measured in $MPa\ m^{1/2}$.

crosslinking: The process whereby chemical links set up between the molecular chains of a polymer bond the chains together, and form a 3D network through the formation of regions from which at least four chains originate. In thermosets, crosslinking makes one infusible super-molecule of all chains contributing to strength, rigidity and high-temperature resistance. Although not typical, thermoplastics can also be crosslinked to produce 3D molecular structures featuring enhanced mechanical performance.

crystalline: (A) Region of a polymer whose molecular chains are aligned in a 3D ordered, compact, and repeated pattern. (B) Polymer containing amorphous and crystalline regions but never completely crystalline, and in fact they are also more properly termed semi-crystalline. Examples of crystalline polymers used as AM commercial feedstocks are: PLA, PA, polyester, PE, and PP.

crystallinity: The presence of crystalline regions in polymers.

crystallization: In polymers the formation of crystalline regions. Crystallization affects optical, mechanical, thermal, and chemical properties of the polymer due to partial alignment of their molecular chains.

curing: Irreversible chemical process of converting a prepolymer or a polymer into a polymer of higher molar mass, and then into a 3D network, changing the properties of the starting material by chemical reaction, and/or by the action of heat, and catalysts, with or without pressure. Humidity and radiation are also used to activate curing. The term *curing* is mostly limited to thermosetting plastics and rubbers. Curing occurs in photopolymers for AM.

cyclic ester: Ester whose atoms are connected forming a ring.

cytotoxicity: Ability of being toxic to cells, like some venoms.

degree of polymerization (DP): The average number of monomeric repeating units in a macromolecule.

dehydration: Removal of water from a substance through various methods: drying, heating, chemical reaction, etc.

derivative: Substance derived from a similar substance by means of a chemical reaction.

design allowables: Materials property values statistically determined and derived from test data. They are relative to strength, strain, stiffness, are lower than their corresponding test values, and enable to design force-bearing components with margin of safety.

design of experiments (DOE): A methodology to maximize the number of output information by running the smallest number of experiments. It consists in planning, conducting, analyzing, and interpreting test results to evaluate at the same time and cost-effectively how a number of factors (input variables like printing speed and temperature) affect the values of output variables (like tensile strength and modulus of a printed part).

dielectric constant or **permittivity:** The ability of a material in an electric field to store electrical energy. It is the ratio of the capacitance formed by two plates with a material between them to the capacitance of the same plates with air as the dielectric. The higher the dielectric constant, the better a material functions as an electrical insulator, and the greater its ability to store electric charge.

differential scanning calorimetry (DSC): Instrumental technique to characterize polymers. It consists in heating or cooling a sample and a reference compound at a controlled rate, monitoring the difference in temperature between the sample and the compound, and measuring the amount of heat emitted or absorbed by the sample as a function of temperature. It enables to study phase transitions, thermal stabilities, sample composition, kinetics, etc. and measure, among others, degree of crystallization, degree of cure, melting temperature, crystallization temperature, and glass transition temperature.

digester: Container where chemical or biological reactions are conducted in laboratories and process industries.

diisocyanate: Compound containing two isocyanate groups –NCO.

dimensional resolution: In 3D printers is the minimum size of the geometric features of the printed part.

dissipation factor (DF): Quantifies the inefficiency of an insulating material by measuring the electrical energy absorbed and dissipated (mostly as heat) when electrical current is applied to the insulating material. The lower DF, the more efficient is the insulating material.

disubstituted: Referred to a molecule having two atoms or groups of atoms taking the place of another atom or group.

ductility: Property of a solid material that displays considerable strain when subjected to stress, and is expressed by the amount of strain at failure.

dynamic mechanical analysis (DMA): Instrumental technique to characterize polymers in the solid state that consists in applying an oscillating (sinusoidal) stress, in tension, bending, or shear, over a range of temperature, time, or frequency, and recording the polymer's elastic and viscous response and amorphous phase transitions (including glass transition temperature) in terms of strain, stiffness, and viscosity of the polymer.

elastic deformation: Dimensional change – in response to an applied force – of a component that returns to its original shape after the force is removed. Measured in mm.

elastic limit: The maximum stress (typically reported in tension) applicable to a test sample that produces nopermanent deformation after stress removal. Measured in MPa.

elastic modulus, modulus of elasticity, Young's modulus: See **tensile modulus.**

elasticity: Property of a material that is substantially deformed by force, and fast recovers to its approximate initial shape and dimensions after subsequent release of the force.

elastomer: Polymer that under stress is substantially deformed (at least twice its initial length) by an applied force, and fast recovers its initial approximate shape and dimensions after the force is released.

electrical capacitance: Ratio of the amount of electric charge stored on a conductor to a difference in electric potential applied to the conductor.

electrolyte: Substance that produces an electrically conducting solution when dissolved in a polar solvent such as water.

emulsifier: Substance that stabilizes an emulsion by keeping the dispersed phase in suspension. Emulsifiers are used as food additives to stabilize processed foods, and in food as feedstock for AM.

emulsion: Permanent mixture of two or more incompletely miscible liquids (phases) one of which is dispersed as globules in the other.

enantiomer: One of a pair of molecular entities which are mirror images of each other and do not exactly coincide when superimposed on one another. A polymer with enantiomer is PLA.

end-group: Terminal group at the end of a molecular chain. Although end-groups constitute a minuscule portion of the entire polymer, they may significantly affect the polymer's properties.

engineering strain: Ratio of initial gage length to final gage length.

engineering stress: Ratio of load to initial cross-sectional area where the load is applied.

enthalpy: Thermodynamic quantity of a material in a system that is equal to the sum of its internal energy in form of heat plus the product of pressure and volume added or removed from the system. Enthalpy is measured in joules, and on polymers through DSC.

enzymatic hydrolysis: Process in which enzymes facilitate the splitting of chemical bonds in molecules with the addition of water.

enzyme: Substance produced by a living organism and acting as a catalyst to cause a specific biochemical reaction.

epoxide: Subgroup of epoxy compounds whose molecule contains a triangular ring featuring an O atom and two C atoms at its vertices.

epoxy: Thermosetting liquid or solid compound in which an O atom is directly attached to two adjacent or non-adjacent C atoms forming a three-atom ring. Epoxy compounds are resins in form of prepolymer (lower MW) or polymer (higher MW). Epoxy is a commercial feedstock for VP.

ester: Organic compound produced by reacting an alcohol with an organic or inorganic acid, or exchanging an H atom of an acid with an –O– alkyl group. When the alcohol and the acid contain two or more reactive groups, ester is polymerized into polyester, a feedstock for AM.

eukaryote: Organism whose cells feature a nucleus enclosed by membranes, unlike bacteria.

excipient: Inactive substance serving as a carrier or medium for drugs, and other active substances. Present in 3D printed drugs.

extracellular matrix (ECM): 3D network of extracellular macromolecules, such as collagen, enzymes, and proteins that provide biochemical and structural support to the surrounding cells.

extrusion: Continuous process of forming articles by forcing a molten plastic through a die.

fatigue: (A) Subjecting a material or component to stress or strain variable and fluctuating (alternating) in amount and type (f.e. tension and compression together). (B) Progressive and permanent structural changes caused on a test coupon or component by fatigue stress or strain applied to it.

fatigue life: Number of cycles endured by the test coupon or component subjected to fatigue stress or strain before failure occurs, either in form of cracking or complete break.

fatty acid: Carboxylic acid featuring mostly long hydrocarbon unbranched aliphatic chains ranging from 4 to 36 carbons. Fatty acids are structural components of cells and important dietary sources of fuel for animals.

fibroblast: Type of biological cell that are present in connective tissue and secrete collagen proteins forming the structural framework of many tissues. Fibroblasts are the most common cells of connective tissue in animals.

filler: Relatively inert material in particulate form that is added to a plastic to change its properties and processability, and/or reduce its cost. Fillers differ from reinforcements because the latter ones are fibrous and improve mechanical strength.

flexural modulus: Resistance to elastic deformation in bending, and calculated as the ratio – within the elastic limit – of bending stress to the corresponding strain. Measured in MPa.

flexural strength: Value of bending stress at failure, or at a specific strain value if no failure occurs. Measured in MPa.

flexural stress: Stress applied to a material in bending, according to three-point or four-point bending loading scheme. Measured in MPa.

Fourier transform infrared spectroscopy (FTIR): Preferred method of infrared (IR) spectroscopy and utilized for chemical analysis. When a sample is hit by an IR radiation, part of the radiation is absorbed by it, and part passes through (*transmission*) and is detected. The detected signal is a spectrum that represents a molecular *fingerprint* of the sample, because different compounds and their relative molecules produce different spectral fingerprints. Hence, FTIR enables to identify and characterize unknown materials (through the molecular structure and molecular bonds), contaminations in a polymer, and additives after extraction from a polymer.

fracture toughness: Ability of a material to resist fracture in the presence of cracks, and manifested as its ability to absorb strain energy prior to fracture. Fracture toughness is expressed by two material properties: *critical strain energy release rate G_c*, and *critical stress intensity factor K_c*. Polymers fail in fatigue in two ways: (a) thermal failure caused by softening and melting from heating developing from cyclic loading; (b) mechanical failure from initiation and propagation of crack.

free radical: Atom, molecule, or ion that has an unpaired valence electron, and is hence highly chemically reactive.

functional group: Atom, or a group of atoms possessing similar chemical properties whenever they are present in different compounds.

functionally gradient materials: Engineering composites made of two or more constituent phases featuring continuous and smooth gradients of composition, structure, and particular properties within their volume. They can be obtained with high spatial control via AM.

furan: Colorless, flammable, highly volatile liquid organic compound that consists in a five-membered aromatic ring with four C atoms and one O atom. It can be derived from lignocellulosic biomass.

furfural: Organic compound with the formula C_4H_3OCHO, and a colorless or brown liquid. Obtained by dehydration of sugars from sawdust and agricultural by-products: wheat bran, oat, and corncobs.

gel: In mers "Semisolid system consisting of a network of solid aggregates in which liquid is held" (ASTM D883), or a soft, solid, or solid-like materials consisting in two or more components one of which is a liquid, present in substantial quantity (Almdal et al. 1993).

glass transition: The change in an amorphous polymer, or in the amorphous regions of a crystalline polymer, that, upon being heated, transitions from being (relatively) hard, stiff, and strong to becoming (relatively) soft, flexible, and weak. In brief, it is transitioning from a glassy state to a rubbery state. The change is reversible, and can take place from rubbery to glassy state upon cooling. The change is due to the degree of molecular motion in the polymer: lower degree when transitioning from glassy to rubbery state, and greater degree when transitioning from rubbery to glassy state.

glass transition temperature (T_g)**:** The approximate midpoint of the temperature range over which the glass transition of a polymer occurs. Since typically polymers have functional properties below their T_g and rubbers above their T_g, T_g is a critical parameters in assessing the utility of a polymer, and widely mentioned in this book and the technical literature cited in it. Amorphous polymers feature T_g but lack melting temperature. Crystalline polymers possess both temperatures, because they are partially amorphous, and their T_g is always below their melting temperature. T_g value varies (about 10%) depending upon the measuring method.

glucose (or **dextrose**)**:** Simple sugar with the molecular formula $C_6H_{12}O_6$, and the most abundant monosaccharide.

glycerol: Carbohydrate that is a colorless, odorless, hygroscopic, and sweet tasting viscous liquid. Used as lubricant in pharmaceutical and hygiene products, like toothpaste and soap, as food additive for sweetening and preserving foods, and as car antifreeze.

glycol: Alcohol containing two hydroxyl groups on adjacent C atoms.

glycosaminoglycan: Compound occurring mainly as a component of connective tissue, and used in the body as a lubricant or shock absorber.

glycosidic bond: Type of covalent bond joining a carbohydrate molecule to another group, which may or may not be another carbohydrate.

grafting: Reaction in which one or more portions of a polymer are linked to the main chain of another polymer as side chains and differ in some features from the main chain.

graphene: One atomic thick sheet of C atoms linked in a hexagonal arrangement.

greenhouse effect: The trapping of the sun's warmth in the earth's lower atmosphere by some gases, like carbon dioxide (CO_2), methane (CH_4), etc. that, due to their infrared absorption, are more transparent to the shorter wavelength radiation from the sun to earth than to longer wavelength infrared radiation emitted from the earth's surface, and hence impair the outward dissipation of the latter radiation.

greenhouse gas: Any of the atmospheric gases that contribute to the greenhouse effect.

group: See **functional group**.

halloysite nanotubes: Naturally occurring alumina-silicate clay mined from natural deposits and featuring a tubular structure with microscale length and nanoscale inner and outer diameter.

heat distortion temperature (HDT): Maximum temperature at which a polymer can be utilized as a rigid material. It is measured through three-point bending, by applying a steady bending weight to a bar-shaped test coupon that results in a constant flexural stress of 0.45 MPa or 1.82 MPa, while the coupon is immersed in a liquid bath at raising temperature. The HDT value is the temperature at which the coupon deflects 0.25 mm. It is also termed *deflection temperature under load* (DTUL).

heat resistance: Ability of materials to maintain specific mechanical properties at a desired level at the maximum service temperature.

hemicellulose: Carbohydrate present in the cell walls of plants and possessing a simpler structure than cellulose.

homopolymer: Polymer formed of identical monomer units, such PP and PE, two feedstock for AM.

hydride: Anion of hydrogen H^-, and binary compound containing H atoms covalently bound to a more electropositive element or group.

hydrocarbon: Compound consisting only of C and H.

hydrocolloid: Long-chain polymer (polysaccharides and proteins) that forms a gel in the presence of water. Some hydrocolloids are used in surgical dressings.

hydrogel: Gel in which the liquid component and swelling agent is water, and the network component is usually a hydrophilic polymer that is crosslinked to form a 3D network. Hydrogel is a soft, highly porous material.

hydrolysis: The chemical breakdown of a compound by reacting with water.

hydrophobic: Material tending to repel water, and typically exhibiting a low surface energy, expressed as a high contact angle.

hydroxyl group: Functional group –OH in which a hydrogen atom is covalently bonded to an oxygen atom.

immunogenic: Related to immunogenicity.

immunogenicity: Ability of a substance foreign to a human or animal, such as an antigen, to provoke an immune response in the body of the human or animal where it is introduced.

impact strength: Resistance of materials to fracture when subjected to loads applied at high speed. It is measured by the *crack propagation energy* and *crack initiation energy* on notched and unnotched test coupons, respectively. Two types of notched and unnotched impact tests are conducted: in the Izod test the sample is held in upright, with its length oriented vertically, whereas in the Charpy test the sample is held in flat position, with its length oriented horizontally.

indentation hardness: Hardness evaluated from measuring the area or depth of the indentation caused by pressing a specific indenter with a specific force into the material's surface.

infill: A 3D printed part can be inside completely filled or porous, even if its outer surface is non-porous. Infill is the part's volume occupied by its outer surface minus the total porosity volume, or, in brief, the part's volume occupied by material. Infill is expressed on a scale from 0 (theoretical) to 100% (a part without voids).

inhibition: Decrease in the rate of a chemical reaction produced by the addition of a substance (*inhibitor*) that diminishes that rate.

interface: In a matrix-filled composite, the "two-dimensional boundary between the matrix and the filler" (Mozafari et al. 2019), with the filler being particles, fibers, platelets, or whiskers.

interlaminar shear strength: The maximum shear stress recorded between layers of a laminated material subjected to three-point bending as a coupon featuring length = thickness × 6, and width = thickness × 2. Measured in MPa.

interphase: In a matrix-filled composite, the "three-dimensional region that includes the interface plus a zone of finite thickness on both sides of the interface. The interphase boundaries are generally defined from the point in the matrix where the local properties start to deviate from the bulk properties in the direction of the polymer/particle interface" (Mozafari et al. 2019).

ionotropic: Affecting cell membrane channels that open or close in the presence of ions.

irradiance: At a point of a surface it is the radiant power of all wavelengths incident from all upward directions on a small element of surface containing the point. It is expressed in radiant power divided by the area of the element (W/m^2). Radiant refers to *radiation*, a term indicating electromagnetic waves in this case.

isocyanate: Compound containing the isocyanate group –NCO linked to H or an organic radical.

isomer: Substance made of molecules that contains the same number of atoms of the same elements (same molecular formula) but linked differently, and hence different in structure and properties.

isotropic: Material whose physical and mechanical properties are constant regardless of the direction in which they are measured.

lactic acid: Organic compound with the formula $CH_3CH(OH)COOH$. In its solid state it is white and water-soluble, and in its liquid state it is colorless. It is produced both naturally (f.e. by bacteria in our guts and yogurt) and synthetically.

lactide: The cyclic carboxylic di-ester of lactic acid.

lactonitrile: Organic compound with formula $CH_3CH(OH)CN$. A colorless liquid utilized in the industrial production of lactic acid.

lignin: Organic, amorphous, hydrophobic polymer that provides structural integrity to tissues of vascular plants and some algae, and binds individual fiber cells together. It has aliphatic and aromatic constituents.

line chemical formula: 2D representation of molecules in which atoms are shown joined by lines representing single or multiple bonds, without additional information on the spatial direction of bonds.

lipid: Biological substance that is soluble in nonpolar solvents (such as benzene, hexane, chloroform, etc.). Examples include fats and oils.

liquid-crystalline polymer (LCP): Polymer that spontaneously orders itself when melted, allowing relatively easy processing at relatively high temperatures. It feature a highly crystalline molecular chain compared to most common polymers. Commercial examples of LCPs are Kevlar® (DuPont) and Vectra® (Celanese).

loss modulus: Measure of the mechanical energy received by a polymer subjected to sinusoidal stress, and not returned when the stress is removed but dissipated as heat. It represents the purely viscous behavior of the polymer in which the resulting strain lags the applied stress, and is also a measure of *damping*. Measured in MPa.

loss tangent: See **tanδ**.

lyotropic: A liquid-crystalline mesophase that is formed by dissolving in a suitable solvent a compound that displays liquid-crystalline properties and is also hydrophilic (prone to be wetted by, mix with, and dissolve in water) and lipophilic (prone to combine with or dissolve in fats and lipids).

macromolecule: Molecule featuring high molecular mass and a structure that comprises multiple repetition of units derived from molecules of low molecular mass.

magnetostrictive: Capable to change shape and dimensions, and expand and contract in response to a magnetic field.

melt flow index (MFI) or **melt flow rate (MFR):** It indicates how fast a thermoplastic polymer flows in its molten state, and is inversely proportional to its viscosity. It is expressed in the quantity in g of polymer that is extruded through a die in 10 min. MFI affects the polymer processability during 3D printing, and the interlayer cohesion in articles printed via extrusion-based AM.

meso: Scale between microscale and macroscale, and ranging from nanometers to microns.

metamaterial: Man-made material with properties usually absent in natural materials, such as negative Poisson's ratio.

methacrylic: A plastic or an acrylic resin made from a derivative of methacrylic acid.

methoxy group: Functional group consisting of a methyl group CH_3 bound to an O atom.

methoxylated: Modified by the addition of one or more methoxy groups.

mixture: Combination of two or more substances that are intermingled in which each substance maintains its original properties.

modulus of elasticity: Ratio of applied stress (tension, compression, shear, flexure) to the resulting strain within the linear portion of the stress-strain curve below the proportional limit. For polymers lacking a linear region, it is calculated as the slope of: (a) a secant line from the origin to a specified point on the stress-strain curve; (b) a line tangent to stress-strain curve at a specific point. Measured in MPa.

modulus of toughness: Amount of strain energy per unit volume that a material can absorb until the moment it fractures. The modulus of toughness is calculated as the area under the stress-strain curve up to the fracture point. See also **toughness**.

moiety: Part of a molecule but not a small fragment of it.

mole: The amount of a substance containing 6.022×10^{23} atoms of that substance.

molecular weight (MW): In a molecule, it is the sum of the atomic weight of all the atoms in the molecule. In a polymer, it expresses as the average size of the all molecules in the polymer, and is calculated as number-average MW and weight-average MW according to different formulas (Mathias 2021) and measured as g/mol. In almost all technical literature cited in this book, the number-average MW was reported. The polydispersity index (PDI) is the ratio of the the mass-averaged MW and the number-averaged MW, and represents the degree of "non-uniformity" of a distribution of MW values across the polymer chains: the closer PDI is to 1 (its minimum value) the more uniform is the distribution.

molecule: The smallest amount of a substance that can exist by itself and retain all the properties of the original substance (Whittington 1978).

monomer: Molecule that links to copies of itself forming long chain-like larger molecules that in turn constitute a polymer.

monosaccharide or **simple sugar:** It is the simplest form of sugar and the most basic units of carbohydrates. It has chemical formula $(CH_2O)_n$, with n equal or greater than 3.

motif: Pattern of amino acids in a protein sequence thatperforms a specific function.

nanocellulose: Cellulose with at least one dimension in the nanometer range.

nanoclay: Material usually made of layered silicates or clay minerals with traces of metal oxides and organic matter.

nanomaterial: Material engineered in form of particles featuring at least one dimension in the nanoscale, and capable to exploit the unique chemical and physical properties existing at the nanoscale.

neoalkoxy: Alkoxy with a neopositioned quaternary C (Monte 2002).

nuclear magnetic resonance spectroscopy (NMR): Analytical technique consisting in taking measurements relative to nuclei exposed to an external magnetic field. NMR is mostly applied to determine the structure of pure compounds, and the quantitative determination of mixtures. Monitoring the progress of chemical reactions is also an application. For practical purposes, it is used for solutions only.

number of contours (contour width): In extrusion-based AM it is the number of outer and, in hollow objects, inner perimeters of each layer that make up the edge of the printed part.

nylon: Generic name of commercial derivation for all synthetic PAs. See **polyamides**.

oligomer: Polymer consisting of a small number of monomers.

open-source: See **RepRap**.

organic: Referred to substances containing C atoms, regardless whether the substances come from living organisms (animals and plants) or are manmade.

osteoblast: Cell that synthesizes bone.

oxygen inhibition: Inhibition of chemical reactions caused by oxygen.

part or build orientation: Inclination of the part on the build platform with respect to the printers' Cartesian axes X, Y, Z with XY the plane of the print platform, and Z perpendicular to XY (Figure 2.5).

particle size distribution d(10), d(50), d(90): Typical measures to characterize a powder, and equal to percent of powder particles not exceeding a certain diameter. F.e. d10 = 25 µm means that 10% of the powder sample has a diameter not exceeding 25 µm, d50 = 45 µm means that 50% of the powder sample has a diameter not above 45 µm, and so on. They are reported for powder feedstocks for AM.

path or toolpath: Sequence traced by the nozzle's positions during 3D printing in extrusion-based AM.

permittivity: See **dielectric constant.**

phenol: Aromatic organic compound with molecular formula C_6H_5OH. It is a crystalline, volatile solid, soluble in water, and mildly acidic. Along with its derivatives, it is employed to produce feedstocks processed in AM, such as PC and epoxy.

phenolic resins: Synthetic thermosetting resins obtained by reacting a phenol with an aldehyde through condensation polymerization. Utilized as feedstocks for AM, such as UV-curable acrylates, and binders for binder jetting.

photoinitiator: Molecule that serves as a catalyst and initiates a chemical chain reaction of polymerization by forming reactive species (ions, free radicals) when exposed to UV or visible radiation. Photoinitiators are combined with photopolymers compatible with VP. Examples of photoinitiators for VP are: benzophenone, and diphenyl (2,4,6-trimethylbenzoyl)-phosphine oxide for UV light, and camphorquinone and bis(4-methoxybenzoyl)diethylgermanium for visible light.

photon: Particle of zero charge, zero mass, carrier of electromagnetic force.

photopatterning: Process to form patterns using light and photocrosslinkable hydrogels. In its simplest version, the material to be patterned is photosensitive, and its surface is covered with a photomask featuring a pattern that locally shields the material from light exposure and subsequent chemical modification.

photopolymer: Light-sensitive resin that changes its physical, chemical, and mechanical properties when exposed to UV or visible light. The photopolymers for VP fall in two basic groups: (a) acrylic- and methacrylic-based resins, and (b) resins based on epoxy or vinyl ether.

photopolymerization: Reaction of polymerization consisting in exposing monomers or oligomers to visible or UV lights in presence or absence of a catalyst.

piezoelectric material: Material developing electric charge in response to an applied mechanical stress and vice versa. Examples of piezoelectric materials can be naturally occurring like quartz, and manmade like barium titanate $BaTiO_3$.

piezoelectric strain coefficients: They express the electrical charge developed in a material in response to stress applied to it, or the strain in the material upon applying an electrical field to it. They are also called *strain constants*, or *d* coefficients. Piezoelectric coefficients with double subscripts connect mechanical and electrical quantities. The first subscript indicates the direction of the electric field associated with the applied voltage or the charge produced, whereas the second subscript specifies the direction of the mechanical stress or strain. The greater d_{ij} constants are, the larger the mechanical displacements are, which is generally desirable in motional transducer devices. d_{ij} are measured in pC/N (picoCoulomb/Newton) or m/V.

piezoelectric voltage coefficient: It measures of the electric field developed in a material subjected to stress, and is also called *g coefficient*, and measured as volt meter/Newton. The subscript 33 in g_{33} indicates that the electric field and the mechanical stress are both along the polarization axis.

plastic deformation: Dimensional change – in response to an applied force – of a component that does not return to its original shape after the force is removed. Measured in mm.

plasticizer: Liquid of low MW added to a polymer to increase its flexibility and workability, by placing itself among the polymer's molecular chains, reducing the polymer's intermolecular forces, and hence allowing the chains to slide over one another more easily. Plasticizers lower viscosity, glass transition temperature, elastic modulus, and brittleness, and increase elongation at break and toughness. Examples of plasticizers selected for AM feedstocks are glycerol and tributyl citrate (TBC) for PLA/poplar wood flour composite.

plastics: Materials composed of additives and mostly one or more organic polymers of large MW, solid in their finished state, and shaped by flow into a finished product. Essentially, plastics are the final product started from polymers.

plastisol: Liquid substance that consists of particles of PVC or another polymer dispersed in liquid plasticizer upon heating, and is converted into a more or less flexible solid plastic upon cooling.

pluripotency: It indicates: (a) in biological compounds their ability to produce distinct biological responses; (b) in stem cells their capability to differentiate into other cell types.

Poisson's ratio: The absolute value of the ratio of the transversal strain to the relative axial strain below the proportional limit of a material measured in a coupon undergoing axial tensile or compressive stress. It is dimensionless.

polyacid or **polyelectrolyte:** Polymer whose repeating units include an electrolyte group.

polyacrylate: Polymer formed from acrylates.

polyamide (PA): Organic, thermoplastic, linear, highly crystalline polymer whose chemical formula contains repeating amide links –CO–NH– in its molecular chain. Present in nature in proteins and peptides. The following grades of PA are commercially available for AM: PA6, PA11 (biobased), PA 12.

polyesters: Family of polymers in which its main backbone contains ester formed through a condensation polymerization of a polyfunctional alcohol and a polyfunctional acid. The most common polyester is polyethylene terephthalate (PET), available also in a biobased version that is biodegradable, compostable, and recyclable. Various commercial filaments of glycol modified PET (PETG) are sold for extrusion-based AM.

polyfunctional: Substance possessing two or more different functional groupss.

polyglycolic acid (PGA): Thermoplastic biodegradable polymer, and the simplest aliphatic polyester, with chemical formula $(C_2H_2O_2)_n$.

polyimide (PI): Aromatic high performance polymer featuring rings of four C atoms tightly bound together. PI features remarkable properties: outstanding mechanical properties and thermal stability, low thermal expansion, excellent chemical resistance, and favorablel electrical properties, such as low dielectric constant and high breakdown voltage.

polymer blend: Mixture of two or more polymers that is macroscopically homogeneous, separable by physical means, and miscible or immiscible.

Polymer nanocomposites: mixture of two or more materials, in which the matrix is a polymer and the dispersed phase *(filler)* has at least one dimension smaller than 100 nm. Since nanoscale fillers greatly exceed fillers of conventional size in surface area, the interfacial area between matrix and nanofiller is enormous, which (even at marginal amounts of nanofiller uni-formely distributed) significantly enhances mechanical strength and stiffness, impact strength, thermal and electrical conductivity, barrier to moisture and gases, flame retardancy, etc. in com-parison to conventional composites.

polymethylmethacrylate (PMMA): Amorphous, acrylic polymer that features outstanding values of optical clarity and resistance to weathering and sun (UV) rays, rigidity, hardness, resistance to scratching. It possesses impact strength, but it is also penalized by poor resistance to fatigue and solvents. Its chemical formula is $(C_5O_2H_8)_n$. PMMA is available as a filament for extrusion-based AM from the printing service company Craftcloud®.

polyolefin: Polymer obtained from monomers that are *olefin* (or *alkene*) and have chemical formula C_nH_{2n}. The most known polyolefins are PE and PP produced from ethylene and propylene, respectively. Commercial PP and high density PE filaments are sold for extrusion-based AM.

polysaccharide or **polycarbohydrate:** Long chain polymeric carbohydrate composed of monosaccharide units bound together by glycosidic bonds. It is the most abundant carbohydrate found in food.

polystyrene (PS): Aromatic thermoplastic polymer of styrene. General purpose PS is transparent, hard, brittle, thermally and dimensionally stable, and inexpensive. It is commercially available as a filament for extrusion-based AM.

polyvinyl chloride (PVC): Thermoplastics polymer with chemical formula $(C_2H_3Cl)_n$, available in two basic forms: rigid (bottles, pipes, windows, etc.) and flexible (electrical cable insulation, fake leather, flooring, plumbing, etc.). PVC is hard, strong, rigid, and tough (except in its flexible grades), resistant to water, chemicals, and abrasion, and is an excellent electrical insulator. The commercial PVC filament *Vinyl 303* (Fillamentum®) is available for extrusion-based AM.

porosity: Ratio of total volume of void or air contained within a material sample to the total (solid material plus void or air) volume of that sample. Typically expressed as a percentage. Porosity in a printed object is detrimental because it reduces its density, matrix-filler interfacial bonding, and ultimately its mechanical properties.

prepolymer: Polymer or oligomer whose molecules are capable of achieving, through reactive groups, further polymerization.

proportional limit: It is the largest stress that can be applied to a material and be proportional to the corresponding strain, following Hooke's law. Measured in MPa.

pyrolysis: Process that uses heat and absence of air (to prevent formation of CO_2) and converts a biomass into solids, low MW liquids, and gaseous products that can be utilized as fuels.

racemic: Relative to a *racemate*, that is an equimolar mixture of two enantiomers.

radiation: Term indicating electromagnetic waves and fast moving particles .

radical: Group of atoms in a molecule that does not change upon undergoing chemical reactions, and is denoted with *R*.

raster: A rectangular pattern of parallel lines traced by the bead that fills in the area within the part's contour.

raster angle: Direction of raster relative to the X axis of build platform.

raster width (road width): Width of deposited filament inside the part's outline.

reactive group: A linked collection of atoms or a single atom in a molecule undergoing change in a chemical reaction.

recyclable material: Material used to make objects that can be reprocessed and reused to make new objects. Examples of recyclable materials are cardboard, metals, and bottle glass.

recycled plastics: Plastics "composed of postconsumer material or recovered material only, or both, that may or may not have been subjected to additional processing steps of the types used to make products such as recycled-regrind or reprocessed or reconstituted plastics" (ASTM D883).

recycling: "Reprocessing in a production process of the waste materials for the original purpose or for other purposes including organic recycling but excluding energy recover" (DIN EN 13437). Organic recycling is the recycling of materials derived from animal and plants.

renewable energy: Energy derived from resources that are practically inexhaustible in duration but limited in amount available per unit of time, such as biomass (wood and wood waste, ethanol, biogas, etc.), solar, wind, geothermal, and hydropower.

renewable material: Material that can be fully replenished on a timescale for practical utility without degradation or loss in its quality.

repeat: In statistics and experimentation (including material testing) it is the repetition of an experiment conducted under all the same conditions during the same experimental event or consecutive events.

replicate: In statistics and experimentation (including material testing) it is the repetition of an experiment conducted under all same conditions but during experimental events that occur at different and not successive times, and often randomized.

RepRap: Project started in UK in 2005 to develop low-cost 3D printers capable of printing most of their own components, and later supported by many collaborators worldwide. RepRap is short for *replicating rapid prototyper*. All the 3D printer designs developed by RepRap are open-source, that is are printers for which the design of their hardware, software, and firmware is not a company's intellectual property, but freely available without requiring a license.

resin: Soft solid or highly viscous substance that typically contains prepolymers of thermosetting polymers with reactive groups.

resolution: In AM is the size of the smallest feature that can be printed in an article.

responsive: Materials that react to external stimuli, such as electric and magnetic fields, light, moisture, pH, stress, substances, temperature, and water, and change one or more of their properties in a predictable way.

Reynolds number Re: Determines whether the flow of a liquid is streamlined or turbulent.

rheology: Study of flow and deformation of matter when subjected to an external force.

rubber: A type of elastomers, namely a polymer that can rapidly recover from large deformations and insoluble (but it can swell) in boiling solvent, such as benzene (ASTM D1566).

saccharide: Synonym of carbohydrate.

salt: Compound consisting of a combination of organic and inorganic positive and negative ions. .

selectively: In AM it is a key term, and means "in selected locations that lay in the XY plane of the 3D printer", and are defined according to the geometry of the 3D digital model of the part to be 3D printed.

shape fidelity: In AM the closeness in dimensions between a printed article and its 3D digital model.

shape-memory polymer: Polymer capable to return from a deformed state (temporary shape) to its initial (permanent) shape when exposed to an external stimulus (trigger), such as electricity, light, magnetism, pH, pressure, temperature, and water.

shear thinning (see **thixotropy**): Non-Newtonian behavior of fluids whose viscosity decreases with raising shear stress, and increases with diminishing shear stress. Examples of shear thinning fluids are blood, ketchup, and paint. Shear thinning is a property of polymers well suited for extrusion-based AM.

shell: The outmost layer or layers of a 3D printed article that define its volume and shape, and are printed at 100% infill.

silane: Biocompatible silicon-based organic-inorganic material, used in composites as a coupling agent to promote adhesion between matrix and filler (particles, platelets, and fibers).

silicone: "Semi-organic polymer comprising chains of alternating silicon and oxygen atoms, modified with various organic groups attached to the silicon atoms" (Whittington's 1978).

sol: A fluid colloidal system of two or more components, or a type of colloid in which solid particles are suspended in a liquid.

sol–gel: Process whereby microparticles or molecules present in a colloidal solution (*sol*) agglomerate and, under controlled conditions, link together and form a network (*gel*).

solvent: (A) Any substance, typically liquid, dissolving other substances. Examples are alcohols and esters. (B) The constituent of a solution that is present is larger amount than the other constituent(s).

stereochemistry: Study of the 3D arrangement of atoms in their molecules and stereoisomers.

stereoisomers: Isomers that possess identical molecular formula and sequence of bonded atoms but differ in the arrangement of their atoms in space.

stiffness: In materials, the resistance to deformation in response to a force in tension, compression, shear, and flexure, and expressed by the value of elastic modulus in tension, compression, shear, and flexure, respectively. However, the stiffness of 3D printed articles depends not only on the material's stiffness but other variables such porosity, printing orientation, infill, and so on.

storage modulus: A measure of the elastic mechanical energy that is stored by a polymer under stress and is returned when the stress is removed. It represents the purely elastic behavior of the polymer. It is calculated from DMA data as the ratio of stress in phase with the resulting strain to the strain. Although it is conceptually and numerically different from the modulus of elasticity, when the value of the latter is unknown, it can serve as a good approximation of it.

strain hardening: Strengthening of a polymer undergoing great deformation, caused by large-scale orientation of chain molecules and lamellar crystals. It typically occurs in polymers when they are extended beyond their yield point.

strain: In an object, the change of dimension – as a result of an applied stress – relative to the same dimension before the stress application. F.e. the tensile strain is the ratio of elongation caused by a stretching stress divided by the length prior to the stress. Measured in mm/mm and %.

strength: In material properties, the maximum value of stress sustained by the test coupon during the test before failure. It may be equal or higher than the value of stress at failure. As with stiffness, the strength of printed articles depends not only on the material's strength but other variables such porosity, printing orientation, infill, and so on. Measured in MPa.

stress: It is the ratio of the force over the area where it is applied. It has the dimension of pressure, and is measured in MPa.

stress relaxation: The time-dependent decay of stress in a polymer subjected to constant strain over time.

superposable: Two molecules are superposable if one can be an exact replica of the other only applying translation and rigid rotation.

support material: In AM it is processed by the 3D printer to oppose gravity and prevent overhangs to sag, beams to bend, circular holes to become oval, etc., i.e. it prevents distortions and undesired deformations before the article fully solidifies. It is removed before the article is finished.

surface tension: It is the property of the surface of a liquid that enables the liquid to resist an external force. It is caused by the fact that the cohesive forces present in the top layer of water molecules are stronger than the same forces acting on the molecules below that layer.

suspension: A liquid in which solid particles are dispersed.

tanδ, loss factor, loss tangent: A measure of the polymer's ability to dissipate (as heat) the mechanical energy received through the stress applied to the polymer. It is adimensional, and calculated from DMA data as the ratio of the loss modulus to the storage modulus in presence of tensile, compressive, shear, and flexural stress. The lower tanδ, the stiffer the polymer, the higher tanδ, the more energy-dissipating the polymer.

tensegrity structure: 3D deployable structure consisting of a continuous network of tension members (cables) supported and prestressed by discontinuous compression members (struts). *Tensegrity* is contraction of *tens*ile int*egrity*.

tensile modulus, tensile modulus of elasticity: Resistance of a material to its elastic deformation while being extended, and calculated as the slope of the tangent to the initial linear portion of the tensile stress-strain curve. Measured in MPa.

tensile strength: Maximum stress that a material can withstand while being extended before breaking. Measured in MPa.

thermogravimetric analysis (TGA): An experimental technique consisting in continuously measuring the mass of a sample as a function of its temperature in air or controlled atmosphere. It provides information on polymerization reactions, thermal stability, and degradation of materials under specific conditions. It also enables identification and quantification of individual compounds in mixtures, such as water and organic matter.

thermomechanical analysis (TMA): An experimental technique in which the dimension of a sample is monitored against temperature or time while a very small, constant compressive force is applied to the sample. Used to determine the coefficient of thermal expansion and T_g, study resin cure, perform penetration experiments, measure expansion due to moisture absorption (by measuring shrinkage due to moisture loss), etc.

thermoplastic (TP): It indicates plastics and polymers that can be repeatedly softened and melted by heating, and hardened and solidified by cooling. Once it is solid, a TP typically assumes two molecular arrangements: amorphous or crystalline, each with its own set of properties.

thermosetting, thermoset (TS): Referred to plastics and polymers that, when solid, are permanently crosslinked (hindering molecular bending and rotations), and, if heated from solid to above a certain temperature, do not soften but irreversibly degrade below functionality. TSs are harder, more dimensionally stable, and more brittle than TPs.

thiol: Any organosulfur compound in which SH is attached to a C atom of any aliphatic or aromatic molecule, being S sulfur.

thixotropy: Sometimes considered the same as shear thinning. Barnes (1997) argues that there are various definitions of thixotropy falling in two "families": (a) conferring to a liquid gel-like properties that disappear upon shaking/shearing but reappear upon standing; (b) "time response of the microstructure brought about by shearing or resting, respectively, and the rheological effects arising therefrom".

tissue engineering: Field that applies biology, chemistry, materials science, and engineering to develop functional substitutes for damaged or missing tissue.

titanate: Salt, ester, or anion of titanic acid.

ton: Equal to 907 kg and 2,000 lbs.

tonne, metric ton: Equal to 1,000 kg and 2,205 lbs.

toughness: Energy required to break a material upon impact or in tension. In the latter case it is equal to the area (also called *modulus of toughness*) under the tensile stress-strain curve up to failure, and hence it depends on elongation at break and tensile strength, and is measured in J/m^3. It also measures the material's capability to withstand plastic stress before failure (*ductility*). Material with higher toughness are more suited for applications in which accidental stresses exceed elastic stresses.

transesterification: Exchanging the organic group R'' of an ester with the organic group R' of an alcohol.

triglyceride: A lipid molecule made up of one unit of glycerol (head) and three units of saturated and/or unsaturated fatty acids. The fatty acids of a triglyceride are what make this compound useful as an energy source. One end of triglyceride is made of long chains (tails) of C atoms. When the tails are broken down, they release energy. Triglycerides are the main constituents of body fat in humans and animals and vegetable fat.

true strain: Instantaneous increase of the instantaneous gauge length, and equal to $\ln(1 + \varepsilon)$, with ε the engineering strain.

true stress: The ratio of the instantaneous load acting on the instantaneous cross-sectional area, and equal to $\sigma(1 + \varepsilon)$, with σ the engineering stress, and ε the engineering strain

ultimate tensile strength: See **tensile strength**.

unsaturated: Having double or triple C–C bonds.

vinyl ether: Colorless, very reactive monomer with formula CH_2CHOR, with R being CH_3, C_2H_5, etc., and a commercial feedstock for VP.

viscosity: A measure of the internal friction of a fluid whose layers are in reciprocal motion, and fluid's tendency to resist flow at a specific shear rate. Measured in Pa s, mPa s, and centipoise (cP). More exactly termed as *dynamic viscosity*. It is the opposite of fluidity.

viscous: Generic term indicating a fluid thick and slow in flowing rather than thin and fast-flowing.

Weber number We: Dimensionless number to analyze the flowing behavior of a liquid at the interface with its substrate.

whiskers: Monocrystalline, short fibers possessing exceptionally high strength due to the absence of crystalline imperfections.

yield point: The point at the lowest stress on the stress-strain curve at which the strain increases without an increase in stress. Typically measured in tension.

yield strength: The stress at the yield point. When designing load bearing parts out of materials featuring tensile yield strength, the latter is preferred in the calculations to its tensile strength.

Young's modulus: See tensile modulus.

zeolites: Minerals consisting of hydrated aluminosilicates (minerals mainly composed of aluminum, silicon, and oxygen) of barium, calcium, potassium, and sodium that are employed as catalysts or catalyst supports in refining and petrochemical industries.

zero-shear viscosity: The viscosity of a material when it is at rest, and is measured a very low shear rate.

REFERENCES

Almdal, K., Dyre, J., Hvidt, S., Kramer, O. 1993. Towards a phenomenological definition of the term 'gel'. *Polym. Gels Netw.* 1 (1): 5–17.

ASTM D883. 2017. Standard Terminology Relating to Plastics. 100 Barr Harbor Drive, PO Box C700, West Conshohocken, PA, USA: ASTM International.

ASTM D1566. 2011. Standard Terminology Relating to Rubber. 100 Barr Harbor Drive, PO Box C700, West Conshohocken, PA, USA: ASTM International.

ASTM D6400. 2012. Standard Specification for Labeling of Plastics Designed to be Aerobically Composted in Municipal or Industrial Facilities. 100 Barr Harbor Drive, PO Box C700, West Conshohocken, PA, USA: ASTM International.

Barnes, H. A., 1997. Thixotropy – a review. *J. Non-Newtonian Fluid Mech.* 70:1–33.

DIN EN 13437 2004. *Packaging and Material Recycling – Criteria for Recycling Methods.* Germany: Deutsches Institut fur Normung.

Groll, J., Burdick, J. A., Cho, D.-W. 2019. A definition of bioinks and their distinction from biomaterial inks. *Biofabrication* 11:013001.

Harper, C. ed. 1996. *Handbook of Plastics, Elastomers, and Composites*. New York: McGraw-Hill. 3rd edition.

Huang, J. 2017. Design and development of ceramics and glasses. In *Biology and Engineering of Stem Cell Niches*, ed. A. Vishwakarma and J. M. Karp, 315–330. London: Elsevier.

IUPAC 2014. Compendium of Chemical Terminology - Gold Book. Version 2.3.3. http://goldbook.iupac.org/pdf/goldbook.pdf (accessed January 16, 2021).

Mathias, L. J. 2021. Calculating Molecular Weights. https://pslc.ws/macrog/average.htm (accessed January 16, 2021).

Monte, S. J. 2002. Neoalkoxy titanate and zirconate coupling agent additives in thermoplastics. *Polym. Compos.* 10 (2): 121–172.

Mozafari, H., Dong, P., Hadidi, H., Sealy, M. P., Gu, L. 2019. Mechanical characterizations of 3D-printed PLLA/steel particle composites. *Materials* 12:1. doi:10.3390/ma12010001

Osswald, T. A. and G. Menges 2012. *Material Science of Polymers for Engineers*, p. 49. Munich: Hanser. 3rd edition.

Vert, M., Doi, Y., Hellwich, K.-H., et al. 2012. Terminology for biorelated polymers and applications (IUPAC Recommendations 2012). *Pure Appl. Chem.* 84 (2): 377–410.

Whittington, L. R. 1978. *Whittington's Dictionary of Plastics* Westport: Technomic. 2nd edition.

1 Sustainable Polymers for Additive Manufacturing

Sustainable development is development that meets the needs of the present without compromising the ability of future generations to meet their own needs.

– United Nations 1987, Our Common Future

1.1 INTRODUCTION

Currently, there is consensus among governments, corporations, and public opinions in developing and industrialized countries worldwide on the necessity to reduce pollution and waste, and efficiently use the natural resources, in order to protect health and the natural environment, and achieve and maintain a sustainable standard of living and economic development. A material is *sustainable* if it is derived from renewable feedstocks, and can be recycled and disposed of in ways harmless to the environment. *Sustainable polymers* (SPs) are polymers derived from renewable feedstocks, such as plants, animals, and microorganisms. Their utilization enables to achieve the above objectives, and is growing, thanks to the development of cost-effective industrial production routes, consumers' demand for eco-friendly products, and pro-environment policies and legislations at national and local level, and international agreements that have been implemented for decades, among other factors. SPs also critically contribute to transitioning from the current economy towards the *circular economy* that is an economic model (described later in this chapter) minimizing the use of natural resources and energy, and generation of waste. Furthermore, when SPs are processed through specific fabrication methods termed *additive manufacturing* (AM) or *3D printing* (simply referred to as *printing* hereafter), SPs reduce feedstock waste, and at the same time are turned into innovative products with novel and advanced properties and functionalities.

Although the figures may vary depending on the information sources, experts agree that the global AM market will significantly grow, as it has steadily and rapidly done in the last 30 years. The global AM market (including printers, materials, software, and services) is expected to grow from USD 8.4 billion in 2018 to USD 36.6 billion by 2027 at an impressive CAGR of 17.7% (Report Buyer 2019). According to another source, the AM global market will steadily increase from USD 9.3 billion in 2018 to USD 41.6 billion in 2027 (Sher 2018). A 3D or AM printer (referred to as *printer* hereafter) is a machine fabricating a physical object by adding material selectively (that is in selected locations) and layer upon layer to reproduce a 3D digital model.

SPs for AM are benefitting from the following factors: fast diffusion of printers among industrial, business, educational, institutional (libraries) and personal users; dropping price of personal printers; and commercial availability of several SPs for printers, varying in appearance, properties, and composition. Numerous SPs are also being developed and investigated for a wide range of applications, ranging from the Internet of Things (IoT) to tissue engineering to fabricating buildings, and they greatly contribute to creative solutions driving leading-edge innovation.

Currently, SPs for AM are preferred for non-functional, do-it-yourself items instead of engineering and load-bearing applications, although reinforced SPs have been chosen for architectural-size load-bearing structures, such as the Leaf Bridge (Relander-Koivisto 2018) made of materials supplied by UPM Formi 3D (Finland), and the pavilion components designed by SHoP Architects (USA) for

DOI: 10.1201/9781003221210-1

Design Miami 2016 event (Cascone 2016). If SPs for AM are to transition from household and display items to engineering and functional components, they have to combine a competitive price and adequate and consistent values of physical and mechanical (static, dynamic, long-term, above and below room temperature, etc.) properties under various more or less demanding service conditions associated with load-bearing applications.

This book describes a multitude of experimental and commercial SPs for AM, but only those that meet a combination of the following requirements will be commercially successful: possessing a ratio performance/cost competitive with or superior to existing feedstocks, filling a specific need currently unfilled, and being processable through a commercial or cost-effective (f.e. featuring high-rate output) version of an AM process.

1.2 POLYMERS

Basically *polymers* are substances consisting in very long chains comprising many of the same molecule linked together in a sequence. All linked molecules are called *repeating units*. Strong forces (*covalent* chemical bonds) hold the polymer molecule's atoms together, and weaker intermolecular forces (Van der Waals, hydrogen chemical bonding, etc.) link groups of polymer chains together, preventing them from disentangling (Jansen 2016), and form the polymer. The International Union of Pure and Applied Chemistry (IUPAC) defines a polymer as "a substance composed of macro-molecules" (IUPAC 2014), with *macromolecule* being a molecule composed of many atoms. Each polymer's repeating unit is called *monomer*, meaning *one part* in Greek, whereas *polymer* means *many parts*. Polymers, such as DNA, skin, hair, starch, cellulose, cotton, spider silk, proteins, and sugar, are produced by living organisms and called *natural* polymers. Cellulose, starch, and sugar are examples of natural polymers processed by printers. Natural polymers were used thousands of years ago when plant gum was utilized to bond wood pieces together to build houses, and as a waterproof coating for boats (Sin et al. 2012). Man-made polymers are termed *synthetic*, and constitute the basic ingredient of synthetic plastics, rubber, fibers, adhesives, and coatings. Most commercial polymers processed in AM are formulated from fossil feedstocks such as crude oil and natural gas, with a few commercial polymers derived from plants, such as poly(lactic acid) (PLA). Figure 1.1 is a schematic representation of the molecule of the common sugar, or *sucrose*, and having chemical formula $(C_{12}H_{22}O_{11})_n$, with n (*degree of polymerization*, DP) referring to the average number of monomeric repeating units $C_{12}H_{22}O_{11}$ in the sugar's macromolecule, and capable of reaching many thousands. F.e., in high density polyethylene (HDPE), a most widespread polymer, n ranges from 10,000 to 100,000 (Michigan State University 2013). *Oligomers* are essentially polymers consisting of relatively short chains of molecules, and feature small DP values. Collagen is an example of a natural oligomer. DP is a key property of polymers, because it affects their physical properties (including f.e. their melting temperature T_m), and in most polymers must reach several thousand to result in useful physical properties. DP of sucrose is 1–2 (Cummings and Stephen 2007). Figure 1.2 is a graphic representation of the

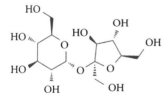

FIGURE 1.1 Schematic representation of the molecule of common sugar or sucrose $(C_{12}H_{22}O_{11})_n$. The solid and dashed wedges represent atoms located towards and away from the viewer, respectively.

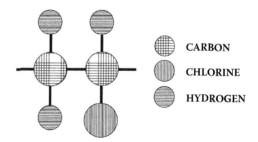

FIGURE 1.2 Graphic representation of the molecule of polyvinyl chloride (PVC) C_2H_3Cl.

molecule of polyvinyl chloride (PVC), a widely popular synthetic polymer, featuring chemical formula $(C_2H_3Cl)_n$. PVC is the feedstock of the commercial filament 3D Vinyl (Chemson Pacific, AUS) for AM.

Polymers are produced from the starting molecules (monomers or oligomers) through a process called *polymerization*, which is basically a chemical reaction in which the same monomers or different types of monomers link together to form macromolecules. There are two major types of polymerization reaction: *addition* and *condensation*. In *addition* (or *chain growth*) *polymerization* the same molecules are added to each other by breaking the double bonds between carbon atoms, and allowing them to link to neighboring carbon atoms to form long chains (Osswald and Menges 2012). As a result, there is no by-product, and the total number of atoms in the formed polymer is a multiple of the numbers of atoms of the starting monomeric units (Mahanta and Mahanta 2016). Many common synthetic polymers are produced following this route, including polypropylene (PP) and high impact polystyrene (HIPS), employed also as AM feedstocks in form of filaments commercialized by Verbatim (PP), Ultimaker (PP), and MatterHackers (HIPS). In *condensation* (or *stepwise* or *radical initiated*) *polymerization* molecules of two or more types with end-groups are mixed and react with each other, and the end-groups link the molecules to one another forming chains, and leave a by-product such as water, ammonia, or CO_2 (Osswald and Menges 2012). This polymerization route applies to most natural polymers (Mahanta and Mahanta 2016), such as cellulose (Dai et al. 2019) and proteins (Derby 2008), both AM feedstocks, and only a few synthetic polymers, including polycarbonate (PC) and polyamide (PA), also ingredients of commercial filaments and powders for AM, respectively. Figure 1.3 schematically illustrates the two type of polymerization, citing as examples polystyrene (PS) and PA 12, two polymers processed via AM, with PA even available in a sustainable version (PA 11) from major suppliers of AM polymers, such as EOS, ALM, and 3D Systems.

Once polymerization forms the macromolecules, the arithmetic product of DP and the *molecular weight* (MW) of the repeating unit is the polymer's own MW, which is a key parameter controlling the polymer's properties. In fact, a large part of the tensile stress-strain curve of a polymer can be expressed by the following equation, in which A and B are two constants, and σ is the stress (Sperling 2006):

$$\sigma = A - \frac{B}{MW} \tag{1.1}$$

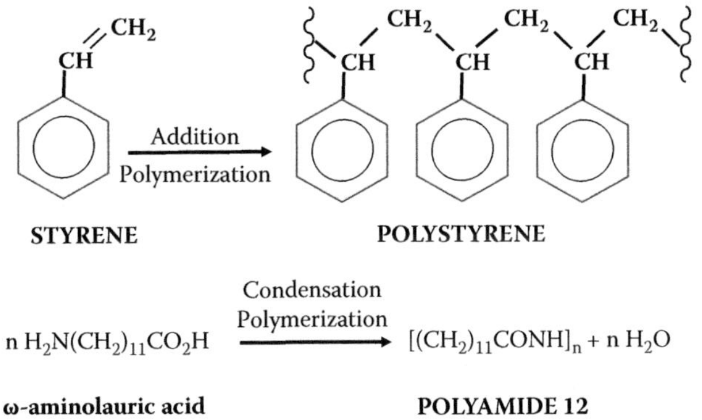

FIGURE 1.3 Examples of addition (upper reaction) polymerization and condensation (lower reaction) polymerization of two polymers for AM: polystyrene and polyamide 12.

If a polymer with a specific chemical formula is available in versions characterized by different values of MW, its version featuring higher MW outperfroms its version possessing lower MW in mechanical and thermal properties and chemical resistance. Most commercial polymers feature an average MW between 10,000 and 500,000 g/mol. When formulating polymeric feedstocks for AM filaments, the feedstock's MW must be considered, because it affects not only various types of properties, such as strength, stiffness, impact resistance, thermal behavior, chemical resistance, weatherability (Shrivastava 2018), etc., but also melt viscosity, and hence processing conditions, fabrication and quality of the printed article. Formulators of polymers for conventional fabrication processes strive to reach the optimal value of MW that balances functional properties with ease and competitive cost of processing, and so should formulators of SPs for AM.

1.3 PLASTICS

Polymers differ from *plastics* or *plastic materials* because polymers constitute the main and not only ingredient making up plastics. In fact, the American Society for Testing and Materials (ASTM) defines *plastic* as "a material that contains as an essential ingredient one or more organic polymeric substances of large molecular weight, is solid in its finished state, and, at some stage in its manufacture or processing into finished articles, can be shaped by flow" (ASTM D883-17). The International Standard Organization (ISO) provides a similar definition (ISO 472). Both definitions imply that other ingredients, called *additives,* are added to the polymer in varying combinations and amounts when fabricating plastic objects, in order to achieve any combination of the following: enhancing or tailoring aesthetic, physical, mechanical, thermal, optical, electrical, etc. properties, reducing the material and processing cost, improving quality, and facilitating processing by improving flow, flexibility, workability, compatibility with blended polymer or filler, etc. These additives includes several type of compounds, such as fillers, plasticizers, reinforcing agents, flame retardants, stabilizers (against heat, light, oxidation), impact modifiers, antioxidants, pigments, lubricants, and antistatic agents. Formulators of new plastics for AM should be familiar with additives, and their major aspects: benefits and drawbacks, chemistry, cost, processability, etc. Basically, polymers are the starting ingredient, and plastics are the final product. Similarly, plastics for AM contain one or more polymers and several ingredients. F.e. an AM filament may incorporate, among other additives: (a) a lubricant to facilitate mixing the feedstocks in form of pellets and extrude them into the filament that will be successfully run on the printer, (b) a plasticizer to increase the filament's flexibility, and prevent its breaking when wound around the spools on which it is sold, (c) a reinforcing agent to increase its strength and/or stiffness, (d) a colorant, and (e) a filler to reduce material cost. *Plasticizers* are liquids with low MW, mainly added to plastic feedstocks for AM to improve processability, lower the *glass transition temperature* (T_g), T_m, crystallinity, increase flexibility, resistance to fracture, tensile strength, and elongation at break (Menčik et al. 2018). Plasticizers also affect printing quality, namely porosity and interlayer adhesion that in turn control the functional performance of the printed part (Wang et al. 2017a). It is worth mentioning that non-AM plastic formulations cannot by default be utilized on printers, because they are formulated for processing conditions different that those associated to AM processes. The Dutch company Proviron supplies a family of plasticizer called Proviplast® with viscosity between water and olive oil, and suited for SPs for AM, such as PLA.

The term *plastic* derives from the Greek πλαστικος (plastikos) meaning "fit for molding" (Liddell and Scottl. n.d.). It is easy to recognize what plastic material makes up a consumer product by checking the plastics identification code printed on the product itself. Figure 1.4 displays the *resin identification codes* of the most sold polymers, with the code "other" encompassing acrylic, PC, PA, and fiberglass.

Since between 35,000 (Rosato and Rosato 2003) and 60,000 (Middlecamp et al. 2015) synthetic polymers have been formulated so far, they enable to produce an enormous number of plastics that feature a wide range of functional (physical, mechanical, chemical, thermal, electrical, etc.) and

FIGURE 1.4 Resin identification codes of the most sold polymers.

aesthetic properties meeting demanding application requirements at a competitive cost, and hence they have become indispensable in every aspect of our lives. Their consumer and industrial applications widely range from inexpensive (f.e. single use shopping bags) and mundane (f.e. clothing items), to very costly and high-performance products, such as airplane structural components and transplant organs. Thanks to their high ratio of strength/density and stiffness/density, currently plastics make up about 50% and 10% of a vehicle's volume and weight, respectively (American Chemistry Council 2019), and, in form of composite materials made of glass- and carbon-reinforced resins, plastics constitute 50% of the weight of the Boeing 787 (Anrady and Neal 2009), far exceeding aluminum (20%), titanium (15%), steel (10%), and the remaining materials (5%) (Giurgiutiu 2016).

In 2018, world and European production of plastics were 359 and 61.8 million tonnes respectively (PlasticsEurope 2019), excluding fibers made of polyethylene terephthalate (PET or PETE), PA, and acrylic plastics. The most sold synthetic polymers worldwide are: polyethylene (PE), which includes linear low (LLDPE), low (LDPE), medium (MDPE), and high density (HDPE) grades, PP, PET, PS and expanded polystyrene (EPS), PVC, and polyurethane (PUR or PU). They are listed in Table 1.1, along with their chemical formula, typical applications, and share (% of total tonnes) of all world and European production in 2015 and 2018, which corresponds to a total of 77.7% and 81%, respectively. PE, PP, and PVC (in decreasing order) led both lists, and together amounted to 55.4% and 59.3% of all plastics produced worldwide and in Europe, respectively. Table 1.1 contains examples of applications that may inspire possible applications of AM grades of the same plastics. However, many examples in Table 1.1 are mass-produced articles that instead would be too expensive if produced via current AM. However, AM is rapidly advancing in terms of technology and capabilities, and can achieve mass production and be profitable under specific conditions, as demonstrated by two companies: Align Technology Inc. that operates over 60 printers to produce 8 million customized orthodontics a year under the trade name of Invisalign® (Haria 2017), and Carbon® Inc. that announced printing sneakers for Adidas (Zahnd 2018). The announcement in 2018 by Chanel® to manufacture 1 million mascara brushes a month through AM demonstrates that AM enables concurrently fine geometric features and high-volume production: namely, AM will form microcavities in the core of the brush, allowing to load a high volume of mascara, and form high-surface areas improving adhesion to eyelashes (Zahnd 2018). Finally, in the middle of COVID-19 crisis, a consortium including commercial, academia and medical enterprises was formed to print clinically tested, FDA-approved COVID-19 nasopharyngeal test swabs up to 4 million per week (Arnold 2020). Excluded from Table 1.1 are acrylonitrile butadiene styrene (ABS) and PLA, the most popular polymers for AM, with the latter one being central in this book, because it is derived from corn starch and sugar cane, and not from fossil sources, and hence is sustainable.

Table 1.2 lists all the polymers in Table 1.1, and the website of the suppliers of their commercial grades for AM, except for LDPE and LLDPE for which we found no AM commercial grade. However, Baechler et al. (2013) and Hamod (2014) experimentally demonstrated that recycled HDPE (a major ingredient in household waste stream) could be extruded into an AM filament. AM polymers are currently adopted in markets such as aerospace, automotive, consumer goods, household items, electronics, and medical, and in a broad range of end products and experimental applications, comprising aircrafts heating and cooling ducts, structures of unmanned aircrafts, scaffolds for tissue engineering, sneakers, tooling to make parts out of composite materials, food, pedestrian bridges, buildings, microfluidic analytical devices, miniaturized inductor and capacitors, sensors, medicine tablets, miniature forceps, miniature lithium battery, aerogels, orthopedics, and orthodontics.

1.4 IMPORTANT PROPERTIES OF PLASTICS

Throughout this book the values of physical and mechanical properties of polymers (especially SPs for AM) are emphasized, because those properties are the most critical in practical applications.

TABLE 1.1
Polymers Most Sold: Chemical Formula, Applications, and Market Shares

Polymer	Chemical Formula	Applications	World Production Share in 2015 (% of 403 million tonnes)[a]	European Production Share in 2018 (% of 51.2 million metric tonnes)[b]
LDPE, LLDPE	$(C_2H_4)_n$	Bags, films, sheets, toys, wire insulation trays, containers, agricultural film, food packaging film.	15.8	17.4
MDPE, HDPE	$(C_2H_4)_n$	Containers of detergent, milk, juice, and shampoo,buckets, crates, fencing.	13.5	12.4
PP	$(C_3H_6)_n$	Automotive parts, bank notes, car parts, carpet yarn, containers, fiber-reinforced items, food packaging, hinged caps, housing for electrical appliances, laboratory equipment, microwave-proof containers, pipes, ropes, suitcases, sweet and snack wrappers.	17.1	19.4
PVC	$(C_2H_3Cl)_n$	Garden hoses, house sidings, intra-vein tubing, plumbing pipes, shoes, shower curtains, toys, window frames, floor and wall covering, cable insulation, inflatable pools.	9.0	10.0
PET/PETE	$(C_2H_8O_4)_n$	Carpet yarn, clothing fabric, food trays, soft drink bottles, films.	8.3	7.9
PUR	$(C_2H_4)_n$	Automotive instrument panels, building insulation, hoses, insulating foams for fridges, mattresses and pillows, performance films, tubes, wire and cable jacketing.	7.2	7.9
PS, EPS	$(C_8H_8)_n$	Transparent and thermally insulating articles: eyeglasses frames, CD cases, insulated containers, egg cartoons, food packaging trays, transparent cups, packaging items, foam cups, thermally and water insulating sheets for building construction, disposable razors.	6.8	6.2

Sources: [a]Data calculated from plot in Geyer, R., Jambeck, J. R., Lavender Law, K. 2017. Supplementary Materials for Production, use, and fate of all plastics ever made. *Sci. Adv.* 3, e1700782. [b]PlasticsEurope 2019. Plastics – the Facts 2019 file:///C:/Users/us899c/AppData/Local/Temp/AF_Plastics_the_facts-WEB-2020-ING_FINAL-1.pdf (accessed January 16, 2021).

TABLE 1.2

Commercial AM Grades of Polymers Listed in Table 1.1 and Their Suppliers' Website

Polymer	Website of Producers of AM Grade Polymers
EPS, PS	https://www.matterhackers.com, https://www.imakr.com/,https://www.lightinthebox.com/en/p/ kcamel-3d-printer-filament-carbon-fiber-eps-17-5-mm-1-kg-for-3d-printer_p7043717.html
LDPE, LLDPE	NA
MDPE, HDPE	https://filaments.ca, https://www.filaments.directory/en/filaments/3r3dtm/hdpe/hdpe-lupolen
PET/PETE	PETG (PET glycol modified): https://www.kodak.com/US/en/Consumer/Products/3d-printing/3d-filament/petg/default.htm, www.innofil3D.com,https://www.matterhackers.com, https://designbox3d.com/products/smartfil-petg,https://www.formfutura.com/shop/product/arnite-id-3040-black-1192
PP	https://ultimaker.com, https://www.verbatim.com/
PUR	https://airwolf3d.com, https://tractus3d.com, https://forward-am.com/find-material/powder-bed-fusion/tpu01/, https://www.lubrizol.com/Engineered-Polymers/Products/Estane-TPU/Estane-3D
PVC	https://filaments.ca/products/vinyl-303-pvc-filament-black-1-75mm

Besides the chemical composition, two sets of factors affect physical and mechanical properties of polymers (Nielsen and Landel 1994): (a) factors relative to polymer's structure and molecule, such as MW, crosslinking and branching, amount of crystallinity, crystal morphology, copolymerization, plasticizers, fillers, and so on; (b) environmental factors such as temperature, moisture content, duration (time), frequency and rate of applied stress and strain, amplitude of stress and strain, heat treatment, etc. Polymers are *viscoelastic* because they behave at the same time like elastic solids, in which the deformation is proportional to the applied force, and viscous liquids in which their rate of deformation is proportional to the applied force. Examples of viscous liquids are pitch and honey, and of elastic solids are metal springs and rubber bands.

Differently than metals, mechanical properties of polymers strongly depend on temperature and duration (time) of the applied stress, because the ways polymeric molecules respond to stress are affected by time and temperature. The most followed test methods to measure physical, thermal, and mechanical properties of polymers are released by two organizations: U.S.-based ASTM and Europe-based ISO. The tests methods by the two organizations are very similar. For those not familiar with those test methods, the Glossary included in this book explains some basics terms and concepts relative to properties of polymers. Additional information on polymer properties and how to measure them is provided by the books listed under Polymers in the section Further Readings of this chapter.

Since polymers are viscoelastic, and their mechanical properties are sensitive to several factors such as temperature, time, and stress rate application, their tensile stress-strain curve only provides a basic indication of the polymer's performance in the finished article, and more comprehensive information is required by conducting tests under different types of stress (shear, compression, etc.), at various values of temperatures and humidity, rate of load application, etc.

Properties necessary for structural design and stress analysis are those in tension, shear, and compression if the material behavior in tension differs from that in compression. In plastics, elongation at break can be a proxy for impact resistance, because higher elongation at break indicates more ductility and hence stronger impact resistance. Stress-strain curves of polymers in tensile, compression, and flexure are quite different from one another not only overall but also in their initial portion, where the respective modulus of elasticity is measured. Commonly, the behavior of plastics in compression differs from that in tension, with the elastic modulus in compression generally higher than that in tension, likely because different molecular and small-scale processes take place between tension and compression (Nielsen and Landel 1994). Brischetto et al. (2020) printed test coupon made of a commercial PLA (Shenzhen Eryone Technology Co., China), and their tensile and compressive moduli were 2,549 MPa and 2,035 MPa, respectively. Compressive strength in brittle polymers is typically higher than tensile strength because the latter is controlled by flaws and submicroscopic cracks that open up under stress and reduce the resisting cross section, and locally raise the stress, whereas in compression the applied stress closes the cracks, so that in brittle polymers compressive strength is predicted to be 1.5 to 4 times higher than tensile strength (Dukes 1966). PLA is brittle, and we found compressive strength reported only for two commercial PLA filaments for AM, made by MakerBot (USA) and REC3D (Russia), and their ratio compressive strength (MPa)/tensile strength (MPa) is 93.7/65.7 = 1.4, and 77.4/34.8 = 2.2 (Koslow and de Valensart 2017), respectively. Song et al. (2017) tested coupons printed from commercial PLA (IMAKR, USA) and recorded a ratio tensile strength/compressive strength equal to 89.8 MPa/46.5 MPa = 1.9. Forces in pure shear are infrequently applied to structural articles, and, usually, shear stresses develop as a by-product of other types of stresses or where transverse forces are present, such as bending forces. Exact ultimate shear stress is not easy to determine, but it can be approximated as 0.75 of the ultimate tensile stress of the material (Tres 2014). Flexural strength and modulus are useful to predict the behavior of components in actual service conditions, and as a metric for quality control. Moreover, across plastics, there is a positive and rather strong linear correlation between flexural and tensile strengths, and between tensile and flexural moduli (Paloheimo n.d.). Particularly, flexural strength tends to exceed tensile strength, and the values of the former generally are not completely correct, because they are calculated from equations assuming a linear correlation between stress and strain that does not fully exist in practice. Plotting the values of flexural and tensile strengths of 15 commercial PLA filaments for AM, the former was on average 54% higher than the latter. Flexural strength can be either lower or higher than compressive strength (Heger et al. 1978). Flexural modulus is nominally the average between the tensile and compressive moduli of elasticity, when they differ. The following relationship relates tensile, compressive, and flexural modulus, assuming E_{ten}, E_{com}, and E_{fle} are the elastic modulus in tension, compression, and flexure, respectively (Crawford 1998):

$$E_{fle} = E_{ten} \left(\frac{2\sqrt{(E_{com}/E_{ten})}}{1 + \sqrt{(E_{com}/E_{ten})}} \right)^2 \qquad (1.2)$$

However, when this formula was applied on PLA, and ABS filaments manufactured by REC 3D (Koslow and de Valensart 2017), the only supplier we found disclosing tensile, compressive, and flexural modulus of the same filament, the calculated E_{fle} was about 30–40% smaller than the values reported. In the case of the 13 commercial PLA filaments for AM we selected, their flexural modulus was on average 12% higher than their tensile modulus, although in a few cases the former modulus lagged the latter. Some suppliers of SPs for AM report strength and modulus of elasticity in bending but not in tension, likely because deflection in bending is easier to measure accurately than elongation in tension.

About the performance of polymeric feedstocks for AM subjected to fatigue, the review by Safai et al. (2019) is recommended because it covers fatigue behavior of ABS, PLA, PA 12, PA 6, PA 11, PCL, PU, PC/PEI, and acrylic-based photopolymers processed by means of three major AM methods for polymers: material extrusion, powder bed fusion, and material jetting.

It is worth mentioning here a critical property for polymers: the *glass transition temperature* or T_g, loosely defined, from an applicative standpoint, as the temperature about which polymers (not rubbers) are not functional because they transition from being hard, rigid, and strong to being soft, flexible, and weak. A different definition of T_g related to the polymer molecular structure is offered in the next section. The T_g values of SPs for AM are often reported throughout this book. In polymers possessing T_g and T_m, an empirical formula correlating T_g and T_m in many polymers is the following, in which the temperature values are expressed in K (Nielsen, Landel 1994):

$$T_g/T_{m=} = 0.63 - 0.71 \tag{1.3}$$

The ratio T_g/T_m of the following SPs for AM is: 0.74 in Ingeo™ 3D850 PLA, 0.65 for Rilsan PA 11 (Di Lorenzo et al. 2019), 0.62 in polybutylene succinate (Succinity PBS), and 0.64 for poly (3-hydroxyalkanoate) (PHA) (Czerniecka-Kubicka et al. 2017).

A property of plastics that is critical to fabrication of filaments for AM is viscosity. *Viscosity* is a measure of the internal friction of a fluid whose layers are in reciprocal motion, and the fluid's tendency to resist flow at a specific shear rate. Viscosity is measured in Pa s, mPa s, and centipoise (cP). Fluids whose viscosity does not depend on the shear rate are called *Newtonian*, and examples of them are water, honey, and organic solvents. Denoting shear stress and shear rate as τ and $\dot{\gamma}$, respectively, their viscosity η is calculated as in equation (1.4):

$$\eta = \tau/\dot{\gamma} \tag{1.4}$$

In *non-Newtonian* fluids, as the shear rate raises, the shear stress increases not linearly but at either increasing rate (*shear-thickening* or *dilatant fluids*) or decreasing rate (*shear-thinning fluids*). A widely adopted model for shear-thinning and shear-thickening fluids is the power law in equation (1.5), where the *consistency index K* is the shear stress at a shear rate value of 1/s, and the *flow behavior index n* is dimensionless, and <1 for shear-thinning plastics, and >1 for shear-thickening plastics.

$$\tau = K\,(\dot{\gamma})^n \tag{1.5}$$

For $n = 1$, K is equal to viscosity. Shear thinning in polymer solutions and melts is caused by the disentanglement of polymer chains during flow, and the degree of disentanglement depends on the shear rate. An example of shear thickening fluid is a colloidal suspension transitioning from a stable state to a state of flocculation, a state in which colloidal particles come out of suspension to sediment.

1.5 THERMOSETS, THERMOPLASTICS, ELASTOMERS

AM and non-AM polymers are classified in *thermosets* (TSs), *thermoplastics* (TPs), and *elastomers*.

TSs or *thermosetting (TS) polymers* are characterized by the fact that they are shaped into final products starting from a liquid that *irreversibly* turns into a solid material through a chemical process (*curing*) triggered by any combination of heat, UV light, and a catalyst, whereby the atomic double bonds within the molecules break, and allow the molecules to link with their neighbors, and *crosslink* to one another, i.e. the molecules get reciprocally interconnected by physical not chemical links that, like bridges, impede the molecules to slide past each other, and form a 3D molecular network. The crosslinking of TSs mostly involves strong covalent bonding and sometimes hydrogen bonding which is a stronger form of a weaker atomic bond. Typically TSs are stiff and strong but brittle, and, once cured, if heated above a certain temperature, do not melt but degrade and char. TSs must be shaped before they crosslink, and can be shaped only once. Common examples of TSs are phenolic resins, PUR, unsaturated polyester, urea-formaldehyde, silicone, epoxy, and vinyl ester. Examples of TSs for AM are acrylate, acrylic resins, epoxy resins, vinyl ether, and thiol monomers, which are suited for the AM processes falling in the *vat*

TABLE 1.3

Commercial TP and TS Polymers for AM

Polymer	Amorphous TP	Semi-Crystalline TP	TS
ABS	x	N/A	N/A
Acrylics	N/A	N/A	x
Acrylates	N/A	N/A	x
Epoxies	N/A	N/A	x
PA 11, PA 12, PA 6, PA 6/PA 66			
Unfilled	N/A	x	N/A
Glass filled	N/A	x	N/A
Carbon filled	N/A	x	N/A
Aluminum filled	N/A	x	N/A
Polymer bound	x	N/A	N/A
PC	x	N/A	N/A
PC/ABS	x	N/A	N/A
PC/PEI	N/A	x	N/A
PLA	x	N/A	N/A
PE	N/A	x	N/A
PEI	x	N/A	N/A
PET	N/A	x	N/A
PEEK			
Unfilled	N/A	x	N/A
Carbon filled	N/A	x	N/A
PEKK			
Unfilled	x	x	N/A
Carbon filled	x	x	N/A
PP	N/A	x	N/A
PPSF	x	N/A	N/A
PS	x	N/A	N/A
PSU	x	N/A	N/A
PVC	x	N/A	N/A
TPU	N/A	N/A	N/A

photopolymerization (VP) family. TSs represent less than 20% of plastic production (Auvergne et al. 2014).

TPs comprise long molecules bonded to each other by chemical bonds far weaker than covalent bond, and do not crosslink when heated and shaped. After being cooled they solidify, and can melt again without degrading, since their non-crosslinked molecules are free to move again once heated. In other words, TPs can be heated and shaped repeatedly. Some TPs for AM are listed in Table 1.2. Other TPs for AM are listed in Table 1.3, and include polymers that are more costly than those in Table 1.2 but also feature superior mechanical properties at room and high temperature, such as polyetherimide (PEI), polyetherketoneketone (PEKK), and polyether ether ketone (PEEK). Recent and broad overviews of polymers for AM recommended are those published by Bourell et al. (2017) and Ligon et al. (2017).

Elastomers are polymers exhibiting rapid recovery to their approximate initial shape and dimensions after substantial deformation by an applied force and following release of that force (ASTM D1566-20). *Rubbers* are a type of elastomers, namely polymers that can rapidly recover from large deformations and are insoluble (but can swell) in boiling solvent, such as benzene (ASTM D1566-20). Examples of common elastomers are: isoprene rubber, fluoroelastomers, natural rubber (a polymer derived from the *Hevea brasiliensis* tree), neoprene, nitrile rubber, silicone, styrene-butadiene rubber, and urethane. Elastomers can be (lightly) crosslinked and left uncrosslinked: the former ones are called *thermosetting* or *vulcanized elastomers* (from the heat-driven crosslinking process called *vulcanization*), and the latter ones are termed *thermoplastic* (TP) *elastomers*. Examples of elastomers for AM include: BASF Ultrasint® TPU01, ESTANE® family of TP polyurethane (TPU) grades by Lubrizol varying in hardness (Lubrizol n.d.), TPU 92A (Stratasys 2021a), PUR called EPU (Carbon® Inc. 2021a), and a high performance TP called PEBA 2301 composed of polyether block amide (EOS n.d.). Sustainable elastomers for AM are also available, the TP urethane WillowFlex FX1504 (BioInspiration 2015), and ESTANE® 3D TPU F95A-030 BR ECO PL (Lubrizol n.d.).

Polymers have a structure whose order is intermediate between ideally crystalline solids and liquid-like amorphous materials. Hence, a critical and functional distinction within TPs is between *amorphous* and *crystalline* polymers. In amorphous polymers, as they cool, the molecular chains randomly coil and entangle, and form a disordered molecular structure, like that of glass. In crystalline polymers, upon cooling, the molecular chains fold multiples times, and form ordered, laminar, highly oriented structures called *crystallites*. *Crystallinity* or *degree of crystallinity* is measured as the ratio of crystalline regions over the entire polymer expressed in percent. Although some polymers can be made highly crystalline (like PE prepared under high pressure and featuring 95–99% crystallinity), almost no polymers are 100% crystalline, but contain some extent of amorphous regions. Most polymers do not exceed 30% in crystallinity, hence crystalline almost always means *partially crystalline* or *semi-crystalline*. PLA is a semi-crystalline polymer, whose crystallinity ranges from about to 10 to above 30%. Polymer chains can exceed the length of a crystallite, and be part of crystallites and amorphous regions. Crystallites act as crosslinks, and impart the polymer values of tensile strength and modulus higher than the values in the disordered regions. Figure 1.5 is a schematic representation of the molecular structures of crystalline (left) and amorphous (right)

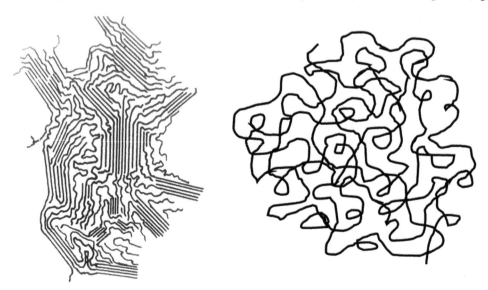

FIGURE 1.5 Schematic representation of semi-crystalline (left) and amorphous (right) molecular structure. In the left picture, the parallel lines represent the *crystallites*.

Sources: (left picture) Bryant, W. M. D. 1947. Polythene fine structure. *J. Polymer Science* 2, 547. Reproduced with permission from J. Wiley and Sons; (right picture) courtesy of Mario Paesano.

polymers: the crystallites are the sets of parallel lines on the left, and are mixed with amorphous regions. This distinction in the molecular structure and the degree of crystallinity is another key molecular parameter in polymers (Nielsen and Landel 1994), besides MW, because, generally speaking, the following differences in material properties (Jansen 2016; Crawford 1998) exist between amorphous and crystalline polymers:

- Crystalline polymers are mostly opaque, and, compared to amorphous polymers, feature superior density, tensile strength and modulus, creep, wear, and fatigue resistance, resistance to higher temperatures and chemicals, greater shrinkage, and T_m of the crystalline regions, along with the T_g, that is the midpoint of the temperature range over which the amorphous regions transition from being hard, rigid, and strong to being soft, flexible, and weak. They also feature T_m exceeding their T_g.
- Amorphous polymers are usually transparent, and, compared to crystalline polymers, display lower density, higher ductility and toughness, weaker chemical resistance, smaller shrinkage and T_g but no T_m, poor resistance to wear and fatigue, and broad softening temperature range.

Table 1.3 includes the types of polymers suitable for AM processes, some of which require TPs, whereas others need TS. For example, *material extrusion* (ME) processes rely on amorphous polymers, because these polymers do not have a distinct melting temperature, but upon raising temperature they increasingly soften and their viscosity decreases, and they turn into a viscous paste that is fluid enough to be extruded through an 0.2–0.5 mm diameter nozzle, but viscous enough to maintain its shape immediately after deposition (Gibson et al. 2015). It is worth noting that ME include two processes frequently encountered in the technical literature that are essentially the same: one is *fused deposition modeling*™ or FDM™ and was patented by Stratasys, the other is *fused filament fabrication* (FFF), and is derived from FDM™. Semi-crystalline TPs are fitting for *powder bed fusion* (PBF) processes (also known as *sintering-based processes*), because they have a distinctive melting temperature which is adequately higher than their crystallization temperature, and between them a sintering temperature can be set, at which the TP powder melts and re-crystallization (i.e. new formation of crystalline regions) does not occur in the immediately deposited layers and hence prevents shrinkage and warping resulting from residual stresses (Drummer et al. 2010). Finally, photosensitive TSs are suited for VP processes such as *stereolithography* (SL), because they are available as liquid monomers and oligomers, capable to be cured by light into solid parts. PEKK constitutes a special case, because it is formulated by Arkema (France) (under the trade name of Kepstan®) from two monomers, and, by varying the ratio of them, it is supplied in two forms: one featuring amorphous structure for ME processes (Thermax™ and CarbonX™ filaments by 3DXTech® (3DXTech® n.d.)), and the other featuring semi-crystalline structure for PBF processes (HexPEKK™ and HT 23). Polyphenylsulfone (PPSF or PPSU) is a high-performance amorphous TP exhibiting outstanding thermal and chemical resistance. As already mentioned, PLA and PA 11 are SPs, commercialized in grades derived from crops instead of fossil sources.

All polymers in Table 1.1 are TPs. TSs are also widely employed as engineering materials. A large share of TSs are *resins*, which are soft solid or highly viscous substances at room temperature, usually containing prepolymers with reactive groups of atoms, being *prepolymer* a "polymer or oligomer whose molecules can undergo further polymerization" (IUPAC 2014). TSs include the following family of polymers, listed below along with examples of their applications, spanning from ordinary and economical ones to demanding and very expensive ones:

- *Acrylic resin*: binder in dental fillers, coatings, lighting fixtures, top coat, and varnish
- *Epoxy resin*: structural parts for aircrafts, adhesives, electronics encapsulation, construction, protective coatings, and printed circuit boards
- *Phenolic resin*: paints, varnishes, and adhesives for plywood

TABLE 1.4

Commercial Non-AM and AM Grades of TPs: Their Features and Suppliers of AM Grades

Thermoplastic Classification	Features	Non-AM Grades	AM Grades	Suppliers of AM Grades
Commodity	Lowest performance (values of physical, mechanical, and thermal properties), lowest price, and largest use	PE, PP, PVC, PS, EPS, bottle grade PET, PUR, PVC	PE, PET, PP, PS, PUR, PVC	Airwolf 3D, BASF, DSM Somos®, Kodak, Innofil3D, Stratasys, Ultimaker, iMakr, 3D Lubrizol, Vinyl™
Engineering	Intermediate performance and price.	PET, PA, PC, PI	ABS, acrylics, PA 11, PA 12. PA 6, PA 6/PA 66, PC, PC/ABS PLA	3D4Makers, 3D-Fuel, 3D Systems, EOS, Airwolf 3D, colorFabb, Innofil3D, Stratasys, DSM Somos®, Ultimaker, ECOMAX®, EcoTough™
High-Performance	Suited for continuous use above 150°C, strongest performance, highest price, and very limited use	PEEK, PEK, PEKK, PES, PAI, PEI, PPSU, PSU, PPS, LCP	PC/PEI (Ultem™), PEEK, PEKK, PPSU, PEI, PSU	Stratasys, ALM, Hexcel, Thermax™, CarbonX™

- *Polyester resin*: repair compounds, protective coatings, toner of laser printers
- *Polyurethane*: bedding, furniture, automotive interiors and bumpers, carpet underlay, packaging, coatings, adhesives, sealants, and elastomers
- *Vinyl ester resin*: flooring, tanks, vessels, parts for cars and other vehicles, fasciae for buildings, and reinforcements for bridges.

Since the early 1980s R&D activity has been conducted to formulate TSs (phenolic, epoxy, PUR, and polyester) from the same renewable resources currently utilized to produce SPs for AM, such as castor oil, soy bean, starch, and wood (Raquez et al. 2010; Auvergne et al. 2014; Kumar et al. 2018). The experience gained in this field may provide guidelines for developing sustainable TSs for AM processes based on VP, such as SL and CLIP™ process (Carbon® Inc. 2021b).

TPs are also split in three groups, based on their functional properties: *commodity, engineering,* and *high-performance* or *specialty* polymers. Table 1.4 lists features and examples of commercial grades of TPs for AM and conventional fabrication processes, along with suppliers of the AM grades. It is worth noting that the blend PC/PEI is commercialized by SABIC under the name of Ultem™ 9085 and Ultem™ 1010, and the two grades offer an outstanding combination of physical and mechanical properties at room and high temperature.

Obviously, the price of non-AM polymers increases going from standard to high-performance grades, and varies also based on the quantity purchased. The price values publicly available depend on the information source. Table 1.5 lists the price in USD/kg of widely used TP and TS commodities and engineering plastics. The price of engineering polymers can be 10–100 times that of commodity polymers depending on the specific polymer. Density, tensile, and impact properties and heat deflection temperature (HDT) of most common commodities, engineering, and specialty plastics are reported in Table 1.6.

A similar price distinction based on material performance holds for AM polymers, as illustrated in Figures 1.6 and 1.7, showing 2019 prices of commercial polymeric filaments for ME and available from two suppliers: Stratasys (Stratasys 2021b), a world leader manufacturer of ME printers, and 3DXTech®, a U.S. material manufacturer and distributor. We selected polymers for ME because this

TABLE 1.5

Prices of Widely Sold TP and TS Commodities and Engineering Plastics

Plastic	Price Range (July 2020)	
	USD/kg	USD/kg
ABS	2.14	3.57
Acrylates	2.75	2.86
Cellulosics	3.85	3.92
Epoxy	2.20	6.54
HDPE	1.32	1.83
LDPE	1.43	1.72
LLDPE	1.12	1.87
PA 6,6, PA 6	2.86	3.96
PC	2.71	4.63
PET	1.30	2.11
Phenolic	1.65	1.87
Polyamideimide (30% glass)	39.65	57.27
Polyester	2.60	3.85
Polyester Unsaturated	3.77	5.84
PEEK	99.1	99.1
PEI	17.62	19.38
Polyphenylene sulphide (PPS) (30% glass)	13.11	16.85
PP	1.10	2.11
PS	2.00	3.11
PU	2.36	3.33
PVC	1.85	2.71
Ultra high molecular weight PE (UHMWPE)	2.64	3.22
Vinyl Ester	4.49	5.22

Source: Plastics News. https://www.plasticsnews.com/resin. Reproduced with permission from Plastics News. https://everchem.com/european-pu-overview-april-2020/ (for PU only, European prices).

family of processes is a most popular type of AM processes. In Figures 1.6 and 1.7 the black and grey bars refer to engineering and high-performance polymers, respectively. Since SL (or SLA) is another popular AM process, Figure 1.8 includes the 2019 current prices of commercial TSs formulated for SL by two major suppliers, 3D Systems, and Formlabs. The price of AM plastics is one order of magnitude or more higher that of the same non-AM plastics, which can be in part explained by the fact that AM plastics are supplied in far smaller batches than those of plastics for conventional fabrication processes, and hence the suppliers cannot leverage economy of scale to reduce cost.

Figure 1.9 reports the price in USD/kg of several commercial AM polymers: those next to the grey bars are SPs, and those next to the black bars are engineering fossil-based polymers featuring physical and mechanical properties similar to those of SPs. The fossil-based polymers have a different chemical formula than that of SPs, except for Rilsan® Invent, which is a PA like PA 650.

The large majority of commercial feedstocks for AM plastics are fossil-derived polymers, and their tensile strength, tensile modulus, Izod notched impact strength, and HDT at 1.82 MPa are plotted in Figures 1.10–1.13, respectively. These property values provide a benchmark in formulating SPs for AM that would be competitive with fossil-based AM polymers not only in terms of price but also performance. The polymers plotted are a sample of the commercial products, and: (a) are mostly offered by major suppliers, and include, labeled by grey bars, also Duraform EX

TABLE 1.6

Properties of Main Commodities, Engineering, and Specialty Plastics for Conventional Manufacturing

Polymer	Density	Ultimate Tensile Strength	Elongation at Break	Tensile Modulus	Izod Notched Impact Strength	HDT at 1.8/ 0.45 MPa
	g/cm³	MPa	%	MPa	J/m	°C
ABS	1.03–1.07	32–45	15–30	1,900–3,000	133–641	80/120
Epoxy[a]	1.04–1.40	28–90	3–6	2,400	11–53	46–2
HDPE	0.94–0.96	18–35	100–1,000	600–1,400	21–214	50/–
LDPE	0.91–0.93	8–23	300–1,000	200–500	No Break	35/–
PA12	1.02–1.04	56–65	300	570–1,600	107–133	140/150
PC	1.20–1.24	56–67	100–130	2,100–2,400	640–961	130/145
PEI	1.27	96	60	2,963	53–64	199/209
PEKK[b]	1.27	88 (yield)	>80	2,900	5.0–5.5[d]	139/–
PEEK[c]	1.30	105	20	4,100	187	156/–
PET	1.37	47	50–300	3,100	43–53	21–66/75[a]
PP	0.90	21–37	20–800	1,100–1,300	27–1,068	45/120
PS	1.05	45–65	3–4	3,200–3,250	13–32	80/110
PVC	1.2–1.4	10–75	10–400	1,000–3,500	21–1,068	60/82

Source: Osswald, T. A. and G. Menges 2012. *Material Science of Polymers for Engineers*. Appendix I, Table I, 3rd ed. Munich: Hanser.

Notes:

[a]*Handbook of Plastics, Elastomers and Composites*, ed. Harper. New York: McGraw-Hill 1996.

[b]Kepstan™ 6000.

[c]Victrex™ 90G.

[d]Charpy method, kJ/m².

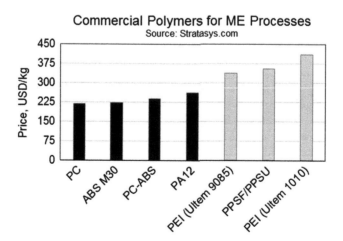

FIGURE 1.6 Prices of commercial polymers for material extrusion (ME) processes (2019 data). The black and grey bars refer to engineering and high-performance polymers, respectively.

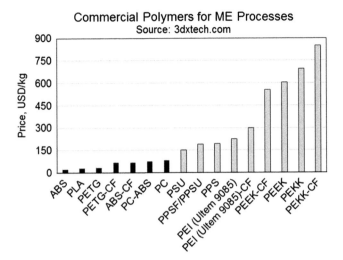

FIGURE 1.7 Prices of commercial polymers for material extrusion (ME) process (2019 data). The black and grey bars refer to engineering and high-performance polymers, respectively.

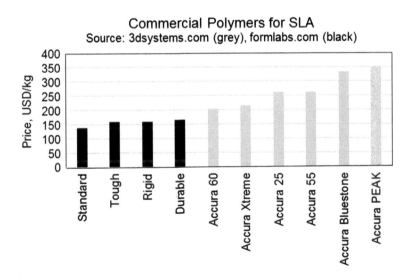

FIGURE 1.8 Prices of commercial polymers for stereolithography (2019 data).

(composed of PA formulated from plants), PLA, and carbon-reinforced (CF) PLA; (b) are processed through the three most popular AM technologies for polymers: ME (crosshatched bars), PBF/LS (black bars), and VP (striped bars); (c) span from commodity (like PP) to engineering (like PA) to high-performance polymers (like PEKK). LS stands for *laser sintering*, a subset of PBF. Fossil-based polymers and SPs for AM are sold by material suppliers and printer manufacturers, and both groups of companies release a list of material properties for each feedstock that depends on the specific supplier and feedstock. Even long lists of released properties often omit properties for designing engineering components, such as tensile yield strength, values measured above and below room temperature, shear modulus, creep, fatigue life, and so on, making difficult to initially select the candidate material. Furthermore, comparing materials is complicated by the fact that a specific property is measured according to different test standards, and reported in different units, such as impact strength measured following Izod or Charpy test methods, and reported in J/m or kJ/m^2, respectively. Other times a property is measured in different conditions across suppliers,

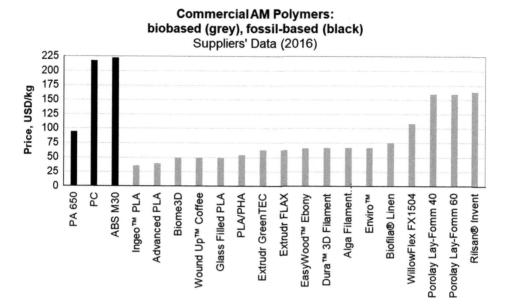

FIGURE 1.9 Price of commercial AM polymers: polymers derived from renewable feedstocks (grey), and fossil-based polymers (black).

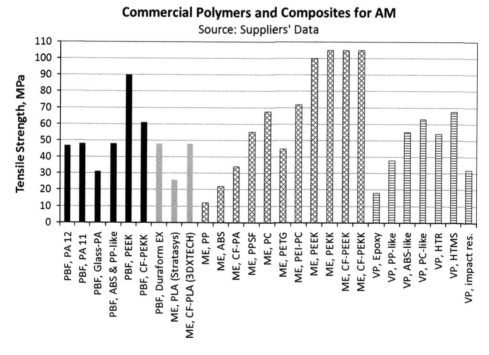

FIGURE 1.10 Tensile strength of commercial polymers and composites for AM processes: PBF, ME, and VP. The polymers indicated with solid grey bars are biobased.

such as HDT that can be measured at 0.45 and 1.82 MPa of applied stress. The values of properties in Figures 1.10–1.13 span a broad range, and provide designers with options, but, when designing critical load-bearing articles operating over temperature ranges and time-dependent stress or strain, a list of property values is required that is rarely available for AM polymers and composites.

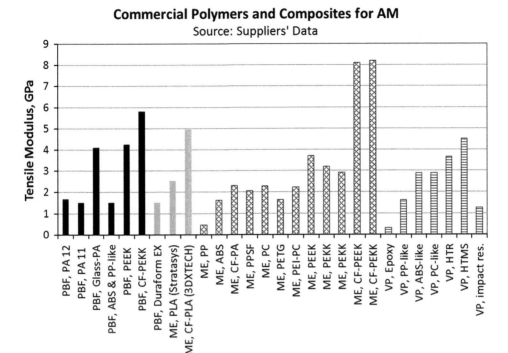

FIGURE 1.11 Tensile modulus of commercial polymers and composites for AM processes: PBF, ME, and VP. The polymers indicated with solid grey bars are biobased.

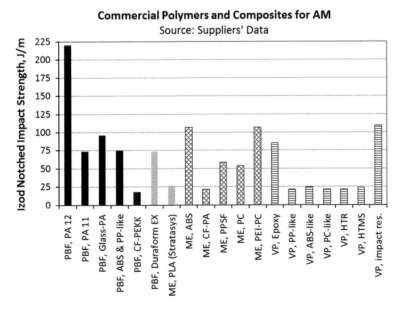

FIGURE 1.12 Izod notched impact strength of commercial polymers and composites for AM processes: PBF, ME, and VP. The polymers indicated with solid grey bars are biobased.

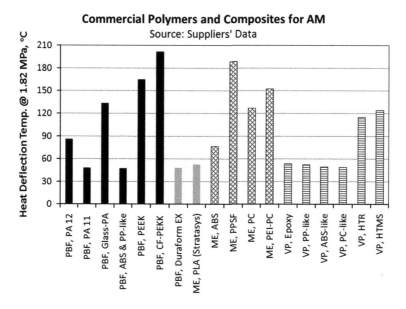

FIGURE 1.13 Heat deflection temperature at 1.82 MPa of commercial polymers and composites for AM processes: PBF, ME, and VP . The polymers indicated with solid grey bars are biobased.

Most polymers for PBF are based on unfilled and filled PA, because currently only this polymer best meets the material requirements imposed by the process equipment, and hence the range of property values in designing a part for PBF is limited. VP employs TSs that typically are brittle, and hence penalized by low impact strength, and articles printed via VP typically possess a maximum service temperature lower than that of heat-resistant TPs. The ME process is compatible with a broad range of polymers and composites that span from commodities to specialty polymers, and reach top values in the four properties plotted in Figures 1.10–1.13 relative to the plastics considered, only outperformed by PBF PA 12 in impact strength. The plot of tensile modulus vs. tensile strength (Figure 1.14) of commercial AM polymer and polymeric composites shows that ME, VP, and PBF are mostly equivalent in those properties, with CF-PEKK and CF-PEEK top performers in both properties. It also appears that there is some correlation between the two properties across all materials plotted, although strength and modulus are measured at high and low

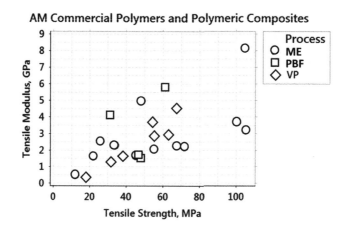

FIGURE 1.14 Tensile modulus vs. tensile strength of commercial polymers and composites for AM processes: PBF, ME, and VP.

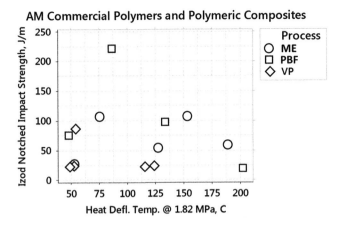

FIGURE 1.15 Izod notched impact strength vs. heat deflection temperature at 1.82 MPa of commercial polymers and composites for AM processes: PBF, ME, and VP.

strain, respectively, and hence express a material's behavior under different stress levels. The plot of impact strength vs. HDT (Figure 1.15) does not include all the materials plotted in Figure 1.14, due to the lack of property values available for some materials in Figure 1.14. In impact strength PBF PA 12 far outperforms the rest of plotted materials, whereas the VP materials stand in the lower half of the plotted values for both properties.

1.6 LIQUID-CRYSTALLINE POLYMERS (LCPs)

LCPs (or *liquid crystal polymers*) form a class of aromatic polyester TP polymers, inert, and resistant to high temperature that can be melted and molded. LCPs are characterized by a highly ordered crystalline structure in liquid and solid states featuring rod-like stiff segments, and, when melt, they feature properties between those of conventional liquids and those of solid crystals. The polymer chains are oriented in loose parallel lines along the direction of the flow when they are molded and drawn, and acquire anisotropic physical, mechanical, and optical properties, and superior mechanical properties in the flow direction when they melt upon being spun into fibers or molding, because, after they are cured, their anisotropic structure is "locked in." By varying the alignment of layers, properties such as thermal expansion coefficient can be optimized.

When LCPs undergo stress, their molecular chains slide over one another but retain their ordered molecular structure that impart their excellent properties. The molecular structure of LCPs imparts them outstanding strength at extreme temperatures, dimensional stability, and resistance to weathering, radiation, burning, and many chemicals, and enhances properties of the materials to which LCPs are added.

LCPs were first developed in the 1960s, and have been investigated as AM feedstocks for more than two decades, demonstrating that AM leverages scientific and technological innovation. LCPs were initially studied as a SL resin for high temperatures (Chartoff et al. 1998; Ullett et al. 2000), and currently are being evaluated as a feedstock for AM and specific applications, such as: (a) elastomer for a cutting edge version of AM termed *4D printing* (Ambulo et al. 2017) described in Section 2.11; (b) high density, compact graphene microlattices for an AM method called *direct ink writing* (Wang et al. 2020) presented in Section 2.10; (c) recyclable lightweight structures featuring remarkable stiffness and toughness, and complex geometries (Gantenbein et al. 2018) and studied for a monopulse antenna (Tamayo-Dominguez et al. 2018). The Swiss company NematX has achieved a precise control of molecular orientation of LCPs by means of self-assembly of LCP molecules into highly oriented domains during extrusion of the molten feedstock, and has commercialized filaments for FFF that are 10 times stronger than PEEK, resistant to temperatures above 250 °C, and apt for end-use parts for aerospace, medical, and special industrial applications. In stiffness, strength, and toughness, these filaments outperform state-of-the-art printed polymers by an order of magnitude, and are comparable with top performing lightweight composites.

1.7 POLYMER MATRIX COMPOSITE MATERIALS (PMCs)

1.7.1 INTRODUCTION

A material is defined as *composite* if it combines two or more constituent materials (*phases*) being present on a macroscopic scale and distinct in the fabricated objects, and features properties superior to those of its constituents individually considered, which is the reason for combining the two phases. The stiffer and stronger phase is called *reinforcement*, the less stiff and strong phase is called *matrix*, and incorporates, binds, and protects the reinforcement that is in a loose form, such as particles and fibers. The region between matrix and reinforcement, called *interphase*, is important because it affects the interactions between the two constituents, and ultimately the performance of the ensuing composite material. Reinforcements are in the form of natural (bamboo, coconut, cotton, flax, hemp, jute, kenaf, sisal, wheat straw, wood, etc.) and man-made (alumina, aramid, boron, carbon/graphite, glass, oriented PE, silicon carbide, etc.) fibers, flakes, and particles. Matrices are ceramic, metallic, and polymeric, with the last type including TPs (ABS, PA, PAI, PC, polyester, PEEK, PE, polyphenylene oxide or PPO, poly(phenylene sulfide) or PPS, PP, PVC, etc.) and TSs (bismaleimide, cyanate esters, epoxy, phenolic, polyester, polyimide or PI, PU, silicone, vinyl ester, etc.). Fibers are: (a) continuous and oriented randomly (mat, cloth), or orderly, in specific directions to form unidirectional and woven layers; (b) discontinuous (chopped, milled). Recent and very small-sized reinforcements with extraordinary properties are: (a) *graphene* that is a 2D matrix of carbon atoms arranged in a honeycomb lattice, and the thinnest existing material (Mohan et al. 2018), and (b) *carbon nanotubes* (CNTs) that are hollow cylinders of graphene.

The strongest and stiffest *polymer matrix composite materials* (PMCs) are those composed of continuous and oriented high-strength and high-modulus fibers (aramid, carbon, glass) embedded in, mostly, epoxy matrix. They belong to the family of *advanced composite materials*. Table 1.7

TABLE 1.7

Properties of Epoxy, Metals, and Structural PMCs

Material	Specific Gravity, SG	Tensile Modulus, E	Ultimate Tensile Strength, UTS	Specific Modulus, E/SG	Specific Strength, UTS/SG
	N/A	GPa	MPa	GPa	MPa
Quasi-isotropic carbon/epoxy composite[a]	1.6	70	276	44	173
Quasi-isotropic glass/epoxy composite[a]	1.8	19	73	10.6	41
Epoxy[b]	1.25	3.25	87.5	2.6	70
Stainless steel 304	7.8	200	505	26	65
Aluminum 2024 T4, T351	2.8	73	469	26	167

Sources: [a]A. Kaw 2006. *Composite Materials*. Boca Raton: Taylor and Francis, 2nd ed. p. 6. [b]K. K. Chawla 1998, *Composite Materials*, New York: Springer, 2nd ed., p. 80.

Note: A plate made of quasi-isotropic composite material features the same values of mechanical properties in all directions within its plane but not across its thickness.

TABLE 1.8

Advantages and Disadvantages of Structural PMCs vs. Metals

Advantages	Disadvantages
Weight reduction: high strength/density, high stiffness/density ratios	No yield strength
Controlled or zero thermal expansion	Weak properties perpendicular to fiber length
Custom-designed bending and twisting response to loading	Brittle matrices, low to medium resistance to impact
Fatigue resistance	Aging/degradation of matrix and interphase due to humidity
Vibration damping	Mechanical properties strongly depend on temperature, humidity
Resistance to water and corrosion	Corrosion upon aluminum contacting carbon fibers
Environmental resistance	Paint thinners attack epoxy
Radiation transparent	Costly raw materials and autoclave fabrication
Reduction of part count	Repair process is complicated
Low inertia, fast start-ups	Smoke emitted upon fire can be toxic

lists room temperature values of density, absolute and specific tensile strength and modulus of structural PMCs, epoxy, steel, and aluminum. Specific properties are property values divided by specific gravity, and matter in structural applications where reducing weight is critical, such as in aerospace, transportation, and sport equipment. The PMCs in Table 1.7 feature specific strength and specific modulus exceeding those of steel and aluminum. The advantages and disadvantages of structural PMCs vs. metals are listed in Table 1.8, and serve as a benchmark for material developers and suppliers of commercial PMCs for AM.

Major markets of structural PMCs are aerospace, automotive, batteries and fuel cells, boatbuilding and marine, civil infrastructure, construction, electrical and electronics, industrial applications, oil and gas, pressure vessels, renewable energy, recreation and sports, and utility infrastructure (Sloan 2020). These markets offer opportunities for PMCs for AM, such as tooling for molding parts in composite materials, although some particular applications of PMCs featuring any combination of large size, high throughput, and outstanding mechanical properties (f.e. wind turbine blades, aircraft wing boxes and floor beams, yacht hulls, and composite pipes for energy industry) have cost and performance requirements that cannot typically be met by current AM processes employing PMCs. Moreover, a set of material property data as large and dependable as that of PMCs is not yet available for PMCs for AM, and will require time and money to be gathered. However, these limitations will not prevent spreading the use of PMCs for AM. In fact, the scale of printable parts is increasing, with AM technologies such as *big area additive manufacturing* or BAAM (BAAM 2020), and *contour crafting* (Contour Crafting 2017) enabling to print a maximum volume of $6.1 \times 2.3 \times 1.8$ m, and buildings, respectively. One main distinction is that advanced PMCs are typically employed as TS resin-impregnated sheets (*prepregs*) of unidirectional or bidirectional fibers that are stacked and cured, whereas feedstocks of most utilized AM technologies are in form of powder, filament, and liquid. However, some less popular AM methods utilize feedstocks in form of sheet, such as *composite-based additive manufacturing method* or CBAM (Impossible Objects 2018), and *selective lamination composite object manufacturing* or SLCOM (EnvisionTEC 2017), both described in Chapter 2.

The global market of composite materials is expected to grow to USD 132 billion by 2024 (MarketsandMarkets™ 2020a), and to USD 182 billion by 2026 (Globenewswire 2020), according to consultancies. Hence, even a small portion of this rising market captured by PMCs for AM will generate a sizeable and attractive amount of business.

Many of the polymer matrix composites described in the following chapters are filled with rigid particles. These fillers increase the tensile, flexural, and shear modulus, but they typically reduce the elongation at break, and often the tensile strength as well. Hence, when considering fillers, the material formulators must find a balance among the material property values they intend to achieve. Main factors affecting the modulus include filler content, particles shape, relative modulus of the component, and spatial distribution of the particles, f.e. close vs. loose packing, and ag-glomerated vs. non-agglomerated. The nature of the filler/matrix interface (except its effect on particle packing), and the amount of interface adhesion are not critical to the modulus, but very critically affect the strength and the stress-strain behavior of composites (Nielsen, Landel 1994).

Since a significant number of particle-filled composite materials for AM is commercialized, a few equations are reported below that predict the tensile strength σ_c of particulate-filled composite materials, based on the tensile strength of their polymeric matrix σ_m, and the degree of matrix/filler interfacial adhesion.

In the case of absent or poor interfacial adhesion, σ_c drops as the volume fraction of the filler V_f raises, as indicated by equations (1.6) (Danusso and Tieghi 1986; Levita et al. 1989), (1.7) (Nicolais and Narkis 1971; Nicolais and Nicodemo 1974) and (1.8) (Nicolais and Narkis 1971; Nicolais and Nicodemo 1973), in which a and b are constants:

$$\sigma_c = \sigma_m (1 - V_f) \tag{1.6}$$

$$\sigma_c = \sigma_m (1 - aV_f^b) \tag{1.7}$$

$$\sigma_c = \sigma_m (1 - 1.21V_f^{2/3}) \tag{1.8}$$

In case of partial interfacial adhesion, the penalizing effect of V_f is mitigated, in comparison to absent or poor interfacial adhesion, and the filler may even improve σ_c, as predicted by 1.9 (Bigg 1987) and (1.10) (Lu et al. 1992), in which c and d are constants:

$$\sigma_c = \sigma_m (1 - aV_f^b + c\, V_f^d) \tag{1.9}$$

$$\sigma_c = \sigma_m (1 - 1.07\, V_f^{2/3}) \tag{1.10}$$

Fu et al. (2008) authored a comprehensive review of models predicting tensile strength and modulus, and fracture toughness of particulate-filled polymer composites. Among those models, some preferred equations to estimate the tensile modulus E_c from the matrix' tensile modulus E_m are the Lewis-Nielsen equation (1.11–1.13), in which A value depends on the filler's aspect ratio, and φ_m is the packing density of the filler (Nielsen, Landel 1994):

$$E_c = E_m \left(\frac{1 + ABV_f}{1 - \psi BV_f} \right) \tag{1.11}$$

$$B = \frac{(E_f/E_m) - 1}{(E_f/E_m) + A} \tag{1.12}$$

$$\psi \approx 1 + \left[V_f (1 - \varphi_m)/\varphi_m^2 \right] \tag{1.13}$$

Another established model to estimate E_c is the Halpin-Tsai equation (1.14), in which $\zeta \approx 2$ for particulate-filled polymer composites, and η is calculated from equation (1.15) (Halpin and Kardos 1976; Halpin 1992):

$$E_c = E_m \frac{(1 + \zeta\eta V_f)}{(1 - \eta V_f)} \tag{1.14}$$

$$\eta = (E_f/E_m - 1)/(E_f/E_m + \zeta) \tag{1.15}$$

1.7.2 PMCs FOR AM

PMCs for AM, or printable PMCs, are commercially available as feedstocks in the form of filament, powder, liquid, and sheet in order to be compatible with AM processes that are extrusion-, liquid-, powder-, and lamination-based (Chapter 2), respectively. Recent and extensive reviews on PMCs for AM by El Moumen et al. (2019), Goh et al. (2019), Parandoush and Lin (2017), Saroia et al. (2020), Singh et al. (2019), and Wang et al. (2017b) are recommended to learn about performance, advantages, limitations, applications, R&D conducted, and areas of improvements. Hereafter, we focus on *fiber-reinforced polymer composites* (FRPCs) for AM, because they are the most popular commercial printable PMCs, and hold potential for engineering applications, but in the following chapters PMCs filled with particles made of natural materials, such as cellulose (Wang et al. 2017b) and wood, are discussed. For a review of particle-filled PMCs for AM the reviews by El Moumen et al. (2019), and Saroia et al. (2020) are suggested. Information on polymer-based nanocomposites for AM is contained in the comprehensive review by Khan et al. (2019) that includes the mostly investigated nanoparticles for printable PMCs: CNTs and nanoclays.

Commercial FRPCs for AM are formulated to be processed by the following families of AM processes: ME, VP, *sheet lamination* (ShL), and PBF. These processes are described in detail in Chapter 2, but are very briefly described in the following sub-sections, where the term *selectively* means *in selected regions*. Tables 1.9 and 1.10 includes examples of commercial PMCs for PBF and ME, respectively, and values of their major properties.

1.7.3 FRPCs FOR ME

In ME the printed material is mostly heated in the printer and selectively dispensed through a scanning nozzle. We add *mostly* because several versions of ME are available, and different forms of FRPCs associated to them exist as filaments (mostly), paste-like, and fluids.

Filaments consist of TP matrices reinforced with discontinuous and continuous fibers of aramid, carbon, and glass. Matrices range from commodities (PP, polyethylene terephthalate glycol-modified or PETG) to engineering (ABS, PA, PC, PLA) to high-performance polymers (PC/PEI or Ultem™, PEEK, PEI, PEKK). Discontinuous fibers vary in size, ranging from nanoscale CNTs and graphene, to micrometer-sized metal powders of copper and iron, to millimeter long chopped fibers. Fibers in FRPs for AM can also be natural, such as hemp and harakeke (Goh et al. 2019). Printing FRPCs with continuous fibers is challenging, and hence this type of material is available at the experimental stage. Methods developed to print continuous fiber PMCs are (Goh et al. 2019): (a) in situ coating of dry fiber with liquid TP before extrusion; (b) in situ fusion of molten TP matrix and impregnation of dry fibers in the printer's extruder; (c) extrusion of preimpregnated fibers. Experiments have been conducted with aramid, carbon, and glass, combined with ABS, PA, PLA, and Ultem™. Voids, fiber-to-matrix bonding, and alignment, amount, and length (in case of

TABLE 1.9
Properties of Commercial PMCs for PBF

Material Name	Supplier	Composition	Density	Tensile Strength	Tensile Modulus	Elongation at Break	Flexural Strength	Flexural Modulus	HDT at 0.45 MPa	HDT at 1.82 MPa	Izod Notched Impact	Izod Unnotched Impact
			g/cm³	MPa	MPa	%	MPa	MPa	°C	°C	J/m	J/m
Duraform® ProX AF	3D Systems	PA 12 + aluminum	1.31	37	4,340	3	64	3,710	182	174	54	255
Duraform® ProX HST	3D Systems	PA 12 + mineral fiber	1.12	44	4,123	4.3	75	3,430	183	171	55	307
Duraform® ProX GF	3D Systems	PA 12 + glass fiber	1.33	45	3,720	2.8	60	3,120	180	129	48	207
Duraform® HST Composite	3D Systems	N/A	1.2	48–51	3,475–5,725	4.5	83–89	4,400–4,550	184	179	37	310
Duraform® GF	3D Systems	PA 12 + glass fiber	1.49	26	4,068	1.4	37	3,106	179	134	41	123
PA 3200 GF	EOS	PA 12 + glass fiber	1.22	47–51	2,500–3,200	5.5–9	73	2,900	96	157	43	214
PA12-GFX 2550	Prodways	PA 12, glass fiber, aluminum	1.35	30	2,550	8	NA	2,275	NA	116	NA	814

Source: Online material data posted by suppliers.

TABLE 1.10

Properties of Commercial PMCs for ME

Material Name	Composition	Density	Tensile Strength	Tensile Modulus	Elongation at Break	Flexural Strength	Flexural Modulus	HDT at 0.45 MPa
		g/cm³	MPa	MPa	%	MPa	MPa	°C
CarbonX™ Carbon Fiber PEEK	Carbon fiber, PEEK	1.39	105	8,100	3	136	8,300	265
CarbonX™ Carbon Fiber PEKK (aerospace grade)	Carbon fiber, PEKK	1.38	115	9,560	3	135	9,850	280
CarbonX™ Carbon Fiber Ultem™ PEI	Carbon fiber, PEI	1.31	145	7,700	1.5	120	7,500	205
MAX G™ Glass Fiber PETG	Glass fiber, PETG	1.41	55	3,285	4	77	3,320	70
CarbonX™ Carbon Fiber PC	Carbon fiber, PC	1.36	70	6,200	2	90	5,890	135
CarbonX™ Carbon Fiber NYLON Gen3	Carbon fiber, PA	1.17	63	3,800	3	84	3,750	147
CarbonX™ Carbon Fiber PETG	Carbon fiber, PETG	1.34	56	5,230	3	80	5,740	77
CarbonX™ Carbon Fiber ABS	Carbon fiber, ABS	1.11	46	5,210	2	76	5,260	76
CarbonX™ Carbon Fiber PLA	Carbon fiber, PLA	1.29	48	4,950	2	89	6,320	91
AmideX™ PA6-GF30 Glass Fiber Nylon	Glass fiber, PA	1.35	62.8	4,261	6	72	3,600	186

Source: https://www.3dxtech.com/tech-data-sheets-safety-data-sheets/.

discontinuous fibers) of fibers impact physical and mechanical properties of printed parts and processability in opposite ways, and hence their optimization is not straightforward. F.e. greater fiber amount increases strength and stiffness but also viscosity, hence complicating processability.

In paste-like and fluid feedstocks the reinforcement is mixed with liquid resin and other ingredients, and only discontinuous fibers have been utilized, namely carbon, CNTs, glass, and silicon carbide whiskers. Matrices are PLA and TS acrylic- and epoxy-based resins. Since extrusion occurs at room temperature, TS resins are preferred to TPs because it is easier for the former ones to be liquid at room temperature than for the latter ones.

1.7.4 FRPCs for VP

In VP a liquid photopolymer kept in a vat is selectively cured by light-activated polymerization. FRPCs for VP include polyacrylate, (mostly) polyester, and epoxy resins combined with discontinuous or continuous fibers. The former fibers range from nanoscale fibers, such as carbon black, graphene oxide, and SiO_2 nanoparticles, to micrometer-scale fibers of alpha-alumina (Al_2O_3), bioglass, ferromagnetic fibers, SiC, and TiC, to millimeter-scale glass fibers (Goh et al. 2019). The composite materials are prepared by premixing resin and reinforcement, or dispersing the fiber on the resin's surface. Continuous fibers have been investigated in form of bundles and mats of glass and carbon fiber. Continuous fibers have been added to SL polymers by manual laying or incorporating fiber laying. Recently, Sano et al. (2018) disclosed a successful SL-based method to print composite materials made of a UV-cured epoxy resin filled with glass in form of powder, short fibers, or continuous fibers, and reported that tensile strength and modulus were respectively 7.2 and 11.5 times higher than those of the resin only.

1.7.5 FRPCs for ShL

In ShL sheets of material are laid, bonded, and cut to form an object. FRPCs are compatible with two versions of ShL: *laminated object manufacturing* (LOM), and CBAM. LOM is compatible with feedstocks in form of sheet material made of any commercial prepreg (f.e. glass-epoxy) or any fiber (*preform*). Since ShL focuses on load-bearing applications, it processes engineering-grade TPs such as PA and PEEK. In CBAM essentially a PEEK or PA 12 powder is sprinkled on a carbon or glass fiber sheet of oriented or randomly placed fibers, later stacked and fused together. Under development are composite materials with PA 6 matrix.

1.7.6 FRPCs for PBF

In PBF thermal energy selectively fuses and binds the very top layers of powder that is spread on a platform. The fact that the feedstock is in form of particles limits the reinforcement's length in the composite, and hence reinforcements of PMCs for PBF are discontinuous fibers, namely nanoscale and microscale fibers and particles, such as carbon black, carbon nanofiber, CNTs, glass beads, nanosilica, SiC, and yttrium-stabilized zirconia. As shown in Table 1.9, in commercial PMCs for PBF the reinforcements are milled carbon and glass fibers, while matrices are PA 12 (mostly), and PS.

1.8 NATURAL POLYMERS

Besides synthetic polymers, natural polymers are the other family of polymers, when considering their source. *Natural* polymers are derived from biomass, that is animals, plants, and microorganisms, and include cellulose, some resins, starch, etc.. Natural polymers can be grouped based on their source in the following types: lignin, polyesters, polyisoprene, polynucleotides, polysaccharides, proteins (Atkins 1987). Some natural polymers such as cellulose and lignin are

TABLE 1.11

Natural Polymers for AM, and References to Their AM Applications

Natural Polymer	Reference to AM Application
Alginate or alginic acid	Ozbolat and Hospodiuk (2016)
Cellulose	Dai et al. (2019) and Wang et al. (2018)
Chitin	Ozbolat and Hospodiuk (2016)
Hyaluronate	Gopinathan and Noh (2018)
Poly(3-hydroxyalkonate)	Pereira et al. (2012)
Polypeptides	Raphael et al. (2017)
Polysaccharide	Voisin et al. (2018) and Kim et al. (2019)
Proteins	Brindha et al. (2016), Derby (2008), and Melchels et al. (2012)
Starch	Chen et al. (2019), Kuo et al. (2016), and Laird (2015)

Note: Full references are at the end of this chapter.

abundant and economical. Natural polymers are present in several markets, such as food industry, pharmacy and biomedicine, and studied to be leveraged in other markets, such as electronics and constructions. A natural version of PET is PHA, produced by microorganisms (bacteria), and studied (Kim and Lenz 2001), and commercialized as a SP for AM by colorFabb (The Netherlands). Table 1.11 lists examples of natural polymers employed as AM feedstocks, and the corresponding technical literature describing their utilization in AM, which currently is more experimental than commercial, and underscores the remarkable flexibility of AM printers that can be adapted to different materials affordably and quickly to carry out experimentation.

The polymers in Table 1.11 are briefly described here. A *polysaccharide* is a carbohydrate macromolecule composed of molecules called *monosaccharides* that are simple sugars, such as glucose, and are linked together by glycosidic bonds. Starch and cellulose are polysaccharides. *Starch* is the end product of photosynthesis in plants, and a plant polysaccharide that is present in root, seeds, and tubers, and one of most abundant carbohydrates. It is mainly composed of two molecules: linear amylose, and highly branched amylopectin. *Cellulose* is the main constituent of cell walls of all plants, and the most abundant organic polymer on Earth. It features a crystalline morphology consisting of a linear chain of several hundred to many thousands of glucose units. It is found in wood, green plants, algae, bacteria, and oomycetes, a type of fungi (Mahanta and Mahanta 2016). Because it combines low price and functional physical and mechanical properties, cellulose is extensively studied as a feedstock for AM, as described in Chapter 7. *Alginate* is a linear polysaccharide, synthetized by soil bacteria and brown seaweeds of the class Phaeophyceae in which it has a structural function (Draget et al. 1997). *Chitin* is structurally similar to cellulose, derived from glucose, and ubiquitous in nature: it is the main component of the cell walls of fungi, and the exoskeletons of arthropods such as crabs, insects, lobsters, and shrimps. *Hyaluronate* is the salt of hyaluronic acid, and a glycosaminoglycan present in the extracellular matrix of mammalian connective, epithelial, and neural tissues, and corneal endothelium. *Proteins* and *polypeptides* are composed of α-amino acids that chemically are carboxyl amine or amino carboxylate, both containing amino and carboxyl groups covalently attached to a single carbon atom. PHA is one of the polyesters produced in nature by numerous microorganisms through bacterial fermentation of sugar and lipids.

1.9 MARKET DATA FOR POLYMERS

Our current life, especially in the most industrialized countries, relies on synthetic polymers, because they possess the following features in combinations that are unmatched by other engineering materials (Mülhaupt 2013):

- Very attractive cost/performance ratio
- Design flexibility
- High versatility in properties and applications
- Low density and high ratio mechanical properties/density
- Outstanding corrosion resistance
- Processes to fabricate articles require short cycle times
- Enabling highly cost-, resource-, eco-, and energy-effective mass production. Producing plastics requires less energy than that for producing metals and ceramics
- Diversified base of raw materials (crude oil, coal, gas, and biomass)
- High energy content similar to that of crude oil and superior to wood
- Being recyclable as chemical feedstocks and source of energy
- Substantially contributing to energy savings in applications, such as lower fuel consumptions in vehicles due to polymers' low density, and less energy to heat up and cool down houses due to polymers' efficient thermal insulation.

Although, as stated earlier, printers are not ready to fully replace conventional fabrication processes of plastics parts such as compression molding and injection molding, the following market data relative to plastics in general are also relevant to AM plastics, because each plastic available as an AM grade and non-AM grade has the same main ingredient and similar properties, and these market data can guide in formulating, developing, and marketing SPs for AM.

The world production of plastics (including TPs, PUs, TSs, adhesives, coatings, sealants, but not acrylic, PET, PA, and PP fibers) reached 348 million tonnes in 2017 (PlasticsEurope 2019). From 1950 to 2017 the production rate has considerably increased (PEMRG et al. n.d.), propelled by new chemical formulations and applications ranging from consumer goods to aerospace, and fueling expectations of further growth in the near future. Plastics are so pervasive, that in 2015 the global output of plastics in volume was 1.5 times that of steel, being however the density of steel about eight times that of plastics (PEMRG et al. n.d.). As anticipated in Table 1.1, in 2015 the most produced polymers were in decreasing order PE, PP, PVC, PET, PS, and PUR (Geyer et al. 2017). PLA and ABS (the most popular AM polymers) are not among them because they are more costly than the polymers in Table 1.1. As mentioned in Table 1.2, PE, PP, PVC, PET, PS, and PUR are also available in AM grades. The world distribution of plastics (TP, PUR, TS, adhesives, coating, sealants, but not acrylic, PET, PA, and PP fibers) in geographic areas in 2018 saw China as the world leader with 30%, obviously also due to its large workforce and low labor cost, with the latter factor making Chinese plastics competitively priced. China was followed by NAFTA (18%), Europe (17%), Asia excluding China and Japan (17%), Middle East and Africa (7%), Latin America (4%), Japan (4%), and Commonwealth of Independent States (CIS) comprising Russia and its neighboring states (3%). Europe included the 28 countries members of the European Union plus Norway and Switzerland. NAFTA comprised Canada, Mexico, and USA. In 2017, the U.S. plastics industry reported USD 432 billion in shipments (an increase of 6.9% vs. 2016) and 989,000 jobs, which became 1.81 million when jobs at suppliers were included. In the period 2012–2017 in U.S. employments growth was 1.6% for plastics manufacturing, and 0.9% for all manufacturing (Plastics Industry Association 2018).

In 2010, China surpassed Europe in production of plastics (The European House – Ambrosetti 2013), and increased its world share of plastics production from 21% in 2006 to 28% in 2015 (PEMRG et al. n.d.). Between 2000 and 2017, while plastic production in the world doubled, in China it grew by seven times, exceeding 80 million tonnes.

In 2017, the worldwide revenues from plastics amounted to USD 523 billion. Packaging, construction, consumer goods, and automotive in decreasing order made up together 75.1% of the revenues, with electrical, agriculture, medical devices, etc. trailing (Grand View Research 2019). In 2018, 51.2 million tonnes of plastics were produced in Europe. The largest markets were packaging, building/construction, automotive, and electrical/electronic in decreasing order, that generated 76% of the revenues, followed by household/leisure/sport, agriculture, and others (PlasticsEurope 2019). Packaging and construction/buildings topped both lists. Plastics for packaging typically comprise PP, PE, and PS, which are very inexpensive, and profitable when sold in large quantities, and include articles for food and shipping, and many other products that are single-use and inexpensive. Packaging is not a profitable application for printed parts, because packaging is mass-produced at high speed and low cost, which are requirements unmet by most current AM processes. Construction is not a sizeable market currently for AM, but, driven by R&D and newly formed companies worldwide, it could become a profitable niche market. Currently AM polymers are preferably targeting prototypes, models, tooling, hobby and display items, and engineering articles in order to meet any one or any combination of the following requirements: making home-designed items, cutting lead time, accelerating path from idea to market, no mold/tooling, low production volumes, highly complex shape, user-tailored design (f.e. medical implants), less constrained geometries, reducing part count in assemblies, making legacy parts with blueprint or mold no longer available, minimizing weight without sacrificing structural performance, multifunctionality, and specific surface finish. However, AM technology and industry are evolving so rapidly that even AM experts find difficult to predict the near future of AM, and hence applications now deemed improbable or even excluded for AM may become reality.

Of the 322 million tonnes of all plastics produced in 2015 worldwide, TPs amounted to 269 million (83.5%), and the remaining 54 million tonnes (16.5%) comprised TSs, elastomers, adhesives, coatings, sealants, and PP fibers, and did not include acrylic, PET, and PA fibers (PlasticsEuropeMarket Research Group n.d.).

Plastics are the prevalent AM materials, due to the fact that plastics are affordable and were the first successful AM feedstocks, and printers for plastics are far less costly than those for metals. The price of industrial printers for plastics and industrial printers for metals starts around USD 0.3 million and USD 1.5 million, respectively. Metals and ceramics, in decreasing order, lag behind plastics in utilization. The global AM plastics market was estimated at USD 616 million in 2018 and projected to jump to USD 1.97 billion by 2023, at a remarkable CAGR of 26.1% between 2018 and 2023 (Research and Markets 2018), while, according to another source, it will grow less rapidly and exceed USD 1.8 billion in 2025 (Grand View Research 2018). Use of AM polymers will spread also thanks to larger demands (f.e. from the medical sector), and the expansion of the global AM market, expected to grow from USD 8.4 billion in 2018 to USD 36.6 billion by 2027 at an outstanding CAGR of 17.7% (Report Buyer 2019). Presently, ME filaments and hence TPs dominate the global market of AM materials in terms of revenue. In fact, a survey of nearly 1,000 respondents conducted in 2017 by Sculpteo (Core-Ballais et al. 2017), a French company specialized in 3D printing in the cloud, concluded that TPs were used by 88% of the AM users pooled, followed by TS resins (35%), and metals (28%). Polymers are compatible with the most popular AM processes ME (by far the most preferred on desktop printers), PBF, and VP, which will keep polymers in their leading role in the near future. ME will generate the most revenues among AM printers for professional and industrial use through at least 2024, and, by "2027, the polymer material extrusion market of non-hobbyist and consumer users will generate USD 2.2 billion in printer and hardware sales" (Manolis Sherman 2018). Switching to PBF processes, sales of polymers for PBF have leaped from about USD 50 million in 2008 to about USD 400 million in 2018, with the steeper rise recorded from 2015 to 2018, leading to an optimistic forecast for the near future, according to the *Wohlers Report 2019* by Wohlers Associates Inc., a well-known consultancy specialized in AM (Wohlers Associates 2019). In summary, polymers are expected to stay dominant among AM materials in the near future.

We conclude cautioning that figures relative to production of non-AM and AM plastics in USA and globally depend on the information source, and hence may differ for the same time period and geographic region, even when provided by specialized sources such as consultancies and market research companies. Similarly, forecast data for AM markets and materials have to be considered with greater caution, because they are predictions about the AM industry that has a significantly shorter record than other industries, and, as mentioned earlier, is changing very rapidly.

1.10 BIOBASED POLYMERS

In 2016 it was reported that crude oil and natural gas made up more than 90% of the feedstocks for the plastics industry (World Economic Forum 2016), and this results in the following drawbacks impacting the plastics manufacturers as well the entire population (Amass et al. 1998; Chandra and Rustgi 1998; Mohanty et al. 2000; Siracusa et al. 2008; NatureWorks 2021a; Islam et al. 2014):

- Crude oil and natural gas are fossil-based and hence finite resources, whose supply is declining with time. In 2008 the company BP estimated that the proven oil reserves were sufficient for 41 years, or 80 years according to others' estimate (British Plastics Federation 2014)
- Price of crude oil and natural gas raises (and drops) over time and is subjected to geopolitical conditions that can be unpredictable
- Lack of or slow biodegradability of crude oil-based plastics
- Significant amount of greenhouse gas emissions, which, according to most experts, contribute to global warming
- Large heat required for processing crude oil and natural gas
- High cost and cross-contaminations associated to recycling
- Risks of rendering toxic the food in contact with recycled plastics.

In 2009 it was reported that about 8% of global crude oil and natural gas production was utilized as feedstocks (4%) for plastics and energy (3–4%) for their manufacture (Hopewell et al. 2009). The rising demand for plastics in emerging economies makes it likely that the share of crude oil and natural gas for plastic production will raise in the future (Ghaddar and Bousso 2018).

One way to address the shortcomings of fossil-derived polymers is to turn to polymers derived from *biomass* that is "material produced by the growth of microorganisms, plants or animals" (IUPAC 2014), or "material of biological origin excluding material embedded in geological formations and/or fossilized" (CEN/TR 16208). Examples of biomass (and their extracted ingredients to derive industrial polymers) are: wood (cellulose, lignin), plants, agricultural crops (beet, cassava, corn, sugar cane), fuels from alcohol, animals (proteins), and microorganisms. "Polymers derived from biomass" are also termed *bio-based* polymers (CEN/TR 16208) (or *biobased*, according to ASTM D6866-18). A more specific definition of biobased polymer is a polymer containing at least in part carbon of renewable origin such as microorganisms, plant, animal, fungi, agricultural, forestry, and marine products living in a natural environment in equilibrium with the atmosphere. Similarly, *biobased plastics* refer to plastics that contain ingredients wholly or partly derived from biomass.

Selecting plants as renewable feedstocks for polymers is even more beneficial, because it not only addresses the above shortcomings, but also offers the following advantages:

- Supporting rural economy, especially in agricultural areas that may otherwise decline economically
- Supporting areas not industrially developed
- Attracting customers: 78% percent of consumers say they want to buy more sustainable products, although only 43% will pay more. Particularly, millennials, who are the largest demographic group in USA, prefer materials converted into products more natural and durable than current products (Muenchinger 2017). Hence, the ideal situation would be a material or product that is sustainable and affordable at the same time.
- Environmental benefits: governments, corporations, and public opinions in industrialized and developing countries worldwide agree on the necessity to preserve now and in the future the natural environment from pollution and waste, and efficiently using the natural resources, in order to achieve and maintain a sustainable economy.

One advantage in preferring biobased polymers as feedstocks for AM to synthetic polymers is that, as studies have shown, during printing synthetic polymers release volatile harmful substances, such as aldehydes, benzene, ethylbenzene, phthalates, toluene, and so on (Horst et al. 2016).

Synthetic polymers have been produced from natural feedstocks already in the past. In fact, the first synthetic polymer, *celluloid*, was invented in 1869 by J. W. Hyatt, who formulated it from the cellulose present in cotton, and, therefore, at the same time he introduced the first man-made polymer from natural feedstocks, preceding by 38 years Bakelite, a TS resin invented in 1907 that was the first man-made polymer from synthetic ingredients. Since the onset of polymer science and engineering, about 100 years ago, polymers have been derived from natural feedstocks, such as cellulose, casein, shellac, gum, and natural rubber, chemically modified and converted into materials featuring new and specific properties (Mülhaupt 2013). A well-known example of crops converted into plastics comes from the past, and is the *soybean car* built and unveiled by H. Ford in 1941: it consisted in a tubular steel frame covered by a body comprising 14 panels in plastic formulated from soybeans, wheat, hemp, flax, and ramie (The Henry Ford 2020). Some of Ford's goals were utilizing surpluses of crops such as soybeans, and fabricating a lighter car with improved fuel economy, and are still relevant today. Unfortunately for H. Ford, after World War II, an abundant and inexpensive supply of crude oil made fossil-based plastics products less expensive to fabricate, and more durable and weather-resistant than the plant-based versions of them (Science History Institute 2021).

Fast forward to 75 years later, and again Ford Motor Co. has chosen plant-derived polymers for its vehicles. In a 2016 interview about use of plant-derived plastics on cars (CGTN America 2016), Dr. Mielewski, Technical Fellow at Ford, reported that Ford had successfully researched, developed, and launched eight plant-based plastics on its vehicles, including wheat straw in bins on Ford Flex, and kenaf fiber in arm rests on Ford Escape. She mentioned the following reasons from switching from fossil-based plastics to plant-derived plastics: oil price fluctuation, lightweight of plant-derived-materials, providing farmers with source of revenue, protecting environment, and cost reduction. She also predicted that in the future almost all fossil-based plastics could be replaced with sustainable feedstocks from CO_2, algae, bamboo, etc. H. Ford's vision of plant-based plastics on cars was proved right again in 2017, when Goodyear Tire and Rubber Company announced the release of Assurance® WeatherReady® tires, equipped with a unique soy-based tread with enhanced traction, with soybean oil replacing fossil-based mineral oil, and enabling the rubber to remain more flexible at lower temperatures than oil-based rubber compounds, ultimately improving traction in different weather conditions (United Soybean Board 2019). Another recent application of soybean oil was developed by WCCO Belting Inc. that replaced synthetic aromatic oil with soybean oil in their conveyor belting products, which increased abrasion resistance (the most critical performance parameter in conveyor belts) while maintaining all other physical properties, including tensile and tear strength, and puncture resistance (Anon. 2018). SPs in form of natural fibers (bamboo, banana, bast, flax, hemp, kenaf, pineapple leaf, sisal) present in composite materials are also being considered for aircraft components (Arockiam et al. 2018), such as radome (Haris et al. 2011), wing boxes (Boegler et al. 2014), and cabin interior panels (Anandjiwala et al. 2008).

The case of Ford and Goodyear underscores the current benefits of adopting biobased polymers. Car interiors offer most opportunities for biocomposite. International Automotive Components Group (USA) has produced trunk trim, map boxes, rear seat liners, and door components made of natural fiber composites, and in 2017 launched the first natural fiber composite sunroof frame, mounted on all Mercedes-Benz E-Class and A-Class models: it was made from hemp and kenaf fiber (70%), and was 50% lighter than its metal version.

A most sold biobased polymer is natural rubber, which is derived from plants, and has been around since the invention of the automobile in the late 19th century. In 2017 the global production of natural rubber from the *Hevea brasiliensis* tree was 13.6 million tonnes, quite close to the world output of synthetic rubber that reached 15 million tonnes in the same year (Market Publishers 2018).

A biobased polymer is also a *renewable* material, because it is a "material used to produce a product and replenished by natural processes at a rate comparable to its exploitation rate" (CEN/ TR 16208). Examples of renewable raw materials are wood, plants, agricultural crops, and animal fur and leather.

It is worth pointing out that polymers from biomass alone will not reduce the amount of solid plastic waste, which will decrease also by recycling, reusing, composting, and selecting biode-grading plastics.

1.11 SUSTAINABLE POLYMERS

A material is *sustainable* if it is produced from a natural resource that is generated indefinitely by nature, and, in order to produce the material, is consumed at the same or lower rate at which it is naturally renewed. SPs are defined in different ways: (a) polymers "derived from renewable feedstocks that are safe in both production and use and can be recycled or disposed of in ways that are environmentally innocuous" (Schneiderman and Hillmyer 2017); (b) polymers "that are de-rived from renewable feedstocks and exhibit closed loop life cycles" (Zhang et al. 2018); (c) polymers meeting the needs of consumers and companies without damaging health, environ-ment, and economy. Specifically, SPs are polymers made from renewable feedstocks such as vegetable oils and plants, feature a smaller carbon footprint, and, when produced, use less net water and non-renewable energy, emit less greenhouse gases, and generate less waste in com-parison to fossil-derived polymers (Center for Sustainable Polymers 2020). The main feature of SPs is that they are biobased. Hong and Chen (2019) divide SPs in two families:

- *Plant-based* polymers: they are polymerized from monomers derived from biomasses, and feature inherent negative carbon footprint, since plants absorb CO_2. PLA, polycaprolactone (PCL), bioderived PET (Nakajima and Kimura 2013), sugar-based poly(ethylene furanoate), and renewable PC belong to this family.
- *Degradable* polymers: polymers formulated to degrade into environmentally harmless substances. Polyglycolide, polytrimethylene carbonate, and PHAs are examples of them.

More broadly a material is sustainable if it enables any type of development (economic, urban, social, etc.) that is sustainable, as defined in 1987 by the World Commission on Environment and Development of the United Nations in its report on environment and development. In the same report, the United Nations argue that a development is sustainable if "it meets the needs of the present without compromising the ability of future generations to meet their own needs" (United Nations 1987). That report strongly urged the industrial and developing nations to achieve a sustainable development that is to protect the environment while pursuing economic development, and proves that the importance and need for sustainable materials was recognized at the highest international level more than 30 years ago. In 2005, a new step towards a truly sustainable future was taken when the Kyoto Protocol, ratified by 145 nations, mandated emission cuts for in-dustrialized nations (United Nations Climate Change 2020). Lately, the United Nations have set sustainability development goals, some of which can be pursued also by leveraging sustainable materials (United Nations n.d.).

Consensus has been reached among governments worldwide on protecting the environment, as demonstrated by several policies regarding the transition from traditional, non-renewable energy sources, such as crude oil, coal, and natural gas, toward renewable energy sources, such as wind, solar energy, and biomass. Since the early 2000s policies, laws, incentives, and bans promoting the use of biobased materials, and reducing waste and littering relative to fossil-based plastics have been enacted worldwide, and summarized by Hermann et al. (2011). F.e. a mandate common to USA and many European countries states to give preference to biobased/environmentally friendly products over traditional ones when purchasing for public administrations. Moreover, numerous countries across Africa, Asia, Europe, and U.S. cities enforce bans of plastic carrier bags, and some bans exclude plastic bags made from renewable resources, and/or being biodegradable, or compostable. In January 2020, the government of China announced a plan to cut plastic use by 2025 in order to contain pollution (State Council of China 2020).

Corporations also have agreed to promote sustainability. An example of such intent is the joint statement signed by seven of the world's largest aviation manufacturers, in which CTOs of Airbus, Boeing, Dassault Aviation, GE Aviation, Rolls-Royce, Safran, and United Technology Corporation committed themselves "to driving the sustainability of aviation," and "supporting the commercialization of sustainable, alternate aviation fuels" in order to protect the planet (UTC 2019). Namely, in 2021 Boeing announced that its commercial airplanes will be capable and certified to fly on 100% sustainable aviation fuels by 2030 (Il Bioeconomista 2021).

However, we must recognize that all current plastics, even fossil-based, already contribute positively to environmental, economic, and social sustainability, as illustrated by the following examples (British Plastics Federation 2019):

- *Environmental sustainability*
  - Packaging: (a) food packaged in plastics lasts longer than unpackaged food, reducing waste and transport energy; (b) some packaging materials such as EPS are recyclable and recycled.
  - Construction: (a) plastics such as PUR and EPS provide high thermal insulation and are lightweight, reducing heating and cooling energy; (b) plastic pipes require less energy to be produced than that for concrete and iron pipes.
  - Transportation: being lightweight enables plastic components to save fuel and electrical power when inserted on road vehicles, watercrafts, aircrafts, spacecrafts, and rail cars. Saving weight in cars also reduces emissions. In a car 100 kg of plastics can replace 200–300 kg of heavier materials, and every 100 kg of plastics in a car will reduce fuel consumption of the vehicle by 750 liters over the car's useful life (150,000 km).
  - Waste management. Plastics can be recycled, and converted into: (a) granules to make new products; (b) their polymer constituents that in turn are utilized to make new plastics and chemicals; (c) thermal energy recovered through waste incineration, with an average value per polymer of 38 MJ/kg versus 31 MJ/kg for coal.
- *Economic sustainability*. Plastic industry contributes to jobs, exports, and supply chain, not only in the manufacturing sector, but also in R&D, and industries serving the plastics industry, such as construction, logistics, safety, engineering, power supply, maintenance, etc. According to data published in October 2019, the following statistics were reported for the plastic industry active in the 28 countries members of the European Union: 1.6 million jobs, close to 60,000 companies, turnover exceeding EUR 360 billion in 2018, and 7th industry in Europe in industrial value-added contribution (PlasticsEurope 2019).
- *Social sustainability*. Plastics industry contributes to jobs, and training, and funds university activities. Generally, it is geographically distributed across an entire country, and follows procedures and regulations to protect the health and safety of its employees. It makes products for personal well-being (medicine, medical supplies, artificial organs, stents, prostheses) and products for safety at home, work, and elsewhere.

SPs are getting more attention by policy makers and citizens worldwide in recent years also because they constitute a pillar of the *circular economy* (Zhang et al. 2018) mentioned in Section 1.1 that aims at making most of all technical (metals, minerals, fossil resources) and biological (food, plants, water, etc.) resources, while minimizing waste and pollution (Ellen MacArthur Foundation 2013). Circular economy has numerous definitions, and mostly is described as a combination of reducing, reusing, recycling (Kirchherr et al. 2017), and (less frequently mentioned) recovering. Circular economy is based on three principles (Ellen MacArthur Foundation n.d.):

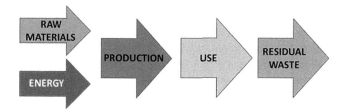

FIGURE 1.16 Schematic visualization of the model of linear economy.

- Design out waste and pollution, i.e. minimize them at the product design stage
- Keep products and materials in use through reusing, repairing, refurbishing, re-manufacturing, and recycling
- Regenerate natural systems, f.e. returning valuable nutrients to the soil and other ecosystems.

Because it is based on the above tenets, and employs sustainable materials and resources (like renewable energy), the circular economy differs from the current main economic models of *linear economy* (Figure 1.16) which is the least sustainable, and *economy with feedback loops* which involves recycling (van der Heijden et al. 2017).

In December 2015, the European Commission adopted an E.U. Action Plan for a circular economy, and in January 2018 published its communication *A European Strategy for Plastics in a Circular Economy* that represented a step towards making the European plastics system more resource-efficient, and driving the transformation from a linear to a circular economic system (European Commission 2018).

Sustainable feedstocks, such as dedicated crops f.e., are a key element of the *green chemistry*, a concept introduced in the mid-1990s that refers to the design of chemical products and processes that reduce or eliminate the use and generation of hazardous substances by achieving maximum conversion of reactants into a determined product, and minimum waste production through enhanced reaction design. Namely, green chemistry affirms that "a raw material or feedstock should be renewable rather than depleting whenever technically and economically practicable" (Pfaltzgraff and Clark 2014).

Achieving sustainable products has been already pursued to a certain degree in the last two decades by (Graedel and Allenby 1995; Lewis and Gertsakis 2001):

- Selecting materials that are abundant, nontoxic, natural, minimally processed
- Minimizing waste by choosing recyclable materials, and extending the product's life
- Minimizing the impact of disposal.

According to Faludi et al. (2017), SPs will be adopted in AM if they meet the following requirements: compatibility with AM processes, commercial availability (distributors) at competitive price and adequate scale to meet demand, print quality (look, surface finish, etc.), expected performance, and acceptance by industry. A market where printed SPs can be successful is biomedical, namely bone scaffolds, because SPs can: (a) be biocompatible in terms of cell attachment and proliferation as well as lack of toxicity and inflammatory reactions; (b) be biodegradable, enabling substitution of the scaffold material with bone, and possibly bioresorbable, meaning totally eliminated or bioassimilated by the hosting organism (Vert et al. 1992); (c) possess adequate load-bearing mechanical properties; (d) feature proper architecture in terms of porosity and pore sizes for cell penetration, transfer of nutrients and waste, and development of new blood vessels; (e) be sterile and bioactive; and (f) allow controlled deliverability of bioactive molecules or drugs (Hutmacher 2000; Ghassemi 2018; Brown et al. 2009; Hutmacher 2000; Porter et al.

2009; Ghassemi et al. 2018). Another incentive to use SPs in AM would be their ability to: (a) provide tunable physical-mechanical properties (such as strength and stiffness) whose values are adjusted within the volume of the same article during printing, or changed in response to an external stimulus such as temperature or pH; (b) generate multifunctional components, such as, f.e., those combining structural and electrical functions and fully printed in one session.

1.12 DEGRADABLE, BIODEGRADABLE, RECYCLABLE, COMPOSTABLE POLYMERS

The terms *biobased*, *biopolymer*, *degradable*, *biodegradable*, and *compostable* have multiple and overlapping meaning, and there is no consensus in technical literature and patents over their exact definition (Niaounakis 2015).

Biobased polymers differ from *biopolymers* that are defined as "macromolecules (including proteins, nucleic acids and polysaccharides) formed by living organisms" (IUPAC 2014). However, Niaounakis (2015) defines a *biopolymer* as a polymer derived from:

- Biodegradable renewable resources
- Non-biodegradable renewable resources
- Fossil-based biodegradable polymers.

We must clarify that a biopolymer is not a biomaterial. A most accepted definition of biomaterials is provided by the American National Institute of Health that identifies a *biomaterial* as "any substance or combination of substances, other than drugs, synthetic or natural in origin, which can be used for any period of time, which augments or replaces partially or totally any tissue, organ or function of the body, in order to maintain or improve the quality of life of the individual" (Bergmann and Stumpf 2013). Biomaterials comprise metals, ceramics, polymers, and their combinations. Selected sustainable and non-sustainable polymers employed as biomaterials are listed in Table 1.12, along with their applications as biomaterial. Some of the listed polymers, such

TABLE 1.12
Sustainable and Non-Sustainable Polymers Applied as Biomaterials

Polymer	Sustainable	Applications as Biomaterial
Acetal	No	Heart pacemaker
PA	Yes, No	Surgical sutures, tracheal tubes, gastrointestinal segments
PE	No	Blood vessels
PLA	Yes	Bone scaffolds, ureteral stents, drug delivery systems, bone screw and plates, surgical sutures and meshes
PGA	Yes	Surgical sutures, fiber meshes, scaffolds for tendon and cardiac tissue, pins[a]
PMMA	No	Bone cement
PTFE	No	Blood vessels
PUR	No	Blood vessels
PVC	No	Blood vessels, gastrointestinal segments, facial prostheses
Silicone	No	Finger joints
UHMWPE	No	Knee, hip, and shoulder joints

Source: Adapted from ASM International 2003. Reproduced with permission from AM International.
Note:
[a]www.sciencedirect.com: polyglycolide.

as polyglycolic acid or polyglycolide (PGA) (Göktürk et al. 2015), and PLA are sustainable, being formulated from natural and renewable resources; others such as PA, PVC and PTFE are derived from fossil feedstocks. PA is sustainable when formulated from corn, and not sustainable when derived from crude oil. Since PLA, PA, and PE listed in Table 1.12 are also feedstocks for AM and derived from plants (PLA from sugar cane and corn, PA from castor oil, and PE from corn), in principle SPs for AM may also be utilized as biomaterials.

Based on a definition of degradation (Whittington 1978), a polymer or plastic material is *degradable* if it undergoes a damaging change in its chemical structure, physical properties or appearance caused by exposure to heat, light, oxygen, or weathering. According to ASTM D5488-94d and CEN EN 13432, *biodegradable* means "capable of undergoing decomposition into carbon dioxide, methane, water, inorganic compounds, or biomass in which the predominant mechanism is the enzymatic action of micro-organisms." A *biodegradable plastic* is "a degradable plastic in which the degradation results from the action of naturally occurring microorganisms such as bacteria, fungi, and algae" (ASTM D6400-19). The rate of biodegradation is affected by various factors, such as the medium of biodegradation, its temperature and humidity, and chemical features of the degrading polymer, such as composition and MW. Biodegradable polymers are cellulose, chitin, starch, PHAs, PCL, and collagen. Not all biobased polymers are biodegradable. F.e. PET is biobased and not biodegradable. Some SPs are biodegradable, and other SPs do not biodegrade in landfill, because the amount of heat and/or oxygen is not adequate for these plastics to break down. Further information on biodegradable polymers is found in Averous and Pollet (2012). Polyvinyl alcohol (PVA) is a synthetic polymer that is water soluble.

The durability of plastic products is advantageous for consumers but harmful to the environment, because discarded plastic products take years to degrade under natural conditions, and can survive even hundreds of years as solid waste before they dissolve. In fact, according to the U.S. National Oceanic and Atmospheric Administration, plastic products ended up in the oceans as marine debris can take up to 600 years before they biodegrade (Table 1.13), depending on the polymers of which they are made (NOAA n.d.). Table 1.13 includes non-plastic products serving as a benchmark along with some most common plastic items. The time taken by plastic items before they degrade span from 20 years for PE grocery bags to 600 years for monofilament fishing lines made of PA, PVDF, PET, PE, and the durability of plastic articles is 1,000 to 10,000 times

TABLE 1.13
Duration of Solid Waste in the Marine Environment

Product	Biodegradation Time
Paper towel	4 weeks
Newspaper	6 weeks
Apple core	2 months
Cotton shirt	5 months
Plastic grocery bag (HDPE, LDPE, LLDPE)	20 years
Foam drinking cup (PS)	50 years
Aluminum can	200 years
6-pack beer plastic ring (HDPE)	400 years
Disposable diaper (PE + polyacrylate + PET or PP)	450 years
Plastic bottle (PET, HDPE, LDPE)	450 years
Monofilament fishing line (PA, PVDF, PET, PE)	600 years

Source: National Oceanic and Atmospheric Administration (NOAA).

superior to items made of paper and cotton. Switching to plastics buried in land, their degradation time is 500 years for HDPE bottles 5,000 years for PET bottles, PS insulating packaging and PVC pipes, and 10,000 years for HDPE pipes (Chamas et al. 2020).

Another consideration about biodegradation is that natural environments feature wide diversity in humidity, microorganisms, oxygen, sunlight, and temperature, and other variables, making it very difficult, if not impossible, to control and ensure the complete degradation of even potentially degradable plastics. Instead, in biomedical applications, degradable polymers such as PGA, PLA, and PCL have been successfully employed for decades, because the design of their molecular architecture results in a specific degradation rate, due to the rather controlled conditions (temperature, composition, pH, etc.) of the human body (Albertsson and Hakkarainen 2017). Biodegradable is different than compostable. A *compostable* plastic is "a plastic that undergoes degradation by biological processes during composting to yield CO_2, water, inorganic compounds, and biomass at a rate consistent with other known compostable materials and leave[s] no visible, distinguishable or toxic residue" (ASTM D6400-19). *Composting* is "a managed process that controls the biological decomposition and transformation of biodegradable materials into a humus-like substance called compost" and "results in the production of carbon dioxide, water, minerals" (ASTM D6400-19). The primary difference between biodegradable and compostable is the biomass or stabilized organic matter (*compost* or *humus*) left after composting that provides valuable nutrients to the soil, and is 100% able to reintegrate with the natural cycle. Composting is beneficial because it reduces the amount of solid waste, but not all SPs can be composted, and only SPs marked with a seal from the U.S. Composting Council can be composted in industrial compost facilities (Center for Sustainable Polymers 2020). An example of sustainable and compostable material is flax, which is added as a reinforcement in PLA filament formulated for ME (Makershop n.d.).

Recycling plastics is an effective way to reduce solid waste: it is cheaper in comparison to recycling other materials, and can be accomplished through common equipment such as extruders and injection molding presses, and in fact it has been applied to plastic waste for many years (Żenkiewicz et al. 2009). CEN/TR 16208 defines *recycling* as reprocessing the waste materials in a production process "for the original purpose or for other purposes including organic recycling but excluding energy recover." Organic recycling is the recycling of materials from animal and plants. ASTM D883-17 defines *recycled plastics* as "those plastics composed of postconsumer material or recovered material only, or both, that may or may not have been subjected to additional processing steps of the types used to make products such as recycled-regrind or reprocessed or reconstituted plastics" (ASTM D883-17).

GreenGate3D (USA) sells filaments and pellets for AM in various colors made of recycled post-industrial PETG that is high grade and clean, and converted in a feedstock featuring low shrinkage, high strength, and a glossy sheen. GreenGate3D plans to add filaments made of recycled HIPS, PA, and TPU (Hendrixson 2020). Żenkiewicz et al. (2009) demonstrated that recycling multiple times an extrusion-grade commercial PLA (by repeatedly extruding it) does not significantly impact its performance in terms of tensile strength and strain at break, impact strength, T_g, and so on. Sanchez Cruz et al. (2015) recycled PLA filaments for ME five times and measured a reduction of its MW that caused a slight reduction of its mechanical properties and a reduction in viscosity that benefited flowability and processability. The Dutch company Refil developed a process to recycle plastics into filament for ME that consists in drying and shredding plastics waste, and loading the resulting granules in an extruder where they are melted and finally extruded as a filament (Refil n.d.). Refil supplies the following filaments: ABS from car dashboards, HIPS from inside of refrigerators, PET from blue bottles, and PLA from yogurt cups. Oak Ridge National Laboratory (USA) is developing an effective process to take carbon and glass fibers isolated from polymer-matrix composite parts disposed of (such as wind turbine blades), and process them on very large printers (Donaldson 2020).

Not all SPs are recyclable, while some SPs can be recycled through chemical processes. Salt modified corn starch is an example of sustainable, recyclable, and compostable polymer that is similar in mechanical properties to fossil-derived plastics, and similar in mechanical strength to PP and PE (Abbott et al. 2012). PLA is a widespread SP, but is currently not recyclable, although Ingeo™ PLA can be identified in the stream of mixed waste plastics with accuracy higher than 90%, and sorted out

(NatureWorks 2021b). Using recyclable materials will obviously reduce the plastic solid waste, and environmental pollution. Every year, at least 8 million tonnes of plastics leak into the ocean, which is equivalent to dumping the contents of one garbage truck into the ocean per minute (Jambeck et al. 2015). Without any action taken, the contents will increase to two tracks per minute by 2030 and four tracks per minute by 2050 (Ocean Conservancy, McKinsey Center for Business and Environment 2015). According to the U.S. Environment Protection Agency, in 2015 the plastics contained in the U.S. municipal solid waste amounted to 34.5 million ton (13.1%) of the total municipal waste, of which 3.1 million recycled, 5.4 million combusted with energy recover, 26 million landfilled, and nothing composted. Within the municipal waste, plastics were the fourth largest material group, after paper and paperboard (25.9%), food (15.1%) and yard trimming (13.2%) (EPA 2019). In 2016, it was reported that 72% of plastic packaging consumed every year was not recovered at all: 40% was landfilled, and 32% escaped the collection system, or was illegally dumped (World Economic Forum 2016).

1.13   PROPERTIES AND MARKET OF BIOBASED POLYMERS

PLA, PHA, polyhydroxybutyrate (PHB), and PBS are biobased polymers utilized as feedstocks for AM materials, and below briefly introduced (Babu et al. 2013), since they are individually described in details in the following chapters.

PLA is an aliphatic polyester whose monomer, lactic acid, is the hydroxyl carboxylic acid obtained via bacterial fermentation from sugar and corn starch. Leading PLA producers are NatureWorks (USA), and Total Corbion (The Netherlands).

PHA indicates a family of polyesters obtained by bacterial fermentation and accumulated in granules inside bacterial cells. PHA is produced from several renewable waste feedstocks, such as derivatives of cellulose, fatty acids, municipal solid waste, organic waste, and vegetable oils. It is not only biodegradable but also biocompatible. Some major PHA suppliers are BioMatera (Canada), Danimer Scientific (USA), and Yield10 (USA).

PHB is the simplest PHA. PHB producers are TianAn Biologic Materials, and Tianjin GreenBio Materials, both in China.

PBS is an aliphatic highly crystalline TP polyester with properties similar to those of PET, and T_m higher than that of PLA. It is obtained by fermenting renewable resources into sugar that in turn produces the basic ingredients of PBS: succinic acid and 1,4-butanediol. It features high temperature resistance, and a wide processing window which makes it suited for many fabrication processes. PBS producers are PTTMCC that is a joint venture of Mitsubishi Chemical Corp. (Japan) and PTT Global Chemical (Thailand), Anqing Hexing Chemicals (China), and Zhejiang Hangzhou Xinfu Pharmaceutical (China). Commercial (3D Printlife DURA™) and experimental (Ou-Yang et al. 2018) filaments made from PBS are available for ME.

Currently, PLA is the most popular commercial biobased feedstock for AM, supplied as unfilled and composite filament, reinforced with sustainable materials such as bamboo, and cork, and non-sustainable materials such as aramid, carbon, glass, and metals.

Epoxy resins are hard, stiff, resistant to many solvents and alkali solutions, and possess excellent dielectric properties, low shrinkage, strong adhesion to metals and non-metals. They are widely consumed as adhesives, coatings, composites, and lamination materials, and their demand is growing from aerospace, construction, automotive, paints and coatings, and wind energy. Epoxy is also a feedstock for two families of AM processes, binder jetting and VP, and along with its curing agents can be formulated in biobased versions besides those from fossil sources. Reviews of biobased epoxy were authored by Auvergne et al. (2014) and Kumar et al. (2018). The list of natural raw materials to formulate biobased epoxy monomers is long, and includes soybean oil, starch, sugar, lignin, vanillin, cardanol, rosin (Wang et al. 2017c), to name only those most known. Some of those natural compounds should be considered by formulators of epoxy for AM. Here we touch on a few of them.

Soybean oil is a triglyceride molecule derived from unsaturated acids, and an inexpensive and abundant, renewable resource available worldwide.

Cardanol is an alkylphenolic compound, distilled from the liquid extracted from a soft hon-eycomb structure inside the cashew nutshell. This liquid contains phenols from which the epoxy monomer is formulated (Auvergne et al. 2014). Cardanol-based epoxy resins, curing agents, and modifiers are sold by Cardolite Corporation (USA), the world's largest producer of chemical products from cashew nutshell liquid.

Furan is a compound derived from cellulose and hemicelluloses, and whose chemical structure contains a five-member ring. Furan resins derive from furfural and furfuryl alcohol which is produced from agricultural by-products such as bagasse, corn cobs, oat, and hulls rice. Furan compounds have been investigated for over a century (van Es et al. 2014).

Rosin is a solid resin obtained from pines and some other coniferous plants by heating fresh liquid resin to vaporize the volatile liquid terpene components. It is semi-transparent and varies in color from yellow to black.

According to nova-Institute, a German research institute and consultancy with more than two decades of experience with biobased materials and products, in 2019 the worldwide actual production volume of biobased polymers was 3.8 million tonnes, equal to 1% of the production volume of fossil-based polymers and about 3% greater than in 2018 (Skoczinski et al. 2020).

Table 1.14 (Skoczinski et al. 2020) indicates the distribution of the worldwide installed production capacity of biobased polymers across formulations, and that the global installed capacity

TABLE 1.14
Worldwide Production Capacity of Biobased Polymers in Volume

Material	2019	2019	2024	2024
	Actual Data		**Forecast**	
	%	Mt[a]	%	Mt[a]
Epoxy	23.8	1.02	22.8	1.13
CA[b]	22.5	0.96	20.4	1.01
Starch	9.2	0.40	8.0	0.40
PUR	7.1	0.31	8.4	0.41
PLA	7.1	0.31	6.6	0.32
PA	5.9	0.25	6.6	0.32
PBAT[c]	5.9	0.25	5.1	0.25
PE	5.0	0.22	4.7	0.23
PET	4.6	0.20	2.9	0.14
PTT[g]	4.6	0.20	3.6	0.18
PBS and copolymers	2.5	0.11	2.2	0.11
PHA, APC[d], EPDM[e], PEF[f], PP	1.6	0.07	8.5	0.42
Total	100.0	4.28	100.0	4.93

Source: Values from Skoczinski, P., Chinthapalli, R., Carus, M. et al. 2019. Bio-based Building Blocks and Polymers – Global Capacities, Production and Trends 2019–2024. http://bio-based.eu/reports/ (accessed May 1, 2019). Reproduced with kind permission from nova-Institute.

Notes:

[a]Million tonnes

[b]Cellulose acetate

[c]Polybutylene adipate-co-terephthalate

[d]Aliphatic polycarbonates

[e]Ethylene propylene diene monomer rubber

[f]Polyethylene furanoate

[g]Polytrimethylene terephthalate

TABLE 1.15

Geographic Past (2019) and Future (2024) Distribution of Global Production Capacities of Biobased Polymers

Region	2019 %	2024 %
Asia	45	43
Europe	26	31
North America	18	16
South America	10	9
Australia/Oceania	1	1

Source: Skoczinski, P., Chinthapalli, R., Carus, M. et al. 2020. Bio-based Building Blocks and Polymers – Global Capacities, Production and Trends 2019–2024. Nova-Institute. http://bio-based.eu/reports/ (accessed January 16, 2021). Reproduced with kind permission from nova-Institute.

will grow from 4.3 million tonnes in 2019 to about 4.9 million tonnes in 2024, with epoxy and cellulose acetate (CA) the dominant biobased polymers in 2019 and 2024, compared to the rest of the list. PLA is expected to increase, growing from 0.31 to 0.32 million tonnes, but accounts for only 4% of the total.

According to nova-Institute, the global production capacities of biobased polymers by geographic region will change between 2019 and 2024, as illustrated in Table 1.15, with the European share growing, and shares by all other regions except Australia decreasing. The present dominance of Asia stems also from the large workforce, low labor cost, and vast amount of biomasses available in China and India, and is predicted to continue in the near future. The increment of the European share is chiefly related to polyethylene furanoate (PEF), PHA, PLA, starch blend, new production of PP, and increase in capacity of PE, PA, and polybutylene adipate-co-terephthalate (PBAT). Biobased polymers are currently present in almost all market segments and applications, and their applications vary depending on the polymer. The market segments of biobased polymers in 2019 and 2024 are listed in Table 1.16, and no significant difference in shares of market segments between the two periods is expected.

1.14 BIOPLASTICS

It is generally undisputed that our society needs alternatives for innumerable fossil-based products such as plastics (Koller 2017), and bioplastics represent such an alternative. The Plastics Industry Trade Association define *bioplastics* as a type of plastic partially or fully biobased and/or biodegradable. *Bioplastics* include (Figure 1.17):

- Partially or fully biobased biodegradable plastics, in the upper-right quadrant
- Partially or fully biobased non-biodegradable plastics, in the upper-left quadrant
- Any biodegradable plastic, even if not biobased, in lower-right quadrant.

Table 1.17 lists major bioplastics and shows that not all of them are biodegradable.

Clarinval and Halleux (2005) compared T_g, T_m, and tensile properties of sustainable and non-sustainable biodegradable polymers to those of three commodity fossil-based polymers featuring different performance and serving as baselines, namely: LDPE, a flexible material with medium

TABLE 1.16

Market Segments of Biobased Polymers in 2019 (Actuals) and 2024 (Forecast)

Market Segment	2019 %	2024 %
Textiles	22	20
Automotive, transportation	15	15
Consumer goods	13	15
Building, constructions	13	14
Packaging, flexible	13	13
Packaging, rigid	11	10
Electric, electronics	5	5
Agriculture, horticulture	3	3
Functional (adhesives, coatings, cosmetics)	3	3
Others	2	2

Source: Skoczinski, P., Chinthapalli, R., Carus, M. et al. 2020. Bio-based Building Blocks and Polymers – Global Capacities, Production and Trends 2019–2024. Nova-Institute. http://bio-based.eu/reports/ (accessed January 16, 2021). Reproduced with kind permission from nova-Institute.

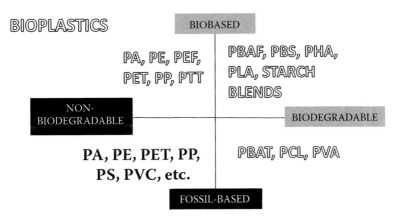

FIGURE 1.17 Bioplastics, shown in whitefonts. All abbreviations are in the chapter text, except PBAF standing for poly(butylene adipate-co-butylene 2,5 furandicarboxylate).

service temperature, PS, a relatively stiff polymer with medium service temperature, and PET, a stiff material with high service temperature (Table 1.18). The SPs are biodegradable and biobased, and labeled as BDB, and fare well against LDPE and PS in all five properties, with starch exceeding LDPE in tensile modulus and strength, and cellulose far outperforming PS in all tensile properties. However, drawbacks of BDB polymers vs. PS is that they have to be processed at significantly higher T_m, and their maximum service temperature is limited by a substantially lower T_g.

Bioplastics have been available since the 1950s. Initially they derived from traditional agricultural and renewable feedstocks, such as corn, soybeans, and sugar cane. The second-generation bioplastics was formulated from non-food renewable sources, such as castor beans, hemp, sawdust, switch grass, and the by-products of first-generation sources, such as husks and peels. For the third-generation bioplastics, new feedstocks have been developed, such as algae and modified methanobacteria (SPI 2016). Today, bioplastics are mostly composed of agro-based (mainly) and lignocellulosic feedstocks. The former ones are plants that are rich in carbohydrates, such as corn and sugar cane, and are the most

TABLE 1.17

Biodegradability of Main Bioplastics

Bioplastic	Biodegradable
PA	No
PBAT	Yes
PBS	Yes
PE	No
PET	No
PHA	Yes
PLA	Yes
PTT	No
Starch Blends	Yes

efficient and profitable option, since these plants produce the highest yields, and withstand pests and severe weather conditions. Namely, bio-PE produced by Braskem (Brazil) is derived from sugar cane, and waste fats and oils. Lignocellulosic feedstocks include plants that are not suitable for food or feed production, such as wood, agricultural or forestry wastes, and consist of cellulose, hemicellulose, and

TABLE 1.18

Properties of Biodegradable, and Non-Biodegradable Polymers. Top Group (LDPE to Starch): Flexible, Medium Service Temperature (ST). Middle Group (PS to PVA): Stiff, Medium ST. Bottom Group (PET to PGA): Stiff, High ST

Polymer	Type	T_g	T_m	Tensile Strength	Tensile Modulus	Elongation at Break
		°C	°C	MPa	MPa	%
LDPE (baseline)	NBDF	−100	98–115	8–20	300–500	100–1,000
PBAT	BDF	−30	110–115	34–40	N/A	500–800
PCL	BDF	−60	59–64	4–28	390–470	700–1,000
PTMAT	BDF	−30	108–110	22	100	700
Starch	BDB	NA	110–115	35–80	600–850	580–820
PS (baseline)	NBDF	70-115	100	34–50	2,300–3,300	1.2–2.5
Cellulose	BDB	N/A	NA	55–120	3,000–5,000	18–55
Cellulose acetate	BDB	N/A	115	10	460	13–15
PHA	BDB	−30 to 10	70–170	18–24	700–1,800	3–25
PHB	BDB	0	140–180	25–40	3500	5–8
PHB-PHV	BDB	0-30	100–190	25–30	600–1,000	7–15
PLA	BDB	40-70	130–180	48–53	3,500	30–240
PVA	BDF	58-85	180–230	28–46	380–530	N/A
PET (baseline)	NBDF	73-80	245–265	48–72	200–4,100	30–300
PEA	BDF	−20	125–190	25	180–220	400
PGA	BDF	35-40	225–230	890	7,000–8,400	30

Source: Adapted from Clarinval, A. M., Halleux, J. 2005. Classification of biodegradable polymers. In *Biodegradable polymers for industrial applications*, ed. R. Smith, 3–31. Boca Raton: CRC.

Legend: NBDF non-biodegradable fossil-based (non-sustainable), BDB biodegradable biobased (sustainable), BDF biodegradable fossil-based (non-sustainable).

lignin (Dotan 2014). Under development are technologies to turn waste materials from the mentioned feedstocks into feedstocks for bioplastics. According to a wide-ranging review of Brodin et al. (2017), lignocellulosic resources have a potential to replace plastics derived from fossil resources, and contribute to the growth of the biobased industry, but production of bioplastics from forestry biomass currently requires steps that are technologically difficult and not profitable. Therefore, innovative and economically viable technologies to produce bioplastics from renewable sources have to be developed.

The land allocated to grow the renewable feedstocks for producing bioplastics amounted to about 0.81 million hectares in 2018, corresponding to 0.016% of the global agricultural area of 4.9 billion hectares, 97% of which were used for pasture, feed, and food. Although the bioplastics market is predicted to grow from 2019 to 2024, the share of land use for bioplastics will exceed 1 million hectares equal to 0.02% of the global agricultural area and hence it will not hinder farming for food and feed.

European Bioplastics, the association representing the interests of the bioplastics industry in Europe, released market data for 2020 (European Bioplastics n.d.): the global production of bioplastics was 2.1 million tonnes, comprising 41.9% of biobased/non-biodegradable grades (PE, PA, PET, polytrimethylene terephthalate (PTT), others, and PP in decreasing order), and 58.1% of biodegradable grades (starch blends, PLA, PBAT, PBS, others, and PHA in decreasing order). The main markets were packaging, textiles, consumer goods, agriculture/horticulture, and automotive/transport in decreasing order. Some materials are preferred for specific applications: f.e. PET and PTT are chosen for rigid packaging (like the Coca-Cola's PlantBottle™), and textile respectively. Consumer goods for expensive interior components of costly cars may represent a profitable and technically suitable opportunity for AM, because these components are made not in large numbers, and sold with a substantial mark up. If that is the case, SPs for AM have to compete with the following bioplastics:

- Automotive/transport: PA, PBS, PET, PTT
- Consumer goods: PA, PBS, PE, PLA, starch blends.

The distribution of bioplastics is dominated by the Asia region with 45% of the total volume, leading Europe (25%), North America (18%), and South America (10%) (European Bioplastics 2020).

In Figure 1.18 the prices in 2009 of SPs (PLA, starch based, cellulose, poly(hydroxybutyrate–co-hydroxyvalerate or PHBV, PHB, PBT, and PBS) and commodity fossil-based polymers (HDPE, PP, PS, PVC, poly(vinyl alcohol) or PVOH, PC, PET, ethylene vinyl acetate or EVA) are plotted. Although the values are not recent, they offer an indicative comparison between the two sets of polymers, and show that SPs cost from about 2 to 5 times the price of the most inexpensive fossil-based counterparts, and, not surprisingly given its leading share of use among SPs, PLA was the least expensive among sold SPs, and the closest in price to the non-sustainable commodity polymers. This confirms the need, mentioned earlier, for production routes generating to more economical SPs. For those involved in R&D in SPs for AM, Table 1.19 lists the price of small quantities of bioplastics sold online in granules.

Table 1.20 lists the bioplastics commercially available or in development in 2016, and the relative reference to their use in AM, namely paper author and date (in case of journal papers), supplier, supplier's website, etc.

The near future of bioplastics looks favorable: their global capacity is expected to steadily and slowly expand from 2.01 billion tonnes in 2018 to 2.43 billion tonnes in 2024 (European Bioplastics 2020).

Spierling et al. (2018) reviewed the assessments of environmental, social, and economic impact caused by biobased plastics, and concluded that they could potentially save 241 to 316 million tonnes of CO_2 equivalent annually, assuming that biobased plastics can substitute fossil-based plastics, and will approximately substitute 2/3 of the global plastics demand, which will not be achieved soon in our opinion.

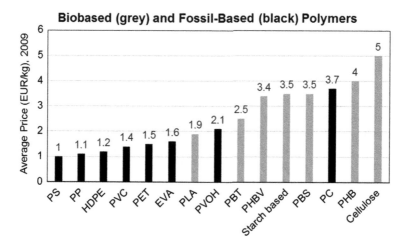

FIGURE 1.18   Average price of biobased (grey bars) and fossil-based (black bars) polymers in 2009.

Source: Sin, L. T., Rahmat, A. R., Rahman, W. A., 2012. Overview of poly(lactic acid) in *Handbook of Biopolymers and Biodegradable Plastics*. ed. S. Ebnesajjad, 11–54. Chadds Ford: William Andrew. Adapted and reproduced with permission from Elsevier.

Another benefit of bioplastics is that they critically contribute to the U.S. industry of biobased products from American farms and forests that posted the following achievements in 2013: USD 369 billion revenues, 4 million jobs, 300 million gallons of crude oil replaced by biobased products, and reduction of greenhouse emissions equivalent to 200,000 cars taken off the road (Golden et al. 2015).

Lists of producers of biobased building blocks and polymers, and relative semi-finished products are publicly available. Particularly, in its February 2019 report, nova-Institute (Chinthapalli et al. 2019) listed 19 biobased building blocks, 18 biobased polymers, and the profile of 173 producers of biobased building blocks and polymers worldwide. Moreover, a list of 87 major suppliers of raw materials and semi-finished products for bioplastics was included in a 2015 report on bioplastics and bioelastomers by the Istituto Italiano di Tecnologia (ITT) (Fondazione Istituto Italiano di Tecnologia 2015). Most companies on the two previous lists overlap, such as: Arkema, BASF, Biome Bioplastics, Biomer, Braskem, Corbion, DuPont, FKuR, Innovia Films, Kaneka, Mitsubishi, NatureWorks, Novamont, Purac, Solvay, Tianjin GreenBio Materials, Yunan Fuji Bio-Material

TABLE 1.19

Price of Bioplastics in Granule Form and Relative Quantity Sold

Bioplastic	Price (July 2020)	Quantity Sold	Price (July 2020)	Quantity Sold
	USD/kg	kg	USD/kg	kg
Cellulose acetate	3.13–3.80	25	3.50–7	1,000
Corn starch	1.70–2.20	250	1.59–2.47	1,134
Ethanol	0.50–1.50	1	0.8–1.50	1,000
Lactic acid	N/A	N/A	1–1.20	1,000
PBAT	5.70–6	100	2.7–3.2	1,000
PBS	5.29–5.66	200	3–3.80	1,000
PHA	5.3–5.4	100	1–3	1,000
PLA	3.45–3.95	200	0.7–0.98	1,000

Source: www.alibaba.com (accessed July 14, 2020).

TABLE 1.20

Bioplastics Commercially Available or in Development in 2016

Polymer	Biobased	Biodegradable	Reference on the use of the bioplastic in AM
PA 10	Partially	No	N/A
PA 1010	Yes	No	N/A
PA 11	Partially	No	Bai et al. (2017). Arkema (Rilsan®)
PA 410	Partially	No	N/A
PA 610	Partially	No	N/A
PBAS	In development	IC, SB	N/A
PBAT	Partially	IC, HC, SB	Singamneni et al. (2018)
PBS	Yes	IC, SB, MB	Patent CN10511699A. Ou-Yang et al. (2018)
PE	Yes	No	https://filaments.ca/
PEET	Partially	No	N/A
PEF	Yes	No	N/A
PES	Partially and fully bio-based. In development	No	Topolkaraev et al. (2014)
PET	Partially	No	MatterHackers. Verbatim. MadeSolid (www.matterhackers.com). www.innofil3d.com. https://designbox3d.com/products/smartfil-petg
PHA	Yes	IC, HC, SB, MB, AD	https://colorfabb.com
PLA	Yes	IC	https://www.3dsystems.com/materials/cubepro-pla-plastic/tech-specs, https://ultimaker.com/en/products/materials/pp, https://www.3dfuel.com/pages/pla-comparison Stratasys
PP	In development	No	https://www.simplify3d.com/support/materials-guide/polypropylene/ Gizmodorks.
PPA	Partially	No	https://www.3dxtech.com/ppa-filament/
PTT	Partially	No	Yu et al. (2017)
TPC-ET	Partially	No	N/A
TPS	Yes	IC, HC, SB, MB, AD	Kuo et al. (2016). https://www.filaments.directory/en/superpowers/starch
TPU	Partially	No	https://airwolf3d.com/shop/platinum-series-abs-filament-2–88mm-2-2lbs-123/, https://tractus3d.com/3d-printing-materials/tpu-material/

Source: Adapted from "Plastic Market Watch – Bioplastics," Summer 2016, Issue VI, copyright 2016 The Society of Plastics Industries SPI 2016. The Plastics Industry Trade Association. Reproduced with permission from The Society of Plastics Industries.

Legend: AD anaerobically digestible, HC home compostable, IC industrially compostable, MB marine degradable, SB soil degradable.

Technology, and Zhejiang Hisun Biomaterials. Although the above lists are extensive, we found some companies not listed, such as Matrìca and Green Dot Bioplastics. In the list by ITT, most companies are located in Europe (54), followed by USA (14), Japan (6), and China (5), confirming the pro-environment policies and public opinion in many European countries, such as Germany, Switzerland, and The Netherlands, and their preference for eco-friendly materials and products.

Formulators of SPs for AM may appreciate Table 1.21 that provides a sample of biobased building blocks and polymers, and name and country of their producers. The fact that some companies have been established as joint ventures may reflect that the business of producing

TABLE 1.21

Producers and Joint Ventures (JV) of Biobased Building Blocks and Polymers

Biobased Materials	Company	Country
Butadiene	Lanzatech	USA
BDO	Genomatica	USA
Caprolactam	Genomatica	USA
Dicarboxylic acids	Matrica	JV of Novamont (IT) and Versalis (IT)
DDDA	Verdezyne	USA
Enzymes	Novozymes	Denmark
Esters	Matrica	JV of Novamont (IT) and Versalis (IT)
FDCA	Synvina	Owned by Avantium (NL)
Glycerol ECH for bio-epoxy	Solvay	Belgium
PA 10	Arkema	France
PA 10	Dupont and Tate and Lyle	USA, UK
PA 11	Arkema	France
PHA	Yield10 Bioscience	USA
PLA	NatureWorks	JV of Cargill (USA) and PHH (Thailand)
PLA	Total Corbion	JV of Total (FR) and Corbion (NL)
PBS	PTT MCC Biochem	JV of Mitsubishi Chemical Corp. (Japan), PTT Global Chemical (Thailand)
PEF	Synvina	Owned by Avantium (NL)
PTT	Dupont and Tate and Lyle	USA, UK
PDO	Dupont and Tate and Lyle	USA, UK
Starch based resins	Novamont	Italy
Succinic acid	Reverdia	JV of DSM (Holland) and Roquette (France)
Succinic acid	Succinity GmbH	JV of BASF (Germany) and Corbion (NL)
Succinic acid	Bioamber	Canada

Source: Data from https://bioplasticsnews.com/top-bioplastics-producers/ (accessed January 16, 2021).

bioplastics is not well established, and hence some companies prefer sharing initial capital and operational costs with a partner, instead of assuming all the business risks alone.

Two commercial products composed of bioplastics are pictured in Figure 1.19: on the left, a ski boot using bio-based Rilsan® (PA 11); on the right, a compostable grocery bag from biodegradable and compostable MATER-BI bioplastic by Novamont (Italy). Coca-Cola also uses the PlantBottle™ that is the first ever fully recyclable PET plastic beverage bottle made partially (30%) from plants, namely sugar cane (Coca-Cola 2018). The PlantBottle™ packaging is found in several Coca-Cola's brands (Dasani, Minute Maid, Smartwater, etc.), and it is estimated that over 40 million bottles have been sold in more than 40 countries, since PlantBottle™ was introduced in 2009 (SPI 2016). If standard PET is not recycled and left in sea water, it takes about fifteen years before it starts degrading (Ioakeimidis et al. 2016).

It is important to stress that not all biobased polymers are 100% derived from biomass. In fact, Table 1.22 contains a list of biobased structural polymers, their biomass content and number of producing companies in 2016 (Aeschelmann et al. 2017). This content varies from 10% to 100%, being 100% for PA, PE, PLA, and PUR. From the number of producers for each material it appears that in 2016 the polymers most successful commercially were PLA, PHA, CA, and starch blends in decreasing order. As elucidated in Chapter 7, cellulose has been and is being studied in its various forms as an AM feedstock for various applications. The high number of PLA producers may partly derive from the demand for AM PLA.

FIGURE 1.19 Commercial products made of biobased plastics: left, ski boot using bio-based Rilsan® (PA 11); right, compostable grocery bag. Reproduced with permission from Arkema S.A. (left), and Novamont S.p.A. (right).

TABLE 1.22

Information on Biobased Structural Polymers

Biobased Structural Polymer	Average Biomass Content	Suppliers in 2016
	Weight %	Number
CA	50	16
Epoxies	30	N/A
EPDM	50–70	1
PA	40–100	9
PBAT	Up to 50	4
PBS	Up to 80	8
PE	100	1
PET	20	4
PHA	100	19
PLA	100	26
PTT	27	2
PU	10–100	N/A
Starch Blends	25–10	16

Source: Aeschelmann, F., Carus, M., Baltus, W., et al. 2017. Bio-based building blocks and polymers. http://www.bio-based.eu/reports/ (accessed April 25, 2019). Reproduced with permission from nova-Institute.

Vegetable or *plant oils* are oils extracted mainly from the seeds of the following plants (Islam et al. 2014): castor, coconut, corn, cottonseed, linseed, nahar seed, olive, palm, rapeseed, safflower, sesame, soybean, and sunflower. The main constituents of plant oils are triglycerides that are esters 95 wt% composed of three molecules of saturated and/or unsaturated fatty acids that can be identical or different, and are attached to a glycerol backbone. Triglycerides are also in the people's blood, and skin oils. Vegetable oils are ingredients in formulations of paints, coatings, composites, adhesives, varnishes, and biomedical products such as wound healing devices, surgical sealants and glues, pharmacological patches, and scaffolds for tissue engineering (Islam et al. 2014; Lligadas et al. 2013).

Vegetable oil-based polymers are attractive as bioplastics because of their sustainability, biodegradability, low price, low toxicity towards humans and environment, and universal availability. They are: (a) precursors for monomer chains selected to synthesize polymers widely utilized and mentioned later; (b) suitable for producing monomers similar in structure and properties to crude oil-based monomers; (c) apt for synthesis of hydrophobic polymers that complement biobased hydrophilic polymers such as carbohydrates and protein. Polyester, polyether, polyolefin, and PU are the four most important families of vegetable oil-based polymers, and possess excellent biocompatibility and distinctive properties, and therefore are possible candidates as bioplastics, also because they comprise crosslinked TSs and linear TPs, and hence can meet quite a range of performance requirements and end-use applications (Miao et al. 2014). Other fossil-based plastics targeted to be replaced by sustainable monomers from plant oils are (Islam et al. 2014): alkyds (available in three forms: high-solid-content, liquid-crystalline, waterborne), epoxies, interpenetrating polymer networks, PA, poly(ester amid) (PEA), PUR (organic-solvent-based and water-based grades), and vinyl polymers. Interestingly, from the above list, epoxy, PA, polyester, TPU, and vinyl polymers are already feedstocks for the following experimental and commercial AM polymers:

- PET: Innofil3D's filaments for ME manufactured in two grades (BASF n.d.): (a) EPR InnoPET, a premium, food approved PET, present in food and beverage containers and bottles, and 100% recyclable; (b) rPET made from recycled PET
- PA: biobased PA 11 is sold by Arkema (Rilsan® Invent), and EOS (Shapeways 2019) as a powder for LS process.
- Epoxies are commercially available for two AM families: VP and material jetting.
- TPU is supplied by Airwolf 3D (Airwolf 3D 2019), and Tractus3D (Tractus3D n.d.)
- Vinyl: filament for AM produced by Chemson Pacific (Australia).

Given their advantages mentioned above, polymers based on vegetable oils are also considered attractive sustainable sources for AM polymers, and their performance/price ratio will be critical for their commercial success. In fact, f.e., soybean oil was converted in epoxidized acrylate, a photocurable, biocompatible, liquid resin processed through SL into highly biocompatible porous scaffolds supporting growth of stem cells (Miao et al. 2016). Advancements in processes and technologies for the synthesis of polymers from vegetable oils are summarized by Islam et al. (2014). Raquez et al. (2010) published an extensive critical review of TSs (epoxy, phenolic, polyester, and PU) derived from renewable resources comprising the same biomasses currently selected to formulate AM materials, such as castor oil, soybean, starch (Laird 2015), and wood. Among these TSs, specific SPs may be chosen as candidates for AM processes based on VP, such as SL and CLIP™ (Carbon® Inc. 2021b), if their performance/price ratio is competitive with that of current feedstocks.

In conclusion, based on documented results, it is reasonable to expect that any fossil-based monomer can be derived from a sustainable resource at least in laboratory, which, however, does not guarantee success in producing the same monomers on an industrial scale (Schneiderman and Hillmyer 2017).

An interesting biobased plastic was formulated by Bayer et al. (2014), who directly converted inedible wastes from edible vegetable and cereal into plastics with a wide range of mechanical properties that are suitable for functional applications. Namely, they took wastes of parsley and spinach stems, rice hulls, and cocoa pod husks, and, through digestion in trifluoroacetic acid, casting, and evaporation, synthetized amorphous sustainable, biodegradable, cellulose-based plastics, whose mechanical properties are equivalent to those of crude oil-based synthetic polymers. This research is twofold remarkable because it used sustainable resources and reduced solid waste.

1.15 NEAR FUTURE OF SUSTAINABLE POLYMERS

There is consensus among experts in predicting that the industry of SPs will grow in the near future in terms of output quantity and innovative materials. The market growth of SPs depends on several factors: stronger consumers' demand stemming from awareness of their environmental benefits, and esthetic and functional appeal, government policies and legislations favorable to SPs, feedstocks' price competitive with that of fossil-derived feedstocks, ability to meet the expected performance at a competitive cost, not competing for resources with food crops in specific locations, and being processed through routes compatible with existing industrial infrastructures and supply chains to produce and sell monomers and polymers (Zhu et al. 2016). In order for biobased polymers to be transformed into additional commercial products beyond those currently prevalent in packaging, construction, films, tableware, and fibers, biobased polymers have to demonstrate that they are equivalent or better than fossil-based polymers in terms of physical, mechanical, thermal and long-term properties, and display characteristics adequate to target high-value products, instead of competing with fossil-based polymers for low-value products. It is reasonable to expect that more pro-environment policies will be implemented at national and local level, driving up the use of SPs, and providing an economic incentive to producers of SPs. This may be countered by a low price of crude oil and natural gas that makes the price of SPs typically not competitive with the price of fossil-based polymers.

Technologies to produce SPs from waste of wood and paper, food industries, stems and leaves, and solid municipal waste streams will be further introduced or improved, in order to fabricate products out of SPs more efficiently and cost-effectively. Another type of waste is *electronic and electric waste* or *e-waste*, a term comprising "items of all types of electrical and electronic equipment (EEE) and its parts that have been discarded by the owner as waste without the intention of re-use." (StEP 2019). Some examples of e-waste are: refrigerators, air conditioners, heat pumps, screens and monitors on televisions, laptops, tablets, lamps, large equipment (washing machines, electric stoves, etc.), small equipment (microwaves, electric shavers, scales, calculators, radio sets, electrical and electronic toys, etc.), small IT and telecommunication equipment (computers, printers, mobile phones, etc.). In 2017 e-waste reached 44.7 million tonnes annually, but only 20% of it was collected and recycled, and the rest was likely dumped, traded or recycled under inferior conditions (StEP 2019). Gaikwad et al. (2018) transformed PC collected from computer printers, and converted it into AM filaments that exhibited tensile strength and modulus 83% and 86% of those of virgin ABS filament, respectively.

In 2013, producing biobased monomers and polymers economically viable was considered dependent on developing: logistics for biomass feedstocks, new high-yield manufacturing routes, new microbial strains/enzymes, and efficient downstream processing methods to recover biobased products (Babu et al. 2013).

The market of biobased chemicals and polymers is predicted to grow in the period 2018–2026, mainly because of the preference of the consumers for eco-friendly products, growing demand of ethanol and methanol by the pharmaceutical industry, and companies' adoption of cost-effective green materials and products. Europe will dominate the market of biobased chemicals, followed by North America, while Asia Pacific will experience enormous growth, also due to its vast amount of plant feedstocks (Coherent Market Insights 2021).

In 2018, the bioplastics market reached USD 6 billion globally and USD 1.5 billion in North America. In 2020, it was predicted that by 2026 the global and North America markets would raise to USD 19.9 billion and USD 4.3 billion, respectively, driven by growing demand for eco-friendly packaging materials (Fortune Business Insight n.d.).

The diffusion of SPs will benefit from the support for the environment by plastics producers. PlasticsEurope, the Association of Plastics Manufacturers in Europe, has voluntarily established *Plastics 2030*, a series of targets and initiatives aimed at 2030 that are focused on preventing leakage of plastics into the environment, improving resource efficiency, and increasing recycling and reuse rates (PlasticsEurope 2019). One activity of Plastics 2030 is to "accelerate research of alternative feedstocks," which obviously include SPs. Similar environmental concerns guide the World Plastics Council, an organization whose members are executives from some of the world's largest plastics producers, which states that it "will work to promote the ethic of sustainability and the responsible use of plastics" (World Plastics Council n.d.).

Hong and Chen (2019) recognized the following emerging families of SPs:

- *Chemically recyclable polymers.* They are completely recyclable chemically because they are fully depolymerized into their monomers, and repolymerized into virgin-quality polymers, with this process in theory repeatable endlessly (Tang and Chen 2019). Examples of such polymers are polymers containing the inherently recyclable γ-butyrolactone ring: they perform like common plastics and can be chemically recycled repeatedly (Hong and Chen 2016; Zhu et al. 2018), and thus implement a circular economy.
- *Upcycling* or *repurposing post-consumer polymers.* They are waste made of post-consumer polymers and mixed polymers that are upcycled and repurposed into useful materials through innovative and economical routes. A recent example is an ethylene-propylene copolymer developed to make PE and PP compatible with each other and mixed together to be recycled into equal- or higher-value materials (Eagan et al. 2017).
- *Reprocessable* and/or *recyclable crosslinked polymers.* TSs are strong, stiff, and durable but, being crosslinked, they cannot be melted and recycled like TPs. However, a new family of TSs has been formulated by condensation of paraformaldehyde with bisanilines ($C_{24}H_{28}N_2$): these TSs are stronger than bone, available also as self-healing gels, and completely recyclable back to their starting material, employing acid digestion to recover the bisaniline monomers (Garcia et al. 2014).

The above families of SPs may successfully generate polymers for AM, if the resulting polymers are processable in a cost-effective way and with adequate and consistent quality, and possess functional and cost-competitive properties.

Switching to the near future of specific feedstock for SPs, carbohydrates and lignin are sustainable and very abundant feedstocks for SPs. High-value products can be prepared from carbohydrate-based polymers by exploiting the rigid ring structures, and the non-covalent interactions and stereoregularity of carbohydrate (Haba et al. 2005; Mikami et al. 2013; Haba 2005).

Raising the yield and reducing the cost of monomers from lignocellulosic biomass will drive the conversion of biomass into SPs. Optimizing the yield and cost of the monomers for SPs will boost the use of lignocellulosic biomass (Zhu et al. 2016). New methods to selectively transform lignin into monomers are necessary, and can leverage catalysis (Wang and Rinaldi 2013; Rahimi et al. 2014), and application of lignin-derived ferulic acid to formulate polyesters and resins for practical applications (Maiorana 2016; Mialon et al. 2010; Maiorana et al. 2016).

The future of PLA's market looks auspicious with opportunities in agriculture, bio-medical, electronics, packaging, and textile industries. The global PLA market is expected to reach USD 3.7 billion by 2024 with a CAGR of 13.8% from 2019 to 2024. Improved mechanical properties, heat resistance, and production of PLA from second-generation feedstock will drive PLA's diffusion (Business Wire 2019). One route to enhancePLA performance is to combine it with PGA via copolymerization, physical blending, and (although not suitable for filaments and powder for AM) multilayer lamination, but the production cost of PGA has to be reduced to make the resulting PLA-PGA composite more competitive with other polymers (Jem et al. 2020).

PHA is a promising biomaterial because of its physicochemical properties, biodegradability, and biocompatibility, especially as a candidate to replace traditional plastics in biomedical, packaging, and other market sectors (Zinn et al. 2001; Chen 2009; Keshavarz and Roy 2010; Tan et al. 2014; Zinn 2001; Chen 2009). High-MW PHA is synthesized and stored in the cell cytoplasm as water-insoluble inclusions by various microorganisms (Sudesh et al. 2000), and currently is industrially produced from natural cultures of isolated microorganisms, and engineered strains on pure substrates. Wide production, commercialization, and application of PHA as a biodegradable alternative to fossil-based plastics is still limited due to high production cost (mainly caused by expensive fermentation of carbon sources), making its price twice or more that of PP and PE. The increasing availability of raw renewable materials, and raising demand and use of biodegradable polymers for biomedical, packaging, and food applications, along with "green" procurement policies are expected to benefit PHA's market growth. In 2019 the global PHA market reached USD 57 million, and is expected to reach USD 98 million by 2024 at a CAGR of 11.2%, with Europe predicted to account for the largest market share in value during the forecast period. Key companies in the PHA market are Kaneka (Japan), Danimer Scientific (USA), Shenzhen Ecomann Biotechnology (China), Newlight Technologies (USA), and TianAn Biological Materials Co. Ltd. (China) (MarketsandMarkets™ 2020b). Another information source was more optimistic about growth of PHA, and in 2019 expected that its production capacity would more than triple in the next five years (European Bioplastics 2020). A way to diminish the production cost of PHA and increase its utilization would be leveraging mixed culture biotechnology, which employs mixed microbial consortia under non-sterile conditions and ecological selection principles, where microorganisms able to accumulate PHA are selected based on the operational conditions imposed on the biological system (Kourmentza et al. 2017). Another way to cut PHA's cost is to increase its derivation from the many waste sources being investigated: domestic wastewater, food waste, molasses, olive oil mill effluents, palm oil mill effluents, tomato cannery water, lignocellulosic biomass, coffee waste, starch, biodiesel industry waste, used cooking oil, pea-shells, paper mill wastewater, bio-oil from the fast-pyrolysis of chicken beds, and cheese whey (Amaro et al. 2019).

The global demand for biobased PUR is forecast to swell between 2019 and 2025 at higher rate than that recorded in the previous five years, propelled by consumption in major emerging markets, and featuring increasing domestic and export-oriented revenues for key players (Research and Markets 2019).

Moving to PBS, on one hand some factors boosts its consumption, such as using its blends as feedstocks for ME filaments and the demand for heat resistance plastics and biodegradable polymers, but on the other hand its complex synthesis, and large emission of greenhouse gas associated with its production are restrictive factors. The global market of PBS was estimated to be USD 97 million in 2018 and expected to grow at a CAGR of 10.5% during the period 2018–2023 (Orion Market Research 2019).

Overall, the growth of the market of bioplastics has the potential to trigger a significant increase in job opportunities. According to European Bioplastics, in 2013 the bioplastics industry in Europe accounted for about 23,000 jobs that could jump to 300,000 by 2030 with "the right framework conditions in place."

Shifting to SPs for AM, their near future market may be promising for the following reasons: the growth of the worldwide market of AM itself; small size of investments required for developing and launching new feedstocks for AM compared to investments for the same goal in conventional fabrication processes; many customers of personal printers prefer green products, and like experimenting with new feedstocks; current policies and legislations at national and local level, international agreements, and public opinion delineate a pro-environment trend; growing production of SPs by world leading chemical corporations, deriving from the availability of additional cost-effective production routes.

FURTHER READINGS

Additive Manufacturing:

Gibson, I., Rosen, D., Stucker, B. 2015. *Additive Manufacturing Technologies*. New York: Springer, 2nd edition.
Wohlers Associates (USA) https://wohlersassociates.com/

Bioplastics:

nova-Institut GmbH (Germany) http://nova-institute.eu/
Kabasci, S. ed. 2014. *Bio-Based Plastics – Materials and Applications.* London: Wiley.

Polymers:

Harper, C. ed. 1996. *Handbook of Plastics, Elastomers, and Composites.* New York: McGraw-Hill. 3rd edition.
McCrum, N. G., Buckley, C. P., Bucknall, C. B. 1997. *Principles of Polymer Engineering,* Oxford: Oxford University Press. 2nd edition.
Osswald, T. A. and G. Menges 2012. *Material Science of Polymers for Engineers.* Munich: Hanser. 3rd edition.
Osswald, T. A., Bauer, E. 2019. *Plastics Handbook: The Resource for Plastics Engineers.* Munich: Hanser. 5th edition.
Whittington, L. R. 1978. *Whittington's Dictionary of Plastics.* Westport: Technomic.
Ligon, S. C., Liska, R., Stampfl, J., Gurr, M., Mülhaupt, R. 2017. Polymers for 3D printing and customized additive manufacturing. *Chem. Rev.* 117:10212–10290.
Bourell, D., Kruth, J. P., Leu, M., et al. 2017. Materials for additive manufacturing CIRP Annals. *Manuf. Technol.* 66:659–681.
Chawla, K. K. 1998. *Composite Materials – Science and Engineering.* New York: Springer. 2nd edition.
Kelly, A. ed. 1989. *Concise Encyclopedia of Composite Materials.* Oxford: Pergamon Press. 1st edition.

REFERENCES

3DXTech® 2021. https://www.3dxtech.com/ (accessed January 16, 2021).
Abbott, A. P., Ballantyne, A. D., Conde, J. P., Ryder, K. S., Wise, W. R. 2012. Salt modified starch: sustainable, recyclable plastics. *Green Chem.* 14:1302–1307.
Aeschelmann, F., Carus, M., Baltus, W., et al. 2017. *Bio-Based Building Blocks and Polymers.* https://www.bio-based.eu/reports/ (accessed April 26, 2019).
Airwolf 3D. 2019. *WOLFBEND TPU Flexible Filament.* https://airwolf3d.com/shop/platinum-series-abs-filament-2–88mm-2-2lbs-123/ (accessed August 31, 2019).
Albertsson, A.-C., Hakkarainen, M. 2017. Designed to degrade. *Science* 358 (6365): 872–873.
Amaro, T. M. M. M., Rosa, D., Comi, G., Iacumin, L. 2019. Prospects for the use of whey for polyhydroxyalkanoate (PHA). *Prod. Front. Microbiol.* doi:10.3389/fmicb.2019.00992.
Amass, W., Amass, A., Tighe, B. 1998. A review of biodegradable polymers: uses, current developments in the synthesis and characterization of biodegradable polyesters, blends of biodegradable polymers and recent advances in biodegradation studies. *Polym. Int.* 47:89–144.
Ambulo, C. P., Burroughs, J. J., Boothby, J. M., et al. 2017. Four-dimensional printing of liquid crystal elastomers. *ACS Appl. Mater. Interf.* 9 (42): 37332–37339. doi:10.1021/acsami.7b11851.
American Chemistry Council. 2019. *Plastics – Major Market.* https://plastics.americanchemistry.com/Automotive/ (accessed April 26, 2019).
Anandjiwala, R. D., John, M. J., Wambua, P. et al. 2008. Bio-based structural composite materials for aerospace applications. *2nd South African International Aerospace Symposium,* Cape Town, South Africa, 14–16 September 2008. http://researchspace.csir.co.za/dspace/handle/10204/2645.
Anon. 2018. WCCO belting launches patented soybean oil rubber belt technology. *Press Release.* https://www.wccobelt.com/latest-news/wcco-belting-launches-patented-soybean-oil-rubber-belt-technology-press-release/ (accessed April 26, 2019).
Anrady, A., Neal, M. 2009. Applications and societal benefits of plastics *Phil. Trans. R. Soc. B* 364(1526). doi:10.1098/rstb.2008.0304.
Arnold, K. 2020. Consortium aims to print COVID-19 test swabs at rate of millions per week. *Addit. Manuf.* 9 (3): 8. Cincinnati: Gardner Business Media Inc.
Arockiam, N. J., Jawaid, M., Saba, N. 2018. Sustainable bio composites for aircraft components. In *Sustainable Composites for Aerospace Applications,* ed. M. Jawaid, M. Thariq, 109–123. Oxford: Elsevier. doi:10.1016/B978-0-08-102131-6.00006-2.
ASM International. 2003. Overview of biomaterials and their use in medical devices. In *Handbook of Materials for Medical Devices,* ed. J. R. Davis, 1–11. https://www.asminternational.org/documents/10192/1849770/06974G_Chapter_1.pdf (accessed May 31, 2018).

ASTM D883-17. *Standard Terminology Relating to Plastics. ASTM D1566-19 Standard Terminology Relating to Rubber.* West Conshohocken, PA USA: ASTM International. www.astm.org.

ASTM D5488-94d. *Environmental Labeling of Packaging Materials and Packages ASTM International.* West Conshohocken, PA USA: ASTM. www.astm.org.

ASTM D6400-19. *Standard Specification for Labeling of Plastics Designed to be Aerobically Composted in Municipal or Industrial Facilities.* West Conshohocken, PA USA: ASTM International. www.astm.org.

ASTM D6866-18. *Standard Test Methods for Determining the Biobased Content of Solid, Liquid, and Gaseous Samples Using Radiocarbon Analysis.* West Conshohocken, PA USA: ASTM International. www.astm.org.

Atkins, W. P. 1987. *Molecules.* Scientific American Library (Book 21). New York: W.H. Freeman.

Auvergne, R., Caillol, S., David, G., Boutevin, B., Pascault, J. P. 2014. Biobased thermosetting epoxy: present and future *Chem. Rev.* 114 (2): 1082–1115.

Averous, L., Pollet, E. 2012. Biodegradable polymers. In *Environmental silicate nano-biocomposites*, ed.L. Avérous, E. Pollet, 13–39. Heidelberg: Springer.

BAAM 2020. *Big Area Additive Manufacturing 3D Printer.* https://www.e-ci.com/baam (accessed March 21, 2020).

Babu, R. P., O'Connor, K., Seeram, R. 2013. Current progress on bio-based polymers and their future trends. *Prog. Biomater.* 2:8.

Baechler, C., DeVuono, M., Pearce, J. M. 2013. Distributed recycling of waste polymer into RepRap feedstock. *Rapid Prototyp. J.* 19 (2): 118–125.

Bai, J., Yuan, S., Shen, F., Zhang, B., Chua, C. K., Zhou, K. 2017. Toughening of polyamide 11 with carbon nanotubes for additive manufacturing. *Virt. Phys. Prototyp.* 12 (3): 235–240. doi:10.1080/17452759. 2017.1315146.

BASF. n.d. *Sustainable Filaments – Technical Data.* https://www.ultrafusefff.com/material-data/sustainable-technical-data/ (accessed June 2, 2019).

Bayer, I. S., Guzman-Puyol, S., Heredia-Guerrero, J., et al. 2014. Direct transformation of edible vegetable waste into bioplastics. *Macromolecules* 47:5135–5143.

Bergmann, C.P., Stumpf, A. 2013. Biomaterials. In *Dental Ceramics. Microstructure, Properties and Degradation.* Berlin: Springer.

Bigg, D. M. 1987. Mechanical properties of particulate filled polymers. *Polym. Compos.* 8:115–122.

BioInspiration. 2015. *WillowFlex FX1504.* https://bioinspiration.eu/wp-content/uploads/2016/03/BioInspiration_ FX1504_Pellet_TDS.pdf, https://bioinspiration.eu/wp-content/uploads/2016/03/WillowFlex_FX1504_MSDS_ eng.pdf (accessed July 4, 2019).

Boegler, O., Kling, U., Empl, D., Isikveren, A. 2014. Potential of sustainable materials in wing structural design. *Presented at congress: Deutscher Luft- und Raumfahrtkongress* 2014. http://www.dglr.de/ publikationen/2015/340188.pdf (accessed November 15, 2019).

Bourell, D., Kruth J. P., Leu, M., et al. 2017. Materials for additive manufacturing. *CIRP Ann. Manuf. Technol.* 66:659–681.

Brindha, J., Privita, Edwina, R. A. G., Rajesh, P. K., Rani, P. 2016. Influence of rheological properties of protein bio-inks on printability: A simulation and validation study. *Mater. Today: Proc.* 3:3285–3295.

Brischetto, S., Torre, R. 2020. Tensile and compressive behavior in the experimental tests for PLA specimens produced via fused deposition modelling technique. *J. Compos. Sci.* 4:140.

British Plastics Federation. 2014. *Oil Consumption.* https://www.bpf.co.uk/Press/Oil_Consumption.aspx (accessed July 27, 2019).

British Plastics Federation. 2019. *Plastics: Recycling and Sustainability.* https://www.bpf.co.uk/sustainability/ Plastics_and_Sustainability.aspx (accessed August 28, 2019).

Brodin, M., Vallejos, M., Tanase Opedal, M. et al. 2017. Lignocellulosics as sustainable resources for production of bioplastics – a review. *J. Cleaner Prod.* 162:646–664.

Brown, B. N., Valentin, J. E., Stewart-Akers, A. M., McCabe, G. P., Badylak, S. F. 2009. Macrophage phenotype and remodeling outcomes in response to biologic scaffolds with and without a cellular component. *Biomaterials* 30 (8): 1482–1491.

Business Wire. 2019. *Global Polylactic Acid Market Study, 2019–2024.* https://www.businesswire.com/news/ home/20190416005498/en/Global-Polylactic-Acid-Market-Study-2019–2024-Future (accessed August 28, 2019).

Carbon® Inc. 2021a. *3D Printing Materials for Real-World Applications.* https://www.carbon3d.com/ materials/epu-elastomeric-polyurethane/ (accessed June 2, 2019).

Carbon® Inc. 2021b. https://www.carbon3d.com/our-technology/ (accessed June 2, 2019).

Cascone, S. 2016. *Design Miami Honors SHoP Architects, Taps 35 Dealers for 2016 Edition.* https:// news.artnet.com/market/design-miami-exhibitors-architects-687941.

CEN EN 13432. 2011. *Packaging – Requirements for Packaging Recoverable Through Composting and Biodegradation – Test Scheme and Evaluation Criteria for the Final Acceptance of Packaging.* Brussels, Belgium: European Committee for Standardization.

CEN/TR 16208. 2011. *Biobased Products – Overview of Standards.* Brussels, Belgium: European Committee for Standardization.

Center for Sustainable Polymers. 2020. *Sustainable Polymers 101.* https://csp.umn.edu/sustainable-polymers-101/ (accessed August 28, 2019).

CGTN America 2016. Debbie Mielewski on Ford's agave project. https://www.youtube.com/watch?v= fY5FT0KM-jU&t=70s (accessed May 31, 2019).

Chamas, A., Moon, H., Zheng, J. et al. 2020 Degradation Rates of Plastics in the Environment ACS Sustainable Chem. Eng. 8: 3494-3511.

Chandra, R, Rustgi, R. 1998. Biodegradable polymers. *Prog. Polym. Sci.* 23:1273–1335.

Chartoff, R. P., Schultz, J. W., Bhatt, J., Ullett, S. 1998. *Properties of a High Temperature Liquid Crystal Stereolithography Resin.* https://repositories.lib.utexas.edu/handle/2152/73521 (accessed August 28, 2019).

Chen, G. Q. 2009. A microbial polyhydroxyalkanoates (PHA) based bio-and materials industry. *Chem. Soc. Rev.* 38:2434–2446.

Chen, H., Xie F., Chen, L., Zheng, B. 2019. Effect of rheological properties of potato, rice and corn starches on their hot-extrusion 3D printing behaviors. *J. Food Eng.* 244:150–158. doi: 10.1016/j.jfoodeng.2018. 09.011.

Chinthapalli, R., Skoczinski, P., Carus, M. et al. 2019. *Bio-based Building Blocks and Polymers – Global Capacities, Production, And Trends 2018–2023.* http://bio-based.eu/reports/ (accessed May 28, 2019).

Clarinval, A. M., Halleux, J. 2005. Classification of biodegradable polymers. In *Biodegradable polymers for industrial applications*, ed.R. Smith, 3–31. Boca Raton: CRC.

Coca-Cola. 2018. https://www.coca-colacompany.com/our-company/plantbottle (accessed April 26, 2019).

Coherent Market Insights. 2021. *Renewable Chemicals Market Analysis.* https://www.coherentmarketinsights.com/ ongoing-insight/renewable-chemicals-market-262 (accessed August 31, 2019).

Contour Crafting. 2017. *Introducing Contour Crafting Technology.* http://contourcrafting.com/ (accessed March 6, 2020).

Core-Ballais, M., Bensoussan, H., Richardot, A., Kusnadi, H. 2017. *The State of 3D Printing.* https:// www.sculpteo.com/media/ebook/State%20of%203DP%202017_1.pdf (accessed May 28, 2019).

Crawford, R. J. 1998. *Plastics Engineering.* Oxford: Butterworth Heinemann. 3rd edition.

Cummings, J. H., Stephen, A. M. 2007. Carbohydrate terminology and classification. *Eur. J. Clin. Nutr.* 61 (Suppl. 1): S5–S18.

Czerniecka-Kubicka, A., Frącz, W., Jasiorski, M. et al. 2017. Thermal properties of poly(3-hydroxybutyrate) modified by nanoclay. *J. Therm. Anal. Calorim.* 128:513–1526.

Dai, L., Cheng, T., Duan, C. et al. 2019. 3D printing using plant-derived cellulose and its derivatives: areview. *Carbohydr. Polymers* 203:71–86.

Danusso, F, Tieghi, G. 1986. Strength versus composition of rigid matrix particulate composites. *Polymer* 27:1385–1390.

Derby, B. 2008. Bioprinting: inkjet printing proteins and hybrid cell-containing materials and structures. *J. Mater. Chem.* 18:5717–5721.

Di Lorenzo, M. L., Longo, A., Androsch, R. 2019. Polyamide 11/poly(butylene succinate) bio-based polymer blends. *Materials* 12:2833. doi: 10.3390/ma12172833.

Donaldson, S. 2020. Why "recycled" doesn't mean inferior for 3D printing filament. *Additive Manufacturing.* 9 (4): 32–34.

Dotan, A. 2014. Biobased thermosets. In *Handbook of Thermoset Plastics*, ed.H. Dodiuk, S. H. Goodman, 577–621. Oxford: Elsevier.

Draget, K. I., Skjåk-Braek, G., Smidsrød, O. 1997. Alginate based new materials. *Int. J. Biol. Macromol.* 21:47–55.

Drummer, D., Rietzel, D., Kühnlein, F. 2010. Development of a characterization approach for the sintering behavior of new thermoplastics for selective laser sintering. *Phys. Proc.* 5:533–542.

Dukes, W. H. 1966. *Unsolved Problems in Brittle Material Design.* US Govt. Report AD 654119.

Eagan, J. M., Xu J., Di Girolamo, R., et al. 2017. Combining polyethylene and polypropylene: enhanced performance with PE/iPP multiblock polymers. *Science* 355:814–816.

El Moumen, A., Tarfaoui, M., Lafdi, K. 2019. Additive manufacturing of polymer composites: Processing and modeling approaches. *Composit. B* 171:166–182.

Ellen MacArthur Foundation. 2013. *Towards the Circular Economy: an economic and business rationale for an accelerated transition.* https://www.ellenmacarthurfoundation.org/assets/downloads/publications/Ellen-MacArthur-Foundation-Towards-the-Circular-Economy-vol.1.pdf (accessed May 28, 2019).

Ellen MacArthur Foundation. n.d. *Circular Economy.* https://www.ellenmacarthurfoundation.org/circular-economy/concept (accessed May 28, 2019).

EnvisionTEC. 2017. *Large Format 3D Printer for Industrial Composites.* https://envisiontec.com/3d-printers/slcom-1/ (accessed March 21, 2020).

EOS. n.d. *PrimePart® ST PEBA 2301 TPA.* https://eos.materialdatacenter.com/eo/en (accessed January 16, 2021).

EPA. 2019. *Facts and Figures about Materials, Waste and Recycling.* https://www.epa.gov/facts-and-figures-about-materials-waste-and-recycling/plastics-material-specific-data#PlasticsTableandGraph (accessed July 7, 2019).

European Bioplastics. 2020. *Bioplastics Market Data 2019.* https://docs.european-bioplastics.org/publications/market_data/Report_Bioplastics_Market_Data_2019.pdf (accessed July 4, 2020).

European Bioplastics. n.d. *Welcome to European Bioplastics.* https://www.european-bioplastics.org/ (accessed January 16, 2021).

European Commission. 2018. *A European Strategy for Plastics in a Circular Economy.* EC: Brussels, Belgium. https://ec.europa.eu/environment/circular-economy/pdf/plastics-strategy-brochure.pdf (accessed July 4, 2020).

Faludi, J., Cline-Thomas, N., Agrawala, S. 2017. *3D Printing and Its Environmental Implications in Next Production Revolution – Implications for Governments and Business*, 171–213. OECD Library. doi:1 0.1787/9789264271036-en.

Fondazione Istituto Italiano di Tecnologia. 2015. *Bioplastics and Bioelastomers.* https://iit.it/technology-transfer-docs/505-new-technology-teaser-bioplastics-and-bioelastomers/file (accessed March 6, 2019).

Fortune Business Insight. n.d. *Bioplastics Market Size, Share & Industry Analysis.* https://www.fortunebusinessinsights.com/industry-reports/bioplastics-market-101940 (accessed July 4, 2020).

Fu, S.-Y., Feng, X.-Q. Lauke, B., Mai, Y.-W. 2008. Effects of particle size, particle/matrix interface adhesion and particle loading on mechanical properties of particulate–polymer composites. *Composites: Part B* 39:933–961.

Gaikwad, V., Ghose, A., Cholake, S. et al. 2018. Transformation of e-waste plastics into sustainable filaments for 3D printing. *ACS Sustain. Chem. Eng.* 6:14432–14440.

Gantenbein, S., Masania, K., Woigk, W. et al. 2018. Three-dimensional printing of hierarchical liquid-crystal-polymer structures. *Nature* 561:226–230. doi:10.1038/s41586-018-0474-7.

Garcia, J.M., Jones, G. O., Virwani, K. et al. 2014. Recyclable, strong thermosets and organogels via paraformaldehyde condensation with diamines. *Science* 344:732–735.

Geyer, R., Jambeck, J. R., Lavender Law, K. 2017. Supplementary materials for production, use, and fate of all plastics ever made. *Sci. Adv.* 3:e1700782.

Ghaddar, A., Bousso, B. 2018. *Rising Use of Plastics to Drive Oil Demand to 2050: IEA.* October 4, 2018. https://www.reuters.com/article/us-petrochemicals-iea-idUSKCN1ME2QD (accessed August 28, 2019).

Ghassemi, T., Shahroodi, A., Ebrahimzadeh, M. H., et al. 2018. Current concepts in scaffolding for bone tissue engineering. *Arch. Bone Joint Surg.* 6 (2): 90–99.

Gibson, I., D. Rosen, B. Stucker, 2015. *Additive Manufacturing Technologies,* p. 164. New York: Springer. 2nd edition.

Giurgiutiu, V. 2016. *Structural Health Monitoring of Aerospace Composites.* Oxford: Elsevier.

Globenewswire. 2020. *Global-Composites-Market-is-Expected-to-Reach-USD-181-49-Billion-by-2026.* https://www.globenewswire.com/news-release/2020/01/17/1971822/0/en/Global-Composites-Market-is-Expected-to-Reach-USD-181-49-Billion-by-2026-Fior-Markets.html Posted January 17, 2020 (accessed March 21, 2020).

Goh, G. D., Yap, Y. L., Agarwala, S., Yeong, W. Y. 2019. Recent progress in additive manufacturing of fiber reinforced polymer composite. *Adv. Mater. Technol.* 4:1800271.

Göktürk, E., Pemba, A. G., Miller, S. A. 2015. Polyglycolic acid from the direct polymerization of renewable C1 feedstocks. *Polymer Chem.* 21:3918–3925.

Golden, J. S., Handfield, R. B., Daystar, J., McConnell, T. E. 2015. An economic impact analysis of the U.S. biobased products industry. *Indus. Biotechno.* 11 (4): 201–209.

Gopinathan, J., Noh I. 2018. Recent trends in bioinks for 3D printing. *Biomater. Res.* 22:11. https://www.ncbi.nlm.nih.gov/pmc/articles/PMC5889544/ (accessed July 4, 2019).

Graedel, T. T. and B. R. Allenby. 1995. *Industrial Ecology.* Englewood Cliffs: Prentice-Hall.

Grand View Research. 2018. *3D Printing Plastics Market Size, Share & Trends Analysis Report.* https://www.grandviewresearch.com/industry-analysis/3d-printing-plastics-market/segmentation (accessed August 28, 2019).

Grand View Research. 2019. *Plastics Market Size*. https://www.grandviewresearch.com/industry-analysis/global-plastics-market (accessed July 4, 2019).

Haba, O., Tomizuka, H., Endo, T. 2005. Anionic ring-opening polymerization of methyl 4,6-*O*-benzylidene-2,3-*O*-carbonyl-α-d-glucopyranoside: a first example of anionic ring-opening polymerization of five-membered cyclic carbonate without elimination of CO_2. *Macromolecules* 38:3562–3563.

Halpin, J. C. 1992. *Primer on Composite Materials Analysis*. Lancaster: Technomic.

Halpin, J. C., Kardos, J. L. 1976. The Halpin-Tsai equations: A review. *Polym. Eng. Sci.* 16 (5): 344–352.

Hamod, H. 2014. *Suitability of Recycled HDPE for 3D Printing Filament*. Master Thesis, University of Helsinki. https://www.theseus.fi/bitstream/handle/10024/86198/Thesis%20final.pdf?sequence=1&isAllowed=y (accessed August 28, 2019).

Haria, R. 2017. *How 3D Printing Has Changed Dentistry, a Billion Dollar Opportunity*. https://3dprintingindustry.com/news/3d-printing-impact-on-dentistry-121284/ (accessed June 2, 2019).

Haris, M. Y., Laila, D., Zainudin, E., et al. 2011. Preliminary review of biocomposites materials for aircraft radome application. *Key Eng. Mater. Trans. Tech. Publ.* 471–472:563–567.

Heger, F., Chambers, R., Dietz, A. 1978. *Structural Plastics Design Manual*. New York, NY: American Society of Civil Engineers.

Hendrixson, S. 2020. Why "recycled" doesn't mean inferior for 3D printing filament. *Additive Manufacturing* 9 (4): 12.

Hermann, B., Carus, M., Patel, M., Blok, K. 2011. Current policies affecting the market penetration of biomaterials. *Biofuels, Bioprod. Bioref.* 5:708–719. doi:10.1002/bbb.327.

Hong, M., Chen, E. Y.-X. 2019. Future directions for sustainable polymers. *Trend. Chem.* 1 (2): 148–151.

Hong, M., Chen, E.Y.-X. 2016. Completely recyclable biopolymers with linear and cyclic topologies via ring opening polymerization of γ-butyrolactone. *Nat. Chem.* 8:42–49.

Hopewell, J., Dvorak, R., Kosior, E. 2009. Plastics recycling: Challenges and opportunities. *Philos. Trans. R. Soc. B* 364:2115–2126.

Horst, D. J., Tebcherani, S. M., Kubaski, E. T. 2016. Thermo-mechanical evaluation of novel plant resin filaments intended for 3D printing. *Int. J. Eng. Res. Technol. (IJERT)* 5 (10): 582–586.

Hutmacher, D. W. 2000. Scaffolds in tissue engineering bone and cartilage. *Biomaterials* 21 (24): 2529–2543.

Il Bioeconomista. 2021. *Boeing Goes Green: 100% Sustainable Aviation Fuels by 2030*. https://ilbioeconomista.com/2021/01/26/boeing-goes-green-100-sustainable-aviation-fuels-by-2030/ (accessed January 16, 2021).

Impossible Objects. 2018. *3D FROM 2D*. https://www.impossible-objects.com/process/ (accessed April 3, 2020).

Ioakeimidis, C., Fotopoulou, K. N., Karapanagioti, H. K., et al. 2016. The degradation potential of PET bottles in the marine environment: An ATR-FTIR based approach. *Sci. Rep.* 6:23501.

Islam, R. M., Beg, M. D. H., Jamari, S. S. 2014. Development of vegetable-oil-based polymers. *J. Appl. Polym. Sci.* doi:10.1002/app.40787.

ISO 472:2013. *(E/F) Plastics – Vocabulary*. Geneva, Switzerland: International Organization for Standardization. www.iso.org.

IUPAC. 2104. *Compendium of Chemical Terminology – Gold Book*. Accessed at http://goldbook.iupac.org/pdf/goldbook.pdf (accessed August 28, 2018).

Jambeck, J. R., Geyer, R., Wilcox, C. et al. 2015. Plastic waste inputs from land into the ocean. *Science* 5 (10): 582–586.

Jansen, A. J. 2016. Plastics – it's all about molecular structure. In *Plastics Engineering*, p. 28–32. https://www.madisongroup.com/ (accessed April 26, 2019).

Jem, K. J., Tan, B. 2020. The development and challenges of poly (lactic acid) and poly (glycolic acid). *Adv. Indus. Eng. Polym. Res.* 3:60–70.

Keshavarz, T., Roy, I. 2010. Polyhydroxyalkanoates: bioplastics with a green agenda. *Curr. Opin. Microbiol.* 13:321–326.

Khan, I., Kamma-Lorger, C. S., Mohan, S. D., et al. 2019. The exploitation of polymer based nanocomposites for additive manufacturing: a prospective review. *Appl. Mech. Mater.* 890:113–145.

Kim, S. W., Kim, D. Y., Roh, H. H. et al. 2019. Three-dimensional bioprinting of cell-laden constructs using polysaccharide-based self-healing hydrogels. *Biomacromolecules* 20 (5): 1860–1866.

Kim, Y. B., Lenz, R. W. 2001. Polyesters from microorganisms. *Adv. Biochem. Eng. Biotechnol.* 71:51–79.

Kirchherr, J., Reike, D., Hekkert, M. 2017. Conceptualizing the circular economy: An analysis of 114 definitions. *Resour. Conserv. Recycl.* 127:221–232.

Koller, M. 2017. *Preface in Advances in Polyhydroxyalkanoate (PHA) Production,* ed. M. Koller, special issue published in Bioengineering, Basel: MDPI. 1st edition. www.mdpi.com/journal/bioengineering (accessed June 13, 2019).

Koslow, T., de Valensart, G. 2017. *REC 3D Releases Comprehensive Stress Test for 3D Printing Materials Filaments Directory.* https://www.filaments.directory/en/blog/2017/01/27/rec-3d-releases-comprehensive-stress-test-for-3d-printing-materials (accessed August 31, 2020).

Kourmentza, C., Plácido, J., Venetsaneas, N. et al. 2017. Recent advances and challenges towards sustainable polyhydroxyalkanoate (PHA). *Prod. Bioeng.* 4 (2): 55.

Kumar, S., Samal, S. K., Mohanty, S., Nayak, S. K. 2018. Recent development of biobased epoxy resins: a review. *Polymer-Plastics Technol. Eng.* 57 (3): 133–155. doi:10.1080/03602559.2016.1253742.

Kuo, C.-C., Liu, L.-C., Teng, W.-F., et al. 2016. Preparation of starch/acrylonitrile-butadiene-styrene copolymers (ABS) biomass alloys and their feasible evaluation for 3D printing applications. *Compos. B: Eng.* 86:36–39.

Laird, K. 2015. Starch-based Biome3D filament now available on US 3D printing market. *Plastics Today.* https://www.plasticstoday.com/content/starch-based-biome3d-filament-now-available-on-us-3d-printing-market/4029890922524 (accessed May 28, 2019).

Levita, G, Marchetti, A, Lazzeri, A. 1989. Fracture of ultrafine calcium carbonate/polypropylene composites. *Polym. Compos.* 10:39–43.

Lewis, H., Gertsakis, J. 2001. *Design + Environment: A Global Guide to Designing Greener Goods.* Sheffield: Greenleaf Publishing Ltd.

Liddell, H. G., Scott, R. n.d. *A Greek-English Lexicon.* http://www.perseus.tufts.edu (accessed April 26, 2018).

Ligon, S. C., Liska, R., Stampfl, J., et al. 2017. Polymers for 3D printing and customized additive manufacturing. *Chem. Rev.* 117:10212–10290.

Lligadas, G., Ronda, J. C., Galia, M., Cadiz, V. 2013. Renewable polymeric materials from vegetable oils: a perspective. *Materials Today* 16 (9): 337–343. doi:10.1016/j.mattod.2013.08.016.

Lu, S, Yan, L, Zhu, X, Qi, Z. 1992. Microdamage and interfacial adhesion in glass bead-filled high-density polyethylene. *J. Mater. Sci.* 27:4633–4638.

Lubrizol n.d. *ESTANE® 3D TPU (Thermoplastic Polyurethane).* https://www.lubrizol.com/Engineered-Polymers/Products/Estane-TPU/Estane-3D (accessed May 31, 2020).

Mahanta, R., Mahanta, R. 2016. Sustainable polymers and applications. In *Handbook of Sustainable Polymers: Processing and Applications*, ed. V. K. Thakur, M.K. Thakur, 1–57. Singapore: Pan Stanford Publishing Pte. Ltd.

Maiorana, A., Reano, A. F., Centore, R., et al. 2016. Structure property relationships of biobased n-alkyl bisferulate epoxy resins. *Green Chem.* 18:4961–4973. doi:10.1039/C6GC01308B.

Makershop. n.d. *PLA 1.75mm flax fibre.* https://www.makershop3d.com/expert-filament/743-pla-fibre-flax.html
(accessed July 4, 2019).

Manolis Sherman, L. 2018. *Market-Trends-in-Polymer-Additive-Manufacturing.* https://www.ptonline.com/blog/post/market-trends-in-polymer-additive-manufacturing (accessed June 4, 2019).

Market Publishers. 2018. *Natural vs. Synthetic Rubber: Key Market Trends & Statistics.* https://marketpublishers.com/lists/23821/news.html (accessed May 31, 2019).

MarketsandMarkets™. 2020a. *Composites Market – Global Forecast to 2024.* https://www.marketsandmarkets.com/ (accessed April 26 2020).

MarketsandMarkets™. 2020b. *Polyhydroxyalkanoate (PHA) Global Forecast to 2021.* https://www.marketsandmarkets.com/Market-Reports/pha-market-395.html (accessed April 3, 2020).

Melchels, F. P. W., Domingos, M. A. N., Klein, T. J., et al. 2012. Additive manufacturing of tissues and organs. *Prog. Polym. Sci.* 37:1079–1104.

Menčik, P., Prikryl, R., Stehnová, I., et al. 2018. Effect of selected commercial plasticizers on mechanical, thermal, and morphological properties of poly(3-hydroxybutyrate)/poly(lactic acid)/plasticizer biodegradable blends for three-dimensional (3D) print. *Materials* 11:1893. doi:10.3390/ma11101893.

Mialon, L., Pemba, A. G. & Miller, S. A. 2010. Biorenewable polyethylene terephthalate mimics derived from lignin and acetic acid. *Green Chem.* 12:1704–1706.

Miao, S., Wang, P., Su, Z., Zhang, S. 2014. Vegetable-oil-based polymers as future polymeric biomaterials *Acta Biomater.* 10:1692–1704.

Miao, S., Zhu, W., Castro, N. J. et al. 2016. 4D printing smart biomedical scaffolds with novel soybean oil epoxidized acrylate. *Scientific Rep.* 6:27226.

Michigan State University. 2013. *Polymers.* https://www2.chemistry.msu.edu/faculty/reusch/VirtTxtJml/polymers.htm (accessed May 28, 2019).

Middlecamp, C. H., M. T. Mury, K. L. Anderson et al. 2015. *Chemistry in context.* New York: McGraw Hill.

Mikami, K. et al. 2013. Polycarbonates derived from glucose via an organocatalytic approach. *J. Am. Chem. Soc.* 135:6826–6829.

Mohan, V. B., Lau, K., Hui, D., Bhattacharyya, D. 2018. Graphene-based materials and their composites: A review on production, applications and product limitations. *Compos. B: Eng.* 142:200–220.

Mohanty, A. K., Misra, M., Hinrichsen, G. 2000. Biofibres, biodegradable polymers and biocomposites: an overview. *Macromol. Mater. Eng.* 276–277 (1): 1–24.

Muenchinger, K. 2017. Research indicates millennials want durable, natural products. *Plastics Eng.* 73 (9): 40–43.

Mülhaupt, R. 2013. Green polymer chemistry and bio-based plastics: dreams and reality. *Macromol. Chem. Phys.* 214:159–174. doi:10.1002/macp.201200439/full.

Nakajima, H., Kimura, Y. 2013. General introduction: Overview of the current development of biobased polymers. In *Bio-Based Polymers,* ed. Y. Kimura, 1–23. Tokyo: CMC Publishing.

NatureWorks. 2021a. *What-is-Ingeo/Why-it-Matters/Eco-Profile.* https://www.natureworksllc.com/What-is-Ingeo (accessed July 4, 2019).

NatureWorks. 2021b. *Recycling.* https://www.natureworksllc.com/What-is-Ingeo/Where-it-Goes/Recycling.

Niaounakis, M. 2015. *Biopolymers: Processing and Products.* Oxford: Elsevier, William Andrew Publishing.

Nicolais, L, Narkis, M. 1971. Stress–strain behavior of styrene–acrylonitrile/glass bead composites in the glassy region. *Polym. Eng. Sci.* 11:194–199.

Nicolais L., Nicodemo, L. 1973. Strength of particulate composites. *Polym. Eng. Sci.1*, 13, 469.

Nicolais, L, Nicodemo, L. 1974. Effect of particles shape on tensile properties of glassy thermoplastic composites. *Int. J. Polym. Mater.* 3:229.

Nielsen, L. E., Landel, R. F. 1994. *Mechanical Properties of Polymers and Composites.* New York: Dekker. 2nd edition.

NOAA. n.d. *Marine Debris Is Everybody's Problem.* https://www.whoi.edu/fileserver.do?id=107364&pt=2&p=88817 (accessed May 31, 2019).

Ocean Conservancy, McKinsey Center for Business and Environment. 2015. *Stemming the Tide: Land-Based Strategies for a Plastic-Free Ocean.* https://oceanconservancy.org/wp-content/uploads/2017/04/full-report-stemming-the.pdf (accessed June 13, 2018).

Orion Market Research. 2019. *Polybutylene Succinate (PBS) Market.* https://www.omrglobal.com/industry-reports/polybutylene-succinate-market. Published July 2019 (accessed July 4, 2020).

Ou-Yang, Q., Guo, B., Xu, J. 2018. Preparation and characterization of poly(butylene succinate)/polylactide blends for fused deposition modeling 3D printing. *ACS Omega* 3:14309–14317.

Ozbolat, I. T., Hospodiuk, M. 2016. Current advances and future perspectives in extrusion-based bioprinting. *Biomaterials* 76:321–343.

Paloheimo, M. n.d. *Tensile or Flexural Strength/Stiffness – Is There Really a Difference?* https://www.plasticprop.com/articles/tensile-or-flexural-strengthstiffness-there-really-difference/ (accessed November 15, 2019).

Parandoush, P., Lin, D. 2017. A review on additive manufacturing of polymer-fiber composites. *Compos. Struct.* 182:36–53.

Patent CN10511699A. *Polybutylene Succinate 3D Printing Wire and Preparation Method thereof.* In Chinese. https://patents.google.com/patent/CN105111699A/en (accessed May 28, 2019).

PEMRG/Consultic Marketing & Industrieberatung. n.d. *World Plastics Production 1950–2015.* https://committee.iso.org/files/live/sites/tc61/files/The%20Plastic%20Industry%20Berlin%20Aug%202016%20-%20Copy.pdf (accessed May 31, 2019).

Pereira, T. F., Oliveira, M. F., Maia, I. A., et al. 2012. 3D printing of poly(3-hydroxybutyrate) porous structures using selective laser sintering. *Macromol. Symp.* 319:64–73.

Pfaltzgraff, L. A., Clark J. H. 2014. Green chemistry, biorefineries and second generation strategies for re-use of waste: an overview. In *Advances in Biorefineries,* ed. Keith Waldron, 3–33. Swaston: Woodhead Publishing.

Plastics Industry Association 2018. 2018. *2018 Size & Impact Report.* Washington, DC. https://www.plasticsindustry.org/sites/default/files/SizeAndImpactReport_Summary.pdf (accessed May 28, 2019).

PlasticsEurope. 2019. *Plastics – the Facts 2018.* https://www.plasticseurope.org/application/files/6315/4510/9658/Plastics_the_facts_2018_AF_web.pdf (accessed August 28, 2019).

PlasticsEurope. 2019. *Plastics – the Facts 2019.* https://www.plasticseurope.org/application/files/9715/7129/9584/FINAL_web_version_Plastics_the_facts2019_14102019.pdf (accessed August 28, 2019).

Porter, J. R., Ruckh, T. T., Popat, K. C. 2009. Bone tissue engineering: a review in bone biomimetics and drug delivery strategies. *Biotechnol. Prog.* 25 (6): 1539–1560.

Rahimi, A., Ulbrich, A., Coon, J. J., Stahl, S. S. 2014. Formic-acid-induced depolymerization of oxidized lignin to aromatics. *Nature* 515:249–252.

Raphael, B., Khalil, T., Workman, V. L., et al. 2017. 3D cell bioprinting of self-assembling peptide-based hydrogels. *Mater. Lett.* 190:103–106.

Raquez, J.-M., Deléglise M., Lacrampe M.-F., Krawczak P. 2010. Thermosetting (bio)materials derived from renewable resources: a critical review. *Prog. Polym. Sci.* 35:487–509.

Refil. n.d. *Refilament.* https://www.re-filament.com/ (accessed January 16, 2021).

Relander-Koivisto, M. 2018. *A 3D printed bridge: Naturally with UPM Formi.* https://www.upm.com/news-and-stories/articles/2018/04/a-3d-printed-bridge-naturally-with-upm-formi/built (accessed August 28, 2019).

Report Buyer. 2019. https://www.reportbuyer.com/product/5751917/additive-manufacturing-market-to-2027-global-analysis-and-forecasts-by-material-technology-and-end-user.html (accessed June 13, 2019).

Research and Markets. 2018. *3D Printing Plastics Market by Type, Form, End-Use Industry, Application, and Region – Global Forecast to 2023 Report.* https://www.researchandmarkets.com/research/mzf7vv/3d_printing?w=12 (accessed June 13, 2019).

Research and Markets. 2019. *Future of Global Bio-Based Polyurethane Market to 2025.* https://www.researchandmarkets.com/reports/4618654/2019-future-of-global-bio-based-polyurethane (accessed April 26, 2019).

Rosato, D. and D. Rosato 2003. *Plastics Engineered Product Design.* New York: Elsevier.

Safai, L., Cuellar, J. S., Smit, G., Zadpoor, A. A. 2019. A review of the fatigue behavior of 3D printed polymers. *Addit. Manuf,* 28:87–97.

Sanchez Cruz, F., Lanza, S., Boudaoud, H., Hoppe, S., Camargo, M. 2015. Polymer recycling and additive manufacturing in an open source context: optimization of processes and methods. In *Annual International Solid Freeform Fabrication Symposium – An Additive Manufacturing Conference,* Austin, Texas, 10–12 August 2015.

Sano, Y., Matsuzaki, R., Ueda, M. 2018. 3D printing of discontinuous and continuous fiber composites using stereolithography. *Addit. Manuf.* 24:521–527.

Saroia, J., Wang, Y., Wei, Q. et al. 2020. A review on 3D printed matrix polymer composites: its potential and future challenges. *Int. J. Adv. Manuf. Technol.* 106:1695–1721.

Schneiderman, D. K., Hillmyer, M. A. 2017. 50th anniversary perspective: There is a great future in sustainable polymers. *Macromolecules* 50:3733–3749.

Science History Institute. 2021. *Conflicts in Chemistry: The Case of Plastics.* https://www.sciencehistory.org/case-study-for-sustainability-the-future-of-plastics (accessed May 28, 2019).

Shapeways. 2019. *New Material Launch: EOS Biobased Polymer PA11.* https://www.shapeways.com/blog/archives/39563-new-material-launch-eos-biobased-polymer-pa11.html (accessed June 2, 2019).

Sher, D. 2018. *3D Printing Media Network The Global Additive Manufacturing Market 2018 Is Worth $9.3 Billion.* https://www.3dprintingmedia.network/the-global-additive-manufacturing-market-2018-is-worth-9-3-billion/ (accessed August 28, 2019).

Shrivastava, A. 2018. Polymerization. In *Introduction to Plastics Engineering,* 17–48. Oxford: William Andrew..

Sin, L. T., Rahmat, A. R., Rahman, W. A.W.A. 2012. Overview of poly(lactic acid). In *Handbook of Biopolymers and Biodegradable Plastics,* ed.S. Ebnesajjad, 11–54. Chadds Ford: William Andrew. doi:10.1016/B978-1-4557-2834-3.00002-1.

Singamneni, S., Dawn Smith, D., LeGuen, M.-J., Truong, D. 2018. Extrusion 3D printing of polybutyrate-adipate-terephthalate-polymer composites in the pellet form. *Polymers* 10:922.

Singh, S., Ramakrishna, S., Berto, F. 2019. 3D Printing of polymer composites: a short review. *Mat. Design Process Comm.* 2:97.

Siracusa, V., Rocculi, P., Romani, S., Rosa, M. D. 2008. Biodegradable polymers food packaging: a review. *Trends Food Sci. Technol.* 19:634–643.

Skoczinski, P., Chinthapalli, R., Carus, M. et al. 2020. *Bio-based Building Blocks and Polymers – Global Capacities, Production and Trends 2019–2024.* Nova-Institute. https://www.bio-based.eu/reports/ (accessed January 16, 2021).

Sloan, J. 2020. *Composites 2020: A multitude of markets Composites World.* https://www.compositesworld.com/blog/post/composites-a-multitude-of-markets (accessed March 6, 2020).

Song, Y., Li, Y., Song, W., et al. 2017. Measurements of the mechanical response of unidirectional 3D-printed PLA. *Mater. Des.* 123:154–164.

Sperling, L. H. 2006. *Introduction to Physical Polymer Science*, p. 6. Hoboken: J. Wiley and Sons. 4th edition.

SPI. 2016. *The Plastics Industry Trade Association. Plastic Market Watch – Bioplastics.* Summer 2016 Issue VI. https://www.plasticsindustry.org/sites/default/files/2016PMWBioplasticsIA.pdf (accessed August 28, 2019).

Spierling, S., Knüpffer, E., Behnsen, H., et al. 2018. Bio-based plastics – a review of environmental, social and economic impact assessments. *J. Cleaner Prod.* 185:476–491.

State Council of China. 2020. *China Reveals Plan to Cut Plastic Use by 2025.* http://english.www.gov.cn/statecouncil/ministries/202001/19/content_WS5e243ea1c6d0db64b784ccd1.html (accessed July 4, 2020).

StEP. 2019. *What Is E-Waste?* https://www.step-initiative.org/e-waste-challenge.html (accessed January 16, 2021).

Stratasys. 2021a. *TPU 92A Elastomer.* https://www.stratasys.com/elastomers-material (accessed May 28, 2018).

Stratasys. 2021b. *Our Materials.* https://www.stratasys.com/materials/search (accessed May 31, 2018).

Sudesh, K., Abe, H., Doi, Y. 2000. Synthesis, structure and properties of polyhydroxyalkanoates: biological polyesters. *Prog. Polym. Sci.* 25 (10): 1503–1555.

Tamayo-Dominguez, A., Fernández-González J., Sierra-Castañer, M. 2018. Additive manufacturing and liquid crystals for new millimeter-wave devices. *IEEE MTT-S Latin America Microwave Conference (LAMC 2018)*, Arequipa, Peru, pp. 1–3. doi:10.1109/LAMC.2018.8699025.

Tan, G. Y. A., Chen, C. L., Li, L. et al. 2014. Start a research on biopolymer polyhydroxyalkanoate (PHA): A review. *Polymers* 6:706–754.

Tang, X.-Y., Chen, E.Y.-X. 2019. Toward infinitely recyclable plastics derived from renewable cyclic esters. *Chem.* 5:284–312.

The European House – Ambrosetti. 2013. *L'eccellenza della filiera della plastica per il rilancio industriale dell'Italia e dell'Europa.* Page 25. In Italian. https://www.ambrosetti.eu/wp-content/uploads/130907_Filiera_della_plastica_Ricerca_completa_HD.pdf (accessed August 28, 2019).

The Henry Ford. 2020. *Popular Research Topics – Soybean Car.* https://www.thehenryford.org/collections-and-research/digital-resources/popular-topics/soy-bean-car/ (accessed June 13, 2019).

Topolkaraev, V. A., McEneany, R. J., Scholl, N. T. 2014. Polymeric Material for Three-Dimensional Printing. Patent Application Publication US 2016/0185050 A1.

Tractus3D. n.d. *TPU Material.* https://tractus3d.com/materials/tpu (accessed July 4, 2020).

Tres, P. 2014. *Designing Plastic Parts for Assembly,* p. 51. Munich: Hanser Gardner.

Ullett, J. S., Schultz, J. W., Chartoff, R. P. 2000. Novel liquid crystal resins for stereolithography – processing parameters and mechanical analysis, *Rapid Prototyp. J.* 6 (1): 8–17.

United Nations. 2020. *Climate Change.* https://unfccc.int/kyoto_protocol (accessed July 4, 2020).

United Nations. 1987. *Report of the World Commission on Environment and Development (WCED): Our Common Future.* https://www.are.admin.ch/are/en/home/sustainable-development/international-cooperation/2030agenda/un-_-milestones-in-sustainable-development/1987--brundtland-report.html (accessed March 21, 2020).

United Nations. n.d. *About the Sustainable Development Goals.* https://www.un.org/sustainabledevelopment/sustainable-development-goals/ (accessed on May 31, 2019).

United Soybean Board. 2019. *Goodyear Tires Gain Traction with Soy.* https://soynewuses.org/case-study/goodyear-tires-gain-traction-with-soy/ (accessed on May 31, 2019).

UTC. 2019. *The Sustainability of Aviation: A Joint Statement by Seven of the World's Major Aviation Manufacturers.* https://www.utc.com/news/2019/06/18/the-sustainability-of-aviation (accessed June 13, 2019).

van der Heijden, R., Coenen, J., van Riel, A. 2017. Transitioning from a linear economy towards a circular economy: the case of the apparel industry. In *Constructing A Green Circular Society*, ed. M. E. Moula, J. Sorvari, P. Oinas, 14–38. Helsinki: Unigrafia Oy. https://www.researchgate.net/publication/325417234_CONSTRUCTING_A_GREEN_CIRCULAR_SOCIETY (accessed May 28, 2019).

van Es, D. S., van der Klis, F., Knoop, R. J. I. et al. 2014. Other polymers for biomass derived monomers. In *Bio-Based Plastics – Materials and Applications*, ed.S. Kabasci, 241–274. Chichester: Wiley.

Vert, M. et al. 1992. Bioresorbability and biocompatibility of aliphatic polyesters. *J. Mater. Sci.-Mater. Med.* 3 (6): 432–446.

Voisin, H. P., Gordeyeva, K., Siqueira, G., et al. 2018. 3D printing of strong lightweight cellular structures using polysaccharide-based composite foams. *ACS Sustainable Chem. Eng.* 6 (12): 17160–17167.

Wang, F., Jiang, Y., Liu, Y., et al. 2020. Liquid crystalline 3D printing for superstrong graphene microlattices with high density. *Carbon* 159:166–174.

Wang, J., Liu, S., Yu, J, et al. 2017c. Rosin-derived monomers and their progress in polymer applications. In *Sustainable Polymers from Biomass*, ed. C. Tang, C. Y. Ryu, 103–150. Weinheim: Wiley.

Wang, Q., Sun, J., Yao, Q., et al. 2018. 3D printing with cellulose materials. *Cellulose* 25:4275–4301.

Wang, S., Capoen, L., D'hooge, D. R., Cardon, L. 2017a. Can the melt flow index be used to predict the success of fused deposition modelling of commercial poly(lactic acid) filaments into 3D printed materials? *Plast. Rubber Compos.* doi:10.1080/14658011.2017.1397308.

Wang, X., Jiang, M., Zhou, Z., Gou, J., Hui, D. 2017b. 3D printing of polymer matrix composites: A review and prospective *Compos. B* 110:442–458.

Wang, X., Rinaldi, R. 2013. A route for lignin and bio-oil conversion: dehydroxylation of phenols into arenes by catalytic tandem reactions. *Angew. Chem. Int. Edn Engl.* 52:11499–11503.

Whittington. 1978. *Whittington's Dictionary of Plastics.* Westport: Technomic.

Wohlers Associates. 2019. *Wohlers Report 2019. Details of Developments in Additive Manufacturing Worldwide.* https://www.sme.org/wohlers-report-2019-additive-manufacturing (accessed August 31, 2019).

World Economic Forum. 2016. The new plastics economy: rethinking the future of plastics. In ed.L. Neufeld, F. Stassen, R. Sheppard, T. Gilman. http://www3.weforum.org/docs/WEF_The_New_Plastics_Economy.pdf (accessed January 16, 2019).

World Plastics Council. n.d. *Overview.* https://www.worldplasticscouncil.org/about/ (accessed June 13, 2019).

Yu, D., Sun, L., Xue, W., Zeng, Z. 2017. A modified poly (trimethylene terephthalate) used for fused deposition modeling: synthesis and application. *Chem. Sci. Int. J.* 18 (3): 1–8.

Zahnd, P. A. 2018. Is 3D printing ready for mass production? *3D printing Industry.* https://3dprintingindustry.com/news/3d-printing-ready-mass-production-132576/ (accessed August 31, 2019).

Żenkiewicz, M., Richert, J., Rytlewski, P. et al. 2009. Characterisation of multi-extruded poly(lactic acid). *Polymer Test.* 28 (4): 412–418.

Zhang, X., Fevre, M., Jones, G. O., Waymouth, R. M. 2018. Catalysis as an enabling science for sustainable polymers. *Chem. Rev.* 118:839–885.

Zhu, J. B., Watson, E. M., Tang, J., Chen, E. Y.-X. 2018. A synthetic polymer system with repeatable chemical recyclability. *Science* 360:398–403.

Zhu, Y., Romain, C., Williams, C. K. 2016. Sustainable polymers from renewable sources. *Nature* 540:354–362.

Zinn, M., Witholt, B., Egli, T. 2001. Occurrence, synthesis and medical application of bacterial poly-hydroxyalkanoate. *Adv. Drug Deliv. Rev.* 53:5–21.

Żenkiewicz, M., Richert, J., Rytlewski, P. et al. 2009. Characterization of multi-extruded poly(lactic acid). *Polym. Test.* 28 (4): 412–418.

2 Additive Manufacturing and Its Polymeric Feedstocks

A nation that destroys its soils destroys itself. Forests are the lungs of our land, purifying the air and giving fresh strength to our people.

Franklin D. Roosevelt

2.1 INTRODUCTION TO ADDITIVE MANUFACTURING (AM)

AM is the most recent class of manufacturing processes, and consists in "joining materials to make parts from 3D model data, usually layer upon layer, as opposed to subtractive manufacturing and formative manufacturing methodologies" (ISO/ASTM 52900 2015(E)). All AM processes comprise the following main steps: generating a computer-aided design (CAD) or 3D digital model of the article to build, converting this model into a digital file compatible with AM equipment (*AM printer*), electronically "slicing" this file along the Z axis in very thin cross sections, each representing one layer in the XY plane of the article, and building, on the printer, the article one layer at the time stacked on the previous layer, without performing any operation of machining and molding during the building.

Alternative terms for AM are: *additive fabrication, additive processes, additive techniques, additive layer manufacturing, direct digital manufacturing, layer manufacturing, solid freeform fabrication, freeform fabrication, 3D printing,* and *rapid prototyping. 3D printing* is also a specific AM technique that will be described in Section 2.6, while *rapid prototyping* is a limiting term, because it is a legacy of the initial versions of AM techniques that processed materials adequate for non-functional prototypes but not load-bearing end-use items.

Additive means that the object is shaped by sequentially adding layers of material. In fact "the fundamental principle of AM processes is forming 3D parts by the successive addition of material" (ISO/ASTM 52900 2015(E)). *Formative*, instead, refers to fabrication processes (f.e. forging, bending, casting, compression molding, and injection molding) in which the feedstock material is shaped by applying pressure to it, whereas *subtractive* refers to manufacturing methods (milling, turning, drilling, etc.) in which a block of raw material is shaped by selectively removing some of it. In this book, unless otherwise specified, *to print* and *printer* are synonym of *to fabricate by AM* and *AM printer,* respectively.

AM feedstocks include polymers, metals, ceramics, and any combination (composites) of the above that are processed in the following physical forms (ISO/ASTM 52900 2015(E)):

- Polymers: filament, hydrogel, liquid, molten, pellet, powder, sheet (Ligon et al. 2017)
- Metals: filament/wire, powder, sheet (Frazier 2014; Lewandowski and Seifi 2016; Sames et al. 2016)
- Ceramics: powder and liquid suspension, powder (Zocca et al. 2015; Wang et al. 2019)
- Composites: filament, liquid, powder, sheet (Hofstatter et al. 2017; Parandoush and Lin 2017; Türk et al. 2017; Wang et al. 2017b; El Moumen et al. 2019).

The AM feedstocks most popular are unfilled and filled polymers, metals, and ceramics (Feilden 2017) in decreasing order. In a 2017 market report by Wohlers Associates, 51% of the surveyed providers of AM services printed polymer parts (Wong 2017).

DOI: 10.1201/9781003221210-2

ISO 17296-2 has grouped AM processes in seven categories: *binder jetting*, *direct energy deposition* (DED), *material extrusion* (also known as *fused deposition modeling™* or *FDM™*, and *fused filament fabrication* or FFF), *material jetting* (MJ), *powder bed fusion* (PDF), *sheet lamination* (ShL), and *vat photopolymerization* (VL). In the remainder of the book, when mentioning FFF, we also mean FDM™. They all process polymeric feedstocks (and other materials), except DED. Polymers are mostly processed on printers as filament, hydrogel, liquid, paste, and powder. However, polymeric pellets are gaining ground, because they permit to print large objects at higher speed and lower cost than polymeric filaments (Moreno Nieto et al. 2018). Hereafter we focus especially on AM processes whose feedstocks are sustainable polymers (SPs).

The typical steps to print an object are the following (Figure 2.1):

- Designing the 3D digital model of the object, possibly maximizing its service performance (maximum strength and durability, lowest stress concentrations, etc.) and minimizing its volume, weight, and cost.
- Converting the 3D digital model into a computer file that exists in the STL (*Standard Triangle Language* and *Standard Tessellation Language*) format, and is basically a version of the digital model whose surfaces are replaced with triangles and the 3D coordinates of their vertices.
- Digital modeling the supporting structure that prevents sagging, bending, and other phenomena that cause a mismatch between nominal and actual dimensions.
- Using computer software and digitally "slicing" the STL model into layers stacked on each other, with each layer (*slice*) representing one very thin XY cross section of the article. The 3D coordinates of the layers are sent as instructions to the printer.
- Loading the feedstock material, and setting up the printing parameters and orientation that aim at (a) minimizing printing time and amount of total material (material staying in the final object, and removable material supporting overhanging features, holes, etc.), and (b) maximizing quality (dimensional accuracy and fidelity, surface finish, etc.).
- Printing the object one layer (typically 15 to 500 μm thick) at the time, and each layer on top of the previous one.
- Removing the object from the printer, and, if required, *post-processing* it, which includes heating (*post-curing* for TS polymers, and heat treatment for metals), removing the support material, trimming, boring and reaming holes, sanding, milling, coating, painting, etc.
- Cleaning up the printer.

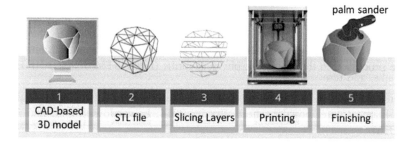

FIGURE 2.1   Typical steps to print an object.

Source: Modified from Cotteleer, M., Holdowsky, H., Mahto, M., 2013. The 3D opportunity primer – The basics of additive manufacturing. https://www2.deloitte.com/us/en/insights/focus/3d-opportunity/the-3d-opportunity-primer-the-basics-of-additive-manufacturing.html (accessed July 24, 2020). Reproduced with permission from Deloitte.

Post-processing addresses some drawbacks of AM, namely the staircase effect (defined in Section 2.3.2), surface quality, and dimensional accuracy, with the latter two being inferior to those in conventional fabrication processes. Since several types of AM processes for polymers exist, post-processing on polymeric printed parts can be performed through the following methods: chemical treatment, electroplating, filling gaps with epoxy resin, hand finishing, hot cutter machining, micromachining, painting, and vibratory bowl abrasion (Kumbhar and Mulay 2018).

AM "is not only a disruptive technology, it has the potential to replace many conventional manufacturing processes, but is also an enabling technology, allowing new business models, new products and new supply chains to flourish" (Mitchell et al. 2018). AM is impacting science and technology (Ligon et al. 2017), and revolutionizing manufacturing across industries (Berman 2012; Kurfess and Cass 2014; Monahan et al. 2017). In fact, AM has been compared to disruptive technologies such as digital books and music downloads. AM is spreading fast, because it enables companies to cut cost and time during development and production phases, and to fabricate new functional designs, which boosts creativity, economic growth, and innovation. In 2016, there were more than 300 companies selling printers of affordable price (Stansbury and Idacavage 2016). AM also makes possible cost-effective mass customization for specific products (Reeves et al. 2011), and localized, and flexible production. AM is not always beneficial and cost-effective, but, when intelligently and properly applied, AM's benefits are significant and listed below, cautioning however that not all of them can be accomplished at the same time:

- Cutting part fabrication cost: AM is a toolless process, meaning it requires no mold (for plastic articles) or die (for metal articles) to fabricate a part, and hence no large capital invested upfront that has to be offset later by making many copies of the same part over time.
- Reducing time from idea to market by cutting lead time between consecutive developmental designs of the same product, when the designs require a die or mold, since there is no need to wait months for a new die and mold to be made, and possibly toss them if their design needs to be modified.
- Shortening overall fabrication time to produce an assembly, by waiting less time for components of that assembly to be made.
- Enabling to fabricate engineering and non-engineering components (*legacy components*) whose production has stopped and whose blueprint, die, or mold are no longer available. In that case, the legacy component is 3D scanned and its resulting 3D digital model is converted into a STL file for a printer.
- Decreasing part's volume and weight by applying *topological optimization*, a computer-assisted technique that minimizes the part's volume and meets (or exceeds) the part's structural and functional requirements, by placing the material at the locations of the (major) stress paths, and removing it from the less stressed locations.
- Unprecedented freedom in geometry design: printers enable to produce novel and extremely complicated geometries impossible or excessively expensive to manufacture through other processes.
- Customizing solutions: starting with 3D digital models, printers can fabricate items with unique shapes meeting the requirements of a specific user, such as prosthetics with shapes perfectly matching a patient's body that was previously 3D scanned and converted into a 3D digital model of the printed item.
- Being the only manufacturing alternative when there is no company bidding to make a part, because the parts to produce are too few and/or too inexpensive to attract a bid.

- Saving on a part's price when the price for the same part fabricated via conventional processes is high because only a very few suppliers exist and they exploit a situation of oligopoly.
- Reducing scrap rates of the final product: printing developmental designs and prototypes enables to fine-tune the final product, and maximize its quality during production.
- Lowering buy-to-fly ratio: when aerospace components are machined from a solid block of metal, such as expensive titanium, the ratio of weight of the initial block over weight of the final component can be as high as 15–20 for flying components (Arcam n.d.). Instead, printers only use the amount of metal corresponding to the component's volume, plus the support material mentioned earlier.
- Decreasing part count: AM facilitates the fabrication of a multi-piece assembly as one single printed part (as in the case of GE's LEAP nozzle), or as an assembly of fewer printed pieces. This reduces the part count, which results in: fewer items to manage at the supply chain level (order, buy, transport, store, and move to the factory floor), fewer component to inspect and assemble (cutting total manufacturing time), possibly lower failure rate (simpler components are less likely to fail than complicated ones), and fewer spares for customers to handle.

AM does not always provide the best value when printing an exact replica of an article conventionally fabricated, unless it is addresses issues related to quality, delivery, and obsolescence, but it is most beneficial when utilized to reshape the component in order to reap some of the above benefits, and to address some shortcomings of the component's current design and performance.

When considering the replacement of an existing part with a printed version of it, one must weigh not only the cost of printing that part (material, time, design optimization, etc.), but also, if required, post processing, such as surface finishing, heat treatment, and machining (f.e. reaming and boring holes not perfectly round or smaller than nominal diameter). Moreover, if the part to be printed is functional and load-bearing, and not all required physical and mechanical properties of the printing material are available, the cost of printing test coupons and at least one test part and testing coupons and part must be included, adding tens of thousands dollar or more to the overall cost, and possibly making the printing of that part too costly. Often, when a company decides to print a load-bearing part selecting a certain printing material and printer, the company *qualifies* both, by collecting test data on the physical and mechanical properties of the material and the performance of the printer, and making sure properties and performance meet the company's requirements. Qualification is expensive and adds more cost to the total printing cost. Generally speaking, AM is economically advantageous more often for new parts designed for AM than in replacing existing parts conventionally made.

An additional advantage of AM is that it can be applied throughout the component's life cycle: development (from idea to market), production, and sustainment (maintenance, spares, retrofit).

However, it is worth stressing that AM is not a one-size-fits-all solution, and also features the following downsides (Oropallo and Piegl 2016; Abdulhameed et al. 2019):

- Metal and polymeric AM feedstocks have high price/kg when compared to feedstocks of the same material formulated for conventional manufacturing, although the comparison is not completely fair, because the price of the non-AM polymers is based on quantities typically sold in tons and more, whereas the AM polymers are sold in kilograms. F.e. in 2020 the price in USD/kg of PC and ABS was 40–300 and 32–250, respectively, for AM filament grades, whereas it was 3–5 and 2–4, respectively, for injection molding grades.
- Limits in printing speed, choice of materials, and maximum size of the printable parts.
- Presence of residual stresses

- Lack of a large material property database for most AM materials, especially for property values measured above and below room temperature, and values of long-term properties, such as fatigue and creep.
- Post-processing is required
- Closed-loop process controls need improvement.
- Variation of material properties among copies of the same article printed on the same printer at (a) different times or (b) at the same time. The issue in (b) is caused by inconsistent process conditions, f.e. printing temperature varying across the build plate.
- Not rarely, printing a functional component requires printing that component in multiple runs, in order to tweak the process parameters, printing orientation, etc. at each run and make an acceptable (surface finish, no distortion, accurate details, etc.) version of it.
- Modeling component's performance from its material properties and process parameters has not been perfected yet.
- Some fixed costs, such as setting the printer up and cleaning it after printing are about the same for one part as for many copies of it individually printed, unless multiple copies are built in one printing session, in which case those fixed costs can be distributed over a number of parts.
- Void formation is present within and between layers, resulting in delamination (Gibbons et al. 2010), porosity, and anisotropy (Chen et al. 2017; Zareiyan and Khoshnevis 2017), with mechanical properties higher in the printer's plane (XY) than normally to it (Z direction)Gibbons 2010.
- Anisotropic mechanical properties for some processes, such as PBF, FFF, and SL.
- Mechanical properties of printed objects may be inferior to those of the same object conventionally fabricated in the same material, which is caused by the AM process conditions, porosity, stronger bonding within each layer (*intralayer*) than between adjacent layers (*interlayer*), low pressure during extrusion, etc.
- Dimensional resolution (that is the minimum size of the printed geometric features) generally worsens as printed parts get bigger, and in XY is typically different than that along Z, where it is controlled by the layer's thickness. In some AM processes resolution is lower than in subtractive processes such as milling and turning, and post-processing is added to achieve the expected dimensional accuracy and surface finish, and meet the required functionality and appearance.
- Limitations in printed overhang surfaces when supporting material is absent.
- Warping and pillowing (Abdulhameed et al. 2019) occurs in some processes.
- In polymeric composites, interfacial adhesion between polymeric matrix and fillers needs improvement.

Besides cost, other factors influencing the final decision of printing an item are its criticality for the functionality or safety of the system to which it belongs, and possibly the safety of individuals depending on that system, like in the case of aircrafts. In the aviation industry, printing can alleviate shortage of spare parts causing the so-called *aircraft-on-the-ground* (AOG) situation, which is very costly for companies operating a grounded aircraft. F.e., in the case of an A380 Airbus grounded for technical reasons the daily cost of AOG was EUR 925,000 (DHL 2020).

Benefits and limitations of AM were extensively discussed by Gao et al. (2015).

However, the above shortcomings are becoming less significant, because they are being addressed by the R&D activity conducted by academia, government, and industry worldwide, and confirmed by the constant growth of the number of AM patent applications worldwide that jumped from 2,355 in 2007 to 24,245 in 2018 (Pohlmann 2019). The R&D activity is also favored by the fact that some printer types (such as the FFF models) are not expensive to modify and run, and make it affordable and very fast to introduce new AM polymeric feedstocks.

Overall, the benefits of AM outweigh its limitations. Since the first commercial printer SLA-1 was introduced in 1987, and patented by Charles Hull as an "Apparatus for production of three-dimensional objects by stereolithography" (Hull 1986), the whole AM business, including printers, materials, software, and services, has been growing steadily and rapidly, and this trend is expected to continue. From about 1978 to about 2013 the AM market has grown at an phenomenal average rate of 20–25% per year (Shortt 2013). In 2018, it was predicted that the world market of AM including hardware, materials, services, and software would develop from USD 9 billion to almost USD 42 billion (Sher 2018). As mentioned in Chapter 1, quantitative forecasting about AM is more complicated than about conventional or older fabrication technologies for which a greater amount of historical data is available, but there is agreement on the near-term proliferation of AM. As to AM polymers consumed by professional users, according to Grand View Research (2018), their U.S. market is expected to steadily rise from USD 0.3 billion in 2017 to USD 1.8 billion in 2025 at a remarkable CAGR of 24%, and their main markets will continue to be medical (Aimar et al. 2019), aerospace (Uriondo et al. 2015) and defense, automotive (Koenig 2019), and consumer goods (Quinlan et al. 2017) in decreasing order. In a 2017 survey, plastics were used by 88% of the AM users pooled, followed by resins (35%) and metals (28%) (Sculpteo 2017).

AM has become pervasive: its applications range from jet engines to human body parts, food, drugs, buildings and infrastructures (Bhardwaj et al. 2018), consumer products, and electronic circuits. In 2016, AM applications were split as follows: functional parts 29%, fit and assembly 17.8%, visual aids 10.3%, patterns for tooling 9.9%, patterns for metal castings 9%, presentation models 8.7%, education and research 8.7%, tooling components 4.7%, and others 1.9% (Wohlers 2016). This distribution indicates that not all AM materials must have functional or load-bearing properties. However, functional uses of AM materials are growing for plastic, composite, and metal feedstocks, and, f.e., reinforced polymer matrix composites such as carbon fiber PEEK, and carbon fiber PEKK are gaining ground in demanding applications where considerable long-term mechanical performance and thermal resistance are required. F.e. the Italian company Roboze prints parts PEEK and carbon fiber reinforced PEEK for high-performance parts that demand low weight, load-bearing properties, and high temperature resistance (Roboze 2021).

AM is also facilitating technological advances, and, f.e., it is a constitutive component of the *Internet of Things* or *Industry 4.0* (Galantucci et al. 2019), which is the network of computers, people, sensors, phones, and devices connected to each other, and transferring information among each other, or, briefly, "the integration of intelligent production systems and advanced information technologies" (Dilberoglu et al. 2017). Printers are already remotely connected to computers and operators who activate the printers, initiate the build, and monitor the results in real time. Factories are being set up with printers connected to robots and other devices feeding the printers with raw feedstocks, collecting the printed parts, and moving them to the next step, such as inspection, packaging, and shipping. A comprehensive review of AM processes and their contributions to Industry 4.0 was published by Dilberoglu et al. (2017). Another cutting-edge application of AM is fabricating *micro-electromechanical systems* or *MEMS* (Aspar et al. 2017). AM is at the juncture of some of the latest advancements in architecture, chemistry, computer science, design, engineering, materials science, medicine, physics, pharmacology, etc. AM is exciting because it is not only poised for the future but also for outer space: U.S. and European space agencies have been evaluating AM for in-space and on-orbit fabrication of components, systems, and structures like self-fabricating satellite, trusses, antenna reflectors, shrouds, etc. (Mitchell et al. 2018).

Table 2.1 provides an overview of the families and versions of AM processes, and ISO 17296-2 contains additional and concise relevant information on them. However, AM users typically choose a process for a specific part to print based initially on the part's size, service performance (strength, stiffness, service temperature range, etc.), and overall cost, hence based on the properties

TABLE 2.1

Families and Versions of AM Processes. Several Process Listed Are Patented

AM Process Category	Description	Binding Mechanism	Process Examples	Feedstocks	Printer Manufacturers	Advantages	Disadvantages	Commercial and Experimental Polymers
Binder Jetting (BJ)	Liquid binder is selectively dispensed onto regions of a thin powder layer	Chemical reaction or thermal reaction	ColorJet Printing (CJP). Multi Jet Fusion (MJF). Three-Dimensional Printing (3DP). HP Metal Jet.	Ceramics (gypsum, sand). Metals. Polymers (adhesives, bonding agents, hydrogels). Composites.	3D Systems, ExOne, VoxelJet, MicroJet Technology. MJF: HP. 3DP: Aprecia Pharmaceuticals, Therics. CJP: 3D Systems. HP Metal Jet (Hewlett Packard).	Fast. Efficient. Low cost materials and printers. No thermal distortions upon printing. Unbound powders are recyclable. No support material needed.	Fragile, non-structural parts. Poor strength. Post-processing required. Maintenance. Rough, grainy appearance. Some shrinking possible. Less accurate than material jetting.	Epoxies, PGA, PCL, PEO, HDPE, PLA, PMMA. starch, waxes, furan, phenolic, acrylates, cellulose.
Direct Energy Deposition (DED)	Concentrated thermal energy melts and fuses the feedstock while it is being deposited	Thermal reaction: melting and solidification	Laser Engineered Net Shaping (LENS). Electron Beam Additive Manufacturing (EBAM). Directed Light Fabrication (DLF). Direct Metal Deposition (DMD). Aerosol Jet. Laser Deposition Welding (LDW). Hybrid Manufacturing. Laser Generation. Laser-Based Metal Deposition (LBMD). 3D Laser Cladding. Laser Freeform Fabrication (LFF). Laser Direct Casting. DM3D Technology.	Metals. Ceramic fillers. Polymers.	RPM Innovations. LENS: Optomec. EBAM: Sciaky. 3D laser cladding: Kuka. DMD: Trumpf. LDW and Hybrid Manufacturing: DMG Mori. Aerosol Jet: Optomec. DMD: Optomec. DM3D: DM3D Technology.	For repairing parts. Multiple axis. Flexible. Fast. High material utilization. Very efficient for repair and add-on features. For large components.	Limited materials. Post processing may be needed. Limited accuracy, geometric complexity. Poor surface finish, resolution.	Thermoplastic matrix composites (AREVO).

(Continued)

TABLE 2.1 (Continued)

Families and Versions of AM Processes. Several Process Listed Are Patented

AM Process Category	Description	Binding Mechanism	Process Examples	Feedstocks	Printer Manufacturers	Advantages	Disadvantages	Commercial and Experimental Polymers
Material Extrusion (ME)	Material is heated and selectively dispensed through a nozzle or orifice	Thermal reaction or chemical reaction	Fused Deposition Modeling (FDM). Fused Filament Fabrication (FFF). Fused Layer Modeling (FLM). Contour Crafting (CC). 3D microextrusion. 3D microfiber extrusion. Continuous Filament Fabrication (CFF).	Polymers. Composites. Elastomers. Ceramic and metal fillers. Concrete. Soil. Living cells.	FDM: Stratasys. FFF: MakerBot, Ultimaker, many others. 3D dispensing: nScrypt. CC: Contour Crafting Corporation. Metal powder and polymeric binder: Markforged, Desktop Metal, AIM3d, Titan Robotics, Pollen AM.	Low price of printers. Low maintenance cost. Simple to use. Large size (CC and open chamber printers). Very low cost of the entry-level printers. Many materials. Broad property range. Versatile, easy to customize. Support materials are easily removed. Sparse fill enables stiff and light-weight parts.	Anisotropy. Voids. Slow. Support material may be needed. Surface featuring grooves and staircase look. No multi-material, multi-color parts. Warping and shrinking possible.	ABS, ASA, blends, HDPE, HIPS, PA, PC, PEI, PEEK, PEKK, PES, PETG, PLA, PP, PPE, PPS, PPSU, PS, PVC, PVDF.
Material Jetting (MJ)	Droplets of build material are selectively deposited, and bonded together	Adhesion by solidification of melted material or chemical reaction	Material Jetting (MJ). Drop on Demand (DOD). NanoParticle Jetting (NPJ). Ink-jet printing. Aerosol Jet Printing (AJP).	Polymers (thermosetting photopolymers). Wax. Ceramics. Composites.	MJ: 3D Systems, Stratasys, Arburg. DOD: Solidscape. NPJ: XJET. AJP: Sirris.	Fast and efficient. Complex shapes. Low waste. High resolution, accuracy. Multiple materials and colors.	Mainly thermosetting photopolymers. Costly maintenance. Post-processing may damage thin and small features Support materials not recyclable, and often required.	Acrylates, ABS, PA 10, PC, PMMA, PP, TPU.

Process	Description	Binding mechanism	Technologies	Materials	Systems	Advantages	Disadvantages	Feedstocks
Powder Bed Fusion (PBF)	Thermal energy selectively fuses together regions of a thin powder layer	Thermal reaction bonding	Selective Laser Melting (SLM). Selective Laser Sintering (SLS). Selective Heat Sintering. (SHS). Electron Beam Melting (EBM). Direct Metal Laser Sintering (DMLS). Laser Melting (LM). LaserCUSING. HP Metal Jet. Hybrid (PBF + CNC). Multi Jet Fusion (MJF).	Thermoplastic polymers. Elastomers. Metals (aluminum, cobalt chrome, copper, stainless steel, steel, titanium). Ceramics. Composites.	DMLS: EOS. LaserCUSING: Concept Laser. SLM: SLM Solutions. LM: Renishaw. SLS: 3D Systems, Sintratec, Sinterit. EBM: Arcam. Hybrid: Lumex.	No support material necessary. Good range of materials. Recyclable polymer and metal powders. Wide range of materials. Tough, and heat and chemical resistant materials. Functional, load-bearing parts. Accurate.	Expensive printers and materials. Slow. High power and ancillary equipment needed. Limited size, surface finish. Long cool down time. Grainy surface.	HDPE, PA 11, PA 12, PA 6, PC, PCL, PEEK, PEK, PEKK, PLA, PMMA, POM, PP, PS, PU, SAN, TPE, UHMWPE
Sheet Lamination (ShL)	Sheets or foils of material are laid and bonded together	Thermal reaction, chemical reaction, ultrasound	Laminated Object Manufacturing (LOM). Ultrasonic Additive Manufacturing (UAM). Selective Lamination Composite Object Manufacturing (SLCOM). Composite Based Additive Manufacturing (CBAM).	Paper. Polymers. Metals. Ceramics.	LOM: EnvisionTEC, Mcor Technologies. UAM: Fabrisonic.	No support structures needed. Low warping and internal stress. Multi-materials and multi-colors available. LOM: high speed, low cost, easy material handling. UAM: low temperature processing, embedded electronics, sensors.	LOM: post processing at times needed, limited materials. UAM: titanium, stainless steel require development. High material waste. Difficult removal of support trapped in internal cavities. Shapes not very complex. Thermal cutting produces harmful fumes. Possible warping.	Adhesives, binders, composites, paper, PVC.
Vat Photopolymerization (VP)	Liquid photopolymer in a vat is selectively cured by light-	Chemical reaction	Stereolithography (SLA or SL). Digital Light Processing (DLP). Continuous Liquid Interface Production or Printing (CLIP). Two-	Thermosetting photopolymers. Ceramic and metal fillers.	SLA: 3D Systems, Formlabs, DWS Systems. DLP: EnvisionTEC, B9Creations. CLIP: Carbon3D. 2PVP:	Excellent accuracy, resolution, surface finish. Complex shapes. Very	Only thermosetting photopolymers. Costly materials. Support required. Post curing needed to enhance	Resins based on acrylic, methacrylic, epoxy, vinyl ether. Cyanate esters.

(Continued)

TABLE 2.1 (Continued)

Families and Versions of AM Processes. Several Process Listed Are Patented

AM Process Category	Description	Binding Mechanism	Process Examples	Feedstocks	Printer Manufacturers	Advantages	Disadvantages	Commercial and Experimental Polymers
	activated polymerization		Photon Vat Photopolymerization (2PVP). Daylight Polymer Printing (DPP). Lithography-based Metal Manufacturing (LMM).		Nanoscribe, Photocentric.	thin walls. Fast. Flexibility in printers (various technologies, sizes, light sources). Broad range of mechanical properties (from stiff to rubbery).	strength, durability. Aging and degradation of mechanical properties over time.	

Sources: ISO 17296-2. ©ISO. This material is reproduced from ISO 17296-2:2015 with permission of the American National Standards Institute (ANSI) on behalf of the International Organization for Standardization. All rights reserved. Mitchell, A., Lafont, U., Hołyńska, M., Semprimoschnig, C. 2018. Additive manufacturing - A review of 4D printing and future applications. *Additive Manufacturing* 24:606–626 (permission to reproduce received from Elsevier). Lee, J.-Y. An, J., Chua, C. K. 2017. Fundamentals and applications of 3D printing for novel materials. *Applied Materials Today* 7:120–133 (permission to reproduce received from Elsevier). Dassault Systèmes 2021. 3D printing – Additive. https://make.3dexperience.3ds.com/processes/directed-energy-deposition (accessed January 16, 2021)

Note: for ease of reading the symbols ™ and ® were omitted when applicable.

of the feedstock material and the maximum printable volume. Secondary factors considered are: range of printable materials, ancillary equipment required, printing time, surface finish, minimum thickness printable, etc. Hence, a basic classification of AM processes more functional to users than that in Table 2.1 would start from the feedstock families: (a) polymers (TSs, TPs) and polymer matrix composites; (b) metals; (c) ceramic and ceramic matrix composites. Each family would be broken up in low-, medium-, and high-performance materials, with performance typically encompassing physical and mechanical properties over a temperature range.

This chapter is an overview of all AM process categories, and their most employed versions, and relative equipment, feedstocks, and applications. It focuses on commercial and experimental filled and unfilled polymers, and their properties, and provides formulators of SPs for AM with benchmarks and guidelines for developing new polymeric feedstocks, and product developers with the performance range of these feedstocks. The basic working equations of the process categories are reported, and for their analytical models we refer to the comprehensive book by Gibson et al. (2015).

2.2 AM FEEDSTOCKS

AM processes all types of engineering materials (ceramics, polymers, composites, metals and metal alloys), non-engineering materials (paper, food, etc.) (Dankar et al. 2018), natural fibers, and agricultural and forestry waste (straw, rice hull, wood, bamboo, corn powders, etc.) (Li et al. 2016; Chinga-Carrasco et al. 2018). Polymers and composites span a very broad range of applications: electronics, biomedical, construction, hobby and DIY items, aviation, etc. Compared to their versions for conventional processes, polymers and composites for AM are penalized by fewer choices of available materials, smaller databases of engineering properties, and higher prices per weight.

Polymers make up the bulk of AM feedstocks. According to Wohlers Associates, TSs constitute nearly 50% of the AM industrial market, followed by TPs (Ligon et al. 2017).

It is unlikely that a plastic grade for conventional processing is suitable for AM without proper modification, because the two families of technologies process their feedstocks at different values of temperature, pressure, time, viscosity, etc.

Currently, filled and unfilled polymers for AM have established themselves in applications with demanding sets of requirements (load-bearing, and temperature and chemical resistance), and have entered greatly dissimilar areas: tissue engineering, electronics, construction, aerospace, etc. However, there is demand for improved polymers, such as polymers that are less expensive, perform better, and enable faster printing, which opens opportunities for material developers.

2.3 MATERIAL EXTRUSION (ME)

2.3.1 PROCESS DESCRIPTION

ME is the preferred technology for inexpensive home printers but also large industrial printers. Essentially, AM processes based on *material extrusion* consist in taking a solid material, initially in form of pellets, powder, or (mostly) filament, liquefying and extruding it from the printer's nozzle, and dispensing it in a molten or semi-molten state on a flat platform (*build bed, print bed, print table*) in stacked layers and *selectively*, that is in specific locations according to the geometry in the 3D digital model of the part to be printed (Figure 2.2). The feedstock filament is available in two standard diameters, 1.75 and 2.85 mm, and is fed to the *printing* or *extrusion head* from reels. When the material leaves the nozzle is semisolid, fully solidifies when deposited, and bonds to the printer's build bed or the disposable *build sheet* laid onto the build bed at the beginning of the print, and to the previously laid layers after the first layer is placed. The thickness of each layer is about equal to the filament's diameter out of the nozzle. Once a complete layer is laid, the build

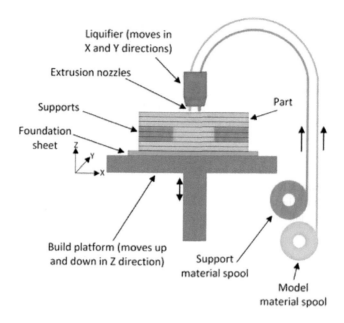

FIGURE 2.2 Schematic of filament-based material extrusion process.

Source: Gebisa, A. W., Lemu, H. G., 2018. Investigating Effects of Fused-Deposition Modeling (FDM) Processing Parameters on Flexural Properties of ULTEM 9085 using Designed Experiment. *Materials* 11, 500; doi:10.3390/ma11040500. Reproduced with kind permission from Prof. Gebisa.

bed is lowered by one layer thickness, the extrusion head deposits the next layer above the previous one, and the process repeats itself. Versions of the ME process are *fused deposition modeling*™ (FDM™), and *fused filament fabrication* (FFF). FDM™ was trademarked by Stratasys (USA), and invented by S. Scott Crump in 1989 (Crump 1992). FFF is basically a copy of FDM™ with a different name, and was implemented on home-grade inexpensive desktop commercial printers flooding the market after the FDM™ patent expired.

ME include the following steps (Gibson et al. 2015):

- Converting the 3D digital model of the article to be printed into a file in STL format and "slicing" it in thin layers
- Loading in the printer the feedstock, which advances driven by gravity (and aid of a plunger, or compressed gas), screw, or pinch rollers depending on its form (filament, pellets, etc.). Typically there are two feedstock filaments: one is the building material (*model material* in Figure 2.2), the other is the support material.
- Liquefying the feedstock in a heated chamber at the optimal temperature to prevent plastic to degrade or burn, and achieve adequate bonding between adjacent layers
- Applying pressure to the feedstock to push it through the nozzle, whose diameter is smaller than the starting filament's diameter, and dictates the minimum geometric feature that can be fabricated (*resolution*). Larger nozzle diameter enables faster printing, but lower shape fidelity to the digital model, and rougher, wavy surface.
- Extruding and depositing selectively the material (*bead, strand,* or *road*) that exits the nozzle as a filament and forms one XY cross section (*slice*) of the article, and, if needed, the support material that prevents distortions such as sagging and undesired deformations before the article fully solidifies. The extrusion head is mounted on a plotting system moving horizontally along two perpendicular axes. In some printers, the extrusion head is stationary and the print table moves in the XY plane.

- Solidifying and full bonding two adjacent layers together. During solidification, the deposited filament may change shape due to gravity and surface tension, and change size (shrinking) due to cooling and drying, when the feedstock gels. Bonding is typically achieved by heat energy that is contained in the filament, leaves the nozzle, and activates the surfaces of the adjacent layers.
- Cooling and removing the article from the printer.
- Post-processing if needed: annealing the article, removing support material, trimming, smoothing surface, reaming and boring holes, painting, installing bushings, etc. The support material can be the same or a different material than the one making up the article.

In ME a polymer behaving as a non-Newtonian fluid inside the printing head is driven through an extrusion tube of length L and radius R if its apparent viscosity η satisfies equation (2.1) in which ΔP and Q are the pressure drop and the volumetric flow rate, respectively (Bikas et al. 2016):

$$\Delta P = (8\eta QL)/(\pi R^4) \tag{2.1}$$

The shear rate $\dot{\gamma}$ (1/s) during extrusion can be calculated using equation (2.2), where v (m/s) is the average velocity of the extruded plastic, D (m) the diameter of the nozzle, Q (m^3/s) is the extrusion rate, and A (m^2) is the nozzle's cross-section area (Le Tohic et al. 2018):

$$\dot{\gamma} = 8v/D = 8(Q/A)/D \tag{2.2}$$

Once outside the extruder, the polymer is deposited, and to prevent buckling its tensile modulus E must satisfy equation (2.3), in which L/R and K are the slenderness ratio of the filament and a scaling factor experimentally determined, respectively:

$$\frac{E}{\eta} < \frac{8Ql\,(L/R)^2}{k\pi^3 R^4} \tag{2.3}$$

Areas of improvement for ME processes are printing speed, and, in printed parts, surface roughness, dimensional accuracy and precision, interlayer adhesion, and residual stresses (Singh et al. 2020).

The filament must feature adequate values of tensile modulus and viscosity: the former must prevent filament buckling under extrusion pressure, and the latter must not offer excessive resistance to extrusion pressure (Gkartzou et al. 2017). Additional details on FFF and relative polymers are discussed in the reviews by Brenken et al. (2018), Goh et al. (2018), Goh et al. (2019), and Ligon et al. (2017).

Viscosity is critical in ME, and must be adequately low to extrude the polymer without "excessive" force, and adequately high so that the extruded thread keeps its shape while standing alone and under the weight of the above layers. Moreover, if the melted polymer is too viscous, the extruded thread may have limited deformation and spread, and not merge thoroughly with the adjacent layers, forming voids. Crystallization rate of the melted deposited polymer can also affect porosity: if the polymer has a fast crystallization rate and a high crystallinity after cooling, the deposited thread exhibits a relatively high shrinkage, and partially merges with the adjacent deposited threads, resulting in pores (Rasselet et al. 2019).

Performing a life-cycle assessment based on many environmental variables, Faludi et al. (2015a) compared FFF to CNC machining in sustainability, and concluded that sustainability depends primarily on the percent utilization of each equipment and secondarily on the specific equipment, and they could not state in general terms that AM was more environmentally friendly than CNC. Higher use decreased idling energy use and amortized the impacts of each process. The dominant environmental impact was caused for FFF always by electricity use, and for CNC at maximum utilization by material waste. At high and low utilization FFF had lower ecological impacts per part than CNC had.

Additional information on process design and modeling, dimensional accuracy, print resolution, and surface roughness in ME and FFF are shared in the reviews by Turner et al. (2014) and Turner et al. (2015).

2.3.2 ME PRINT PARAMETERS

The geometry of the part to be extruded is established in the 3D digital model, but there are several *process* or *build* or *print parameters* that are associated with the printer and influence the internal structure, properties, and surface finish of the part, amount of support material, and printing time, and they have to be possibly set at optimal values before starting printing, in order to maximize the overall outcome, that is fabricating in the shortest time and with the smallest amount of feedstock and support material a high-quality part performing as intended. The major ME build parameters are listed below (Figure 2.3):

- *Air gap* or *raster-to-raster gap*: distance between adjacent deposited beads or *roads*. It can be: (a) zero, if beads touch each other; (b) positive, if there is a gap between adjacent beads; (c) negative, if the beads are overlapped; a negative value can increase strength and stiffness (Ahn et al. 2002).
- *Infill*. A printed part can be inside completely filled or porous, even if its outer surface is non-porous. *Infill* is the part's volume occupied by its outer surface minus the total porosity volume, or, in brief, the part's volume actually occupied by material. Infill is expressed on a scale from 0 (theoretical) to 100% (a part without voids). When infill is <100%, the remainder of the part is voids. Voids may be beneficial when printing a visual model and a non-functional prototype, in order to reduce printing time, material, weight, and cost.
- *Layer thickness* or *slice height*. Obviously greater thickness reduces printing time which is beneficial in prototypes. However, since the object is printed as a series of stacked layers, it features a waviness on flat surfaces, and steps (*stair-stepping* or *staircase effect*) on curved surfaces. Hence, greater thickness induces greater waviness, taller steps, rougher surface, and poorer resolution, which instead must typically be high in end-use articles.
- *Number of contours* (*contour width*): number of outer and, in hollow objects, inner perimeters of each layer that make up the edge of each "slice" and the surface of the part. They

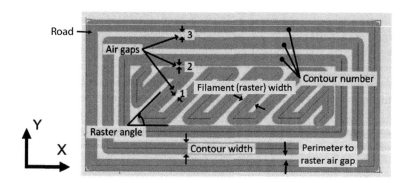

FIGURE 2.3 Material extrusion printing parameters.

Source: Gebisa, A. W., Lemu, H. G., 2018. Investigating Effects of Fused-Deposition Modeling (FDM) Processing Parameters on Flexural Properties of ULTEM 9085 using Designed Experiment. *Materials* 11, 500; doi:10.3390/ma11040500. Reproduced with kind permission from Prof. Gebisa.

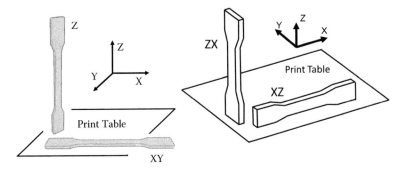

FIGURE 2.4 Printing orientation of test coupons for ME process. Left: Innofil3D printing orientations. Reproduced with permission from Innofil3D BV. Right: Stratasys printing orientations.

are the first portion of the part to be traced in each slice, after which the part's inside is filled with inclined rasters. Then the process is repeated to generate all the remaining layers.

- *Part* or *build orientation*: inclination of the part on the build bed with respect to the printers' Cartesian axes X, Y, Z with XY the plane of the build bed, and Z perpendicular to XY (Figure 2.4). The first decision is how to orient the part on the build bed because the properties of parts printed via ME are anisotropic (typically stronger and stiffer in XY plane than along Z), and the part may be too large to fit in Z direction but not in X and Y.
- *Path* or *toolpath*: sequence traced by the nozzle's positions during printing.
- *Raster*: a rectangular pattern of parallel lines traced by the bead that fills the area within the part's contour.
- *Raster angle*: direction of raster relative to the X axis of the build bed.
- *Raster width* (*road width*): width of deposited filament inside the part's contour.

The effects of the above parameters on mechanical properties of objects printed via ME have been documented, measured, and modeled for at least two decades. Table 2.2 summarizes a survey of the technical literature from 2010 to 2015 on the effect of air gap, filament and layer thickness, part orientation, raster angle and width on the tensile strength of parts printed via FDM™ using two non-SPs: ABS (inexpensive, popular especially for home printing), and Ultem™ 9085, commercial version of PC/PEI (costly, high-performance, for demanding applications). Most studies were conducted on ABS because of its affordable price and diffusion. Ali et al. (2014) reported a comprehensive list of FDM™ process parameters influencing a printed part, and how often they were mentioned in the technical literature: most investigated variables were road width, layer thickness, raster angle, part orientation, and air gap.

Li et al. (2018a) assessed the effects of layer thickness, deposition velocity and infill rate on tensile strength and interlayer bonding degree (observed through a microscope) of coupons printed via FFF from PLA filament. Popescu et al. (2018) published a recent and broad review of technical literature on how different process settings in ME impacted mechanical behavior of built parts made of ABS, PA, PC, PEI, PEEK, and PLA. From this review we extracted a list of mechanical properties measured on PLA coupons to investigate the effect of process parameters on printed parts (Table 2.3). It may surprise that even the feedstock's color influences PLA's strength, because different colors are obtained with additives of dissimilar chemical composition that induces a varying amount of crystallinity in the final filament (Wittbrodt and Pearce Effect of FDM™ Printing Variables on Tensile 2015). Torres et al. (2016) measured the effects of extrusion temperature and speed, infill direction and density, layer thickness, and perimeter on the tensile strength, modulus and strain, and toughness and fracture properties of coupons printed out of PLA via FFF using design of experiment (DOE). The authors identified the combination of the above process parameters that yielded consistently medium to high values for all properties tested. Torres

TABLE 2.2
Effect of FDM™ Printing Variables on Tensile Strength (TS) of ABS and Ultem™ 9085

Printing Variable	Results and Conclusions	Material	Reference
Air Gap	Highest TS values were found with a negative air gap.	Ultem™ 9085	Bagsik and Schöppner (2011) and Bagsik et al. (2010)
Air Gap	Negative air gap significantly increased TS compared to positive air gap.	ABS	Onwubolu and Rayegani (2014)
Air Gap	TS improved with larger air gap and the amount of increase depends on the values of other process parameters.	ABS	Sood et al. (2010)
Filament Thickness	Thicker filaments were better for X and Z build directions, but thinner filament was better for the Y build direction.	Ultem™ 9085	Bagsik and Schöppner (2011) and Bagsik et al. (2010)
Layer Thickness	Higher TS coincides with thicker layer and 0° part orientation.	ABS	Onwubolu and Rayegani (2014)
Layer Thickness	TS was the highest with the thickest of the three layers.	ABS	Sood et al. (2010)
Part Orientation	TS was higher across raster angle values on coupons printed with their thickness contained in the printing platform plane.	ABS	Durgun and Ertan (2014)
Part Orientation	Highest TS was recorded in coupons printed horizontally, that is with their length contained in the printing plane. Lowest TS was measured in coupons printed vertically, that is with their length perpendicular to the printing plane.	ABS	Onwubolu and Rayegani (2014)
Part Orientation	Coupons were printed at 0°, 15°, 30° between their length axis and the printing plane. TS was max at 0° and min at 90°.	ABS	Sood et al. (2010)
Raster Angle	TS increased by 6% with raster angle varying from 0° to 45°.	ABS	Onwubolu and Rayegani (2014)
Raster Angle	TS increased by 14% with raster angle changing from 0° to 60°.	ABS	Sood et al. (2010)
Raster Width	Increasing raster angle by 10% and 20% did not result in a consistent effect on TS which increased and decreased, depending on other variables.	ABS	Sood et al. (2010)
Raster Width	TS was sensitive to raster width but it was also affected by other factors, such as layer thickness and part orientation. The common trend was that at 0° part orientation TS slightly decreased, while at 90° part orientation TS increased.	ABS	Onwubolu and Rayegani (2014)

Sources: Bagsik, A., Schöppner, V., 2011. Mechanical properties of fused deposition modeling part manufactured with Ultem™ 9085 ANTEC 2011, Boston; Bagsik, A., Schöppner, V., Klemp, E. 2010. FDM Part Quality Manufactured with Ultem™ 9085, "Polymeric Materials 2010" Halle (Saale) https://www.semanticscholar.org/paper/FDM-Part-Quality-Manufactured-with-Ultem-*-9085-Bagsik-Sch%C3%B6ppner/63ec4f3fd6e39600080e0a14caaa7754cc653632 (accessed July 4, 2019). Onwubolu, G. C., Rayegani, F. 2014. Characterization and Optimization of Mechanical Properties of ABS Parts Manufactured by the Fused Deposition Modelling Process. *International Journal of Manufacturing Engineering* Article ID 598531. Sood, A. K., Ohdar, R. K., Mahapatra, S. S. 2010. Parametric appraisal of mechanical property of fused deposition modelling processed parts. *Materials and Design* 31:287–295. Durgun, I., Ertan, R., 2014. Experimental investigation of FDM process for improvement of mechanical properties and production cost. *Rapid Prototyping Journal* 20 (3): 228–235.

TABLE 2.3

ME Process Parameters, and Mechanical Properties of PLA on Which the Effect of the Parameters Was Investigated

Process Parameter	Mechanical Properties Measured
Build orientation	Tensile strength and modulus
Filament colors	Tensile strength
Infill percentage	Tensile strength, shear stress
Infill pattern	Tensile strength
Layer thickness	Tensile strength and modulus, shear stress, impact strength, flexural strength
Printing temperature	Tensile strength
Printing speed	Tensile strength
Raster angle	Tensile strength and modulus
Raster gap	Tensile strength, flexural strength, impact strength

Source: Data from Popescu, D., Zapciu, A., Amza, C., F., Marinescu, R. 2018. FDM process parameters influence over the mechanical properties of polymer specimens: A review. *Polymer Testing* 69:157–166. Reproduced with permission from Elsevier.

et al. (2015) also measured the effect of layer thickness, infill density, and post-processing heat-treatment time at 100°C on the shear response and ductility of PLA coupons printed via FFF, with the goal to achieve properties as close as possible to those of bulk PLA. A similar study was conducted by Lanzotti et al. (2015), who quantified the variation of tensile strength, modulus, and strain at break of coupons in PLA printed on a type of low-cost, open-source (Rep-Rap) FFF printer in response to changes in layer thickness, infill orientation, and the number of contours (shell perimeters). The authors elucidated the relationship among the process parameters and stiffness and strength of the printed coupons, and proposed an empirical model relating process parameters and mechanical properties for PLA.

The influence of printing speed on tensile strength of PLA was also assessed by Tsouknidas et al. (2016). They printed test coupons out of an unspecified PLA at 180°C extrusion temperature, 50% infill density, and rectilinear filling, and tested them per ASTM D638 at strain rate of 0.0087/ s. Increasing printing speed from 30 to 220 mm/s resulted in a steady, almost linear decline of tensile strength from 57 to 49 MPa, and rise of elongation at break from 3 to 3.25%.

A review of research and future prospects in optimization of build parameters was published by Mohamed et al. (2015).

Since some process parameters significantly impact the part's physical and mechanical performance, designers of structural components intending to maximize those properties should consult the relevant technical publications and/or experiment with combinations of setting values to assess their optimal combination. Hereafter the effect of some of the above process parameters on part's properties printed via FFF for specific SPs are described.

2.3.3 POLYMERS FOR ME

Commercial and experimental polymers for ME span a wide array: TPs (the most popular), rubbers, PUs, silicones, inorganics and organic pastes, polymer latex, biomaterials, hydrogels, functional polymers, plastisols, biologically active ingredients, and living cells. Table 2.4 lists the main polymeric feedstocks for ME, their applications in non-printed articles, advantages, and disadvantages. Many of these applications are not economically suited for AM, but they can inspire other or similar AM applications.

TABLE 2.4

Main Polymeric Feedstocks for ME: Uses in Non-Printed Articles, Advantages and Disadvantages

Polymer	Applications	Advantages	Disadvantages
Acrylonitrile Butadiene Styrene (ABS)	Domestic appliances, pipes, medical devices, prototypes, tooling, covers, lawn mower covers, safety helmets, luggage shells, fittings, telephone handsets, computer and other office equipment housings.	• Chemically resistant to dilute acids, dilute alkalis, greases, oils • Dimensionally stable • Easy to glue and paint • Easy to machine • Impact resistant • Insoluble in water • Low to intermediate price • Soluble in acetone for achieving smooth finish and joining parts • Tough	• Controlled temperature and humidity during printing is required. • Storing feedstocks in temperature and humidity controlled environment without exposure to sunlight. • Parts exposed to high temperatures or sunlight may deteriorate, discolor, or warp over time. • AM facility should be open and/or ventilated. • Poor chemical resistance to alcohols, aromatic and halogenated hydrocarbons.
High Impact Polystyrene (HIPS)	Displays, scale models, food packaging, cases, computer housings.	• Dimensionally stable • Easy to paint and glue • Fully recyclable (FDA compliant) • Low price • Resistance to heat, impact	• Pure HIPS is very flammable • Poor solvent resistance • AM facility should be open and/or ventilated to expel fumes developed during printing.
Polyamide (PA)	Gears, bushings, plastic bearings, build jigs, fixtures, guards, textiles, carpets, fishing lines, food packaging, card door handles and radiator grilles, low voltage switch gears, ski bindings and in-line skates.	• Durable • Flexible • Low coefficient of friction • Resistant to abrasion, bases, fatigue, heat, solvents • Strong • Wide range of properties due to its many formulations	• Attacked by strong acids, alkalis, alcohols, oxidizing agents, phenols • Dimensionally unstable (shrinkage, warping) • Hygroscopic (must store in dry conditions) • Sensitive to sunlight, requires UV stabilizers • Flammable although slow to ignite
Polycarbonate (PC)	Automotive components (head lamp lenses, dashboards, interior cladding, exterior parts), aerospace, baby bottles, electrical (covers and housings, connectors, household appliances, mobile phones, electrical chargers, lighting, battery boxes), protective eyewear and gear, medical devices, lighting fixtures.	• Chemical resistance to alcohols, dilute acids, aliphatic hydrocarbons • Durable • Impact resistant • Electrical insulator • Stiff • Strong • Thermally stable • Tough • Transparent	• Hygroscopic (must store in dry conditions • Limited chemical (attacked by dilute alkalis, and aromatic hydrocarbons) and scratch resistance • Tendency to shrink and warp • Tendency to yellow upon long-term exposure to UV light

TABLE 2.4 (Continued)
Main Polymeric Feedstocks for ME: Uses in Non-Printed Articles, Advantages and Disadvantages

Polymer	Applications	Advantages	Disadvantages
Polyether ether ketone (PEEK)	Aerospace (engine, exterior, and interior parts), automotive (bearings, seals, washers, components in transmission, braking and air-conditioning systems), healthcare and medical (dental syringes and sterile boxes, load-bearing implants), electrical/electronic (coaxial connector jacks, insulators for connector pins), food contact applications.	• High-performance polymer • Biocompatible: does not cause toxic and mutagenic effects • Compatible with carbon and glass • Continuous use temperature under loading (260°C) • Flow and processing properties • High glass transition (143°C), melting (334°C), and heat distortion temperature (316°C) • Lightweight • Low coefficient of friction and self-lubricating • Resistant to abrasion, chemicals, corrosion, fatigue, flame, heat, hydrolysis, peeling • Stiff • Tough	• Expensive • Attached by some acids (concentrated sulphuric, nitric, chromic), halogens and sodium • High processing temperatures • Low resistance to UV light
Polyether ketone ketone (PEKK)	Similar applications to PEEK. Aerospace, offshore, oil, and gas.	• Flame retardant • High melting temperature (305- 358°C) • Low smoke ad toxicity generation • Resistant to abrasion, chemicals, fire, impact, wear • Stiff • Strong • Strong electrical insulator	• Anisotropic • Expensive • High processing temperatures
Polyetherimide (PEI)	Automotive (transmission components, throttle bodies, ignition components, thermostat housings, bezels, reflectors, lamp sockets, fuses, gears, bearings, solenoid bodies, ignition switches and oil pump drives), aerospace, electrical (housings, switches and controls, motor parts, printed circuit boards, connectors),	• High heat deflection temperature • Strong • Impact resistant • Meets flame, smoke, and toxicity requirements • High strength-to-weight ratio • Resistant to chemicals, UV, weather • Electrical insulator • Stable electrical properties over range of temperatures and frequencies	• Expensive • High processing temperatures • Notch sensitive • Attacked by polar chlorinated solvents • Long drying required before processing

(Continued)

TABLE 2.4 (Continued)

Main Polymeric Feedstocks for ME: Uses in Non-Printed Articles, Advantages and Disadvantages

Polymer	Applications	Advantages	Disadvantages
	household (microwave cookware, steam and curling irons, ovenproof trays for food packaging), medical, tooling.	• Retains mechanical properties at high temperatures • Easy to machine • Biocompatible	
Polyethylene (PE)	Bags, films, sheets, toys, wire insulation trays, containers, agricultural film, food packaging film. Containers of detergent, milk, juice, and shampoo. Buckets, crates, fencing.	• High strength-to-density ratio • Tensile, compressive, impact strength	• Tends to shrink and distort easily when cooled. • Does not stick to itself well.
Polyethylene Terephthalate Glycol-modified (PETG)	Clothing fibers, flexible food packaging, food containers, plastic bottles, space blankets	• Layer-to-layer adhesion • Resistant to impact, heat • Strong	• Easy to scratch • Hygroscopic, requiring storage in a dry conditions • Sensitive to sunlight (UV exposure), requiring UV stabilizers
Poly(lactic) Acid (PLA)	Automotive (floor mats, pillar cover, door trim, front panel and ceiling material) bags, food packaging, drug delivery systems, medical suturing, tissue (bladder, bone, cartilage, liver) remodeling, wound covers, medical devices (fixation rods, plates, screws), textiles (shirts, carpets, bedding, mattress, sportswear)	• Derived from plants • Environmentally friendly • Heated bed optional • Nontoxic • Stiff • Biodegradable and compostable • Low price	• Low glass transition and melting temperatures • Low impact strength • Thermally unstable • Poor gas barrier performance • Slow degradation rate • Not flexible
Polymethyl Methacrylate (PMMA)	Bone cement, dentures, lightweight and shatter resistant replacement for glass such as eyeglass lenses, and vehicle exterior light lenses	• Compatible with human tissue • Impact resistant • Low-cost • Rigid • Stiff • Strong • Tough	• Enclosing the AM chamber may be needed to regulate cooling • High nozzle temperature is required to prevent warping and maximum clarity • Easy to scratch
Polypropylene (PP)	Automotive parts, bank notes, car parts, carpet yarn, containers, fiber-reinforced items, food packaging, hinged caps, housing for electrical appliances, laboratory equipment, microwave-	• Flexible • Food safe • Lightweight • Low-cost material • Resistant to chemicals, corrosion, fatigue and heat • Tough • Can be steam sterilized	• Poor layer-to-layer adhesion • Tends to warp • Poor resistance to UV, impact and scratches • It becomes brittle below −20°C

TABLE 2.4 (Continued)

Main Polymeric Feedstocks for ME: Uses in Non-Printed Articles, Advantages and Disadvantages

Polymer	Applications	Advantages	Disadvantages
	proof containers, pipes, ropes, suitcases, sweet and snack wrappers.		• Low upper service temperature (90–120°C) • Attacked by highly oxidizing acids • Swells in chlorinated solvents and aromatics • Poor paint adhesion
Polystyrene (PS)	Protective packaging, containers, lids, bottles, trays, toys, light diffusers, beakers, cutlery, general household appliances, video/audio cassette cases, electronic housings, refrigerator liners, housewares, glazing.	• Dimensionally stable • Easy to mold and vacuum form • Good electrical properties, low dielectric loss • Hard • Inexpensive • Resistant to gamma radiation • Stiff • Transparent	• Brittle • Degrades under UV • Flammable • Poor barrier to oxygen and water vapor • Poor chemical resistance to organic compounds • Relatively low melting point • Slow to biodegrade
Polyurethane (TPU)	Agriculture, automotive (instrument panels), sheets, building insulation, hoses, insulating foams for fridges, mattresses and pillows, performance films, seals, gaskets, textile coatings, leisure and sport, tubes, wire and cable jacketing. Timing, transmission, and conveyor belts.	• Properties depend on its formulations (polyester, polyether, polycaprolactone) • Can be sterilized • Durable • Flexible, even at low temperatures • Food safe • Resistant to abrasion, chemicals, grease, impact, microbes, oil, tear, UV, solvents, weather • Versatile	• Lower strength than non-elastomeric polymers • Expensive • Drying often required prior to use • Some grades have short shelf life • Clear TPU yellows irreversibly

The basic working principle of ME applies to plastics, metals and composites, as long as the feedstocks are produced in form of filament featuring size, strength, and properties compatible with ME. This section deals with the features that polymers must exhibit to be suitable for ME. Feedstock for ME are fed to the printers in solid form, melted, and go back to solid state without substantial property degradation. As explained in Chapter 1, TPs can be repeatedly melted upon heating and solidified upon cooling without significant property degradation, and hence are well-suited as feedstocks for ME. Furthermore, TPs for ME must meet the following requirements: (a) steadily flowing through a 0.5 mm or smaller diameter nozzle; (b) developing sufficient adhesion to the previously printed layer; (c) being fluid enough to be extruded without a large motor; (d) providing an uninterrupted bead; (e) being viscous enough when out of the nozzle so that the bead maintains its shape after being extruded without spreading; and (f) being adequately stiff outside the nozzle to satisfy equation (2.3) and allow some amount of overhanging. Amorphous polymers are well suited for ME and preferred to crystalline polymers, because the former ones do not have a distinct melting temperature but soften over a wide range of temperature, and their viscosity decreases upon raising temperature, until they turn into a viscous paste that is fluid enough to be extruded through an approximately 0.2–0.5 mm diameter nozzle, and viscous enough to maintain its shape immediately after deposition. Moreover, adjacent printed layers of amorphous polymers bond well to each other (Gibson et al. 2015). Instead, highly crystalline polymers tend to shrink upon cooling, which hinders the fidelity between computer model and actual geometry of the printed article (Chang and Faison 2001). The flowability of a polymer in its molten state is expressed by the *melt flow index* (MFI) or *melt flow rate* (MFR) that is inversely proportional to its dynamic viscosity, and is 6-15 g/10 min in commercial PLA filaments for ME. Wang et al. (2017a) experimented with seven commercial PLA filaments, and concluded that MFI is a key variable to achieve adequate printing quality, along with plasticizer type and degree of crystallinity, and recommended a threshold value of 10 g/min, measured per ISO 1133 (ISO 1133-1:2011).

Commercial filaments for ME keep growing in number, and are, unfilled and filled: ABS, HIPS, PA 12, PLA, PC, PC/ABS, PC/PEI, PEI, PEEK, PEKK, PPSF/PPSU, PPS, PSU, and PVC. The range of property values and prices relative to ME plastics is broad and, along with equipment cost, simplicity of use, low operating cost, and no expensive ancillary equipment needed (except tools to remove support material) partly explains the popularity of ME among corporate, institutional, and personal users. Price and major mechanical properties of ME plastics are plotted in Chapter 1, namely: price (USD/kg) in Figures 1.6 and 1.7, tensile strength in Figure 1.10, tensile modulus in Figure 1.11, Izod notched impact strength in Figure 1.12, and heat deflection temperature at 1.82 MPa in Figure 1.13. The information in Table 2.4 can aid in the preliminary screening of the feedstock for a particular article, and should be followed by consulting suppliers' material property data sheets. Vice versa, Table 2.4 can guide in selecting uses for specific ME materials.

Applications of ME grow in number, and span a very broad range comprising hobby and DIY articles, visual and engineering models, prototypes, tooling, jigs and fixtures, and end-use functional and load-bearing parts for aerospace, automotive, construction, consumer products, and medical.

In general TSs typically outperform TPs in critical areas such as mechanical properties (f.e. tensile modulus), chemical resistance, thermal stability, and overall durability, but only TPs are commercially employed for ME. However, recently two newly developed polymers indicate that it is technically feasible to extrude TSs for ME, and a relative market for TSs may open up in the near future if the ratio performance/price is competitive. In one case, Oak Ridge National Lab (USA) successfully demonstrated a new large-scale, high-rate extrusion-based printing technology using a two-component TS polymer whose individual ingredients were mixed in the deposition nozzle, and underwent a fast crosslinking reaction generating parts with excellent mechanical properties (Rios et al. 2017). TSs offer the following advantages:

- Remaining liquid until deposition allows lightweight pumps and deposition hardware to attain higher deposition rates with greater precision than is currently possible with TPs
- Curing after deposition would form physical crosslinks between printed layers, potentially eliminating the interlayer weakness that penalizes parts printed out of TPs
- The chemistry involved in this study is well established, scalable for manufacturing, and tuned to achieve an extensive range of properties that span from flexible to rigid, and comprise high toughness, impact resistance, and optical transparency.

The other newly experimental TS developed for ME was reported by Yang et al. (2017a), who formulated three TS resins for a version of ME called *Diels–Alder reversible thermoset* (DART) *process*, based on a type of chemical reaction called *furan-maleimide Diels–Alder*. As monomers the authors selected a bismaleimide, two tri-functional furan monomers, and a tetra-functional furan monomer. Bismaleimide and furan are characterized by high temperature resistance. The former one features tensile strength and modulus of 40–80 MPa, and 4.1–4.8 GPa, respectively (Iredale et al. 2017), while furans are mostly obtained from biomass (Kwiecien and Wodnicka 2020), and hence the research by Yang et al. (2017a) is interesting also because it might lead to biobased feedstocks for ME. The monomers were crosslinked, and featured TS properties at service temperatures, very low melt viscosity at print temperatures, smooth surface finish, and isotropic mechanical properties in printed parts, because crosslinks between deposited layers alleviate anisotropy in printed parts. Particularly, in comparison to printed samples from the control materials, in DART polymers the interlayer adhesion in the printed parts was >95% improved and, hence, anisotropy significantly lower. The sequence of DART process is the following: DART polymers are heated in a syringe, extruded through a nozzle and deposited on the platform layer-by-layer with the help of an extra cooling system. The tensile properties measured across the three TS resins and printing orientations were: strength 9–23 MPa, modulus 8–85 MPa, and strain 1.8–2.5%.

Finally, the company Green Cycles® (Spain) sells a PVA filament as that is 100% water soluble, biodegradable, and compostable, and can serve as a building material and (especially) support material. It is offered in three versions that completely dissolve at temperature ranging from 5 to 70°C, and adheres to PLA, PBS, and PET.

2.3.4 ANISOTROPY OF PRINTED PARTS

Anisotropy of articles printed via ME must be accounted for when designing load-bearing articles. In fact, technical data sheets of ME filaments released by suppliers report values of mechanical properties that are measured on test coupons printed in different directions and differ from each other. F.e. Innofil3D and Stratasys report that the PLA's properties measured from coupons printed in two directions (Figure 2.4) are significantly different, as is illustrated in Table 2.5.

Anisotropy in parts printed out of SPs via ME is elucidated in the chapters on PLA and PA. Chacon et al. (2017) extensively described the effect of printing orientation and other process parameters on mechanical properties of PLA coupons printed via ME, and the optimal selection of those parameters.

2.3.5 EFFECT OF BUILD ORIENTATION ON STRENGTH OF PRINTED PARTS

Yao et al. (2019) developed equations (2.4) and (2.5) accurately predicting the tensile strength (*TS*) in PLA coupons printed via ME at various orientations from 0° to 90° with respect to the layers' plane (Figure 2.5):

$$TS_\theta = \left[\left(\frac{cos^4\,\theta}{TS1^2} \right) + \left(\frac{1}{SS_{sf}^2} - \frac{1}{TS1^2} \right) sin^2\,\theta cos^2\,\theta + \frac{sin^4\,\theta}{TS2^2} \right]^{1/2} \tag{2.4}$$

TABLE 2.5
Tensile Properties of Coupons in PLA Filaments Printed in Various Build Orientations

Supplier	Build Orientation	Strength	Yield Strength	Modulus	Elongation at Break
		MPa	MPa	MPa	%
Innofil3D	Z	29	N/A	3,150	1.1
Innofil3D	XY	38	N/A	2,852	2.1
Stratasys	XZ	48	45	3,039	2.5
Stratasys	ZX	26	26	2,539	1.0

Source: Technical data sheets posted online by Innofil3D and Stratasys.

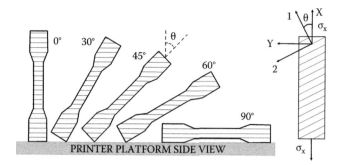

FIGURE 2.5 Printing orientations of tensile coupons defined from the angle between direction of coupon's length and Y axis. Drawing based on Yao, T., Deng, Z., Zhang, K., Li, S., 2019. A method to predict the ultimate tensile strength of 3D printing polylactic acid (PLA) materials with different printing orientations. *Composites Part B* 163:393–402.

$$TS_\theta = \left[\left(\frac{cos^4\,\theta}{TS1^2} \right) + \left(\frac{1}{SS_{HT}^2} - \frac{1}{TS1^2} \right) sin^2\,\theta cos^2\,\theta + \frac{sin^4\,\theta}{TS2^2} \right]^{1/2} \tag{2.5}$$

In equations (2.4) and (2.5) TS_θ is the *TS* value at any angle θ in Figure 2.5, *TS1* and *TS2* are *TS* values in directions 1 and 2, respectively, SS_{sf} and SS_{HT} are the shear strength according to standard formula for unidirectional composites and the Hill-Tsai criterion, respectively, calculated as follows:

$$SS_{sf} = \frac{F\,(45)}{2WT} \tag{2.6}$$

$$SS_{HT} = \left[\left(\frac{4}{TS\,(45)^2} \right) - \left(\frac{1}{TS^2} \right) \right]^{1/2} \tag{2.7}$$

In equations (2.6) and (2.7) $F(45)$ is the maximum tensile load (measured in N) of unidirectional composite coupons with fibers oriented at 45° with respect to the loading direction, W and T are width and thickness of coupon's cross section, respectively, and TS(45) is the TS measured on coupons with layers printed at 45° (Figure 2.5). Equations (2.4) and (2.5) are based on the assumption that the coupons are transversely isotropic, that is they feature isotropic mechanical properties in the printing plane, and symmetric mechanical properties about the axis normal to the printing plane.

2.3.6 PELLET-BASED EXTRUSION

In ME based on pellets (or *granulates*) the feedstock is processed in form of pellets instead of filament. Pellets are less expensive and can be extruded at higher speed than filaments, which is beneficial for large-scale printing. According to UPM Formi (Finland), a supplier of sustainable AM feedstocks, granulate-based printing is on average 37 times faster than filament-based methods, and permits to efficiently construct large objects and structures, such as boats and buildings, respectively. Moreover, AM filaments are available in a limited in the range of materials, whereas all industrial polymer are available as pellets, which opens up a wide range of business opportunities for formulators of AM polymeric feedstocks. However, a plastic formulated as pellets for a non-AM process is not necessarily suitable for ME, and may have to be accordingly modified for ME.

Examples of ME printers processing pellets are those marketed by Pollen AM (France) that employs ABS, PET, and PP, and Titan Robotics (USA) that features a wide choice of feedstocks: (a) unfilled polymers: ABS, PA 6, PA 66, PA 12, PC, PETG, PEKK, PLA, PVC, TPE, TPU; (b) glass or carbon fiber filled polymers with matrix of ABS, PA, PEEK, PEI, PPO, and PPSU.

Pellet-based ME has been described by several authors since 2005 (Moreno Nieto et al. 2018) but it is still less popular than filament-based ME. Other examples of printers suited for plastic pellets are Big Area Additive Manufacturing (BAAM) by Cincinnati Incorporated (USA) that can process up to 36 kg/h, and Gigabot X by re:3D (USA) that is up to 17 times faster than filament printers, and processes pellets and flakes 5–10 times cheaper than filaments.

Moreno Nieto et al. (2018) evaluated the performance of pellets of Ingeo™ PLA 3D850, processed on an experimental large-format polymeric printer, using 5 mm diameter nozzle, 2 mm layer height, and 50 mm/s printing speed. The results were positive: 3D850 showed excellent extrusion and melt flow, without deformation and delamination in the test specimens.

Fused granular fabrication (FGF) is a process employed by BLB Industries (Sweden) compatible with plastic granules that go through a vertical heated extrusion screw that melts and presses them through a replaceable nozzle, featuring variable diameter and shape to control the flow. Greater throughput enables higher building speed but lower surface finish and resolution, which is true for other ME processes. After each layer is deposited, the building platform is lowered, and the next layer is added. BLB Industries claims that "next to all polymers can be used." The maximum printing rate is 35 kg/h when using the 6–14 mm diameter nozzle.

FGF is also adopted on the PrentaRobo, a large scale extrusion-based printer, unveiled in 2019, and consisting of an extrusion head built by Prenta Oy (Finland) mounted on an industrial 6-axis robot by ABB Robotics (Sweden, Switzerland). PrentaRobo is especially interesting, because it prints sustainable composites made of PLA and cellulose formulated by UPM Formi.

2.3.7 BIOEXTRUSION (BE)

BE is a subclass of ME developed for *3D bioprinting* or *additive biofabrication*, an emerging family of technologies to fabricate artificial tissues and organs by printing living cells in a tissue-specific pattern (Ozbolat and Hospodiuk 2016). Many types of cells are printed through BE: vascular tissues, cartilage- and bone-like structures, cardiac tissues, liver tissues, stem cells, cancer cells, adipose tissues, muscle cells, skin tissues, etc. (Lobo and Ginestra 2019).

Additive biofabrication enables to build scaffolds with precise geometries, and control over pore interconnectivity and architectures that is impossible with conventional techniques (Chung et al. 2013). BE comprises AM technologies extruding a biocompatible and/or biodegradable component serving as framework (*scaffold*) for living cells and blood vessels that grow and form various types of tissue. Scaffolds are characterized by high porosity (typically above 66%) consisting in micropores enabling cells and vessels to adhere, and macropores enabling vessels and cells to infiltrate and grow. Scaffolds are a critical element in the field called *tissue engineering* (TE) (Gibson et al. 2015) that applies biology, chemistry, materials science, and engineering to develop functional substitutes for damaged or missing tissue.

TE addresses the issue of increasing need for organ transplants, and insufficient number of donors, and is a growing area of AM application and stimulates a great volume of R&D. Feedstocks for BE are available as liquid, melt, paste, and gel.

A polymer suited for BE of commercial products is PCL, because, although not sustainable, it is biocompatible, bioresorbable, malleable, slow-degrading, and as strong as trabecular (porous) bone. PCL is used by Osteopore International (Singapore), a company printing bioresorbable implants utilized in conjunction with neurosurgery and craniofacial surgery to assist with the natural stages of bone healing.

Hydrogels (HGs) are polymers that can be dispersed in water but not dissolved, and are described in Section 2.11.2.2. HGs made of natural polymers are well-suited for scaffolds, because they are extruded in jelly-like form, followed by water removal, which leaves a porous and solid object. Most HGs for scaffolds are made of natural rather than synthetic polymers, being the former ones more biocompatible than the latter ones. However, scaffolds made of HGs of natural polymers (f.e. gelatin and silk) are mechanically weaker than those made of synthetic polymers (f.e. polyethylene glycol or PEG, and Pluronic®), and hence preferred for supporting soft tissues. Pluronic® is a block copolymers consisting in the sequence PEO-PPO-PEO of poly(ethylene oxide) or PEO and PPO.

BE based on melting solid feedstocks is preferred in printing strong and stiff scaffolds for the growth of bony tissue. Since this type of BE is based on FFF, the feedstock is in form of filament often made of polymers mixed to fillers, such as ceramic. EnvisionTEC (Germany) sells 3D-Bioplotter®, a printer for BE. The feedstock in form of gel, liquid, melt, and paste is fed by a screw extruder or compressed gas for high- and low-temperature polymers respectively, and is dispensed on the build bed from a material cartridge through a needle tip mounted on a three-axis system. Multiple printing heads are included, each with individual material and temperature control, although only one material at the time is dispensed. Table 2.6 lists materials for BE compatible with 3D-Bioplotter®, and that list includes many biobased materials derived from agar, cellulose, chitin, collagen, soy, and sugar.

Noteworthy in this book is *extrusion-based bioprinting* (EBB or EBP), which works with a wide range of sustainable ink (*bioink*) types, such as the HGs, micro-carriers, cell aggregates, and decellularized matrix components described below, and listed by Ozbolat and Hospodiuk (2016) in their comprehensive and well-documented review on EBB. Feedstocks for EBB are:

TABLE 2.6

Sustainable Polymers and Other Materials for Bioextrusion Compatible with EnvisionTEC 3D-Bioplotter®

Polymer/Material	Applications
Agar Hydrogel	Soft tissue, cartilage fabrication, cell and organ printing
Alginate Hydrogel	Soft tissue, cartilage fabrication, cell and organ printing
Cellulose-based	Support materials
Chitosan Hydrogel	Soft tissue, cartilage fabrication, cell and organ printing
Collagen Hydrogel	Soft tissue, cartilage fabrication, cell and organ printing
Fibrin Hydrogel	Soft tissue, cartilage fabrication, cell and organ printing
Gelatin Hydrogel	Soft tissue, cartilage fabrication, cell and organ printing
Gelatin-based hydrogel mix	Soft tissue materials
Hyaluronic Acid (HA)	Bone and cartilage, bone regeneration
HA Hydrogel	Soft tissue, cartilage fabrication, cell and organ printing
PLGA	Bone regeneration, drug release
PLLA	Drug release, bone regeneration
Polycaprolactone (PCL)	Bone and cartilage (PCL MW 45,000–120,000), bone regeneration, drug release
Silicone	Soft tissue materials, support materials
Soy Hydrogel	Soft tissue, cartilage fabrication, cell and organ printing
Sugar-based	Soft tissue materials
Titanium	Bone regeneration
Tricalcium Phosphate	Bone regeneration

Source: Data posted online by EnvisionTEC.

- HGs investigated for EBB are the following ones, containing:
  - *Agarose*, a galactose (sugar)-based polymer extracted from seaweed, and marketed in a few types. The most suitable agarose for EBB is the kind melting and gelling at low temperature.
  - *Alginic acid* or *alginate*, a polysaccharide derived mainly from brown seaweed, and bacteria.
  - *Chitosan* is produced by deacetylation of chitin, is nontoxic biodegradable, antibacterial and antifungal, and selected as wound dressing. Chitosan HGs are widely employed in bone, skin, and cartilage TE, because they have a content of hyaluronic acid and gly-cosaminoglycans (polysaccharides containing amino groups) equivalent to that in native tissues.
  - *Collagen type I* has been widely present in TE as a growth substrate for 3D cell culture, and a scaffold material for cellular therapies.
  - *Fibrin* is a protein developed during the clotting of blood to form a fibrous mesh ob-structing the flow of blood, and has been common in TE thanks to its inherent cell-adhesion capabilities and high cell seeding density.
  - *Hyaluronic acid* (HA) is a natural compound present in almost all connective tissues, has been extensively utilized in medicine as an epidermic filler, and, due to its lubricating properties, as synovial fluid in treating osteoarthritis.
  - *Matrigel* is a gelatinous protein mixture produced by specific mouse cancer cells, pro-motes the differentiation of multiple cell types, and outgrowth from tissue fragments. It is also a candidate for cardiac TE.
  - *Methylcellulose* (MC) is a chemical compound derived from cellulose made of a semi-flexible linear chain of polysaccharide, and features the simplest chemical composition among cellulose products.
  - *Pluronic*®, approved by FDA, is employed as drug delivery carrier and injectable gel in treating burns and healing wounds.
  - *PEG* (and PEO) is a widespread excipient in medicines and non-pharmaceutical products. PEG and PEO are polymers of ethylene oxide, and at times considered the same but they differ in that PEG has MW below 20,000 g/mol, and PEO above 20,000 g/mol. PEG-based HGs are biocompatible, reduce immune response in the body, are FDA approved for internal use, and can be crosslinked using physical, ionic, or covalent crosslinks.
- *Micro-carriers* (M-C) are substances, usually in form of 125–250 μm diameter spheres, into which living cells are inserted, and proliferate effectively in terms of cost and space. Commercial M-C for bone and cartilage regeneration are made out of collagen, dextran, gelatin, glass, and plastic. M-C can be the medium in which to print living cells.
- *Cell aggregates*. Scaffold-free cell aggregates have been considered a promising option in bioprinting because they enable building tissues in a relatively short period of time compared to the commonly used cell-laden hydrogel approach. Cell aggregate-based bioinks have significant advantages: good cellular interactions, close biomimicry, fast tissue formation, and long-term stability of cell traits (*phenotypes*) in space.
- *Decellularized matrix* (dECM) *components*. A dECM component is what is left of a tissue after removing its residing cells, which leaves behind its scaffold. dECM components are important because applied for regeneration of organs such as heart, kidney, liver, cartilage and bone, pancreas, etc. (Ozbolat and Hospodiuk 2016). Pati et al. (2014) took dECM components chopped in small fragments and loaded with cells, and printed tissue analogues supported by a PCL "frame."

Examples of commercial bioinks are Gel4Cell®, CELLINK®, BioInk®, OsteoInk®, Bio127®, and BioGel®. Commercial extrusion-based bioprinters are sold by Advanced Solutions, EnvisionTec,

Organovo, RegenHu, nScypt, Sys+Eng, etc., and print tissue constructs for aortic valves, bone, breast cancer, cartilage, heart, liver, vascularization, etc. Many non-commercial extrusion-based bioprinters have been developed by universities and research centers worldwide, such as Dresden University of Technology, Harvard University, and Korea University.

2.3.8 MICROEXTRUSION PRINTING (μEP)

μEP is a subfamily of BE in which feedstocks contained in cartridges are selectively dispensed through micronozzles or needles (<50 μm diameter) to form 3D objects. Printers with multiple cartridges to print heterogeneous structures, and biomaterials in form of ink are available. For bioprinting cell-laden constructs, cells are blended with bioink, whose function is to encapsulate cells, in order to offer a supportive extracellular matrix (ECM) environment, and safeguard cells from the stresses that cells undergo during printing (Panwar and Tan 2016). In comparison to traditional photolithography, μEP considerably simplifies fabrication of arbitrary structures, and offers a wide range of choices in materials and substrates, according to Udofia (2019), who authored a wide-ranging and insightful review of μEP. Bioink formulators will also benefit from the report of the current status of bioinks for μEP by Panwar and Tan (2016).

Formulating new feedstocks for μEP will further μEP (Udofia 2019). A candidate polymer for bioink is evaluated based on several factors: ease with which it is printed with adequate resolution, stability of its shape after printing, shape fidelity, resolution, biocompatibility, ability to support the cells (Kirchmajer and Gorkin 2015), and viscosity. Various strategies to improve bioink printability and shape fidelity, and ensure good cell viability and function were described by Kirchmajer and Gorkin (2015). Ink viscoelasticity is critical for predicting its flow behavior and achieve fine feature resolution on the printed article. Newtonian and non-Newtonian inks have been employed in μEP, and the latter type features a decrease in viscosity as shear rate increases (*shear thinning*), which is advantageous because it reduces the required amount of driving pressure, and improves the shape retention after the ink is deposited on the substrate, in other words it is conducive to smooth flow from the nozzle and accurate and permanent geometry.

An example of natural polymer for bioink for μEP is *silk*, a high-MW protein produced by silk worms and spiders, selected due to its nontoxic nature, slow degradation, and low immunogenicity.

Ortega et al. (2016) developed flow models and determined the appropriate values of operating conditions (temperature and shear rate) for μEP, experimenting with PLA, and a nozzle 300 μm in diameter, with the goal to obtain an extruded filament with homogeneous morphological characteristics that can be leveraged to print scaffolds for TE via ME.

2.3.9 LIQUID DEPOSITION MODELING (LDM)

LDM has been described by several authors, and essentially consists in extruding from a filament-based printer a paste-like filled and unfilled, made of resins, ceramics, food, and so on.

Compton and Lewis (2014) utilized LDM to print lightweight, strong, and stiff cellular composites imitating natural composites such as wood, and devised an ink containing epoxy resin, 1 nm thick and 100 nm long nanoclay platelets, dimethyl methyl phosphonate (DMMP), silicon carbide whiskers, 10 μm diameter and 220 μm long carbon fibers, and imidazole-based ionic liquid, with each ingredient having a specific function. F.e. the platelets imparted shear thinning and shear strength, and DMMP reduced the resin's initial viscosity to allow higher loading of solids.

Postiglione et al. (2015) employed LDM to fabricate electrically conductive nanocomposite-based microstructures with arbitrary shapes and features as low as 100 μm. On a low-cost, commercial benchtop printer equipped with a dispensing syringe, they deposited layers of a liquid solution of PLA pellets, multi-walled carbon nanotubes (MWCNTs) featuring 9.5 nm diameter and 1.5 μm average length, dispersed in dichloromethane solvent.

Rosenthal et al. (2018) printed via LDM wood-like material in form of a paste-like compound comprising air-dry ground beech sawdust (main ingredient), MC (lubricant and binding agent), and water that was deposited on a standard printer with a self-made extruder. Physical properties were influenced by binder/water ratio and wood particle size: density was 0.33–0.48 g/cm^3, flexural strength and modulus 2.3–7.4 MPa and 285–733 MPa, respectively.

2.3.10 CONTOUR CRAFTING (CC) AND CEMENT AND CONCRETE PRINTING

AM is applied to building and infrastructures (Krimi et al. 2017) through CC, a process patented by B. Khoshnevis (Khoshnevis 2010), and based on a very large robotic version of a ME printer. In CC, a movable computer-controlled crane or gantry robot is mounted to an overhead beam sliding on rails, and is equipped with a nozzle that is fed by a cement truck, and rapidly discharges quick-setting, concrete-like material (Contour Crafting 2017). A critical feature of CC is that, after the material is deposited, a scraping tool smooths the surface, enabling speed and accuracy at the same time. CC is faster, and requires less labor than the conventional building technology. Critical parameters in CC and other AM processes for construction are print quality (measured through surface quality and dimensions of printed layers), shape stability, and printability window. CC is suitable for housing, commercial and government buildings, infrastructures (bridges, foundations, slabs, pylons, etc.), not only on the Earth but even on other planets where their soil can be used as a construction material (Khoshnevis et al. 2016).

Additional details on AM for construction are published by Lim et al. (2012), and Gosselin et al. (2016). Other methods to print concrete and cement are: (a) *concrete printing* developed at Loughborough University (UK) by S. Lim, R. Buswell, and their team; (b) D-Shape®, developed by E. Dini (D-Shape n.d.), employing an inorganic liquid solidifying agent that is poured into the powder mix; (c) the CC-like process impemented by Winsun (China).

Investigating CC, Kazemian et al. (2017) tested four printing mixtures based on Portland cement, and studied the effects of fiber inclusion, nanoclay, silica fume (particles of mainly SiO$_2$ of size 0.1–0.3 µm), finding that the latter two fillers significantly enhanced shape stability of fresh printing mixture.

Geopolymer mixtures for ME of constructions were studied by Hambach and Volkmer (2017), who tested Portland cement paste reinforced by short fibers (carbon, glass and basalt fibers 3–6 mm long), and Panda et al. (2018), who employed geopolymer mixtures based on fly ash (mostly), granulated blast-furnace slag, and undensified micro silica fume.

2.3.11 CONTINUOUS FILAMENT FABRICATION (CFF)

CFF is a technology patented by G. Mark (Mark and Gozdz 2015) founder of Markforged (USA) to print parts out of continuous fiber reinforced polymeric matrix while controlling amount, orientation, and type of the reinforcing fiber. CFF relies on some version of FFF printers equipped with two nozzles: one nozzle dispensing a PA (available in several grades) filament serving as matrix, and another nozzle laying down continuous bundles of aramid (Kevlar®), carbon, or glass (in two grades: standard, and for high strength and temperature) fiber infused with sizing to promote adhesion to the matrix. CFF consists of printing the PA matrix first, and the fiber reinforcement next (Dickson et al. 2017). Markforged has released physical and mechanical properties separately for matrix materials (Onyx, Onyx FR, and Nylon W) and the above individual fibers, but not for the relative composite materials, since the properties of the resulting composites depend on the amount, orientation, and type of reinforcing fiber selected by the users. The tensile yield and ultimate strength, modulus and strain of the matrix materials are in the range of values for PA, and namely: 29–51 MPa, 30–36 MPa, 1.3–1.7 GPa, and 58–150%, respectively. However, Dickson et al. (2017) measured tensile and flexural properties of coupons printed with unfilled PA, and PA filled with fiber oriented in different directions (concentric and bidirectional raster scan)

TABLE 2.7
Tensile and Flexural Properties of Coupons Printed Via CFF Process

Material	Fiber Volume Fraction	Tensile Strength	Tensile Modulus	Elongation at Break	Flexural Strength	Flexural Modulus
	%	MPa	GPa	%	MPa	MPa
PA	0	61	0.53	439	42	1.06
PA-Carbon Fiber (C[a])	11	216	7.73	4.2	250	13.0
PA-Glass Fiber (C[a])	8	194	3.12	9.0	166	3.87
PA-Glass Fiber (BRS[b])	10	206	3.75	8.4	197	4.21
PA-Kevlar® Fiber (C[a])	8	150	4.23	3.6	107	4.61
PA-Kevlar® Fiber (BRS[b])	10	164	4.37	4.5	126	6.65

Source: Dickson, A. N., Barry, J. N., McDonnell, K. A., Dowling, D. P., 2017. Fabrication of continuous carbon, glass and Kevlar fiber reinforced polymer composites using additive manufacturing. *Additive Manufacturing* 16:146–152. Reproduced with permission from Elsevier.
Notes:
[a]Concentric fiber orientation
[b]Bidirectional raster scan fiber orientation

within each deposited layer. These values are included in Table 2.7, and show that even a fiber content of 8–11 vol% improved those properties multiple times.

2.3.12 WATER-BASED ROBOTIC FABRICATION (WBRF)

Mogas-Soldevila et al. (2014) developed WBRF, and extruded biobased and biodegradable composite materials made of natural HGs (chitosan and sodium alginate) and chitin powder. The WBRF equipment consisted of a Kuka 6-axis robotic arm holding at its end a portable three-chamber extrusion system: each chamber was equipped with one 200 cc syringe, and contained the ingredients, and all chambers were connected to a mixing nozzle. Two chambers contained viscous colloids pushed out by plungers controlled by stepper motors, and one chamber contained natural fillers such as cellulose chopped fibers, fine sand, and natural polymer powders. Flow rate, linear motion, and nozzle diameter were 8–4,000 mm³/sec, 10–50 mm/sec, and 0.5–8 mm, respectively. The robotic arm allowed a maximum print area of 1 × 0.5 m. The equipment enabled: consistent volumetric flow rates, features from below 1 mm to macroscale, and graded properties within each article. WBRF targeted biodegradable and fully recyclable products, and architectural parts with graded mechanical, optical, and environmental properties such as those for water storage, hydration-activated shape change, and complete disintegration over time.

2.4 POWDER BED FUSION (PBF)

2.4.1 PROCESS DESCRIPTION

PBF is a family of AM processes in which thermal energy selectively fuses together the feedstock particles of the top layer of a bed covered with a powdered material. The thermal energy is generated by infrared laser (the most common source, and typically a 10.6 μm wavelength CO_2

laser), electron beam, and/or infrared lamps, and the binding mechanism is thermal reaction bonding (ISO 17296-2 2015). PBF encompasses the following similar processes that are protected by patents (Loughborough University 2020):

- *Direct metal laser sintering* (DMLS): metal powders are heated up by a laser until their particles fuse together.
- *Electron beam melting* (EBM): electron beam fuses the particles together.
- *Selective heat sintering* (SHS): it utilizes thermal printheads that are less powerful than laser and electron beam, and heat up and fuse powder particles (Baumers et al. 2015). Patented by Danish company Blueprinter that filed for bankruptcy in 2016.
- *Selective laser melting* (SLM): a laser beam fully liquefies and melts metallic powders into a homogeneous part.
- *Selective laser sintering* (SLS) or *laser sintering* (LS): a laser beam heats up and fuses *polymeric* powder particles.

All PBF processes have in common three steps described later: powder dispensing, energy input, and consolidation (Chatham et al. 2019).

The earliest AM technologies relying on a powder bed were developed in the 1970s (Hopkinson et al. 2006; Bourell et al. 2009a), but the first patent for the modern PBF system was filed in 1994 by C. Deckard (Deckard 1997) under the commercial name of SLS that still represents the most widespread variant of PBF processes and includes HP *multi-jet fusion process* (MJF) (Hewlett-Packard 2020) and *high-speed sintering* (HSS) (Hopkinson and Erasenthiran 2004). Although the term *sintering* is a frequent synonym of PBF, in PBF particles are either partially or fully melt (Kruth et al. 2007; Zarringhalam et al. 2009). PBF is compatible with polymers, metals, ceramics, and composites of all the above. Hereafter, we focus on PBF for polymers and their composites.

In polymeric PBF the feedstocks are typically in form of powder composed of 20–80 μm diameter solid particles (Vock et al. 2019). Figure 2.6 is a schematic of a PBF printer for polymers. The laser (8) sends a laser beam to a mirror (7) which deflects it (6) to the *build area* or *build chamber* (2) where the powder is kept at a temperature (*bed* or *sintering temperature*) slightly below its melting point by the heaters (1) in a closed chamber under inert purging gas. The recoater (5), which can be a counter-rotating roller (pictured) or a rake, pushes some powder from one or two side tanks (3), and spread it forming a thin (typically 100 μm) even layer on (2). The

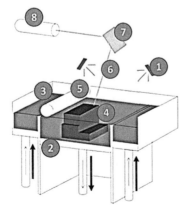

1	Heaters
2	Build Chamber
3	Powder Delivery System
4	Printed Part
5	Recoater
6	Laser Beam
7	X-Y Scanning Mirror
8	Laser

FIGURE 2.6   Schematic of a powder bed fusion printer with two powder tanks and delivery systems.

Source: Adapted from https://formlabs.com/3d-printers/. Reproduced with permission from Formlabs.

laser beam (6) selectively (i.e. only tracing the cross-section of the part (4)) rasters across (2), and fuses together the top particles forming a cohesive layer, and, when full melting is achieved, fuses also this layer to the material layer underneath (Gibson el al. 2015). The laser beam (6) is synchronized with the recoater, and the beam stops when the recoater moves across (2). After one layer is being sintered, (2) is lowered a distance equal to one powder layer, and the cycle iterates.

There are two powder fusion mechanisms for polymers (Gibson et al. 2015):

- *Liquid-phase sintering* (LPS). It consists in melting "smaller powder particles and the outer regions of larger powder particles without melting the entire structure" (Gibson et al. 2015). LPS, also called *partial melting*, applies to amorphous polymers that have no distinct melting point.
- *Full melting* (FM). In this case the whole area of powder targeted by the impinging heat energy is melted to a depth exceeding the top layer's thickness. FM generates well-bonded high-density objects with the highest possible strength, and is applied to PA because, being semi-crystalline, PA has a distinct T_m. Since temperatures associated with FM can cause part growth, process conditions are set at the threshold between FM and LPS to minimize part growth in volume.

The consolidation phase is the last step in PBF, and includes coalescence and cooling, when the molten particles first flow into one another, and join, then cool, solidifying and crystallizing.

The HSS and MJF version of PBF employ an additional material that is "jetted into the powder bed as a processing aid, to enhance optical-to-thermal energy conversion for greater throughput and increased edge-definition and surface finish" (Chatham et al. 2019). HSS and MJF feature lower processing time, and higher process output compared to other versions of PBF.

In PBF, the specific energy input per volume (W) of each scan depends on laser power (P), scan speed (v), scan line spacing (h), and layer thickness (t) as it follows (Simchi and Pohl 2003):

$$W = P/(vht) \qquad (2.8)$$

Powder's properties such as T_g, density ρ (kg/mm^3), specific heat c_p (kJ/kgK), and latent melt energy c_l (kJ/kg) affect the energy E required to fuse a specific powder particle according to (2.9) where V is the volume of the spherical particle (mm^3), and ΔT (K) is the raise from ambient temperature to powder's melting/sintering temperature (Wang and Kruth 2000):

$$E = (c_l + \Delta T c_p)\rho V \qquad (2.9)$$

The description of PBF in this chapter is introductory. For further information on PBF and relative polymers we suggest the relevant papers by J. P. Kruth, and the reviews by Chatham et al. (2019), Goodridge et al. (2012), Ligon et al. (2017), and Schmid et al. (2014).

2.4.2 POLYMERS FOR PBF

Feedstock materials for PBF are in form of polymeric, metal, and ceramic powders, with ceramics the least common material. Polymeric powders for PBF are available as unfilled polymers and composite materials. The polymeric phase is mostly a semi-crystalline TP, but may also be an amorphous TP, a two-component TS, and an elastomeric material (Ligon et al. 2017). Semi-crystalline TPs are best fit for PBF processes, because they have a distinctive T_m which is adequately higher than their crystallization temperature (T_c), and between them a sintering temperature can be set at which the TP powder melts, and recrystallization does not occur in the immediately

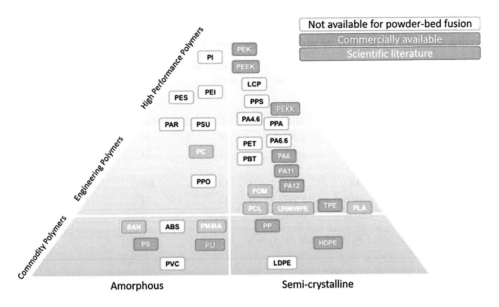

FIGURE 2.7 Commercial and experimental polymers for powder bed fusion.

Source: Yuan S., Shen F., Kai Chua C., Zhou K., 2019. Polymeric composites for powder-based additive manufacturing: Materials and applications. *Progress in Polymer Science* 91:141–168. Reproduced with permission from Elsevier.

deposited layers, preventing shrinkage, curling, and warping resulting from residual stresses (Drummer et al. 2010).

The commercial and experimental polymers for PBF comprise commodity, engineering, and high-performance polymers (Figure 2.7), and provide a broad range of property values and price, which can attract formulators of alternative PBF feedstocks made of SPs. Examples of commercial polymers for PBF are:

- *Commodity polymers*: HDPE, and PP, supplied by Diamond Plastics (Germany), PP sold by EOS (Germany) as PP 1101, PS, which is the only commercial amorphous PBF polymer, commercialized as CastForm PS (3D Systems, USA) and PrimeCast101® (EOS).
- *Engineering polymers*: these are the most preferred, and include mostly PAs, particularly PA 12 (the number 12 refers to the number of carbon atoms included in the monomer), marketed as PA2200 and DuraForm® PA12 by EOS and 3D Systems respectively, and both based on Vestosint® PA 12 produced by Evonik (Germany). Other engineering polymers are PA 6, available from Solvay, and PA 11, sold by ALM (USA), EOS, and Stratasys. Interestingly, Arkema (France) produces biobased PA 11 under commercial name of Rilsan® from castor oil (Arkema 2020) that is 100% renewable, and represents the base material for PBF powders provided by the following suppliers of PBF powders: ALM (PA D80-ST, FR106, PA 850 Nat, PA 850-Black, PA 860), EOS (PA1101, PA1102), 3D Systems (DuraForm® EX, DuraForm® EX Black). Table 2.8 contains the major properties of commercial powders for PBF made from Rilsan®. TP elastomers (TPE), namely DuraForm® Flex (3D Systems), and PrimePart® ST PEBA 2301 (EOS), and urethanes (TPU), namely Luvosint X92A-1, and X92A-2 (Material Data Center n.d.) are also available.
- *High-performance polymers*: PEEK, commercialized by EOS as PEEK HP3, and based on PEEK formulated by Victrex (UK), and Kepstan® PEKK developed by Arkema (Arkema n.d.). PEKK and PEEK are semi-crystalline, expensive TPs combining high mechanical strength and temperature stability for demanding engineering applications.

TABLE 2.8

Properties of Powders for PBF Made from Biobased Arkema PA 11 Rilsan®

Material	Density	Tensile Strength	Tensile Modulus	Elongation at Break	Izod Notched Impact Strength	HDT at 1.8 MPa	HDT at 0.45 MPa
	g/cm³	MPa	GPa	%	MPa	MPa	MPa
ALM PA D80-ST	1.07	46	1,392	38	69	70	186
ALM PA 850-Black	1.03	48	1,475	51	74	48	186
EOS PA 1101	0.99	48	1,200	45	N/A	46	180
3D Systems DuraForm® EX	1.01	48	1,517	47	74	48	188

Source: Data sheets posted online by ALM, EOS, 3D Systems.

The following are non-commercial and/or patented polymers for PBF (Ligon et al. 2017):

- *Polymers*: ultra-high MW PE (UHMWPE), poly(methyl methacrylate) (PMMA), PC, HIPS, poly(lactic-co-glycolic acid) (PLGA), PLA, poly-(hydroxybutyrate-co-hydroxyvalerate) (PHBV), methyl methacrylate-butyl methacrylate copolymers
- *Blends*: PA 12/HDPE, PA 12/PEEK, PS/PA, styrene-acrylonitrile-copolymer (SAN) blends
- *Composites and nanocomposites* based on PA 11, PA 12, PCL, and PLGA.

Further experimental polymers for PBF mentioned in technical literature are: ABS copolymer with starch (Kuo et al. 2016), PCL (Williams et al. 2005; Sabir et al. 2009; Yeong et al. 2010; Mkhabela and Ray 2014), PLA (Bai et al. 2017), poly-L-lactide or PLLA (Duan et al. 2010), PHBV (Duan et al. 2010), PC-epoxy (Shi et al. 2007), poly(butylene terephthalate) (PBT) (Schmidt et al. 2016), poly(ether ketone) (PEK) (Berretta et al. 2016), PE (Bai et al. 2016), poly(phenylene sulfide) (PPS) (Ito and Niino 2016), PP (Wegner 2016), UHMWPE (Goodridge et al. 2010; Khalil et al. 2016), PEO blended with PA12 (Yang et al. 2019). Among the above polymers, PBT, PE, PHBV, PLA, and PLLA are biobased and SPs.

Polymers for PBF are fed solid to the printers, where they are heated, fused, and cooled to their final shape; therefore, PBF can process TPs but not TSs. In PBF printers the bed temperature is kept below the feedstock's T_m but above its T_c, because at T_c the powder shrinks, curls, and warps due to residual stresses. Consequently, candidate polymers for PBF must possess well-defined (that is narrow and repeatable range of) values of T_c and T_m, and a gap between them adequately wide to set up an intermediate value of bed temperature that will keep the powder warm and ready to be fused without crystallizing for as long as possible, and also to enable cohesion among the topmost powder layer and its closest previously laid layers.

We recall that TPs can be amorphous or crystalline: the former ones melt over a wide range of temperature, while the latter ones feature a distinct T_m. Crystalline TPs meet all the above PBF requirements. PA is a crystalline TP and the most prevalent commercial polymer for PBF, in form of PA 11 and PA 12, both filled and unfilled. F.e. T_m of PA 12 is 35°C above its T_c. A representative example of commercial polymers for PBF is the family of powders supplied by 3D Systems, composed of PA 11 and PA 12 unfilled, and filled with glass, carbon, aluminum, and mineral charge, and listed in Table 2.9. Figure 2.8 illustrates the design space for the materials in Table 2.9 defined by their tensile strength and modulus values.

TABLE 2.9

Biobased (PA 11) and not Biobased, Filled and Unfilled, Polymeric Powders Supplied by 3D Systems for SLS

Commercial Name	Ingredients	Density	Tensile Modulus	Tensile Strength	Elongation at Break	Flexural Modulus	Flexural Strength	Notched Izod Impact Strength	Unnotched Izod Impact Strength	HDT at 0.45 MPa	HDT at 1.8 MPa
		kg/dm³	MPa	MPa	%	MPa	MPa	J/m	J/m	°C	°C
DuraForm TPU	TPU	0.78	5.3	2	220	6	NA	NA	NA	NA	NA
DuraForm Flex	TPE	NA	5.9	1.8	110	5.9	48	NA	NA	NA	NA
DuraForm EX	PA 11	1.01	1,517	48	47	1,310	46	74	1,486	188	48
DuraForm PA	PA 12	1.03	1,586	43	14	1,387	48	32	336	180	95
DuraForm GF	Glass filled PA 12	1.49	1,586	43	14	1,387	48	32	336	180	95
DuraForm HST	Mineral fiber filled PA 12	1.2	5,600	49	4.5	4,475	86	37.4	310	184	179
CastForm PS	PS	0.86	1,604	2.84	NA	NA	NA	<11	14	NA	NA
DuraForm FR1200	Fire retardant PA 12	1.02	2,040	41	5.9	1,770	62	25	233	180	94
DuraForm ProX PA	PA12	0.95	1,770	47	22	1,650	63	45	644	182	92
DuraForm ProX GF	Glass filled PA 12	1.33	3,720	45	2.8	3,120	60	48	207	180	129
DuraForm ProX HST	Mineral fiber filled PA 12	1.12	4,123	44	4.3	3,430	75	55	307	183	171
DuraForm ProX EX BLK	PA 12	1.02	1,570	43	60	1,360	51	75	3,336	193	57
DuraForm ProX EX NAT	PA 12	1.02	1,590	51	61	1,436	56	91	no break	192	56
DuraForm ProX AF+	Aluminum filled PA 12	1.31	4,340	37	3	3,710	64	54	255	182	174
DuraForm ProX FR1200	Fire retardant PA 12	1.03	2,010	45	8	1,720	61	24	278	180	94

Source: Data sheets posted online by 3D Systems.

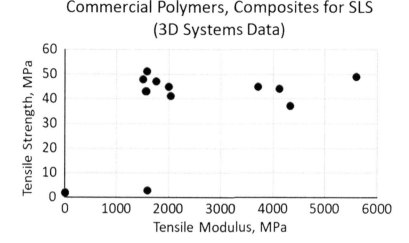

FIGURE 2.8 Tensile strength plotted versus tensile modulus of commercial polymer powders made of PA for laser sintering supplied by 3D Systems. Values from 3D Systems' data sheets.

The first step to evaluate candidate polymers for PBF is to run differential scanning calorimetry (DSC) to ascertain whether T_c and T_m peaks are clearly defined and adequately apart, in order to possibly identify a bed temperature. DSC also measures the amount of energy (specific heat in J/g) required to melt the polymer, and provides the basis to estimate the laser power (in W) required. Thermogravimetric analysis (TGA) determines the thermal stability of a candidate polymer upon processing, and detects its decomposition temperature. When evaluating filled polymers, DSC and TGA also assess the effect of fillers on the composite's properties, while Fourier-transform infrared spectroscopy (FTIR) measures any chemical change due to heating during fusion and/or interaction filler/matrix (Warnakula and Singamneni 2017).

If the bed temperature is too low, a high thermal gradient between the fused and unfused particles arises, resulting in curling. If the bed temperature is too high, a phenomenon called *caking* is observed whereby the powder particle surface starts melting and particles stick together, leading to powder agglomeration (Bai et al. 2017). The correct powder bed temperature for PBF polymers strongly depends on the chosen material, and must be experimentally determined for newly employed materials. The powder bed temperature for PA 12, featuring T_c of 165°C (Zhao et al. 2018), and T_m of 180°C, is set at 170°C (Goodridge et al. 2011). Amorphous polymers lack T_m, which has to be replaced by their T_g.

Particle's size and shape also are critical for successfully dispensing and consolidating the powder during PBF. The optimal particle diameter is 45–90 μm. Particle diameter affects powder's free-flowing behavior: decreasing particle diameter improves powder's rheological properties up to 25–45 μm where the powder's surface energy starts inhibiting flow (Goodridge et al. 2012; Ziegelmeier et al. 2015), which prevents high packing density, low porosity, and high mechanical properties. Particularly, porosity in fused and unfused powder strongly influences the following mechanical properties of parts made via PBF, listed in order of decreasing severity: fracture, fatigue, strength, ductility, and elastic modulus (Hopkinson et al. 2006). Current commercial PBF printers for plastics build parts with about up to 50% porosity, which may seem high, but the density of spherical particles equal in size reaches theoretically a maximum of 74% space filling in a cubic face-centered structure that slightly increases adding smaller particles (Schmid 2018, p. 97). Moreover, particles smaller than 20 μm in diameter are likely to agglomerate during PBF (Schmid et al. 2015). Spherical particles are more suited for PBF than non-spherical particles (Amado et al. 2011), because irregularities in particle shape cause lower packing and part densities (Ziegelmeier et al. 2015). Commercial particles also exhibit potato and irregular shapes. Particle's shape derives from its fabrication process: co-extrusion with soluble and non-soluble material mixtures, precipitation, and cryogenic milling generate round particles, potato-shaped, and irregular particles, respectively (Schmid et al. 2014).

Key properties of candidate powders are zero-shear viscosity of the polymer melt and surface tension: low value of the former combined with high value of the latter are conducive to high rate of coalescence (Verbelen et al. 2016), as established in the Frenkel model (Frenkel 1945), and hence extensive particle cohesion, low porosity, and relatively high mechanical properties.

Since the PBF lasers generate energy in the IR spectrum (10.6 μm wavelength for CO_2 lasers), the candidate polymer for PBF must absorb adequate portions of the energy at 9–11 μm wavelength. This is possible for most polymers as they consist of aliphatic compounds that absorbs well energy at 10.6 μm.

When manufacturing polymer powders for PBF, it is key to consider that the powder's chemical structure affects flow temperature, viscosity, surface energy, and crystallinity, which all in turn influence the powder's printability (Rajan et al. 2001).

Figure 2.9 displays material properties and individual properties of feedstocks for PBF in bulk and powder forms, and the relationship between these properties and the process parameters and effectiveness.

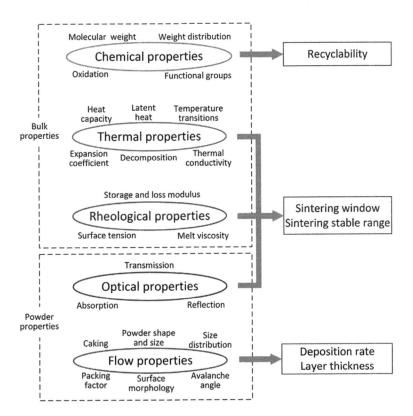

FIGURE 2.9 Material properties of feedstocks for PBF in bulk and powder form, and the relationship between these properties and PBF parameters and effectiveness.

Source: Yuan S., Shen F., Kai Chua C., Zhou K., 2019. Polymeric composites for powder-based additive manufacturing: Materials and applications. *Progress in Polymer Science* 91:141–168. Reproduced with permission from Elsevier.

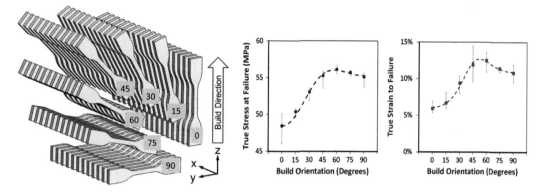

FIGURE 2.10 Variation of tensile properties of PA 12 test coupons printed via selective laser sintering as a function of build orientation.

Source: Oliver M., Xue Z., Abeid B., Brown S. 2017. Testing and modeling anisotropic failure of polymeric SLS materials and structures. SPE ANTEC® Anaheim 2017. https://www.veryst.com/project/Anisotropy-of-3 D-Printed-Polymers. Reproduced with kind permission from M. Oliver, Veryst Engineering LLC.

2.4.3 ANISOTROPY

As with FFF, plastic objects printed via PBF feature anisotropic mechanical properties, as clearly reported in the suppliers' technical data sheets. F.e. tensile strength and modulus in XY plane for ALM PA 850-Black are 48 MPa and 1475 MPa, respectively, whereas in the Z direction they are 42 MPa and 1427 MPa, respectively.

Cooke et al. (2011) quoted technical publications by Gibson and Shi (1997), Gornet et al. (2002), Ho et al. (2003), Ajoku et al. (2006), Caulfield et al. (2007), and Amado-Becker et al. (2008) who had documented how tensile properties of test coupons printed via PBF out of PC and PA had varied depending on their build orientation. A reduction in mechanical properties in the Z direction was attributed to imperfect interlayer bonding (Cooke et al. 2011), previously observed and modeled (Childs et al. 1999; Ho et al. 2003). Veryst Engineering (USA) printed and tested PA 12 tensile coupons with their longitudinal axis aligned in seven build orientation from 0 to 90° relative to the Z axis of the build table, and reported a significant variation of true strength to failure, true strain to failure and energy to break depending on the orientation (Figure 2.10).

2.5 VAT PHOTOPOLYMERIZATION (VP)

2.5.1 PROCESS DESCRIPTION

In VP a liquid TS monomer or oligomer kept in a vat and containing a photoinitiator is *selectively* exposed to UV (in most of the cases) or visible light of specific wavelengths; the photoinitiator acts as a catalyst, and produces a free radical or cation initiating a chain reaction of polymerization (*curing*) that, in turn, converts the monomer and oligomer into a solid object layer by layer. Polymers for VP are resins cured by light and termed *photopolymers*. Common example of photopolymers are the non-AM compounds employed for decades in dentistry as adhesives, sealant composites, and protective coatings. Photopolymers for VP are acrylate, acrylic, epoxy, vinyl ether, and thiol monomers.

VP is also known as *photocuring* and *photocrosslinking*, and is a family of AM processes encompassing *continuous liquid interface production* (CLIP™), *digital light processing* (DLP), and *stereolithography* (commonly SLA, stereolithography apparatus, or SL in this book), the most popular of them, and the first commercially successful AM process to be patented. All the above processes build parts that can feature multifunctionality (Zarek et al. 2016), tunable chemical, electrical, magnetic, mechanical, and optical properties (Ligon et al. 2017; Layani et al. 2018; Taormina et al. 2018), and the highest complexity, accuracy, and resolution (as low as 10 μm); (Wang et al. 2017b) among AM processes (Taormina et al. 2018). VP processes find applications in audiology, biomedical, dentistry, drug delivery devices, education, engineering and product design, entertainment, health care, jewelry, microfluidics, soft robotics, surgery, and TE (Zorlutuna et al. 2011; Wang et al. 2015; Rusling 2018; Formlabs 2020), etc. Additional benefits associated with VP are: isotropic properties, smooth surface finish, watertightness (no voids between layers), and a broad range of material properties. Downsides of VP processes are: support structures are always required; mechanical properties and visual appearance of parts degrade overtime upon exposure to sunlight (spray coating articles with a clear UV acrylic paint before use helps); need to clean the vat after printing; smelly and toxic fumes; safety precautions required when handling the liquid feedstocks; and mandatory post-processing in order to: (a) remove delicate support structures, sanding, and filing; (b) post-cure under UV to fully cure the material, and maximize its strength, stiffness, and temperature resistance.

The monomers and oligomers for VP are formulated from epoxides, polyols, meth(acrylic) acids, and their esters, and diisocyanates (Gibson et al. 2015). Most monomers are multifunctional monomers, and polyol polyacrylates that crosslink upon polymerization. Oligomers are amino acrylates, cycloaliphatic epoxies, and acrylates of epoxy, polyester, and urethane (Dufour 1993).

The price of VP in USD/liter ranges from about \$50 for standard resins to \$400 for the specialty resins, such as castable resins, and dental resins (Varotsis 2020).

Based on the number and types of light source interacting with the resins, there are three main versions of VP, and they are described below, with the first two implemented in commercial printers, and the third in research equipment:

- *Vector scan* or *point-wise*: one laser beam moves in a row by row fashion, traces and cures point-by-point one layer at the time. Implemented in SL.
- *Mask projection* or *layer-wise*: a digital light projector screen flashes at once a single image of each layer to be printed on the resin, and each entire layer is polymerized at once. The image of each layer is composed of square pixels, called *voxels*. It allows faster printing than in the vector scan version, and is applied in DLP and CLIP™.
- *Two-photon*: two scanning laser beams intersect at a point of contact with the resin, and both cure it locally. It is a high-resolution version of the vector scan type, and capable of features finer than those achievable with the other versions.

The VP process parameters are: C_d (mm) cure depth, D_p (mm) depth of penetration of laser at the depth value where irradiance (W/mm^2) is 37% on the resin surface (D_p is a function only of the resin), E_c (mJ/mm^2) *critical exposure*, that is exposure at which resin starts solidifying (a threshold to reach and possibly exceed in order to cure the resin), P (W) laser power, V_s (mm/s) scan speed, and W_0 (mm) radius of laser beam on the resin surface. These parameters are combined together in equation (2.10) (Gibson et al. 2015):

$$C_d = D_p ln\left(\sqrt{\frac{2}{\pi}}\frac{P}{W_0 V_s E_c}\right) \tag{2.10}$$

The standard design of VP printers is based on the empirical formula proposed by Jacobs (Jacobs 1992; Jacobs 1996) and reported in (2.11), where E_{max} (mJ/mm^2) is the peak exposure of the resin to the laser striking the resin surface:

TABLE 2.10

Features of Top-Down and Bottom-Up Models of SL Printers

Features	Top-down (Industrial) SL	Bottom-up (Desktop) SL
Advantages	Very large build size. Faster build times	Lower cost. Widely available
Disadvantages	Higher cost. Require specialist operator. Changing material. Requires emptying the whole tank	Small build size. Smaller material range. Requires more post-processing, due to extensive use of support
Build size	Up to 1500 × 750 × 500 mm	Up to 145 × 145 × 175 mm
Typical layer height	25 to 150 microns	25 to 100 microns
Dimensional accuracy	±0.15% (lower limit ±0.010–0.030 mm)	±0.5% (lower limit: ±0.010–0.250 mm)
Major printer manufacturers	3D Systems	Formlabs

Source: Adapted from Varotsis, A. B. Introduction to SLA 3D Printing. https://www.3dhubs.com/knowledge-base/introduction-sla-3d-printing/ (accessed November 14, 2019).

$$C_d = D_p \ln\left(\frac{E_{max}}{E_c}\right) \tag{2.11}$$

Equation (2.11) indicates that the higher E_{max} is compared to E_c, the higher C_d and hence the faster is the process.

Additional and detailed information is reported in broad and recent reviews by Bagheri and Jin (2019) and Taormina et al. (2018), the books mentioned in the section Further Readings, and those authored by Bartolo (2011) and Gibson et al. (2015).

2.5.1.1   Stereolithography (SL)

Commercial printers for SL come in two configurations with different features, advantages, and disadvantages listed in Table 2.10: top-down and bottom-up orientations.

2.5.1.1.1 Bottom-up SL

The distinctive features of bottom-up SL printers (Figure 2.11) is that the light source is located below the vat, and the object is being built below the platform P to which it is bonded. When the printing session begins, P is positioned inside a vat filled with the liquid photopolymer resin at a distance of one cured layer's height from the vat's bottom, which is transparent and coated with silicone, allowing laser light to pass through and preventing cured resin from sticking to it. The printer's laser sends a single point beam to a scanning mirror that directs it through the bottom of the tank, and point-by-point traces one cross-sectional area (*layer*) of the object to be printed, polymerizing and solidifying the layer that bonds itself to the platform. After that, P edges up by one cured layer's thickness, pulls up with itself the solid layer that detaches from the tank's bottom (*peeling*), and liquid resin fills the space. The laser beam strikes the liquid resin again, and the next layer is cured. This process is repeated layer by layer until the entire solid object is finished. Printed articles are typically post-cured by UV light to increase strength, stiffness, and temperature resistance but, as an unintended consequence, also brittleness. Bottom-up orientation is prevalent in desktop SL printers, because these printers are easier and less expensive to manufacture, and easier to operate than top-down models, and require smaller volumes of resins (De Leon et al. 2016), but their build size is limited, as the forces applied to the object during peeling might cause the cured object to detach from P. A seller of popular bottom-up printers is Formlabs (USA).

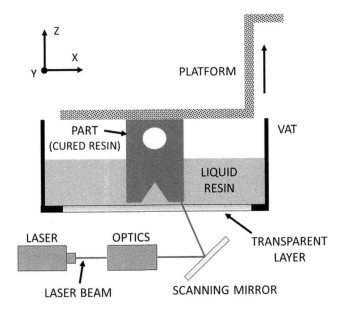

FIGURE 2.11 Schematic illustration of SL printer featuring bottom-up configuration. Courtesy of Agostino Paesano.

2.5.1.1.2 *Top-down SL*

In top-down SL printers (Figure 2.12) the laser source is located above the resin vat, and irradiates the exposed surface of the resin volume, and the part is being built facing up while resting on the build platform that is located at the top of the vat when printing starts, and travels downwards after every layer is cured to let liquid resin to fill the space, hence the term *top-down*. A sweeping blade spreads a thin coat of liquid resin on top of the preceding cured layer (*recoating*) to ensure that the new layer has constant thickness. However, recoating consumes a substantial portion of the total printing time (Pham and Ji 2003). Top-down is preferred in industrial SL printers, since these

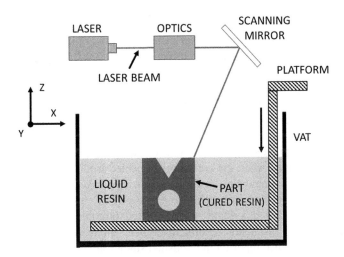

FIGURE 2.12 Schematic illustration of SL printer featuring top-down configuration. Courtesy of Agostino Paesano.

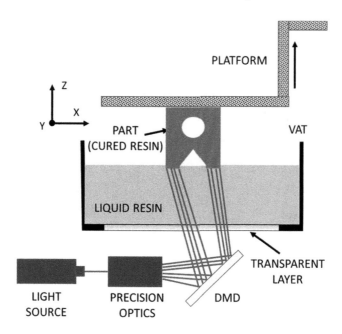

FIGURE 2.13 Schematic illustration of DLP printer.

printers can scale up to very large build sizes without sacrificing accuracy but are sold at a premium. 3D Systems is the largest manufacturer of top-down printers.

2.5.1.2 Digital Light Processing (DLP)

DLP printers (Figure 2.13) are basically bottom-up printers in which the point-wise light beam is replaced by a digital light projector screen that illuminates the clear window at the vat's bottom, and flashes on the resin a digital bitmap (*dynamic mask*) of each entire XY cross section of the part, and cures at once a complete layer in a single exposure, enabling far greater process speed than that with SL, while maintaining high fabrication accuracy (Zhu et al. 2017). The dynamic mask is formed by LCD screen, spatial light modulators, or thousands of moving digital micromirrors (*digital micromirror device* or DMD) that switch between on and off settings to duplicate the object's XY cross section, and reflect and focus the light beam on the resin. The light passing through the vat's bottom cures the bottom layer of resin that is not in direct contact with air, and hence photopolymerization is not inhibited by oxygen (*oxygen inhibition*) (Ligon et al. 2017) and occurs very fast. Over SL, DLP has the further advantages, besides greater process speed: no recoating step is necessary, since gravity lets resin fill the gap between the latest cured layer and the vat's bottom (Gibson et al. 2015); DLP printers are more efficient, operate at a wide range of wavelengths, and are equipped with customized and small resin reservoirs. Shortcomings of DLP versus SL are: (a) DLP is less accurate because of higher speed; (b) in SL the laser spot size is smaller than the minimum pixel size of the DMD, resulting in smaller minimum feature size; (c) fine features may get damaged when the cure layer is peeled from the window.

Most of the polymers in commercial formulations for DLP are acrylic-based, such as most of the following ingredients in feedstocks supplied by EnvisionTEC: unspecified acrylated monomers and oligomers, hexane-1,6-diol diacrylate, methacrylated oligomer, methacrylic oligomer, isobornyl acrylate, ethoxylated bisphenol A, dimethacrylate, epoxy resin, dicyanate, and dimethacrylate. Other polymers for DLP reported in published studies are poly(ethylene glycol) diacrylate (PEGDA), poly(ethylene glycol) dimethacrylate (PEGDMA) (Kadry et al. 2019), and 2-ethylhexyl acrylate (EHA) (Peng et al. 2019). Very recently Shan et al. (2020) investigated an epoxy acrylate-based polymer to print a shape memory polymer via LCD SL. Suppliers of DLP printers are B9Creations (USA) and EnvisionTEC (Germany).

2.5.1.3 Micro VP

In micro VP objects less than 1 mm in size are made out of photopolymers. Versions of micro VP have been around for 30 years (Gibson et al. 2015), and typically employ a stationary light beam focusing to a spot less than 20 μm in diameter, while the vat moves in the X, Y, Z directions.

A typical equipment for point-wise microstereolithography features: maximum size of printed part 10 mm in each direction, minimum size of cured feature $5 \times 5 \times 3$ μm (X, Y, Z), positional accuracy of 0.25 μm in X and Y and 1 μm in Z, and UV beam with 5 μm spot size (Gardner et al. 2001). Micro VP printers are utilized and investigated for fabricating components of microelectromechanical systems (MEMS), chips to analyze fluids in protein synthesis, ultrafiltration membranes, electrical conductors, etc. (Gibson et al. 2015).

A version of micro VP is *microstereo-thermal-lithography* (μSTLG), developed by Bartolo and Mitchell (2003), and offering the following advantages over conventional SL: (a) the equipment is more tunable; (b) printed items are more accurate because the curing reaction is more localized; (c) it combines heat and UV radiation, increasing reaction rates; (d) it works with smaller concentrations of photo and thermal initiators (Bartolo 2011). Interestingly, Gonçalves et al. (2014) formulated new sustainable unsaturated polyesters (UPs) via bulk polycondensation from biobased aliphatic acids (succinic, adipic, and sebacic acid) that were successfully processed through μSTLG, and, through physical-chemical characterization and cell viability essays, demonstrated that μSTLG and the new UPs were suited to prepare biodegradable tailor-made scaffolds potentially applicable in TE.

2.5.1.4 Continuous Liquid Interface Production (CLIP™)

CLIP™ (Tumbleston et al. 2015) was very recently patented (DeSimone et al. 2018), and is basically a far faster version of DLP. In DLP the steps of UV exposure, resin renewal, and raising the platform are separate and discrete, but in CLIP™ they all occur *continuously*, exploiting the fact that oxygen inhibits photopolymerization. In CLIP™ (Figure 2.14), an oxygen-permeable window (made of amorphous fluoropolymer) forms a permanent and thin (tens of μm thick) layer of uncured liquid polymer (*dead zone*) between the UV-clear window and the cured layer: the dead zone enables a very fast photopolymerization, carried out by elevating simultaneously and *without pauses* the platform (*build support plate*) and changing the UV images of the part's XY cross section. CLIP™ can print lattice structures at 500 mm/h in Z. Polymers compatible with CLIP™ are: cyanate ester, epoxy, biocompatible and sterilizable PU, elastomeric and rigid PU, silicone, and urethane methacrylate (Table 2.11), and they offer a wide range of properties. CLIP™ printers are made by Carbon3D (USA) that developed the Digital Light Synthesis™ technology based on CLIP™. The technical data sheets posted by Carbon3D on its website include data typically not disclosed by other suppliers, such as tensile stress-strain curves, dynamic mechanical analysis (DMA) plots, chemical compatibility, UV aging data in terms of tensile strength, modulus and strain at break, and biocompatibility.

2.5.1.5 Daylight Polymer Printing (DPP)

DPP was recently developed by UK resin specialist Photocentric® (Photocentric n.d.), and does not employ a UV laser or projector but light from mass-produced liquid crystal display (LCD) screens to cure liquid photopolymer resins layer by layer. Because DPP screens are the same as those for mobiles, televisions, and tablets, DPP printers are less expensive than SL and DLP printers.

Photocentric® provides printers, and resins compatible with them called Daylight, offered in the grades listed below along with all their ingredients:

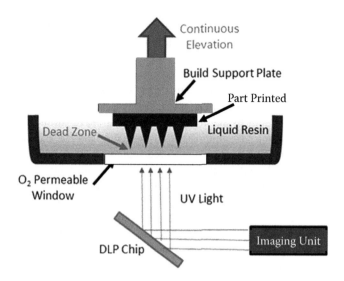

FIGURE 2.14   Schematic illustration of CLIP™. DLP stands for direct laser processing.

Source: Johnson A. R., Caudill C. L., Tumbleston J. R., et al. 2016. Single-Step Fabrication of Computationally Designed Microneedles by Continuous Liquid Interface Production. PLoS ONE 11 (9): e0162518. Reproduced with kind permission from Prof. J. M. DeSimone.

TABLE 2.11

Mechanical Properties of Polymers for CLIP™ Process

Polymer Type, Commercial Name	Tensile Strength	Strain at Break	Tensile Modulus	Izod Notched Impact Strength	HDT at 0.45/ 1.8 MPa
	MPa	%	MPa	J/m	°C
Urethane metacrylate, UMA 90	30	30	1,400	30	113/95
Cyanate ester, CE 221	85	3	3,900	15	230/200
Flexible PU, FPU 50	25	200	700	40	70/45
Epoxy, EPX 82	80	5	2,800	45	130/120
Multipurpose PU, MPU 100	35	15	1,300	30	50/40
Rigid PU, RPU 130	35	100	1,000	75	120/55
Silicone, SIL 30	3.5	350	3	N/A	N/A

Source: www.carbon3d.com (accessed July 25, 2020).

- *Hard* and *High Temperature* grades: bisphenol A ethoxylate dimethacrylate (mostly), acrylate, diacrylate, photoinitiator
- *High Tensile* and *Flexible*: 2-[[(butylamino)carbonyl]oxy]ethyl acrylate (mostly), acrylate, diacrylate, photoinitiator.

2.5.1.6   Two-Photon VP (2PVP)

2PVP was invented in 1975 (Swanson et al. 1978), and its applications included *3D microfabrication*, a term encompassing manufacturing techniques consisting in layering materials to produce 3D structures usually on the scale of μm (Baldacchini 2015) for microelectronics and MEMS.

In 2PVP a tightly focused near-IR laser emits two beams that simultaneously strike a photo-initiator molecule below the resin surface, starting the polymerization that is achieved only at the intersection of the photons. No platform is required, since the part is polymerized inside the resin volume. The lasers for 2PVP are very fast, and called *femtosecond lasers*, because they emit optical pulses whose durations are measured in femtoseconds (fs), with $1 \text{ fs} = 1 \times 10^{-15}$ s, which translates into fast polymerization and high process speed. A typical 2PVP laser is Ti:sapphire (a crystal of sapphire (Al_2O_3) that is doped with Ti3+ ions) laser (Li and Fourkas 2007). In 2PVP the laser beams do not scan but are stationary, and the resin vat is mounted on a precision stage moving in XY. Since polymerization is spatially controlled at extremely low scale, 2PVP offers outstanding resolution, and geometric features smaller than 0.2 μm. Common photopolymers are compatible with 2PVP printers.

Nanoscribe (Germany) supplies 2PVP printers manufacturing nano, micro, and mesoscale structures up to dimensions of millimeters, and relative resins, which include the IP-series, a family of proprietary acrylic resins plus initiator that are biocompatible, and non-cytotoxic. According to Nanoscribe, the applications of 2PVP are many: biomedical engineering, cell scaffolds, diffractive optical elements, integrated photonics, life sciences, mechanical components and metamaterials, micro rapid prototyping, microfluidics, micro-optics, microrobots, photonic metamaterials, photonics, and TE.

For more details on 2PVP the readers are directed to Sun and Kawata (2004), Hayat et al. (2011), Baldacchini (2015), and Gibson et al. (2015).

2.5.1.7 Lithography-Based Metal Manufacturing (LMM)

In 2019, the company Incus (Austria) announced its LMM process developed initially to print high-accuracy dental parts, i.e. patient-specific crowns and bridges (Hendrixson 2019). As feedstocks LMM employs photopolymer resins filled (up to 60 vol%) with metal powder that, unlike most VP resins that are liquid during printing, remain solid but spreadable within the 20°C build chamber. A blade spreads the resin over a heated build plate, and the photopolymer is cured with a digital light projector to form each layer. The polymerized unfinished (*green*) article leaves the printer in a block of material. An IR light liquefies the uncured material, and frees the article that is debonded and sintered in a metal injection molding (MIM) furnace. The LMM is a process tailored for high resolution, small metal objects with smooth finish. The Incus printer, called Hammer Lab 35, features building volume of 90 x 56 x 120 mm, layer thickness 10–100 μm, XY resolution 35 μm, print speed up to 100 cm^3/h. LMM parts are similar in quality to MIM parts, and benefits from the geometric freedom of AM. Printed metals comprise iron-based alloys, titanium, copper, precious metals, and hard metals.

2.5.2 Polymers for VP

2.5.2.1 Introduction

Photopolymerization occurs in VP in two forms, each associated to specific feedstocks:

- *Free-radical*: once the photoinitiator reacts with the emitted light, long polymer chains are formed (by successive addition of free radical building blocks) that are initially linear and later crosslink, when the polymer chains grow to the point of becoming very close to one another. Free radical polymerization applies to acrylates that present the advantage of fast reacting but the disadvantage of shrinking, warping, and curling.
- *Cationic*: when the photoinitiator is hit by light, it produces a cation that reacts with a monomer, and the reaction propagates generating a polymer. Cationic polymerization applies to epoxy and vinyl ether.

The feedstock materials for VP must contain, as mentioned earlier:

- A reactive, UV-curable monomer or oligomer or a blend of them that crosslink and polymerize into a solid object.
- A photoinitiator or a blend of them that is degraded when exposed to the light source, and forms radicals, cations, or carbene-like species that activate the process of polymerization (Bartolo 2011). Bagheri and Jin (2019) published a comprehensive and detailed list of photoinitiators with names, chemical structure, wavelength, and references, including among others:
  - For UV light: benzophenone, diphenyl (2,4,6-trimethylbenzoyl)-phosphine oxide
  - For visible light: camphorquinone, bis(4-methoxybenzoyl)diethylgermanium.

Optionally, other types of materials are added, such as dyes and fillers. Dyes offer the following benefits: controlling layer thickness by controlling light penetration depth, inhibiting resin over-curing and hence preventing a reduction of resolution in all orientations upon printing, and, when a filler is present, tuning the refractive index of the resin closer to that of the filler. Fillers enhance some properties or impart new ones: antibacterial, mechanical (fracture mechanics, stiffness) and optical properties, electrical conductivity, electromagnetic shielding, thermal conductivity, etc. (Taormina et al. 2018).

Ceramic materials were investigated for SL, namely suspension containing alumina powder, UV curable monomer, diluent, photoinitiator and dispersant (Hinczewski et al. 1998). Two diacrylate monomers were alternatively employed: di-ethoxylated bisphenol A dimethacrylate (Akzo Diacryl 101) and 1,6-hexane diol diacrylate (UCB Chemicals).

2.5.2.2 Polymers for VP

There are two main groups of basic polymers for VP, and each group is based on a different polymerization process (Taormina et al. 2018): (a) acrylic- and methacrylic-based resins, and (b) resins based on epoxy or vinyl ether.

Acrylic- and methacrylic-based resins, and other resins modified with acrylate ends are the most popular polymers for VP, and follow free-radical polymerization. Resins' functionality and MW affect the part's reaction time, viscosity, strength, and stiffness. Most common methacrylate monomers and oligomers for VP are: PEGDA (Warner et al. 2016), urethane dimethacrylate (UDMA) (Liska et al. 2007), triethylene glycol dimethacrylate (TEGDMA) (Al Mousawi et al. 2018), bisphenol A-glycidyl methacrylate (Bis-GMA) (Al Mousawi et al. 2017), trimethylolpropane triacrylate (TTA) (Janusziewicz et al. 2016), and bisphenol A ethoxylate diacrylate (Bis-EDA) (Credi et al. 2016). Methacrylate-based resins shrink when polymerized, inducing residual stress, causing in turn curling and deformation during printing. High-MW oligomeric acrylates display less shrinkage.

Resins based on epoxy (more common) or vinyl ether (less common) follow cationic chain-growth polymerization, and are less studied than acrylic and methacrylic resins. Examples of commercially available epoxy monomers are 3,4 epoxycyclohexane)methyl 3,4 epoxycyclohexylcarboxylate (EPOX), and bisphenol A diglycidyl ether (DGEBA). The combination of useful engineering properties and a wide range of processing options has led to many applications of epoxy resins comprising, among others, adhesives, encapsulation for electronics, potting, printed circuit boards, and matrices of aerospace composites. Table 2.12 lists epoxy-based commercial resins for SL and DLP that are sold by Dutch leading supplier DSM Somos®, and exhibit a substantial range of property values: tensile strength and modulus are 31–72 MPa and 1,350–9,653 MPa, respectively, Izod notched impact strength is 20–51 J/m, and heat deflection temperature at 0.46 MPa is 50–268°C. An example of cationic polymerizable vinyl ether monomers commonly utilized in the SL-based systems is 1,4-cyclohexane dimethanol divinyl ether (CDVE). Moreover, disubstituted oxetane monomers are suited for cationic photopolymerization, and, compared to the epoxides, offer high reactivity with comparable low shrinkage rate, and improved water resistance (Bagheri and Jin 2019).

TABLE 2.12

Epoxy-Based Resins for Stereolithography Sold by DSM Somos®

Commercial Name	Polymer	Density	Tensile Strength	Tensile Modulus	Strain at Break	Flexural Strength	Flexural Modulus	Izod Notched Impact Strength	HDT at 0.46/ 1.8 MPa	Features. Applications
		g/cm^3	MPa	MPa	MPa	MPa	MPa	MPa	°C	
9120	Epoxy	1.13	31	1,350	20	45	1,380	51	N/A	Form, fit, function
Element	Epoxy	1.11	53	3,170	2.3	114	3,230	22	58/53	Investment casting
NeXT	Epoxy	1.17	32.8	2,430	9	69.3	2,470	50	56/50	Cracking, fracture resistant
PerFORM	Epoxy	1.61	80	9,800	1.2	146	9,030	20	268/119	Strong, stiff, temp. resistant
PerFORM Reflect	Epoxy	1.61	72.4	9,653	0.96	130	7,722	20	276/122	Strong, stiff, temp. resistant
Taurus	Epoxy	1.13	49 (yield)	2,206	17	62.7	1,724	35.8	91/73	Strong, stiff, temp. resistant
Water Shed XC 11122	Epoxy	1.12	50.4	2,770	15.5	68.7	2,205	25	50/49	Optically clear, moisture resistant, durable

Source: Data posted online by DSM Somos®.

Epoxy and vinyl resins are also mixed together in formulations that limit the disadvantages of both groups (Yugang et al. 2011; Kumar et al. 2012).

LCPs for VP were studied two decades ago to print parts possessing anisotropic properties and service temperatures exceeding 100°C by Ullett et al. (2000). The ensuing printed and post-cured test coupons out of LCPs featured T_g of 75–148°C, depending on the polymer and processing conditions, and values of mechanical moduli along the molecular alignment twice the values measured perpendicularly to it.

Many polymers for VP are brittle once printed because they are crosslinked, and over time yellow at a higher rate if exposed to UV light, and their mechanical properties degrade, because they tend to take up water in humid atmosphere.

The brittleness is addressed with commercial impact resistant formulations performing like ABS, PE, and PP, and fillers that also improve elastic modulus, such as AEROSIL® fumed silica and AEROXIDE® fumed oxides by Evonik. Metal fillers have been investigated for SL more than a decade ago, by combining unsaturated polyester, and epoxy resin with powder of either tungsten carbide or cobalt (Bartolo and Gaspar 2008), and metal-filled resins for VP are only sold by Incus (Austria).

By tuning chemistry and cure, Carbon3D supplies polymers displaying a wide range of properties (from strong, stiff, and brittle to soft, flexible and tough), and targeting performance of ABS, PA, PP, glass filled-PBT, and PP, for applications requiring sterilization and biocompatibility.

A way to stabilize mechanical properties over time is adding ceramic (glass, oxide, etc.) or metal fillers. Ceramic-filled resin are sold by 3D Systems (Accura® CeraMAX™ Composite), Evonik (NANOCRYL® and NANOPOX®), Formlabs (Ceramic), Somos (NanoTool™ and NanoForm™ 15000 Series), and Tethon 3D (Porcelite®).

Yellowing can be controlled by specific photoinitiators, and eliminated by selecting as feedstocks ceramic-filled photopolymers, and acrylated silicones that form a class of polymers for VP exceptionally non-yellowing and possessing resistance to heat, weathering, scratch, abrasion, and chemicals, and a broad service temperature interval from −50 to 260°C.

Table 2.13 lists the main properties of two commercial resins for DLP supplied by B9Creations.

In 2018, the company eSUN (China) launched possibly the first worldwide commercial sustainable resin for VP, called eResin-PLA, made of PLA, specifically designed for LCD/LED light source, and suited for models, prototypes, and functional parts. However, its reported mechanical properties are lower than those of non-PLA resins for VP supplied by eSUN (Table 2.14).

Ding et al. (2019) demonstrated that it is possible formulating sustainable UV-curable acrylates for SL from biobased phenolics. From softwood lignin, they derived sustainable TS resins made of phenolic methacrylates featuring fast reactivity adequate for SL process, and promising thermomechanical properties: tensile strength, modulus, and strain at break of 33–62 MPa, 0.83–1.23 GPa, and 2.8–8.9%, respectively, toughness of 0.5–3.7 MJ/m^3, and temperature at which 5 wt% decomposes equal to 323°C.

TABLE 2.13

Properties of Commercial Resins for DLP Supplied by B9Creations

Material	Tensile Strength	Tensile Modulus	Elongation at Break	Izod Notched Impact Strength	HDT at 1.8 MPa	Hardness
	MPa	MPa	%	J/m	°C	Shore D
Black Resin	36.5	1,750	3.4	11.4	49	83
Grey Resin	44.6	2,110	8.6	20.7	39	87

Source: Data posted online by B9Creations. Reproduced with permission from B9Creations.

TABLE 2.14
Properties of Resins for VP Produced by eSUN

Resin Type	Density	Hardness	Tensile Strength	Elongation at Break	Flexural Strength	Flexural Modulus	Izod Impact Strength
	kg/dm^3	Shore D	MPa	%	MPa	MPa	J/m
eResin-PLA	1.07–1.13	75–80	35–50	20–50	40–60	600–800	15–32
Standard Resin	1.08–1.13	80–82	46–67	28–35	46–72	1,000–1,400	18–40
High-Tough Resin	1.10–1.15	81	55–80	30–50	70–80	1,300–1,400	67–100
Rigid Resin	1.10–1.15	83	60–70	25–35	85–35	1,400–1,500	15–42
Precision Model Resin	1.13–1.15	85	335–410	12.1	77	2,350	44

Source: Online data posted by eSUN at http://www.esun3d.net/products/473.html (accessed April 3, 2021).

2.6 BINDER JETTING (BJ)

2.6.1 PROCESS DESCRIPTION

BJ is a family of AM processes that have in common spreading a layer of powder on a flat platform, and *selectively* jetting on the powder a liquid binder that joins the powder particles into a cohesive slice of the part being built. The first BJ version was patented by researchers at the Massachusetts Institute of Technology (MIT) in 1993 (Sachs et al. 1993; Cima et al. 1995) under the name of *three-dimensional printing*™ (3DP™). Currently 3DP™ is employed on printers offered by ExOne (USA), Voxeljet (Germany), and Z Corporation (USA) that was acquired by 3D Systems in 2012.

BJ starts with a recoating blade or roller spreading a thin layer of powder on a flat platform (*powder bed*), over which the *printer* or *jetting head* sweeps. This is a device that contains many tiny ejection nozzles lined up perpendicularly to the head's moving direction, and selectively dispenses a binder in form of droplets of about 80 μm in diameter. The powder particles in the regions receiving the binder bind together and to the previously spread layer. When the jetting head has run across the whole layer, the powder bed is lowered by one powder layer's thickness, a new powder layer is spread, and the process is repeated until the object is complete. The object is left for some time inside the powder bed among the unbound powder to let the binder fully set and the object gain some cohesive strength, after which the object is extracted, cleaned manually and/or via pressurized air, and coated or infiltrated with adhesives that are cured to maximize its strength and stiffness.

In equation (2.12) the binder droplet's energy E_d is the sum of the droplet's surface energy (the term to the left of + sign) and kinetic energy (the term to the right of + sign), being σ the surface energy (J/m^2), D the diameter, ρ density, V volume, and S speed of the droplet (Chua et al. 1998):

$$E_d = \sigma \pi D^2 + 0.5\rho V S^2 \qquad (2.12)$$

The adhesive energy *BE* binding together the powder is the droplet energy per aggregate volume associated with the droplet, being K a constant, according to equation (2.13):

$$BE = 6E_d/[\pi (KD)^3] \qquad (2.13)$$

Advantages of BJ are: (a) feedstock can be inexpensive; (b) using drop-on-demand with different colored inks, BJ prints objects in multiple colors within individual layers; (c) unprinted powders are fully recyclable; (d) high build rate: a 100-nozzle print head can build at up to 200 cm^3/min (Yun and Williams 2015). Drawbacks of BJ are: (a) printed parts lack the precision of PBF and SL; (b) the mechanical properties and surface roughness of printed polymer objects, even after resin infiltration or sintering during post-processing, often do not meet the service demands of tooling, functional prototyping, and functional parts (Ligon et al. 2017).

2.6.2 Feedstocks for BJ

A wide range of ceramics, metals, and polymers are suited for BJ but only some of them are commercially offered (Gibson et al. 2015). Plaster-based powders and water-based binders are chosen for low-cost printers. PMMA in form of 55 μm and 85 μm particles is the polymeric base material running on Voxeljet printers for investment casting. Sustainable starch-based powders are employed as well. 3D Systems employs a plaster-based powder, and a water-based binder that generate objects that, after infiltration, are stiffer and weaker than objects printed using VP less deformable feedstocks (Gibson et al. 2015).

A few polymers have been processed as powder for BJ, partly because polymers are rarely produced as powders. PGA, PLA, PCL, and PEO have been processed with a suitable solvent (Wu et al. 1996). Powders mostly composed of amorphous cellulose were also successfully utilized when polysaccharides were present in the powder and ink (Holland et al. 2018).

Voxeljet feedstocks include sands based on silica and chromium, and the option of three binders, two of them based on TS high temperature resistant resins: furan and phenolic. SPs serving as alternative to phenolic resins are available. In 2012 a U.S. company converted sustainable feedstocks, such as switch grass, pine, poplar, corn stover, etc., into high-quality phenol materials that can replace fossil-derived phenols present in phenolic resins, and reduce greenhouse gases and air emissions caused by burning fossil fuel (Metrey 2012). In 2018, the Finnish company Storaenso, the largest kraft lignin producer in the world, launched a new biobased lignin called Lineo™ suited to replace fossil-based phenolic resins (Storaenso 2018). Furan is produced from fossil-based 1,3-butadiene or furfural (C$_4$H$_3$OCHO), an organic compound derived from sugar, and hence a SP. Furan resins have also been considered as a biobased alternative to phenolic resins (Ramon et al. 2018).

According to Voxeljet's website, its BJ process generates parts for sandcasting, investment casting, design and functional models, ceramic printing, and concrete formwork, serving a wide spectrum of industries: aerospace, architectural and construction design, art, industrial design, automotive, foundries, film and museum, pump and heavy industry.

Very recently Ziaee and Crane (2019) authored a wide-ranging review of process, materials, and methods for BJ.

2.6.3 Multi Jet Fusion (MJF)

MJF is a version of BJ developed by Hewlett-Packard (HP), based on polymeric powder, fluid agents, and a working principle called *area-based fusion* (Hewlett-Packard 2020). MJF comprises the following main steps (Figure 2.15):

- MJF starts with the HP software preparing the build of the part. If replicates of the same part are printed in one session, air checking and automatic packing inside the build chamber (to minimize distance among replicates on the build platform) are carried out.
- The feedstock material in form of TP powder is contained in a wheeled tank (*build unit*), and is inserted in the printer by an operator, and during printing the powder is automatically transferred from the tank and spread evenly across the build platform, forming a thin layer.

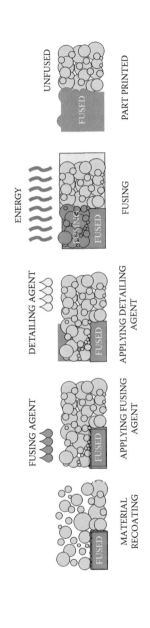

FIGURE 2.15 Schematic representation of the multi jet fusion process. Courtesy by Francis Paesano.

- The *fusing* agent is dispensed on the powder layer as minute droplets. This is a radiation-absorbing pigment whose function is to increase the powder's temperature under radiation, melt and fuse the powder.
- The *detailing* agent is sprinkled in areas of the layer that are extraneous to the part's XY cross section, in order to keep temperature down preventing fusion, and achieve dimensional accuracy and control of boundary and look, shaping the part's outer surface.
- The layer is exposed to radiating energy emitted by lamps, and reacts with the fusing and detailing agents, and only the layer's areas wet with the fusing agent fuse, while the rest of the layer stays loose in powder form.
- The build platform is lowered by one powder layer thickness, and the cycle repeats. The time to print one layer is the same regardless of how many items are printed in one session.

According to HP, MJF enables to fabricate visual aids, presentation models, prototypes, and functional part at higher speed and lower cost than other AM processes for plastics. As to speed, HP maintains that in 3 h MJF, FFF, and LS printers make 1000, 79, and 36 gears of the same size, respectively. Sillani et al. (2019) compared MJF to SLS by analyzing powders and final parts made of virgin and recycled PA 12, and detected differences at the molecular and powder scale and in thermal properties, while flowing properties were similar among the virgin and recycled powders. Against the MJF parts, the SLS parts featured higher tensile modulus but lower tensile strength and strain at break, and out-of-plane unnotched Charpy impact strength.

Feedstocks for MJF have to be TP in form of powder, like in PBF. Commercial polymers compatible with MJF are currently limited to two PA grades, PA 11 and PA 12, whose tensile, impact, and thermal properties are reported in Table 2.15:

- HP 3D High Reusability PA 12
- HP 3D High Reusability PA 12 Glass Beads
- HP 3D High Reusability PA 11
- VESTOSINT® 3D Z2773 PA 12 30L.

Table 2.15 indicates that adding glass beads to PA 12 raises its tensile modulus and elongation at break by 65% and 225%, respectively, but lowers its tensile strength by 38%, possibly indicating that the bonding between filler and polymer can hold at low strains where the modulus is

TABLE 2.15

Properties of Materials for Hewlett-Packard MJF Process

Material	Particle Size	Tensile Strength	Tensile Modulus	Elongation at Break	Notched Izod Impact Strength	HDT at 0.45 MPa	HDT at 1.82 MPa
	μm	MPa	MPa	%	kJ/m²	°C	°C
HP 3D HR PA 12	60	48	1,700	2	3.5	175	106
HP 3D HR PA 12 GB	58	30	2,800	6.5	2.7	173	121
HP 3D HR PA 11	54	50	1,800	50	6	183	50
VESTOSINT® 3D Z2773 PA 12 30L	57	48	1,700	20	N/A	N/A	N/A

Source: Data posted online by Hewlett-Packard 3D at https://www8.hp.com/us/en/printers/3d-printers/materials.html (accessed November 14, 2020).

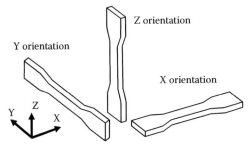

FIGURE 2.16 Printing orientations and tensile coupons in the study of MJF conducted by O'Connor H. J., Dickson A. N., Dowling D. P. 2018. Evaluation of the mechanical performance of polymer parts fabricated using a production scale multi jet fusion printing process. *Additive Manufacturing* 22:381–387.

calculated, but not at high strains where tensile strength is measured. O'Connor et al. (2018) assessed the effect of printing orientation (Figure 2.16) on tensile and flexural properties of test coupons printed out of HP 3D High Reusability PA 12. The test data (Table 2.16) indicated that all properties measured, except tensile strength, were anisotropic, and tensile and flexural moduli and flexural stress were maximum along Z, and this superiority may point out that the interlayer bonding is effective from a mechanical standpoint. The authors also compared tensile properties of HP 3D High Reusability PA 12 to other brands of PA powders for laser sintering: DuraForm® by 3D Systems and PA 2200 by EOS. The HP powder's values fell inside the ranges relative to the other powders that were 35–51 MPa, 1,100–2,171 MPa, and 17–21% for tensile strength, modulus, and strain at break, respectively. The standard deviation values in Table 2.16 are fairly limited for all properties measured.

2.6.4 Three Dimensional Printing™ (3DP™)

3DP™ consists in spreading a layer of powder on a flat platform, and *selectively* jetting on the powder a liquid binding material that joins the targeted powder particles that form a cohesive layer. U.S. patent 5,204,055 (Sachs et al. 1993) on 3DP™ summarizes the process as follows: "A process for making a component by depositing a first layer of a fluent porous material, such as a powder, in a confined region and then depositing a binder material to selected regions of the layer of powder material to produce a layer of bonded powder material at the selected regions. Such steps are repeated a selected number of times to produce successive layers of selected regions of bonded powder material so as to form the desired component. The unbonded powder material is then removed. In some cases the component may be further processed as, for example, by heating it to further strengthen the bonding thereof."

TABLE 2.16

Mechanical Properties of Coupons Made of HP 3D High Reusability PA 12 Printed Via MJF

Build Orientation	Tensile Strength	Tensile Modulus	Elongation at Break	Flexural Stress	Flexural Modulus
	MPa	MPa	%	MPa	MPa
X	47 ± 0.9	1,242 ± 28	19 ± 2.8	50 ± 0.9	1,146 ± 51
Y	48 ± 0.8	1,147 ± 40	27 ± 1.2	66 ± 0.5	1,567 ± 36
Z	49 ± 0.6	1,246 ± 37	16 ± 1.9	70 ± 0.7	1,687 ± 12

Source: Data from O'Connor, H. J., Dickson, A. N., Dowling, D. P. 2018. Evaluation of the mechanical performance of polymer parts fabricated using a production scale multi jet fusion printing process. *Additive Manufacturing* 22:381–387.

In 3DP™ an inkjet-like printing head is mounted on two horizontal perpendicular XY axes, on which it rests, running above and across a flat platform where a thin layer of powder feedstock is spread by the spreading bar. Through many nozzles, that can exceed tens of thousands (Anon. 2018), the printing head dispenses droplets of a liquid binder solution on the areas of the powder layer corresponding to the XY cross section of the article to be printed, and these areas consolidate into one whole layer. The platform is mounted on top of a piston that lowers it by one powder layer thickness, and the process is iterated. Upon printing completion, loose powder is removed by hand and by means of a vacuum hose, the article is extracted from the platform, and infiltrated with a compound (*infiltrant*) to achieve its specific final physical and mechanical properties. Feedstocks for 3DP are ceramics, HGs, metals, and polymers (Katstra et al. 2000).

An example of 3DP printers are the ProJet printers sold by 3D Systems that are based on technology developed by Z Corporation, and print full-color items serving as design models, visual aids, architectural models, prototypes, tooling, dies for forming, and figurines. ProJet feedstock material is gypsum powder, namely calcium sulfate hemihydrate ($CaSO_4 \cdot 0.5H_2O$), featuring density of 2.6–2.7 g/cm^3, and particle size distribution d(10), d(50), and d(90) equal to 2.1 μm, 39.8 μm, and 80.2 μm, respectively (Fonseca Coelho et al. 2019). Table 2.17 lists major mechanical properties of VisiJet® PXL, a feedstock for ProJet printers comprising core and binder. The mechanical properties in the objects printed with PXL vary depending on what infiltrant is selected among the following: (a) one-part, fast-acting adhesive, such as cyanoacrylate (the active ingredient of instant glues such as Krazy Glue®) for strong and color models; (b) two-part, high-strength epoxy for functional parts; (c) eco-friendly and safe salt and water for monochrome models; and (d) eco-friendly and safe wax for color models. Compared to HP's polymers in Table 2.15, VisiJet polymers are stiffer, less strong, and deformable. ProJet's infiltrants can represent an opportunity for SPs. In fact, sustainable epoxies are already commercially available, such as ONE, an epoxy high in biobased content sold by Entropy Resins (Entropy Resins 2020).

Some applications of 3DP are pharmaceutical, and medical. In fact, Aprecia Pharmaceuticals (USA) prints solid, highly porous, high dose (up to 1000 mg of drug) medications that disintegrate in seconds in the mouth in a sip of liquid that breaks up the binder keeping the powder together (Yoo et al. 2014). Aprecia produces Spritam®, a medication that was the first printed oral solid dosage form approved by FDA. Therics LLC (USA) leverages 3DP™ not only for oral medications in solid dosage forms, but also for implants, and tissues (Hsiao et al. 2018).

3DP™ has also been investigated for constructions by Xia and Sanjayan (2016), who studied two powder systems: a slag-based geopolymer powder and a commercially available plaster-based powder.

TABLE 2.17

Mechanical Properties of VisiJet® PXL Combined with Various Infiltrants

Infiltrant	Tensile Strength	Tensile Modulus	Elongation at Break	Flexural Strength	Flexural Modulus
	MPa	GPa	%	MPa	GPa
One-part adhesive	14.2	9.5	0.23	31.1	7.2
Two-part high-strength epoxy	26.4	12.6	0.21	44.1	10.7
Salt and water	2.38	12.9	0.04	13.1	6.4
Wax	9.2	22.6	0.09	11.7	4.8

Source: Data sheets posted online by 3D Systems.

2.7 MATERIAL JETTING (MJ)

2.7.1 PROCESS DESCRIPTION

MJ, or also *3D inkjet printing*, is similar to inkjet document printing. In MJ hundreds of droplets of a liquid photopolymer are simultaneously and *selectively* jetted from multiple printheads (similar to those on standard inkjet document printers) on the build platform, and form one layer at the time, which is then cured and made solid by UV light. Layer is built upon layer until the part is finished. Namely, MJ consist of the following steps: (a) the liquid resin is heated to about 30–60°C to lower its viscosity to an optimal value for printing; (b) the printhead equipped with hundreds of nozzles, and mounted on a carrier, travels in X and Y directions over the stationary build platform, and jets hundreds of minute uniform-size droplets of feedstock material and support material (always required); (c) the UV light source mounted on the carrier very rapidly cures the deposited photopolymer, solidifies it and forms the first layer of the part; (d) the build tray is lowered one layer height (typically 16–32 µm), and the process is repeated until the whole part is complete; (e) the printed part is freed from its support structures (made of wax or water-soluble photopolymer) and cleaned up. In MJ the inkjet printheads are attached side-by-side to the same carrier, and dispense different feedstock materials and support materials on the entire print surface in a single pass and a line-wise fashion. MJ relies on photopolymerization like SL and jetting like BJ, but differently from SL does not require post-curing, and differently from BJ it dispenses not the binder but the *whole* feedstock material. The chief advantage of MJ is printing print multi-material and multi-color parts, because each material can be stored in a subset of theprintheads, and printed simultaneously, allowing different materials and colors to be processed. MJ's drawback is its small build volume and a post-processing step to remove support material and resin residues.

MJ is available in two versions (Gibson et al. 2015):

- *Continuous mode*: a constant pressure is applied to the fluid feedstock reservoir and causes the nozzle to eject a continuous column of fluid that breaks into droplets outside the nozzle before reaching the build platform. After the droplets leave the nozzle, they are charged electrostatically, so they can be steered to the targeted location by an electric field established by electrically charged deflection plates. Droplets that are outside target are captured in a gutter and recirculated.
- *Drop-on-demand* (DOD): individual pressure pulses produced in the nozzle by thermal, electrostatic, piezoelectric (most common), or acoustic actuators (Le 1998) cause the nozzle to expel the feedstock as individual droplets only when needed. Piezoelectric DOD is the most common type of MJ and f.e. is implemented by Stratasys and 3D Systems on their commercial printers PolyJet and ProJet, respectively (Yang et al. 2017b).

MJ printers feature a dimensional accuracy of ±0.1%, with a lower limit typically of ±0.1 mm, and sometimes as low as ±0.02 mm. Accuracy also stems from the fact that, since printing is conducted at near room temperature, warping is infrequent. The build size ranges from about 150 × 150 × 50 mm to 1000 × 800 × 500 mm, while the minimum layer spans from 0.01 to 0.032 mm. The resolution in X and Y directions varies from 600 × 600 to 8,000 × 8,000 dpi (dots per inch), and obviously increases going from larger to smaller build size.

In MJ the image time per layer T depends on the droplet volume V, the droplet rate f, layer area A, and layer thickness L as follows (Chua et al. 1998):

$$T = (AL)/(fV) \qquad\qquad (2.14)$$

Hence, being T_R the reset time, the fabrication speed S for an article is:

$$S = L/(T + T_R) \qquad (2.15)$$

Further growth of MJ hinges on overcoming the following challenges, and perfecting the following features (Gibson et al. 2015): formulating feedstocks in liquid form, forming droplets, controlling droplet deposition, converting liquid droplets into solid form, controlling deposition of droplets on top of deposited layers, monitoring and maintaining nozzle performance while printing, and depositing many small droplets very close together to achieve high resolution. The quality of MJ deposition depends on printhead speed, droplet velocity and frequency.

Advantages of MJ are dimensional accuracy; build speed, unaffected by making multiple articles at the same time; smooth surface finish; full color; printing support materials and printing multiples materials at the same time, and obtaining an object featuring varying properties within its volume (*functionally graded materials* or FGMs); compact hardware configuration, allowing to install printers in an ordinary office space without ancillary equipment and specific environmental requirements. Shortcomings are: expensive printers and feedstocks, limited choice of materials, printed parts are weak, brittle and not durable, and hence not suited for load-bearing applications.

Derby and Reis (2003) formulated a *printability parameter* Z that, when in 1–10 range, ensures stable drops during MJ, and is calculated from fluid density D, surface tension S, and viscosity η, and a characteristic length L traveled by the drop, according to equation (2.16):

$$Z = \frac{\sqrt{SDL}}{\eta} \qquad (2.16)$$

Low values of Z prevent drop ejection, whereas high values cause the primary drop to be accompanied by many satellite droplets. Moreover, the behavior of the ejected droplets is governed by the Reynolds number Re and the Weber number We of the drop's liquid material. Particularly, according to Seerden et al. (2001), the splashing of liquid drops occurs when the value $We Re^{1/4}$ exceeds a critical value, and droplet spreading ε is expressed by equation (2.17), where r and r_{max} are the initial drop radius and the maximum splat radius, respectively, and θ is the equilibrium contact angle between drop and substrate:

$$\varepsilon = \frac{r_{max}}{r} = \left(\frac{W_e^2 + 12}{3(1 - cos\vartheta) + 4\frac{W_e^2}{R_e^{1/2}}} \right)^{1/2} \qquad (2.17)$$

MJ is primarily chosen for visual models and prototypes in architecture, figurative arts, education, engineering, industrial design, medicine, food, etc. rather than for functional articles that are, however, possible with a process called *Arburg plastic freeforming*, described in Section 2.7.2. Other applications are injection molds for a small number of parts. Electronics packaging and system integration are application of MJ being investigated following two main approaches: (a) single process approach, e.g. inkjet, to print conductive and dielectrics multi materials, (b) *hybrid* or *multi-systems* approach combining different digitally driven manufacturing processes to print multi-materials components (Stoyanov et al. 2016). The latter approach seems to be the preferred by the industry.

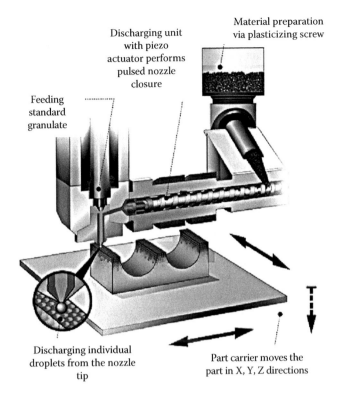

FIGURE 2.17 Schematic illustration of *Arburg plastic preeforming* process and equipment.

Source: Arburg. Reproduced with permission from Arburg.

2.7.2 COMMERCIAL MJ PRINTERS

MJ can be traced back to the U.S. patent 4,665,492 awarded to W. Masters in 1987 (Masters 1987) who described that "mass particles are injected to arrive at predetermined coordinate points in the coordinate system to form an article." Today, major companies selling MJ printers are 3D Systems (ProJet models), Arburg (Freeformer models), Solidscape (DL and S series), and Stratasys (PolyJet models). A newly formed producer of MJ printers is XJET (Israel), which introduced its proprietary *NanoParticle Jetting*™ process in 2015, consisting in printing not with plastic droplets but with liquid droplets loaded with ceramic or metal nanoparticles in order to build metal and ceramic articles (Carlota 2019). Arburg (Germany) developed a version of MJ called *Arburg plastic freeforming* (APF) and noteworthy, because it processes qualified standard plastic granules made of ABS and carbon nanotube-filled ABS, PA 10, PC and aerospace PC, PMMA, medical-grade PLLA, TPU, PP that are converted in functional parts that have movable details, and feature soft and hard regions in the same component. In APF (Figure 2.17), the granules are melted following the same method as in injection molding, and are discharged by a piezoelectric actuator as tiny plastic droplets on the build platform, which moves in the X, Y, Z directions while the printing head is stationary (Arburg n.d.).

2.7.3 REACTIVE INKJET PRINTING (RIJ)

A version of MJ is RIJ (Smith and Morrin 2012) that consists in making an object on an inkjet printer by depositing two or more physical or chemical reactants in form of droplets 20–80 μm in diameter on a substrate according to a pattern, and by controlling size, position, and number of droplets. The reactants are either dispensed simultaneously or sequentially, and mix and react in situ on the substrate. There are two types of RIJ: (a) *single RIJ*, where the inkjet printer dispenses a reactant on top of another reactant placed by another deposition technique; (b) *full RIJ*, in which the inkjet printer dispenses more than one ink.

Applications of RIJ are: (a) synthesis of expensive or hazardous materials; (b) in situ formation of conductive nanoparticles without insulating surfactants, required instead if the nanoparticles are in form of ink; (c) specific articles whose cost is lowered by reducing fabricating steps; (d) patterning alternating layers, and imparting functionally graded features.

Metal (copper, nickel, and silver) and polymeric (PU, and hydrogel arrays) feedstocks have been successfully printed with RIJ. Moreover, SABIC filed a patent in 2016 (Hocker et al. 2019) claiming 3D "printing of reactive polycarbonate precursor compounds onto a substrate to provide for rapid prototyping of one or more polycarbonate layers."

2.7.4 FEEDSTOCKS FOR MJ

Only polymers (waxes and photopolymers) are feedstocks for commercial MJ printers. Feedstocks for building and support materials for MJ must be photocurable (hence liquid) and feature adequate viscosity at the printing temperature, be thermally stable, and quickly cure upon light exposure. Suitable resins are based on urethane acrylate, exhibit viscosity of 10–16 mPa between 70 and 90°C, and contain an inert urethane wax holding the building material in place before it is photocured. Typical inks for MJ have a density close to 1 g/cm^3, viscosity and surface tension of 3–20 mPa, and 20–70 mN/m, respectively (Brindha et al. 2016).

Surface tension influences the adhesion of molecules and spread of ink on the substrate: if it is too low, the fluid drips out of the cartridge and floods the printhead at the orifice; if it is too high, it may cause discontinuous printing. Triton X-100, a non-ionic detergent, derived from polyoxyethylene and containing an alkylphenyl hydrophobic group, is added to adjust bioink's surface tension because of its low impact on biomolecular activity.

Viscosity critically affects ink printability (Brindha et al. 2016). Inks with higher viscosity require a longer application of pressure to dispense the drop, and also form a long filament during printing. Viscosity of bioinks can be changed to acceptable levels by adding viscosity modifiers, such as carboxymethyl cellulose (CMC), a cellulose derivative. Brindha et al. (2016) experimented with ink made of bovine serum albumin featuring density around 1 g/cm^3, viscosity and surface tension of 1–10 mPa and 25–37 mN/m, respectively.

Support materials also must be photocurable but easy to remove. An early support material (Napadensky 2003) patented by Objet included water-soluble monomers and polymers, photocured into weak material easily removed with water: it contained PEG monoacrylate, PEG diacrylate, diacrylates, photoinitiator, stabilizers, and silicone surface additives. Another support material is wax, removed from the printed object by mild heating. A common wax-based support material is made of stearyl alcohol and tall oil rosin.

Polymers for commercial MJ printers are waxy polymers and photopolymers. Commercial examples of the former feedstocks are Midas™ wax (Solidscape n.d.), whose major ingredient is p-ethylbenzenesulfonamide ($C_8H_{11}NO_2S$), and EmeraldCast™ resin, containing wax, acrylic resin trimethylolpropane formal acrylate (CTFA) ($C_{10}H_{16}O_4$), and initiators. Wax is employed to make patterns for jewelry industry, and high precision investment casting. Examples of MJ photopolymers are DiamondCast™ resin and ProtoPearl™ resin, both containing as main ingredients urethane dimethacrylate (UDMA) ($C_{23}H_{38}N_2O_8$) and acrylic resin. Midas™, DiamondCast™, EmeraldCast™, and ProtoPearl™ are processed on Solidscape printers.

MJ feedstocks can be liquid or solid. In the latter case, they are melted. If not solid or liquid, they may be in form of particles suspended in a liquid carrier, or dissolved in a solvent, monomer, or prepolymer mixed with a polymerization initiator.

Ceramics and metals and other polymers are also being investigated. Among the former ones are suspensions of zirconia powder (Tay and Edirisinghe 2001), and alumina particles (Derby and Reis 2003) in a liquid carrier. Printed metals (Gibson et al. 2015) have comprised: alloy of bismuth, lead, and tin; alloy of bismuth, lead, tin, cadmium, and indium; aluminum; copper; mercury; solder; tin.

Auto-generated, melt-away (55–60°C T_m) polymers serve as a support material. They enable very complex shapes with very fine, gravity-defying features, and interlocking components.

2.8 DIRECT ENERGY DEPOSITION (DED)

2.8.1 Process Description and Versions

In DED, focused thermal energy melts and fuses together the feedstock, present in the form of powder (mostly) or wire, as it is being deposited. DED works mostly with many metals, such as aluminum, bronze, copper-nickel alloys, Inconel®, niobium, steel and stainless steel, tantalum, titanium and its alloys, tungsten, zirconium, etc., but also some ceramics with relatively low melting temperature (Balla et al. 2008; Bernard et al. 2010; Niu et al. 2015a; Niu et al. 2015b). However, very recently AREVO (USA) has developed a version of DED suited for carbon reinforced polymers (Zhang et al. 2019), described in Section 2.8.2. In DED, a high-energy beam, generated by an electron beam gun, laser, or plasma arc, is focused onto a spot on a substrate laid on a platform, and generates a molten pool of substrate material. A nozzle mounted on four- or five-axis arm continuously supplies the feedstock to the molten pool where it melts upon deposition, and moves relative to the substrate in the X and Y directions according to a pattern matching the slice in the 3D digital model, and generates a bead of deposited material that quickly solidifies and ultimately forms a layer 0.3–1 mm thick. When the layer is completed, the platform lowers by one layer thickness, and the process is repeated.

Typically DED is conducted in inert atmosphere to avoid oxidation of the melt metal pool. DED is similar to but different from PBF, because PBF only uses powder, and in DED the metal powder is applied only where the energy source is focused and when the energy is provided. The microstructure of articles printed via DED is similar to those made via PBF.

Major suppliers of DED printers are Sciaky (USA) and Optomec (USA), with their proprietary versions of DED called *electron beam additive manufacturing* (EBAM®) and *laser engineered net shaping* (LENS®), respectively. Other versions for DED are *3D laser cladding, laser generation, direct metal deposition* (DMD), *laser-bed metal deposition* (LBMD), *laser cast, laser direct casting,* and *laser preform fabrication* (LPF). All of these versions share the same working principle, but their relative equipment differ in laser type, power, and spot size, and powder delivery method (Gibson et al. 2015). Because in DED the nozzle is not fixed but free to move in various directions, DED is apt for repairing parts such as propellers, transmission shafts, and turbine blades by adding material to the damaged zone.

2.8.2 AREVO® Process

The AREVO® process (Bheda 2018) applies to composite materials in form of filament of TP polymer matrix reinforced with continuous carbon filament fiber embedded in PEEK (CF-PEEK), and relies on the following components: industrial robot, printhead with laser heating, compacting roller, and rotating build platform. The printhead includes the thermal management equipment and vision system for in situ inspection. During the fabrication, the composite filament being laid (along with the layers already placed) is quickly heated by a laser and simultaneously compacted by a roller that compresses the filament, bonds it to the substrate layers, eliminates voids inside and in between the layers, and consolidates the article in situ. The entire deposition system is mounted

TABLE 2.18

Mechanical Properties of CF/PEEK Test Coupons Printed with AREVO® Process

Test	ASTM	Layup	Modulus of Elasticity	Ultimate Strength	Strain at Failure
		degree	GPa	MPa	%
Tension	D3039	0	115 ± 2	1,420 ± 83	1.22 ± 0.06
		[−45/90/45/0]	41 ± 3	479 ± 66	1.18 ± 0.09
Compression	D6641	0	100 ± 2	712 ± 47	N/A
		[−45/90/45/0]	37 ± 2	280 ± 26	N/A
Flexure	D7264	0	104 ± 1	1,173 ± 64	N/A
		[−45/90/45/0]	31 ± 2	492 ± 54	N/A

Source: Zhang, D., Rudolph, N., Woytowitz, P. 2019. Reliable Optimized Structures with High Performance Continuous Fiber Thermoplastic Composites from Additive Manufacturing. SAMPE Conference Proceedings. Charlotte, NC, May 20–23, 2019.

on a six-axis robot laying the filament in any angle and direction. AREVO® reports that PA and glass are also available as matrix and reinforcement, respectively. In the AREVO® process the main material requirement for the matrix is being a TP able to be compacted while maintaining most of its shape, and featuring low void content in the printed article. Property values in tension, compression, and flexure of test coupons in CF-PEEK printed with the AREVO® process are listed in Table 2.18. AREVO® can process carbon fiber volume contents exceeding 50%. Overall, mechanical properties of parts made via AREVO® exceed those in fiber direction exhibited by printed continuous CF reinforced plastics, and typical high-performance carbon fiber TS composite materials (Zhang et al. 2019).

2.9 SHEET LAMINATION (ShL)

2.9.1 PROCESS DESCRIPTION

ShL encompasses AM processes in which sheets of material are stacked, cut, and bonded together to form an object, with each sheet representing a cross-sectional layer of the 3D digital model of that object. Versions of ShL process reciprocally differ in the source material (paper, polymers, composites, metal, and ceramics), process sequence (*bond-then-form, form-then-bond*), and bonding mechanism (adhesive bonding, clamping, thermal bonding, and ultrasonic welding), and feature the following names:

- *Laminated object manufacturing* (LOM)
- *Computer-aided manufacturing of laminated engineering materials* (CAM-LEM)
- *Plastic sheet lamination* (PSL)
- *Selective deposition lamination* (SDL)
- *Ultrasonic additive manufacturing* (UAM) or *ultrasonic object consolidation* (UOC)
- *Composite-based additive manufacturing* (CBAM)
- *Selective lamination composite object manufacturing* (SLCOM).

The above processes are concisely described hereafter. The most popular ShL versions are those utilizing paper sheets bonded with polymer-based adhesives, whose commercial precursor was the process filed for patent by Helisys in 1991 (Feygin 1994), and later termed LOM (Feygin et al.

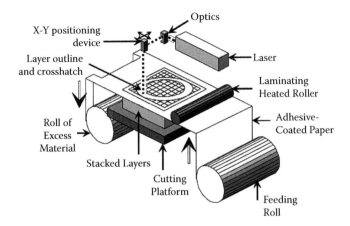

FIGURE 2.18 Schematic illustration of laminated object manufacturing (LOM).

Source: Modified from Wimpenny, D. I., Bryden, B., Pashby, I. R. 2003. Rapid laminated tooling. *Journal of Materials Processing Technology* 138:214–218. Reproduced with permission from Elsevier Science & Technology Journals.

1998; Park et al. 2000). The LOM equipment is schematically illustrated in Figure 2.18, and consists in the following steps: (a) the TP adhesive-backed paper is unwound from the feeding roll and pulled over the cutting (or build) platform, if the first layer is being laid, or, if it is not the first layer, over the top layer currently stacked on the platform; (b) a heated roller runs across the paper, melts and presses the adhesive down, bonding the sheet to what is under it; (c) a computer-controlled laser beam (or mechanical knife in other versions), cutting to a depth of one sheet thickness, traces the contour of a specific cross section of the object on the bonded sheet; (d) the laser beam or knife dices the sheet's portion external to the cross section following a crosshatch pattern, in order to ease the removal of excess material when the object is completed; (e) the platform is lowered by one sheet thickness; (f) the roll advances to locate the next sheet on the previous layer, and the roll's portion previously used is wound around the collected roll; (g) the process is repeated until the object is completed; (h) the object is removed from the platform, the diced material is separated from the object, which is sanded if required, coated (to keep moisture outside, and ensure dimensional stability), and painted as if it were wood. LOM is a bond-then-form process, and there are form-then-bond versions of SL, such as *offset fabbing* by Ennex Corp. (USA), and *computer-aided manufacturing of laminated engineering materials* (CAM-LEM) by CAM-LEM Inc. (USA), in which first the layer's outline and inner cavities are cut, and then the layer is stacked on and bonded to the previous layer. A recent review of LOM was authored by Mekonnen et al. (2016).

If C is the specific heat capacity (J/(kgK), ρ is the sheet paper density (kg/dm^3), and T (K) is the dissolution temperature of the sheet paper adhesive, the specific energy E (J/dm^3) to apply in ShL to bond the layers together is (Chua et al. 1998):

$$E = CT\rho \qquad (2.18)$$

The original LOM process relied on *kraft paper* (or *butcher paper* since used to wrap meat and fish) coated on one side with adhesive because it was a type of paper strong, durable, and tear resistant. Successive ShL processes switched to a different feedstock, as in the case of former Mcor Technologies (Ireland), whose *selective deposition lamination* (SDL) is based on sheets of standard business A4 and letter paper, an adhesive dispensing system, and a tungsten-tip blade to cut the shape. SDL is now applied on printers made by CleanGreen3D Limited (Ireland).

Cubic Technologies (USA), successor of Helisys, in 2013 disclosed a more sophisticated version of LOM, called *laminated object manufacturing with refill* (LOMR), which fabricated complex parts out of thin films and foils, stabilized on removable carrier ribbons.

In UOC, metal foils typically 100–150 μm thick are laid on a heated base, pressed, and welded together by a sonotrode that is a tool producing and applying ultrasonic vibrations to a material. After welding, the foils are shaped by a CNC milling head to their final slice contour, making UOC a hybrid ShL process, because it combines subtractive and additive manufacturing (Gibson et al. 2015). UOC was patented by Solidica (White 2003) that later became Fabrisonic (USA).

Plastic sheet lamination (PSL), patented by Solidimension Ltd. (Israel), replaces paper sheets with PVC sheets.

CBAM by Impossible Objects (USA) is a very recent and interesting version of ShL, because it fabricates articles in fiber-reinforced plastics through the following steps:

- One sheet of long carbon (CF) or glass fibers (GF) is fed to the printer
- Clear aqueous fluid is selectively dispensed on the sheet from conventional thermal inkjet heads (hovering across and above the sheet)
- TP powder is sprinkled onto the sheet, and sticks only to the fluid
- Powder not adhered is removed, and the powder stuck on the sheet has the shape of the object's XY cross section
- The process is repeated for next layer (made of another sheet of long CF or GF) that is stacked on the previous one, and so on for all layers
- The stack of all layers is heated to the melting point of the polymer, and compressed to reach the object's designed height. Heat melts the powder that encases the fibers and fuses the layers together in one cohesive part
- The layers' uncoated fibers are removed mechanically (sandblasting) or chemically, leaving behind the object's final shape.

CBAM processes CF/PEEK, CF/PA 12, GF/PEEK, and GF/PA 12, and under development are CF/PA 6, CF/GF/PA 12, CF/elastomer, and GF/elastomer.

Applications of ShL are full-color fit and visualization models and detailed prototypes for architecture, consumer products, education, geography (3D maps), entertainment, industry and manufacturing, life sciences and medicine (implants, anatomic models, etc.), but also functional items, such as tooling and microfluidic devices.

SLCOM, developed by EnvisionTEC, consists in stacking and cutting sheets from prepregs of TP matrix reinforced by unidirectional or multidirectional (woven fabric) fibers, and building parts fitting in a build envelope of 762 × 610 × 610 mm (EnvisionTEC 2017). SLCOM is compatible with many matrix and fiber materials, and targets applications in aerospace, automotive, consumer products, sporting goods, and medical.

Some companies mentioned above may currently be out of business, but their process was mentioned to provide a comprehensive overview and inspire ideas.

2.9.2 Feedstocks and Biobased Alternatives

2.9.2.1 Current Feedstocks

ShL utilizes paper (LOM, SDL), PVC (PSL), metals (UOC), and ceramics (CAM-LEM), and composites (CBAM). Polymers are present as adhesive coating on specialty paper (LOM), adhesive dispensed on standard paper (SDL), thin sheets of PVC (PSL), matrix for composites

(CBAM), and binders in ceramic-filled tapes. Information on type and properties of polymers for ShL is scarce in the technical literature.

Paper for LOM is coated on one side with TP adhesive that, upon heating and pressure, melts and bonds adjacent sheets. Park et al. (2000) reported that a TP adhesive with a T_m of 65°C formed the coating on 0.11 mm thick paper LPH 042 used by Helisys. The paper featured density of 0.9 g/cm^3, in-plane tensile strength, modulus, and strain at break of 25.6 MPa, 2.52 GPa, and 10.7%, respectively, and degradation temperature of 270°C. The bond strength reported by Helisys was 5 MPa, measured as interlaminar shear strength.

Ceramic tapes for ShL are composites made of microparticles of SiC, TiC-Ni, and alumina bound by a polymer binder (Gibson et al. 2015) such as poly(vinylbutyraldehyde), LDPE, and aqueous styrene-acrylonitrile-copolymer dispersions (Cawley et al. 1998).

CAM-LEM processes tapes of ceramic (alumina, zirconia, zirconia-toughened alumina, silica, silicon nitride), and metal (316L stainless steel) powder bound by an organic binder.

According to its SLCOM 1 printer brochure, SLCOM can process the following materials: (a) matrix: PEEK, PEI, PPS, PP, PE, PC, PET, polyethersulfone (PES), PAs, and PEKK; (b) fiber: aluminum, aramid, carbon, glass, steel, titanium, and polybenzoxazole, a new extremely heat-resistant TP.

2.9.2.2 Sustainable Feedstocks

Base polymers for TP adhesive coatings applicable on paper feedstocks for LOM and SDL include copolyolefins, copolyesters, copolyamides, and TP PUs (Stammen and Dilger 2013). There exist sustainable versions of the above polymers comparable in performance and cost with their fossil-based versions. Biobased adhesives are formulated from sustainable sources such as cellulose, starch, lignin, vegetable oils, and proteins extracted from animals and plants (Magalhães et al. 2019), and offer advantages over their fossil-based counterparts beyond renewability (Heinrich 2019). Biobased TP adhesives for paper are already commercially available, such as that derived by Sealock Adhesives (India) from sugar cane and corn (Barrett 2019). Among sustainable adhesives coated on paper, TPs are very attractive candidates because they exhibit strong bond to paper, low T_m and cost.

As a matrix of composites, sustainable PEEK (Kanetaka et al. 2016) could be evaluated for CBAM.

Biobased PVC being marketed by Solvay (Belgium), and INOVYN (UK) could be studied to derive biobased PVC for PSL.

As to binders, one alternative to fossil-based binders was studied by Cui et al. (2003), who investigated a commercial styrene-acrylic latex binder, and concluded that it was an eligible binder to process an aqueous Al_2O_3 suspensions tape that could be coiled continuously by a roller and serve as a feedstock for LOM.

2.10 DIRECT WRITING (DW)

2.10.1 PROCESS DESCRIPTION

DW or *direct-write* technologies brings AM to the size of the nanoscale. DW are additive processes designed to build without tooling or masks 2D and 3D functional patterned artifacts featuring dimensions of 5 mm or less, and resolution below 50 µm in one or more dimensions by depositing materials on flat and curved substrates. AM DW processes are microdeposition technologies for fabricating functional and geometrically complex structures at meso- and microscale composed of numerous materials of different types, and possess advantages over micromachining and lithographic methods. AM DW is mostly applied to print sensors, electronics, and integrated power sources (Pique and Chrisey 2001), namely to build passive and active microelectronic devices, such as antennas, batteries, capacitors, conductors, insulators, integrated RC filters,

TABLE 2.19

Examples of Polymeric Feedstocks for Non-AM DW

Polymer	Application	Process	Reference
Epoxy Novolak resin polymer	Low-loss polymeric optical waveguides	Electron-beam DW	Wong, W. H., Zhou, J., Pun, E. Y. B., 2001. Low-loss polymeric optical waveguides using electron-beam direct writing. *Appl. Phys. Lett.* 78, 2110 https://doi.org/10.1063/1.1361287
Poly(methyl methacrylate)	Diffractive optical elements for beam shaping, material processing, sensing, optical metrology, and lighting	Femtosecond-laser filamentation	Watanabe, W., Mochizuki, H. 2009. Direct writing of diffractive optical elements in polymer materials. https://spie.org/news/1856-direct-writing-of-diffractive-optical-elements-in-polymer-materials (accessed August 28, 2019)
Poly(thiophene)	Polymer nanostructures: poly (thiophene) nanowires on semiconducting and insulating surfaces	DW	Maynor, B. W., Filocamo, S. F., Grinstaff, M. W., Liu, J., 2002. Direct-Writing of Polymer Nanostructures: Poly(thiophene) Nanowires on Semiconducting and Insulating Surfaces. *J. Am. Chem. Soc.* 124 (4): 522–523

multilayer voltage transformers, strain gages, thermocouples, but also biological scaffolds and porous chemical sensors.

The family of AM DW is the additive subset of the larger family of DW technologies, comprising subtractive processes, and mainly chosen to fabricate microelectronics. Mortara et al. (2009) proposed the following definition of DW technologies, which can also fit AM DW if only additive processes are considered in the definition: "Direct Writing technologies allow the local manipulation of material without the use of interim tools such as molds or masks. They allow the achievement of the desired feature, both by adding material locally or subtracting material from a pre-deposited bulk layer. [...] Most techniques are digitally-driven, work serially and with high resolution building the artefacts layer by layer." Applications, materials, and techniques in using DW technologies are described in details by Pique and Chrisey (2001). Given the scope of this book, we mention only some examples of polymeric feedstocks for non-AM DW in Table 2.19. As to SPs, already in 2001 it was reported that living neural cells, *Escherichia coli* bacteria, and various types of eukaryotic cells, proteins, enzymes, and antibodies had been deposited onto different substrates via DW (Pique and Chrisey 2001).

Examples of AM DW processes are *aerosol jet* (formerly *maskless mesoscale materials deposition* or M3D), *direct write thermal spraying*, *matrix-assisted pulsed laser evaporation* (MAPLE), and *nScrypt 3De*. A recent item made via AM DW is microbatteries with high areal energy and power densities possibly for autonomously powered microdevices (Sun et al. 2013).

Feedstocks for AM DW are ceramics, metals, polymers, and semiconductors. Polymers for AM DW and their relative applications are listed in Table 2.20. Martin et al. (2017) modified an inexpensive desktop FFF printer into a DW printer suited for a solution of epoxy matrix reinforced with ceramic particles that could be spatially oriented by controlled magnetic fields, and hence

TABLE 2.20

Polymers and Relative Applications for AM DW Processes

Polymer	Application
Blend of poly(3,4-ethylendioxythiophene):poly-(styrenesulfonate) (PEDOT:PSS), blend of poly(3-hexylthiophene):C61-butyric acid methyl ester (P3HT:PCBM)	Polymer-based photodetectors, electrochromic devices, electrolyte-gated transistors, transparent conductors, polymer solar cells, organic light emitting diodes
PLGA/TiO$_2$ nanocomposite	Tissue engineering
Polyaniline	Strain sensors
Polyimide	Dielectric layer on circuit board, touch panel display jumpers
Polyvinylpyrrolidone	Adjustment of rheological properties for DW of aqueous dispersions
PMMA-based nanocomposites	Thin film transistors
SU-8 (low molecular weight epoxy)	Photoresist, dielectric material
Teflon	Dielectric materials
UV adhesives	2D and 3D structures on nonplanar surface

Source: Adapted from Ligon, S. C., Liska, R., Stampfl, J., Gurr, M., Mülhaupt, R. 2017. Polymers for 3D printing and customized additive manufacturing. *Chem. Rev.* 117:10212–90. Open access article published under a Creative Commons Attribution (CC-BY) License.

achieve 3D improvement of mechanical properties. Wang et al. (2020) developed a simple procedure to fabricate high-density graphene microlattices with a highly ordered structure via liquid crystalline direct ink writing. They prepared graphene oxide/glycerol inks, featuring a solid content up to 6 wt%, and elevated compressive strength (62.7 MPa at 90% strain), high elasticity (strain up to 15–18%), and high electrical conductivity of 2073 S/m.

Based on their working principle, AM DW technologies are categorized as *beam deposition*, *beam tracing*, *ink-based*, *laser transfer*, *liquid-phase*, and *thermal spray processes* (Gibson et al. 2015).

2.10.2 Ink-Based DW

2.10.2.1 Introduction

In *ink-based DW*, liquid inks containing the building material are deposited as droplets or a continuous bead on a surface, where they solidify due to evaporation, gelation, solvent-driven reactions, or thermal energy, and acquire the intended properties. Inks can be colloidal, filled with nanoparticle, sol-gel, etc. Since a great deal of activity is conducted to develop new and improved DW inks, there is a market opportunity for SPs. Suitable inks are required to freely flow through the deposition equipment, become rigid, set up quickly after deposition, retain shape, and either span voids/gaps or fill voids/gaps (Gibson et al. 2015). The dispensing mechanism in ink-based DW can be: nozzle, quill (with both producing a continuous streak), inkjet printing, and aerosol type (with both generating droplets).

2.10.2.2 Nozzle Ink-Based DW

This method relies on a pump or syringe pushing the ink through an orifice. Example of resolution, accuracy, repeatability in XY, and resolution in Z achievable in this process are 10 nm, ±1.5 μm, ±0.5 μm, and ±0.5 μm, respectively, which are the specification values of 3Dn-DDM-PF printer marketed by nScrypt (USA), and suited for ink-based DW of antennas, batteries, capacitors, electroluminescent lighting, inductors, interconnects resistors, and sensors on flexible, rigid, flat, curved, doubly curved, and even random 3D shapes. With changeable pen tips as small as 10 μm, thousands of materials can be conformally printed, with viscosity from 1 cps to more than 1 million cps. Compatible polymers include, among others, hydrogel, silicone, synthetic and natural polymers, living cells, and TSs such as adhesives and epoxy.

2.10.2.3 Quill Ink-Based DW

In this technology a minute tip is dipped into a reservoir of ink that adheres to the tip's surface, and when the tip is moved near to the substrate, the ink is transferred from the tip to the substrate. The most important example of quill ink-based DW is the technology developed by Quill NanoInk (USA), and called *dip-pen nanolithography* (DPN) in which the tip is part of an atomic force microscope (AFM). The inks are polymers: mercaptohexadecanoic acid, octodecane thiol, glycerol, lipids, nucleic acid, PEG, proteins, silanes, and other thiols.

2.10.2.4 Aerosol Ink-Based DW

This process consists in forming a deposit from inks or ink-like materials suspended in an aerosol mist (Gibson et al. 2015). A commercial version was patented by Optomec as *direct write™ system* (Renn 2006, 2007). It involves atomizing the building material (liquid solutions and dispersions) in an aerosol chamber – through pneumatic atomization or ultrasonification – into a stream of 1–5 μm diameter droplets, transferred by an inert gas stream to the deposition head, where this stream is focused and accelerated through a nozzle, and continuously deposited at 10–100 m/s. Achievable lateral resolutions and layer thicknesses are 10 μm and 100 nm, respectively, hence outperforming traditional inkjet printing. Advantages are high resolution and wide range of applicable feedstocks. It mainly fabricates microelectronic items such as thin film transistors, capacitors, and resistors on different substrates, including flexible polymers and articles made conventionally and additively (Obata et al. 2014).

2.10.2.5 Inkjet Printing DW

This version of DW is similar to MJ described in Section 2.7, with the difference being that the motion control systems and printheads are optimized to print a few layers of electronic traces on flat substrates instead of 3D objects. This process is very popular due to its high speed and accuracy, and low cost. However, its disadvantages are printing on curved surfaces, material continuity impaired by use of droplets, and limited droplet size range (Gibson et al. 2015). Feedstocks suitable as ink work best if their viscosity is low at or near room temperature.

2.10.3 Laser Transfer DW (LTDW)

In LTDW the material is transferred from a foil, tape, or plate onto a substrate by the action of a laser according to two alternative mechanisms. In one mechanism, based on a layer of sacrificial material, the laser beam or pulsed laser strikes and ablates the layer of the sacrificial material, forming a gas or plasma that expands itself, and propels the layer of building material, located adjacent to the sacrificial material, towards the substrate to which it adheres as a coating. The other mechanism is based on *spallation* that is the expulsion of fragments of material from an object caused by impact or stress, and consists of the following steps: the laser ablates a portion of the surface of a foil, which, combined with thermal energy, produces thermal and shock waves in the material that travels through the foil and causes spallation on the surface of the other side of the foil; spallation produces fragments that are propelled toward the substrate, and form a coating on it.

A version of LTDW is the *matrix-assisted pulsed laser evaporation direct write* (MAPLE DW or MDW) process (Fitz-Gerald et al. 2000) developed by the U.S. Navy, which is notable also because it utilizes polymers. In MDW the feedstock is a powder mixed with a polymer binder, and previously deposited as a coating on a laser transparent quartz disc or polymer ribbon. As the laser pulse strikes the coating, the polymer evaporates and the powdered material is deposited on and adheres to the substrate. MWD deposits layers and patterned films with spatial accuracy and re-solution of tens of μm on a number of substrate materials and geometries (Wu et al. 2003).

The U.S. Naval Research Laboratory has developed laser DW processes, and combined them with AM in order to fabricate functional parts, such as 3D electronic circuits and structures capable of detecting, processing, communicating, and interacting with their surroundings in novel ways (Kirleis et al. 2014).

2.10.4 THERMAL SPRAY DW

In *thermal spray DW* the building material in powder or wire form is introduced into a combustion or plasma flame that melts the material, imparts thermal and kinetic energy to it, and converts it into high-velocity droplets that are deposited on the substrate. By controlling the flame char-acteristics and material state (e.g. molten or softened) a wide range of metals, ceramics, polymers, and composites thereof can be deposited. Thermal spray DW can build devices made of multiple stacked layers, such as conductive and insulating layers, and layers of metallic, ceramic, and polymeric material.

A version of thermal spray DW called MesoPlasma™ is commercialized by MesoScribe Technologies Inc. (USA) to manufacture printed electronics (integrated wiring, sensors, heaters, and high-performance UHF, VHF, and microwave antennas) by forming multi-material deposits compatible with most substrate materials for aerospace, power generation, satellites, etc.

2.10.5 BEAM DEPOSITION DW

Beam deposition DW comprises processes in which a solid material is formed by depositing on a substrate a vapor that is primarily derived from the thermal decomposition of precursor gases, and is converted in solid material by condensation, chemical reaction, or conversion. One version of this process is *chemical vapor deposition* (CVD), in which heat converts a gas into a solid layer deposited onto a substrate. CVD enables single-step deposition of 3D structures of metals, semiconductors, and insulators on planar and non-planar substrates (Han and Jeong 2004).

2.11 4D PRINTING (4DP)

2.11.1 PROCESS DESCRIPTION

Four-dimensional printing or *4DP* is printing an item whose shape, property, or functionality change as intended as a function of time from one stable state to another stable state when the item is exposed to a specific external stimulus (Tibbits 2014). In 4DP time is the fourth dimension. 4DP is made possible by combining specific design, *responsive* materials, and AM printers. Materials are termed *responsive, smart, adaptive,* or *intelligent* if they react to external stimuli, such as electric and magnetic fields, light, moisture, pH, stress, substances, temperature, and water, and change one or more of their properties in a predictable way. Similarly, components and structures are designated *smart* if they are designed to respond to external stimuli by varying their shape, property, or functionality. A well-known example of smart component is the car airbag, in which the airbag is initially folded (Laukkonen 2019).

4DP is an emerging technology, introduced in 2013 by S. Tibbits at MIT (Tibbits 2013), and is a time-dependent, printer-independent, and mathematically modeled process that is attracting growing attention because of its potential (Kuang et al. 2019). Smarts materials have been utilized outside AM before 4DP was introduced, but AM was later recognized as a fabrication route more efficient than conventional fabrication processes to build all together adaptive materials, components, and structures, and thus 4DP was born. In fact, AM is similar to fabrication methods for composite materials, because it permits to simultaneously fabricate the intended design, and generate the desired and customized material properties, and multi-material components, that is components featuring material properties varying across their volume in *value* and *location* as designed.

Like other AM processes, 4DP leverages some most innovative materials for advanced applications: anatomical models, drug delivery, medical devices, autonomous robots, sensors, self-assembling structures, smart textiles, soft robotics (a new family of robots primarily made of soft materials usually inspired by biological systems (Mutlu printed object. Examples of stimulus have beenet al. 2015), TE (Sidney Gladman et al. 2016), etc. Self-assembling structures have great potential also for in-space and on-orbit fabrication (Mitchell et al. 2018) of a wide range of parts, included self-fabricating satellites (Hoyt et al. 2013).

The essential elements of 4DP are (Momeni et al. 2017):

- *Printer*: it concurrently generates the object and its constitutive materials featuring the desired combination of design and properties. AM processes leveraged for 4DP listed in decreasing number of publications from 2013 to July 2018 are (Kuang et al. 2019): FDM™/FFF, SLA, inkjet, DW, SLS, and DLP. Inkjet is widely used in 4DP because it enables to print materials with different properties (color, hardness, stiffness, transparency, etc.) simultaneously, and assembled in space. One version of inkjet is implemented on PolyJet printers (Section 2.7.1) that deposit a photocurable polymer through inkjet heads, and cure it using UV light.

- *External stimulus*: it is what triggers the change of shape, property, and functionality of the printed object. Examples of stimulus have been listed earlier. One or more stimuli are applied to the object, and their amount and duration are known or known and controlled.

- *Responsive materials* can be polymers, metals (Ma et al. 2017), and ceramics (Bargardi et al. 2016; Liu et al. 2018). Among responsive materials are piezoelectric materials (that produce a voltage upon stress application), shape-memory polymers, and electroactive polymers, both defined later.

- *Interaction mechanism*: exposing the smart material to the stimulus that is applied under specific and known conditions, such as amounts and duration.

- *Mathematical modeling*: this consists in modeling the final intended shape, material structure (composition and location of the ingredients, volume fraction of fibers and matrix, fiber orientation, etc.), material properties (MW, tensile modulus, swelling ratio, T_g, etc.) of the printed object, and stimulus characteristics (temperature value, light intensity, etc.). Mathematical modeling is necessary not only to obviously predict the final shape, but also to prevent interference and impact between elements of the printed object while its shape changes, and to reduce the number of trial-and-error experiments. There is forward and inverse mathematical modeling: the former type predicts the final shape from the object's material properties and material structure, and stimulus characteristics, whereas the latter type acts backward and models the object's material structure and printing parameters from the object's final shape, material, and stimulus characteristics.

As to shape shifting, changes include bending, folding, linear and nonlinear expansion and contraction, surface curling, twisting, and generation of surface topographical features (buckles, creases, and wrinkles). There are multiple possible shape changes: from 1D to 1D, 2D, and 3D; from 2D to 2D and 3D; and from 3D to 3D (Momeni et al. 2017).

The list of 4D printing applications is long and spans across biology, industrial, medicine, and robotics, and comprises, among others: adaptive joints (Raviv et al. 2014) and scaffolds (Miao et al. 2016), artificial protein structure (Tibbits et al. 2014a, 2014b), bio-origami (Jamal et al. 2013), cell printing (Koch et al. 2016), origami structures (Ge et al. 2014), nanoprinting (Carbonell and Braunschweig 2017), smart gripper (Ge et al. 2016), smart key-lock connector (Kokkinis et al. 2015), smart valve controlling acidic and basic flow (Nadgorny et al. 2016), and stents (Wei et al. 2017). *Bio-origami* refers to: (a) 3D-folded structures composed of biological materials such as proteins, DNA, and cells, and (b) self-folding of polymeric HGs usable to engineer cell-laden HGs for long-term 3D cell culture (Jamal et al. 2013).

Although 4DP is still at a proof of concept stage (Mitchell et al. 2018), its applications are expected initially in aerospace, industrial, and medical sectors if they demonstrate reliable performance at affordable or acceptable cost. 4DP uses in space include responsive visors, wipers, and reversal electrical systems to shed the electrical sticking of the abrasive lunar dust. Table 2.21 lists applications investigated in 4DP and their relative AM process and activating stimulus. The source of Table 2.21 cites the references for each area of research listed.

Broad and detailed reviews on 4DP were published by Momeni et al. (2017) and Mitchell et al. (2018). Polymers for 4DP were described by González-Henriquez et al. (2019).

2.11.2 Polymers for 4DP

2.11.2.1 Introduction

Obviously, polymers for 4DP have to be printable and responsive to external stimuli. Fewer polymers have been developed for 4D printing than for other AM processes. Most polymers for 4D

TABLE 2.21

Applications Investigated in 4DP, Their Relative AM Process and Activating Stimuli

Areas of Research	AM Process	Stimuli
Biomimetic sensors	FFF	Temperature, electricity
Controlled drug delivery systems, diagnostic medicine	SLA	Magnetic field
Flexible electronic, soft robotics	Direct ink writing	Magnetic field
Hinge section	MJ (PolyJet)	Temperature
L-hinge, spiral square, and orchid flower	MJ (PolyJet)	Boiling water
Lightweight, strong armor	SLA	Electric field
Metamaterials	FFF	Thermomechanics
Metamaterials	Direct ink writing	Magnetic field
Morphing structures	SLA	Photo-thermal
Organs	Bioprinting	Simulated body environment
Radio-frequency components	Inkjet	Electricity
Scaffolds	FFF	Temperature
Self-bending structures	FFF	Temperature
Soft robotics such as smart key lock connectors	Direct ink writing	Magnetic field
Solar concentrator	FFF	Hot plate
Textile	FFF	Temperature
Wearable devices	FFF	Heat
Weight sensitive parts	FFF	Temperature

Source: Adapted from Singh, S., Singh, G., Prakash, C., Ramakrishna, S. 2020. Current status and future directions of fused filament fabrication. *Journal of Manufacturing Processes* 55:288–306.

printing are acrylic polymers, elastomers, elastomeric composites, HGs, polymer composites, and shape-memory polymers. Printability varies with the specific AM process enabling 4DP. Next, families of polymers cited as 4DP feedstocks in technical literature and their relative AM technologies are described herein, with the information being derived from a comprehensive and detailed review by Momeni et al. (2017). Until 2019, most widely used materials for 4DP have been shape-memory polymers and HGs (Kuang et al. 2019).

2.11.2.2 Hydrogels (HGs)

According to the International Union of Pure and Applied Chemistry, HGs are gels in which the liquid component and swelling agent is water, and the solid 3D network component is usually a polymer (IUPAC 2014). Gels can be formulated to be sensitive to heat, light, magnetic field, and pH. Polymeric HGs are hydrophilic, and swell and change shape when immersed in water. An example of hydrogel for 4DP is poly(N-isopropylacrylamide) (PNIPA), a temperature-responsive polymer $(C_6H_{11}NO)_n$ that, when heated in water above 32°C, transitions from a swollen hydrated state to a shrunken dehydrated state, losing about 90% of its volume. PNIPA was employed in a bi-layered self-folding stimuli-responsive micro-device, made of two layers of gels that featured a different degree of swelling between them and, upon folding transition, generated a different folded shape between them (Kim et al. 2012).

Acrylamide (C_3H_5NO) matrix reinforced with cellulose fibrils was chosen to fabricate a composite HG ink mimicking the structure of plant cell walls, and was printed combined with other ingredients such as N, N-dimethylacrylamide (C_5H_9NO), to form an aqueous solution and become a printable and UV-curable ink. UV cure resulted in irreversible shape shifting that, however, becomes reversible if N, N-dimethylacrylamide is replaced with thermo-responsive N-isopropylacrylamide $(C_6H_{11}NO)$ (Sidney Gladman et al. 2016).

Examples of artifacts made via HG-based 4DP include: active valves made of thermally sensitive HG, parts made of anisotropic HG composites with cellulose fibrils, complex self-evolving structures actuated by multilayer joints, and transformable patterns out of heat-shrinkable polymer (Ge et al. 2016).

2.11.2.3 Shape-Memory Polymers (SMPs)

SMPs, discovered in 1980s, are polymers capable to return from a deformed state (temporary shape) to their initial (permanent) shape when exposed to an external stimulus (trigger), such as electricity, magnetism, light, pH, pressure, temperature, and water (González-Henríquez et al. 2019). SMPs are a subset of the larger family of *shape-memory materials* that include shape-memory metallic alloys and shape-memory ceramics, but SMPs surpass the other subsets in many properties, such as easy manufacturing, programming, high shape recovery ratio and low cost (Liu et al. 2009).

A typical SMP is that one activated by temperature as follows: the original shape of the object is deformed under an external stress at elevated temperature above its T_g or T_m. If that stress stays constant while the temperature is decreased below T_g or T_m, the SMP will maintain the temporary shape until it is heated above T_g or T_m again. Examples of SMPs utilized for 4DP are mentioned in Yang et al. (2016a) and Yang et al. (2016b).

A SMP with multiple transition temperatures can assume as many shapes (Xie 2010; Zhao et al. 2013). SMPs are applied outside AM for aerospace, automotive, bionic, mechanical, and medical applications, and in products such as heat-shrinkable films for packaging and tubes for electronics, implants for minimally invasive surgery, intelligent medical devices, MEMS, self-deployable sun sails, structures for spacecrafts (Cohee et al. 2003), and self-disassembling mobile phones (Hussein and Harrison 2004). SMPs can be active composites taking multiple shapes: namely, they contain various SMP fibers with different T_g values being activated by increase in temperature and controlling the shaping of the component. Bending deformation is controlled by tuning the volume fraction of the fibers (Wu et al. 2016a). Table 2.22 comprises examples of SMPs for 4DP obtained

TABLE 2.22

SMPs for 4DP, Their 4DP Processes and Applications, Stimuli, and References

Reference (fully listed at end of Chapter 2)	Composition	AM Process Family. Printer	Stimulus	4DP Applications
Bakarich et al. (2015)	Hydrogel ink containing alginate, sodium salt of algin, a polysaccharide present in brown algae. Other ingredients: poly N-isopropylacrylamide (PNIPAAm), α-Keto glutaric acid photoinitiator, acrylamide, calcium chloride, ethylene glycol rheology modifier, N-isopropylacrylamide and N,N'-methylenebisacrylamide crosslinker, and epoxy-based UV-curable adhesive (Emax 904 Gel-SC).	ME. EnvisionTEC D-Bioplotter® coupled with UV-light source for in situ photopolymerization.	Water, temperature	Hydrogel-based valves automatically closing and opening upon exposure to hot and cold water, respectively
Bodaghi et al. (2016)	Photopolymeric ink made of TangoBlackPlus FLX980 as soft and flexible matrix (26–28 Shore A hardness) plus VeroWhitePlus (tensile modulus 2,495 MPa) as stiff fiber.	MJ (PolyJet). Objet Connex 500	Temperature	Self-expanding and self-shrinking structures, employable for: (a) plane actuators serving as structural switches, (b) tubular lattice serving as tubular stents and grippers for bio-medical and piping applications
Ge et al. (2014)	Inks made of photopolymers with different hardness and T_g at room temperature: TangoBlack (26–62 Shore A hardness, T_g −5°C) as elastomeric matrix plus VeroWhite (83–83 hardness Shore D, T_g 47°C) as rigid fiber. Ink for TangoBlack: urethane acrylate oligomer, Exo-1,7,7-trimethylbicyclo [2.2.1] hept-2-yl acrylate, methacrylate oligomer, PU resin, and photo initiator. Ink for VeroWhite: isobornyl acrylate, acrylic monomer, urethane acrylate, epoxy acrylate, acrylic monomer, acrylic oligomer, and photo initiator.	MJ (PolyJet). Objet Connex 260	Temperature	Flat sheets featuring hinges and turning into 3D shapes (box, pyramid, airplane)
Ge et al. (2016)	Photo-curable SMP network combining two categories of materials: (a) benzyl methacrylate (BMA) as the linear chain builder; (b) poly(ethylene glycol) dimethacrylate (PEGDMA), bisphenol A ethoxylate dimethacrylate (BPA) and di	High resolution projection microstereolithography (PµSL). Apparatus comprising LED light, translation stage, stepper motor, and extruder.	Temperature	Grippers, springs, stents

(Continued)

TABLE 2.22 (Continued)
SMPs for 4DP, Their 4DP Processes and Applications, Stimuli, and References

Reference (fully listed at end of Chapter 2)	Composition	AM Process Family. Printer	Stimulus	4DP Applications
Ge et al. (2013)	SMP comprising two materials with different T_g and stiffness at room temperature: elastomeric matrix (T_g −5°C) plus glassy shape memory polymer fibers (T_g 35°C) (ethylene glycol) dimethacrylate (DEGDMA) as crosslinkers.	N/A	Temperature	Sheets assuming 3D forms (bent, coiled, twisted, folded shapes) and returning to flat shape
Jamal 2013	Photopatterned poly(ethylene glycol) (PEG)-based hydrogel bilayers featuring different molecular weights and swelling ratios, with the latter driving self-folding.	Photopatterning (photolithography) by UV exposures. Mercury vapor lamp (Advanced Radiation Corporation, USA) and commercial mask aligner Ultraline (Quintel, USA)	Water	Planar bilayers turning into 3D bio-origami serving as anatomical tissue models and self-folding vascularized tissue constructs
Yu et al. (2015)	Epoxy based UV curable SMP.	MJ (PolyJet). Stratasys Objet Connex 260	Temperature	Multiple hinges, each with different T_g
Kokkinis et al. (2015)	Two inks consisting of magnetically responsive anisotropic stiff particles (alumina platelets) suspended in a light-sensitive liquid resin of tunable composition, containing mainly crosslinking PU acrylate (PUA) oligomers, reactive diluents, photoinitiator, rheology modifier and alumina platelets, present in different concentrations achieving different functionalities. One ink included a soft, highly swellable PUA, the other ink a solid non-swelling PUA.	Magnetically assisted 3D printing (MM-3D printing) based on direct ink-writing system. 3D Discovery (regenHU Ltd) customized to align particles by means of a neodymium magnet.	Magnetic field	Complex adaptive material systems. Reconfigurable key–lock connectors based on shape-changing parts, and possibly used as flexible joints, soft building blocks, selective pick-and-place systems in soft robotics.
Kuksenok and Balazs (2016)	Composite consisting of thermo-responsive polymer gel with poly(N-isopropylacrylamide) (PNIPAAm) serving as matrix, and photo-responsive fibers functionalized with spirobenzopyran chromophores embedded in it.	N/A	Temperature, light	Reconfiguration, actuation

Reference	Material	Printer/Method	Stimulus	Application
Le Duigou et al. (2016)	Hygromorphic biocomposite made of PLA and poly(hydroxyalkanoate) (PHA) matrix reinforced with recycled wood fibers, available as commercial filament called woodFill fine (colorFabb).	FFF. Prusa i3 Rework.	Water	Actuators, hinges.
Liu et al. (2017)	Struts: UV curable liquid photopolymers made of VeroWhitePlus, DM9895 (DM-1) and DM8530 (DM-2) (all by Stratasys). Cables: TPU FilaFlex® (Recreus, Spain).	Struts: material (ink) jetting, printer Objet Connex 260 (Stratasys Cables: FFF, printer Hyrel 3D System 30 M.	Temperature	Tensegrity deployable structures
Mutlu et al. (2015)	Unspecified thermoplastic elastomer.	FDM. UP! Plus 2 (Tiertime, China).	Mechanical force	Fully compliant prosthetic finger usable alone or integrated within a prosthetic hand.
Nadgorny et al. (2016)	pH responsive filaments for material extrusion 3D printing made of poly(2-vinylpyridine) (P2VP) and ABS as reinforcing ingredient.	FFF. MakerBot Replicator 2X.	pH	pH-responsive flow regulating valve that is open at pH 13, closed at pH 2.
Naficy et al. (2016)	Hydrogel-based inks of different formulations to print bilayer composed of poly (N-isopropylacrylamide) (poly(NIPAM)) as smart thermo-responsive polymer, with non-active poly (2-hydroxyethyl methacrylate) (poly(HEMA)). Also added: polyetherbased PU(PEO-PU) as rheology and viscosity modifier, α-ketoglutaric acid as UV initiator. N, N′-methylenebisacrylamide (BIS) as crosslinking agent.	ME. Printer equipped with 300–450 nm UV curing UV spot light (Dymax Bluewave)	Hydration, temperature	Boxes, hinges
Raviv et al. (2014)	UV curable acrylated monomers. Rigid plastic (tensile modulus 2 GPa) plus expandable polymer (tensile modulus 40 MPa) composed of vinyl caprolactam (50 wt%), PE (30 wt%), epoxy diacrylate oligomer (18 wt%), Irgacure 81 (1.9 wt%) photoinitiator, and wetting agent (0.1 wt%).	MJ (PolyJet). Objet Connex 500	Water	Flat sheet bending curling, and folding
Rodriguez et al. (2016)	Inks containing biobased thermoset shape memory composites, made of epoxy resins (epoxidized soybean oil and bisphenol F diglycidyl ether) filled with carbon nanofibers.	MJ/Ink jet printing. Fluid dispenser Ultimus™ V (EFD, Nordson, USA) mounted on a three-axis positioning stage ABL 9000 (Aerotech, USA).	Temperature	Electrically conductive hinges, medical devices
Sidney Gladman et al. (2016)	Hydrogel ink containing soft acrylamide matrix reinforced with anysotropically aligned stiff cellulose fibrils, plus aqueous solution of N,N-dimethylacrylamide, photoinitiator Irgacure	Printing through nozzles mounted on ABG10000 Air-Bearing Direct-Drive Cartesian Gantry System. Curing by OmniCure® Series 2000	Water	Biomimetic devices, soft robots, tissue engineering

(Continued)

TABLE 2.22 (Continued)
SMPs for 4DP, Their 4DP Processes and Applications, Stimuli, and References

Reference (fully listed at end of Chapter 2)	Composition	AM Process Family. Printer	Stimulus	4DP Applications
Tibbits et al. (2014a)	2959 (BASF), nanoclay, glucose oxidase, glucose, and nanofibrillated cellulose (NFC). Hydrophilic acrylated UV curable monomer turning rigid polymer and expanding under water up to 150% of its original volume.	ultraviolet source (Lumen Dynamics). MJ (PolyJet).	Water	Flat sheet curling and folding
Wei et al. (2016) liquid crystal display SL. Their polymer	UV crosslinking PLA-based inks: PLA pellets (NatureWorks, USA), Fe₂O₃ nanoparticles.	DW. Apparatus comprising: ink dispenser EFD HP7X (Nordson, USA) mounted on benchtop gantry robot I&J2200-4 (Fisnar, USA), two 365 nm UV LEDs.	Magnetic field, temperature	Medical devices
Yang et al. (2016a)	TPU, ABS, SMP made of DIAPLEX MM-4520 pellets (SMP Technologies, China).	ME. MakerBot Replicator 2x	Temperature	Foldable flowers, grips
Wu et al. (2016a)	Composite made of TangoBlackPlus, VeroWhite, and two types of fibers: DM8530 and DM9895 (T_g ~38°C).	MJ (Polyjet). Objet 260.	Temperature	Exemplary bending items: self-assembling and disassembling trestle, active helix shape, active "wave" shape, smart hook, insect.
Yu et al. (2017)	Polyacrylate-co-epoxy particles dispersed in epoxy resin.	SLA. iSL200 (ZRapid, China)	Temperature	Actuators, hinges
Zarek et al. (2016)	Semi-crystalline methacrylated PCL with MW 10,000 g/mol, 2,4,6-trimethylbenzoyl-diphenylphosphineoxide (TPO) as photoinitiator, vitamin E to avoid premature cross-linking.	LED-UV stereolithography. Freeform Pico 2	Temperature	Implants, medical devices
Zhang et al. (2016)	PLA (MakerBot) strips, 0.8 mm wide × 0.2 mm thick, printed on a membrane of paper sheet.	FFF. MakerBot Replicator 2X	Temperature	Heat shrinkable pattern transformations

Sources: Momeni, F., Mehdi Hassani, S. M., Liu, X., Ni, J. 2017. A review of 4D printing *Materials and Design* 122:42–79. Mitchell, A., Lafont, U., Holyńska, M., Semprimoschnig, C., 2018. Additive manufacturing – A review of 4D printing and future applications. *Additive Manufacturing* 24:606–626. Reproduced with permission from Elsevier for both papers.

from experimental and commercial materials, and their relative AM processes, and applications as reported in the technical literature.

The experimental procedure to characterize and control the responsive behavior to temperature of PEEK as a SMP described by Wu et al. (2014) can be generalized, and serve as a guideline to tailor the behavior of a SMP. Additional readings on SMPs are the books by Lendlein (2010) and Leng and Shanyi (2010).

Very recently, Shan et al. (2020) investigated a polymer based on epoxy acrylate for 4D printing a SMP via liquid crystal display SL. Their polymer featured high resolution and transparency, tunable properties, high shape recovery properties, and cycling stability.

Electroactive polymers (EAPs), such as poly(vinylidene fluoride-trifluoroethylene)-based fluoroterpolymers (Huang and Zhang 2004; Liu et al. 2009), convert electric charge and voltage into mechanical force and movement. This electromechanical coupling has led to selecting EAPs for actuation and sensing chemical and mechanical stimuli. EAPs are a distinctive family of materials that possess low elastic moduli and high strains, and conform to surfaces of different shapes. These features make them attractive for wearable sensors and interfacing with soft tissues (Wang et al. 2016). Since EAPs exhibit large responsive strains, they have been employed to build actuators that most closely imitate muscles, and movements of humans, animals, and insects in biologically inspired intelligent robots (Bar-Cohen 2005). EAPs are classified into two groups depending on the main type of charge carrier: *electronic* EAPs (dielectric elastomers, liquid-crystal polymers, and piezoelectric polymers) and *ionic* EAPs (conducting polymers and ionic polymer–metal composites). A recent and broad review of electronic and ionic EAPs was authored by Wang et al. (2016).

Dielectric elastomer actuators (DEAs) are a class of EAPs generating strains of 100% and more when electrically stimulated (Bar-Cohen 2010), and studied to make soft dielectric elastomer actuators employed as robot components. An acrylic-based photopolymer and MJ were utilized by Rossiter et al. (2009) to fabricate in one step a complete actuator that consisted in an active 90 µm thick membrane supported by a thick circular collar, with the membrane-changing shape controlled by variation in electrical voltage.

2.11.2.4　Printed Active Composites (PACs)

PACs are printed multi-materials of which some or all materials exhibit some active response, such as a thermally induced shape-memory behavior. PACs can be leveraged for self-actuating and self-assembling structures, as demonstrated by Ge et al. (2014). The authors combined an elastomeric matrix and glass fibers into a SMP, and printed origami with active hinges (a flat polymer sheet folding into a box, a pyramid, and two airplanes), and a 3D box with active composite hinges and programmed to assume a temporary flat shape that successively returned to the 3D box shape on demand.

2.11.2.5　Near Future for 4DP

The near future of 4DP is promising although it will not constitute a large business. In 2016, the total 4DP market was predicted to grow from USD 63 million in 2019 to USD 556 million in 2025 at an impressive CAGR of 43% (Loh 2016).

This market growth will drive the development of (a) smart polymers with properties superior to those of current smart polymers, and (b) new printers suited for smart polymers, and printers performing "better" than current ones, f.e. faster (Momeni et al. 2017). Smart materials are expected to be applied not only in highly technological sectors (f.e. aerospace and automotive), but also in clothing, construction, healthcare, and consumer products (Loh 2016). Higher printing speed will obviously promote the use of smart materials, and may be achieved by planar curing instead of point and linear curing. The penetration of 4DP in medical applications will also be favored by the availability of printers suited for aqueous feedstocks.

Some information below is derived from Kuang et al. (2019) who offered an outlook of 4DP in: (a) smart materials, (b) processes and printers, and (c) design and modeling. Firstly, properties of smart materials affect actuation method, shape-changing speed, and object's strength and stiffness. Multifunctional inks enabling multifunctional materials, strong materials (f.e. through double molecular network), and materials with fast response speed (f.e. through porosity incorporated into HGs) will be valuable for 4DP. SPs for AM have opportunities to be utilized in 4DP as: (a) HGs that are biocompatible, biodegradable, and adaptive to physiological environment and hence suited for tissue and organ regeneration; (b) liquid crystal elastomers, attractive for their reversible shape-changing capability. New 4DP materials may be incompatible with existing AM processes and printers that will require to be accordingly modified.

Mitchell et al. (2018) published a review of 4DP with emphasis on current and upcoming space applications for polymers and metals, and predicted that near-future space applications of AM would proliferate, because AM meets the primary need of in-space manufacturing to reduce weight at takeoff and on resupplying flights from Earth. In fact, U.S. National Aeronautics and Space Administration (NASA) and European Space Agency (ESA) are experimenting with 4DP as an exploration-based manufacturing technology for in-space and on-orbit fabrication of a wide range of parts, including large, complex, 3D structures such as trusses, antennas for radio communication, radar surveillance, instrument booms, solar sails (Dalla Vedova et al. 2014), and shrouds. An AM process enabling 4DP in space is UOC described in Section 2.9.1, because this process works at a relatively low temperature, allowing polymers and electronics to be embedded in metals within the built component. Hence, UOC is suitable to fabricate conductive smart materials that reshape in response to external stimuli. Another future space application of SMPs is to print with them CubeSats, which are inexpensive miniature satellites, weighing typically 1 to 10 kg, and deploying small payloads, radio receivers, and optical cameras. CubeSats are based on rugged, stackable electronic boards housed in 10 cm units, and could be printed in PEEK and other high-grade TPs, since the latter ones are recyclable and biocompatible (ESA 2017).

2.12 3D BIOPRINTING (BP)

2.12.1 INTRODUCTION

Biofabrication is production of complex living and non-living biological products from raw materials such as living cells, molecules, extracellular macromolecules, and biomaterials. A more detailed definition of is "the automated generation of biologically functional products with structural organization from living cells, bioactive molecules, biomaterials, cell aggregates such as micro-tissues, or hybrid cell-material constructs, through Bioprinting or Bioassembly and subsequent tissue maturation processes" (Groll et al. 2016). *3D bioprinting* or *bioprinting* (BP) is a version of biofabrication, and indicates a group of AM processes consisting in printing structures and organs using viable biological molecules, biomaterials, and cells (Murphy and Atala 2014; Moroni et al. 2018). BP is applied in regenerative medicine, TE, organ transplantation (Xia et al. 2018), drug delivery, and cancer studies. BP provides patient-specific spatial geometry, controlled microstructures, and accurate positioning of different cell types to fabricate scaffolds for TE. Particularly, bioprinted scaffolds have the benefit of featuring a suitable microarchitecture imparting mechanical properties and promoting cell ingrowth, while ensuring cell viability. BP lacks the disadvantages of traditional methodologies of scaffold fabrication (fiber bonding; solvent casting and particulate leaching; membrane lamination; melt molding and gas foaming) that include wide use of highly toxic organic solvents, long fabrication times, labor-intensive processes, poor repeatability, irregularly shaped pores, insufficient interconnectivity of pores and thin structures, and restrictions on shape control (Lam et al. 2002).

BP methods are nozzle-based techniques (inkjet and extrusion printing), laser-based techniques (VP), and laser-assisted BP (Kacarevic et al. 2018) Table 2.23 compares inkjet and laser BP in

TABLE 2.23

Comparison between Bioprinting Processes: Inkjet vs. Laser

Method	Inkjet	Laser
Viscosity	Very low (3.5–12 mPa s)	Low (1–300 mPa s)-
Cell viability	>85%	>95%
Cell density	Low, <10^6 cell/ml	Medium, ~10^8 cell/ml
Working principle	Noncontact	Noncontact
Nozzle size	50–300 μm	Nozzle free
Resolution	50–300 μm	20–80 μm
Part thickness	Very thin	Thin
Printer cost	Low	Low

Source: Li, H., Tan, C., Li, L. 2018b. Review of 3D printable hydrogels and constructs. *Materials and Design* 159:20–38. Reproduced with permission from Elsevier.

terms of various features, and performance (cell viability and density). Table 2.24 lists examples of recent biomaterials for in vitro, in vivo, and in situ studies conducted via BP, but omits types of cell (chondrocytes [cartilage cells], stem cells, endothelium [interior surface of blood vessels] cells, etc.) combined with the listed biomaterials.

TABLE 2.24

Examples of Recent Biomaterials for In Vitro, In Vivo, and In Situ Studies Conducted Via Bioprinting (BP)

Process	Biomaterials for In Vitro Studies	Biomaterials for In Vivo Studies	Biomaterials for In Situ Studies
Extrusion-based BP	Alginate PLA fibers, GelMA, SA, SA/AG, SA/collagen	Alginate, alginate/gelatin, alginate/HA, alginate/matrigel, fibrinogen, gelatin, glycerol, human decellularized adipose tissue, HA, laponite XLG, PCL, PCL/Pluronic® F127, PCL/TCP/Pluronic® F127, PEG, PU nanoparticles, PU, sodium alginate	HA-GelMA
Laser-based BP	Human osseous cell sheets	Collagen	nHA
Stereolithography BP	GelMA and graphene nanoplatelets, GelMA and nHA, PEGDA and GelMA	N/A	N/A
Inkjet-based BP	Alginate, cell suspension	Collagen, fibrin, fibrinogen, thrombin	Fibrinogen/collagen PEGDMA

Source: Adapted from Kacarevic Z. P., Rider P. M., Alkildani S., et al. 2018. An Introduction to 3D Bioprinting: Possibilities, Challenges and Future Aspects. *Materials* 11:2199 that include references for each material.

Abbreviations: AG agarose, GelMA gelatin methacryloyl, HA hyaluronic acid, nHA nano-crystalline hydroxyapatite, PCL poly(caprolactone), PEGDA poly(ethylene glycol) diacrylate, PEGDMA poly(ethylene glycol) dimethacrylate, PEG poly (ethylene glycol), PLA polylactide, PU polyurethane, SA sodium alginate, TCP tricalcium phosphates.

Feedstocks for BP are called *bioinks*, and to be suitable for BP they have to meet requirements in terms of printability, mechanical and rheological properties, compatibility with living cells, degradability, etc. The above requirements are interrelated and hence experimentation and analysis is necessary to identify the optimal bioink and printing parameters. Chapter 12 on HGs discusses in detail the above requirements, and how to assess whether they are met. Paxton et al. (2017) proposed a relatively simple and fast two-step screening process for formulating acceptable bioinks at research level that consists in: (1) initial screening to assess: (a) fiber formation versus droplet formation, (b) stability of stacked layers without merging between layers; (2) rheological measurements to characterize: (a) flow initiation properties and yield stress, (b) shear thinning to predict extrusion process and cell survival, and (c) bioink's recovery behavior (that is increased viscosity) after printing. Gatenholm et al. (2016) developed a more articulated process than Paxton's for the same purpose. Ouyang et al. (2016) assessed the printability of bioinks for FFF based on equation (2.19), where L and A are the perimeter and the area of the pore, respectively, and Pr is *printability*, a dimensionless value that characterizes the gelatinization degree of the bioink:

$$Pr = \frac{L^2}{16A} \qquad (2.19)$$

When an extruded bioink possesses an ideal printability, $Pr = 1$, and the printed scaffold displays a clear morphology, i.e. accurate and consistent shape across its volume, fine resolution, smooth surface, regular grid, and pores of constant shape and size with sharp corners (such as in square and diamond shapes). In the case of under gelation, $Pr < 1$, and a filament more fluid and less viscous is printed, resulting in the scaffold's upper and lower layers fusing together, and pores developing a form tending to round. If over gelation occurs, $Pr > 1$, and the printed layers do not fuse together.

In 2015, the global BP market was estimated to be USD 487 million in 2014 and was predicted to reach USD 1.82 billion by 2022 (Kyle et al. 2017).

Some commercial bioinks are mentioned in Section 2.3.7.

2.12.2   Extrusion-Based BP (EBP)

In EBP the bioink is extruded through a nozzle, driven by either pneumatic or mechanical pressure, and deposited to form a specific shape. EBP can replace tissues and organs, and also efficiently fabricate microfluidic chips. EBP is versatile, relatively fast, and generates very high cell densities, but it is penalized by: (a) extruding at pressure levels potentially altering cell's morphology and function, and (b) very limited resolution, with minimum feature size generally >100 μm, which could make it suited for hard tissues with size larger than 10 mm but not for cell positioning, and soft tissue applications requiring small pore sizes. Given the pressure issue, highly viscous HG is recommended to protect the ink cell during extrusion and their survival, achieve adequate mechanical properties and a good geometric definition, although at the expenses of less porous parts with worse cellular infiltration. No chemical additives for curing are necessary.

2.13   FIBER ENCAPSULATION AM (FEAM)

FEAM is a process experimentally developed very recently, and consists in using a single, potentially low-cost printer, and simultaneously extruding a flowable matrix material, such as molten TP, and laying a fiber that is immediately encapsulated by the matrix to form a two-phase composite. FEAM can lend itself also to TS resins and non-wire reinforcements, such as structural fibers (carbon, glass, aramid, etc.), and optical fibers. FEAM targets quite a range of components: 3D circuitry comprising electrical conductors distributed throughout a dielectric matrix and fabricated in multiple layers, moving and sensing soft robotic components, helical 3D coils/inductors, loudspeakers, rheostats, inductive sensors, membrane switch arrays, and a linear variable differential transformer (Saari et al. 2015).

2.14 PRESENT AND FUTURE SUSTAINABILITY OF AM

That the sustainability of an AM process an accurate assessment of sustainability of AM should be comprehensive, specific, and comparative. *Comprehensive* indicates that present and future environment, economy, health, jobs, etc. should be considered. *Specific* means that sustainability should refer to specific families of processes and, even better, specific processes within each family, and specific conditions, such as type of printed item (f.e. one off vs. many production parts), path of going from idea to market, etc. *Comparative* reflects that the sustainability of an AM process for specific materials should be compared to conventional manufacturing technologies for the same materials. The information in this section intends to be as much as possible comprehensive, specific, and comparative within the limits of scope and space of this book, although some broad opinions are included as well in order to provide also an overview of the topic. In brief, AM is more and less sustainable than conventional manufacturing methods, depending on processes, variables, and conditions, as described in the rest of this section.

AM processes offer sustainability benefits, and in the best case they may enable to reduce energy and CO_2 emissions in industrial manufacturing by 5% by 2025 (Gebler et al. 2014), which however seems optimistic. In specific cases, AM technologies make the manufacturing process more efficient, and hence more sustainable, by reducing the amount of manufacturing inputs and outputs, especially in markets of low-volume items (such as aerospace components and high-cost medical prostheses), and lowering energy use, resource demands and related CO_2 emissions over the entire product life cycle. AM fabricate objects without removing material from an initial stock, and therefore there is very little waste of building material. However, some AM processes such as FFF produce waste because they also use support material (Sections 2.3, and 2.7).

Very recently a review by Walter and Marcham (2020) reported that AM processes have demonstrated environmental benefits, being more efficient and sustainable and less wasteful than conventional subtractive manufacturing, and that companies should consider AM in order to increase material and resource efficiency, and gain flexibility in production and part design.

Gebler et al. (2014) estimated that AM has the potential to reduce production costs by USD 170–593 billion, the total primary energy supply by 2.5–9.3 10^{18} J, and CO_2 emissions by 130–525 million tonnes by 2025. The large range of the saving figures was attributed to the uncertainties of predicting market and technology developments associated to a novel family of manufacturing processes with a very short track record. We also note that since 2014 AM has become more popular, and hence the saving figures estimated may be larger than predicted.

Ford and Despeisse (2016), drawing on publically available data, confirmed the favorable implications of AM on sustainability, and found benefits across the life of product and material. They identified an extensive list of benefits and challenges in the following areas associated with adopting AM: product design, material input processing, component and product manufacturing, product use, repair and manufacturing, and recycling. The sustainability benefits in processing material input (feedstocks) include: minimizing use of raw materials, and consequently reducing by-products and waste; reducing toxic chemical associated with processing; decentralized recycling eliminating the transportation costs; diverting by-products from waste stream, and upcycling them; upcycling waste materials into new applications; no need for mold and die. Examples of challenges in input material processing are: current limited knowledge of environmental impact of AM processes; increasing amount of recycled content in printing material; partial recyclability of mixed materials; standardizing materials and processes; qualifying materials for structural applications; building databases of material properties; using support material.

AM's sustainability is not limited to the environment. Taddese et al. (2020) conducted a broad review of the technical literature on sustainability, sustainable design, and the opportunity for AM to manufacture more sustainable products. Sustainability requirements affect all stages of product life cycle, including pre-manufacturing, manufacturing, use, post-use, and logistics, and can be measured in environmental, economic, and social terms. The authors concluded that from a life cycle perspective, AM showed a better trend of addressing all dimensions of sustainability over conventional manufacturing processes.

Agrawal and Vinodh (2019) also analyzed the relevant technical literature on sustainability of AM in regard to economy, environment, society and health. Following are some highlights from their review:

- *Economy.* Woodson (2015) argued against certainty that AM will cause sustainable industrial transformation, because experts had not fully considered the socio-technical system that surrounds AM and is impacted by its adoption. A survey of 105 companies in 23 countries concluded that AM suits only low-volume production, and new product development, while conventional fabrication is preferred for high-volume production (Khorram et al. 2018).
- *Environment.* This aspect was thoroughly addressed in terms of energy consumption, design optimization, and life cycle assessment. AM features a reduced environmental impact and this varies according to the AM process. F.e. life cycle analysis indicated that FFF articles have little environmental impact, whereas MJ parts feature high impact.
- *Society and health.* Social impact of AM is not completely known, and fabrication, customization, sustainability, business models, and work have the most social impact (Matos and Jacinto 2019). Although, the effects of AM on human health have not been fully understood and more relevant information must be collected, some health issues have emerged, such as toxic feedstocks and printers emitting ultrafine particles and toxins.
- *Optimization.* The configuration that minimizes energy consumption and CO_2 emissions is combining AM and conventional manufacturing (Priarone and Ingarao 2017).

Colorado et al. (2020) published a wide-ranging review of the sustainability of AM, spanning from the circular economy and recycling of materials to other environmental challenges involving the safety of materials and manufacturing. They noted that more attention is being paid to sustainability of AM and may lead to a near-term reduction of the limitations of AM relative to the reuse and recycle of materials, health impact, and sustainable processes. AM metals are the most appropriate for an optimal circular economy due to their high recyclability, as demonstrated by the company MolyWorks (USA) turning scrap metal into powder for AM. In conclusion, although a great amount of research is conducted on more efficient use of polymers, ceramics, metals, and composite materials, optimization of materials to minimize energy and waste is still far from a global solution, but AM feedstocks with better recyclability, reuse, or circularity will be more available, as national policies increasingly drive manufacturing toward green materials and processes.

Plastics and polymeric composites form the largest share of AM materials discussed in papers on sustainability (Colorado et al. 2020), because of their predominant use over the rest of AM feedstocks. A way in which plastics can increase sustainability of AM is through FFF printers processing filaments fabricated from recycled polymers. One example of this approach is the Dutch company Refil, supplier of filaments for ME from recycled PLA, ABS, PET, and HIPS, applying a process described in Section 1.12 (Refil n.d.). Behind Refil is Better Future Factory (Better Future Factory n.d.), a "sustainable product design studio" whose clients are global companies (Accenture, Heineken, ING, KLM, Philips, etc.) confirming that improving sustainability is being pursued by large corporations. Another option is to have printers fed by screw extruders that convert pellets obtained from ground items printed in plastics and discarded for any reason into filaments. Beside the cost of buying and running the extruder, this solution has a main disadvantages that a filament of recycled plastics will have properties inferior to those of its virgin grade (Filabot 2020).

If sustainability comprehensively refers to the impact on natural, economic and energetic resources, and society, then sustainability of AM should be evaluated comprehensively through a life cycle assessment, emitted amount of greenhouse gas and other air pollutants, material toxicity, depletion of natural resources, total energy spent (to produce the feedstocks, run the printer, post-process the printed article, etc.), impact on society (health, jobs, skills, and economic development), material recycling and composting, feedstock price, setup cost, etc.

There is a large body of technical literature on sustainability of AM, and some feedstocks, equipment, and procedures currently address this topic. This section is a brief overview of sustainability of AM, its present, future, impact, and limitations due to technical and economic factors. Current AM presents some advantages and disadvantages over conventional manufacturing from a sustainable standpoint, and hence extensive adoption of AM will not necessarily and totally benefit the environment. AM is more sustainable than conventional manufacturing in specific conditions, such as one-off articles, low number of replicates per part, prototyping, and combining components in one, and other cases mentioned in Section 2.1. An effective way to improve sustainability in AM is maximizing printer utilization. Hereafter the focus is mostly on the major AM process for polymers: ME, PBF, VP, and MJ and their feedstocks. J. Faludi has analyzed sustainability of AM broadly and in details, and the following considerations draw on his papers (Faludi et al. 2012, 2014, 2015a, 2015b) that are suggested for further reading.

A great deal of creativity has been driving the formulation of new sustainable polymers for AM, and several experimental and commercial grades are being introduced, and screened by a combination of factors (market, price, performance, etc.) that ultimately declare the winners. AM has adopted some sustainable polymeric feedstocks. FFF converts unfilled PLA, and PLA filled with sustainable fillers such as bamboo, flax, wood, etc., and PBF processes PA 11 derived from castor oil. On the other hand, photopolymers for VP in their liquid form are to some extent toxic, and hazardous, similarly to epoxies, but cease to be so when solidified after printing. Liquid PolyJet material exceeds ABS and PET in toxicity per kilogram. Upon FFF printing, PLA emits smaller amount of toxic fumes than the very popular ABS (Stephens et al. 2013). ABS is also fairly hazardous to produce and is rarely recycled municipally (Rossi and Blake 2014), but ABS for injection molding (IM) has the same downsides. More sustainable polymers can derive from other biobased polymers.

Plastics for AM are at least about 10 times more expensive than their version for conventional plastic manufacturing, which makes printed parts expensive. Customers are willing to pay extra for prototypes and small copies of the same item, because that is offset by shorter lead time and no cost and wait for a mold. Printing multiple materials is already possible, and when it will extend to load-bearing materials it will make AM more cost-competitive for engineering components made of multiple parts, but it will impair the component's recyclability, and in turn sustainability, unless all materials of the component are recyclable or compostable, and require hence no separation after collection.

Sustainability is reduced in AM when specific geometries require support material that is removed after printed, and hence wasted. Support material is typical for FFF and PBF, and can be recycled in the latter case for a few times before its properties overly degrade. Support material is typically not required in SL, and it depends on part geometry in DLP and CLIP™. Machining obviously generates material waste because it starts from a block of material to be shaped into the end use article, but with IM the waste is only the material left in the channels (*sprue* and *runners*) feeding liquid plastic to the cavity (or cavities) forming the article (or articles). Since 80% of plastics is processed via IM and only a small share of them is machined, AM in general does not waste less material than conventional plastic processing.

The effect on sustainability by printing speed and size is elucidated below. The printing processes are slow because parts are built layer by layer. F.e., a C-shaped ABS cover with a swept volume $3.3 \times 3.3 \times 2$ cm and printed volume of 5 cm^3 was built in 3 h on a personal desktop printer, whereas it would take a few seconds to injection mold many of it at once. The higher speed attributed to AM derives from saving time to skip fabrication of molds to build specific parts during their developmental design stage, f.e. molds for casting, IM, and layup of polymeric composite materials. That is why AM is typically cost-effective for prototypes, and producing small number of parts, such as for aerospace and medical, but not for parts built in large numbers. Although printers are getting faster, and in fact CLIP™ claims to be up to 100 times faster than other AM processes, printers still cannot compete in speed with IM and compression molding. This means that the energy required to print one part may be greater than that to injection mold the same part, because, although the printer may require less energy per time unit than that of an injection molder, the latter is orders of magnitude faster. However, in AM the energy cost is lower than other printing costs, longer time is a greater disadvantage than energy, but printers typically run unattended once they start printing, hence one operator can supervise several printers.

Printers are enabling bigger parts, as AM for architecture and construction proves. F.e. BLB Industries (Sweden) can build plastic parts up to $3 \times 3 \times 3$ m, and $1.5 \times 1 \times 5.6$ m. However, most large industrial plastic printers can print up to about $0.6 \times 0.6 \times 0.9$ m. Therefore, parts too large for a printer must be built in sections, and then connected together, which requires additional time and resources, whereas large molds can be as large as airplane wings, and enable one-step fabrication. Shifting to small size manufacturing at micro and nano level, such as for electronics, AM is not yet competitive with current processes in terms of speed and resolution.

The impact of AM on society can be estimated through several parameters. Since AM is not labor-intensive, it reduces the labor cost of articles compared to some conventional processes such as machining or hand lay-up of composite parts. Because AM may require relatively small initial investments, it can be implemented in developing countries and can support growth of local economies, as long as these countries can sustain AM with adequate resources consisting of supply chains of feedstocks, consumable, spares, etc., and a skilled labor force to run, maintain, and repair the printers and their computer hardware and software, ancillary equipment, devices supplying the powering energy, with all the above resources being more available in urban than rural areas. On the other hand, developed countries may benefit from the most technologically sophisticated versions of AM through "reshoring," consisting in manufacturing jobs returning to high-wage countries from low-wage countries. Another advantage of AM for society is the associated demand of jobs requiring new skills, unfortunately countered by a loss of skilled and unskilled jobs in conventional manufacturing due to the reduction of parts conventionally fabricated.

Assessing the environmental impact of AM in general terms is difficult, because it varies for specific AM technologies and printed articles. Similarly, comparing AM versus IM and machining in terms of overall environmental impact is complicated, because that impact derives from many factors: mostly printing energy and material use, but also on feedstock, printer type and setup, part geometry, machine utilization rate, etc. For the specific case of a hollow-shell plastic part without unusual requirements, AM has generally lower impact per part than machining, but higher impact per part than IM making many copies of the same part.

Even if printed parts will replace a sizeable amount of machined parts, it will not have a significant environmental impact, because machining represents an almost negligible niche of the manufacturing market, namely <1% in 2007 (BEA 2014), and is limited to prototyping and parts of complex geometry. This seems to differ with what argued by Gebler et al. (2014).

FFF of ABS, PLA and PET, SLA, PolyJet, and DLP have more negative environmental impact than IM of ABS at maximum utilization (Faludi et al. 2017).

IM of plastics is estimated to account for up to 2% of global greenhouse gas emissions (Hendrickson et al. 1998; Bjorn and MacLean 2003). If AM replaced IM and released a smaller amount of those emissions, that gain would be offset by the fact that AM could not be cost-competitive with and replace IM for large series of parts in the near future.

AM is expected to reduce transportation, because manufacturing will shift from large centralized factories to small regional printing factories and even to warehouses, stores, and customers' houses. However, transportation affects the environment less than manufacturing and energy use during the product's life. Moreover, even assuming that decentralization would be more cost-effective than economies of scale, this model works if the entire product can be printed, but today printers can only build simple items and components that are assembled with other components and shipped to the factories. AM can reduce transportation if, instead of shipping articles in boxes with empty volume, less voluminous feedstocks will be shipped to stores and warehouses where they will be turned into products printed on demand having the right staff and equipment.

In IM material waste is 1-10%, whereas machining a hollow part from plastics usually generates far larger amounts of waste, often over 80%. In comparison to these processes, AM can produce lower and greater material waste depending on the AM specific methods. PBF can reach 44% feedstock waste, PolyJet printers waste almost half of their liquid polymers utilized as building and support material, and some plastic cartridges of filaments for FFF cannot be recycled. On the other hand, FFF of hollow articles certainly wastes less material than when machining them. However, in AM, material waste has lower environmental impact than energy.

TPs for FFF (like ABS) and PBF (like PA 12) are recyclable, but not TSs for VP. However, unsolidified liquid polymer for SL and DLP can be reused a few times. The company RE PET 3D (Czech Republic) sells PET filament for FFF derived from recycled bottles (RE PET 3D, n.d.), featuring: tensile strength and at yield 26 MPa and 53 MPa, respectively, flexural modulus 2150 MPa, notched Izod impact strength 90 J/m, and HDT at 1.8 MPa 70°C. Other recycled filaments are made of ABS, acrylonitrile styrene acrylate (ASA), HIPS, and PLA and supplied by Formfutura (The Netherlands), GreenGate3D (USA), Kimya (France, USA), and Filamentive (UK). Feng et al. (2019) took PA 12 powder heated in a PBF printer but not sintered, and converted it in an extruder into filaments for FFF whose tensile and impact strength, and flexural modulus were 5%, 13%, and 15% lower, respectively than those of filaments made of virgin PA 12 powder. Properties of recycled plastics have to be adequate for functional applications. EOS has established a closed recovery circuit to recycle used plastics, in collaboration with KaJo Plastic that organizes the collection of used powder separated by printer owners in Austria, Germany, and Switzerland, carries it to the recycling company, and pays for the transportation costs. Examples such as RE PET 3D and EOS are initiatives going in the right direction, but obviously the scale of their impact needs to increase orders of magnitude to make a significant change.

Biodegradable plastics, such as PLA, PHA, PBS, and starch, are processed in AM, but cannot be recycled if they are mixed with non-biodegradable plastics, otherwise they can disrupt the operation of recycling systems.

Recyclable polymers for AM should be transported to large current sorting facilities that are efficient for large quantities of plastics, differently than smaller recycling facilities located at printing sites. The option that customers recycle the material of their own printed parts may not be cost- and time-effective for them if virgin feedstocks are affordable, and even if this behavior spreads, it will only represent a sliver of the market of AM plastics. Sorting prior to recycling may result less accurate and costly if the recycled plastics are used as particles bonded together with adhesive. The downside of this approach is that the material properties of the recycled feedstock may degrade and limit their applications. In developing countries, recycling may be affordable in small and decentralized facilities, because the cost of labor to sort plastics is lower than in developed countries, and hence recycling may be less costly than composting.

The growing interest in sustainability by public opinion has extended to AM customers, and some young and small companies, such as those mentioned below, sell desktop equipment to recycle plastics for AM. The energy required by a desktop recycler is about 38 MJ per kg of recycled plastic, not lower than the range of commercial plastic recyclers, and the impact of these firms and similar ones is very limited for now, because their equipment is designed for individual and not industrial use, and AM is still a niche manufacturing sector. However, these firms represent a start, and can scale up, or be acquired by larger companies, if recycling becomes cost-effective and is practiced across companies. ReDeTec (Canada) developed ProtoCycler, a machine converting recycled waste and virgin granules in plastic filaments for FFF. ProtoCycler recycles most commonly used plastics, and is expected to expand its range of materials. Similarly, in 2017 3Devo (The Netherlands) developed: (a) SHR3D IT, a device recycling plastic waste into printable granules at the pace of 5.1 kg of plastics per hour; (b) a machine converting granules into FFF filaments. Precious Plastic (The Netherlands) went farther, and engineered a production system to recycle plastics including a shredder, an extruder to form FFF filaments, and injection and compression molding machines. Francofil is a French startup that has commercialized NaturePlast, a family of FFF filaments based on bioplastics from renewable sources: wheat and coffee waste, and shells of mussels, oysters, and scallops. The Rotterdam-based design studio The New Raw exploits the durability of plastics, and designs outdoor public benches printed from packaging plastics: each bench weighs 50 kg, is 150 cm long and 80 cm wide, and can fit up to four people. Filabot (USA) builds equipment of different sizes converting plastic waste into granules, and granules into standard and customized FFF filaments.

Converting construction waste into feedstock for AM technologies applied to architecture and construction is a way to protect the environment. This is what the firm Winsun (China) does Winsun 3D. 2017, by turning crushed and sifted demolished bricks, concrete and cement blocks, and other construction waste into building materials for road and house construction (Winsun 3D 2017). If this approach spreads where the construction industry is thriving, the environmental will benefit.

Composting has the advantage of being applicable to mixed and inseparable (or costly to separate) plastics, and being less costly per treated unit mass than recycling because less dependent on sorting. However, composting does not allow recyclability, has to be weighed against other options such as incineration, is more efficient when performed in large facilities (f.e. municipal ones) that are not present in most cities, and may be hindered by the future availability of printers processing multiple materials.

Unfilled PLA and PLA mixed with biobased fillers (cellulose, hemp, starch, wood, etc.) can be composted in current facilities. Experimental compostable materials are being investigated, but they will have to pass the market's cost-benefit test to become successful commercially.

2.15   PRINTER SELECTION

This book focuses on feedstocks, but printers are obviously critical for fabricating a successful item, be it a model, prototype, or an end-use product. Given the multitude of brands and models of printers for plastics and polymeric composites, this section lacks the space to discuss printers adequately, and on the other hand it is easy to retrieve online information on printers. However, we propose a method to select among candidate printers the one that meets the specific requirements set by the buyer. This method is math-based in order to rank the candidate printers in an unbiased and quantitative way, and ultimately select the optimal printer as the one featuring the highest score. We illustrate this method by referring to Table 2.25, in which the choice of metrics and their values is arbitrary (although realistic), and can be customized by each reader.

The selection process is the following:

- Select the metrics
- Enter metric values
- Assign priority (PR) to the metrics at your discretion. Highest PR = top priority, lowest PR = lowest priority. The highest PR should not exceed the number of metrics, the lowest PR should be 1 or higher.
- Normalize metric values:
    - If a high metric value (f.e. the number of printable feedstocks) is more desirable than a low value, divide each printer's value by the smallest value of that metric across printers
    - If a low metric value is more desirable than a high value (f.e. printer price), divide each printer's value by the smallest value of that metric across printers, then calculate its inverse.
- Calculate the product of PR and each normalized value

TABLE 2.25

Example of Printer Selection

Metric	Unit	Metric Values			PR	Normalized Metric Values			PR × Normalized Metric Values		
		A	B	C		A	B	C	A	B	C
Sale Price	USD 1000s	275	300	360	5	1	0.92	0.76	5.00	4.58	3.82
Average Feedstock Price	USD/kg	52	68	62	4	1	0.76	0.84	4.00	3.06	3.35
Max Printing Volume	m^3	1.02	1.59	1.19	5	1	1.56	1.17	5.00	7.78	5.83
Number of Feedstocks	NA	5	7	6	4	1	1.40	1.20	4.00	5.60	4.80
Max Printing Speed	cm^3/h	10	15	20	3	1	1.50	2.00	4.00	6.00	8.00
Total									22.00	27.02	25.81

Legend: PR, priority

- Calculate the total of the normalized values of each printer
- Final ranking. The printer with the highest score is the one mostly meeting the requirements. According to Table 2.25, the optimal printer is B, followed by C and A.

2.16 NEAR FUTURE OF AM

2.16.1 INTRODUCTION

An ample variety and amount of R&D is conducted by academia, governments, research institutions and corporations in the fields of materials, technologies, equipment, software, and analytical/predictive models for AM, and is demonstrated by the increasing number of new and relevant technical papers, patents, devices, and companies.

AM progresses very fast because it leverages innovations in materials, equipment, and software, and, given its rapid pace, some credible prediction can only be attempted about its near future that, however, looks promising in terms of diffusion and options. Obviously the market will decide the most successful processes.

AM will be more successful if users see it as a tool (or set of tools) enabling an holistic approach to a challenge, that is a tool permitting to "address" and improve not only a specific component (f.e. a duct) but also the assembly (f.e. the cooling circuit) to which that component belongs, the system (f.e. HVAC system) containing that assembly, and ultimately the final product (f.e. the vehicle) comprising that system.

The Internet of Things (IoT) is being extended to printers through in situ monitoring devices, and process sensors designed to flag quality risks, and will become more employed because it enables real-time feedback compensating for process deviations. The capacity of IoT to improve quality and efficiency by increasing data exchange and automation in manufacturing will automate powder and support removal, and surface finish, and ultimately reduce labor, improve repeatability, and shorten schedule. AM will be increasingly adopted in the near future especially if it is applied in a comprehensive way that includes design to reduce time and/or cost of steps following manufacturing, such as installing, inspecting, servicing, and replacing the component to be printed.

An advance that will increase the benefits of AM is adopting the *digital twin* (Zhang et al. 2020), that is a digital model of a part to be printed that combines in situ sensor data and theoretical, computational predictions, and enables to statistically and closely simulate printing, postprocessing, inspection, and testing with high fidelity. Once the design is finalized and all steps are optimized, the part is printed, but feedback data from the part are still collected and recorded to further optimize the part's design.

In 2016, Wolhers Associates forecast significant investment into the AM market in the following 10 years (Wohlers 2016). As mentioned in Section 2.1, the worldwide market of AM (hardware, materials, services, and software) is expected to grow and reach USD 42 billion in 2027 (Sher 2018). There is consensus among experts, confirmed by indicators such as number of relevant patents and publications over time, on the growing role played globally by AM in academia, arts, education, business, entertainment, households, manufacturing, science, etc., and the consequent growth of worldwide market of AM.

Since traditional manufacturing accounted for 16% of the world's USD 80 trillion economy (Monahan et al. 2017), if AM grasps even a marginal share of that value, it will translate into a considerable amount of business. Monahan et al. (2017) assessed a wide-ranging "performance" in AM of 28 countries in Africa, Asia Pacific, Europe, and North and South America based on 38 quantitative metrics grouped in six categories: printing capabilities (global share of printers sold and installed, AM patent applications, etc.), demand (demand for printers, domestic market size, etc.), trade (export and import of AM goods, foreign investment, etc.), people (workforce skills, support for reskilling, etc.), governance (regulations to protect intellectual property, etc.), and technology (infrastructure and investments in R&D, new business models, etc.). The authors concluded that in 2017 the leading countries were USA, Germany, Republic of Korea, Japan, and UK in decreasing order, and predicted that the countries showing the greatest growth in AM would be Republic of Korea (due to a government-driven R&D road map focused on AM, and possible continuation of growth in patent number), Italy (thanks to multidisciplinary initiatives, and government support of universities and research centers educating and training workforce), UK, Germany (due to its broad Industry 4.0 strategy that includes AM, and focuses on short-term and long-term objectives), and France in decreasing order.

As a small bit of information describing innovation in AM, we mention that the authors of technical papers we found relevant for the present book were mostly affiliated with U.S. universities.

Jiang et al. (2017) predicted the future of AM in 2030 based on interviews with 65 experts, and reported 18 projections in areas of: (a) production, supply chain, and localization; (b) business models and competition; (c) consumer and market trends; (d) intellectual property and policy. Major projections were:

- Market share of printed components and products will exceed 10% of non-printed components and products.
- More than 50% of total industrial AM capacity will be met on in-house printers.
- The majority of private consumers in industrial countries will possess printers at home.
- Across all industries, AM production conducted by chains globally spread will decrease, whereas local AM production will grow.
- A significant amount of printed items will be made of multimaterials, and/or contain embedded electronics.
- Printed human organs will be available and largely substitute donor organs.

Faludi et al. (2017) predicted that medical prosthetics and elite sporting equipment may be the next frontier in low-volume printing.

We expect that companies utilizing AM will make progress in addressing issues stemming from traditional IP laws covering rights and protection for AM (Vogel 2016; Flank 2017). Likely, issues of patent, trademark and copyright infringement will be raised if companies that previously purchased replacement parts print them on their own printers, and, likewise, consumers needing spares for consumer goods and home (HVAC, plumbing, appliances, etc.) print them themselves.

Obviously, AM is not without challenges, but they also represent opportunities for advances in materials, processing, printers, modeling and processing software, functionality, and multifunctionality in the near future.

The growth of AM will be driven by its advantages listed in Section 2.1, but also by achieving improvements such as the following ones:

- Increasing the number of printable polymers and composites and reduce their price compared to their grades for conventional manufacturing.
- Improving the service performance of current printable polymers and composites, f.e. reducing void content in articles printed out of composites, and improving the filler/matrix interfacial adhesion.
- Reducing printing time.
- Developing printers with built-in feedback system to suspend printing if an error occurs, or to respond and adjust to a process change: material temperature too high; viscosity outside the intended range; etc.
- Increasing the ratio geometric resolution/printing time, and improving shape fidelity.

2.16.2 Production

Incorporating AM as one step in a manufacturing process forces to reorganize that process, in terms of scheduling, flow, and layout, and developing in-house expertise. AM will decentralize fabrication of low- and medium-volume items, moving the fabrication possibly closer to their customers, and to areas where the operating cost of manufacturing is lower, and all of this will impact the supply chain. Obviously, if the total cost per article in polymeric AM decreases (due to faster printing and/or less expensive materials), AM will be more cost-competitive in making end-use articles with widespread plastic manufacturing technologies, such as IM, whose initial high cost of tooling and setup is offset by a large number of copies of the same article.

AM will more frequently transition from prototyping to low- to medium-volume production (including repair, overhaul, and maintenance or MRO, where it is already leveraged in aviation), with reductions in inventory and storage costs.

AM is expected to diminish total supply chain cost by 50–90% "as production will move from make-to-stock in offshore/low-cost locations to make-on-demand closer to the final customer with major reductions coming from transportation and inventory costs" (Bhasin and Bodla 2014).

A larger number of AM service companies exist that print prototypes and end-use parts made of polymers, composites, and metals for other firms, and this growing trend will likely continue. These companies are approached not only by businesses not ready to invest in AM, but also firms preferring to test processes and printers before acquiring printers, or having their own printers busy currently and in the near future. Some examples of AM service companies are 3D Hubs (The Netherlands), Protolabs (USA), Sculpteo (France, USA), and Shapeways (USA), and manufacturers of printers such as 3D Systems, EOS, and Stratasys that have opened their own printing service division.

2.16.3 Materials

In general, for new materials to be commercially successful they must provide a performance/cost ratio that is competitive on the market. Ultimately, companies adopt innovation if it is profitable. Material performance is defined not only by adequate values of specific properties, but also the consistency of those values. This is valid especially in engineering applications. Before large and highly technologically companies, such as those in automotive and aerospace, adopt a new material in their products, this undergoes a thorough experimental, lengthy, and costly characterization process (*qualification*) to determine its dependability in service conditions. Qualifying an AM feedstock means also qualifying the printer on which it is processed. A supplier of a new AM feedstock will increase its chances by sharing some of the qualification cost with the customer evaluating that feedstock, f.e. by donating the test coupons. The same applies to a printer manufacturer that can donate printing time for the qualification process. A challenge to the diffusion of AM in structural components is providing designers and stress analysts with *design allowables*, because the latter ones are determined after a costly and long process of testing and analysis that can be only offset by a strong business case.

New materials or same materials with improved properties may require adjustments in printer specifications, and hence progress in feedstocks also depends on process development.

Current efforts in material development will continue to spread across AM processes, unless the demand for some process drops, and diminishes the incentive to conduct specific R&D work. Feedstocks targeting engineering applications will possess more consistent properties and geometry, manifested through smaller batch to batch variations in melting point and ovality (filaments), shape and particle size (powders), viscosity and shelf life (liquids). Other issues to be addressed are: (a) for polymer filaments, interlayer bonding, and shrinkage after printing resulting in part distortion; (b) for photopolymers, longer shelf life, physical ageing, and sustainability (the commercial versions are not recyclable). Higher production rates of feedstocks may drive their price down, and propel their applications. The interfacial adhesion between reinforcement (particles and fibers) and polymeric matrix in filaments of polymeric composites for FFF should be improved in order to drive their engineering uses. We expect that the number of high-performance polymers, f.e. PEEK and PEKK, will continue growing, and they will be considered for components currently made out of metal (Roboze 2021), possibly imitating the application trajectory of non-AM composite materials.

More polymeric feedstocks in granular form will be employed, because they enable high printing speed, and will be more likely adopted if their engineering properties can compete with those of filled polymeric filaments. PI in powder form for PBF is about to be commercialized. High printing speed is particularly beneficial for large objects, such as in construction and mold for parts made of composite materials.

2.16.4 POLYMERS

2.16.4.1 General Requirements

In the near future polymers for AM will provide performance unmet by current polymeric feedstocks but demanded by industrial users, such as filled polymers with high elastic modulus, and fatigue resistant polymers. However, the commercial success of new AM polymers hinges on the ratio benefits/cost, and its value compared to that of its competing materials. Benefits and cost are intended in a broad sense: benefits includes material performance (functional properties, durability, weight saving, etc.), overall cost and time reduction in printing parts (buying and running the printer, etc.), while cost encompasses all costs related to printing the parts (including cost of qualifying material and printer). The expected applications will direct the material R&D, with medicine and military playing a major role due to the number of their costly applications and large funding available, respectively, although the U.S. Department of Defense, considers AM also as a means to reduce the maintenance cost of their current vehicles, ships and airplanes by printing spares. Broadly speaking, the effort to gain new applications will widen the range of AM polymers, and improve their physical-mechanical performance. More competition among materials providers will cause the price of standard AM polymers to drop, and the new feedstocks should be competitively priced, unless their performance justifies a price premium. Reusable materials are already available. In fact, PA powder not melt during printing is already reused, and recently Stratasys patented methods to reuse materials in MJ (Teken and Belocon 2017; Teken and Belocon 2019). More sustainable materials including recycled, reusable, and biodegradable materials will emerge, further decreasing materials costs (Jurrens 2013).

A promising area that will receive increasing attention is design and synthesis of FGMs, which are engineering composites made of two or more constituent phases featuring continuous and smoothly gradients of composition, structure, and particular properties (Aysha et al. 2014; Bhavar et al. 2017) within their volume. FGMs constitute a particularly high-value application for AM, and provide additional functionality to printed components (Bourell et al. 2009b).

In 2014, David Leigh, COO at EOS North America, expected that in the longer term better formulations of PEEK, PEKK, PA, and blends including flame retardants and filled/fiber composites would be available. He also believed that most of the future focus on material properties would be directed to medical grade and high temperature performance, and the work on electrical properties would aim primarily at engineering hybrid printers that would fuse material with a laser and print conductive materials with a print head (Leigh 2014).

Stephen Hanna (SDH Consulting) in 2014 predicted that the focus in AM would continue to be on increasing the functionality of polymers in order to extend the range of manufacturing applications. "To achieve this," he explained, "materials must meet the following design demands":

- *High stability*. Polymers more stable over longer durations will be formulated, pushing the useful life of printed parts further. To realize this, polymers must be environmentally, mechanically, and dimensionally stable over time. Some but not all current polymers feature these properties. Improvements in these areas will greatly increase adoption of AM for functional and engineering parts.
- *Improved mechanical properties*. Enhanced and adequate values of specific properties (ductility, impact resistance, hardness, etc.) will generate more applications.
- *Other properties*. Special properties, be they electrical, medically related, chemical resistance or other, could be important for enabling new and emerging applications. (Hanna 2014).

Many current SL resins have softening temperatures under 100°C, and new SL resins with superior temperature resistance would have a competitive advantage. More formulations will be investigated that are curable at broader UV bandwidths, resulting in a wider range of softer to stiffer materials than the current range.

Common epoxy-based feedstocks for VP are not biocompatible and hence not suitable for medical applications due to irritating and cytotoxic effects on human cells, resulting mainly from uncured epoxies in the cured polymer articles (Bens et al. 2007). Studies on nontoxic materials for VP have started and will continue. F.e. Miao et al. (2016) employed a novel renewable soybean oil–based epoxidized acrylate to print via SL a highly biocompatible scaffolds supporting growth of human bone marrow stem cells and featuring shape memory.

Vegetable oils are a most important classes of biological resources to produce polymeric materials. Polyester, polyether, polyolefin, and PU are the four most important families of vegetable-oil-based polymers (Miao et al. 2014), and are suitable candidates from which to formulate new AM feedstocks for healthcare and medical applications, because many of them feature outstanding biocompatibility and unique properties, including shape memory.

Improvements in specific properties, f.e. tensile strength and elastic modulus, will be achieved not only in single polymers for AM, but also in new combinations, such as the current iron/ABS and copper/ABS for FDM™ (Nikzad et al. 2011), PP/ZrO$_2$ (Shahzad et al. 2014), and Al$_2$O$_3$/PP (Shahzad et al. 2013) for PBF.

There was large and qualified consensus (Shipp et al. 2012) that most likely AM polymers in the near future will comprise: (a) carbon fiber-reinforced polymers, including carbon nanofiber-reinforced materials; (b) liquid-crystalline polymers; (c) biodegradable materials, such as PCL.

2.16.4.2   SPs for AM

The number of SPs available to manufacturers is expected to increase (Faludi et al. 2017). The near future of SPs looks optimistic, according to 3DomFuel (USA), a supplier of many filaments for FFF made of PLA and other sustainable feedstocks (3DomFuel 2016).

In the short term, SPs will benefit from wider applications of AM, and will gain popularity. However, their adoption for functional and engineering components is possible if they possess durability and consistent physical and mechanical performance adequate for load-bearing applications, and are processable with industrial printers. It is expected that SPs will be more present in outdoor and furniture items, architectural components, and medical, pharmaceutical, and dental applications, such as organs for transplant, live tissues for testing during drug development, and printed organs and cadavers for medical training.

The following forecast combines the opinions of the author and several experts from SP suppliers. The global production volume of PLA for conventional and additive manufacturing reached about 190,000 tons in 2019, and its demand has doubled every 3–4 years from 2000 to 2020 (Jem and Tan 2020). The global market of SPs for conventional and additive manufacturing will keep increasing due to current concern of governments, public, and corporations for the environment, especially if the crude oil price will rise.

With the market of SPs growing, major printer suppliers may enter the arena of SPs with their own formulations, possibly compatible with industrial grade printers. R&D will be conducted to file patents on new SP formulations challenging PLA's dominance.

New SPs will focus on:

- Balance between increasing the market of DIY, models, prototypes, and non-engineering applications based on desktop printers, and developing the market of industrial and engineering applications enabled by industrial printers.
- High-value applications, driven by improved material properties or added functionality, leading to printed articles marketable at higher price than that of current articles.
- More stable performance (to widen the targeted market): feedstocks featuring stable physical and mechanical properties in presence of changing service conditions (temperature, load, humidity, chemicals, etc.) and over long durations.

Acceptance of SPs in engineering components is also based on a documented and extensive set of material property data (high and low temperature, humid and dry environments, fatigue, etc.), and low scattering in material property values. Particularly, diffusion of PLA depends on improving its ductility and temperature resistance.

The pace and amount of penetration of SPs in the market (automotive, consumer goods, home, transportation, etc.) is dictated not only by benefits/cost consideration but also the material qualification process, that grows longer and more costly the more demanding are the service requirements associated with each potential application. A penetration route for SPs to build their "credential" among designer and fabricators, is to start with non-critical parts such as models, prototypes, interiors, and aids, jigs, and fixtures for manufacturing shops, instead of end-use parts.

Expectations for the near future of SPs are the following:

- PLA will maintain its preeminence among SPs
- Property ranges will widen and property values will improve
- New materials will include: sustainable ABS; biodegradable PETG; flexible and semi-flexible PA; water soluble, wood filled, metal filled, magnetic, conductive, color-changing materials; liquid UV resins; materials mimicking and inspired by nature
- More materials available filling the 60–110°C maximum service temperature interval
- Printing multi-material parts in one run
- Blends and alloys to tailor properties to special applications
- Improve biodegradability, compostability, lifetime, and recyclability
- Additives, fillers, and reactive extrusion to enhance specific properties
- Price will stay in current USD 40–100/kg range for most materials, unless higher volumes lower it.

Obstacles hindering the diffusion of SPs are:

- The R&D cost, high especially for start-ups. It could be mitigated by financial support: government grants, tax incentives, and scholarships to universities, research institutes, etc.
- Cost of educating customers that commercial biobased materials perform as well as fossil-based materials
- Reliability, build volume, and speed of desktop printers
- Improving brittleness, mechanical and thermal properties, chemical resistance, durability, interface adhesion in composites, and interlayer strength
- Materials' limited performance
- Design software.

Some of the previous issues are being addressed by materials suppliers, and firms providing printers and software for polymers in general that are marketing larger and faster printers, and advanced software packages featuring: (a) design optimization, aimed at minimizing weight while meeting or exceeding the component's structural requirements; (b) optimizing part's orientation inside the build volume; (c) minimizing the support material in case of extrusion-based AM process.

2.16.4.3 Polymer Matrix Composites (PMCs)

PMCs for AM have superior functionality and performance compared to unfilled AM polymers. The number of commercial PMCs for AM has been gradually increasing, but their proliferation is hindered by limitations in polymer selection, material performance, and process (Wang et al. 2017b; Saroia et al. 2019).

There exist more TP matrix (ABS, PA, PC, PEI, PEEK, PEKK, PETG, PLA, etc.) composites commercially available (for ME and PBF) than TS matrix composites, such as ceramic filled resins (for VP). This status is inverse to that in conventional PMCs, where TS matrices (cyanate ester, epoxy, phenolic, unsaturated polyester, vinyl ester, etc.) prevail over TP matrices, and dominate in load-bearing and engineering applications such as aerospace and boating. For PMCs for AM, there are opportunities to synthesize new matrices, introduce new reinforcements, adopt as a matrix other SPs besides PLA, and devise efficient filler-matrix combinations that permit demanding and/ or specific uses.

As to material performance, some issues associated with the presence of filler (f.e. dispersion and porosity) have to be addressed: intralayer porosity must be reduced because it partially offsets the improvement in mechanical properties induced by the filler; filler dispersion has to avoid formation of clusters that form above a certain filler content level and impair mechanical properties. The interfacial bonding between matrix and filler also must be enhanced for more effective load transfer from the former to the latter. The interlayer strength in PBF and ME also needs to improve, in order to enhance properties in Z direction, and consequently reduce anisotropy.

Process challenges are specific to each technology. The following improvements will encourage the adoption of PMCs, if their performance/price ratio is competitive: faster printing (but it will be still slow compared to conventional manufacturing such as casting and IM), and built-in feedback systems that respond to process errors and unexpected changes, reduce variation and anisotropy, and increase resolution without stretching printing time.

Since this section focuses on near-future PMCs, mentioning the following emerging technologies for PMCs for AM can guide in polymer development:

- *Micro automated fiber placement* (μFAP) by Desktop Metal (USA): a desktop printer combines traditional automated fiber placement with ME to print items out of tape of PA, PEEK, and PEKK reinforced by continuous fibers of carbon and glass. Since small continuous fiber PMC parts are often made by hand lay-up and require molds (*tooling*), μFAP reduces the manufacturing cost for the same type of parts.
- In *composite fiber coextrusion* (CFC) by Anisoprint (Luxemburg, Russia) continuous fibers are embedded in a polymeric matrix (PETG, ABS, PC, PLA, PA, etc.) during and not before the printing process.
- In *digital composite manufacturing* (DCM) by Fortify (USA) a projector cures a liquid photosensitive resin mixed with reinforcing additives, such as chopped carbon fibers that are oriented applying a magnetic field during printing, and hence controlling multiple properties (strength, stiffness, and thermal conductivity, etc.) in three dimensions within each voxel. One area of application of DCM is composite tooling.

2.16.4.4 Polymer-Based Nanocomposites (PNCs)

PNCs are already leveraged in VP, PBF, and ME, and their usage reflects a gradual growing trend (Khan et al. 2019) that will continue, because they show benefits, applications, and potential to produce cost-effective, customized, and functional parts. Some challenges of PNCs for AM are the same as in PMCs for AM: slow printing, porosity, and surface quality. Another challenge is making the price of PNCs attractive. In case of VP, the presence of nanofillers must not disturb the UV cure of the resin, and impair its final physical and mechanical properties. As to PBF, adding nanofillers affects the matrix' thermal properties, and requires optimization of powder bed temperature and laser power, in order to prevent curling and caking of the printed part. Porosity is detrimental to the printed object by reducing its density, matrix-filler interfacial bonding, and hence its mechanical properties. When developing a PNC filament for ME, currently this material undergoes two extrusions: the first one in the extruder mixing matrix and filler, the second one upon printing, and the effects of this double processing (such as filler distribution) on the printed part should be assessed and addressed. Alternatively, ME printers could be modified, and permit the formulation of the PNC filament directly in the printer extruder, such as it already takes place in bioextruders. Finally, fully sustainable nontoxic PNCs could be developed by combining SPs as matrix and nano-size sustainable fillers such as chitosan, gelatin, and sodium alginate.

2.16.4.5 Carbon Fiber-Reinforced Polymers (CFRPs)

Adopting nanotechnology offers tremendous potential to improve performance and functionality of polymers formulated for conventional fabrication processes and AM. A class of CFRPs comprises polymers reinforced with carbon nanomaterials, such as carbon nanotubes (CNTs), which are rolled-up one-atom thick sheets (called *graphene*) of carbon that feature aspect ratio greater than 1000, and have demonstrated remarkable mechanical, electrical, and thermal properties. Adding nanomaterials reinforces and stiffens a "matrix" material in a way compatible with the AM layer-by-layer printing pattern that prevents placing a continuous reinforcement in the Z direction across all the object's layers. CNTs constitute an area of development for AM polymers, and broad overviews of CNT applications in AM are authored by Ivanova et al. (2013) and Ghoshal (2017).

CNTs are already added to non-AM polymers to improve their mechanical, thermal, and electrical properties (Coleman et al. 2006; Farzana et al. 2006; Dinesh Kumar et al. 2020). Similarly, when CNTs are combined with SL polymers such as UV-curable epoxy resins, and PBF materials such as PA 12, they improve the mechanical properties of the printed object. Specifically, in the case of SL, CNTs raised tensile strength and fracture stress, but also increased brittleness compared to the unfilled resin (Sandoval and Wickermm 2006; Sandoval et al. 2007), therefore we expect research in the near future directed at reducing the brittleness of CNT-resin printed parts, without penalizing their tensile strength and fracture stress.

CNTs will be further investigated as PBF ingredients, since they have been tested as coating on PA 12 powder particles for LS, and have generated printed parts possessing enhanced tensile, flexural, and impact properties without reduced elongation at break, compared to the laser-sintered PA 12 parts (Bai et al. 2013). CNTs may reduce the gap between properties of parts in PA 12 conventionally made and printed parts in PA 12.

The need to print mechanically strong, thermally stable, and electrically conductive parts will continue pushing research to enhance properties of powders of PA 12 plus carbon black (CB) for PBF, and build on successful formulations (Espera et al. 2019). Topics investigated will be optimizing CB's content and distribution to improve specific properties, and/or printing parameters to reduce the fraction of particles not melt during the process.

Surface quality of printed parts depends on the specific process resolution and dimensions of the raw material, such as filament diameter and particle size. In PBF the typical layer thickness is 60–200 μm, and the average roughness R_a can be as low as 8.5 um (DuraForm PA by 3D Systems). Controlling surface quality is important not only in applications where appearance is a requirement, such as presentation models, but also in others where the surface is functional, such as in the case of aircraft air ducts, part regions mating other parts, and testing models for wind tunnels, where the surface roughness significantly affects the drag coefficient (Daneshmand and Aghanajafi 2012). In medical implants a surface too smooth is detrimental because it impairs attachment of living tissues to the implant. Importantly, surface roughness affects part's fatigue strength that, in general, is lowered by a rougher surface. Studies will be conducted to improve the surface finish of PA 12/CNTs vs. PA 12, focusing also on understanding the influences of parameters affecting surface roughness, and dimensional accuracy during fabrication.

Because varying the loadings of nanomaterials during printing enables to fabricate parts with graded material properties, graded materials are also expected to be further studied.

On the other hand, the diffusion and improvement of nanomaterials in AM applications must overcome the following challenges (Ivanova et al. 2013):

- Enhancing dispersion of CNTs in the matrix has been a challenge for some time (Xie et al. 2005; Ma et al. 2010; Ghoshal 2017) and still is. Agglomeration of the particles makes the nanomaterials behave as a bulk and lose their unique properties. To increase dispersion, nanomaterials can be functionalized with organic linker molecules keeping the particles away from each other, even when embedded into the printing media, ultimately improving the properties of the printed parts.
- Because the cure depth of VP parts can be impaired by the presence of nanomaterials due to their ability to absorb the UV light, selecting the ideal wavelength that will solidify the polymer without being influenced by the nanomaterial will enable to cure the part throughout its volume.
- Reducing the porosity of PBF parts, f.e. by synthesizing core-shell nanostructures, with the core made of nanomaterial, and the shell of the printing polymer, such as PA 12 or PA 11 (Zheng et al. 2006).

2.16.4.6 Liquid-Crystalline Polymers (LCPs)

As explained in Section 1.6, the presence of LCPs effectively enhances mechanical and thermal properties of the materials to which they are added. LCPs have been investigated as AM feedstocks since late 1990s, and been evaluated for AM processes such as VP, ME, and 4DP. The R&D in LCPs for AM will continue in the near term because their generate feedstocks for AM with improved properties, and will also focus on reducing their price that is now higher than that of popular TPs. F.e. *lyotropic liquid crystals* form a class of LCPs for VP and ME that have been studied to control the polymer structure at the nanometer scale, and ultimately enhance material properties (Forney 2013), f.e. by producing hierarchical materials with well-defined anisotropic structure (Rodriguez-Palomo et al. 2021). Houriet (2019) studied and mechanically tested a printable high-strength LCPs, whose anisotropy could be promoted for topology optimization in AM, by orienting the molecules in the printing direction through the printer nozzle.

2.16.4.7 Other Polymers for AM

Bioink is already available, but requires further development to overcome its key, opposing properties that make it complicated to handle it. On one hand, bioink must be adequately fluid to flow through the printing nozzle without damaging cells that are subjected to shear forces as they come out of the printer head. On the other hand, bioink must be stiff enough to hold its shape after printing and prevent the item printed from losing its intended shape.

Elastomers have become AM feedstocks later than TPs and TSs: TPU pellets and filaments were the first AM elastomers to hit the market, and were followed by silicone, such as ACEO® marketed in 2019 by Wacker (Germany) for chemical industry, electrical components, health care, medicine, robotic components, sports gear, and transportation. More AM elastomeric feedstocks may be developed to target applications requiring complex, customized shapes and low- to medium-volume production.

2.16.4.8 Properties of Polymers for AM in Near Future

Experts agree that AM polymers in the near future will widen their range of application and increase their market volume offer by improving: fire resistance, thermal conductivity, electrical conductivity, self-healing properties, and recyclability.

Functionality and multifunctionality will be still pursued. AM enables to build biomimetic materials, cellular materials, deployable structures, new mechanisms, materials and structures featuring multi-scale and multi-resolution, and smart components (Lyons and Devine 2019). AM also provides ways to achieve multiple and cost-effective functionality. F.e. BJ can print multiple materials in one printing session, such as in the case of an aircraft cooling duct made of stiff, tough, and temperature resistant plastics transferring fluids, and two rubbery edges sealing the duct to its mating components. Another example of multiple functionality is 4DP making programmable components changing shape to react to a specific stimulus.

2.16.5 Processes and Printers

The business incentive to develop larger and faster printers is obvious. Commercial polymer-fed, extrusion-based printers having custom size and making very large parts are already available by the companies BLB Industries, Cincinnati Inc. (USA), Thermwood Corp. (USA), and ErectorBot (USA). An obvious challenge is building larger *and* faster printers that process more polymers and feature better surface finish than current models do. The company DMG Mori (Germany, Japan) sells hybrid equipment combining metal powder nozzle-based AM and turn-mill machining in one work area, enabling to print at high speed and then complete the object by removing material and smoothing surfaces. The same type of machine could be engineered for polymeric AM, and merry fast printing to smooth surface finish, if the business case exists. Another way to reach the same goal may be to develop a version of the CC process described in Section 2.3.10, namely combining a large nozzle for rapid printing and a tool for smoothing the outer surface.

Gao et al. (2015) anticipated that more numerous, friendly, and powerful versions of software and hardware than those currently available, such as web-based design software, natural user interfaces, and low-cost optical sensors for 3D scanning, will be introduced for desktop printers.

Simulation and optimization of printing processes to maximize properties and minimize defects in printed parts is currently possible with options such as GENOA 3DP, an AM design tool and software suite for polymers, metals, and ceramics utilized to reduce defect, material waste, and engineering time (GENOA 3DP Simulation n.d.). Efforts to perfect simulation and optimization of AM will continue. In fact universities, organizations such as ADAPT Advanced Characterization Center (USA), and companies such as GE (USA) and its Global Research Center (Zelinski 2018), have been working on computer-developed algorithms capable to fine-tune process parameters and print components with minimal defects and the intended mechanical properties, based on analysis of test data from printed parts conducted multiple times. The above examples reflect the use of data analytics and machine learning to continuously improve quality and consistency of printed objects.

R&D has to develop ways to reduce porosity (in ME and PBF) and anisotropy (in ME, PBF, and VP) that penalize physical and mechanical properties of printed parts.

Besides competing with the strengths of conventional manufacturing (fabrication speed, material range, large data sets, isotropy, etc.), AM should also leverage its own unique strengths, such as handling multiple feedstocks and generate complex shapes, which can result in components with embedded active components providing functionality previously unavailable. "Examples could include arbitrarily shaped electronics with integrated microfluidic thermal management and intelligent prostheses custom-fit to the anatomy of a specific patient" (MacDonald and Wicker 2016).

Another peculiar strength of AM is designing to reduce the number of parts in an assembly, which translates into several savings described in Section 2.1.

A way to reduce overall cost of AM is automating the post-processing step on parts produced in series and relatively high volume, in order to cut time and labor.

2.16.6 EDUCATION AND TRAINING

As AM usage and market grow, so does the need to educate and train technicians, engineers, managers, executives, high school and college students, and their teachers, instructors, and professors about AM, its benefits, limitations, and applications, in order to best leverage AM, and maximize its ratio benefits/cost.

Companies will benefit from investing in educating technicians, engineers, supply management professionals, managers, etc. in AM. Design engineers need to learn new design practices specific for AM and collectively labeled as *design for AM* (DfAM), in order to exploit topological optimization (Section 2.1), and reduce weight, part count, stress concentrations, and service life, in other words in order to leverage as many advantages of AM as possible. As size and complexity of printed objects increase, designers who generate multi-materials artifacts featuring complex shapes and multiple functionality will need more efficient and powerful computational and modeling software optimizing material support, simulating, and stress-analyzing their designs. As to AM processes themselves, they will become more automated and integrated, f.e. including real-time and feedback-based controls monitoring and adjusting the process as needed. Loading feedstocks, unloading printed items, post-processing, quality inspection, and moving printed items to next steps (f.e. packaging and shipping) will be automated.

Companies and governments aware of DfAM are releasing internal design guidelines to achieve the strongest, stiffest, lightest, most durable parts featuring the best quality.

2.16.7 AREAS OF APPLICATIONS

2.16.7.1 Introduction

So far, AM is studied and applied across a wide range of areas, but the scope of the present book limits the description of future applications to some major areas. However, throughout the book current and future applications cited in technical literature are mentioned for all the SPs discussed. An advanced and most recent application area for AM is shape-morphing items that are utilized in smart textiles, autonomous robotics, biomedical devices, drug delivery, and TE. In their functionality shape-morphing items imitate plants, whose leaves, flowers, tendrils, etc., respond to environmental stimuli (humidity, light, etc.) by altering their own shape (Sidney Gladman et al. 2016). Another promising area is electronics where AM is not typically applied but is technical progressing: in 2020 Nano Dimension (USA) used dielectric polymeric ink and printed a 10-layer circuit board on which high-performance electronic devices were soldered for the first time on both sides. AM offers the military the opportunity to print spare parts quickly (and possibly performing "better" if they are redesigned) and even in a theater of war, ensuring minimal downtime and continuous operation. Ultimately, every potential application will be successful if it has an attractive benefits/cost ratio.

2.16.7.2 Aerospace

AM has been steadily advancing in aerospace, not only in models, prototypes, production aids, and tooling, but also in parts flying on airplanes, drones, helicopters, and satellites. These parts must comply with airworthiness and air transport safety, and regulations issued by civil aviation organizations such as the European Aviation Safety Agency (EASA) and the Federal Aviation Administration (USA). In the case of military aircrafts, printed parts must meet the requirements established by the specific customer, such as f.e. Royal Air Force (UK) and U.S. Air Force.

Moreover, when it comes to flying parts, new polymeric and metallic feedstocks and processes must undergo costly, and lengthy test programs to prove that they can meet the demanding service requirements that include static, dynamic, fatigue, bearing, environmental and chemical resistance, and other properties measured from below to above room temperature, and, in case of interiors, fire, smoke, toxicity (FST) requirements. The above qualification process represents a steep entry barrier for materials and processes/printers.

AM is well poised for aerospace and will continue to gradually spread, because, among its other advantages, its processes are tailored for low to medium production volumes that are typical of aerospace, and aerospace can absorb the cost of AM parts more than other industries, and obviously greatly benefits from the weight reduction generated by topological optimization. For decades polymeric feedstocks for AM have been successful in flying parts that were not flight critical and not structural: air ducts, vents, panels, housings, mounts, plugs, bumpers, fairing collars, louvers, plenums, wire shrouds, etc. Structural components require materials ensuring steady mechanical performance over time, across loading modes (static, fatigue, impact, vibrations), and in harsh environmental conditions, and some PMCs for AM such as CF-PEKK are suited for that.

PMCs for AM are expected to penetrate aerospace because of their high ratios strength-to-weight and stiffness-to-weight, and high resistance to corrosion and fatigue, especially if the fiber-matrix interface adhesion is improved and boosts the mechanical properties of PMCs.

Unmanned aerial vehicles (UAVs) are becoming more frequent in the military, private, and public sector, and possess less demanding requirements than those for manned aircrafts, and represent a potential area of applications of newly developed materials.

SPs for AM have potential use in aviation. In fact, biobased polymers have been considered for aerospace applications for some time. John et al. (2008) studied a sandwich panel made of Nomex® honeycomb and flax fiber fabric treated with flame retardant and impregnated with phenolic resin, and tested it as a candidate material for aerospace structures. Researchers funded by European Union and Chinese researchers are developing eco-friendly biobased materials for aircrafts with the goal to replace fossil-based carbon- and glass fiber-reinforced plastics. Namely, they are aiming at recycled carbon fibers and bioresins for secondary structures and interiors of aircrafts (Anon. 2020). However, "green" aircrafts would still require high-performance resins to substitute the traditional epoxies in structural applications, therefore epoxy resins for composites have been formulated starting from biobased rosin acid ($C_{20}H_{30}O_2$) and itaconic acid ($C_5H_6O_4$) and through genetic engineering technology. A review of recent research on biobased epoxies derived from natural oils, natural polyphenols, saccharides, natural rubber and rosin, and their potential in aviation has been published by Ramon et al. (2018), who reported several examples of biobased epoxies whose mechanical properties and viscosity were adequate for applications in fiber composites and made them promising candidates for aircraft interior components. Even with significant progress in biobased epoxies, further optimization of their performance to make them competitive for applications in the aviation sector is still needed. The current strongest and stiffest epoxies for AM are those processed through VP.

2.16.7.3 Automotive

AM supports the automotive industry in production and R&D by providing visual models and functional prototypes, robot and machine components, tools, fixtures, jigs, and part count reduction. Old parts no longer sold can be duplicated via AM for car restoration. Automotive tooling already constitutes about 18% of the AM market (Faludi et al. 2017). Leveraging AM in automotive must be conducive to maintain costs very low, and production rates very high, two most critical factors for automotive firms to competitively operate.

In 2016, in a white paper entitled "Five Ways 3D Printing Is Transforming the Automotive Industry," Stratasys argued that these ways were: flexible optimized design, rapid tooling, fast customization, production aids, and real-world testing. Hereafter we touch on some of these "ways."

Printing new and spare parts for low-volume production vehicles has been already adopted. In a 2016 press release, Daimler mentioned printing plastic spare parts for Mercedes-Benz trucks, explaining that "30 genuine spare parts can be ordered and supplied at the press of a button from the 3D printer, quickly, economically, in any quantity, and always in consistent genuine manufacturer's quality […] Today at Daimler, more than 100,000 printed prototype parts are manufactured for the individual company divisions every year" (Lecklider 2016).

AM can be advantageous for spare parts, because single car models have thousands of spare parts which is problematic for: (a) automotive manufacturers that have to predict what parts to produce and keep them in inventory, (b) car owners, and (c) intermediaries. One option is to set a cloud-based two-sided platform with the purpose of automating orders of printed automotive spare parts (Mansourkhaki 2019). It may be speculated that eventually car and spare part manufacturers sell not only built spare parts but also wire electronic copies of 3D digital models of them to stores equipped with printers that will manufacture them for customers, and, in the case of body parts and accessories, customers can even personalize the design (color, finish, size, etc.). Another use of AM is for rare spare parts and bespoke car designs, with the latter ones being more prevalent and profitable for luxury cars (Nichols 2019).

An example of production aids is the orthotic device shaped as thumb cast printed for BMW assembly line workers who would push a huge number of rubber plugs into holes on the assembly line, which resulted in many workers suffering from an injury due to a repetitive strain. Each thumb cast is custom-sized for the specific worker based on a 3D scan of their hand (Lecklider 2016).

Design and production of tools, dies, and molds are two critical steps in developing new car models, because they determine the lead time and amount of the investments required to start the production. Asnafi et al. (2019) demonstrated that laser-based PBF printed forming tools and trimming/cutting/blanking tools and dies out of maraging steel 1.2709 that met the performance requirement of automotive production.

Messina (2020) focused on robots operating in shops for sheet metal working installed at the FIAT plant in Mirafiori (Italy), and improved the robots' grippers by redesigning them for AM.

Reducing weight of automotive components is favored by environment-conscious consumers and driven by government regulations, and Schmitt et al. (2020) concentrated on this aspect, by printing ABS components, and, instead of solid components, printed ABS components with partially filled cross sections, whose internal structure and void content were tailored to the stress levels experienced by the specific parts.

The BARBARA project is an effort funded by the European Union that aims at developing novel sustainable engineering plastics with innovative functionalities that are available as filaments for AM and possess properties suitable for automotive components such as door handles and dashboard fascia (Anon. 2017).

2.16.7.4   Biomedical and Pharmaceutical

Biomedical and pharmaceutical applications of AM cover a broad range. They relate to exoskeleton devices, prosthetics (including exoprostheses), dentistry, scaffolds, tissue modelling and engineering (Javaid and Haleem 2018; Durfee and Iaizzo 2019), and controlled drug release. Particularly, in TE, AM offers unparalleled versatility in terms of possible topographies, and defect/patient-specific geometries, and generates asymmetric and/or anisotropic geometries that mimic natural tissues. Mota et al. (2015) authored an extensive overview of AM techniques for tissue-engineered constructs out of polymers and other feedstocks.

AM is also being exploited in research and practice of orthopedic, urogenital, cardiovascular, and neurological surgery (Puppi and Chiellini 2020), including biodegradable and non-biodegradable surgical implants, instruments, educational and training tools, and items for preoperative planning.

Polymers have been evaluated in combination with the following AM methods to produce medicines: PBF, SL, ME, and less familiar techniques such as *drop on solid*, *drop on drop*, and *microneedles* (Jamroz et al. 2018). The global pharmaceutical market is expected to grow at 3–6% rate from USD 1.2 trillion of 2018 to USD 1.5 trillion by 2023, with the U.S. market predicted to advance faster (4–7%) than the top five European countries (1–4%) and Japan, and emerging countries predicted to play a significant role (5–8%) (Miglierini 2019). Therefore, pharmaceutical applications of AM can be lucrative even if SPs capture a minimal share of that market. Through AM, personalized medicines can be prepared, by producing medications to exact specifications tailored to the needs of individual patients. Dosage forms, even very small, can be accurately printed in shapes, sizes and textures that are difficult with conventional manufacturing techniques, and production waste is decreased. The potential of AM for the pharmaceutical industry has been recognized (Lamichhane et al. 2019), but so far the above benefits have not offset the higher cost of AM versus that of conventional drug production routes. In fact, until 2020, despite of the large volume of R&D conducted, there has been only one practical application of a printed drug, namely Spritam, produced for epileptic seizure using the technology ZipDose® patented by Aprecia Pharmaceuticals (USA), and the first FDA-approved (in 2015) printed drug. Other challenges to the success of pharmaceutical AM are: concerns regarding quality control in printing medications, regulatory issues, and scaling-up AM. An additional barrier is the limited number of biodegradable, biocompatible, and physically and chemically stable feedstocks suited for AM processes (including VP), such as nontoxic solvent for BJ, active pharmaceutical ingredient and excipients that are laser curable in PBF, and thermally stable polymers other than TP polymers for ME. Formulators of new AM materials for pharmaceutical uses should read the document "Technical Considerations for Additive Manufactured Medical Devices" published by the U.S. Food and Drug Administration (FDA 2017). Experts argue that AM could expedite early-phase drug development and production of formulations on demand (with excellent dose flexibility at low cost), and accelerate formulation development through rapid product iterations for testing (Trenfield et al. 2018).

Likely uses of AM will comprise: implantable medical devices, new in vitro technological tools required for studies in advanced diagnostics, pharmaceutical development, and tissue physiology (Puppi and Chiellini 2020).

Future and/or improved SPs for medical applications of AM may likely be developed within the following families, with their relative AM process in parentheses, and SE-AM standing for *solution extrusion AM*: alginate (SE-AM, SL), cellulose (SE-AM, ME, SL), chitosan (SE-AM, SL), collagen (SE-AM, BJ), fibrin (BP), gelatin (SE-AM, SL), hyaluronic acid (SE-AM, SL), and PHA (PBF, ME, SE-AM) (Puppi and Chiellini 2020). In *solution* (or *slurry*) *extrusion* a solution or slurry is continuously extruded according to a specific pattern onto a construction platform (Mota et al. 2015).

In the case of AM relative to cells, issues to be addressed are cytotoxicity due to the reactive nature of some AM polymers, and interaction among living cells, feedstocks' mechanical properties, and cell growth factors. Bioplotting (extrusion-based cell printing) has been developed for cases in which living cells and growth factors are processed with HGs, but bioplotting multimaterials requires accurate tuning of processing parameters (Deshmukh et al. 2020).

A cutting-edge application of AM involving living cells possibly more utilized in the near future is *integrated organ-on-a-chip*, a device that is an artificial living organ mimicking the complex and physiological responses of a real organ in order to test drugs (Yi et al. 2017).

SPs for AM will be continued to be pursued for biomedical applications (Singh and Thomas 2019) due to their biocompatibility.

A promising and valuable medical area for which developing new materials will continue to be printing is personalized organs: they will considerably reduce the risk of graft rejection, and the need to take anti-rejection medicines for life that weaken the patients' immune system, and possibly reduce their life after transplant compared to what it could be (Yao et al. 2016).

New AM feedstocks for the medical sector will keep targeting educational and training tools, preoperative planning, surgical instruments, and BP.

AM polymers reinforced with graphene and carbon fibers, and electroconductive polymers for devices with connectivity allowing remote and in vivo monitoring will be developed for health care. Emerging application fields within TE and regenerative medicine for BP are: (a) printing organ-on-a-chip for medical research; (b) replicating diseased tissues, to investigate the disease process; (c) regenerating human tissues for transplant; and (d) conducting stem-cell research (Lyons and Devine 2019).

Several customized SL and SE-AM methods are being industrially implemented and some of them will probably be integrated in the near future in approved diagnostic and therapeutic procedures.

Conducive to progress in biomedical AM will be also advances in automation, 4DP, nano/ microfabrication, and digital control of the design and fabrication processes.

2.16.7.5 Architecture, Buildings, and Construction

Remarkably, AM spans from the micron scale of biomedical applications to the meter scale of buildings. Obviously, the market of architecture, buildings, and construction offers enormous business potential to feedstock suppliers because of the involved quantities of materials. Detailed and recent reviews of use and future of AM in the construction industry are offered by Craveiro et al. (2019). Wang et al. (2018) discussed the potential contribution of AM to the advancement of construction processes by employing mechanization and automation, and included a systematic treview of the relevant technical literature.

Currently, BJ, ME, PBF, and DED are applied in the architecture and construction industry, and process cementitious (the most), polymeric (PP, polyester, PU, flax, cork, etc.), and metallic (the least) feedstocks in three types of equipment: gantry (mostly), robotic, and other. Examples of firms already applying AM to building and constructions are the following:

- Mighty Buildings (USA) builds homes by combining advanced materials, patented AM technology, and robotic automation.
- Winsun printed parts for office buildings later shipped and assembled in Dubai, cutting labor costs by 50–80% and construction waste by 30–60% (MacRae 2016).
- D-Shape® (Italy) builds artificial coral reef for fish for underwater ecological park, coffee tables, and houses.
- Aectual (The Netherlands) is pioneering large architectural components such as floors, wall panels, stairs, and facades that are tailored to specific projects, printed in a factory, and delivered to the construction site for installation, enabling a faster, more affordable and efficient building process than the conventional one (Autodesk University 2020).
- Apis Cor (Russia) built in Moscow a 400 square foot home from scratch in 24 hours costing USD 10,000, and entirely on site using only a mobile 3D printer (Sakin and Kiroglu 2017).
- Holcim (Switzerland), the world's largest producer of cement, formed a joint venture with CDC Group (UK) that can print a house for a less than USD 12,000 in 12 hours.

AM polymers that are suited for aesthetic purposes and combined with strength- and rigidity-enhancing ingredients meet structural requirements of construction (Camacho et al. 2018). Moreover, since polymers are lightweight, in order to be transported to the construction site they do not need large machines that would in turn require wide streets and large areas. In Amsterdam, entire houses were printed out of SPs (Allen 2018). Construction items printed out of SPs are the support legs, seating areas, and counter space of the pavilions installed at the 2016 Design Miami Exhibition: they were built with 100% biobased and biodegradable bamboo fiber and PLA (Branch Technology n.d.). Italian printer manufacturer WASP (WASP n.d.) has integrated AM, construction, and sustainable materials by developing Crane WASP, a modular machine that prints houses from recyclable and reusable materials, such as soil from construction sites, straw, natural waste (rice fibers) from rice production, and compost. Crane WASP comprises a main printer unit that can be assembled in different configurations, depending on the printing area and dimensions of the architectural structure to build, and can reach very large dimensions by adding ties and printing arms to the module. The print area of the single module is 6.6 m in diameter and 3 m in height.

Demonstrating to be ready for the future, AM has successfully integrated itself with *building information modeling* (BIM), an emergent approach to design, construction, and management of facilities that includes a digital representation of physical and functional characteristics of a facility, its geometry information, material properties, spatial relationships, and manufacture information (Eastman et al. 2011).

The future of AM in construction is promising, because, compared to traditional building processes (prefabrication and casting on site), AM offers the following benefits (Hager et al. 2016): (a) it accelerates building schedules by integrating automation in the building process and reducing labor; (b) it lacks the cost associated with a change in geometry that instead penalizes those processes; (c) it provides more freedom in architectural design and fabrication of complicated shapes (Teizer et al. 2016), f.e. internal passageways, undercuts, and other features problematic or even impossible to manufacture with conventional techniques (Bogue 2013); (d) it cuts the number of injuries and fatalities onsite by using printers to take away the most hazardous and dangerous work from personnel; (e) it reduces wet construction processes, hence generating less material waste and dust; (f) it limits material transportation and storage on site; (g) it utilizes "green" feedstock, such as biobased, recycled, and waste materials. However, Krimi et al. (2017) cautioned that the comparison in overall cost (labor, time, materials, equipment, etc.) among AM and other building processes should not be made in general, but between a specific AM method and a specific non-AM building process under specific operating conditions.

SPs for AM may partake in the near future of construction. A goal of the BARBARA project mentioned in Section 2.16.7.4 is to formulate sustainable plastics for AM that withstand 140°C, and produce components for construction such as molds to build truss joints (Anon. 2017)

According to Faludi et al. (2017), penetration of AM in constructions may be slowed down by: (a) the fact that the building industry is reluctant to change; (b) lack of incentives that may be politically unpopular to issue, because of the number of jobs that would be lost by including automation in construction.

Another clue about future expansion of AM in construction is that new forms of AM are being developed. Recently, Asprone et al. (2018) illustrated a novel method for fabricating reinforced concrete (RC) members based on concrete AM, and consisting in partitioning a RC member into different concrete segments printed separately, and then assembled into a unique element along with the steel reinforcement system. This method should facilitate the production of free-form structurally optimized RC elements, save concrete material, and build lighter structures.

Potential applications of AM in construction comprise: selective utilization of multi-materials, namely high-performance feedstocks selected only for areas where they are required, in situ repair in locations difficult or dangerous for people to access, buildings for disaster relief in areas with limited construction workforce and material resources, structural and non-structural elements with optimized topologies, and customized elements of high value. Expansion of AM will be accelerated by new materials, new processes, faster printing (also to avoid solidifying prematurely the feedstock and clogging the printer), quality assurance, and availability of wide-ranging databases of mechanical properties (Camacho et al. 2018).

Furthermore, AM for construction has to address the following challenges (Wu et al. 2016b; Craveiro et al. 2019):

- *Printing in large scale.* This is already being addressed in some ways (Liang and Liang 2014): large printers (f.e. Winsun has employed a printer featuring dimension of 150 × 10 × 6.6 m), modular printers such as Crane WASP, and, in case of high-rise building, by printing structural components piece-by-piece and then assembling them together as a real-scale building.

- *Materials.* They have to be adequately strong and stiff, harden quickly, show limited shrinking, and result in a product stable in properties and shape. Feedstocks for extrusion-based processes must easily extrude, possess strong interlayer adhesion, and form a strongly cohesive structure.
- *Mass customization.* The construction industry has a low degree of customization, and AM might at the same time meet more affordably and promote the demand for higher degree of customization. We see potential in customization that targets luxury homes by realizing original architectural designs fulfilling customers' wishes.
- *Cost.* For specific projects (houses and office buildings), firms employing AM have reported lower cost of fabrication vs. conventional processes. Indeed, fabrication cost can be potentially reduced through AM by cutting labor cost via automation, but AM feedstocks must be competitive in price with current ones. Moreover, additional information has to be gathered to assess the cost of a printed construction not only in its building phase but also over its entire life cycle, but these data lack due to the short history of AM versions employed for buildings and constructions.
- *Standards.* Establishing standards for materials testing and characterization, and construction. Developing building codes and standards, and releasing structural design requirements.
- *Surface finishing.* Improving it to eliminate rough appearance and attract more residential customers.
- *Real-time feedback* during printing to correct geometry and appearance.

FURTHER READINGS

Journal papers with additional information on AM and its feedstocks are the recent and comprehensive review by Bourell et al. (2017), Ligon et al. (2017), Ngo et al. (2018), and Molitch-Hou (2018). Recent books on AM processes and relative feedstocks that we recommend are:

Bandyopadhyay, A., Bose, S. 2016. *Additive Manufacturing*. Boca Raton: CRC Press.
Chua, K., Leong, K. F. 2017. *3D Printing and Additive Manufacturing. Principles and Applications.* Singapore: World Scientific Publishing Co.
Gebhardt, A., Hötter, J.-S. 2016. *Additive Manufacturing: 3D Printing for Prototyping and Manufacturing.* Cincinnati: Hanser Publications.
Gibson, I., D. Rosen, B. Stucker. 2015. *Additive Manufacturing Technologies.* New York: Springer.
Milewski, J. O. 2017. *Additive Manufacturing of Metals.* Cham: Springer.
Ovsianikov, A., J. Yoo, V. Mironov ed. 2018. *3D Printing and Biofabrication.* Cham: Springer.
Ozbolat, I. T. 2016. *3D Bioprinting: Fundamentals, Principles and Applications.* Academic Press.
Redwood, B., Schöffer, F., Garret, B. 2017. *The 3D Printing Handbook. Technologies, design and applications.* Amsterdam: 3D Hubs B.V.
Schmid, M. 2018. *Laser Sintering with Plastics Technology, Processes, and Materials*, Munich: Hanser Verlag.
Shimamura, K., Kirihara, S., Akedo, J., Ohji, T., Naito, M. ed. 2016. *Additive Manufacturing and Strategic Technologies in Advanced Ceramics.* Hoboken: John Wiley & Sons.

REFERENCES

3DomFuel 2016, personal communication.
Abdulhameed, O., Al-Ahmari, A., Ameen, W., Mian, S. H. 2019. Additive manufacturing: Challenges, trends, and applications. *Adv. Mech. Eng.* doi:10.1177/1687814018822880.
Agrawal R., Vinodh, S. 2019. State of art review on sustainable additive manufacturing. *Rapid .Prototyp. J.* 25 (6): 1045–1060. doi 10.1108/RPJ-04-2018-0085.
Ahn, S.-H., Montero, M., Odell, D., et al., 2002. Anisotropic material properties of fused deposition modeling ABS. *Rapid Prototyp. J.* 8 (4): 248–257.
Aimar, A., Palermo, A., Innocenti, B. 2019. The role of 3D printing in medical applications: A state of the art. *J. Healthcare Eng.* Article ID 5340616.

Ajoku, U., Saleh, N., Hopkinson, N., Hague, R., Erasenthiran, P. 2006. Investigating mechanical anisotropy and end-of-vector effect in laser-sintered nylon parts. *Pro. Inst. Mech. Eng. B: J. Eng. Manuf.* 220 (7): 1077–1086.

Al Mousawi, A., Dumur, F., Garra, P., et al. 2017. Carbazole scaffold based photoinitiator/photoredox catalysts: toward new high performance photoinitiating systems and application in LED projector 3D printing resins. *Macromolecules*, 50:2747–2758.

Al Mousawi, A., Garra, P., Schmitt, M., et al., 2018. 3-hydroxyflavone and *N*-phenylglycine in high performance photoinitiating systems for 3D printing and photocomposites synthesis. *Macromolecules* 51:4633–4641.

Ali, F., Chowdary, B. V., Maharaj, J., 2014. Influence of some process parameters on build time, material consumption, and surface roughness of FDM processed parts: inferences based on the Taguchi design of experiments. *Proceedings of The 2014 IAJC/ISAM Joint International Conference* ISBN 978-1-60643-379-9.

Allen, M., 2018. 3D printed bioplastic: the future of construction? http://www.allthings.bio/3d-printed-bioplastic-the-future-of-construction/ (accessed January 16, 2021).

Amado, A., Schmid, M., Levy, G., Wegener, K., 2011. Advances in SLS powder characterization. *Solid Free Fabr. Symp. Proc.* Austin, TX. pp. 438–452.

Amado-Becker, A., Ramos-Grez, J., Yanez, M. J., Vargas, Y., Gaete, L. 2008. Elastic tensor stiffness coefficients for SLS Nylon 12 under different degrees of densification as measured by ultrasonic technique. *Rapid Prototyp. J.* 14 (5): 260–270.

Anon. 2017. *BARBARA Project.* https://www.barbaraproject.eu/ (accessed January 16, 2021).

Anon. 2018. *3D Metal Printing: How Classic Foundries Remain Competitive.* https://www.voxeljet.com/company/news/3d-metal-printing-is-gaining-ground/ (accessed February 14, 2020).

Anon. 2020. *Bio-Based Materials for Aircraft.* Published: 26 April 2018 https://ec.europa.eu/research/infocentre/article_en.cfm?id=/research/headlines/news/article_18_04_26_en.html%3Finfocentre&amp%3Bitem=Infocentre&amp%3Bartid=48256 (accessed July 28, 2020).

Arburg. n.d. https://www.arburg.com/en/products-and-services/additive-manufacturing/freeformer-system/ (accessed January16, 2020).

Arcam. n.d. Just Add. http://www.arcam.com/solutions/aerospace-ebm/ (accessed January16, 2020).

Arkema. 2020. *Rilsan® Polyamide 11 Resin.* https://www.arkema.com/en/products/product-finder/product-viewer/Rilsan-Polyamide-11-Resin/ (accessed April 26, 2020).

Arkema. n.d. *Kepstan® PEKK Resins for Extremely Demanding Applications.* https://www.extrememterials-arkema.com/en/product-families/kepstan-pekk-polymer-range/ (accessed May 25, 2020).

Asnafi, N., Rajalampi, J., Aspenberg, D. 2019. Design and validation of 3D-printed tools for stamping of DP600. *IOP Conf. Series: Materials Science and Engineering* 651 012010. doi:10.1088/1757-899X/651/1/012010.

Aspar, G., Goubault B., Lebaigue O., et al. 2017. 3D printing as a new packaging approach for MEMS and electronic devices. *2017 IEEE 67th Electronic Components and Technology Conference (ECTC).*

Asprone D., Auricchio, F., Menna, C., Mercuri, V. 2018. 3D printing of reinforced concrete elements: Technology and design approach. *Construct. Build. Mater.* 165:218–231.

ASTM D638. *Standard Test Method for Tensile Properties of Plastics.* West Conshohocken, PA, USA: ASTM International.

Autodesk University. 2020. *Large-Scale 3D Printing for Architecture – Hedwig Heinsman.* https://www.autodesk.com/autodesk-university/content/large-scale-3d-printing-for-architecture (accessed July 4, 2020).

Aysha, S., Varghese, B., Baby, A., 2014. A review on functionally graded materials. *Int. J. Eng. Sci. (IJES)* 3 (6): 90–101.

Bagheri, A., Jin, J. 2019. Photopolymerization in 3D. *ACS Appl. Polym. Mater.* 1 (4): 593–611.

Bagsik, A., Schöppner, V. 2011. *Mechanical Properties of Fused Deposition Modeling Part Manufactured With Ultem™ 9085.* Boston: ANTEC.

Bagsik, A., Schöppner, V., Klemp, E., 2010. *FDM Part Quality Manufactured with Ultem™ 9085, "Polymeric Materials 2010" Halle (Saale).* https://www.semanticscholar.org/paper/FDM-Part-Quality-Manufactured-with-Ultem-*-9085-Bagsik-Sch%C3%B6ppner/63ec4f3fd6e39600080e0a14caaa7754cc653632 (accessed July 4, 2019).

Bai, J., Zhang, B., Song, J., et al. 2016. The effect of processing conditions on the mechanical properties of polyethylene produced by selective lasers sintering. *Polym Test.* 52:89–93.

Bai, J., Goodridge, R. D., Hague, R. J. M., Okamoto, M. 2017. Processing and characterization of a polylactic acid/nanoclay composite for laser sintering. *Polym. Compos.* 38:2570–2576. doi:10.1002/pc.23848.

Bai, J., Goodridge, R. D., Hague, R. J. M., Song, M. 2013. Improving the mechanical properties of laser-sintered polyamide 12 through incorporation of carbon nanotubes. *Polym. Eng. Sci.* 53 (9): 1937–1946

Bakarich, S. E., Gorkin, III R., in het Panhuis, M., Spinks, G. 2015. 4D printing with mechanically robust, thermally actuating hydrogels. *Macromol. Rapid Commun.* 36:1211–1217.

Baldacchini, T., 2015. *Three-Dimensional Microfabrication Using Two-Photon Polymerization: Fundamentals, Technology, and Applications.* New York: William Andrew.

Balla, V. K., Bose, S., Bandyopadhyay, A. 2008. Processing of bulk alumina ceramics using laser engineered net shaping. *Int. J. Appl. Ceram. Technol.* 5:234–242.

Bar-Cohen, Y. 2010. Electroactive polymer as actuators. In *Advances in piezoelectric materials* ed.K. Uchino, 287–317. Woodhead Publishing.

Bar-Cohen, Yoseph. 2005. *Biomimetics Biologically Inspired Technologies.* Boca Raton: CRC Press.

Bargardi, F. L., Le Ferrand, H., Libanori, R., Studart, A. R. 2016. Bio-inspired self-shaping ceramics. *Nat. Commun.* 7, 13912.

Barrett, A. 2019. *New Bio-Based Thermoplastic Adhesives.* August 10, 2019 https://bioplasticsnews.com/201 9/08/10/new-bio-based-thermoplastic-adhesive (accessed January 16, 2020).

Bartolo, P. J. 2011. *Stereolithography: Materials, Processes and Applications.* New York: Springer.

Bartolo, P. J., Gaspar, J. 2008. Metal filled resin for stereolithography metal part. *CIRP Ann. Manuf. Technol.* 57:235–238. 10.1016/j.cirp.2008.03.124.

Bartolo, P. J., Mitchell, G., 2003. Stereo-thermal-lithography: a new principle for rapid prototyping. *Rapid Prototyp. J.* 9:150–156.

Baumers, M., Tuck, C., Hague, R. 2015. Selective heat sintering versus laser sintering: comparison of deposition rate, process, energy consumption, and cost performance. *2015 Solid Freeform Symposium*, University of Texas.

BEA. 2014. *Use of Commodities by Industry Valued at Producers' Prices, 2007 (Dataset), Input-Output Accounts Data.* Bureau of Economic Analysis, www.bea.gov/industry/io_annual.htm (accessed May 28, 2020).

Bens, A., Seitz, H., Bermes, et al. 2007. Non-toxic flexible photopolymers for medical stereolithography technology. *Rapid Prototyp. J.* 13 (1): 38–47. doi:10.1108/13552540710719208.

Berman, B., 2012. 3-D printing: the new industrial revolution. *Business Horiz.* 55 (2): 155–162.

Bernard, S. A., Balla, V. K., Bose, S., Bandyopadhyay, A. 2010. Direct laser processing of bulk lead zirconate titanate ceramics. *Mater. Sci. Eng. B Solid-State Mater. Adv. Technol.* 172:85–88.

Berretta, S., Evans, K. E., Ghita, O. 2016. Predicting processing parameters in high temperature laser sintering (HT-LS) from powder properties. *Mater. Des.* 105:301–314.

Better Future Factory. n.d. https://www.betterfuturefactory.com/ (accessed May 31, 2019).

Bhardwaj, A., Zou, N., Pei, Z. J. 2018. Additive manufacturing for civil infrastructure design and construction: current state and gaps. *ASME 2018 13th International Manufacturing Science and Engineering Conference.* doi:10.1115/MSEC2018–6688.

Bhasin, V., Bodla, M. R. 2014. *Impact of 3D Printing on Global Supply Chains by 2020.* PhD diss. Massachusetts Institute of Technology. https://ctl.mit.edu/sites/ctl.mit.edu/files/library/public/2014 ExecSummary-BhasinBodla.pdf (accessed June 13, 2020).

Bhavar, V., Kattire, P., Thakare, S., et al. 2017. A review on functionally gradient materials (fgms) and their applications. *IOP Conf. Series: Materials Science and Engineering* 229:012021.

Bheda, Hemant 2018. Reinforced fused-deposition modeling. US Pat 10011073B2 filed Dec. 12, 2016, issued July 3, 2018.

Bikas, H., Stavropoulos, P., Chryssolouris, G., 2016. Additive manufacturing methods and modelling approaches: a critical review. *Int. J. Adv. Manuf. Technol.* 83:389–405.

Bjorn, A., MacLean, H., 2003. A comparison of US and Canadian industry environmental performance using EIO-LCA models. *InLCA/LCM Conference*, Seattle, 27 September.

Bodaghi, M., Damanpack, A., Liao, W. 2016. Self-expanding/shrinking structures by 4D printing. *Smart Mater. Struct.* 25:105034.

Bogue, R., 2013. 3D printing: the dawn of a new era in manufacturing. *Assem. Autom.* 33 (4): 307–311.

Bourell, D. L., Beaman, J. J., Leu, M. C., Rosen, D. W. 2009a. A brief history of additive manufacturing and the 2009 roadmap for additive manufacturing: looking back and looking ahead. *RapidTech 2009: US-Turkey Workshop on Rapid Technologies*, 2009, p. 5–11.

Bourell, D. L., Rosen, D. W., Leu, M. C., et al. 2009b. The roadmap for additive manufacturing and its impact. doi:10.1089/3dp.2013.0002.

Bourell, D., Kruth, J. P., Leu, M. et al. 2017. Materials for additive manufacturing. *CIRP Annals - Manufacturing Technology* 66:659–681.

Branch Technology n.d. *Design Miami Pavilions.* https://www.branch.technology/projects-1/2017/6/9/shop (accessed August 28, 2020).

Brenken, B., Barocio, E., Favaloro, A., Kunc, V., Byron Pipes, R., 2018. Fused filament fabrication of fiber-reinforced polymers: a review. *Addit. Manuf.* 21:1–16.

Brindha, J., Privita Edwina, G., Rajesh, P. K., Rani, P. 2016. Influence of rheological properties of protein bioinks on printability: a simulation and validation study. *International Conference on Advances in Bioprocess Engineering and Technology* 2016 (ICABET 2016).

Camacho, D. D., Clayton, P., J. O'Brien, W. J. 2018. Applications of additive manufacturing in the construction industry – A forward-looking review. *Automat. Construct.* 89:110–119.

Carbonell, C., Braunschweig, A. B. 2017. Toward 4D nanoprinting with tip-induced organic surface reactions. *Acc. Chem. Res.* 50 (2): 190–198.

Carlota, V., 2019. *The Faces of Additive Manufacturing: Hanan Gothait.* Posted on May 13, 2019. https://www.3dnatives.com/en/hanan-gothait-130520194/ (accessed April 26, 2020).

Caulfield, B., McHugh, P. E., Lohfeld, S. 2007. Dependence of mechanical properties of polyamide components on build parameters in the SLS process. *J. Mater. Process. Technol.* 182 (1-3): 477–488.

Cawley, J. D., Liu, Z., Mou, J., Heuer, A. H. 1998. Materials issues in laminated object manufacturing of powder-based systems. *SFF Symp. Proc.* 503–510.

Chacon, J. M., Caminero, M. A., García-Plaza, E., Nunez, P. J. 2017. Additive manufacturing of PLA structures using fused deposition modelling: effect of process parameters on mechanical properties and their optimal selection. *Mater. Des.* 124:143–157.

Chang, T. C., Faison, E. I. 2001. Shrinkage behavior and optimization of injection molded parts studies by the Taguchi method. *Polym. Eng. Sci.* 41:703–710.

Chatham, C. A., Long, T. E., Williams, C. B. 2019. A review of the process physics and material screening methods for polymer powder bed fusion additive manufacturing. *Prog. Polym. Sci.* 93:68–95.

Chen, L., He, Y., Yang, Y. 2017. The research status and development trend of additive manufacturing technology. *Int. J. Adv. Manuf. Technol.* 89:3651–3660. doi:10.1007/s00170-016-9335-4.

Childs, T. H. C., Berzins, M., Ryder, G. R., Tontowi, A. 1999. Selective laser sintering of an amorphous polymer – simulations and experiments. *Proc. Inst. Mech. Eng. B: J. Eng. Manuf.* 213 (4): 333–349. doi:10.1243/0954405991516822.

Chinga-Carrasco, G., Ehman, N. V., Pettersson, J., et al. 2018. Pulping and pretreatment affect the characteristics of bagasse inks for three-dimensional printing. *ACS Sust. Chem. Eng.* 6 (3): 4068–4075.

Chua, C. K., Chou, S. M., T. S. Wong. 1998. A study of the state-of-the-art rapid prototyping technologies. *Int. J. Adv. Manuf. Technol.* 14:146–152.

Chung, J. H. Y., Naficy, S., Yue, Z., et al. 2013. Bio-ink properties and printability for extrusion printing living cells. *Biomater. Sci.* 7:763.

Cima, M., Sachs, E., Fan, T., et al. 1995. US patent 5,387,380 Three-dimensional printing techniques filed Jun. 5, 1992 granted Feb. 7, 1995.

Cohee, D., Paesano, A., Palmese, G. R. 2003. Carbon-fiber reinforced thermoplastic materials for rigidizable space systems. *J. Thermoplast. Compos. Mater.* 16:139.

Coleman, J. N., Khan, U., Gun'ko, Y. K., 2006. Mechanical reinforcement of polymers using carbon nanotubes. *Adv. Mater.* 18 (6): 689–706.

Colorado, H. A., Gutiérrez Velásquez, E. I., Monteiro, S. N. 2020. Sustainability of additive manufacturing: the circular economy of materials and environmental perspectives. *J. Mater. Res. Technol.* 9 (4): 8221–8234.

Compton, B. G., Lewis, J. A. 2014. 3D-printing of lightweight cellular composites. *Adv. Mater.* 26:5930–5935.

Contour Crafting. 2017. *Introducing Contour Crafting Technology.* http://contourcrafting.com/ (accessed June 13, 2020).

Cooke, W., Tomlinson, R. A., Burguete, R. et al. 2011. Anisotropy, homogeneity and ageing in an SLS polymer. *Rapid Prototyp. J.* 1 7(4): 269–279.

Craveiro, F., Duarte, J. P., Bartolo, H., Bartolo, P. J. 2019. Additive manufacturing as an enabling technology for digital construction: a perspective on Construction 4.0. *Automa. Construct.* 103:251–267.

Credi, C., Fiorese, A., Tironi, M., et al. 2016. 3D printing of cantilever-type microstructures by stereolithography of ferromagnetic photopolymers. *ACS Appl. Mater. Interfaces* 8:26332–26342.

Crump, S. S. 1992. Apparatus and method for creating three-dimensional objects. US Patent 5,121,329, filed October 30, 1989, and issued June 9, 1992.

Cui, X., Ouyang, S., Yu, Z., Wang, C., Huang, Y., 2003. A study on green tapes for LOM with water-based tape casting processing. *Materials Lett.* 57 (7): 1300–1304.

Dalla Vedova, F., De Wilde, D., Sempriomoschnig, C. et al. 2014. The solar sail materials projects: results of activities. In *Advances in Solar Sailing*, ed.M. MacDonald, 509–524. Berlin: Springer.

Daneshmand, S., Aghanajafi, C. 2012. Description and modeling of the additive manufacturing technology for aerodynamic coefficients measurement. *J. Mech. Eng.* 58 (2): 125–133.

Dankar, I., Haddarah, A., Omar, F. E. L., et al. 2018. 3D printing technology: The new era for food customization and elaboration. *Trends Food Sci. Technol.* 75:231–242.

De Leon, A. C., Chen, Q., Palaganas, et al. 2016. High performance polymer nanocomposites for additive manufacturing applications. *React. Funct. Polym.* 103:141–155.

Deckard, C. R. 1997. Apparatus for producing parts by selective sintering. US5,597,589, filed May 31, and 1994 issued Jan. 28, 1997.

Derby, B., Reis, N. 2003. Inkjet printing of highly loaded particulate suspensions. *MRS Bull.* 28 (11): 815–818.

Deshmukh, K., Muzaffar, A., Kovarik, T., et al. 2020. AM, Fundamentals and applications of 3D and 4D printing of polymers: Challenges in polymer processing and prospects of future research. In *3D and 4D Printing of Polymer Nanocomposite Materials*, ed. K. K. Sadasivuni, K. Deshmukh, M. Al Ali Al Maadeed, 527–560. Amsterdam: Elsevier.

DeSimone, J. M., Ermoshkin, A., Ermoshkin, N., Samulski, E. T. 2018. Continuous liquid interphase printing US Pat. 10144181, filed November 10, 2015, and issued December 4, 2018.

DHL. 2020. *Keeping the Aircraft flying – Aircraft On Ground (AOG) Services.* http://www.africa.dhl.com/en/logistics/industry_sector_solutions/aerospace_logistics/aircraft_on_ground.html (accessed November 15, 2020).

Dickson, A. N., Barry, J. N., McDonnell, K. A., Dowling, D. P. 2017. Fabrication of continuous carbon, glass and Kevlar fiber reinforced polymer composites using additive manufacturing. *Addit. Manuf.* 16:146–152.

Dilberoglu, U. M., Gharehpapagh B., Yaman U., Dolen M. 2017. The role of additive manufacturing in the era of Industry 4.0. *Proc. Manuf.* 11:545–554.

Dinesh Kumar, S., Ravichandran, M., Alagarsamy, S. V., et al. 2020. Processing and properties of carbon nanotube reinforced composites: a review. *Mater. Today: Proc.* doi:10.1016/j.matpr.2020.02.006.

Ding, R., Du, Y., Goncalves, R. B., et al. 2019. Sustainable near UV-curable acrylates based on natural phenolics for stereolithography 3D printing. *Polym. Chem.* 10:1067–1077.

Drummer, D., Rietzel, D., Kühnlein, F. 2010. Development of a characterization approach for the sintering behavior of new thermoplastics for selective laser sintering. *Phys. Proc.* 5:533–542.

D-Shape® n.d. www.d-shape.com (accessed May 25, 2020).

Duan, B., Wang, M., Zhou, W. Y., et al. 2010. Three-dimensional nanocomposite scaffolds fabricated via selective laser sintering for bone tissue engineering. *Acta Biomater.* 6:4495–4505.

Dufour, P. 1993. State-of-the-art and trends in radiation curing. In *Radiation Curing in Polymer Science and Technology*, ed.J. P. Fourassier, J. F. Rabek, vol. 1. London: Elsevier Applied Science.

Durfee, W. K., Iaizzo, P. A. 2019. Medical applications of 3D printing. In *Engineering in Medicine*, ed. P. A. Iaizzo, 527–543. Academic Press, Elsevier.

Durgun, I., Ertan, R., 2014. Experimental investigation of FDM process for improvement of mechanical properties and production cost. *Rapid Prototyp. J.* 20 (3): 228–235.

Eastman, C., P. Teicholz, R. Sacks, K. Liston, 2011. *BIM Handbook: A Guide to Building Information Modeling for Owners, Managers, Designers, Engineers and Contractors*, Hoboken: John Wiley & Sons.

El Moumen, A., Tarfaoui, M., Lafdi, K. 2019. Additive manufacturing of polymer composites: processing and modeling approaches. *Compos. B* 171:166–182.

Entropy Resins. 2020. *Epoxy Resins.* https://entropyresins.com/product/super-sap-one/ (accessed January 7, 2020).

EnvisionTEC. 2017. https://envisiontec.com/wp-content/uploads/2016/09/2017-SLCOM1.pdf (accessed March 6, 2020).

ESA. 2017. *3D Printing CubeSat Bodies for Cheaper, Faster Missions.* http://www.esa.int/Our_Activities/Space_Engineering_Technology/3D_printing_CubeSat_bodies_for_cheaper_faster_missions.

Espera, A., Valino, A., Palaganas, J., et al. 2019. 3D printing of a robust polyamide-12-carbon black composite via selective laser sintering: thermal and electrical conductivity. *Macromol. Mater. Eng.* 1800718. doi:10.1002/mame.201800718.

Faludi, J., Bayley, C., Bhogal, S., Iribarne, M. 2015a. Comparing environmental impacts of additive manufacturing vs traditional machining via life-cycle assessment. *Rapid Prototyp. J.* 21 (1): 14–33.

Faludi, J., Cline-Thomas, N., Agrawala, S. 2017. 3D printing and its environmental implications. In *Next Production Revolution – Implications for Governments and Business*, 171–213. OECD Library. doi:1 0.1787/9789264271036-en.

Faludi, J., Ganeriwala, R., Kelly, B., et al. 2014. Sustainability of 3D printing vs. machining: Do machine type & size matter? *Proceedings of the 2014 EcoBalance Conference*, Tsukuba, Japan.

Faludi, J., Hu, Z., Alrashed, S., et al. 2015b. Does material choice drive sustainability of 3D printing? *In. J. Mech. Aerosp. Indus. Mech. Manuf. Eng.* 9 (2): 216–223.

Faludi, J., Lepech, M. D., Loisos, G. 2012. Using life cycle assessment methods to guide architectural decision-making for sustainable prefabricated modular buildings. *J. Green Build.* 7:151–170.

Farzana, H., Mehdi, H., Masami, O., Russell, E. G. 2006. Polymer-matrix nanocomposites, processing, manufacturing, and application: an overview. *J. Compos. Mater.* 40:1511–1575.

FDA 2017. *Technical Considerations for Additive Manufactured Medical Devices Guidance for Industry and Food and Drug Administration Staff.* Issued on December 5, 2017 Available at https://www.fda.gov/ media/97633/download.

Feilden, E. 2017. *Additive Manufacturing of Ceramics and Ceramic Composites via Robocasting*, Ph.D. diss., Imperial College London. https://spiral.imperial.ac.uk/handle/10044/1/55940.

Feng, L., Wang, Y., Wei, Q. 2019. PA12 powder recycled from SLS for FDM. *Polymers* 11 (4): 727.

Feygin, M. 1994. Apparatus and method for forming an integral object from laminations. US Patent 5,354,414 filed April 4, 1991, granted Nov. 11, 1994.

Feygin, M., Shkolnik, A., Diamond, M. N., Dvorskiy, E. 1998. Laminated object manufacturing system. US Patent 5730817A, filed April 22, 1996, and issued March 24, 1998.

Filabot. 2020. *Machines.* https://www.filabot.com/collections/filabot-core (accessed August 28, 2020).

Fitz-Gerald, J., Chrisey, D., Pique, A., et al. 2000. Matrix assisted pulsed laser evaporation direct write (MAPLE DW): a new method to rapidly prototype active and passive electronic circuit elements. *MRS Proc.* 625:99. doi:10.1557/PROC-625-99.

Flank, S. 2017. Legal issues in IP protection for additive manufacturing. *Tex. A&M J. Prop. L.* 1(2017–2018). HeinOnline.

Fonseca Coelho, A. W., da Silva Moreira Thiré, R. M., Araujo, A. C. 2019. Manufacturing of gypsum–sisal fiber composites using binder jetting.*Add. Manuf.* 29:100789.

Ford, S., Despeisse, M. 2016. Additive manufacturing and sustainability: an exploratory study of the advantages and challenges. *J. Clean. Prod.* 137:1573–1587.

Formlabs. 2020. *The Ultimate Guide to Stereolithography (SLA) 3D Printing.* https://formlabs.com/blog/ ultimate-guide-to-stereolithography-sla-3d-printing/ (accessed January 16, 2020).

Forney, B. S. 2013. *Controlled Polymer Nanostructure and Properties Through Photopolymerization in Lyotropic Liquid Crystal Template*, Ph. D. Diss., University of Iowa. https://ir.uiowa.edu/cgi/ viewcontent.cgi?article=4623&context=etd (accessed May 31, 2020).

Frazier, W. E. 2014. Metal additive manufacturing: a review. *J. Mater. Eng. Perf.* 23 (6): 1917–1928.

Frenkel, J. 1945. Viscous flow of crystalline bodies under the action of surface tension. *J. Phys.* 9 (5): 358–391.

Galantucci, L. M., Guerra, M. G., Dassisti, M., Lavecchia, F. 2019. Additive manufacturing: new trends in the 4th industrial revolution. In *Proceedings of the 4th International Conference on the Industry 4.0 Model for Advanced Manufacturing.*, ed.L. Monostori, V. Majstorovic, S. Hu, D. Djurdjanovic, AMP 2019. Lecture Notes in Mechanical Engineering. Cham: Springer.

Gao, W., Zhang, Y., Ramanujan, D. 2015. The status, challenges, and future of additive manufacturing in engineering. *Comput.-Aided Des.* 69:65–89.

Gardner, J. W., Varadan V. K., Awadelkarim O. O. 2001. *Microsensors, MEMS, and Smart Devices.* Chichester: Wiley.

Gatenholm, P., Martinez, H., Karabulut, E., et al. 2016. Development of nanocellulose-based bioinks for 3D bioprinting of soft tissue. In *3D Printing and Biofabrication*, ed.A. Ovsianikov, J. Yoo, V. Mironov, 1–23. Cham: Springer.

Ge, Q., Dunn, C. K., Qi, H. J., Dunn, M. L. 2014. Active origami by 4D printing. *Smart Mater. Struct.* 23(9) IOP Publishing.

Ge, Q., Qi, H. J., Dunn, M. L. 2013. Active materials by four-dimension printing. *Appl. Phys. Lett.* 103:131901.

Ge, Q., Sakhaei, A. H., Leem, H., et al. 2016. Multimaterial 4D printing with tailorable shape memory polymers. *Sci. Rep.* 6, 31110.

Gebler, M., Schoot Uiterkamp, A. J. M., Visser, C. 2014. A global sustainability perspective on 3D printing technologies. *Energy Policy* 74:158–167.

GENOA 3DP Simulation. n.d. http://www.alphastarcorp.com/products/genoa-3dp-simulation/ (accessed May 28, 2020).

Ghoshal, S. 2017. Polymer/carbon nanotubes (CNT) nanocomposites processing using additive manufacturing (three-dimensional printing) technique: an overview. *Fibers* 5 (4): 40.

Ghoshal, S. 2017. Polymer/carbon nanotubes (CNT) nanocomposites processing using additive manufacturing (three-dimensional printing) technique: an overview. *Fibers* 5(40) doi:10.3390/fib5040040.

Gibbons, G. J., Williams, R., Purnell, P. 2010. 3D printing of cement composites. *Adv. Appl. Ceram.* 109:287–290.

Gibson, I., D. Rosen and B. Stucker. 2015. *Additive Manufacturing Technologies*. New York: Springer.

Gibson, I., Shi, D. 1997. Material properties and fabrication parameters in selective laser sintering process. *Rapid Prototyp. J.* 3 (4): 129–136.

Gkartzou, E., Koumoulos, E. P., Charitidis, C. A. 2017. Production and 3D printing processing of bio-based thermoplastic filament. *Manuf. Rev.* 4(1).

Goh, G. D., Yap, Y. L., Tan, H. K. J., et al. 2019. Process–structure–properties in polymer additive manufacturing via material extrusion: a review. *Crit. Rev. Solid State Mater. Sci.* ISSN: 1040-8436 (Print) 1547–1561.

Goh, G. D., Yap, Y. L., Agarwal, S., Yeong, W. Y. 2018. Recent progress in additive manufacturing of fiber reinforced polymer composite. *Adv. Mater. Technol.* 4:1800271.

Gonçalves, F., Costa, C., Fabela, I., et al. 2014. 3D printing of new biobased unsaturated polyesters by microstereo-thermal-lithography. *Biofabrication* 6(3).

González-Henríquez, C. M., Sarabia-Vallejos, M. A., Rodriguez-Hernandez, J. 2019. Polymers for additive manufacturing and 4D-printing: Materials, methodologies, and biomedical applications. *Prog. Polym. Sci.* 94:57–116.

Goodridge, R. D., Hague, R. J. M., Tuck, C. J. 2010. An empirical study into laser sintering of ultra-high molecular weight polyethylene (UHMWPE). *J. Mater. Process Technol.* 210:72–80.

Goodridge, R. D., Tuck, C. J., Hague, R. J. M. 2012. Laser sintering of polyamides and other polymers. *Prog. Mater. Sci.* 57:229.

Goodridge, R. D., Shofner, M. L., Hague, R. J. M., et al. 2011. Processing of a Polyamide-12/carbon nanofibre composite by laser sintering. *Polym. Test.* 30:94–100.

Gornet, T. J., Davis, K. R., Starr, T. L. and Mulloy, K. M. 2002. Characterization of selective laser sintering materials to determine process stability. *Proceedings of the Solid Freeform Fabrication Symposium on Characterization of Selective Laser Sintering Materials to Determine Process Stability*, The University of Texas, Austin, TX, USA, pp. 546–553.

Gosselin, C., Duballet, R., Roux, P., et al. 2016. Large-scale 3D printing of ultra-high performance concrete - a new processing route for architects and builders. *Mater. Des.* 100:102–109.

Grand View Research. 2018. *3D Printing Plastics Market Size, Share & Trends Analysis Report*. https://www.grandviewresearch.com/industry-analysis/3d-printing-plastics-market (accessed July 4, 2020).

Groll, J., Boland, T., Torsten Blunk, T. 2016. Biofabrication: reappraising the definition of an evolving field. *Biofabrication* 8 (1): 013001.

Hager, I., Golonka, A., Putanowicz, R. 2016. 3D printing of buildings and building components as the future of sustainable construction? *Proc. Eng.* 151:292–299.

Hambach, M., Volkmer, D. 2017. Properties of 3D-printed fiber-reinforced Portland cement paste. *Cem. Concr. Compos.* 79:62–70.

Han, S. I., Jeong, S. H. 2004. Laser-assisted chemical vapor deposition to directly write three-dimensional microstructures. *J. Laser Appl.* 16 (3): 154.

Hanna, S. 2014. Personal communication.

Hayat, A., Nevet, A., Ginzburg, P., Orenstein, M., 2011. Applications of two-photon processes in semiconductor photonic devices: Invited review. *Semicond. Sci. Technol.* 26 (8): 083001. doi:10.1088/02 68-1242/26/8/083001.

Heinrich, L. A. 2019. Future opportunities for bio-based adhesives - advantages beyond renewability. *Green Chem.* 21:1866–1888.

Hendrickson, C., Horvath, A., Joshi, S., Lester Lave, L. 1998. Economic input-output models for environmental life-cycle assessment. *Environ. Sci. Technol.* 32:184A–191A, American Chemical Society.

Hendrixson, S. 2019. Stereolithography for metal produces detailed parts. Gardner Business Media Inc. Published Dec. 20, 2019 (accessed July 4, 2020) https://www.moldmakingtechnology.com/articles/stereolithography-for-metals-produces-detailed-parts.

Hewlett-Packard. 2020. *3D Printing Solutions*. https://www.8hp.com/us/en/printers/3d-printers.html (accessed April 3, 2020).

Hinczewski, C., Corbel, S., Chartier, T. 1998. Stereolithography for the fabrication of ceramic three-dimensional parts. *Rapid Prototyp. J.* 4 (3): 104–111.

Ho, H. C. H., Cheung, W. L., Gibson, I. 2003. Morphology and properties of selective laser sintered bisphenol A polycarbonate. *Indus. Eng. Chem. Res.* 42 (9): 1850–1862.

Hocker, T., Gu, H., Kamps, J. H., Noordegraaf, D. 2019. System and Method for reactive inkjet printing of polycarbonate. European Patent EP 3291970B1 filed May 6 2016 issued Dec. 18, 2019.

Hofstatter, T., Pedersen, B., Tosello, G., Hansen, H. N. 2017. Applications of fiber-reinforced polymers in additive manufacturing. *1st Cirp Conference on Composite Materials Parts Manufacturing, cirp-ccmpm2017*, Procedia CIRP 66:312–316.

Holland S., Tuck C., Foster T. 2018.Selective recrystallization of cellulose composite powders and microstructure creation through 3D binder jetting. *Carbohydr. Polym.* 200:229–238.

Hopkinson, N., Erasenthiran, P. 2004. High speed sintering – early research into anew rapid manufacturing process. *Solid Free Fabr. Symp. Proc.* 2004:312–320.

Hopkinson, N., Hague, R. J. M., Dickens, P. M. 2006. *Rapid Manufacturing: An Industrial Revolution for the Digital Age*, p. 293. Hoboken: John Wiley & Sons.

Houriet, C. 2019. Additive Manufacturing of Liquid Crystal Polymers Interlayer features: formation & impact on interlaminar shear strength. Master Thesis diss. Delft University of Technology. https://repository.tudelft.nl/islandora/object/uuid:1104f824-a4a7-4cf6-bdae-fac9441b5b24

Hoyt, R. P., Cushing, J. I., Slostad, J. T. et al. 2013. SpiderFab: an architecture for self-fabricating space systems. *AIAA Sp. Conf. Expo.*, pp. 1–17.

Hsiao, W.-K., Lorber, B., Reitsamer, H., Khinast, J. 2018. 3D printing of oral drugs: a new reality or hype? *Expert Opin. Drug Deliv.* 5 (1): 1–4. doi:10.1080/17425247.2017.1371698.

Huang, C., Zhang, Q. 2004. Electroactive polymer (EAP)-based deformable micromirrors and light-valve technology for MOEMS display and imaging systems. *Proceedings Volume 5389, Smart Structures and Materials 2004: Smart Electronics, MEMS, BioMEMS, and Nanotechnology*. doi.org/10.1117/12.540191.

Hull, Charles 1986. Apparatus for production of three-dimensional objects by stereolithography. US Patent 4575330, filed August 8, 1984, and issued March 11, 1986.

Hussein H., Harrison D. 2004. Investigation into the use of engineering polymers as actuators to produce 'automatic disassembly' of electronic products. In *Design and Manufacture for Sustainable Development*, ed.T. Bhamra, and B. Hon, 36–49.Weinheim: Wiley-VCH.

Iredale, R. J., Ward, C., Hamerto, I. 2017. Modern advances in bismaleimide resin technology: A 21st century perspective on the chemistry of addition polyimides. *Prog. Polym. Sci.* 69:1–21.

ISO 1133-1:2011. *Plastics – Determination of the Melt Mass-Flow Rate (MFR) and Melt Volume-Flow Rate (MVR) of Thermoplastics. Part 1: Standard Method.* West Conshohocken, PA, USA: ASTM International.

ISO 17296-1. 2015. *Additive Manufacturing – General Principles – Part 1: Terminology*. Geneva, Switzerland: ISO.

ISO 17296-2. 2015. *Additive Manufacturing – General Principles – Part 2: Overview of Process Categories and Feedstock*. Geneva, Switzerland: ISO.

ISO/ASTM 52900, 2015(E). *Standard Terminology for Additive Manufacturing – General Principles – Terminology*. Geneva, Switzerland: ISO. ASTM International, West Conshohocken, PA, United States.

Ito, F., Niino, T. 2016. Implementation of top hat profile laser into low temperature process of poly phenylene sulfide. *Solid Free Fabr. Symp. Proc.*, Austin, TX, pp. 2194–2203.

IUPAC (International Union of Pure and Applied Chemistry). 2014. *Compendium of Chemical Terminology – Gold Book*, Version 2.3.3.

Ivanova, O., Williams, C. and Campbell, T. 2013. Additive manufacturing (AM) and nanotechnology: promises and challenges. *Rapid Prototyp. J.* 19 (5): 353–364. doi:10.1108/RPJ-12-2011-0127.

Jacobs, P. F. 1992. *Rapid Prototyping and Manufacturing*. Dearborn, MI: Society of Manufacturing Engineers.

Jacobs, P. F. 1996. *Stereolithography and Other RP&M Technologies*. Dearborn, MI: Society of Manufacturing Engineers.

Jamal, M., Kadam, S. S., Xiao, R., et al. 2013. Bio-origami hydrogel scaffolds composed of photocrosslinked PEG bilayers. *Adv. Healthc. Mater.* 2:1142–1150.

Jamroz, W., Szafraniec, J., Kurek, M., Jachowicz, R., 2018. 3D printing in pharmaceutical and medical applications – recent achievements and challenges. *Pharm. Res.* 35:176. doi:10.1007/s11095-018-2454-x.

Janusziewicz, R., Tumbleston, J. R., Quintanilla, A. L., Mecham, S. J., DeSimone, J. M. 2016. Layerless fabrication with continuous liquid interface production. *Proc. Natl. Acad. Sci. U.S.A.*, 113:11703–11708.

Javaid, M., Haleem, A., 2018. Additive manufacturing applications in medical cases: a literature based review. *Alexandria J. Med.* 54:411–422.

Jem, J., Tan, B. 2020. The development and challenges of poly (lactic acid) and poly (glycolic acid). *Adv. Indus. Eng. Polym. Res.* 3 (2): 60–70.

Jiang, R., Kleer, R., Piller, F. T. 2017. Predicting the future of additive manufacturing: A Delphi study on economic and societal implications of 3D printing for 2030. *Technol. Forecast. Social Change* 117:84–97.

John, M. J., Anandjiwala, R. D., Wambua, P., et al. 2008. Bio-based structural composite materials for aerospace applications. *2nd South African International Aerospace Symposium*, Cape Town, South Africa, 14–16 September 2008.

Jurrens, K., 2013. NIST measurement science for additive manufacturing engineering laboratory. In C. Brown, J. Lubell, R. Lipman, *Additive Manufacturing Technical Workshop Summary Report, NIST, Technical Note 1823.* https://dokumen.tips/documents/additive-manufacturing-technical-workshop-summary-this-report-summarizes-the.html (accessed November 15, 2020).

Kacarevic, Z. P., Rider, P. M., Alkildani, S., et al. 2018. An introduction to 3D bioprinting: possibilities, challenges and future aspects. *Materials* 11:2199.

Kadry, H., Wadnap, S., Xu, C., Ahsan, F. 2019. Digital light processing (DLP) 3D-printing technology and photoreactive polymers in fabrication of modified-release tablets. *Eur. J. Pharm. Sci.* 135:60–67.

Kanetaka, Y., Yamazaki, S., Kimura, K. 2016. Preparation of poly(ether ketone)s derived from 2,5-furandicarboxylic acid via nucleophilic aromatic substitution polymerization. *J. Polym. Sci. Part A Polym. Chem.* 54:3094–3101.

Katstra, W. E., Palazzolo, R. D., Rowe, C. W. et al. 2000. Oral dosage forms fabricated by three dimensional printing™. *J. Controll. Release* 66:1–9.

Kazemian, A., Yuan, X., Cochran, E., Khoshnevis, B., 2017. Cementitious materials for construction-scale 3D printing: Laboratory testing of fresh printing mixture. *Constr. Build. Mater.* 145:639–647.

Khalil, Y., Kowalski, A., Hopkinson, N. 2016. Influence of energy density on flexural properties of laser-sintered UHMWPE. *Addit. Manuf.* 10:67–75.

Khan, I., Kamma-Lorger, C., Mohan, S., Mateus, A., Mitchell, G. 2019. The exploitation of polymer based nanocomposites for additive manufacturing: a prospective review. *Appl. Mech. Mater.* 890:113–145.

Khorram, N. M., Nonino, F., Palombi, G.,Torabi, S. A. 2018. Economic sustainability of additive manufacturing: contextual factors driving its performance in rapid prototyping. *J. Manuf. Technol. Manag.* 30 (2): 353–365.

Khoshnevis, B. 2010. Robotic system for automated construction US Pat 7,641,461 B2 filed Jan. 21, 2005 awarded Jan. 5, 2010.

Khoshnevis, B., Yuan, X., Zahiri, B., Zhang, J., Xia, B. 2016. Construction by Contour Crafting using sulfur concrete with planetary applications. *Rapid Prot. J.* 22 (5): 848–856.

Kim, J., Hanna, J. A., Hayward, R. C., Santangelo, C. D. 2012. Thermally responsive rolling of thin gel strips with discrete variations in swelling. *Soft Matter.* 8:2375–2381.

Kirchmajer, D. M., Gorkin, III, R. 2015. An overview of the suitability of hydrogel-forming polymers for extrusion-based 3D-printing. *J. Mater. Chem. B* 3:4105–4117.

Kirleis, M., Simonson, D., Charipar, N., et al. 2014. Laser embedding electronics on 3D printed objects. *Proc. SPIE 8970, Laser 3D Manufacturing* 897004 (6 March 2014). doi:10.1117/12.2044222.

Koch, L., Deiwick, A., Chichkov, B. 2016. Laser-based cell printing. In *3D Printing and Biofabrication*, ed. A. Ovsianikov, J. Yoo, V. Mironov, 303–329. Reference Series in Biomedical Engineering. Cham: Springer.

Koenig, B. 2019. *Automotive Industry Warms to 3D Printing.* https://www.sme.org/technologies/articles/2019/august/automotive-industry-warms-to-3d-printing/.

Kokkinis, D., Schaffner, M., Studart, A. R., 2015. Multimaterial magnetically assisted 3D printing of composite materials. *Nat. Commun.* 6, Article number 8643.

Krimi, I., Lafhaj, Z., Ducoulombier, L. 2017. Prospective study on the integration of additive manufacturing to building industry – case of a French construction company. *Addit Manuf.* 16:107–114.

Kruth, J.-P., Levy, G., Klocke, F., Childs, T. H. C. 2007. Consolidation phenomena in laser and powder-bed based layered manufacturing. *CIRP Ann. Manuf. Technol.* 56:730–759.

Kuang, X., Roach, D. J., Wu, J., et al. 2019. Advances in 4D printing: materials and applications. *Adv. Funct. Mater.* 29, 1805290.

Kuksenok, O., Balazs, A. C. 2016. Stimuli-responsive behavior of composites integrating thermo-responsive gels with photo-responsive fibers. *Mater. Horiz.* 3:53–62.

Kumar, S., Hofmann, M., Steinmann, B., et al. 2012. Reinforcement of stereolithographic resins for rapid prototyping with cellulose nanocrystals. *Appl. Mater. Interfaces*, 4:5399–5407.

Kumbhar, N. N., Mulay, A. V. 2018. Post processing methods used to improve surface finish of products which are manufactured by additive manufacturing technologies: a review. *J. Inst. Eng. C* 99:481–487.

Kuo, C.-C., Liu, L.-C., Teng, W.-F., et al. 2016. Preparation of starch/acrylonitrile-butadiene-styrene copolymers (ABS) biomass alloys and their feasible evaluation for 3D printing applications. *Compos. B.* 86:36–39.

Kurfess, T., Cass, W. J. 2014. Rethinking additive manufacturing and intellectual property protection. *Res.-Technol. Manag.* 57 (5): 35–42.

Kwiecien, H., Wodnicka, A., 2020. Five-membered ring systems: furans and benzofurans. *Prog. Heterocyclic Chem.* 5.3 (31): 281–323. doi:10.1016/B978-0-12-819962-6.00007-5.

Kyle, S., Jessop, Z. M., Al-Sabah, A., Whitaker, I. S. 2017. 'Printability' of candidate biomaterials for extrusion based 3D printing: state-of-the-art. *Adv. Healthcare Mater.* 6 (10): 1700264.

Lam, C. X. F., Mo, X. M., Teoh, S. H., Hutmacher, D. W. 2002. Scaffold development using 3D printing with a starch-based polymer. *Mater. Sci. Eng. C* 20 (1–2): 49–56.

Lamichhane, S., Bashyal, S., Keum, T., et al. 2019. Complex formulations, simple techniques: Can 3D printing technology be the Midas touch in pharmaceutical industry? *Asian J. Pharm. Sci.* 14:465–479.

Lanzotti, A., Grasso, M., Staiano, G., Martorelli, M. 2015. The impact of process parameters on mechanical properties of parts fabricated in PLA with an open-source 3-D printer. *Rapid Prototyp. J.* 21:604–617.

Laukkonen, J. 2019. How smart airbag systems save lives. *Lifewire.* Posted March 14, 2019 https://www.lifewire.com/how-smart-airbags-save-lives-534818 (accessed April 26, 2020).

Layani, M., Wang, X., Magdassi, S. 2018. Novel materials for 3D printing by photopolymerization. *Adv. Mater.* 30:1–7.

Le, H. P., 1998. Progress and trends in ink-jet printing technology. *J. Imaging Sci. Technol.* 42 (1): 49–62.

Lecklider, T., 2016. 3D printing drives automotive innovation. *Eval. Eng.* Published Dec 21, 2016. https://www.evaluationengineering.com/test-issues-techniques/technology/3d-printing/article/13014908/3d-printing-drives-automotive-innovation (accessed July 4, 2020).

Le Duigou, A., Castro, M., Bevanc, R., Martin, N. 2016. 3D printing of wood fibre biocomposites: From mechanical to actuation functionality. *Mater. Des.* 96:106–114.

Leigh, D. 2014. personal communication.

Lendlein, A. ed. 2010. *Shape-Memory Polymers.* Berlin: Springer-Verlag.

Leng, J., Shanyi, D. 2010. *Shape-Memory Polymers and Multifunctional Composites.* London: CRC Press.

Le Tohic, C., J. O'Sullivan, J. J., Drapala, K. P. et al. 2018. Effect of 3D printing on the structure and textural properties of processed cheese. *J. Food Eng.* 220:56–64.

Lewandowski, J. L., Seifi, M. 2016. Metal additive manufacturing: a review of mechanical properties. *Ann. Rev. Mater. Res.* 46 (1): 151–186.

Li, H., Tan, C., Li, L. 2018b. Review of 3D printable hydrogels and constructs. *Mater. Des.* 159:20–38.

Li, H., Wang, T., Sun, J., Yu, Z. 2018a. The effect of process parameters in fused deposition modelling on bonding degree and mechanical properties. *Rapid Prototyp. J.* 24 (1): 80–92. doi:10.1108/RPJ-06-201 6-0090.

Li, L., Fourkas, J. T. 2007. Multiphoton polymerization. *Mater. Today* 10 (6): 30–37.

Li, T., Aspler, J., Kingsland, A., Cormier, L. M., Zou, X. 2016. 3D printing – a review of technologies, markets, and opportunities for the forest industry. *J. Sci. Technol. For Prod. Process* 5 (2): 30–37.

Liang, F., Liang, Y. 2014. Study on the status quo and problems of 3D printed buildings in China. *Glob. J. Hum. Soc. Sci.* 14 (5): 7–10.

Ligon, S. C., Liska, R., Stampfl, J., Gurr, M., Mülhaupt, R. 2017. Polymers for 3D printing and customized additive manufacturing. *Chem. Rev.* 117:10212–10290.

Lim, S., Buswell, R. A., Le, T. T., et al. 2012. Developments in construction scale additive manufacturing processes *Autom. Constr.* 21:262–268.

Liska, R., Schuster, M., Inführ, R. et al. 2007. Photopolymers for rapid prototyping. *J. Coatings Technol. Res.* 4:505–510.

Liu, K., Wu, J., Paulino, G. H., Qi, H. J. 2017. Programmable deployment of Tensegrity structures by stimulus-responsive polymers. *Sci. Rep.* 7 (1): 3511.

Liu, G., Zhao, Y., Wu, G., Lu, J. 2018. Origami and 4D printing of elastomer-derived ceramic structures. *Sci. Adv.* 4(8) eaat0641. doi:10.1126/sciadv.aat0641.

Liu, Y., Lv, H., Lan, X. et al. 2009. Review of electro-active shape-memory polymer composite. *Compos. Sci. Technol.* 69:2064–2068.

Lobo, D. A., Ginestra, P. 2019. Cell bioprinting: The 3D-bioplotter™ case. *Mater.* 12:4005. doi:10.3390/ma12234005.

Loh, X. J. 2016. Four-dimensional (4D) printing in consumer applications. In *Polymers for Personal Care Products and Cosmetics*, ed.X. J. Loh, 108–116. Cambridge: Royal Society of Chemistry.

Loughborough University. 2020. *About Additive Manufacturing – Powder Bed Fusion.* https://www.lboro.ac.uk/research/amrg/about/the7categoriesofadditivemanufacturing/powderbedfusion/.

Lyons, J. G., Devine, D. M., 2019. Additive manufacturing: future challenges. In *Polymer-Based Additive Manufacturing*, ed.D. M. Devine, 255–264. Cham: Springer Nature.

Ma, J., Franco, B., Tapia, G., et al. 2017. Spatial control of functional response in 4D-printed active metallic structures. *Sci. Rep.* 7, 46707.

Ma, P.-J., Siddiqui, N. A., Marom, G., Kim, J.-K. 2010. Dispersion and functionalization of carbon nanotubes for polymer-based nanocomposites: a review. *Compos. A: Appl. Sci. Manuf.* 41 (10): 1345–1367.

MacDonald, E., Wicker, R. 2016. Multiprocess 3D printing for increasing component functionality. *Science* 353(6307). doi:10.1126/science.aaf2093.

MacRae, M. 2016. The 3D printed office of the future, *ASME.* Published Oct. 7, 2016. https://www.asme.org/topics-resources/content/3d-printed-office-the-future (accessed July 31, 2020.)

Magalhães, S., Alves, L., Medronho, B. et al. 2019. Brief overview on bio-based adhesives and sealants. *Polymers* 11:1685.

Mansourkhaki, S. 2019. *Can Additive Manufacturing Change the Market of Spare Parts in Automotive?* Degree thesis, Politecnico Milano, https://www.politesi.polimi.it/handle/10589/149635 (accessed July 31, 2020).

Mark Gregory, and Antoni Gozdz. 2015. Three dimensional printer for fiber reinforced composite filament fabrication US 9,126,367 B1, filed Dec. 18, 2014, issued Sep. 8, 2015.

Martin, J. J., Caunter, A., Dendulk, A., et al. 2017. Direct-write 3D printing of composite materials with magnetically aligned discontinuous reinforcement. *Proc. SPIE 10194, Micro- and Nanotechnology Sensors, Systems, and Applications IX.*

Masters, William 1987. Computer automated manufacturing process and system. US Patent 4,665,492, filed July 2, 1984, and issued May 12, 1987.

Material, Data Center. n.d. https://www.materialdatacenter.com/ms/en/Luvosint/LEHVOSS+Group/7437 (accessed June 13, 2020).

Matos, F., Jacinto, C. 2019. Additive manufacturing technology: mapping social impacts. *J. Manuf. Technol. Manag.* 30 (1): 70–97, doi:10.1108/JMTM-12-2017-0263.

Mekonnen, B. G., Glen Bright, G., Walker, A. 2016. A study on state of the art technology of laminated object manufacturing (LOM). In *CAD/CAM, Robotics and Factories of the Future, Lecture Notes in Mechanical Engineering*, ed.D. K. Mandal and C. S. Syan, doi:10.1007/978-81-322-2740-3_21.

Messina, I., 2020. *Additive Manufacturing in BiW: Opportunities and Developments for the Design and Manufacturing in Automotive Industry.* University thesis. Politecnico di Torino, https://webthesis.biblio.polito.it/14326/ (accessed July 31, 2020).

Miao, S., Wang, P., Su, Z., Zhang, S. 2014. Vegetable-oil-based polymers as future polymeric biomaterials. *Acta Biomater.* 10 (4): 1692–1704.

Metrey, D. 2012. *Non-Formaldehyde Biobased Phenolic Resins.* https://cfpub.epa.gov/ncer_abstracts/index.cfm/fuseaction/display.abstractDetail/abstract/9666/report/F (accessed January 16, 2020).

Miao, S., Zhu, W., Castro, N. J., et al. 2016. 4D printing smart biomedical scaffolds with novel soybean oil epoxidized acrylate. *Sci. Rep.* 6:27226.

Miglierini, G., 2019. *Emerging Trends for the Pharmaceutical Market.* Published February 20, 2019. https://www.pharmaworldmagazine.com/emerging-trends-for-the-pharmaceutical-market/ (accessed April 26, 2020).

Mitchell, A., Lafont, U., Hołyńska, M., Semprimoschnig C. 2018. Additive manufacturing - A review of 4D printing and future applications. *Addit. Manuf.* 24:606–626.

Mkhabela, V. J., Ray, S. S. 2014. Poly (ε-caprolactone) nanocomposite scaffolds for tissue engineering: a brief overview. *J. Nanosci. Nanotechnol.* 14:535–545.

Mogas-Soldevila, L., Duro-Royo, J., Oxman, N., 2014. Water-based robotic fabrication: large - scale additive manufacturing of functionally graded hydrogel composites via multichamber extrusion. *Mary Ann Liebert Inc.* 1 (3): 141–150.

Mohamed, O. A., Masood, S. H., Bhowmik, J. L. 2015. Optimization of fused deposition modeling process parameters: a review of current research and future prospects. *Adv. Manuf.* 3:42–53. doi:10.1007/s4043 6-014-0097-7.

Molitch-Hou, M. 2018. Overview of additive manufacturing process. In *Additive Manufacturing – Materials, Processes, Quantifications and Applications*, ed. J. Zhang, Yeon-G. Jung, 1–38. Amsterdam: Elsevier.

Momeni, F., Mehdi Hassani, S. M., Liu, X., Ni, J. 2017. A review of 4D printing. *Mater. Des.* 122:42–79.

Monahan, S., Brannen, S., Kurdys, A., Angelo, R. 2017. *3D Printing: Disrupting the $12 Trillion Manufacturing Sector.* https://www.kearney.com/operations-performance-transformation/article?/a/3d-printing-disrupting-the-12-trillion-manufacturing-sector (accessed January 16, 2021).

Moreno Nieto, D., Casal López V., Molina, S. I. 2018. Large-format polymeric pellet-based additive manufacturing for the naval industry. *Addit. Manuf.* 23:79–85.

Moroni, L., Boland, T., Burdick, et al. 2018. Biofabrication: a guide to technology and terminology. *Trends Biotechnol.* 36:384–402.

Mortara, L., Hughes, J., Ramsundar, P. S., Livesey, F., Probert D. R. 2009. Proposed classification scheme for direct writing technologies. *Rapid Prototyping J.* 15 (4): 299–309.

Mota, C., Puppi D., Chiellini, F., Chiellini, E. 2015. Additive manufacturing techniques for the production of tissue engineering constructs. *J. Tissue Eng. Regen. Med.* 9:174–190.

Murphy, S. V., Atala, A. 2014. 3D bioprinting of tissues and organs. *Nat. Biotechnol.* 32:773–785.

Mutlu, R., Alici, G., in het Panhuis, M., Spinks, G.2015. Effect of flexure hinge type on a 3D printed fully compliant prosthetic finger. *2015 IEEE International Conference on Advanced Intelligent Mechatronics (AIM)* July 7–11, 2015. Busan, Korea.

Nadgorny, M., Xiao, Z., Chen, C., Connal, L. A. 2016. Three-dimensional printing of pH-responsive and functional polymers on an affordable desktop printer. *ACS Appl. Mater. Interfaces* 8:28946–28954.

Naficy, S., Gately, R., Gorkin, R., Xin, H., Spinks, G. M. 2016. 4D printing of reversible shape morphing hydrogel structures. *Macromol. Mater. Eng.* 302 (1): 1–9.

Napadensky Eduard o 2003. Compositions and methods for use in three dimensional model printing. US Patent 6569373, filed March 12, 2001, and issued May 27, 2003.

Ngo, T. D., Kashani, A., Imbalzano, G., Nguyen, K. T. Q., Hui, D. 2018. Additive manufacturing (3D printing): A review of materials, methods, applications and challenges. *Compos. B: Eng.* 143:172–196.

Nichols, M. R. 2019. How does the automotive industry benefit from 3D metal printing? *Metal Powder Rep.* 74 (5): 257–258.

Nikzad, M., Masood, S. H., Sbarski, I. 2011. Thermo-mechanical properties of a highly filled polymeric composite for fused deposition modeling. *Mater. Des.* 32:3448–3456.

Niu, F., Wu, D., Ma, G., et al. 2015a. Nanosized microstructure of Al_2O_3–ZrO_2 (Y_2O_3) eutectics fabricated by laser engineered net shaping. *Scr. Mater.* 95:39–41.

Niu, F., Wu, D., Ma, G. et al. 2015b. Effect of second phase doping on laser deposited Al_2O_3 ceramics. *Rapid Prototyp. J.* 21 (2): 201–206.

O'Connor, H. J., Dickson, A. N., Dowling, D. P. 2018. Evaluation of the mechanical performance of polymer parts fabricated using a production scale multi jet fusion printing process. *Addit. Manuf.* 22:381–387.

Obata, K., Klug, U., Koch, J., Suttmann, O., Overmeyer, L. 2014. Hybrid micro-stereo-lithography by means of aerosol jet printing technology. *J. Laser Micro/Nanoeng.* 9:242–247.

Onwubolu, G. C., Rayegani, F. 2014. Characterization and optimization of mechanical properties of ABS parts manufactured by the fused deposition modelling process. *Int. J. Manuf. Eng.* Article ID 598531.

Oropallo, W., Piegl, L. A. 2016. Ten challenges in 3D printing. *Eng. Comput.* 32:135–148.

Ortega Z., Aleman, M. E., Benitez, A. N., Monzon, M. D. 2016. Theoretical-experimental evaluation of different biomaterials for parts obtaining by fused deposition modeling. *Measurements* 89:137–144.

Ouyang, L., Yao, R., Zhao, Y., Sun, W. 2016. Effect of bioink properties on printability and cell viability for 3D bioplotting of embryonic stem cells. *Biofabrication* 8 (3): 035020.

Ozbolat, I. T., Hospodiuk, M. 2016. Current advances and future perspectives in extrusion-based bioprinting. *Biomaterials* 76:321–343.

Panda, B., Unluer, C., Tan, M. J. 2018. Investigation of the rheology and strength of geopolymer mixtures for extrusion-based 3D printing. *Cement Concr. Compos.* 94:307–314.

Panwar, A., Tan, L. P. 2016. Current status of bioinks for micro-extrusion-based 3D bioprinting. *Molecules* 21:685. doi:10.3390/molecules21060685.

Parandoush, P., Lin, D. 2017. A review on additive manufacturing of polymer-fiber composites. *Compos. Struct.* 182:36–53.

Park, J., Tari, M. J., H. Hahn, H. T. 2000. Characterization of the laminated object manufacturing (LOM) process. *Rapid Prototyp. J.* 6 (1): 36–49.

Pati, F., Jang, J., Ha D.-H., et al. 2014. Printing three-dimensional tissue analogues with decellularized extracellular matrix bioink. *Nat. Commun.* 5:3935. doi:10.1038/ncomms4935.

Paxton, N., Smolan, W., Bock, T., et al. 2017. Proposal to assess printability of bioinks for extrusion-based bioprinting and evaluation of rheological properties governing bioprintability. *Biofabrication* 9 (4): 044107.

Peng, B., Yang, Y., Gu, K., Amis, E. J., Cavicchi, K. A. 2019. Digital light processing 3D printing of triple shape memory polymer for sequential shape shifting. *ACS Mater. Lett.* 1 (4): 410–417.

Pham, D. T., Ji, C. 2003. A study of recoating in stereolithography. *Proc. Inst. Mech. Eng. C*217 (1): 105–117.

Photocentric. n.d. https://photocentricgroup.us/ (accessed July 25, 2020).

Pique, A., Chrisey, D. B. 2001. *Direct Write Technologies for Rapid Prototyping Applications. Sensors, Electronics and Integrated Power Sources.* Boston: Academic Press.

Pohlmann T. 2019. *Patent and Litigation Trends for 3D Printing Technologies.* https://www.iam-media.com/patent-and-litigation-trends-3d-printing-technologies posted March 12, 2019 (accessed April 26, 2020).

Popescu, D., Zapciu, A., Amza C., Baciu, F., Marinescu, R. 2018. FDM process parameters influence over the mechanical properties of polymer specimens: A review. *Polym. Testing* 69:157–166.

Postiglione, G., Natale, G., Griffini, G., Levi, M., Turri, S. 2015. Conductive 3D microstructures by direct 3D printing of polymer/carbon nanotube nanocomposites via liquid deposition modeling. *Compos. A: Appl. Sci. Manuf.* 76:110–114.

Priarone, P. C., Ingarao, G. 2017. Towards criteria for sustainable process selection: on the modelling of pure subtractive versus additive/subtractive integrated manufacturing approaches. *J. Cleaner Prod.* 144:57–68.

Puppi, D., Chiellini, F. 2020. Biodegradable polymers for biomedical additive manufacturing. *Appl. Mater. Today* 20:100700.

Quinlan, H. E., Hasan, T., Jaddou, J., Hart J. A. 2017. Industrial and consumer uses of additive manufacturing. *J. Indus. Ecol.* 21:S1.

Rajan, J., Wood, K., Malkovich, N. 2001. Experimental study of selective laser sintering of Parmax®. In *Solid Freeform Fabr. Symp. Proc.* 242–255.

Ramon, E., Sguazzo, C., Moreira, P. M. G. P. 2018. A review of recent research on bio-based epoxy systems for engineering applications and potentialities in the aviation sector. *Aerospace* 5(4):110. doi:10.3390/aerospace5040110.

Rasselet, D., Caro-Bretelle, A.-S., Taguet, A., Lopez-Cuesta, J.-M. 2019. Reactive compatibilization of PLA/PA11 blends and their application. *Addit. Manuf. Mater.* 12:485.

Raviv, D., Zhao, W., McKnelly, C., et al. 2014. Active printed materials for complex self-evolving deformations. *Scientific Rep.* 4:7422.

RE PET 3D, n.d. *Recycled Filaments.* https://re-pet3d.com/product-category/recycled-filaments/ (accessed August 28, 2020).

Reeves, P., Tuck, C., Hague, R., 2011. Additive manufacturing for mass customization. In *Mass Customization*, ed.F. Fogliatto, G. da Silveira, 275–289. Springer Series in Advanced Manufacturing. London: Springer.

Refil. n.d. https://www.re-filament.com/ (accessed January 16, 2021).

Renn, Michael J. 2006. Direct Write™ System. US Patent 7,108,894, filed Feb. 5, 2002, and issued Sep. 19, 2006.

Renn, Michael J. 2007. Direct Write™ System US Patent 7,270,844, filed Sep. 20, 2004, and issued Sep. 18, 2007.

Rios, O., Carter, W., Kutchko, C., Fenn, D., Olson, K. 2017. Evaluation of advanced polymers for additive manufacturing. ORNL/TM-2017/509.

Roboze 2021. *Design of 3D Printing Solutions.* https://www.roboze.com/en/ (accessed April 3, 2021).

Rodriguez, J. N., Zhu, C., Duoss, E. B., et al., 2016. Shape-morphing composites with designed microarchitectures. *Sci. Rep.* 6:27933.

Rodriguez-Palomo, A., Lutz-Bueno, V., Guizar-Sicairos, M. et al. 2021. Nanostructure and anisotropy of 3D printed lyotropic liquid crystals studied by scattering and birefringence imaging. *Additive Manufacturing* 47:102289. https://doi.org/10.1016/j.addma.2021.102289

Rosenthal, M., Henneberger, C., Gutkes, A. et al. 2018. Liquid deposition modeling: a promising approach for 3D printing of wood. *Eur. J. Wood Prod.* 76:797–799.

Rossi, M., Blake, A. 2014. The Plastics Scorecard: Evaluating the chemical footprint of plastics. *Clean Production Action*, Somerville, Mass., www.bizngo.org/sustainable-materials/plastics-scorecard.

Rossiter, J., Walters, P., Stoimenov, B. 2009. Printing 3D dielectric elastomer actuators for soft robotics. *Proc. SPIE 7287, Electroactive Polymer Actuators and Devices (EAPAD) 2009*, 72870H (6 April 2009). doi.org/10.1117/12.815746.

Rusling, J. F. 2018. developing microfluidic sensing devices using 3D printing. *ACS Sens.* 3:522–526.

Saari, M., Cox, B., Richer, E., Krueger, P. S., Cohen, A. L. 2015. Fiber encapsulation additive manufacturing: an enabling technology for 3D printing of electromechanical devices and robotic components. *3D Print.* 2 (1): 32–39.

Sabir, M. I., Xu, X. and Li L. 2009. A review on biodegradable polymeric materials for bone tissue engineering applications. *J. Mater. Sci.* 44:5713–5724.

Sachs, E. M., J. S. Haggerty, M. J. Cima, P. A. Williams. 1993. Three-dimensional printing techniques. US patent 5,204,055, filed Dec. 8, 1989, and issued Apr. 20, 1993.

Sakin, M., Kiroglu, Y. C. 2017. 3D printing of buildings: construction of the sustainable houses of the future by BIM. *Energy Proc.*134:702–711.

Sames, W. J., List, F. A., Pannala, S., Dehoff R. R., Babu S. S. 2016. The metallurgy and processing science of metal additive manufacturing. *Inst. Mater. Miner. Mining SM Int.* 61 (5): 315–360. doi:10.1080/095 06608.1116649.

Sandoval, J. H., Soto, K. F., Murr, L. E. et al. 2007. Nanotailoring photocrosslinkable epoxy resins with multi-walled carbon nanotubes for stereolithography layered manufacturing. *J. Mater. Sci.* 42:156–165.

Sandoval, J. H., Wickermm, R. B. 2006. Functionalizing stereolithography resins: effects of dispersed multi-walled carbon nanotubes on physical properties. *Rapid Prototyping J.* 12 (5): 292–303.

Saroia, J., Wang, Y., Wei, Q., et al. 2019. A review on 3D printed matrix polymer composites: its potential and future challenges. *Int. J. Adv. Manuf. Technol.* 106:1695–1721.

Schmid, M., 2018. *Laser Sintering with Plastics - Technology, Processes, and Materials*. Munich: Hanser.

Schmid, M., Amado A., Wegener, K. 2014. Materials perspective of polymers for additive manufacturing with selective laser sintering. *J. Mater. Res.* 29:1824–1832.

Schmid, M., Amado, A., Wegener, K. 2015. Polymer powders for selective laser sintering (SLS). *AIP Conf Proc* 1664:160009.

Schmidt, J., Sachs, M., Fanselow, S., et al. 2016. Optimized polybutylene terephthalate powders for selective laser beam melting. *Chem. Eng. Sci.* 156:1–10.

Schmitt, M., Mehta, R. M., Kim, I. Y. 2020. Additive manufacturing infill optimization for automotive 3D-printed ABS components. *Rapid Protot. J.* 26(1):89–99. doi:10.1108/RPJ-01-2019-0007.

Sculpteo. 2017. *The State of 3D Printing*. https://www.sculpteo.com/media/ebook/State%20of%203DP%202 017_1.pdf (accessed January 16, 2020).

Seerden, K. A. M., Reis, N., Evans, J. R. G., Grant, P. S., Halloran, J. W., Derby, B., 2001. Ink-jet printing of wax-based alumina suspensions. *J. Am. Ceram. Soc.* 84:2514–2520.

Shahzad, K., Deckers, J., Kruth, J. P., Vleugelfor, J. 2013. Additive manufacturing of alumina parts by indirect selective laser sintering and post processing. *J. Mater. Process. Technol.* 213:1484–1494.

Shahzad, K., Deckers, J., Zhang, Z., Kruth, J. P., Vleugelfor, J. 2014. Additive manufacturing of zirconia parts by indirect selective laser sintering. *J. Eur. Ceramic Soc.*34:81–89.

Shan, W., Chen, Y., Mo, H., et al. 2020. 4D printing of shape memory polymer via liquid crystal display stereolithography. *Preprints.* 2020070163.

Sher, D. 2018. *The Global Additive Manufacturing Market 2018 Is Worth $9.3 Billion*. December 14, 2018. https://www.3dprintingmedia.network/the-global-additive-manufacturing-market-2018-is-worth-9-3-billion/ (accessed June 13, 2020).

Shi, Y., Chen, J., Wang, Y., Li, Z., Huang, S. 2007. Study of the selective laser sintering of polycarbonate and postprocess for parts reinforcement. *Proc. Inst. Mech. Eng. L*221 (1): 37–42. doi:10.1243/146442 07JMDA65.

Shipp, S. S., Gupta, N., Lal, B. et al. 2012. *Emerging Global Trends in Advanced Manufacturing, by the Institute for Defense Analysis*. https://apps.dtic.mil/dtic/tr/fulltext/u2/a558616.pdf (accessed February 14, 2020).

Shortt, M., 2013. Will 3D printing ever replace conventional manufacturing? *Design-2-Part Magazine* 30(8), Eastern Edition, October 2013:4. The Job Shop Company Inc.

Sidney Gladman, A., Matsumoto, E. A., Nuzzo, R. G., Mahadevan, L., Lewis, J. A. 2016. Biomimetic 4D printing. *Nature Materials* 15:413–419.

Sillani, F., Kleijnen, R. G., Vetterli, M. et al. 2019. Selective laser sintering and multi jet fusion: process-induced modification of the raw materials and analyses of parts performance. *Additive Manufacturing* 27:32–41.

Simchi, A., Pohl, H. 2003. Effects of laser sintering processing parameters on the microstructure and densification of iron powder. *Mater. Sci. Eng. A* 359:119–128.

Singh, D., Thomas, D. 2019. Advances in medical polymer technology towards the panacea of complex 3D tissue and organ manufacture. *Am. J. Surg.* 217:807–808.

Singh, S., Singh, G., Prakash, C., Ramakrishna, S. 2020. Current status and future directions of fused filament fabrication. *J. Manuf. Proces.* 55:288–306.

Smith, P. J., Morrin, A., 2012. Reactive inkjet printing. *J. Mater. Chem.* 22:10965.

Solidscape. n.d. *3D Printer Materials Castable Wax and Resin.* https://www.solidscape.com/products/3d-printing-materials/ (accessed July 31, 2020).

Sood, A. K., Ohdar, R. K., Mahapatra, S. S. 2010. Parametric appraisal of mechanical property of fused deposition modelling processed parts. *Mater. Des.* 31:287–295.

Stammen, E., Dilger, K. 2013. Adhesive bonding of textiles: applications. In *Joining Textiles*, ed.I. Jones, G. K. Stylios, 275–308. Woodhead Publishing Ltd.

Stansbury, J. W., Idacavage, M. J. 2016. 3D printing with polymers: Challenges among expanding options and opportunities. *Dental Mater.*32:54–64.

Stephens, B., Azimi, P., El Orch, Z., et al. 2013. Ultrafine particle emissions from desktop 3D printers. *Atmos. Environ.* 79:334–339.

Storaenso. 2018. *Lineo™ by Stora Enso A Future of Opportunities.* https://www.storaenso.com/en/products/lignin (accessed January 16, 2020).

Stoyanov, S., Bailey, C., Tourloukis, G. 2016. Modelling the 3D-printing process for electronics packaging. *6th Electronic System-Integration Technology Conference (ESTC), Grenoble*, pp. 1–6. doi:10.1109/ESTC.2016.7764481.

Sun, H. B., Kawata, S. 2004. Two-photon photopolymerization and 3D lithographic microfabrication. In *NMR – 3D Analysis – Photopolymerization. Advances in Polymer Science*, vol. 170, 169–273. Berlin: Springer. doi:10.1007/b94405.

Sun, K., Wei, T.-S., Ahn, B. Y. et al. 2013. 3D printing of interdigitated Li-Ion, microbattery architectures. *Adv. Mater.* 25:4539–4543.

Swanson, Wyn K., and D. Stephen 1978. Three dimensional systems. US Pat. 4078229, filed Jan. 27, 1975, and issued Mar. 7, 1978.

Taddese, G., Durieux, S., Duc, E. 2020. Sustainability performance indicators for additive manufacturing: a literature review based on product life cycle studies. *Int. J. Adv. Manuf. Technol.* 107:3109–3134.

Taormina, G., Sciancalepore, C., Messori, M., et al. 2018. 3D printing processes for photocurable polymeric materials: technologies, materials, and future trends. *J. Appl. Biomater. Funct. Mater.* 16 (3): 151–160.

Tay, B. Y., Edirisinghe, M. J. 2001. Investigation of some phenomena occurring during continuous ink-jet printing of ceramics. *J. Mater. Res.* 16 (2): 373–384.

Teizer, J., A. Blickle, T. King, et al. 2016. Large scale 3D printing of complex geometric shapes in construction. In *International Symposium on Automation and Robotics in Construction*, 2016.

Teken, Avraham, and Boris Belocon, 2017, Method and system for reuse of materials in additive manufacturing systems. US Patent 9,688,020, filed Dec. 21, 2011, and issued June 27, 2017.

Teken, Avraham, and Boris Belocon, 2019, Method and system for reuse of materials in additive manufacturing systems. US Patent 10,245,784, filed Jun. 26, 2017, and issued Apr. 2, 2019.

Tibbits, S. 2013. The emergence of "4D printing". *TED Conference.*

Tibbits, S., McKnelly, C., Olguin, C., Dikovsky D., Hirsch S., 2014a. *4D Printing and Universal Transformation.* http://papers.cumincad.org/data/works/att/acadia14_539.content.pdf (accessed January 16, 2020).

Tibbits, S. 2014b. 4D Printing: multi-material shape change. *Archit. Des.* 84:116–121.

Torres, J., Cotelo, J., Karl, J., Gordon, A. P. 2015. Mechanical property optimization of FDM PLA in shear with multiple objectives. *JOM* 67 (5): 1183–1193.

Torres, J., Cole, M., Owji, A., DeMastry, Z., Gordon, A. P. 2016. An approach for mechanical property optimization of fused deposition modeling with polylactic acid via design of experiments. *Rapid Prototyp. J.* 22 (2): 387–404. doi:10.1108/RPJ-07-2014-0083.

Trenfield, S. J., Awad, A., Goyanes, A., Gaisford, S., Basit, A. W. 2018. 3D printing pharmaceuticals: drug development to frontline care. *Trends Pharmacol. Sci.* 39 (5): 440–451.

Tsouknidas, A., Pantazopoulos, M., Katsoulis, I., et al. 2016. Impact absorption capacity of 3D-printed components fabricated by fused deposition modelling. *Mater Des.* 102:41–44.

Tumbleston, J. R., Shirvanyants, D., Ermoshkin, N. et al. 2015. Continuous liquid interface production of 3D objects. *Science* 347 (6228): 1349.

Türk, D.-A., Kussmaul, R., Zogg, M. et al. 2017. Composites part production with additive manufacturing technologies. *1st Cirp Conference on Composite Materials Parts Manufacturing Procedia CIRP* 66:306–311.

Turner, B. N., Strong, R., Gold, S. A. 2014. A review of melt extrusion additive manufacturing processes: I. Process design and modeling. *Rapid Prototyp. J.* 20 (3): 192–204.

Turner, B. N., Strong, R., Gold, S. A. 2015. A review of melt extrusion additive manufacturing processes: II. Materials, dimensional accuracy, and surface roughness. *Rapid Prototyp. J.* 21 (3): 250–261.

Udofia, E. N. 2019. *Microextrusion 3D Printing of Optical Waveguides and Microheaters*. Theses and Dissertations Retrieved from https://scholarworks.uark.edu/etd/3386.

Ullett, J. S., Schultz, J. W., Chartoff, R. P. 2000. Novel liquid crystal resins for stereolithography - Processing parameters and mechanical analysis. *Rapid Prototyp. J.* 6 (1): 8–17.

Uriondo, A., Esperon-Miguez, M., Perinpanayagam, S. 2015. The present and future of additive manufacturing in the aerospace sector: A review of important aspects. *Proc. Inst. Mech. Eng. G* 229(11): 2132–2147.

Varotsis, A. B. 2020. *Introduction to SLA 3D Printing.* https://www.3dhubs.com/knowledge-base/introduction-sla-3d-printing/ (accessed January 16, 2020).

Verbelen, L., Dadbakhsh, S., Van den Eynde M., et al. 2016. Characterization of polyamide powders for determination of laser sintering processability. *Eur. Polym. J.* 75:163–174.

Vock, S., Klöden B., Kirchner, A., Weißgärber, T., Kieback, B. 2019. Powders for powder bed fusion: a review. *Prog. Addit. Manuf.* 4:383–397.

Vogel, B. J. 2016. Intellectual property and additive manufacturing/3D printing: strategies and challenges of applying traditional IP Laws to a transformative technology. *Minn. J. L. Sci. & Tech.* HeinOnline. 881.

Walter, A., Marcham, C. 2020. Environmental advantages in additive manufacturing. *Profes. Safety/ASSP* 65(01). https://commons.erau.edu/publication/1365 (accessed April 3, 2020).

Wang, F., Jiang, Y., Liu, Y., et al. 2020. Liquid crystalline 3D printing for superstrong graphene microlattices with high density. *Carbon* 159:166–174.

Wang, J.-C., Dommati, H., Hsieh, S.-J. 2019. Review of additive manufacturing methods for high-performance ceramic materials. *Int. J. Adv. Manuf. Technol.* 103:2627–2647.

Wang, M. O., Vorwald, C. E., Dreher, M. L. 2015. Evaluating 3d-printed biomaterials as scaffolds for vascularized bone tissue engineering. *Adv. Mater.* 27:138–144.

Wang, S., Capoen, L., D'hooge, D., Cardon, L., 2017a. Can the melt flow index be used to predict the success of fused deposition modelling of commercial poly(lactic acid) filaments into 3D printed materials? *Plastics Rub. Compos.* doi:10.1080/14658011.2017.1397308 (accessed November 15, 2020).

Wang, T., Farajollahi, M., Choi, Y. S., et al. 2016. Electroactive polymers for sensing. *Interf. Focus* 6:20160026. doi:10.1098/rsfs.2016.0026.

Wang, X., Jiang, M., Zhou, Z., Gou, J., Hui, D. 2017b. 3D printing of polymer matrix composites: a review and prospective. *Compos. B* 110:442–458.

Wang, X. C., Kruth, J. P., 2000. A simulation model for direct selective laser sintering of metal powders. In *Computational Techniques for Materials, Composites and Composite Structures*, ed.B. H. V. Topping, 57–71. Edinburgh: Civil-Comp Press. doi:10.4203/ccp.67.1.7.

Wang, Q., Zhang, S., Wei, D. et al. 2018. Additive Manufacturing: A Revolutionized Power for Construction Industrialization. Int. Conf. on Construction and Real Estate Management 2018. https://doi.org/10.1061/9780784481721.010.

Warnakula, A., Singamneni, S. 2017. Selective laser sintering of nano Al_2O_3 infused polyamide. *Materials* 10: 864.

Warner, J., Soman, P., Zhu, W., Tom, M., Chen, S. 2016. Design and 3D printing of hydrogel scaffolds with fractal geometries. *ACS Biomater. Sci. Eng.* 2:1763–1770.

WASP n.d. *WASP Crane*. https://www.3dwasp.com/en/3d-printer-house-crane-wasp/ (accessed July31, 2020).

Wegner, A. 2016. New polymer materials for the laser sintering process: polypropylene and others. *Phys. Procedia* 83:1003–1012.

Wei, H., Zhang, Q., Yao, Y., et al. 2016. Direct-write fabrication of 4D active shape-changing structures based on a shape memory polymer and its nanocomposite. *ACS Appl. Mater. Interfaces* 9(1): 876–883.

Wei, H., Zhang, Q., Yao, Y., et al. 2017. Direct-write fabrication of 4D active shape-changing structures based on a shape memory polymer and its nanocomposite. *ACS Appl. Mater. Interf.* 9(1):876–883.

White, Dawn 2003. Ultrasonic object consolidation. US Patent 6,519,500, filed March 23, 2000, and issued February 11, 2003.

Williams, J. M., Adewunmi, A., Schek, R. M., et al. 2005. Bone tissue engineering using polycaprolactone scaffolds fabricated via selective laser sintering. *Biomaterials* 26:4817–4827.

Winsun 3D. 2017. *Yingchuang 3D Printing Architecture Makes Building Solid Waste Recycled.* http://www.winsun3d.com/En/News/news_inner/id/474 (accessed August 28, 2020).

Wittbrodt, B., Pearce, J. M. 2015. The effects of PLA color on material properties of 3-D printed components. *Addit. Manuf.* 8:110–116.

Wohlers T. 2016. *Wohlers Report 2016. 3D Printing and Additive Manufacturing State of the Industry Annual Worldwide Progress Report.* Fort Collins: Wohlers Associates.

Wong, K. 2017. *Wohlers 2017. Report on 3D Printing Industry Points to Softened Growth.* https://www.digitalengineering247.com/article/wohlers-2017-report-on-3d-printing-industry-points-to-softened-growth/ (accessed January 6, 2020).

Woodson, T. S. 2015. 3D printing for sustainable industrial transformation. *Development* 58 (4): 571–576.

Wu, B. M., Borland, S. W., Giordano, R. A., et al. 1996. Solid free-form fabrication of drug delivery devices. *J. Control. Release* 40 (1): 77–87.

Wu, J., Yuan, C., Ding, Z., et al. 2016a. Multi-shape active composites by 3D printing of digital shape memory polymers. *Sci. Rep.* 6:24224. doi:10.1038/srep24224.

Wu, P., Ringeisen, B., Krizman, D. et al. 2003. Laser transfer of biomaterials: Matrix-assisted pulsed laser evaporation (MAPLE) and MAPLE Direct Write. *Rev. Sci. Instrum.* 74:2546–2557.

Wu, P., Wang, J., Wang, X. 2016b. A critical review of the use of 3-D printing in the construction industry. *Autom. Construct.* 68:21–31.

Wu, X. L., Huanh W. M., Ding, Z. et al. 2014. Characterization of the thermoresponsive shape-memory effect in poly(ether ether ketone) (PEEK). *J. Applied Polym. Sci.* 131:39844.

Xia, M., Sanjayan, J. 2016. Method of formulating geopolymer for 3D printing for construction applications. *Mater. Des.* 110:382–390.

Xia, Z., Jin, S., Ye, K. 2018. Tissue and organ 3D bioprinting. *SLAS Technol.* 23(4):301–314.

Xie, T. 2010. Tunable polymer multi-shape memory effect. *Nature* 464:267–270.

Xie, X., Mai, Y., Zhou, X. 2005. Dispersion and alignment of carbon nanotubes in polymer matrix: A review. *Mater. Sci. Eng. Rep.* 49 (4): 89–112.

Yang, H., Lim, J. Charlotte, Liu, Y. C., et al. 2017b. Performance evaluation of ProJet multi-material jetting 3D printer. *Virtual Phys. Prototyp.* 12 (1): 95–103.

Yang, K., Grant, J. C. Grant, Lamey P., et al. 2017a. Diels–alder reversible thermoset 3D printing: isotropic thermoset polymers via fused filament fabrication. *Adv. Funct. Mater.* 27:1700318.

Yang, X., Wei, Y., Xi, S., Huang, Y., Kong, M., Li, G. 2019. Preparation of spherical polymer powders for selective laser sintering from immiscible PA12/PEO blends with high viscosity ratios. *Polymer* 172:58–65.

Yang, Y., Chen, Y., Li, Y., Chen, M. Z. 2016a. 3D printing of variable stiffness hyper -redundant robotic arm. *2016 IEEE International Conference on Robotics and Automation (ICRA)* pp. 3871–3877.

Yang, Y., Chen, Y., Wei, Y., Li, Y. 2016b. 3D printing of shape memory polymer for functional part fabrication. *Int. J. Adv. Manuf. Technol.* 84 (9–12): 2079–2095.

Yao, R., Xu, G., Mao, S. S., et al. 2016. Three-dimensional printing: review of application in medicine and hepatic surgery. *Cancer Biol. Med.* 13 (4): 443–451.

Yao, T., Deng, Z., Zhang, K., Li, S. 2019. A method to predict the ultimate tensile strength of 3D printing polylactic acid (PLA) materials with different printing orientations. *Composites B* 163:393–402.

Yeong, W., Sudarmadji, N., Yu, H., et al. 2010. Porous polycaprolactone scaffold for cardiac tissue engineering fabricated by selective laser sintering. *Acta Biomater.* 6:2028–2034.

Yi, H. G., Lee, H., Cho, D. W. 2017. 3D printing of organs-on-chips. *Bioengineering* 4(1):10. doi:10.3390/bioengineering4010010.

Yoo, J., T. J. Bradbury, B. T. Jebb, J. Iskra, H. L. Surprenant, and T. G. West. 2014. Three-dimensional printing system and equipment assembly. US Patent 65194A1 filed Sep. 3, 2013, and issued Mar. 6, 2014.

Yu, K., Ritchie, A., Mao, Y., et al. 2015. Controlled sequential shape changing components by 3D printing of shape memory polymer multimaterials. *Procedia IUTAM* 12:193–203.

Yu, R., Yang, X., Zhang, Y., et al. 2017. Three-dimensional printing of shape memory composites with epoxy-acrylate hybrid photopolymer. *ACS Appl. Mater. Interf.* 9 (2): 1820–1829.

Yuan, S., Shen, F., Kai, Chua, C., Zhou, K. 2019. Polymeric composites for powder-based additive manufacturing: Materials and applications. *Prog. Polym. Sc.* 91:141–168.

Yugang, D., Yuan, Z., Yiping, T., et al. 2011. Nano-TiO$_2$-modified photosensitive resin for RP. *Rapid Prototyp. J.* 17 (4): 247–252.

Yun, B., Williams, C. B. 2015. An exploration of binder jetting of copper. *Rapid Prototyp. J.* 21 (2): 177–185.

Zareiyan, B., Khoshnevis, B. 2017. Interlayer adhesion and strength of structures in Contour crafting—effects of aggregate size, extrusion rate, and layer thickness. *Automat. Constr.* 81:112–121.

Zarek, M., Layani, M., Cooperstein, I., et al. 2016. 3D printing of shape memory polymers for flexible electronic devices. *Adv. Mater.* 28:4449–4454.

Zarringhalam, H., Majewski, C., Hopkinson, N. 2009. Degree of particle melt in Nylon-12 selective laser-sintered parts. *Rapid Prototyp. J.* 15:126–132.

Zelinski, P. 2018. Where AM meets AI. *Addit. Manuf.* 7 (1): 22–26.

Zhang, Q., Zhang, K., Hu, G. 2016. Smart three-dimensional lightweight structure triggered from a thin composite sheet via 3D printing technique. *Sci. Rep.* 6: (22431): 1–8.

Zhang, D., Rudolph, N., Woytowitz, P. 2019. Reliable optimized structures with high performance continuous fiber thermoplastic composites from additive manufacturing (AM). *SAMPE Conference Proceedings.* Charlotte, NC, May 20–23, 2019.

Zhao, M., Wudy, K., Drummer, D. 2018. Crystallization kinetics of polyamide 12 during selective laser sintering. *Polymers* 10:168. doi:10.3390/polym10020168.

Zhang, L., Chen, X., Wei, Z., et al. 2020. Digital Twins for Additive Manufacturing: A State-of-the-Art Review. *Appl. Sci.* 10:8350.

Zhao, Q., Behl, M., Lendlein, A. 2013. Shape-memory polymers with multiple transitions: complex actively moving polymers. *Soft Matter* 9:1744–1755.

Zheng, H., et al. 2006. Effect of core-shell compositecah particles on the sintering behavior and properties of nano-Al_2O_3/polystyrene composite prepared by SLS. *Mater. Lett.* 60:1219–1223.

Zhu, W., Qu, X., Zhu, J., et al. 2017. Direct 3D bioprinting of prevascularized tissue constructs with complex microarchitecture. *Biomaterials* 124:106–115.

Ziaee, M., Crane, M. 2019. Binder jetting: A review of process, materials, and methods. *Addit. Manuf.* 28:781–801.

Ziegelmeier, S., Christou, P., Wöllecke, F. et al. 2015. An experimental study into the effects of bulk and flow behaviour of laser sintering polymer powders on resulting part properties. *J. Mater. Process Technol.* 215:239–250.

Zocca, A., Colombo, P., Gomes, C. M., Günster, J. 2015. Additive manufacturing of ceramics: issues, potentialities, and opportunities. *J. Am. Ceram. Soc.* 98 (7): 1983–2001.

Zorlutuna, P., Jeong, J. H., Kong, H., Bashir, R. 2011. Stereolithography-based hydrogel microenvironments to examine cellular interactions. *Adv. Funct. Mater.* 21:3642–3651.

3 Poly(Lactic Acid)

We won't have a society if we destroy the environment.

Margaret Mead

3.1 OVERVIEW OF POLY(LACTIC ACID) (PLA)

3.1.1 INTRODUCTION

Part of the information included in this section derives from comprehensive and detailed reviews on PLA published by Henton et al. (2005), Lim et al. (2008), Farah et al. (2016), and Murariu and Dubois (2016), Henton Mohanty 2005 that are recommended for additional details. PLA is a sustainable, biodegradable, bioabsorbable (Farah et al. 2016), recyclable, and compostable (Hartmann and Whiteman 2000; Auras et al. 2004; Lim et al. 2008) polymer. In its solid state, PLA can be amorphous or semi-crystalline depending on the spatial arrangement of its atoms, and thermal history (Loureiro and Esteves 2019). It is also the first commodity polymer that was produced from 100% annually renewable resources (Henton et al. 2005). PLA is a linear rigid TP polymer (Van den Eynde and Van Puyyelde 2018), whose chemical formula is $(C_3H_4O_2)_n$, and whose chemical structure is shown in Figure 3.1. PLA is derived not from fossil sources but renewable sources, that is by fermentation of polysaccharides and sugar contained in corn, sugarcane, sugar beet, rice, wheat, cassava and tapioca (Lovett and de Bie 2016), wheat, maize, and cellulose (Futerro 2019). PLA belongs to the family of aliphatic polyesters. Polyesters occupy a key role as biodegradable polymers, because their ester chemical bond is potentially broken down upon reacting with water, and hence they are biodegradable (Nampoothiri et al. 2010).

PLA is a very popular SP because: (a) it possesses values of barrier, thermal, mechanical, and optical properties that are comparable to those of some common fossil-based polymers (Auras et al. 2003) and high vs. those of other commercial SPs (Table 3.1); (b) it can be specifically formulated to provide enhanced mechanical and thermal properties that are adequate for some engineering applications (Tsuji 2005; Tan et al. 2016); (c) its production technology is mature for mass production (Sin et al. 2012), and compatible with conventional polymer processing equipment only slight modified (Lim et al. 2008); (d) its price per kg is low among SPs: in 2009 it was 1.9 EUR/kg vs. 2.5–5 EUR/kg for other SPs (Sin et al. 2012); (e) it is classified as *generally recognized as safe* (GRAS) by the U.S. Food and Drug Administration (FDA), and is safe for all food packaging applications (Jamshidian et al. 2010; FDA n.d.); (f) it can be converted in films, fibers, and non-wovens. For the same reasons, PLA is a most studied biobased polymer, and a very promising candidate to extensively replace conventional fossil-based polymers (Lasprilla et al. 2012; Saba et al. 2017). Along with ABS, PLA is also the most popular feedstock for FFF filaments for the following reasons: (a) it is easy to print with PLA because it has a lower printing temperature than that of ABS; (b) it does not warp as easily as ABS, and hence it does not always require a heated bed; (c) it is odorless when printing, at most emitting fumes smelling like sweet candy instead of an unpleasant smell like, f.e., that of ABS; (d) it is more environmentally friendly than most types of FFF filaments, being made from annually renewable resources such as corn starch, and sugar cane. PLA is also being investigated in powder and liquid form for other AM processes, namely PBF and VP (Grunewald 2015Van den Eynde and Van Puyyelde 2018), which can represent a profitable opportunity for material formulators.

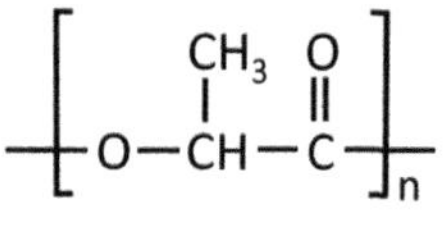

FIGURE 3.1 Chemical structure of PLA.

TABLE 3.1

Properties of PLA, Biodegradable and Non-Biodegradable Polymers

Polymer	Biodegradable	T_g	T_m	Tensile Strength	Tensile Modulus	Elongation at Break
		°C	°C	MPa	MPa	%
Cellulose	Yes	N/A	N/A	55–120	3,000–5,000	18–55
Cellulose Acetate	Yes	N/A	115	10	460	13–15
PBAT	Yes	–30	110–115	34–40	N/A	500–800
PCL	Yes	–60	59–64	4–28	390–470	700–1,000
PEA[a]	Yes	–20	125–190	25	180–220	400
PGA	Yes	35–40	225–230	890	7,000–8,400	30
PHA	Yes	–30 to 10	70–170	18–24	700–1,800	3–25
PHB	Yes	0	140–180	25–40	3,500	5–8
PHB-PHV	Yes	0–30	100–190	25–30	600–1,000	7–15
PLA	Yes	40–70	130–180	48–53	3,500	30–240
PTMAT	Yes	–30	108–110	22	100	700
PVA	Yes	58–85	180–230	28–46	380–530	N/A
Starch	Yes	N/A	110–115	35–80	600–850	580–820
LDPE	No	–100	98–115	8–20	300–500	100–1,000
PET	No	73–80	245–265	48–72	200–4,100	30–300
PS	No	70–115	100	34–50	2,300–3,300	1.2–2.5

Source: Clarinval, A. M., Halleux, J. 2005. Classification of biodegradable polymers. In *Biodegradable polymers for industrial applications*, ed. R. Smith, 3–31. Boca Raton: CRC. Reproduced with permission from CRC.
Note:
[a]Polyester amide.

PLA was discovered in 1932 by W. H. Carothers at DuPont (Jamshidian et al. 2010), but its low MW limited its physical and mechanical properties (as explained in Chapter 1), and hence its diffusion until 1954, when DuPont patented a high-MW version of it obtained by ring-opening polymerization of *lactide*, the cyclic carboxylic di-ester derived from lactic acid (Futerro n.d.). PLA was first sold in 1974 as a material for sutures produced by combining it with PGA (Mehta et al. 2005). Initially, PLA was confined to medical applications, namely internal sutures, tissue scaffolds, mesh, and medical implants, such as anchors, screws, plates, pins, and rods (Auras et al. 2010). These limited applications resulted not only from PLA being biocompatible due to its non-toxic feedstocks (Conn et al. 1995; Farah et al. 2016), and bioabsorbable due to its low MW (less than 100,000 g/mol), but also because of its considerable cost, and low availability. Starting around 1990, advances in PLA's formulation leading to economical production of high-MW grades (Lim et al. 2008), and new production routes, combined with FDA's approval, have increased its production and applications (Lunt 1998; Mohanty et al. 2000; Sodergard and Stolt 2002; Weber et al. 2002; Anonymous 2006; Siebott 2007). Figure 3.2 shows: (a) on the left, PLA pellets, processed as feedstock to manufacture end-use parts, such as drinking cups, tableware and cutlery, water bottles, medical screws, and phone case; (b) on the right water bottles made of PLA.

FIGURE 3.2   Left: pellets of Ingeo^{TM} PLA. Right: water bottles made of PLA.

Sources: Left: NatureWorks. Reproduced with permission from NatureWorks. Right: Sant'Anna Water SpA. Reproduced with permission from Sant'Anna Water SpA.

3.1.2 APPLICATIONS

PLA objects not printed have the following numerous uses: food (chip bags, trays for frozen cheese, fish, meat) and beverage packaging (water, yogurt and juice bottles, yogurt cups); diapers; disposable cutlery and tableware; drinking cups (due to PLA's properties similar to those of PS); shampoo bottles; clamshells; cosmetic packaging; shrink tunnels; tea bags; mulch films; films; fibers for nonwovens; textiles and carpets; engineering component for automotive: door trim ornamentations, scuff plates, cowl side trim, floor finish plate, toolbox, spare tire covers, floor mats, dashboard, door trim, etc. on Hyundai, Mazda, Mitsubishi, and Toyota cars (Bouzouita et al. 2017a); engineering components for electronics; medical products, such as dental materials, drug delivery systems, guided tissue and bone regeneration platforms, porous scaffolds for tissues growth and fracture fixation devices (Auras et al. 2004; Gupta et al. 2007; Jamshidian et al. 2010; Nampoothiri et al. 2010; Sin et al. 2012; Babu et al. 2013; Raquez et al. 2013; Murariu et al. 2014; Wertz et al. 2014; Cui et al. 2015; Hamad et al. 2015; Bayer 2017; Mallegni et al. 2018). Further products are listed in Jamshidian et al. (2010). The above non-AM applications are intended as a benchmark, indication, guide, and suggestion for PLA designers, researchers, and product developers, but most of current PLA products will not be profitable if they were currently printed, because they are being produced in large scale, at high rates, according to simple designs, and in one or a few versions of the same product. An exception is medical products such as implants, because they are manufactured in low volumes, and feature custom and complex designs, and, sometimes, multifunctionality. Moreover, some of the products listed above could be produced in a customized versions on demand, such as personalized cell phone cases. In conclusion, the diffusion of PLA in AM parts depends on developing applications that leverage the unique capabilities and economics of AM.

3.1.3 CURRENT AND FUTURE MARKET

The world production of PLA reached 0.31 million tonnes in 2018 and is predicted to reach 0.40 million tonnes in 2023, corresponding to 3.9% and 4.2% of the world production of biobased polymers in the same years (Chinthapalli et al. 2019). According to Jem et al. (2020) in 2018–2020 leading suppliers of PLA were NatureWorks USA), Total Corbion (Thailand and The

Netherlands), Hisun (China), BBCA & Galactic/Futerro (Belgium, China), and COFO (China) in decreasing order. The global market of PLA alone is growing, and expected to reach USD 3.7B by 2024 with a notable CAGR of 13.8% from 2019 to 2024 (Research and Markets 2019). The U.S. PLA market will maintain its growth and reach USD 1.9 billion in 2025 with packaging, textiles, transportation, agriculture, and electronics the largest markets in decreasing order. The European market of PLA considerably increased from USD 302 million in 2013 to USD 395 million in 2015, and in 2016 was predicted to grow at a high rate, and remarkably reach USD 1610 million in 2024, with the largest markets being packaging, textiles, transportation, and electronics in reducing order (Global Market Insights 2016). All of the above market forecast data obviously contain some uncertainty and, as mentioned in Chapter 1, vary with their source, but sources agree in predicting a worldwide expanding production of PLA, which should reflects well on the PLA feedstocks for AM by lowering their price and increase their availability, possibly resulting also in PLA versions that offer enhanced performance and are more competitive with other polymers. Currently, we expect that packaging and textiles will not likely be lucrative applications for printed PLA articles.

In October 2017, PLA's price in USD/ton in China, Germany, Japan, and USA was 4,770, 2,642, 2,160, and 1,910, respectively (Plastics Insights 2019). According to Bouzouita et al. (2017a), the price of PLA (2 USD/kg) is competitive with ABS (2.1–2.5 USD/kg), PA (3.3–3.6 USD/kg), PC (3.7–4 USD/kg), and PU (4.1–5.6 USD/kg), but not with PVC (0.93–1 USD/kg) and PP (1.2–1.3 USD/kg). PLA is mostly produced and utilized across North America, Asia, and Europe which are the world's geographic areas most productive overall. In Europe, Dutch companies and government have taken a leading role in the areas of sustainable business and economy (The Guardian 2013; Government of the Netherlands 2016), which is remarkable since The Netherlands' population was 17 million in 2017. In fact, in The Netherlands are headquartered not only suppliers of PLA such as Corbion, and of PLA filaments for AM such as Innofil3D BV, 3D4MAKERS BV, and Formfutura BV to name a few, but also manufacturers of 3D printers such as Ultimaker, Tractus, BLACKBELT 3D, and atum3D.

Other producers of PLA are, besides those mentioned: BASF, Bayer, FkuR, Uhde Inventa-Fischer (all in Germany); Danimer Scientific, Dow-DuPont, Eastman Chemical (USA); Mitsubishi Chemical, Mitsui, Teijin, and Toray (Japan); Purac-Sulzer-Synbra (The Netherlands and Switzerland); and Wei Mon Industry (Taiwan). Although this list is a sample, it includes some of the largest chemical companies, and may reveal that PLA is a profitable business with a bright future. In fact, according to Jem et al. (2020), the demand of PLA has doubled every 3–4 years since 2000.

3.1.4 Advantages of PLA

PLA features several advantages (Jamshidian et al. 2010) and disadvantages compared to other SPs, of which designers, material formulators, product developers, and investors need to be aware. Benefits of PLA include:

- *Versatile processability*. As anticipated, PLA can be processed leveraging several standard technologies, that is processes for fossil-based plastics: blending, blow molding, film and sheet casting, electrospinning, fiber spinning, film extrusion, film forming, foaming, injection molding, nanocompositing, thermoforming (Auras et al. 2004; Clarinval and Halleux 2005; Auras et al. 2010; Aldana et al. 2014; Farah et al. 2016; Garrison et al. 2016; Bayer 2017; Muller et al. 2017; Siracusa et al. 2017).
- *Rigidity superior to and clarity similar to PS and PET*. Tensile modulus (GPa) of PLA, PS, and PTE is 3.4, 2.9, and 2.8, respectively (Cygan 2009)
- *Environmental friendliness*. PLA comes from renewable feedstocks, and, as already mentioned, is biodegradable, recyclable, and compostable. PLA also reduces greenhouse gases, since its production consumes CO_2 (Dorgan et al. 2001). In fact, the amount of CO_2 absorbed

TABLE 3.2
Net Emissions of PLA Grades by NatureWorks®
and Fossil-Based Polymers

Polymer	Net Emissions
	kg CO_2 eq./kg polymer
PLA/NG	−0.7
PLA6	0.3
PP	1.9
PLA5	2
LDPE	2.1
PVC Suspension	2.7
PET am	3.2
GPPS	3.4
HIPS	3.4
PC	7.6
PA 6	7.9
PA 6,6	7.9

Source: Data from Jamshidian, M., Arab-Tehrany, E.A., Imran, M., Jacquot, M., Desobry, S. 2010. Poly-lactic acid: production, applications, nanocomposites, and release studies. *Comprehensive Reviews in Food Science and Food Safety* 9:552–571.
Abbreviations: PLA/NG: next generation. PLA5: PLA 2005 formulation. PLA6: PLA 2006 formulation. HIPS: high impact PS. GPPS: general purpose PS. PET am: amorphous, PVC Suspension is a type made by suspension method.

from air when corn is grown exceeds the amount of CO_2 emitted during PLA production (Bogaert and Coszach 2000). *Net* emissions are calculated as total emissions from cradle to factory gate minus CO_2 uptake occurring during corn production. When the total CO_2 consumption from cradle to factory is more than its emission to the environment, net emissions are negative. Table 3.2 (Vink et al. 2003) includes the net emission of fossil-based commodity polymers, and NatureWorks® PLA polymers, that is PLA/NG (Next Generation), PLA5 (2005 formulation) and PLA6 (2006 formulation). NatureWorks® PLA's net CO_2 emissions not only are the lowest except for those of PP, but also decreased from 2 in 2003 to 0.3 kg CO_2 equivalent per kg of polymer in 2006, and it was estimated that they will further shrink to −0.7 kg of CO_2 for the next PLA generation processed with electricity generated by wind energy. Another critical environmental advantage is that PLA requires smaller amount of fossil energy per kg of polymer produced than what needed for fossil-based commodities and engineering polymers, and will require even smaller energy in the near future using wind power (Table 3.3).

- *Biocompatibility*. PLA does not cause toxic or carcinogenic effects in host tissues, its degradation products do not impede tissue healing, and it has been approved by FDA for direct contact with biological fluids (Gupta et al. 2007). Its biodegradation products are nontoxic (Rasal et al. 2010).
- *Tunable performance*. By properly selecting the polymer architecture, the mechanical and physical properties can be partially tailored to meet specific application requirements, widening the PLA market (Migliaresi et al. 1991; Miyata and Masuku 1998).

TABLE 3.3

Amount of Fossil Energy Required to Produce 1 Kg of Polymer

Polymer	Fossil Energy for Production
	MJ/kg of polymer
PA 6,6	144
PA 6	120
PC	117
HIPS	92
General purpose PS	87
LDPE	82
PET, bottle grade	80
PP	78
PET, amorphous	78
First generation polylactide	55
Polylactide from biomass, using wind power	8

TABLE 3.4

Energy and Water Required for Polymer Production

Polymer	Water	Energy
1 lb.	gal	kWh
PLA (from corn)	8.29	7.39
PP	5.16	9.25
PET	7.44	10.17
EPS	20.53	11.17

Source: Shanghai Duxia Industry and Trade Co. Innovative biodegradable plastic manufacturing film blowing machine. https://bagmakingmachine-china.com/news-and-ustomer-case/biodegradable-plastic-film-blowing-machine/ (accessed January 16, 2021).

- *Energy savings*. Compared to PP, PET, and EPS, PLA production requires the least amount of energy, although more water than for PP and PET and less than for EPS (Table 3.4). Energy and water affect the polymer's price.

3.1.5 DISADVANTAGES OF PLA

PLA has the following drawbacks:

- *Low T_g and heat resistance*. T_g determines the maximum service temperature of polymers. PLA's T_g ranges between 40 and 70°C (Clarinval and Halleux 2005) depending on crystallinity, isomeric composition (Van de Velde and Kiekens 2002), MW, absorbed water, polydispersity (ratio of weight average MW to number average MW), amount of oligomers (Vert et al. 1981; Jamshidi et al. 1988), and thermal history (Farah et al. 2016). When PLA approaches its T_g, its molecular chains in the crystalline regions are unlikely to move but not the chains in the amorphous regions, comprising more rigid and less rigid segments Upon increasing temperature, crystalline and rigid amorphous regions hinder the mobility of the rest of PLA chains, and the resulting deformations under load, hence low crystallinity in PLA penalizes its heat resistance. Routes to improve PLA's heat resistance are (Nagarajan et al. 2016) adding compounds that provide nucleation sites (*nucleating agents*) around which polymer chains crystallize, blending PLA with heat-resistant polymers and nanofillers, annealing, and high temperature for injection molding followed by annealing. Yang et al. (2016) extensively discussed nucleating agents for PLA. Nucleating agents can be inorganic (talc, clay, $CaCO_3$, TiO_2, etc.), organic (f.e. aliphatic amide), and biobased (f.e. cellulose nanocrystals and starch). Inorganic agents are easily available at low cost. Organic compounds have the advantages to induce nucleation of PLA even at high temperatures, and are very finely dispersed in molten PLA. Biobased agents are melt-blended with molten PLA, and even at 1 wt% starch and cellulose nanocrystals were effective in increasing crystallization rate. Among blends of PLA, notable is the blend with PHBV and PBS because all three ingredients are biobased and biodegradable, and possess a HDT of 72°C vs. a typical value around 55°C (Zhang et al. 2012). Nagarajan et al. (2016) reported a long list of commercial PLA formulations featuring relatively high temperature resistance and toughness, and the values of their HDT, impact strength, and relative references. This list can include PLA candidates for AM.

- *Brittleness.* PLA is a brittle polymer because in tension it breaks without significant plastic deformation, featuring mostly an elongation at break of 2–10%, depending on its isomeric composition (Van de Velde and Kiekens 2002).

- *Low-impact resistance.* Impact resistance (*toughness*) depends on multiple factors including polymer chain structure, entanglement density, and morphology of PLA phase in case of multiphase PLA blends. PLA displays low crack initiation energy (measured by unnotched impact test) and low crack propagation energy (measured by notched impact test), and fails by crazing (Nagarajan et al. 2016). Notched Izod impact strength of PLA is 9%, 47%, 10%, and 12% that of PET, PS, HIPS, and PP, respectively (Cygan 2009), limiting the number of PLA's applications. Research was successfully conducted to toughen PLA: (a) at the expenses of the tensile modulus and yield strength (no free lunch), adding crosslinked core-shell impact modifiers, consisting in low T_g rubbery core encapsulated by a glassy shell featuring good interfacial adhesion with PLA (NatureWorks LLC 2007; Cygan 2013); (b) by blending PLA and formulating composites of PLA, which resulted in several commercial formulations (Nagarajan et al. 2016), including those for FFF filaments, such as Terraloy 3D-40040 (Grunewald 2015) featuring an outstanding unnotched Izod impact strength of 267 J/m. Glycidyl methacrylate (GMA) in varying forms, acrylic impact modifiers, random aliphatic copolyesters, PUs, and other flexible polymers have been blended with PLA to improve its toughness (Nagarajan et al. 2016).

- *Slow degradation rate.* PLA degrades primarily by hydrolysis, after several months of exposure to moisture, with degradation affected by its MW. Medical implants are an important application of PLA in AM (Yan et al. 2018). Medical implants in high-MW PLA showed complete resorption in 2–8 years, and extended permanence in vivo may lead to inflammation and infection in some organs (Bergsma et al. 1995). Therefore, implants made of low-MW (f.e. 60,000 g/mol) PLA are preferred because they degrade faster than implants in high-MW PLA, but they must degrade after bone healing is completed, otherwise they cannot provide adequate mechanical support (Mainil-Varlet 1998).

- *Thermal instability and degradation.* "PLA is thermally unstable and exhibits rapid loss of MW as the result of thermal treatment at processing temperatures. The ester linkages [i.e. bonds] of PLA tend to degrade during thermal processing or under hydrolytic conditions" (Jamshidian et al. 2010). PLA's degradation during processing is related to the processing temperature and the time spent in the extruder, and, in case of injection molding, in contact with the mold's heated elements (Taubner and Shishoo 2001). According to Farah et al. (2016), PLA thermally degrades above 200°C, whereas according to Jamshidian et al. (2010) PLA undergoes thermal degradation at temperatures lower than its "melting point," and its "degradation rate rapidly increases above the melting point." PLA's T_m varies depending on the reporting source that probably refers to some specific grade: it is 130–180°C for Clarinval and Halleux (2005), and 150–162°C for Van de Velde and Kiekens (2002).

- *Low crystallization rate.* Since PLA's semi-crystalline regions feature higher tensile modulus and strength than those in amorphous regions (Farah et al. 2016), low crystallization rate results in small percent (*degree*) of crystallinity, and hence low tensile strength and modulus. As mentioned, solid PLA can be amorphous or semi-crystalline, depending on its thermal history and the spatial position of its atoms (Farah et al. 2016). Semi-crystalline PLA is preferred over its amorphous version when higher mechanical properties are required.

- *Cost.* PLA is more costly than fossil-based polymers comparable in properties to it, which limits its diffusion. A strategy to reduce its cost and maintain a complete biodegradability is to blend it with low-cost biobased fillers, such as cellulose, natural fibers, and starch. Since these fillers contain a large amount of hydroxyl groups and are hydrophilic, whereas PLA is typically hydrophobic, when fillers and PLA are blended together, their interfacial adhesion is poor (originating brittle composites, especially at high filling concentrations), and needs to be enhanced. Other chapters of this book describe the properties of PLA blended with sustainable fillers, such as cellulose, natural fibers, and starch, hence here it is only mentioned that: (a) alkali treatment is an efficient way to improve the interface adhesion between PLA and natural fibers by removing some non-cellulosic components (hemicelluloses and lignin) and increasing the surface roughness of the fibers; (b) as to PLA/cellulose blend, surface modification of cellulose and adding compatibilizers increases tensile strength and modulus of PLA/cellulose composites, but it also penalizes their flexibility; (c) starch is inexpensive but far more hydrophilic than PLA, hence the PLA/starch weak interface bonding is a main obstacle to achieve high mechanical properties.

3.2 PRODUCTION OF PLA

PLA derives from the polymerization of *lactic acid* (LA) or (2-hydroxypropionic acid) ($C_3H_6O_3$) (Figure 3.3) that is an organic acid produced by chemical synthesis or, in case of LA specific for formulatingPLA, fermentation of 100% renewable sources (Murariu and Dubois 2016). LA is also present in the human body, where it is mainly produced in muscle cells and red blood cells, and forms when the body breaks down carbohydrates to utilize them as energy when oxygen levels are inadequate (MedlinePlus n.d.). *Lactide* is a natural and renewable compound produced from LA by fermentation of sucrose or glucose (Futerro 2019), and that is why PLA is also called *polylactide*. LA's molecule exists as two enantiomers, L- and D-lactic acid (Figure 3.4) that differ in the way they turn polarized light.

Currently, PLA is industrially produced through two major routes (Murariu et al. 2016):

- Direct polymerization by condensation (*polycondensation*) of LA
- Ring-opening polymerization (ROP) of lactide.

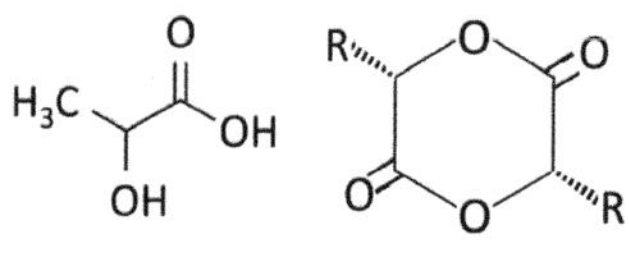

FIGURE 3.3 Left: chemical structure of lactic acid. Right: chemical structure of lactide. The dashed wedges represent bonds extending away from the viewer.

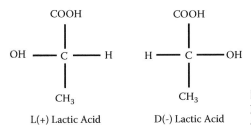

FIGURE 3.4 Molecules of lactic acid for formulating PLA: they are mirror images of each other and non-superimposable, and called *optical isomers*.

Direct polycondensation of LA requires removing water by condensation and employing solvent under high vacuum and temperature. Because eliminating water and impurities thoroughly is difficult, this method only generated PLA with low to intermediate MW (<100,000 g/mol) that can be increased by adding a step consisting in extending the molecular chains (Murariu and Dubois 2016).

ROP of lactide was patented in the 1990s by NatureWorks (patents by Gruber, Hall, Kolstad, Iwen, Benson, and Borchardt), the world's leading producer of PLA, and is based on a combination of agricultural (crop growing), biological (fermentation), and chemical (polymerization) routes. ROP consists of the following steps (Vink et al. 2003; Gruber 2005; Henton Mohanty 2005):

- Starch is extracted exclusively from corn and converted (through enzymatic hydrolysis) into glucose ($C_6H_{12}O_6$), a natural sugar, also known as dextrose.
- Through microbial fermentation, sugar is converted into LA, composed of 99.5% L-enantiomer.
- LA is polymerized through a chemical reaction of condensation into low MW (1,000–5,000 g/mol) prepolymer.
- The prepolymer is depolymerized, and converted into a mixture of monomers, namely L,L-lactide, D,D-lactide, and L,D-lactide or *meso-lactide*.
- The mixture is purified by means of vacuum distillation.
- Through a solvent-free ring-opening reaction in presence of a tin-based catalyst, the lactide is finally polymerized into PLA featuring high MW.
- After the polymerization is complete, any remaining monomer is removed under vacuum, and recycled to the beginning of the process.

Depending on the starting amount of L- and D-lactide chosen, three versions (*isomers*) of PLA are produced: poly(L-lactide) or PLLA, poly(D-lactide) or PDLA, and poly(D,L-lactide) or PDLLA, and all reciprocally differ in properties. Corn is the feedstock preferred by NatureWorks because it is the cheapest, starch-rich, and most widely available raw material in the USA, but in other countries crops locally grown, such as wheat, sweet potatoes, sugarcane, sugar beets, rice, and possibly others, can replace corn as feedstocks.

3.3 PROPERTIES OF PLA

Table 3.5 lists values of price, and physical, mechanical, and thermal properties of PLA and fossil-based polymers competing with PLA in some of the same applications: HIPS, PET, PP, and PS. The property values are only indicative, because the above polymers are commercialized in multiple formulations with specific tailored properties, such as high-impact strength or temperature resistance that may differ from the values in Table 3.5. Hence, selecting the actual material for engineering components should be based on the property values of the particular commercial formulations being considered, and possibly testing a prototype of the printed component. Nevertheless, values in Table 3.5 confirm what already mentioned, that is, in terms of basic formulations, PLA leads PET, PS, HIPS, and PP in tensile modulus and strength, but trails them in

TABLE 3.5

Properties and Price of PLA, PET, PS, HIPS, PP

Polymer	Price in 2019[e]	Density	Tensile Strength	Tensile Modulus	Elongation at Break	Notched Izod Impact Strength	Gardner Impact	Glass Transition Temperature	Heat Deflection Temperature at 0.46 MPa[c]	Heat Capacity	Thermal Conductivity
	EUR/kg	g/cm³	MPa	MPa	%	J/m	J	°C	°C	kJ/kgK	W/mK
PLA	2	1.24	53	3.4	6	12.8	0.06	55	53–56	1.63	0.13
PET	0.98	1.33	54	2.8	130	138.7	0.32	75	75–115	1.84	0.18
PS	1.27	1.05	45	2.9	7	26.7	0.51	105	75–100	2.26	0.21
HIPS	1.36	1.03–1.08[a]	23	2.1	45	122.7	11.3	93–105[b]	80–100	1.4[d]	0.16–0.22
PP	1.13	1.09	31	0.9	120	106.7	0.79	−10	100–120	1.26	0.12

Source: Cygan, Z. 2009. Improving Processing and Properties of Polylactic Acid. http://plasticstrends.com/index.php/last-months-mainmenu-28/218-improving-processing-and-properties-of-polylactic-acid (accessed May 31, 2018).

Notes:

[a] Polymer Database High Impact Polystyrene. https://polymerdatabase.com/Commercial%20Polymers/PS2.html.

[b] Harper (1996).

[c] Omnexus HDT @ 0.46 MPa https://omnexus.specialchem.com/polymer-properties/properties/hdt-0-46-mpa-67-psi.

[d] https://www.makeitfrom.com/material-properties/High-Impact-Polystyrene-HIPS.

[e] Vraag en Aanbod. 2019. Plastics Prices. In Dutch. https://www.vraagenaanbod.nl/vraag-aanbod-kunststofprijzen-week-36-2019/.

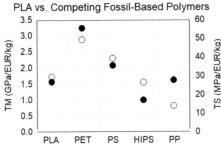

FIGURE 3.5 Ratio tensile modulus/price and tensile strength/price of PLA and fossil-based polymers competing with PLA in similar applications.

elongation at break, impact strength, T_g, and HDT. The higher the T_g, and HDT, the higher the material's maximum service temperature. In Figure 3.5 PLA is compared to PET, PS, HIPS, and PP in terms of tensile strength/price, and tensile modulus/price, measured in MPa/EUR/kg and GPa/EUR/kg, respectively, based on September 2019 prices. PLA is in the middle of the range of both metrics, lagging behind PET and PS, and competing in tensile strength/price with PP, and in tensile modulus/price with HIPs. In Table 3.6, using value ranges, PLA is compared to other biodegradable, biobased polymers: cellulose, cellulose acetate, PHA, PHB, PHB-PHV, and starch. Given the ranges of tensile strength (10–120 MPa) and modulus (460–5,000 MPa), and elongation at break (1–820%), PLA exhibits average strength, upper stiffness, and bottom ductility. T_g and T_m have practical implications. The lower the T_m the less energy is required to melt the polymer during printing, and the more economical is the process. The higher the T_g the higher the material's service temperature, since above T_g polymer chains are mobile, making

TABLE 3.6
Price and Properties of Biodegradable Biobased Polymers

Polymer	Tensile Strength	Tensile Modulus	Elongation to Break	T_g	T_m	Price, 2016[b]	Price, 2009[c]
	MPa	MPa	%	°C	°C	EUR/kg	EUR/kg
Cellulose	55–120	3,000–5,000	18–55	N/A	N/A	N/A	5
Cellulose Acetate	10	460	13–15	N/A	115	5	N/A
PHA	18–24	700–1,800	3–25	30–10	70–170	5	N/A
PHB	25–40	3,500	5–8	5–15[d]	140–180	N/A	4
PHB-PHV	25–30	600–1,000	7–15	0–30	100–190	N/A	3.4
PLA[a]	45–61	3,600–4,700	1–6	55–60	145–180	2	1.9
Starch	35–80	600–850	580–820	N/A	110–115	2–4	3.5

Sources: Adapted from Clarinval, A.-M., Halleux, J. 2005. Classification of biodegradable polymers 2005. In *Biodegradable polymers for industrial applications*, ed. R. Smith, 3–31 Woodhead Publishing.

Notes:

[a]NatureWorks Ingeo Resin Product Guide 2017. https://www.natureworksllc.com/~/media/Files/NatureWorks/Technical-Documents/One-Pagers/ingeo-resin-grades-brochure_pdf.pdf.

[b]Data from Van den Oever M., Molenveld K., van der Zee M., Bos H. 2017. Bio-based and biodegradable plastics – Facts and Figures. http://dx.doi.org/10.18174/408350.

[c]Data from Sin L. T., Rahmat A. R., Rahman W. A. 2012. Overview of poly(lactic acid). In *Handbook of biopolymers and Biodegradable Plastics*, ed. Sina Ebnesajjad. Elsevier.

[d]Van de Velde K., Kiekens P. 2002. *Polymer Testing* 21 (4): 433–442.

polymers soft, weak, flexible, and rubbery, whereas below the T_g there is no polymer chain mobility, rendering polymers hard, strong, stiff, and brittle.

In high-MW PLA the ratio of L- and D-isomer and MW affect its rate of crystallization, and hence degree of crystallinity, mechanical properties, and processing temperatures (Auras et al. 2004). Particularly, MW affects mechanical strength, degradation, solubility (Farah et al. 2016), and T_g. F.e., increasing PLLA's MW from 50,000 to 100,000 g/mol doubles its tensile modulus, and increasing MW from 50,000 to 150,000 to 200,000 g/mol raises its tensile strength from 15.5 to 80 to 150 MPa respectively (Van de Velde and Kiekens 2002). Crystallinity influences tensile modulus and strength, hardness, and T_m (Farah et al. 2016). All commercial PLAs consist of L-, and D,L-lactide monomers (Ajiro et al. 2012), with varying ratios of L- and D-lactide. PLA derived from >93% L-lactide is semi-crystalline, while PLA produced from 50 to 93% L-lactide is amorphous. D,L-lactide may feature an equimolar (or *racemic*) mixture (*rac-lactide*), and non-equimolar mixture (*meso-lactide*) of D- and L-enantiomer (Auras et al. 2004). T_m and T_g of PLA decrease with decreasing L-isomer content (Tsuji and Ikada 1996; Dorgan et al. 2001; Urayama et al. 2003). PLA's T_g is dependent on both MW and ratio of L- and D-isomer (Dorgan et al. 2001). T_g increases with MW, and reaches maximum values of 60.2, 56.4, and 54.6°C in PLA consisting of 100, 80, and 50% L-lactide contents, respectively (Lim et al. 2008). Processing of PLA may affect its MW and cause degradation, as mentioned, and ultimately lower its properties. F.e., Perego et al. (1996) reported that injection molding decreased MW by 14–40%, and combining extrusion and injection molding also reduced MW (Farah et al. 2016). Table 3.7 illustrates how D- and L-isomer affected properties of PLA, and annealed PLLA surpassed PDLLA and not annealed PLLA in all properties reported, except elongation at break. Particularly, PDLLA is more rigid and brittle than PLLA, whether the latter is stronger and more ductile. Table 3.8 compares PLA, PDLLA, and PLLA to other biodegradable polymers PHB, PCL, and PGA in terms of major properties, realistically expressed as ranges of values. Comsidering all its formulations, PLA fares well: in density it is similar to PHB, and between PCL and PGA, its strength spans the range of that of PCL, PHB and PGA, and in modulus outperforms PCL but not PGA, and overlaps with PHB.

When PLA is extruded into a filament for AM, its flow behavior is key, because it influences the processability, printing quality, and mechanical properties of the printed article. The flow behavior in turn is affected by branching degree and molar mass distribution, configuration of L and D enantiomers in PLA, and additives (Dorgan et al. 2000; Mohanty et al. 2005; Silva et al. 2014). Critical PLA's properties to control printability and final product quality are T_g, T_m, heat capacity, and shear stress (Wang et al. 2017a).

TABLE 3.7
Properties of PLA Types

PLA Composition	Tensile Strength	Tensile Modulus	Elongation at Break	Flexural Strength	Unnotched Izod Impact	Notched Izod Impact	Heat Deflection Temperature
	MPa	MPa	%	MPa	J/m	J/m	°C
D,L-PLA (PDLLA)	44	3,900	5.4	88	150	18	50
L-PLA (PLLA)	59	3,750	7	106	195	26	55
Annealed L-PLA (PLLA)	66	4,150	4	119	350	66	61

Source: Adapted from Perego, G., Cella, G. D., Bastioli, C. 1996. Effect of molecular weight and crystallinity on poly (lactic acid) mechanical properties, *J. Appl. Polym. Sci.* 59:37–43.

TABLE 3.8

Properties of PLA Grades and Other Biodegradable Polymers

PLA Composition	Density	Tensile Strength	Tensile Modulus	Elongation at Break	T_g	Molding Temperature
	g/cm^3	MPa	GPa	%	°C	°C
PLA	1.21–1.25	21–60	0.35–3.5	2.5–6	45–60	150–162
PDLLA	1.25–1.27	28–50	1–3.45	2.0–10	50–60	NA (amorphous)
PLLA	1.24–1.30	16–150	2.7–4.14	3.0–10	55–65	170–200
PCL	1.06–1.13[a]	14–27[b]	0.25–0.43[b]	20–120	-60[a]	60[a]
PHB	1.2–1.3[c]	14–43[c]	0.8–3.5[c]	4–15	5–15	168–182
PGA	1.50–1.71	60–100	6.0–7.0	1.5–20	35–45	220–233

Sources: Van de Velde, K., Kiekens, P. 2002. *Polymer Testing* 21 (4): 433–442.

Notes:

[a]Tsuji, H. 2002. Polylactide in *Polyesters 3* (Biopolymers, vol. 4), eds. Y. Doi and A. Steinbuchel, 129–177, Wiley-VCH, Weinheim.

[b]Eshraghi, S., Das, S. 2010. Mechanical and microstructural properties of polycaprolactone scaffolds with one-dimensional, two-dimensional, and three-dimensional orthogonally oriented porous architectures produced by selective laser sintering. *Acta Biomaterialia* 6 (7): 2467–2476. https://doi.org/10.1016/j.actbio.2010.02.002.

[c]Values from commercial grades.

3.4 PLA FEEDSTOCKS FOR FFF

This section describes physical and mechanical properties (Table 3.9) and processing conditions (Table 3.10) of examples of commercial PLA polymers that are produced by world's leading companies, and are feedstocks in form of pellets for extruding AM filaments. These examples are BioFlex® F 7510 by FkuR (Germany); General Purpose PLA by Futerro® (Belgium); Ingeo™ 3D850, 3D870, 3D860, and 4043D by NatureWorks (USA); and Luminy® L175 and LX175 and Compound C by Total Corbion (The Netherlands). These suppliers confirm the leading role of Europe and USA in production and utilization of sustainable materials. More suppliers are present in China and India. The information in this section is intended for developers of AM filaments and pellets, and more details are published on the PLA suppliers' websites. However, the available data sheets of the above mentioned suppliers omit property values required to design load-bearing components, such as shear strength and modulus, Poisson's ratio, creep, and properties measured above and below room temperature. On the other hand, the property values that most matter to designers and fabricators of AM parts are those of the filaments and pellets feeding the printers, and the latter values are not necessarily the same values grouped in Table 3.9, because of the additives and processing involved in converting the base ingredients in Table 3.9 into filaments and pellets for AM. The technical data sheets of the PLA feedstocks in Table 3.9 do not list all the same properties, or use the same units (such as for impact strength), and, hence it is not possible to comprehensively compare all materials to each other. However, some observations can be still drawn. Although these polymers are based on the same chemical formula, their property values span a wide range (f.e. tensile strength spans from 24 to 55 MPa, and tensile modulus ranges from 2,315 to 4,000 MPa), and provide choices to the designers and material users. Table 3.9 also shows that material producers were able to overcome some limitations of basic PLA, by developing grades with improved HDT and impact strength, such as L175 for the former and 3D860 for the latter. No material outperforms all the others in all the property reported. For example, 3D860 has the largest notched Izod impact strength, but its tensile strength is the lowest, whereas Compound C leads in tensile modulus, elongation at break, and notched Charpy impact strength, but features

TABLE 3.9

Properties of Commercial PLAs Converted into AM Feedstocks

Property	Melt Flow Index	Specific Gravity	T_g	T_m	Tensile Strength	Tensile Strength at Yield	Tensile Modulus	Elongation at Break	Flexural Strength	Flexural Modulus	Izod Notched Impact Strength	Charpy Notched Impact Strength	HDT at 0.45 MPa
Unit	g/min	–	°C	°C	MPa	MPa	MPa	%	MPa	MPa	J/m	kJ/m²	°C
Range	6–15	1.22–1.27	52–60	145–220	24–55	33–60	2,315–4,000	1.6–47[a]	46–83	1,979–3,824	16–233	4.2–18	55–110

Source: Online data sheet of: Ingeo™ 4043D, 3D850, 3D870, and 3D860 (NatureWorks), Luminy® L175, LX175, Compound C (Total Corbion), BioFlex® F 7510 (FkuR), and General Purpose PLA (Futerro).

Notes:

[a] 47% is an outlier. Most values collected do not exceed 6%.

TABLE 3.10

Extrusion Parameters of PLA Formulations for AM Feedstocks

Parameter	Length/ Diameter Ratio	Compression Ratio	Temperature					Screw Speed
			Melt	Feed Section	Compression Section	Metering Section	Die	
Unit			°C	°C	°C	°C	°C	RPM
Value	24:1–32:1	2.5:1–3.0:1	150–210	45–200	190–200	200–210	190–210	20–150

Source: Online data sheet of: Ingeo™ 4043D, 3D850, 3D870, and 3D860 (NatureWorks), Luminy® L175, LX175, Compound C (Total Corbion), BioFlex® F 7510 (FkuR), and General Purpose PLA (Futerro).

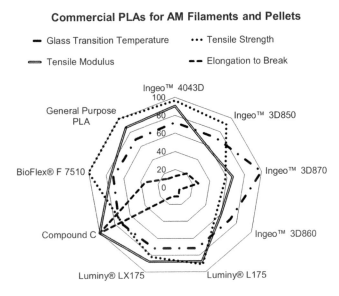

FIGURE 3.6 Normalized values of selected properties of commercial PLA for AM filaments and pellets.

Source: Data sheets posted online by suppliers.

intermediate tensile strength. In Figure 3.6, all materials are visually compared to each other in terms of all the properties common to all of them. As to the Ingeo™ polymers, 3D850, 3D870, and 3D860 surpass 4043D in impact strength, HDT, and melt flow index, but trail in tensile strength and modulus. The melt flow index (MFI) or melt flow rate (MFR) is expressed in g/(10 min), and indicates how fast a polymer flows in its molten state. MFI affects polymer processability and is related to the interlayer cohesion in printed articles. In case of FFF, higher MFI reduces risk of clogging printer nozzle. In the four Ingeo™ PLAs increasing the impact strength is accomplished at the expenses of their tensile strength (Figure 3.7). The Ingeo™ formulations show that, as expected for polymers, there is some positive correlation between tensile modulus and flexural modulus (Figure 3.8). According to NatureWorks, 3D870 offers excellent printing characteristics, such as precise details, good adhesion to build plates, low warping, curling, and odor. Ingeo™ 4043D and 3D850 received FDA approval to be utilized in food packaging materials. The Ingeo™ polymers are sold to the following manufacturers of FFF filament that mix them with several

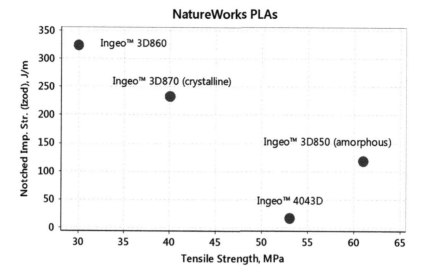

FIGURE 3.7 Impact strength and tensile strengths of Ingeo™ PLA grades.

Source: Data sheets posted online by the supplier.

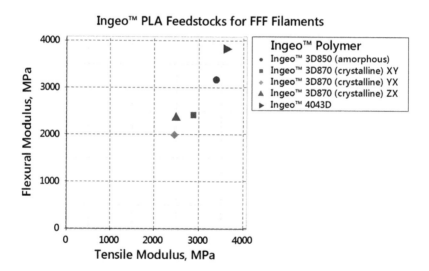

FIGURE 3.8 Flexural and tensile moduli of Ingeo™ PLA grades for AM.

Source: Data sheets posted online by the supplier.

ingredients (pigments, plasticizer, fillers, etc.) to obtain the final product: 3DomFuel (USA, Ireland), Diamond Age (New Zealand), Gehr (Germany), Kingfa (China), MakerBot (USA), Plastrude (Australia), Polymaker (China), TLC Korea (South Korea), and Unitika (Japan).

Since PLA is considered a potential replacement for fossil-based plastics, Figure 3.9 the PLAs included in Table 3.9 have been compared in Figure 3.9 in terms of tensile strength and modulus to fossil-based TPs ABS, HIPS, PP, PET, and PS selected as benchmarks because ABS is the typical alternative to PLA for AM (at least for desktop printers) and the others are commodities with price similar to PLA's. The values for non-PLA polymers are averages of values reported by Harper

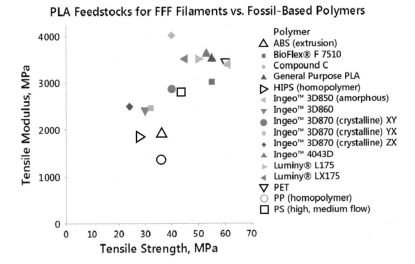

FIGURE 3.9 Comparison between polymers in Table 3.9 and fossil-based commodity polymers.

(1996). In Figure 3.9, most of the PLA grades are located in the upper-right corner, outperforming the basic version of the fossil-based polymers plotted. However, obviously, price is key in the material selection process by designers and fabricators. Assuming that the prices in Table 3.5 apply to all materials in Figure 3.9, PLA's price is about 0.5 to 2 times that of the fossil-based polymers in Figure 3.9. Figure 3.9 also illustrates that, within the range of materials considered, higher strength accompanies higher modulus, as it is the case for polymers broadly speaking (Granta 2010). However, if we consider the specific product lines by NatureWorks and Total Corbion, their PLA versions with highest tensile modulus (4043D and Compound C, respectively) do not lead in tensile strength as well.

The MFI in Table 3.9 has practical importance. A high value of it often indicates a lower coefficient of friction, which offers multiple advantages: (a) the extruder's motor on FFF printers does less work to extrude the same length of filament, and can then print faster; (b) extruder's cooler operating temperatures, and less wear on extruder's parts; (c) a lower probability of extruder gear slippage or stripping of the filament when back pressure occurs in the hot end (Bouthillier 2016). T_g is also significant to users, because it dictates the polymer's service temperature range. Below T_g there is no polymer chain mobility, and polymers are glassy, that is hard, strong, stiff, and brittle, whereas above it the polymer chains are mobile, and polymers are rubbery, namely soft, weak, and flexible. Figure 3.10 represents the variation of the dynamic storage modulus E' of PLA with temperature, with E' abruptly dropping above 55°C to a value inadequate for practical applications.

As mentioned earlier, the PLAs in Table 3.9 are sold in pellets, and converted into filaments for FFF by being processed in extruders. About 1/3 of all plastics products made worldwide out of TPs are fabricated through extrusion (Rosato et al. 2004), and span from simple shapes like films, rods, strips, and sheets up to complex profiles. The single screw extruder is the most common equipment in the polymer industry (Osswald, Menges 2012), and is schematically shown in Figure 3.11. Extrusion is a continuous process, starting by pouring a mixture of pellets, fillers, and additives (such as PLA, metal powder, pigments, plasticizers) in the hopper, from which the mixture enters a heated barrel, where the temperature is gradually raised from the feeding end (location of the hopper) to the discharging end (where the filament exits) of the extruder. In the barrel the mixture is melted, and, by a long rotating shearing screw, compressed and pumped forward, and finally forced through a metal plate (*die*), featuring an orifice whose diameter dictates the filament's

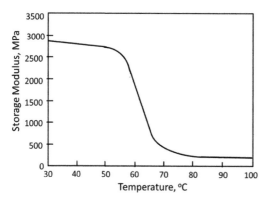

FIGURE 3.10 Dynamic storage modulus of PLA measured using DMA.

Source: Adapted from Silverajah, V. S. G., Ibrahim, N. A., Yunus, W. M. Z. W., Hassan, H. A., Woei, C. B. 2012. A Comparative Study on the Mechanical, Thermal and Morphological Characterization of Poly (lactic acid)/Epoxidized Palm Oil Blend. *Int. J. Mol. Sci.* 13:5878–5898. Reproduced under the terms and conditions of the Creative Commons Attribution license.

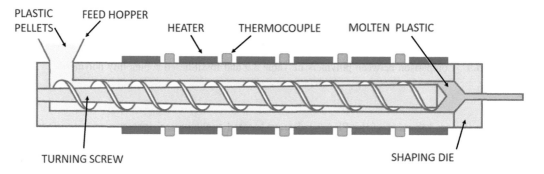

FIGURE 3.11 Schematic representation of single-screw extruder.

Source: Courtesy of Francis Paesano.

diameter. Outside the die, the FFF filament is drawn by a pulling mechanism, calibrated, cooled, wound up into a spool, and cut. Table 3.10 lists the value ranges of extrusion parameters relative to the PLA feedstocks reported in Table 3.9. Feed, compression, and metering are three sections of the extruder screw that perform the corresponding functions. Key extruding parameters are temperature profile along the screw length, screw speed and torque, meeting rate, and feedstock temperature. PLA feedstock suppliers recommend optimal values of temperature for printing, print bed, and annealing of printed parts.

Total Corbion is a 50/50 joint venture between French multinational Total, and Dutch company Corbion, and formulates three grades of PLA for AM: standard grade Luminy® LX175, high-heat Luminy® L175 homopolymer, and high-impact Compound C. Properties of Total Corbion's PLAs were measured following the European ISO standard, whereas the Ingeo™ polymers were tested using the American ASTM standards, however, the ISO and ASTM test methods for the same property practically coincide. Compound C stands out, featuring the highest tensile modulus (4,000 MPa), elongation at break (47%, three times the second highest value 16%), and Charpy impact strength (18 kJ/m²) among the polymers in Table 3.9. Luminy® PLA polymers are compostable and compliant with the EN-13432 specification for compostability. They are food-safe, since they comply with European Union regulation 10/2011 on plastic materials and articles intended to come into contact with food. The extrusion parameters for LX175 and L175 are: length/diameter ratio 24:1–32:1, feed zone temperature 20–40°C, mixing and conveying temperature 190–210°C, die head temperature 190–210°C. The differences are in pre-drying temperature and melt zone temperature. Drying before extrusion is critical, because it removes the moisture, and improves the resulting filament's overall properties, and functional performance.

NatureWorks, maker of Ingeo™ PLAs, is jointly owned by PTT Global Chemical, a Thai state-owned company, and Cargill. NatureWorks offers the option to improve the mechanical properties of their PLA through *annealing*, which consists of heating at a controlled rate a plastic component up for a specific period of time, before cooling the component down. Annealing is performed to relieve the internal stresses that are generated into the parts during their fabrication, and can cause dimensional distortion and cracking of the part. Printed parts in PLA are annealed at about half of T_m, in order to increase PLA's crystallinity, which in turn improves its mechanical and thermal properties. F.e. annealing for Ingeo™ 3D850 is performed at 110°C for 15 min.

FKuR was founded in Germany in 1992 as Foundation of the Research Institute for Plastics and Recycling, and is currently a global company with operations in several continents, reaching a production capacity of 20,000 tonnes per year in 2012. FKuR supplies Bio-Flex® F 7510, a bio-degradable batch based on PLA, also containing copolyester and additives. FKuR does not disclose the processing conditions for extruding Bio-Flex® F 7510.

Futerro is a Belgian company established in 2007 that sells three versions of PLA: injection grade, sheet grade, and extrusion grade. The extrusion grade is made from LA sourced from Galactic (Belgium), and can be converted into AM filament, besides being suited for a wide range of other products, such as dairy containers, food serviceware, transparent food containers, blister packaging, cold drink cups, candy twist wrap, salad and vegetable bags, envelope window film, label film, and lidding film.

3.5 COMMERCIAL UNFILLED PLA FILAMENTS FOR FFF

3.5.1 INTRODUCTION

The following subsections elucidate key aspects of the most important material properties of unfilled PLA filaments for FFF by focusing on a representative sample of commercial filaments, whose full list is extensive and constantly growing.

This chapter focuses on (a) filaments that are supplied by the largest and best-known manufacturers, and feature broader lists of material property data, and (b) filaments with high values of specific properties, like impact strength. Suppliers are located mostly across Asia (China, Japan, South Korea), Europe (especially The Netherlands), and USA, and commercialize numerous versions of PLA and PLA-based filaments that reciprocally differ in performance (high temperature resistance, impact resistance, tensile modulus, etc.), filler (metals, natural ingredients such as wood, cellulose, fibers, etc.), and look (color, finish). Based on the property values, the short list of properties disclosed, and the wide range of finish and (especially) colors, it seems that most filament suppliers target hobbyists and do-it-yourself users rather than companies. Many filaments suppliers are small companies recently established, creative and fast in innovating their offering. Some small suppliers have been acquired by a large company, as in the case of Innofil3D, formed in 2014 and fully purchased in 2017 by the multinational corporation BASF. This section is also intended for designers and fabricators of AM parts, and illustrates the options for them offered by commercial PLA filaments for FFF, whose material property data sheets are not as extensive as those of established plastics formulated for conventional fabrication processes (such as injection molding and compression molding), and engineering applications, and, normally, they lack property values outside room temperature, and long-term properties, such as creep and fatigue curves. This data scarcity is explained by the following: (a) PLA filaments for FFF are newer than conventional plastics, and mostly employed for models and prototypes for room temperature and no or low load-bearing applications, and (b) testing is time consuming and costly, particularly for small companies like many PLA filament producers. Obviously, PLA filaments suited for engineering applications must feature a wide-ranging data set of material properties, measured under several conditions (static, impact, fatigue, varying temperature and humidity conditions, etc.). Even when all the material properties necessary for structural designing are available, also required are the *design allowables*, statistically-based strength values that are calculated from test values, and enable to design load-bearing parts with safety margin. Filament suppliers report material properties measured according to different test standards (ASTM and ISO), different test methods (f.e. impact strength assessed according to Izod or Charpy method), and different test conditions (f.e. HDT quantified under stress of 0.45 MPa vs. 1.8 MPa), and hence it is not always possible comparing filaments across all their released properties.

As explained in Section 2.3.2, properties of articles printed via FFF depend on multiple printing settings. Torres et al. (2016) measured the influence of extrusion temperature and speed, infill direction and density, layer thickness, and perimeter layers on the tensile (strength, modulus and strain, toughness) and fracture (critical stress intensity factor) properties of coupons printed out of PLA via FFF using the statistical technique of design of experiment (DoE) that maximizes the output of information while minimizing the number of tests. The authors identified the combination of process parameters that consistently yielded medium to high values for all properties tested, whereas other combinations led to higher values for either fracture or tensile properties.

This section contains values in form of tables and plots of major material properties of selected PLA filaments that are produced by the largest suppliers. Being PLA the most popular SP for AM, it exceeds all other SPs for AM in the amount of disclosed material property data, although the values are relative almost exclusively to tensile, flexural, thermal, and impact properties.

Before delving in specific aspects of the mechanical properties of commercial PLA filaments, in Figure 3.12 the tensile strength and modulus values of commercial PLA filaments of major suppliers are compared to those of ABS, HIPS, PET PP, PS that are fossil-based benchmarks for PLA. Only PLA filaments whose tensile modulus and strength are measured from dumbbell-shaped coupons (not films) are included. Figure 3.12 shows that in those properties the PLA filaments are comparable to ABS, HIPS, PET PP, and PS, with most PLA filaments being stiffer than PP, HIPS, and ABS. The range of values for the PLA filaments are quite wide: 20 to 60 MPa for tensile strength, and 1,820 to 4,000 MPa for tensile modulus. Only a few suppliers disclosed the standard deviation values, and therefore it is unknown how significantly the plotted property values differ among most of the materials. It is worthwhile noting that filaments' material properties: (a) differ from those of the resulting printed part, which in turn depend on the printing parameters, void content, etc. (b) are anisotropic and affected by the direction in which they are measured with respect to the deposition direction of the filament, and hence,

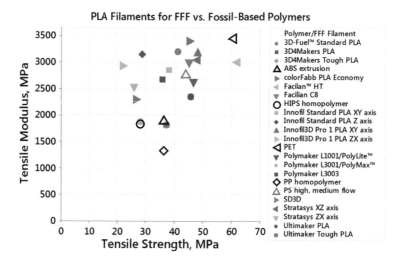

FIGURE 3.12   Tensile modulus and strength values of major PLA filaments for FFF (solid symbols) and commodity fossil-based polymers ABS, HIPS, PET PP, PS (hollow symbols).

Source: Data sheets posted online by PLA suppliers; Harper C. ed. 1996. *Handbook of Plastics, Elastomers, and Composites*. New York: McGraw-Hill. 3rd edition for ABS, HIPS, PET PP, and PS.

TABLE 3.11
Material Properties of Common Commercial PLA Filaments for FFF

Property	Density	Tensile Strength	Tensile Modulus	Elongation at Break	Flexural Strength	Flexural Modulus	Izod Notched Impact Strength	HDT at 0.45 MPa	Vicat Softening Temperature
Unit	g/cm^3	MPa	MPa	%	MPa	MPa	J/m	°C	°C
Range	0.8–1.4	21.8–71.0	1,890–3,400	1–47	45–99	2,119–4,451	16–323	45–144	60–95

Sources: Online data sheets of 3D4Makers, 3D-Fuel, colorFabb, Formfutura, Innofil3D, MakerBot, Polymaker, SD3D, Stratasys, and Ultimaker.

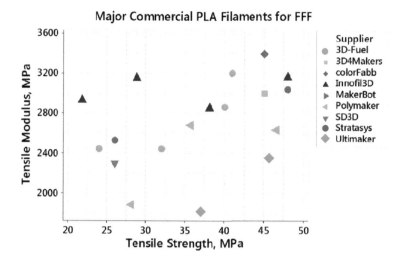

FIGURE 3.13   Tensile modulus and strength of major PLA filaments for FFF.

Source: Data sheets posted online by suppliers.

to be conservative, the lowest values of mechanical properties should be chosen when designing load-bearing parts. Finally, tensile properties are only one metric to characterize the mechanical performance of plastics, thus Figure 3.12 represents one, although critical, detail of the whole picture.

Table 3.11 includes a summary of material properties of commercial PLA filaments produced by major suppliers worldwide. The reader is referred to technical data sheets published online by each supplier for more details and their latest versions available. Some of the major producers in terms of size and longevity are 3D4MAKERS, 3D-Fuel®, 3DXTECH, colorFabb, Formfutura, Innofil3D, MakerBot, Polymaker, SD3D, Stratasys, and Ultimaker. The range of tensile properties is quite wide, with the maximum values of tensile strength and modulus being about three times and almost two times their minimum values, respectively. Obviously, since no PLA outperforms all the others in every property, in selecting a material, trade-offs in its property values are necessary, and the performance metrics of the application being evaluated must be prioritized. Figure 3.13 delimitates the design space of the commercial PLA filaments in Table 3.11, whose tensile strength and modulus were disclosed by their suppliers. Since the data sheets of some commercial PLA filaments exclude tensile modulus but include flexural modulus, Figure 3.14 shows tensile modulus versus flexural modulus of some commercial PLA filaments Figure 3.14, and some positive correlation seems to exist. Finally, there is evidence that passing through the printer's extruder alters the tensile properties of PLA filaments. Grasso et al. (2018) tested an unspecified PLA filament, and measured that its tensile modulus after extrusion from a printer decreased by 30% and 16% when the extrusion temperature was 40°C and 50°C, respectively.

3.5.2   ANISOTROPY

As described in Chapter 2, the layer-by-layer nature of the FFF process imparts anisotropic properties to printed articles depending on the orientation in which they are printed that can be one of the three orientations illustrated in Figure 3.15. Anisotropy extends to articles made of PLA, and in fact some PLA suppliers provide values of tensile, flexural, and impact properties collected on test coupons printed in multiple orientations. When property values are reported without specifying their test orientation, those designing load-bearing articles should measure properties in different

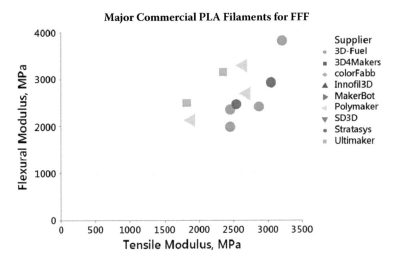

FIGURE 3.14 Tensile and flexural modulus of major PLA filaments for FFF.

Source: Data sheets posted online by suppliers.

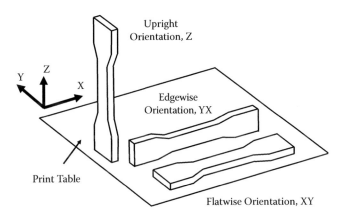

FIGURE 3.15 Printing orientations for properties reported by some suppliers of PLA filaments. Layers of material are stacked along the Z direction. Tensile coupons shown.

orientations, use the lowest values for their modeling, and possibly test the printed article under its expected service loads.

Anisotropy can be significant in elastic properties (tensile and flexural moduli) and ultimate properties (tensile, flexural, and impact strengths, and elongation at break). Figure 3.16 exemplifies the differences in tensile strength and modulus reported for test coupons of three common PLA filaments depending on their printing orientation indicated in Figure 3.15, and shows that, across the three filaments, the effect of anisotropy is superior on strength than modulus. Figure 3.17 illustrates the effect of printing orientation on flexural strength and modulus, and impact strength.

3.5.3 INFILL

A printed part can be inside completely filled or porous, even if its outer surface is non-porous. Infill is the part's volume occupied by its outer surface minus the total porosity volume, or, in brief,

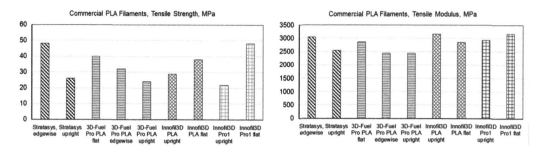

FIGURE 3.16 Effect of printing orientation on tensile properties of major PLA filaments for FFF.

Source: Data sheets posted online by suppliers.

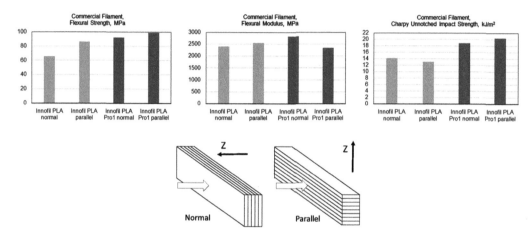

FIGURE 3.17 Upper plots: variation of flexural modulus and strength and impact strength values of PLA filaments for FFF with printing orientation. Lower drawings: test set up to measure impact and flexural properties. White arrows represent the load direction during bending impact. During printing the strands are laid in the XY plane perpendicularly to to Z.

Source: Values in plots from Innofil3D's online data sheet, samples drawings adapted from Innofil3D's online data sheet.

the part's volume occupied by material. Infill is expressed on a scale from 0 (theoretical) to 100% (a part without voids). Infill percentage and printing pattern influence weight, and functional properties of the printed object. Lowering infill decreases the amount of time and material spent to print a part and its overall cost, but it also penalizes its physical and mechanical properties. The effect of infill is illustrated in Table 3.12 by the values of the tensile properties of two Innofil3D filaments: a 50% versus 100% infill caused a reduction in strength and modulus of 37–39% depending on filament and measured property. It is worth noting that: (a) 50% infill means <50% of total amount of voids because the shell's layers are printed with 100% infill, (b) the direction of infill with respect to the testing direction also affects the measured property's value. The value of tensile modulus as a function of infill can be closely predicted by applying the equation (3.1) (Paul 1960) with E_p and E_0 being tensile modulus at P volumetric porosity and tensile modulus at zero volumetric porosity, respectively, and P expressed as a value from 0 to 1:

TABLE 3.12

Effect of Infill on Tensile Properties of Innofil3D PLA Filaments (Flat Orientation)

Filament	Infill	Strength	Modulus	Strain at Break
	%	MPa	MPa	%
PLA	50	24.1	1,760	2.2
PLA	100	38.1	2,852	2.1
Pro1	50	29.3	1,993	8.7
Pro1	100	48.0	3,166	21.9

Source: Innofil3D's online data sheets.

TABLE 3.13

Experimental and Predicted Values of Tensile Modulus of Innofil3D PLA Filaments at 50% Infill

Filament	Unit	Upright Orientation		Flat Orientation	
		Experimental	Predicted	Experimental	Predicted
PLA	MPa	2,028	2,022	1,760	1,831
Pro1	MPa	2,111	1,881	1,993	2,033

Source: Experimental values from Innofil3D's online data sheets.

$$E_p = E_0 \frac{1 - P\left(\frac{2}{3}\right)}{1 - P\left(\frac{2}{3}\right) + P} \tag{3.1}$$

Assuming that the tensile coupons complied with ISO 527 and hence featured a cross section of 10×40 mm, and two outer layers with no infill, and each layer 0.14 mm thick, then 50% infill in the cross section corresponded to 0.406% volumetric porosity. Table 3.13 lists the experimental and predicted tensile modulus values of PLA and Pro1 at 50% infill applying equation (3.1), and plugging in it $P = 0.406$. The predicted values are very close (2–4% smaller) to the experimental values.

3.5.4 Films vs. Dumbbells

We caution that tensile properties measured by suppliers on films according to ASTM D882, as in the case of 3D-Fuel®, 3D4MAKERS, and Formfutura, cannot be compared to tensile properties collected on dumbbell-shaped coupons complying with ASTM D638, because the latter ones, not the former ones, represent the tensile behavior of printed parts shaped far differently than sheets and films. Furthermore, when mechanical properties are measured on films and sheets, they ordinarily display differing values in machine and transverse directions due to the anisotropy induced in film and sheets during their fabrication.

3.5.5 Precision of Property Values

Most supplier of PLA filaments publish only average values of their material properties, but scattering of property values is important, because it is an indicator of consistency in material performance, stemming from consistency and quality control in material composition and

TABLE 3.14

Tensile Properties of Innofil3D PLA Filament

Property	Unit	Upright Orientation		Flat Orientation	
		Avg. ± Stand. Dev.	CV[a]	Avg. ± Stand. Dev	CV[a]
Strength	MPa	28.8 ± 4.2	14.6	38.1 ± 0.9	2.4
Modulus	MPa	3,150 ± 54	1.7	2,852 ± 88	3.1
Strain at Break	%	1.1 ± 0.3	27.3	2.8 ± 0.2	7.1

Source: Data sheets posted online by Innofil3D.
Note:
[a]Coefficient of variation.

TABLE 3.15

Flexural and Impact Properties of Innofil3D PLA Filament

Property	Unit	Normal Orientation		Parallel Orientation	
		Avg. ± Stand. Dev.	CV	Avg. ± Stand. Dev.	CV
Flexural Strength	MPa	65.7 ± 5.3	8.1	86.2 ± 3.2	3.7
Flexural Modulus	%	2,409 ± 206	8.6	2,551 ± 100	4.0
Flexural Strain	MPa	4.1 ± 0.2	4.9	3.8 ± 0.2	5.3
Charpy Unnotched Impact Strength	kJ/m^2	14.2 ± 0.7	4.9	13.1 ± 0.7	5.3

Source: Data sheets posted online by Innofil3D.
Note:
[a]Coefficient of variation.

fabrication process. Scattering in material properties also derives from accuracy and precision of test equipment and in machining and measuring coupons, accuracy in performing tests, and repeatability of the printing process to generate the test coupons.

Innofil3D and Polymaker are the only suppliers we found to release standard deviations of their property values. The values of average, standard deviation, and coefficient of variation (CV) of tensile and flexural properties of Innofil3D PLA filament are listed in Tables 3.14 and 3.15, respectively. The CV values of tensile strength and elongation at break in upright coupons are higher than those in flat coupons (Figure 3.15). This is due to the fact that in the former case the test force is perpendicular to numerous layers and their interlayer bonding (with the latter being weaker than in-layer or intralayer cohesion), and location of failure. On the other hand, coupons printed flat are made of far fewer layers, and the test force is directed along and not normal to the layers. The fact that the interlayer adhesion weakens the coupon's response to load seems confirmed by the fact that elongation at break in Innofil3D PLA upright coupons is about half that in flat coupons. Based on data sheets and CVs released by Innofil3D, Pro1 seems overall a more repeatable material than PLA, because the average CVs of all its properties in XY and Z directions are lower than those for PLA. PLA features higher CVs in Z direction for all properties than those in the XY direction, while the CVs for Pro1 are almost the same across directions. This may be due to the fact that, being the melt flow rate of Pro1 about three times that of PLA, the latter one is three times more viscous, and hence may generate coupons with rougher surface due to the layer interface recesses (Figure 3.18) that act as stress raisers when load is applied normally to the layers' plane.

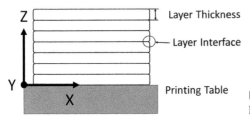

FIGURE 3.18 Schematic of section of printed part using FFF process.

Polymaker sells four PLA filaments, and the average CV for their tensile properties (calculated from their standard deviation values) is: 8.3% for L1001/PolyLite™, 10.6% for L3001/PolyMax™, 11.5% for L3003, and 6.4% for L7001.

3.5.6 Electrical Properties

PLA is an electric insulator (or *dielectric*) material. However, when it approaches its T_g (60–65°C), it quickly starts losing its electrically insulating properties, and hence it behaves as an ineffective insulator in electrical applications generating high heat or featuring high voltage. PLA can become electrically conductive by adding a carbon filler, and extruding carbon-PLA composite filaments with varying carbon content. The most prominent use of conductive PLA is in resistors of electric circuits, electric shielding for sensitive equipment, capacitive touch-sensitive buttons, and ultra-ergonomic wearable electronic parts (Vihaan 2020). Although most applications of printed parts are not electrical, Table 3.16 lists the only values of all electrical properties disclosed by filament suppliers, namely Stratasys and Ultimaker, and likely indicative of the same properties for other PLA filaments. Ultimaker sells PLA and Tough PLA: the dielectric constant and dissipation factor at 1 MHz are 2.62 and 0.014, respectively. Table 3.16 also includes values of the same properties for widely sold non-AM TPs. Since the lower is the dissipation factor, the more efficient is the insulating material, Ultimaker's filaments are slightly less efficient insulators than Stratasys' filaments. Being the average dielectric constant of Stratasys's PLA lower than that of Ultimaker's PLA and Tough PLA, Ultimaker's filaments can store more electric charge. Some PLA filaments are made electrically conductive because they incorporate conductive fillers like carbon, and are illustrated in Sections 3.7.3.1 and 3.7.4.1.

The electrical conductivity of PLA articles can be enhanced by coating them with a layer of materials possessing strong electrical conductivity, as demonstrated by Kumar et al. (2015), who coated coupons of PLA printed via FFF with a layer of copper onto which nickel was deposited in the following steps:

TABLE 3.16
Electrical Properties of Commercial PLA Filaments and Common TPs

Property	Unit	ASTM	Stratasys		Ultimaker	PE[a]	PS[a]	PVC[a]	PA[a]
			Edgewise	Upright					
Volume Resistivity	Ohm-cm	D257-07	2.9×10^{15}	3.2×10^{15}	N/A	$>10^{16}$	10^{18}	10^{12}	10^{15}
Dielectric Constant at 1 MHz	N/A	D150-98	1.51[b]	2.33[b]	2.70	2.3	2.5	3.2	3.6
Dissipation Factor at 1 MHz	N/A	D150-98	0.003[b]	0.005[b]	0.008	N/A	N/A	N/A	N/A
Dielectric Strength	MV/m	D149-09	6.06	11.5	N/A	20	20	40	20

Source: Data sheets posted online by the suppliers.
Notes:
[a]Askeland, D. R. 1994. *The Science and Engineering of Materials*. PWS Publishing Co. 3rd ed.
[b]Not specified at what MHz value.

TABLE 3.17

Comparison among Innofil3D Filaments across Printing Orientations (Figures 3.15, 3.17)

Filament	Tensile Strength (ISO 527)		Flexural Strength (ISO 178)		Unnotched Charpy Impact Strength (ISO 179)	
	MPa	MPa	MPa	MPa	kJ/m^2	kJ/m^2
	Flat	Upright	Normal	Parallel	Normal	Parallel
PLA	37	29	89	67	14	13
PLA Pro1	48	21	99	92	18	20
ABS	29	7	72	66	39	35
InnoPET	41	22	94	75	5	12

Source: Data sheets posted online by Innofil3D.

- Surface preparation of printed coupons through mechanical scoring with 425 μm particle size sand paper to improve adhesion strength
- Rinse in alkaline solution (magnesia 50 g/l, sodium phosphate 10 g/l) for 10 min
- Painting with electrically conductive paint containing silver-coated copper particles
- Electrolytic deposition of 5 μm of copper, followed by 10 μm, 30 μm, or 50 μm of nickel.

However, the authors did not focus on electrical properties, but on improving strength and rigidity of PLA components, reducing absorption of water, moisture and light, and hence aging, therefore they did not take electrical measurements. However, the coatings improved tensile strength and modulus of PLA vs. those of uncoated PLA: a total coating thickness of 55 μm (copper plus nickel) raised strength and modulus from 17 MPa and 3.5 GPa, respectively, to 26 MPa and 7.1 GPa, respectively.

3.5.7 COMPARISON OF PLA TO FOSSIL-BASED POLYMERS

One condition for greater utilization of PLA is that PLA becomes competitive in mechanical performance vs. filaments made of fossil-based plastics. A comparison between the former one and the latter ones is offered by Innofil3D that has published tensile, flexural, and impact strength values across its own AM filaments made of ABS, PET, and two PLA grades, listed in Table 3.17. Assuming for structural design the lower value of each property between the two sets of directions (flat vs. upright, normal vs. parallel), PLA is the strongest in tension, Pro1 the strongest in bending, and ABS is the most impact resistant. The degree of anisotropy in tensile strength varies with the material, and is the least and most manifest in PLA and ABS, respectively.

3.5.8 RECYCLED PLA

The following commercial PLA filaments are twice environmentally friendly because they are made of recycled PLA: rPLA (RDJake, Austria), (RDJAKE 2021), RePLAy (3R Recycling, USA), and rPLA (Filamentive, UK). The second and the third on the list are manufactured from 100% recycled PLA and 55% recycled factory waste streams, respectively. Properties of RDJake rPLA are tensile yield strength 63 MPa, tensile modulus 3,252 MPa, and elongation at break 4%. No material properties are disclosed for RePLAy, whereas Filamentive rPLA features tensile modulus of 3,120 MPa, and high tensile yield strength (70 MPa) and elongation at break (19.5%). Formfutura also manufactures ReForm™ rPLA filament that is based on exactly the same formulation as Formfutura's EasyFil™, but is composed of residual extrusion waste streams that are re-compounded and homogenized into a high-end and easy-to-print upcycled PLA filament with significant less environmental impact. ReForm™ rPLA features the same material property values as EasyFil™.

3.5.9 Major Suppliers of PLA Filaments and Their Products

Stratasys (USA, Israel) is a world leader of AM equipment and materials, reporting USD 668 million in revenues in 2017, and accounting for 20% of world sales of printers in the first quarter of 2018 (Nadel 2019). Stratasys supplies a PLA filament for their printers, compatible with its patented FDM™ process. The material property data sheet of Stratasys PLA is the most comprehensive we found for PLA filaments, and includes even electrical properties. Stratasys markets its PLA as a low-cost option for "fast-draft part iterations," i.e. fast printing of conceptual models and prototypes during product development. Stratasys PLA is compatible with Stratasys F123 series of office printers.

3D-Fuel® runs manufacturing facilities in USA and Ireland, and from starch-rich feedstocks derives three grades of PLA filament for FFF, all available in a range of colors:

- Standard PLA: basic version from Ingeo™ PLA, recommended for initial prototypes
- Workday PLA or APLA: high-temperature version
- Pro PLA or APLA+: version with improved temperature resistance and impact strength.

Tensile properties were measured per ASTM D638 (dumbbell-shaped coupons) for Standard PLA and per ASTM D882 (films) for Workday PLA, hence they cannot be reciprocally compared. However, flexural properties were quantified across filaments according to the same standard, and are plotted in Figure 3.19. Since generally with plastics, flexural properties are a proxy for tensile properties (Paloheimo n.d.), tensile strength and modulus may increase going from Pro to Standard to Workday. Workday PLA stands out due to its high HDT at 0.45 MPa of 144°C, almost three times the value of 53°C for Stratasys PLA. 3D-Fuel® recommends that parts printed in Workday PLA be annealed at 80-130°C in order to increase their crystallinity and maximize their strength and HDT. The properties on Pro PLA were measured on crystalline injection molded bars not printed coupons, which means that the properties of the printed parts may differ due to process differences between injection molding and FFF. Since Workday's flexural modulus is 49–76% higher than that of Stratasys PLA (depending on the printing orientation of the latter one), Workday is expected to be significantly stiffer in tension than Stratasys PLA. Pro was formulated to compete with ABS in terms of impact resistance, and in fact its Izod notched impact resistance of 233 J/m is almost twice 128 J/m of ABS-M30 filament by Stratasys, and nine and six times that of Standard (26 J/m) and Workday (40 J/m), respectively. The tradeoff is that Workday displays higher flexural strength (1.8 times) and modulus (0.7 times) than those of Pro.

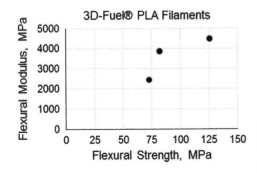

FIGURE 3.19 Flexural properties of 3D-Fuel® PLA filaments.

Source: Data sheets posted online by the supplier.

SD3D company was founded in 2013, is located in California and Texas, and offers product design, CAD, casting, and printing services, along with a range of many printing materials, including PLA unfilled and filled with carbon fibers. The latter version is sold in a range of colors (black, blue, green, grey, natural, orange, red, white, and yellow). Unusually for a supplier of AM feedstocks, SD3D discloses its PLA filament *toughness*, a term that, when applied in tension, quantifies the energy required to break a test coupon, and is numerically the area under the tensile stress–strain curve. The reported T_g values vary (about 10%) depending upon the measuring method; this is a known fact, and it is advised to measure it via TMA (Nielsen, Landel 1994). SD3D also releases the values of heat capacity, thermal conductivity, and coefficient of thermal expansion (CTE) of its PLA. The CTE is almost 60% lower than that of Stratasys PLA, which is advantageous, because smaller expansion upon heating leads to lower internal stresses developed, if the expanding component is rigidly connected to components made of other materials.

3D4MAKERS (The Netherlands) was formed in 2014, and sells grades of FFF filaments, from PLA to high-performance plastics PEEK and PPSU. PLA is offered in colors, and suited for applications such as medical, automotive, and machinery. The tensile properties have been collected on sheets not dumbbell-shaped coupons. 3D4MAKERS also sells a PLA grade called Facilan™ C8, which features a density of 1.4 g/cm^3, higher than the typical 1.2–1.26 g/cm^3 for PLA, meaning that Facilan™ C8 would require more energy than other PLAs in applications relative to transportation. 3D4Makers includes end-use industrial parts among the applications of Facilan™ C8, which implies that the supplier must trust its performance and durability in service. 3D4MAKERS offers a PLLA filament featuring high crystallinity (60–70%), MW of 60,000–80,000 g/mol, viscosity of 300,000 cP, and the highest tensile strength of 71 MPa among the PLA filaments we reviewed. PLLA is mostly present in food packaging, since it meets the European regulations EC No. 1935/2004, EC No. 2023/2006, and EC No. 10/2011 concerning plastics and articles coming into contact with food. Being a version of PLA, PLLA filament is biodegradable and biocompatible, releases nontoxic degradation products, and is also bioactive, meaning it stimulates a biological response from the body, such as bonding to bone and promoting bone growth along the surface of the bioactive material (Jones 2005). Therefore, PLLA is a strong candidate for medical implants. Jabłoński et al. (2018) leveraged these features in the reconstruction of nasal cartilage, and built, via FFF on a Zortrax M200 (Poland) desktop printer, scaffolds made of 3D4MAKERS PLLA that provided structural support to growing tissue, and guide to tissue regeneration, by enabling formation of bone-like apatite on the scaffold's surface when immersed in simulated body fluid.

Innofil3D (The Netherlands) is a supplier of AM materials, established in 2014 and acquired by BASF in 2017, proving that the size and growth expectation of the AM business are appealing to a major material supplier. Innofil3D manufactures filaments in ABS, PET, PP, PVA, and two grades of PLA: standard grade PLA, and improved grade Pro1.

Ultimaker (The Netherlands) is a printer manufacturing company started in 2011, with locations also in Singapore and USA. Ultimaker sells a range of plastic filaments, included PLA and Tough PLA, and recommends its PLA for household tools, toys, educational projects, models, prototyping, architectural displays, and lost casting methods to make metal parts, and excludes applications in contact with food, in vivo, and long-term outdoors above 50°C for PLA and 60°C for Tough PLA. Compared to its standard version, Tough PLA is almost 80% superior in notched impact strength, earning its name, but exhibits 5% lower hardness, 22% lower tensile modulus, and 19% lower tensile strength, confirming that often enhancing some material properties in polymers comes at expense of others, such as in the case of impact strength and tensile modulus that are often negatively correlated. In both formulations tensile and flexural moduli appear positively correlated, as expected, and negatively correlated to impact strength. Tough PLA, although almost twice impact resistant as PLA, exhibits lower tensile yield strength and elongation at break than PLA's.

colorFabb extends the list of Dutch companies active in AM and renewable materials, demonstrating the prominent role of The Netherlands and Europe in implementing a sustainable economy. Opened in 2013, colorFabb supplies an array of printers, mostly desktop models, and filaments for FFF made of PLA (called Economy), PLA/PHA, and PA (for use up to 120°C).

Polymaker is a global company founded in 2012, and headquartered and manufacturing in China, and is also present in Japan, The Netherlands, and USA. Polymaker offers FFF filaments made of ABS, PA, PC, flexible TPU, and the following family of PLA filaments:

- L1001/PolyLite™: featuring patented Jam-Free™ technology
- L3001/PolyMax™: with improved fracture toughness
- L3003: high in toughness and stiffness
- L7001: closed-cell foam, with 30–40% lower density than that of standard PLA.

According to Polymaker, the Jam-Free™ feature comes from its PLA's superior heat resistance due to its elevated softening point (>140°C), and a sharp melting profile, enabling PLA to liquefy only in the heating zone, and preventing extruder jams, which is particularly beneficial for large and time-consuming prints, and dual-extruder printing. Polymaker claims that PLA filaments by other brands tend to soften in the filament barrel due to their low softening point (~60°C), leading to inconsistent feeding and possibly printer jams.

MakerBot is an American company, headquartered in New York City, formed in 2009, and acquired by Stratasys in 2013. MakerBot manufactures printers and two filament grades: PLA and Tough. MakerBot only discloses strength values in tension, bending (both higher than most PLA's), compression (the only value available for a commercial PLA), and impact. The only mechanical properties disclosed for Tough are the remarkably high values of notched and un-notched impact strength, 384 and 2550 J/m, respectively: the former value is 64% higher than the second largest value among PLAs reported in this chapter (234 J/m of 3D-Fuel® Pro PLA), and the latter is one order of magnitude superior to that of other PLA filaments. An example of application of Tough is a simulator (duplicate) to train surgeons in endovascular aneurysm repair (Torres and De Luccia 2017).

Formfutura is a Dutch company established in 2012, producing a wide portfolio of FFF filaments made of TPs, included the following PLA grades:

- EasyFil™ PLA
- Premium PLA
- Volcano PLA
- ReForm™ rPLA.

Formfutura recommends Premium PLA for larger scale prototyping on industrial-sized printers, and Volcano for industrial applications benefiting from high printing speeds, and high-temperature resistance.

MatterHackers is a USA-based company started in 2012 that provides services and products in AM, and distributes a wide range of printers and FFF filaments, including PLA filaments available in 1.75 and 3 mm diameter and displaying very high tensile strength at yield (65 MPa), flexural strength (97 MPa), and modulus (3,600 MPa).

3DXTECH, headquartered in USA, sells Ecomax PLA 3D, a filament possessing high-tensile strength and the highest flexural strength we found reported (115 MPa).

3.6 EXPERIMENTAL PLA POWDER

We found no commercial PLA powders for AM, but two experimental versions of it, one unfilled, and another loaded with 5 wt% nanoclay, both disclosed by Bai et al. (2017), who proved that LS can print articles out of PLA and PLA/nanoclay powder, although the flexural modulus of coupons printed with both powders was one order of magnitude smaller than that of PLA filaments. In Bai's

TABLE 3.18

Effect of Laser Power on Flexural Modulus of Coupons Printed Via LS

Property	Laser Power, 15 W		Laser Power, 16 W		Laser Power, 17 W	
	PLA	PLA/Nanoclay	PLA	PLA/Nanoclay	PLA	PLA/Nanoclay
Flexural Modulus, MPa	102	132	191	198	182	257

Source: Bai, J., Goodridge, R. D., Hague, R. J. M., Okamoto, M. 2017. Processing and Characterization of a Polylactic Acid/Nanoclay Composite for Laser Sintering. *Polymer Composites* 38:2570–2576.

study PLA was supplied by Toyota Technological Institute (Japan), and the nanoclay was organically modified montmorillonite supplied by Nanocor (Japan). Since this was a producibility study, test coupons were printed at different values of laser power (15, 16, 17 W), powder bed temperature (60, 70, 80°C), and laser scan (single and double). The highest flexural modulus for PLA unfilled powder was reached at 80°C, value at which the PLA powder became too hard or agglomerated. Higher powder bed temperature raising the coupons' stiffness was consistent with what reported in the technical literature (Caulfield et al. 2007). Since the optimal bed temperature for PLA/nanoclay was 60°C, PLA was compared to PLA/nanoclay at 60°C, and the results are shown in Table 3.18. The highest value of flexural modulus for PLA and PLA/nanoclay sintered applying single laser scan were about 283 MPa and 258 MPa, respectively, far below the range of 1,979-4,451 MPa of values for commercial PLA filaments reported in this chapter. Switching from single to double (0°–90° pattern) laser scan at 17 W boosted the flexural modulus for PLA powder from 185 to 549 MPa, and for PLA/nanoclay from 258 to 666 MPa. Obviously, PLA powder for AM requires further study to become competitive with PLA filaments for FFF in terms of stiffness, and one step to make progress is reducing porosity of sintered parts. It would be interesting to assess whether adding clay to PLA would also improve flexural strength.

Another non-commercial PLA powder for AM has been recently reported by Gayer et al. (2018), and their paper is mentioned here because it not only describes a version of PLA differing from the PLA in Bai et al. (2017), but also proves again that AM, namely LS, and its sustainable materials are suited for cutting-edge applications like medical implants. PLA has been studied as a feedstocks for medical implants, because it enables a complete regeneration of large bone defects, and is resorbed by the human body, rendering surgical removal unnecessary (Martina and Hutmacher, 2007). F.e. PLA has been employed in low-load-bearing parts of the maxillofacial area (Meara et al. 2012). Implants are manufactured out of PLA or PLA-based composites mostly by injection molding (Eppley 2005), and hence are not as customizable in terms of shape and porosity as if they were built via LS and FFF. Gayer et al. (2018) developed a PLA composite powder for LS to print scaffolds with adequate mechanical strength and tailored properties. The authors synthesized two batches of PDLLA/β-tricalcium phosphate (β-TCP) composite powder with 50 wt % of PDLLA and 50 wt% of β-TCP, in order to achieve a balance between melting behavior, and bone-like properties, provided by PDLLA and β-TCP, respectively. The two batches differed in polymer particle size, filler particle size, and polymer MW. Amorphous PDLLA (Evonik, Germany) containing 50% D-lactide monomers and 50% L-lactide monomers, and β-TCP (Biovision, Germany) were the raw materials. LS requires particles <150 μm in diameter to achieve a layer thickness adequately small to be sintered upon laser exposure on the millisecond timescale. Since PDLLA and β-TCP originally procured were up to 500 μm in size, they were mixed in a 50/50 wt% and milled to smaller size. Two batches of composite powders were processed: one (batch 1) with particle diameter <125 μm, the other (batch 2) with particles diameter <100 μm. Both batches were processed with CO_2 laser (FEHA 600 SM-E, Germany), featuring

0.20–0.50 W power, and 290–360 μm diameter laser beam. During LS, batch 2 showed improved melting behavior due to its smaller polymer particle size and lower melt viscosity vs. batch 1, which resulted in specimens less porous than batch 1, and ultimately to biaxial bending strength (62 MPa) almost triple that of batch 1 specimens (23 MPa), measured conducting the *ball on three balls* test, a new method for biaxial strength testing of brittle materials (Danzer et al. 2007) conducted on test specimens 10 mm in diameter and 1 mm in thickness.

PLA in powder form has been investigated by some researchers as a feedstock for a recent AM process called *selective vacuum manufacturing* (SVM) developed by T. Phattanaphibul and his team at the Asian Institute of Technology in Thailand (Phattanaphibul et al. 2014). SVM starts with placing a layer of support material onto a platform that has been lowered by the thickness of one powder layer. A roller travels over to level and pack the support material. A vacuum nozzle *selectively* removes the support material and forms a cavity in which the building material is deposited, packed, and sintered to form a solid layer. These steps are repeated until the last layer is sintered. Phattanaphibul and Koomsap (2012) printed PLA scaffolds for TE, featuring porosity of 72%, pore size of 20-90 μm and compressive modulus of elasticity of 2.1 MPa, matching the lower range of mechanical properties reported for soft tissue applications. Cell viability was 76%. The authors developed a technique to convert PLA pellets into a powder different from the typical mechanical crushing in dry ice. The new technique overcame the tacky nature of PLA and yielded high particle size (>200 μm), and consisted in: (a) dissolving PLA pellets in dichloromethane, (b) spraying the solution into water mixed with PVA surfactant added to improve the dispersion of PLA solution droplets, (c) filtering and drying the powder precipitant, and (d) disintegrating the stuck powder. Depending on the values of process parameters to make the powder, the particle size ranged from 71 to 416 μm, and was <200 μm under most process conditions.

3.7 COMMERCIAL COMPOSITE PLA FILAMENTS

3.7.1 INTRODUCTION

As mentioned earlier, PLA suffers from some shortcomings, such as brittleness, lack of ductility, and low maximum temperature of service, hence PLA composites have been developed and commercialized in which a PLA matrix is filled with materials enhancing its physical and mechanical performance. However, some commercial PLA composites do not feature improved functional properties vs. those of unfilled PLA, but simply exhibit a different look. The PLA fillers are sustainable materials such as wood, bamboo, cork (Silva et al. (2005), cellulose (Graupner et al. 2009), algae, and hemp (Baghaei et al. 2013; Baghaei and Skrifvars 2016), and non-sustainable materials such as glass, metals, and carbon. In the following subsections PLA composites with the latter type of fillers will be described, whereas PLA composites containing sustainable fillers will be discussed in individual chapters focused on a single filler. Siakeng et al. (2019) published a review of natural fiber-reinforced PLA composites

3.7.2 GLASS-FILLED PLA

3D-Fuel® sells a glass fiber (GF) filled PLA filament, called Glass Filled PLA™, with significantly superior mechanical properties in comparison to those of unfilled Standard PLA by 3D-Fuel®, as shown in Table 3.19, evidently in order to target engineering and functional applications. The filler consists of recycled glass particles. The filament's materials safety data sheet does not disclose the particle size, and the type of glass. E and S glass have been feedstocks for plastic composites for decades, with E less expensive than S, but S outperforming E in tensile strength and modulus. Glass Filled PLA™ features a significant enhancement vs. unfilled Standard PLA, ranging from 12% to 89% across all its properties disclosed. From the density of the composite filament reported

TABLE 3.19

Comparison between Unfilled and Glass-Filled Feedstocks

Property	Unit	AM Filaments by 3D-Fuel®			Non-AM Pellets by RTP		AM Filaments by BASF
		Glass-Filled PLA (GP)	Unfilled Standard PLA (SP)	GP Value/ SP Value	PLA	PLA/10 wt% GF (RTP 2099 X 121249 A)	PP/glass 70/30 wt% (Ultrafuse PP GF30)
Density	g/cm³	1.40	N/A	N/A	N/A	1.32	1.07
Tensile Strength at Break	MPa	57	41	1.39	62	79	15.9–41.7
Tensile Strength at Yield	MPa	46	37	1.24	N/A	N/A	N/A
Tensile Modulus	GPa	4	3.2	1.25	N/A	6.9	2.24–2.63
Elongation at Break	%	3.4	1.8	1.89	N/A	1.5	0.8–4.4
Flexural Strength	MPa	N/A	N/A	N/A	108	93	19.3–76.8
Flexural Modulus	GPa	N/A	N/A	N/A	3.83	7	1.67–3.51
Notched Izod Impact Strength	J/m	29	26	1.12	16	43	14.3–63.2

Source: Online data sheets posted by 3D-Fuel®, RTP Co., BASF. Reproduced with permission from 3D-Fuel®.
Note: Value ranges are the effect of variations in properties due to printing orientations.

in the material's safety data sheet and equal to 1.40 g/cm³, the fraction of glass V_f was calculated as 12.5 vol%. Selecting 72 GPa and 3,200 MPa as the tensile modulus of E glass (the most economical among the glass grades) and PLA, respectively, we calculated that the nominal tensile modulus of Glass Filled PLA™ was 4,357 MPa, based on the Lewis-Nielsen equations (1.11–1.13) for particulate-filled composites, in which $A = 1.5$ in the case of Glass Filled PLA™. The fact that the tensile modulus reported by the supplier (4 GPa) trails the predicted value may be due to an unsatisfactory adhesion between glass and PLA, and porosity, not atypical in composite filaments for FFF.

Table 3.19 also includes, as a benchmark, two PLA produced by RTP Company (USA) as pellets not for AM: one is unfilled, and the other (RTP 2099) is filled with 10 wt% GF (RTP Company 2011). Higher filler content translates in greater strength and modulus if the filler is stronger and stiffer than the matrix and load transfer between matrix and filler is effective. GF-filled PLA filaments have potential to be stronger and stiffer than Glass Filled PLA™, since RTP sells glass-PLA with 40 wt% GF that features a tensile strength and modulus of 110 MPa and 13.8 GPa, respectively, while for commercial unfilled PLA filaments the values of these properties are 22–71 MPa and 2–4 GPa, respectively. In order to offer some perspective, we note that Glass Filled PLA™ surpasses in tensile strength and modulus Ultrafuse PP GF30 (Table 3.19), a filament for FFF comprising 65–75 wt% PP, 25–35 wt% of glass fiber, and 0–2 wt% additives.

In adding filler to PLA, one must consider that filler content affects the processing of the resulting compound, since the higher the content the lower the MFI. The strength of composites is governed not only by the filler content but also the filler-matrix interface adhesion that controls the load transfer between matrix and filler, stress concentrations, and size/spatial distribution of defects (Fu et al. 2008).

3.7.3 METAL-FILLED PLA

3.7.3.1 Comparison

Metal-filled PLA filaments are potentially conducive to functional and engineering applications for AM, and may open up new markets to PLA. Possible applications comprise magnetic devices, and 3D, embedded, conformal, and flexible electronics (such as pressure and strain sensors (Yang et al. 2014), electronic components (resistors, capacitors, inductors, radio frequency filters, horn and cavity antennas, etc.), batteries, and circuits for LED signs, LED matrix displays, electronic games, gaming controllers, and digital music devices (keyboards, drum machines, MIDI controllers), touchpads, etc. A list of technical papers on AM of electrical and electronic articles is available at Multi 3D (2017).

The suppliers colorFabb, Formfutura, MCCP (The Netherlands), and Protoplant (USA) offer commercial filaments for FFF made of PLA filled with powders of brass, bronze, copper, magnetic iron, and stainless steel, with powder content of about 40–80 wt% equal to about 11–40 vol%. These PLA-metal filaments seem to be formulated for aesthetic prints, to impart a metallic look to DIY items, models, and decorative articles, rather than structural uses, since the list of properties disclosed by each supplier is short, and these metal-filled PLA filaments do not outperform neat unfilled PLA filaments in mechanical properties. F.e. the tensile strength of unfilled PLA filament and metal-PLA filament by colorFabb is 45 MPa and 23–30 MPa, respectively. However, metal-PLA materials may be fit for functional application requiring thermal and electrical conductivity. Table 3.20 is based on suppliers' data and experimental readings published by researchers, in order to report the most comprehensive data available. Values of density, weight and volume fraction varied according to the measuring methods. Volume fraction matters because material properties, such as strength and stiffness, of particulate composites, like these PLA-metal filaments, depend on the properties of their constituents multiplied by their volume fractions. Density is a key metric is applications related to transportation. The particles in BronzeFill and CopperFill are spherical and uniformly distributed, whereas the particles in Magnetic Steel and Stainless Steel are similar to flakes. The particle shape is important because, along with size and distribution of the inclusions, it governs the macroscopic behavior of particulate composites (Ahmed and Jones 1990). Moreover, the particle shape also affects the choice of the model followed to predict the tensile modulus of the relative composite material (Fu et al. 2008). The tensile stress-strain curves of BronzeFill, CopperFill, and Magnetic Steel (Kuentz et al. 2016) showed yield, differently than those of Stainless Steel. Elongation at break of BronzeFill and CopperFill is about two and three times that of Magnetic Iron and Stainless Steel, respectively, which may translate in higher impact strength for BronzeFill and CopperFill. The four metal-PLA filaments were stiffer but weaker that an unfilled PLA filament tested: the tensile modulus and strength of the latter were 3,600 MPa and 52.5 MPa, respectively, whereas tensile modulus and strength of the filled filaments were 3,637–5,400 MPa, and 15–38.5 MPa, respectively. Increasing the volume fraction was positively and negatively correlated with tensile modulus and strength, respectively. In Figure 3.20 the tensile and flexural properties of commercial metal-PLA filaments released by suppliers are plotted.

The experimental values of tensile modulus of BronzeFill and CopperFill measured by Kuentz et al. (2016) were 5,400 MPa and 4,494 MPa, respectively. Based on the tensile modulus of the matrix and the filler E_m and E_f, respectively, we predicted the tensile modulus of these composite filaments E_c by applying the Halpin-Tsai equations (1.14–1.15), with $\zeta = 2$ in the case of both materials. Choosing 103 GPa and 117 GPa as E_f for bronze and copper, respectively, and as E_m of PLA the value of 3,145 MPa experimentally measured by Kuentz et al. (2016), we calculated a tensile modulus E_c of 7,525 MPa and 7613 MPa for BronzeFill and CopperFill, respectively that

TABLE 3.20

Properties of Metal-Filled Commercial PLA Filaments for FFF, from Values Reported by Suppliers or Measured by Researchers

Property	Density	Filler Weight Fraction	Filler Volume Fraction	Tensile Strength	Tensile Modulus	Elongation at Break	Flexural Strength	Flexural Modulus	Thermal Conductivity	Charpy Notched Impact Strength
Unit	g/cm^3	%	%	MPa	MPa	%	MPa	MPa	W/(mK)	kJ/m^2
Value	1.68–4	43–80.6	11.1–40.2	15–52.5	3,637–5,400	1.3–8	26–82	4,300–9,000	0.18–0.55	9.3–11.3

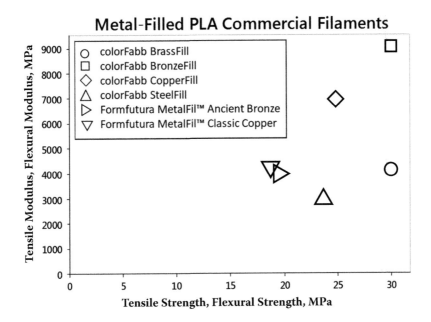

FIGURE 3.20 Commercial metal-PLA filaments whose tensile and flexural properties are reported by suppliers.

Source: Data sheets posted online by suppliers.

far exceeds the experimental values of 5,400 MPa and 4,494 MPa, respectively. This large gap may be due to the substantial porosity experimentally detected and a poor interfacial adhesion between metals and PLA.

Laureto et al. (2017) investigated not only filament density, filler Vf, and particle size distribution of BronzeFill, CopperFill, Magnetic Steel, and Stainless Steel, but also their thermal conductivity measured at 55°C per ASTM F433 by means of a Holometrix TCA300 Through-Plane Thermal Conductivity Tester. The measured values were 0.29–0.55 W/mK for the composites vs. 0.18 W/mK for PLA. Theoretically, the thermal conductivity of a particulate composite can be increased at exponential rate by raising the filler content, and almost linearly by decreasing the void content.

The values of mechanical properties of colorFabb composite PLA filaments were normalized by equating the value of the highest property value among the filaments to 100, and plotted in Figure 3.21.

Protoplant sells five metal-PLA filaments, described henceforth. Composite PLA – Polishable Stainless Steel is recommended for jewelry, costumes, props, figurines, crafting, and robots, and its mechanical properties are not disclosed. Composite PLA – Rustable Magnetic Iron is a ferromagnetic material, which means it forms permanent magnets when exposed to a magnetic field, and can be rusted by abrasion (filing, sanding, or wire brushing) to expose the metal, followed by immersion in a 50/50 vinegar/hydrogen peroxide mixture saturated with salt. Copper Metal Composite HTPLA, Bronze Metal Composite HTPLA, and Brass Metal Composite HTPLA are made of heat treatable PLA, and particles of diameter <250 μm. Heat treating the filament at 110°C for 10 min before printing increases its crystallinity, and raises the printed part's maximum service temperature from 50-60°C to 150–175°C which is a remarkable enhancement for PLA, and makes these materials competitive in thermal resistance with ABS and PETG filament, whose T_g is 108°C (Stratasys) and 88°C (Airwolf), respectively. The typical dimensional changes after heat treatment are 1.5% shrinkage in the X and Y directions, and 1% growth in the Z direction. Applications

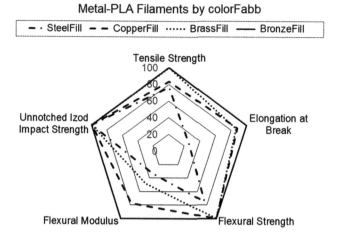

FIGURE 3.21 Normalized values of mechanical properties of metal-filled PLA filaments in Table 3.20.

suggested by Protoplant for the HTPLA materials above are ornamental (faux brick and clay, fixtures, hardware, fine art, sculpture, emblems, signage, trophies, jewelry, game pieces, and figurines), and functional (knobs, buttons, heat sinks, and exchangers). Printed articles in Copper Metal Composite HTPLA look like terra-cotta but can be polished to a metal finish. Printed items in Brass and Bronze Metal Composite HTPLA look like brick, but can also be polished to a metal finish. The polishing process for Copper, Brass, and Bronze Metal Composite HTPLA is very similar to that for Composite PLA – Rustable Magnetic Iron. Copper, Brass, and Bronze Metal Composite HTPLA can as well be painted and coated. Protoplant also manufactures Proto-Pasta Conductive PLA, an electrically conductive filament based on NatureWorks 4043D PLA, and suited to print simple circuits like LED, touch sensitive devices, and items with embedded electrical conductivity, such a 9V battery flashlight, and a controller kit featuring touch buttons.

Formfutura markets the metal-filled PLA filaments MetalFil™ Ancient Bronze, MetalFil™ Classic Copper, and MetalFil™ Brass, and is the only supplier of metal-PLA filaments disclosing impact strength, which was measured according to the Charpy notched method, and equal to 11.3 kJ/m^2 for Ancient Bronze and 9.3 kJ/m^2 for Classic Copper.

3.7.3.2 Analysis of Commercial Metal-Filled PLA Filaments

Mozafari et al. (2019) demonstrated the influence of the interphase on the mechanical properties of PLLA-stainless steel composite material for FFF. Combining 1.75 mm diameter PLLA from 3D4MAKERS, and 420 stainless steel 45–105 μm diameter powder, 25-layer coupons were printed, using 0.2 mm layer thickness, 90% infill, 45° raster angle, 220°C nozzle temperature, and 50 mm/s deposition speed. Applying nanoindentation the authors measured the tensile modulus of the matrix, particles, and interphase layer, and plugged their values in a micromechanical representative volume element model of the composite filament designed by using Digimat (e-Xstream engineering, Luxemburg), a software program modeling composite materials. They also distinguished between *interface*, defined as the two-dimensional boundary between the matrix and the particles, and *interphase*, defined as "the three-dimensional region that includes the interface plus a zone of finite thickness on both sides of the interface. The interphase boundaries are generally defined from the point in the matrix where the local properties start to deviate from the bulk properties in the direction of the polymer/particle interface." In their computer model, the authors simulated perfect and imperfect bonding, with the latter resulting from voids at the fiber matrix interface during composite fabrication or starting with matrix degradation, leading to deterioration of interphase, ultimately leading to imperfect bonding. The model predicted that raising filler volume fraction from 0 to 10%

would result in increasing (because the filler typically carries larger share of load) or decreasing tensile, compressive, and shear modulus if the bonding is perfect or imperfect, respectively.

Fafenrot et al. (2017) investigated the effect of printing settings, namely infill amount (20, 60, 100%) and orientation (45°, 90°) on the tensile and flexural strength of coupons printed in Magnetic Iron PLA and BronzeFill. The coupons' shell was printed at 100% infill. For both filaments, the variation was greater in flexural than tensile strength: the former was 26–46 MPa and 50–82 MPa for Magnetic Iron PLA and BronzeFill, respectively, whereas the latter was 14–15 MPa and 36–42 MPa for Magnetic Iron PLA and BronzeFill, respectively. The tensile and flexural strength values were calculated from the test force data accounting for the internal voids due to infill <100%, and the non-porous portion of the coupons cross section.

3.7.4 CARBON-FILLED PLA FILAMENTS

3.7.4.1 Commercial Carbon-Filled PLA Filaments

Carbon fibers (CFs) are an attractive material to add to PLA because of their combination of high-tensile strength and modulus, temperature resistance, electrical conductivity, low weight, and thermal expansion, partly offset by being more expensive than glass fibers that are in fact widespread for engineering applications. Commercial PLA filaments with CFs are: (a) Carbon Fiber HTPLA (milled CFs), for high temperatures, Original Carbon Fiber PLA (milled CFs), Conductive PLA (carbon black) by Proto-pasta; (b) CarbonX™ CF-PLA (3DXTECH, USA) (fiber, not powder or milled); (c) CF-PLA by SD3D; (d) graphene-PLA Conductive Graphene Filament (Black Magic 3D, USA). As an example, the values of volume resistivity of Conductive PLA reported by its supplier are 30 ohm-cm in the XY plane and 115 ohm-cm in the Z direction. Ibrahim et al. (2016) measured the volume resistivity of 1.75 mm diameter Proto-pasta Conductive PLA to be 200–1,500 ohm-cm depending on the length (1 to 10 cm) of the tested filaments, whereas volume resistivity of Stratasys PLA is about 3×10^{15} ohm-cm.

Applications of CF/PLA comprise those listed in Section 3.7.3.1, and electromagnetic and radio frequency shielding for automotive, aerospace, telecommunications, hospital equipment, medical devices, enclosures, and packaging. CF-filled filaments require hardened nozzles on printers, because they are more abrasive than unfilled PLA filaments.

Ferreira et al. (2017) accurately and broadly measured the elastic properties and strengths (Table 3.21) of coupons made of Proto-pasta Carbon Fiber PLA and its unfilled PLA (equivalent to

TABLE 3.21
Properties of Proto-Pasta Carbon Fiber PLA (CF/PLA), and Its Unfilled PLA

Property	Direction	Unit	ASTM	PLA	CF/PLA
Tensile Strength, S_1	0°	MPa	D638	54.7 ± 1.9	53.4 ± 0.2
Tensile Strength, S_2	90°	MPa	D638	37.1 ± 3.5	35.4 ± 1.5
Tensile Modulus, E_1	0°	MPa	D638	3,376 ± 212	7541 ± 96
Tensile Modulus, E_2	90°	MPa	D638	3,125 ± 148	3,920 ± 167
In-plane Shear Modulus, G_{12}	±45°	MPa	D3518	1,092 ± 3 6	1,268 ± 5
Poisson's Coefficient, v_{12}	0°	N/A	D638	0.331 ± 0 .011	0.400 ± 0.012
Poisson's Coefficient, v_{21}	90°	N/A	D638	0.325 ± 0.014	0.150 ± 0.008
In-plane Shear Strength, S_{12}	±45°	MPa	D3518	18.0 ± 0.8	18.9 ± 0.8

Source: Ferreira, R. T. L., Cardoso Amatte, I., Dutra, T. A. 2017. Experimental characterization and micrography of 3D printed PLA and PLA reinforced with short carbon fibers. *Composites Part B* 124:88–100. Reproduced with permission from Elsevier.

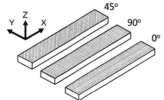

FIGURE 3.22   Test coupons printed in different orientations. The thinner lines represent the orientation of the deposited filament.

NatureWorks Ingeo™ 4043D) via FFF according to the printing orientations 0°, 90°, and 45° shown in Figure 3.22. According to scanning electron microscopy (SEM) analysis, the length of CFs was estimated at 60 µm. The presence of CFs raised tensile modulus at 0° (by 123%) and 90° (by 25%) and in-plane shear modulus (16%) vs. unfilled PLA. Instead, the tensile strength decreased by 2% to 5% (depending on printing orientation), while shear strength rose by 5%. This overall effect indicated that the fibers were effective at low strains where the elastic constants are measured but not at high strains, and this can be due to ineffective load transfer between fiber and matrix, resulting from weak fiber-matrix interface adhesion and/or low fiber's aspect ratio (lenght/ diameter). The CV values were contained, which typically derives from a combination of good quality control of feedstock and fabrication process, and sound testing and analysis practices. The property values reported are adequate for structurally designing load-bearing parts, made of layers printed in multiple orientations. The stress-strain curves CF/PLA displayed that: (a) yield was present at 0° and 45°, and in one out of five coupons at 90°, and (b) that curves at 0° and 45° were linear up to half of of tensile strength, and at 90° up to 80% of tensile strength.

CarbonX™ CF-PLA comprises about 80/20 wt% of PLA/high-modulus CFs (of undisclosed length), and features the properties in Table 3.22. CarbonX™ CF-PLA features higher tensile (+73%) and flexural (+98%) moduli but lower tensile (-14%) and flexural (-23%) strengths and elongation at break (-75%) compared to Ecomax® PLA sold by 3DXTECH.

Since the ranges of tensile strength and modulus we collected for commercial PLA filaments for FFF are 22–71 MPa and 1,820–4,000 MPa, respectively, the tensile strength of CarbonX™ Carbon Fiber PLA, and Proto-pasta Carbon Fiber PLA are about in the middle of the range, whereas their values of tensile modulus exceed that range, indicating that the specific CF used have more a stiffening than a reinforcing effect. Routes to enhance the strength of CF-PLA filaments are improving fiber-matrix interface adhesion, increasing fiber's aspect ratio, and lowering porosity.

TABLE 3.22

Properties of CarbonX™ CF-PLA

Property	Unit	Value
Density	g/cm^3	1.29
T_g	°C	60
Tensile Strength	MPa	48
Tensile Modulus	MPa	4,950
Elongation at Break	%	2
Flexural Strength	MPa	89
Flexural Modulus	MPa	6,320
HDT at 0.45 MPa	°C	91

Source: Data sheets posted online by 3DXTech. Reproduced with permission from 3DXTech.

TABLE 3.23
Properties of PLA and Experimental Continuous CF/PLA

Property	Unit	PLA	CF/PLA	Sized CF/PLA
Tensile Strength	MPa	28	80	91
Flexural Strength	MPa	53	59	156
Storage Modulus	GPa	1.22	0.72	3.25
Loss tangent	N/A	2.32	1.52	1.32
T_g	°C	63.6	65.2	66.8

Source: Adapted from Li, N., Li, Y., Liu, S. 2016. Rapid prototyping of continuous carbon fiber reinforced polylactic acid composites by 3D printing. *Journal of Materials Processing Technology* 238:218–225. Reproduced with permission from Elsevier.

3.7.4.2 Experimental Carbon-Filled PLA Filaments

Investigators have recognized the potential of continuous CFs to improve mechanical performance of PLA filaments, and by leveraging continuous CFs have reached values of tensile strength and modulus one order of magnitude superior to those of milled CFs.

Namiki et al. (2014) manufactured a filament for FFF made of PLA combined with continuous CFs in a heater before extrusion from the printer. Test values of tensile strength and modulus in the composite filament surpassed those in PLA, although they were slightly lower than the values predicted by the rule of mixture. However, examination of the fracture surface revealed some interlayer gaps, which prevented further improvement of tensile properties.

Li et al. (2016) developed a PLA filament for FFF containing continuous CFs, by combining and extruding a bundle comprising 1000 single CFs (HtA40 H13, Toho Tenax, Japan), and PLA (NatureWorks). In order to improve the weak CF/PLA adhesion (Yu et al. 2010), they manufactured a second version of the CF/PLA filament, by modifying the fiber surface with a sizing based on a solution of PLA particles and methylene dichloride. This sizing was effective in improving the PLA filament's tensile and flexural strengths and storage modulus by 71%, 64%, and 51%, respectively (Table 3.23). SEM analysis of the sized filament revealed homogeneous distribution of fibers within PLA, and a microstructure almost void-free, which indicates a more intimate wetting and stronger fiber-matrix adhesion in the surface-modified filament vs. the filament with non-sized surface.

Yao et al. (2017) embedded continuous CFs into PLA structures built via FFF to improve mechanical properties, and also leverage the piezoresistive behavior of CFs to self-monitor the structural health of printed structures: variation in strain is manifested by change in electrical resistivity, namely the resistivity decreases reversibly upon compression, increases reversibly upon tension, and irreversibly upon damage (Wen and Chung 2006). The authors experimented with CF bundles composed of 3,000 (3K), 6,000 (6K), and 12,000 (12K) filaments, and CF/PLA specimens with 12K bundle showed an increase in tensile and flexural strength up to 70% and 19%, respectively vs. unfilled PLA specimens. TORAY fibers from Torayca (Japan), and an unspecified PLA were utilized. The test specimen fabrication process basically consisted in: (a) printing the specimen's lower layers out of unfilled PLA filament; (b) manually placing on the laid layers 250 mm long CFs previously impregnated with epoxy resin adhesive, and loaded with a 2N tensile force to remain straight; (c) printing the upper layers with unfilled PLA. The slope of fractional change in electric resistance with strain served as an accurate indicator of strain measurement within the elastic region, and damage detection in the yield region. Additionally, weight and printing time were reduced up to 26% and 11%, respectively, by decreasing the fill density without penalizing structural strength.

The applicability of self-monitoring printed CF/PLA components was demonstrated in a cutting edge application by printing an artificial hand, where CFs were placed in fingers during the fabrication process, and proved effectively in accurately sensing the load applied to the fingers.

Intimate impregnation of CFs by PLA boosts the mechanical properties of the resulting CF/PLA filament, as Tian et al. (2016) demonstrated by experimenting with their newly developed FFF process to build parts. They combined PLA (Flashforge, China) and a continuous bundle of 1K HtA40 H13 carbon filaments (Toho Tenax[TM], Japan) by simultaneously feeding them in a nozzle, where the PLA filament was liquefied impregnating the continuous CFs, and subsequently deposited under pressure on the print table along with the CFs. Raising the temperature in the nozzle liquefier from 180 to 240°C spiked PLA's flowability from 1.98 to 35.59 g/min, enabled the resin to penetrate inside the fiber bundle, and effectively transfer the load between fibers and matrix. Ultimately, flexural strength and modulus increased from 112 MPa to 155 MPa, and from 5.5 MPa to 8.7 MPa, respectively. Microscopic examination of broken test specimens showed that at 180°C fibers were pulled out and stripped from the matrix, whereas at 240°C fiber bundle breakage occurred. The increase of mechanical properties in a composite material due to intimate impregnation of the reinforcement by the matrix applies to composite materials in general, regardless of the reinforcement's form that is, mostly, in form of particle, flake, and fiber.

Tian et al. (2017) made CF/PLA for AM even more sustainable: they demonstrated a new method to produce continuous CF/PLA filament by recycling 100% of printed parts made of continuous CF/PLA composites, and employing the recycled filament as AM feedstock. Remarkably, recycling did not degrade the mechanical properties of the recycled material vs. the original CF/PLA (Table 3.24). The recycling method recovered 73% of PLA matrix, and consisted in: (a) heating the scrapped printed parts made of continuous CF/PLA filament with a heat gun at 240–300°C to melt the PLA, and debond the layers from each other; (b) pull the debonded CF/PLA filament at 200 mm/min through a 0.8 mm diameter nozzle (*remolding nozzle*) at 240°C to remold the filament and smooth its surface by removing some PLA; (c) roll up the filament around a spool, or pull it through a heating chamber, where it gets impregnated by a PLA filament supplying the polymer removed by the remolding nozzle; (d) the recycled filament is deposited by a printer (*remanufacturing*). The starting filament was 1.75 mm diameter PLA by Flashforge containing continuous 1K CF tows by Toho Tenax[TM] (Japan). The recycled filament showed negligible improvement in tensile strength (264 vs. 256 MPa), tensile modulus (20.8 vs. 20.6 GPa), and 12% rise in unnotched impact strength. The property improvement was attributed to deeper impregnation of CF by PLA, and more effective load transfer from matrix to fiber. The multiple heating steps in the recycling process lowered the MW and increased the flowability of PLA, improving the CF impregnation, and the bonding at CF/PLA interface, although it decomposed and aged the

TABLE 3.24

Properties of PLA, Continuous Virgin, and Recycled CF/PLA Filaments for FFF

Filament	Tensile Strength	Tensile Modulus	Flexural Strength	Flexural Modulus	Charpy Unnotched Impact Strength	Interlaminar Shear Strength
	MPa	GPa	MPa	GPa	kJ/m^2	MPa
PLA	62	4.2	99	3.5	20.0	5.69
CF/PLA	256	20.6	211	14.5	34.5	20.3
Recycled CF/PLA	264	20.8	264	13.3	38.7	22.7

Source: Tian, X, Liu, T, Wang, Q, Dilmurat, A, Li D, Ziegmann, G. 2017. Recycling and remanufacturing of 3D printed continuous carbon fiber reinforced PLA composites. *J. Clean Prod.* 142:1609–1618.

PLA. Combining the recycled filament with virgin PLA filament in step (c) of the recycling process compensated for the drop in MW. However, the energy consumption of 68 and 66 MJ/kg for recycling and remanufacturing, respectively exceeds that in conventional methods for manufacturing composite materials (autoclave molding, RTM, filament winding, pultrusion, injection molding) and recycling (grounding, thermo-chemical recycling).

Flowers et al. (2017) demonstrated that FFF was suited to fabricate individually three of the most basic electronic components, that is resistors, capacitors, and inductors, and, in one step, a complete circuit, namely a high-pass filter made of capacitor, inductor, conductive traces, PLA dielectric, and screw terminal breakouts. The authors chose as conductive materials: (a) PET-based Electrifi (Multi3D LLC, USA), (b) a graphene-based Conductive PLA (Black Magic 3D, USA), and (c) carbon black-based Conductive PLA (Proto-pasta). Proto-pasta and Black Magic 3D filament were not sufficiently flexible to perform as conductors, because they fractured during the 180° bending test (respectively at first and second bending). The resistivity of the printed filaments was measured to be 12 and 0.78 ohm-cm for Conductive PLA and Electrifi, respectively.

A challenge for printing electronic components and circuits is to become cost-competitive with current manufacturing techniques to produce the same products.

3.7.5 ARAMID-PLA FILAMENT

Bettini et al. (2017) developed an experimental composite filament for FFF combining PLA and continuous aramid fiber, by simultaneously inserting fibers and solid polymer in the extrusion head of a commercial FFF printer. A commercial 1.75 mm diameter PLA filament was combined with continuous Technora® HFYT-240 220 dtex (Teijin Aramid, The Netherlands) aramid fiber in form of a yarn of about 150 fibers with 12 μm diameter each. Aramid was selected to avoid potential drop in toughness and elongation at break vs. unfilled PLA. The printer's extruder head was modified to concurrently accept aramid fed from a spool and PLA extruded through a 1 mm die. Low deposition speed and thin thickness of the deposited layer enabled to achieve the required solidification rate, and optimal fiber impregnation, in turn resulting in a proper deposition. The measured fiber content was 8.6 vol% (9.5 wt%). Tensile, compressive, and shear properties were measured on unidirectional coupons. Switching from unfilled PLA to PLA/aramid, tensile strength and modulus in fiber direction jumped from 34 MPa and 3,260 MPa, respectively, to 203 MPa and 9,340 MPa, respectively, whereas the tensile modulus perpendicular to fibers (*transverse*) dropped from 3,260 MPa to 1,530 MPa, and compressive strength improved from 50 MPa to 94 MPa. The tensile strength of PLA/aramid was lower than expected theoretically, possibly due to an uneven distribution of stress across the fibers leading to premature failure. The large reduction in transverse tensile modulus may indicate inadequate bonding at the PLA/aramid interface, which is not uncommon in PLA composites for AM.

3.8   EXPERIMENTAL PLA COMPOSITE FILAMENTS

The advanced experimental PLA composite filaments described hereafter demonstrate that AM blends with cutting-edge materials and drives innovation.

Paspali et al. (2018) formulated a PLA/nanoclay filament, experimenting with Ingeo™ 2003D PLA, and three sodium bentonite nanoclays from BYK-Chemie (Germany), two organically modified, and one natural. Adding 5 wt% organomodified clay to PLA enhanced the elastic and flexural moduli by 10% and 14%, respectively, but penalized tensile and bending strength and strains. The porosity of the tensile test coupons was in the same range for unfilled PLA and filled PLA. Importantly, applying the Halpin-Tsai equation to model the tensile modulus showed that increasing the amount of clay distributed across PLA (clay *intercalation*) was more effective in raising the composite's tensile modulus than raising the clay content.

Coppola et al. (2017) also developed and studied a PLA/nanoclay filament made of Ingeo™ 2003D PLA filled with organically modified Cloisite 30B nanoclay. PLA/nanoclay was stiffer than unfilled PLA: the tensile modulus of the former and latter was 2,584 MPa and 2,254 MPa, respectively, confirmed by the storage modulus that was about 15% higher in PLA/nanoclay between 35°C and 55°C. T_g stayed about unchanged. The loss modulus was lower in PLA/nanoclay, and indicated that the latter was less ductile than unfilled PLA. Printed samples of filled PLA displayed a better shape stability, featuring sharper edges than in unfilled PLA printed samples.

Coppola et al. (2018) investigated the fabrication optimization of a PLA/nanoclay filament, by assessing the effect of printing temperature (185, 200, 215°C) on the filament's properties. They selected Ingeo™ 4032D and 2003D PLA grades, and compounded both withCloisite 30B silicate in an extruder to produce a 1.75 mm diameter filament. First, across printing temperature the presence of clay raised the tensile modulus but lowered tensile strength and strain at break for 4032D-30B and 2003D-30B. Particularly, in the former, tensile modulus improved with raising printing temperature, whereas for the latter it slightly decreased. This opposite behavior was explained considering the difference in polymer macromolecular structure and nanocomposite morphology (that was exfoliated in 4032D and intercalated in 2003D) between the two PLAs: exfoliation permitted orientation of the filler in the flow direction upon filament extrusion from the printer orifice, while intercalation did not. One downside of the two composite filaments is that their T_g lagged that of 4032D and 2003D at all tested printing temperatures.

Lamberti et al. (2018) focused on electrical and thermal applications of filaments fabricated by combining a PLA matrix, and two forms of CFs: multi-walled carbon nanotubes (CNTs), and graphene nanoplates or nanoplatelets (GNPs). Both forms are very effective in changing electrical and thermal properties of polymers but are expensive: Sigma-Aldrich sells 10 g of 99% pure multi-walled CNTs for USD 115–125, whereas industrial grade GNPs cost 50–75 USD/kg. Ingeo™ 3D850 PLA formulated for FFF, and CNTs and GNPs from Times Nano (China) were combined by melt extrusion with a filler content of 1.5, 3, 6, 9, and 12 wt%. The CNTs had 10–30 nm outer diameter, 10–30 μm length, and aspect ratio of about 1000. The GNPs had <30 layers, 5–7 μm median size, and aspect ratio of about 240. Two types of composite filaments were extruded: CNT/PLA and GNP/PLA. Thermal conductivity and DC electrical conductivity were measured. The thermal conductivity of PLA was 0.183 W/mK, and increased by 2.6 times in GNP-PLA linearly, and only 0.8 times in CNT/PLA vs. that of 3D850. This occurred because PLA wetted more thoroughly the plates in GNPs than the entangled CNTs, and the *interfacial thermal resistance* (different than *contact thermal resistance*) was lower in the former than in the latter, resulting in a more efficient heat flow. The bulk electrical conductivity in 3D850 was around 1×10^{-12} S/m, and reached about 1 S/m in both composites at 12 wt% of each filler, but 1.5 wt% of CNTs was adequate to jump electrical conductivity to 1×10^{-8} S/m, while it took 6 wt% of GNPs to improve the same property vs. unfilled PLA. The authors concluded that the bidimensional structure of GNPs was more suited for heat conduction, while the monodimensional form of CNTs was more effective in conducting electricity.

3.9 PROPERTIES OF PLA FEEDSTOCKS FOR AM

3.9.1 Introduction

In tensile and flexural moduli PLA outperforms PS, PP, and HDPE. On the other hand, PLA is brittle, featuring lower notched Izod impact strength than that of PS, PP, and HDPE, and elongation at break of only 4%, just higher than that of PS. Physical and mechanical properties of printed PLA parts depend not only on the filament properties, but other factors, including the printing parameters. Moreover, printing in itself affects PLA's properties: a strong relationship was measured between printed PLA's percent crystallinity and the extruder temperature, and between tensile strength and percent crystallinity of printed samples. Although standard PLA has a T_g of

40–70°C, heat-resistant PLA withstanding temperature of 110°C is available. PLA's applications are limited by its low impact resistance, sensitivity to high relative humidity, and low softening temperature, which however can be raised from 60°C for PLLA to up to 190°C by blending PLLA and PDLA. An example of heat-resistant AM-grade PLA is the filament Workday whose HDT at 0.45 MPa is 144°C, which is a remarkable increase in comparison to HDT of typical PLA filaments, such as 53°C for Stratasys PLA filament.

The next subsections illustrate what factors influence what properties of PLA filaments and how. The property values of PLA filaments and printed articles mentioned in the rest of the chapter are relative to specific settings chosen by the investigators, and each functional article printed should be tested to measure its actual performance prior to its introduction to service.

Unless otherwise mentioned, all data reported hereafter are relative to coupons printed via FFF.

3.9.2 Porosity

Porosity is defined as the ratio of voided area to total area of a sample's cross section, and is present even in functional and structural polymers such as composite materials (Mehdikhani et al. 2019).

Choren et al. (2013) studied how porosity in printed parts influenced their tensile modulus, and how controlling porosity during printing could generate parts with desired properties, such as orthopedic implants with specific flexibility. The authors reviewed the equations predicting the actual tensile modulus in porous materials, and concluded that equations that are provable and provide consistent and reproducible results related to pore structures are not available yet.

Song et al. (2017) showed that porosity of PLA printed blocks is strongly affected by layer height, and extruder temperature and speed. They printed blocks from IMAKR (USA) PLA filament, and varied layer height (0.1–0.4 mm), and extruder temperature (200–240°C) and speed (45–150 mm/s). The porosity was 1.5–13.5%, and consisted in sporadic spherical pores of 100 μm average diameter. The smallest porosity was relative to 0.2 mm layer height, 220°C extruder temperature, and 60 mm/s extruder speed. A statistical analysis indicated that the layer height had by far the greatest influence among the three independent variables. As a benchmark, the porosity of composite materials must be <1% for aerospace structural components (Ghiorse 1993) and up to 2–5% for other applications, depending on the applications.

3.9.3 Moisture

Moisture absorption notoriously adversely affects the mechanical performance of printed articles in PLA. In fact, it is recommended that, once the vacuum-sealed package containing the PLA spool is opened, the spool should be stored in a dry box in order to minimize its moisture absorption.

Kakanaru et al. (2020) measured water uptake of MakerBot PLA filament and relative printed parts immersed in distilled water. Maximum moisture absorption was measured as weight gain, and for PLA filament and printed parts was 1.3% after 15 days, and 2% after 67 days. After those intervals, the weight diminished due to degradation of the material that fractured and dissolved. The degradation was attributed to the formation of lactic acid oligomers due to chain scission (Elsawy et al. 2017).

3.9.4 Tensile properties

3.9.4.1 Effect of Build Orientation

Tensile strength and modulus and elongation at break of PLA printed parts via FFF are influenced by process parameters such as build orientation (BO), layer thickness (LT), raster angle, and printing speed because of the nature of the FFF processes, and this influence extends to other plastics for FFF such as ABS and PC, as reported by several studies going back to almost 20 years (Ahn et al. 2002; Sood et al. 2010; Sood et al. 2011; Ziemian 2012; Domingo et al. 2015).

Chacón et al. (2017) conducted a wide-ranging experimental study on the effect of BO, LT, and feed rate (FR) on tensile strength, modulus, and elongation at break of PLA unidirectional coupons fabricated on a low-cost FFF desktop printer, and demonstrated that filament's properties are not sufficient to predict the performance of the printed part. LT is of practical interest because smaller LT results in more layers, and finer geometrical resolution, making the part's actual dimensions closer to their nominal values, but of course it increases printing time and cost. FR obviously affects time and cost as well. LT and FR ranged from 0.06 mm to 0.24 mm and 20 mm/s to 80 mm/s, respectively. Temperature and air gap were fixed at 210°C and 0 mm, respectively. Test coupons were built out of commercial 1.75 mm diameter SMARTFIL® PLA filament (Smart Materials 3D n.d.) on WitBox desktop printer (BQ, Spain). Tensile strength, strain, and modulus varied as a function of BO, LT, and FR, and the variation with building orientation confirmed the isotropy associated to the FFF process, and described in the technical literature. Anisotropy has practical importance, because it must be accounted for especially when designing load-bearing articles, which can be accomplished conservatively by adopting the lowest values of each anisotropic property as design values. Test coupons were designed and tested according to ASTM D638, and printed according to Figure 3.15. Figure 3.23 graphically summarizes the effect on tensile strength by BO, FR, and LT. At constant feed rate (Figure 3.23, upper plots) flatwise and edgewise coupons featured similar strength values that were about two to four times those of upright coupons, because the upright coupons consisted of a high number of interlayer interfaces whose cohesive strength is lower than the intralayer strength. In fact, upright samples exhibited interlayer failures. Although LT was less influential than BO on strength, the amount of its effect varied depending on BO and FR. The effect of LT and FR was not substantial within each BO (Figure 3.23 lower plots). Raising LT increased and decreased tensile strength in upright and flatwise coupons respectively, whereas it had a variable effect on edgewise coupons. The authors derived the following well-fitting regression quadratic models of tensile strength (TS) for each BO:

$$\text{Upright coupons:} \quad TS = 403 \times LT - 0.726 \times LT \times FR - 828 \times LT^2 \tag{3.2}$$

$$\text{Edgewise coupons:} \quad TS = 257 \times LT + 2.55 \times FR - 4.74 \times LT \times FR - 0.018 \times FR^2 \tag{3.3}$$

$$\text{Flatwise coupons:} \quad TS = 240 \times LT + 2.68 \times FR - 5.30 \times LT \times FR - 0.018 \times FR^2 \tag{3.4}$$

The Young's or tensile modulus value is required in designing structural components (such as aircraft fairings) that do not exceed a certain elastic (that is, recoverable) deformations under load. Figure 3.24 illustrates the variation of the Young's modulus as a function of BO, FR, and LT, and that this property was less sensitive than the tensile strength to the changes in the process parameters. This is because the Young's modulus is measured at the beginning of the stress-strain curve under low load and deformation, and is less sensitive to material and geometric features and defects that may lead to failure at high loads. In fact, the tensile strength values ranged from 20.2 to 89.1 MPa, corresponding to 3.4 times increase, whereas the Young's modulus values spanned from 3.27 to 4.41 GPa, equivalent to 35% rise. From a practical standpoint, if a load-bearing component is printed with filament deposited along X and Y, at FR of 80 mm/s and LT of 0.24 mm (the two values minimizing time), the tensile strength and modulus will be at the lower end of 39.5–71.9 MPa and 3.47–3.93 GPa, respectively. From the stress-

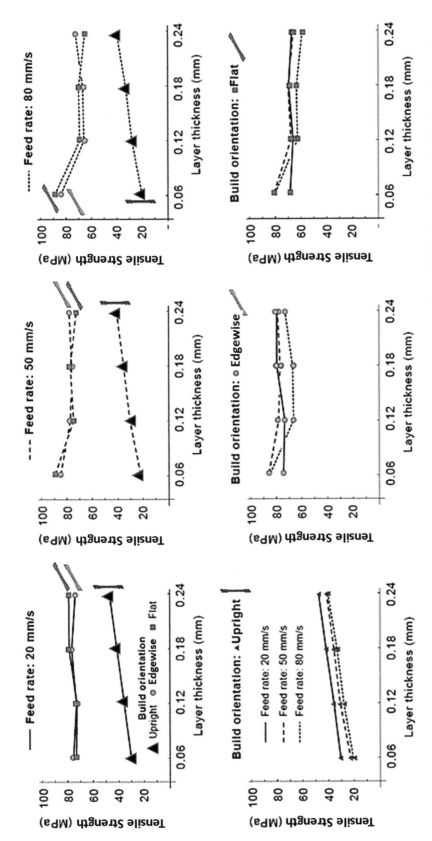

FIGURE 3.23 Average tensile strength of PLA specimens as a function of layer thickness. Upper plots: effect of build orientation at a fixed feed rate. Lower plots: effect of feed rate at a fixed build orientation.

Source: modified from Chacón, J. M., Caminero, M. A., García-Plaza, E., Núñez, P. J. 2017. Additive manufacturing of PLA structures using fused deposition modelling: Effect of process parameters on mechanical properties and their optimal selection. *Materials and Design* 124:143–157. Reproduced with permission from Elsevier.

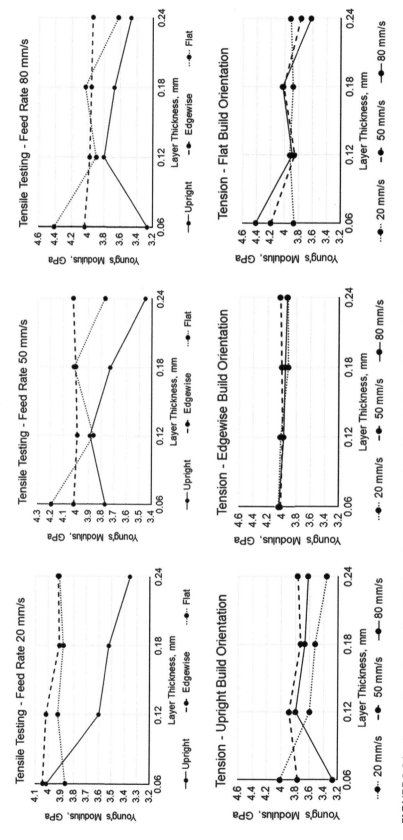

FIGURE 3.24 Average Young's modulus of PLA specimens as a function of layer thickness. Upper plots: effect of build orientation at a fixed feed rate. Lower plots: effect of feed rate at a fixed build orientation.

Source: Data from Chacón, J. M., Caminero, M. A., García-Plaza, E., Núñez, P. J. 2017. Additive manufacturing of PLA structures using fused deposition modelling: Effect of process parameters on mechanical properties and their optimal selection. *Materials and Design* 124:143–157.

strain plots, the failure mode appears dependent not on FR and LT but BO. In fact, the upright coupons exhibited brittle failure, characterized by relatively small strain at failure and no yield strength. Inversely, the edgewise coupons featured relatively high strain to failure and yield, signs of ductile failure. Flatwise coupons displayed intermediate failure mode between ductile and brittle, featuring yield but lower elongation at break. The same stress-strain curves showed that tensile strength and strain at break were positively correlated. When structural parts are not loaded in one direction in service, they should be printed by depositing material in perpendicular orientations along X and Y or at +45/−45° to X (Figure 3.15) in a ratio dictated by the amount and direction of the service loads. In that case, the effects of the printing parameters on mechanical properties are expected to be less significant.

Song et al. (2017) tested unidirectional PLA coupons, and recorded that: (a) the material mechanical response was elastoplastic, orthotropic, and exhibited a strong tension-compression asymmetry, and (b) PLA was tougher when loaded in the extrusion direction than in the transverse direction. The authors measured ultimate strength, yield strength, and modulus in tension, and Poisson's ratio on unidirectional coupons printed at 0°, 90°, and 45° (Figure 3.22). Testing load was applied always in the X direction at two strain rates: 1.25×10^{-4} and 2.5×10^{-4} 1/s. A 1.75 mm diameter IMAKR PLA filament was extruded on Flashforge Replicator printer at 220°C, 0.2 mm layer height, and 60 mm/s deposition speed. Test results are summarized in Table 3.25. As expected with polymers, test strain rate affected the results: ultimate, yield, and modulus values at a high rate were 40%, 10%, and 34%, respectively, higher than those at a low rate. Printing orientation influenced the measured properties, and, given their CV, modulus was the least anisotropic property at both strain rates, as also reported by Chacón et al. (2017). However, since each value in Table 3.25 is relative to one coupon, these results should be considered as preliminary and indicative, and do not permit to quantify how statistically significant the differences among values of the same property (especially of modulus) are.

TABLE 3.25
Tensile Test Results on Unidirectional PLA Coupons Built Via FFF

Test Strain Rate	Orientation	Ultimate Strength	Elastic Modulus	Yield Strength	Statistics
1/s	degrees	MPa	GPa	MPa	
2.5×10^{-4}	0	54.9	3.98	54.8	
2.5×10^{-4}	45	61.4	3.96	54.7	
2.5×10^{-4}	90	46.2	4.04	45.2	
2.5×10^{-4}	0, 45, 90	54.2	3.99	51.6	Avg[a]
2.5×10^{-4}	0, 45, 90	7.61	0.04	5.55	CV[b]
1.25×10^{-4}	0	34.6	3.75	34.3	
1.25×10^{-4}	45	43.3	3.56	42.5	
1.25×10^{-4}	90	38.5	3.58	38.5	
1.25×10^{-4}	0, 45, 90	38.8	3.63	38.4	Avg[a]
1.25×10^{-4}	0, 45, 90	4.36	0.10	4.13	CV[b]
1.25×10^{-4}, 2.5×10^{-4}	0, 45, 90	46.5	3.81	45.0	Overall Avg[a]

Source: adapted from Song, Y., Li, Y., Song, W., et al. 2017. Measurements of the mechanical response of unidirectional 3D-printed PLA. *Materials and Design* 123:154–164. Reproduced with permission from Elsevier.

Notes:

[a]Average.

[b]Coefficient of variation.

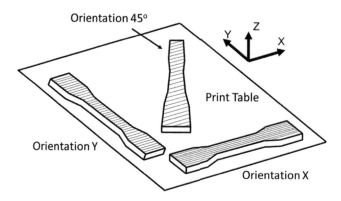

FIGURE 3.25   Build orientations of test coupons.

Statisticians recommend to collect at least 15 data points for each average to make that assessment. Moreover, ASTM D638 test method for tensile properties of plastics states to test "ten specimens, five normal to, and five parallel with, the principle axis of anisotropy, for each sample in the case of anisotropic materials." The overall averages of 46.5 MPa and 3.81 GPa for tensile strength and modulus fall in the range of values reported for most commercial PLA filaments that is 22–71 MPa and 1.82–4 GPa, respectively. The reported tensile stress-strain curves documented the elastoplastic behavior of PLA coupons, and the yield point at about 1.6% true strain on coupons loaded along X and Y, but not on those loaded along Z. The method in the paper by Song et al. (2017) can be followed to generate values of tensile modulus and Poisson's ratio along X, Y, and Z that will serve as material elastic constants and strength values to: (a) calculate in-plane (in XY) and out-of-plane (along Z) shear modulus; (b) apply elastoplastic models such as the yield criterion for anisotropic materials (Hill 1948).

Afrose et al. (2014) also investigated the effect of BO on tensile strength, modulus, and strain of flatwise dumbbell-shaped coupons with their length aligned to the X and Y axis, and at 45° (Figure 3.25). They experimented with the standard PLA filament running on the Cube-2 printer by Longer3D (China), and printed and tested per ASTM D638 three coupons for each orientation. The BO affected all measured properties, although it is unknown how significantly the properties differ across orientation, since standard deviation was not reported. Tensile strength and modulus are highest (38.7 MPa and 1,568 MPa, respectively) in the X direction, and lowest (31.4 MPa and 1,246 MPa, respectively) in the Y direction, because in the X and Y directions the filament are parallel and perpendicular to the loading direction, respectively. This result is expected because the printed coupons are similar to unidirectional fiber composites, whose matrix is reinforced by continuous fibers. Yield was detected in stress-strain curves at all orientations, and was most and least pronounced at 45° and X, respectively.

In comparing results reported in the papers above by Song and Afrose, we have to consider, besides testing different PLA filaments, the following differences: (a) test coupons were printed flatwise in Afrose, and edgewise in Song; (b) Song tested one coupon per orientation, Afrose three; (c) Afrose reported the crosshead speed (50 mm/min) not the strain rate which we calculated to be approximately 0.008/s, and hence higher than 1.25×10^{-4} and 2.5×10^{-4} 1/s; (d) in Afrose tensile strength and modulus had the same degree of anisotropy (expressed as CV), whereas in Song, overall, modulus was 40 times less anisotropic than strength. In Afrose, tensile strength was at the lower end of the range of values of commercial PLA filaments (22–71 MPa), and the tensile modulus below the range of values of commercial PLA filaments (1.82–4 GPa). The values of modulus were collected by measuring the slope of the elastic portion of the stress-strain curve, not by applying strain gages, extensometer, and optical devices that are more accurate techniques than reading the slope. Since there is no mention of subtracting the compliance of testing equipment from the coupon deformation in the above papers, the modulus values recorded may be likely lower than its actual values.

3.9.4.2 Effect of Printers

The tensile and other mechanical properties of printed PLA parts not only depend on the printing settings on the same printer, but also vary across printers. Currently PLA is mostly employed not on professional and industrial printers but low-cost desktop printers typically to build visual models, DIY, hobby, household and ornamental articles, toys, research equipment (Pearce 2012), and scientific instruments (Pearce 2014). A type of low-cost printer is an *open source printer*, like RepRap printers (RepRap 2019), whose hardware, firmware, and software information are available to the public (differently from a proprietary printer), typically under a license, and can be utilized to build, modify, and improve the printer by yourself.

Since open-source low-cost (below $500) desktop printers are being considered for manufacturing functional, load-bearing, and durable components, it would be useful to assess whether the mechanical properties of PLA items built on such printers are comparable to those of PLA articles built on commercial printers. Such assessment was experimentally conducted by Tymrak et al. (2014) by: (a) measuring the variation of tensile strength, modulus, and strain corresponding to the tensile strength across four open-source inexpensive desktop printers operating at 100% infill and varying pattern orientation (0/90°, +45/−45°), and layer height (0.2, 0.3, 0.4 mm); (b) comparing such tensile properties to those measured on commercial printers. The following printing settings were not kept constant across the printers, but on each printer they were equal to the values generating the best print: extruder temperature, print bed temperature, nozzle diameter, cooling, print speed, air gap, and number of extruded perimeters composing the shell of the test coupons. PLA filaments specific to each printer tested were utilized. All specimens were printed flatwise, aligning the filament with the direction of test loading, since varying the build orientation significantly affect tensile properties, as described in Section 3.9.4.1. Open-source printers and filaments evaluated are listed in Table 3.26. The plot of tensile strength vs. strain at tensile strength for PLA printed specimens in Figure 3.26 illustrates that results depended on the printer: printer P1 yielded all the highest values of strength and strain in the upper right hand cluster, whereas the lowest values of both properties were associated to P4 printer. Table 3.27 illustrates the variation of tensile strength, elastic modulus, and strain at tensile strength quantified through various metrics, indicating as *printer subset* the data relative to each individual printer. The authors concluded

---

TABLE 3.26

Filaments and Open-Source Printers Involved in the Study by Tymrak et al. (2014)

Printer ID	Printer	Filament
P1	MOST RepRap	Clear PLA
P2	Lulzbot Prusa Mendel RepRap	Purple PLA, White PLA
P3	Prusa Mendel RepRap	Black PLA
P4	Original Mendel RepRap	Natural PLA

Source: Adapted from Tymrak, B. M., Kreiger, M., Pearce, J. M. 2014. Mechanical properties of components fabricated with open-source 3-D printers under realistic environmental conditions. *Materials and Design* 58:242–246. Reproduced with permission from Elsevier.

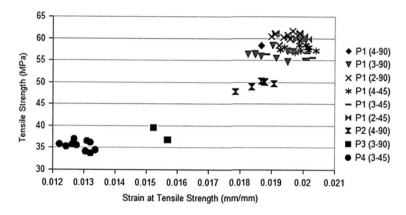

FIGURE 3.26 Tensile strength vs. strain at tensile strength for PLA coupons fabricated on open source printers.

Legend: printer ID in parentheses (layer height in mm × 100 – orientation angle).

Source: Tymrak, B. M., Kreiger, M., Pearce, J. M. 2014. Mechanical properties of components fabricated with open-source 3-D printers under realistic environmental conditions. *Materials and Design* 58:242–246. Reproduced with permission from Elsevier.

TABLE 3.27
Tensile Property Values of PLA Coupons Built on Open-Source Printers

Property	Unit	Average of All Values	Minimum Average of Printer Subset	Maximum Average of Printer Subset	Max Average/Min Average
Strength	MPa	56.6	48.5	60.4	1.25
Modulus	MPa	3,368	3,286	3,480	1.06
Strain at Tensile Strength	%	1.93	1.71	1.96	1.13

Source: Data from Tymrak, B. M., Kreiger, M., Pearce, J. M. 2014. Mechanical properties of components fabricated with open-source 3-D printers under realistic environmental conditions. *Materials and Design* 58:242–246.

that the components built on RepRap printers are comparable in tensile strength and elastic modulus to the components built on commercial 3D printers. Their conclusion seems validated by the data in Table 3.28, comparing the PLA's properties measured by Tymrak to those of coupons out of commercial PLAs most likely printed on professional printers. However, we think that test coupons printed on commercial printers from the same PLA processed on RepRap printers should have been added to this study. PLA filaments of different colors were tested, and it has been documented that filament color in PLA filaments affects their crystallinity, and hence their tensile properties (Wittbrodt and Pearce 2015), therefore this was another variable affecting the results.

3.9.4.3 Effect of Interfacial Bonding Strength between Adjacent Filaments

Li et al. (2018) experimented with PLA filaments, and studied the quantitative relationship between tensile strength of printed part and interfacial bonding strength between adjacent filaments, and through testing learned how, in turn, the latter was affected by - in decreasing order - layer thickness, deposition velocity and infill rate in decreasing order. The bonding degree was measured

TABLE 3.28

Tensile Properties of PLA Coupons Made on Professional Printers

PLA Supplier	Strength	Modulus	Strain at Break
	MPa	MPa	%
Unknown (Tymrak et al. 2014)	48.5–60.4	3,286–3,480	1.7–1.9
Stratasys[a]	26–48	2,539–3,039	1.0–2.5
3D Fuel (Standard PLA)	41	3,200	1.8
Ultimaker	45.6	2,347	5.2
colorFabb	45	3,400	6

Source: website of PLA suppliers, except for the data in "unknown" row.
Notes:
[a]Values depend on build orientation.

visually through a super-high magnification 3D microscope. The bonding strength of filaments is positively associated with the extruding temperature of the filament: the filament at higher temperature forms a stronger bond than the filament at lower temperature does. The authors also shared the analytical models that they proposed to accurately predict the tensile strength from layer thickness, deposition velocity, and infill rate.

3.9.4.4 Mechanical Models of Strength and Modulus of Printed PLA

Zhao et al. (2019) introduced two theoretical models predicting the tensile strength and modulus of PLA coupons printed at multiple angles (Figure 3.27) and layer thicknesses. The strength model was based on the transversely isotropic material hypothesis and Tsai-Hill strength criterion (Azzi and Tsai 1965), whereas the modulus model was based on the orthotropic material hypothesis under plane stress state. Experimental results show that the tensile strength and modulus increased with greater printing angle or smaller layer thickness. The models for the tensile strength σ_x and modulus E_x at the generic printing angle x were expressed by equations (3.5) and (3.6), respectively, with α the orientation angle (Figure 3.27), μ_{12} and G_{12} the Poisson's ratio and the shear modulus in the plane 12, respectively, T_1 and T_2 the tensile strength in direction 1 (laid filament) and 2 (perpendicular to 2), respectively, and S_2 shear strength in plane 12:

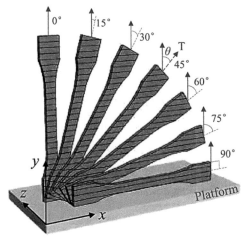

FIGURE 3.27 Printing orientation of test coupons.

Source: Zhao, Y., Chen, Y., Zhou, Y. 2019. Novel mechanical models of tensile strength and elastic property of FDM AM PLA materials: Experimental and theoretical analyses, *Materials & Design* 181, 108089. Reproduced with permission from Elsevier.

$$\sigma_x = \left[\frac{cos^4 \alpha}{T_1^2} + \left(\frac{1}{S_2^2} - \frac{1}{T_1^2} \right) \ sin^2 \alpha \cdot cos^2 \alpha + \frac{sin^4 \alpha}{T_2^2} \right]^{-\frac{1}{2}} \quad (3.5)$$

$$E_x = \left[\frac{1}{E_1} cos^4 \alpha + \left(\frac{1}{G_{12}} - \frac{2\mu_{12}}{E_1} \right) \ sin^2 \alpha \cdot cos^2 \alpha + \frac{sin^4 \alpha}{E_2} \right]^{-1} \quad (3.6)$$

S_2 and G_{12} in the above equations are calculated as follows, being P_{45} the bearing capacity of off-axis specimens printed at 45° angle, b and t the width and thickness of the off-axis specimens, respectively:

$$S_2 = \frac{P_{45}}{2bt} \quad (3.7)$$

$$G_{12} = \left(\frac{4}{E_{45}} - \frac{1}{E_1} - \frac{1}{E_2} + \frac{2\mu_{12}}{E_1} \right)^{-1} \quad (3.8)$$

The models above fit well the experimental data.

3.9.5 COMPRESSIVE PROPERTIES

Compressive properties are not released by suppliers of commercial PLA filaments. Two exceptions are MakerBot and REC3D (Russia) that reported for their PLA filaments a compressive strength of 93.7 MPa at 100% infill, and 77.4 MPa (Koslow and de Valensart 2017), respectively. In polymers, tensile and compressive elastic and strength values do not necessarily coincide, hence in structural design values of compressive properties should not be assumed equal to the values of tensile properties. Song et al. (2017) measured strength, yield stress, and modulus in compression, and compared them to the same properties in tension for the same material. Unidirectional blocks were machined from bars printed out of IMAKR PLA at 0°, 90°, and 45° (Figure 3.28), and tested

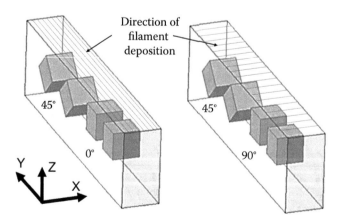

FIGURE 3.28 Schematic of compressive coupons printed at 0° (left), 90° (right), and 45° (left and right).

Source: Modified from Song, Y., Li, Y., Song, W., et al. 2017. Measurements of the mechanical response of unidirectional 3D-printed PLA. *Materials and Design* 123:154–164. Reproduced with permission from Elsevier.

in compression at two strain rates: 1.25×10^{-4}/s and 2.5×10^{-4}/s. Table 3.29 lists the compressive test values: *yield strength* is the stress at 0.2% strain, and *strength* corresponds to the peak stress recorded in the stress-strain curves. Like in tension, strain rate affected results: higher strain rate resulted in 10%, 6%, and 13% higher average compressive strength, modulus, and yield strength, respectively, compared to the same properties at lower strain rate. At both strain rates, strength and elastic modulus were the least and the most anisotropic property, respectively, whereas it was the opposite in tension. However, this analysis is only based on one coupon tested per strain rate and orientation. ASTM D695 standard test method for compressive properties of rigid plastics requires, like for tension, to test "ten specimens, five normal to and five parallel with the principal axis of anisotropy [...] for each sample in the case of anisotropic materials". Strong compression-tension asymmetry was detected: in terms of overall averages, compressive strength, yield stress, and elastic modulus were 93%, 80%, and 5% higher than tensile strength, yield, and modulus, respectively. The overall average strength of 89.8 MPa was 4% lower and 16% higher than that of MakerBot PLA and REC3D, respectively.

Brischetto et al. (2020) measured the compressive modulus of black PLA sold by Shenzhen Eryone Technology (China) on coupons with a square cross section of 12.7 × 12.7 x 40 (H) mm meeting requirements of ASTM D695. Coupons were printed flatwise with their height parallel to X, and cross section perpendicular to the XY plane. The raster angle orientation was ±45° in order to induce a quasi-isotropic response. Adopting a testing strain rate of 0.045/s, the compressive modulus was 2,035 ± 26 MPa, and compressive strength was not calculated because no yield or failure occurred before the maximum applicable load was reached. Stress-strain curves were also reported.

Perkowski (2017) tested tensile (ASTM D638) and compressive (ASTM D695) coupons printed out of Ultimaker PLA filament in several orientations from flatwise to upright (Figure 3.15), and documented property values and stress-strain curves for each orientation. The results summarized across all the printing orientations are the following: (a) tension: strenght and modulus 2.33–39.7

TABLE 3.29
Compression Test Results on Unidirectional PLA Coupons Built Via FFF

Test Strain Rate	Orientation	Strength	Elastic Modulus	Yield Strength	Value Type
1/s	degrees	MPa	GPa	MPa	
2.5×10^{-4}	0	98.4	4.73	95.8	
2.5×10^{-4}	45	99.7	4.25	98.3	
2.5×10^{-4}	90	98.1	3.73	94.9	
2.5×10^{-4}		98.7	4.24	96.3	Avg[a]
2.5×10^{-4}		0.87	11.8	1.9	CV[b]
1.25×10^{-4}	0	78.5	3.19	76.5	
1.25×10^{-4}	45	84.2	3.42	78.2	
1.25×10^{-4}	90	79.8	4.65	65.9	
1.25×10^{-4}		80.8	3.75	73.5	Avg[a]
1.25×10^{-4}		3.7	20.9	9.0	CV[b]
	0, 45, 90	89.8	4.00	84.9	Overall Avg[a]

Source: Adapted from Song, Y., Li, Y., Song, W., et al. 2017. Measurements of the mechanical response of unidirectional 3D-printed PLA. *Materials and Design* 123:154–164. Reproduced with permission from Elsevier.

Notes:

[a]Average.

[b]Coefficient of variation.

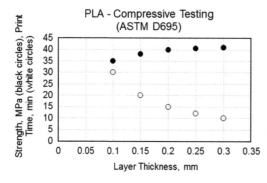

FIGURE 3.29 Compressive testing of PLA coupons.

Source: Data from Abbas, T. F., Othman, F. M., Ali, H. B. 2017. Influence of Layer Thickness on Compression Property of 3D-Printed PLA. *Al-Muhandis Journal JMISE* 154 (4): 41–48.

MPa, and 0.64–3.15 GPa, respectively; (b) compression: strenght and modulus 21-63.6 MPa, and 0.65–1.64 GPa, respectively.

Abbas et al. (2017) experimentally showed that the compressive strength of PLA coupons steadily improved by increasing the layer thickness, when this printing setting was the only one varying. The authors did not specify the PLA filament and printer selected, but the printing parameters were: printing speed 100 mm/s, infill density 80%, shell thickness 1.6 mm, build orientation 0–90°. One coupon per layer thickness was printed per ASTM D695, and tested at 2 mm/min. The results are illustrated in Figure 3.29. Higher thickness resulted in lower printing time, but also smaller fidelity to nominal dimensions, "staircase effect" on curved surfaces, and higher surface roughness, with the latter two phenomena in turn potentially reducing fatigue life, and requiring surface finishing to increase fatigue life or meet surface finish requirements. We speculate that higher thickness translates in fewer interlayer interfaces featuring lower cohesion than intralayer cohesion. The strength values reported trailed those in Table 3.29, but this comparison is not fitting because (a) Abbas built coupons at 0–90° instead of unidirectional ones, (b) the two PLAs are unlikely the same; (c) Abbas built coupons with 80% infill whereas no infill is mentioned for the coupons of Table 3.29, although they contained porosity.

Tsouknidas et al. (2016) assessed the variation of compressive yield strength in PLA as a function of infill density (25, 50, 100%), layer thickness (0.1, 0.2, 0.3 mm), and filling pattern (rectilinear, octagonal, concentric). The results in Figure 3.30 illustrate that, across layer height and infill patterns (the dominant variable), yield strength steadily increased with declining infill density, although this finding is counterintuitive.

Yamamura et al. (2009) measured the compressive behavior of an unspecified PLA, and its experimental compressive stress-strain curve, derived according to the Japanese standard test

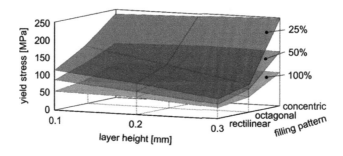

FIGURE 3.30 Compressive yield stress of PLA coupons as a function of printing settings.

Source: Tsouknidas, A., Pantazopoulos, M., Katsoulis, I. et al. 2016. Impact absorption capacity of 3D-printed components fabricated by fused deposition modelling. *Mater Des.* 102:41–44. Reproduced with permission from Elsevier.

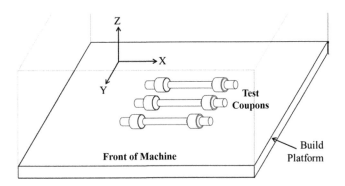

FIGURE 3.31 Test coupons for shear measurements in torsion of PLA.

Source: Torres, J., Cotelo, J., Karl, J., Gordon, A. P. 2015. Mechanical property optimization of FDM PLA in shear with multiple objectives, *JOM* 67 (5): 1183–1193. Reproduced with permission from Springer Nature.

method JIS K7181, showed the following values: ultimate strength and strain at failure 90 MPa and 0.5%, respectively, strength and strain at yield 120 MPa and 0.06%, respectively, and tangent modulus 2 GPa.

3.9.6 SHEAR PROPERTIES

Shear properties are not disclosed for commercial PLA feedstocks for AM, hence the paper by Torres et al. (2015) is particularly notable, because the authors measured the following shear properties in torsion of coupons printed out of an unspecified PLA (likely supplied by MakerBot): ultimate strength, 0.2% yield strength, proportional limit, shear modulus, and fracture strain, and how they were affected by layer thickness (0.1, 0.2, 0.3 mm), infill density (20, 60, 100%), and duration (0, 5, 20 min) of post-processing heat-treatment at 100°C. Stress-stain test curves were also plotted. Coupons were cylindrical, built in 0–90° orientation, with their longitudinal axis parallel to the printer's X axis (Figure 3.31). For each set of parameters three coupons were tested according to ASTM E143, and following DOE. Across the combination of printing parameters, the following range of averages in shear were measured: ultimate strength 20–53 MPa, 0.2% yield strength 15–42 MPa, proportional limit 12–36 MPa, fracture strain (individual values) 0.04–3.2 rad (not %, being the test coupons cylindrical), and shear modulus 644–1,223 MPa. Infill density and layer thickness had the most effect on all properties (specifically in the combination of 100% and 0.1 mm, respectively) except fracture strain. The 20 min duration of heat treatment maximized all properties. Assuming that the tested PLA was supplied by MakerBot, the ratio ultimate shear strength/ultimate tensile strength at 100% infill and without heat treatment was 39.5/65.7 = 0.60 instead of the generic estimate of 0.75 for polymers (Tres 2014).

In comparison, in-plane and cross-sectional shear at failure measured on printed ABS coupons were 32.1 and 30.8 MPa, respectively, and on printed HIPS coupons were 30.4 and 25.1 MPa, respectively (Edwards et al. 2017).

3.9.7 IMPACT PROPERTIES

The brittle fracture behavior of unfilled PLA under impact (and tension) has been attributed to the rigidity of its chains that favors crazing deformation over shear yielding (Wu 1990), with crazing consisting in tiny cracks formed perpendicular to the applied stress. The intrinsic low impact strength of PLA is a reason impeding its diffusion in functional applications. The Izod impact strength of PLA depends on fracture morphology, MW, crystallinity, and morphology, size, and orientation of crystals, and interaction among crystals (Perego et al. 1996; Yang et al. 2012; Bai et al. 2014).

Liu and Zhang (2011) and Yang et al. (2016) published an extensive and detailed review on the progress in toughening PLA and the routes to achieve it: (a) plasticization, (b) copolymerization, (c) melt blending with flexible polymers, and (d) reactive blending. The review also contained numerous property data and material examples. Plasticizers improve not only the processability of polymers but also their flexibility and ductility (elongation at break). However, an increase in elongation usually is offset by substantial decrease in tensile strength and elastic modulus. A suitable plasticizer for PLA is biodegradable, nonvolatile, and nontoxic, and exhibits minimal leaching and migration during aging, but unfortunately lowers its T_g, precluding high temperature uses. Many low-MW compounds have already been employed as potential plasticizers for PLA, such as PEG, poly(propylene glycol), polyadipates, epoxidized soybean oil, citrate esters, and poly (butylacrylate). PEG and citrate esters have been very often investigated as plasticizers. Toughening through copolymerization alters the composition and architecture of the molecule and the sequence of its monomers. PLA is copolymerized through polycondensation with other polymers, and, especially, ring-opening copolymerization. Melt blending with flexible polymers is far less expensive than synthesizing new polymers to enhance the properties of current polymers. Suitable compounds for melt blending include non-biodegradable (PE, ABS, etc.) and biodegradable polymers. Examples of the latter ones are: (a) polyesters and copolyesters: PCL, PBS, PHA, PBAT; (b) elastomers: TP poly(ether)urethane, and TP polyamide elastomer (PAE); (c) soybean oil derivatives. Examples of compounds for reactive blending are (Liu and Zhang 2011): triphenyl phosphate (TPP), dicumyl peroxide (DCP), isocyanate compounds, random terpolymer of ethylene, acrylate ester and glycidyl methacrylate, maleimide-terminated PLA (HEMI-PLLA), styrene/acrylonitrile/GMA copolymer (SAN-GMA), and poly(ethylene-coglycidyl methacrylate) (EGMA). Examples of commercial impact modifiers for PLA are Biomax® Strong 100 and 120 (DuPont), Biostrength™ (Arkema), OnCap™ BIO Impact (PolyOne), and Paraloid BPM (Dow).

Abbas et al. (2018) run a basic experiment, and demonstrated that Izod impact strength of PLA coupons steadily declined upon increasing the coupon's layer thickness, with the latter being the only varying printing setting. An unspecified PLA filament was processed on an Ultimaker 2+ printer running at the following settings: extrusion temperature 235°C, print speed 100 mm/s, infill density 80%, shell thickness 1.6 mm, and build orientation 0–90°. One unnotched coupon per layer thickness value was printed per ISO 180, and flatwise that is with its length and width laying on the printer table. The Izod unnotched impact strength was 13.9–16.7 kJ/m² (Figure 3.32), higher than for metal-filled PLA filament by colorFabb, and Clariant Natural Color HIPS filament (MatWeb 2020), whose values are 10 kJ/m² and 12 kJ/m², respectively. The impact strength linearly diminished with raising layer thickness, and, a fourfold increase in layer thickness resulted in 17% reduction of impact strength, but also a fourfold decrease in printing time.

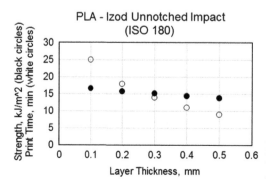

FIGURE 3.32 Impact testing of PLA coupons.

Source: Data from Abbas, T. F., Othman, F. M., Ali, H. B. 2018. Influence of Layer Thickness on Impact Property of 3D-Printed PLA. *International Research Journal of Engineering and Technology (IRJET)* 5, 2.

TABLE 3.30

Notched Izod Impact Strength of Coupons Made of Ingeo™ 4032D PLA Printed Via FFF at 100% Infill

Coupon	Layer Thickness	Printer Table Temperature	Crystallinity	Notched Izod Impact Strength
	mm	°C	vol%	J/m
A	0.2	30	20.5	37.6
B	0.2	160	44.4	95.7
C	0.4	30	18.9	36.5
D	0.4	160	34.3	75.8

Source: Data from Wang, L., Gramlich, W. M., Gardner, D. J. 2017. Improving the impact strength of Poly(lactic acid) (PLA) in fused layer modeling (FLM). *Polymer* 114:242–248.

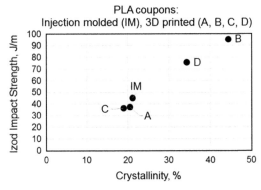

FIGURE 3.33 Variation of Izod impact strength of printed and injection molded PLA with crystallinity.

Source: Data from Wang, L., Gramlich, W. M., Gardner, D. J. 2017b. Improving the impact strength of Poly(lactic acid) (PLA) in fused layer modeling (FLM). *Polymer* 114:242–248.

Wang et al. (2017b) experimentally measured how impact strength of PLA was affected by layer thickness (0.2, 0.4 mm), and platform temperature (30, 160°C). Other printing parameters were fixed: extruder temperature (200°C) and speed (90 mm/s), infill density (100%), layer width (0.4 mm), and build orientation (unidirectional along the coupon length). Coupons were printed flatwise. The authors tested, per ASTM D256-10, Izod notched coupons printed out of Ingeo™ 4032D PLA that features Izod impact strength of 16 J/m, according to its data sheet. Table 3.30 reports the printing settings and test results as average of 10 coupons. Coupons printed at 160°C outperformed the others. After applying microscopy, chromatography, DSC, and X-ray diffraction, the increase in impact strength was attributed to higher crystallinity (Figure 3.33) that was induced by annealing (Harris and Lee 2008; Vadori et al. 2013) at 160°C. Coupons made with 0.2 mm layers at 160°C were annealed for double time than 0.4 mm coupons (because printing the former coupons took double time), and reached 26% higher impact strength than the latter ones. This may also explain the results in Abbas et al. (2018): printed layers were repeatedly subjected to very fast heating and cooling, coinciding with approach and withdrawal of the printing nozzle, respectively, and the oscillating thermal cycling interrupted crystal growth, and produced small crystals, previously recognized to significantly raise impact strength. Hence, the thinner layers led to more thermal cycles and a greater amount of small. The 160°C temperature also resulted in smaller interlayer voids, and greater number of molecules diffusing at the filament interfaces. Higher MW raises impact strength in PLA (Perego et al. 1996), but in this case MW was lower in the 160°C samples than in the 30°C samples, since high temperatures degrade and lower MW, and hence MW did not boost impact strength.

TABLE 3.31
Properties of Printed Coupons of PLA and PLA/CSR

Material	Impact Strength	Tensile Modulus	Tensile Strength	Elongation at Break	tanδ at 1 Hz, 10 Hz
	kJ/m^2	GPa	MPa	%	No unit
PLA	10	3.0	60	4.5	0.010, 0.007
PLA/CSR (90/10 wt%)	23	3.2	49	4.4	0.018, 0.015
ABS	22	2.5	38	5.1	N/A

Source: Data from Slapnik, J., Bobovnik, R., Mešl, M., Bolka, S. 2016. Modified polylactide filaments for 3D printing with improved mechanical properties. *Contemporary Materials* VII–2. doi:10.7251/COMEN1602142S.

Slapnik et al. (2016) improved impact strength of PLA by following the typical route of mixing an impact modifier to PLA, and experimentally investigated the fracture toughness of printed samples of unfilled PLA, PLA blended with core-shell rubber (CSR), and ABS (as an example of impact resistant plastic and best seller AM feedstock). Charpy impact strength was measured according to ISO 179. CSR comprised particles featuring a rubber core inside a shell of a different material ensuring good dispersion in a specific matrix to which it is also compatible. The blend PLA/CSR 90/10 wt% featured twice the impact strength of unfilled PLA, and about the same as ABS. Specifically, test coupons made of Ingeo™ 2003D PLA, LG Chem (South Korea) HF-380 ABS, and an unspecified (trade secret) commercial CSR were printed at 0.3 mm layer height, 20% infill, 230°C nozzle temperature, and 55°C and 110°C platform temperature for PLA and ABS, respectively. Beside impact and tensile testing, DMA was also performed to measure *tanδ* (or *loss factor*) that quantifies the material's energy dissipation, and mechanical damping: the higher tanδ, the more energy a material absorbs without breaking. The main results are summarized in Table 3.31. The PLA blend more than doubled its impact strength vs. unfilled PLA, and out-performed ABS. The improvement in impact strength was accompanied by a greater mechanical damping expressed by a rise of tanδ.

Tsouknidas et al. (2016) assessed the impact absorption capacity of PLA. Cylindrical test coupon, 30 mm high and 20 mm in diameter, were printed at 100 mm/s deposition speed, 180°C extrusion temperature at the following settings: (a) filling patterns: rectilinear, octagonal and concentric; (b) layer height: 0.1, 0.2, 0.3 mm; (c) infill densities: 25, 50, 100%. All cylinders had two solid layers, one at the top and another at the bottom surface. A 0.4 mm thick shell covered the cylindrical surface, completely encapsulating the diverse internal structures. Impact testing was conducted at 8.3 m/s speed, 5.6 kg impact load, and 194 J impact energy, with three replicates for each process variation. The absorbed impact force (N) raised with increasing infill, but the absorbed energy per effective volume of material (J/cm^3) increased as infill decreased, demonstrating that porous structures absorb energy more effectively than solid structures. In fact, Figure 3.34 illustrates that at 0.1 mm layer height the absorbed energy per effective volume was maximum and reached 7.35, 25.5, and 43.6 J/cm^3 for 100, 50, and 25% infill, respectively. Figure 3.34 also shows the effect of filling pattern and infill density on energy per volume.

3.9.8 FRACTURE TOUGHNESS

We recall that *fracture toughness* describes the damage tolerance of a material, namely its resistance against propagation of a preexisting crack, manifested as its ability to absorb strain energy

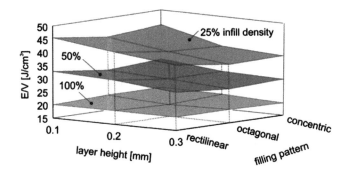

FIGURE 3.34 Impact energy absorbed per effective volume of PLA coupons as a function of printing settings.

Source: Tsouknidas, A., Pantazopoulos, M., Katsoulis, I. 2016. Impact absorption capacity of 3D-printed components fabricated by fused deposition modelling. *Mater. Des.* 102:41–44. Reproduced with permission from Elsevier.

prior to fracture. Fracture toughness is expressed by two material properties: *critical stress intensity factor K_c* and *critical strain energy release rate Gc*. The former is associated with failure due to crack propagation. In fact, when K_c is equal to the *stress intensity factor K* of a notched and stressed component, the crack propagates and the component breaks. For a sharp elastic crack in an infinitely wide plate, K is calculated as it follows, with a half the length of the crack and σ the tensile stress applied to the cracked plate perpendicularly to a:

$$K = \sigma \, (\pi a)^{1/2} \tag{3.9}$$

G_c quantifies the energy needed to increase a crack by one unit length in a component of one unit width, and depends on amount and type of stress applied to the material around the crack, crack length, and the tensile modulus of the material.

Song et al. (2017) measured the K_{Ic} (which is K_c calculated in presence of plain strain) in tensile loading mode of coupons made of IMAKR PLA, and printed at 0° and 90° orientation (Figure 3.28), at 220°C extrusion temperature, 0.2 mm layer height, 0.4 mm nozzle diameter, and 60 mm/s deposition speed. Coupons were tested in single-edge-notch bending per ASTM D5045-14 (Figure 3.35). K_{Ic} was 5.05 and 4.06 MPa m$^{1/2}$ at 0° and 90°, respectively. Analysis of fracture surface through SEM revealed a difference between 0°, and 90° coupons: the former ones displayed a smooth fracture surface, and the crack advanced mainly in the single plane YZ (Figure 3.28), inducing delamination in the relatively weak interlayer bonding, whereas the fracture surface of 90° samples had a coarse texture with features aligned to the layers, and the fracture progressed irregularly.

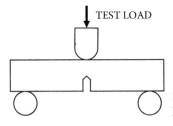

FIGURE 3.35 Schematic of single-edge-notch bending test per ASTM D5045-14.

Yao et al. (2016) followed a route to improve fracture toughness of PLA in general that may be suited to PLA for AM, and consisted in adding 1 wt% of SiC whiskers, which increased the fracture toughness of PLA by 153%. Namely, PLA pellets with a 200,000 g/mol MW were mixed and stirred in dichloromethane (CH_2Cl_2) with SiC whiskers featuring 0.05–2.50 µm diameter and length/diameter >20, and compression-molded into test samples. K_{Ic} was measured via single-edge notched testing in three-point bending performed per ASTM E399. SiC whiskers were added at 0.5, 1, 1.5, and 2 wt%, with 1 wt% producing the most improvement. Since commercial SiC whiskers (Tateho 2020) are far shorter than the diameter of PLA filaments, they can be incorporated into PLA filaments and retain their length, which is what produces the strengthening effect. However, since the cost of SiC whiskers is about USD 1–2/g or less for "large" quantities, performance benefits should offset the material cost.

3.9.9 FATIGUE PROPERTIES

Only scarce data on the fatigue performance of PLA in general is released in the technical literature, but some data have been recently published, indicating that the interest in PLA for durable load-bearing components is relatively new. A very recent and extensive review of the technical publications on fatigue behavior of PLA (and other polymers for AM) has been published by Safai et al. (2019).

Afrose et al. (2016) measured the effect of the build orientation X, Y, and 45° (Figure 3.25) on the fatigue behavior of PLA coupons printed and subjected to tensile fatigue (featuring R = min stress/max stress = 0). The maximum load at each orientation was set at 80, 70, 60, and 50% of the tensile strength for each orientation that was 31.0, 24.9, and 26.9 MPa at X, Y, and 45°, respectively, hence the stress levels were not the same for all orientations. Coupons dumbbells compliant with ASTM D638, and only one specimen was tested for each orientation and maximum load with the intent to detect possible trends in the fatigue behavior across build orientations. In Figure 3.36 the S–N (stress vs. number of cycles survived) curves of the printed coupons are plotted.

Ezeh and Susmel (2018) took the values of applied fatigue stress vs. number of cycles at failure for X (0°), Y (90°), and 45° coupons reported by Afrose et al. (2016), plotted them on an S-N chart (Figure 3.37), and extrapolated the stress values up to 2 million cycles. Figure 3.37 shows that: (a) under tensile cyclic loading condition, the 45° samples displayed the longest fatigue life, and those at X the shortest one, and (b) the orientation did not substantially affect the fatigue behavior. The S–N plot also included three straight lines marking the stress values that the PLA will survive for a specific number of cycles with a probability (P_s) of 10, 50, and 90%.

Letcher and Waytashek (2014) also assessed the influence of build orientations (0°, 45° and 90°, in Figure 3.25) on fatigue life of PLA loaded in tension-compression. Coupons were printed and tested per ASTM D638. Their perimeter was made of two shells and their inside was printed with 100% infill at each specific orientation. The PLA was extruded at 230°C at 100 mm/s on bed surface at 65°C. The coupons were fatigued undergoing a sinusoidal loading waveform at 2 Hz up

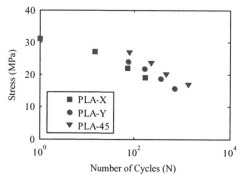

FIGURE 3.36 S–N curves of tensile coupons made of PLA filament and printed at X, Y and 45° orientations.

Source: Afrose, M. F., Masood, S. H., Iovenitti, P. et al. 2016. Effects of part build orientations on fatigue behaviour of FDM-processed PLA material. *Prog. Addit. Manuf.* 1:21–28. Reproduced with permission from Springer.

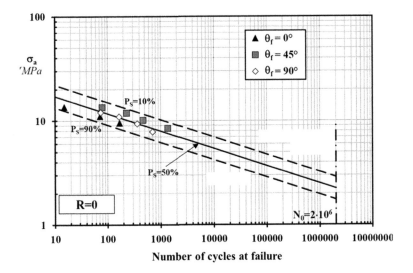

FIGURE 3.37 Extrapolated curve of applied tensile stress vs. number of cycles at failure (S-N curve) plotted with test results reported by Afrose et al. (2016).

Source: Ezeh, O. H., Susmel, L. 2018. Fatigue behavior of additively manufactured polylactide (PLA). *Procedia Structural Integrity* 13:728–734. Reproduced with permission from Elsevier.

to 1,000 cycles, then 5 Hz up to 10,000 cycles, and finally 20 Hz until failure. All testing was conducted at a stress ratio of $R = -1$ that is in tension and compression at the same absolute value of stress. The S-N curve (Figure 3.38) of the test values of 0°, 90°, and 45° coupons was plotted by Ezeh and Susmel (2018), and displays not a large difference in fatigue performance across the build orientations. It also contains the same 10, 50, and 90% probability lines traced in Figure 3.37.

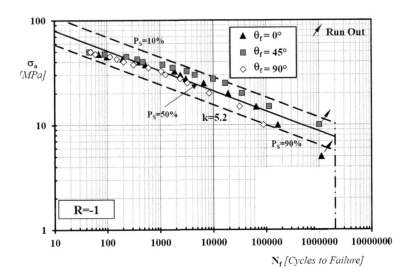

FIGURE 3.38 Applied tensile stress vs. number of cycles at failure (S-N curve) plotted with test results reported by Letcher and Waytashek (2014).

Source: Ezeh, O. H., Susmel, L. 2018. Fatigue behavior of additively manufactured polylactide (PLA). *Procedia Structural Integrity* 13:728–734. Reproduced with permission from Elsevier.

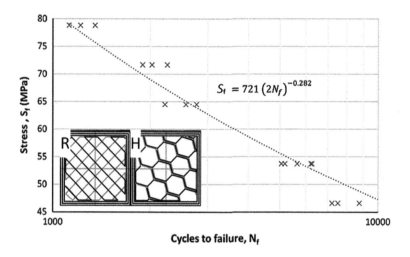

FIGURE 3.39 Fatigue (S–N) curve for coupons loaded in flexure, and printed with honeycomb infill, 75% infill density, 0.5-mm diameter nozzle and 0.3 mm layer height. Inset: R is rectilinear infill, H is honeycomb infill. The equation is relative to the curve best fitting the data.

Source: Gomez-Gras, G., Jerez-Mesa, R., Travieso-Rodriguez, J.A., Lluma-Fuentes, J. 2018. Fatigue performance of fused filament fabrication PLA specimens. *Mater. Des.* 140:278–285. Reproduced with permission from Elsevier.

Gomez-Gras et al. (2018) studied the effect of infill density (25, 50, 75%), pattern (rectilinear and honeycomb, as shown in Figure 3.39) Figure 3.39, nozzle diameter (0.3, 0.4, 0.5 mm), layer thickness (0.1, 0.2, 0.3 mm), and printing speed (25, 30, 35 mm/min) on fatigue performance of specimens subjected to rotating bending fatigue. Test coupons were cylindrical and clamped at one end to the chuck of the GUNTWP140 rotating bending stress machine and spindled at 2,800 rpm/min. The fluctuating stress derived from applying a concentrated load at the free end of the beam through a load cell, and the number of cycles survived by each coupon was registered by a digital revolutions counter. The test results (Figure 3.39) led to the following conclusions: (a) fill density was the dominant variable on fatigue performance (as expected), followed by nozzle diameter and layer thickness in about the same amount, whereas printing speed was irrelevant; the influence of infill density was not linear across the three values, with the jump from 50 to 75% density having most impact; (b) the honeycomb infill pattern resulted in a longer fatigue life than with a rectilinear infill, and required about the same printing time; (c) the combination of 75% infill density, 0.5 mm nozzle diameter and 0.3 mm layer height provided the longest fatigue life; (d) the nozzle diameter should be at least 50% greater than the layer thickness to ensure proper cohesion between filaments, and part integrity; (e) in all coupons brittle and ductile fracture patterns coexisted: the fracture crack started in the outer layers at the locations of concentrations; this finding indicated that surface finishing can extend the fatigue life in printed coupons, as it is already known for non-printed coupons.

Arbeiter et al. (2018) studied how build orientation affected crack initiation and fracture in PLA under tensile fatigue. Specimens in compact tension (CT) configuration were printed per ASTM D5045 from PLA filament supplied by Prirevo e.U. (Austria) in three build orientations, 0°, 90°, and 0/90° (Figure 3.40), with high print infill and 250°C nozzle temperature to ensure best possible interlayer and intralayer cohesion. We recall that the stress intensity factor K_c is a material parameter predicting failure due to crack propagation, and the greater K_c the more resistant to crack is the material. The cycles to failure were measured as a function of build orientation and initial stress intensity factor range $\Delta K = K_{max} - K_{min}$, with K_{max} and K_{min} corresponding to maximum and

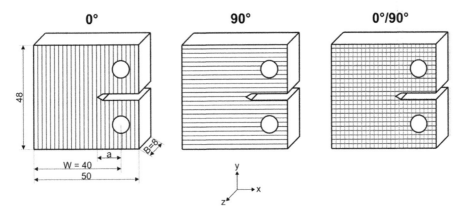

FIGURE 3.40 Compact tension (CT) coupons, with dimensions in mm, displaying build orientations: 0°, 90°, and 0°/90°. All the coupons were printed by adding layers in Z direction, and with 48 × 50 mm area parallel to XY the plane of printer's platform.

Source: Arbeiter, F., Spoerk, M., Wiener, J., Gosch, A., Pinter, G. 2018. Fracture mechanical characterization and lifetime estimation of near-homogeneous components produced by fused filament fabrication. *Polym. Test.* 66:105–113. Reproduced with permission from Elsevier.

minimum fatigue loads, respectively. Figure 3.41 illustrates that the build orientation was irrelevant to the number of cycles to propagate the crack, and fracture the coupons. It was also demonstrated that the well-known Paris law for fatigue (equation (4.2)) closely predicted the crack initiation and propagation, and was insensitive to print orientation.

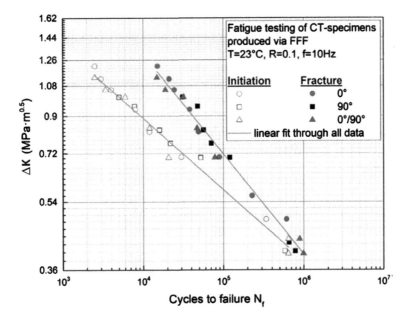

FIGURE 3.41 Crack growth initiation and fatigue fracture plotted as a function of the build orientation and applied initial stress intensity factor range, and linear fits.

Source: Arbeiter, F., Spoerk, M., Wiener, J., Gosch, A., Pinter, G. 2018. Fracture mechanical characterization and lifetime estimation of near-homogeneous components produced by fused filament fabrication. *Polym. Test.* 66:105–113. Reproduced with permission from Elsevier.

Gong al. (2017) targeted applications of porous PLA scaffolds in bone repair (maxillofacial, tubular bones, and so on) where the mechanical stability of the scaffold is critical, and studied the effect of pore shape on the behavior of PLA scaffolds undergoing low-cycle (10,000 cycles) compression fatigue in strain-controlled mode. Starting from PLA pellets of 60,000 g/mol MW (Sigma Aldrich), dried and fed to a HORI Z500D (China) printer with 0.4 mm nozzle diameter, two types of porous scaffolds were printed at 60% porosity: one featuring holes shaped as a 3.3 mm side equilateral triangle, and the other containing 3 mm diameter circular holes. The cyclic compression tests were conducted at strain levels (input) of 0.7–3.0% at 0.2 Hz frequency. The fatigue life was quantified as the number of cycles associated to 70% decline of the resulting peak load (output), and the resulting curves of strain amplitude vs. cycle number for both pore geometries were plotted and fitted by the Coffin-Manson equation (typically chosen for low-cycle fatigue) with good approximation. The pore geometry significantly affected the resistance to fatigue damage: circular pores were more resistant to fatigue than triangular pores, because the former ones produced a homogeneous distribution of the applied mechanical stress, and hence lower stress concentrations in the scaffolds. Under uniaxial static compression, the scaffold with circular pores was crushed in buckling, with the crush band perpendicular to the loading direction, whereas the scaffold with triangular pores failed in shear, with the crush band oriented at 45° to the loading direction. The tensile modulus of both scaffold types marginally increased during the first 10% of the total test cycles, but, after that, it reached a plateau in case of circular pores, whereas it dropped and cancelled the initial gain in case of triangular pores. Both pore structures accumulated almost the same inelastic deformation. Strain hardening occurred for both scaffolds after early cycles, induced by the reorientation of the molecular chains under cyclic strain.

3.9.10 FLEXURAL PROPERTIES

We recall that flexural strength and modulus are not required for structural design. However, they can predict the behavior of components in actual service conditions, and serve as a metric for quality control by assessing the effect of changes in composition, service temperature, print settings, etc. on the material's mechanical performance. They are also a proxy for tensile strength and modulus of plastics in general (Paloheimo n.d.), although the stress state in the cross section of tensile coupons and components is only tension, whereas it is tension and compression in the cross section of flexural coupons and components. Some suppliers of PLA filaments disclose flexural but not tensile properties, possibly because the flexural modulus is easier to accurately measure than the tensile modulus. Given the above uses of flexural properties, some investigations on them relative to PLA filaments are hereafter summarized.

Jaya Christiyan et al. published two experimental studies measuring the effect of printing parameters on flexural properties. In one study (Jaya Christiyan et al. 2018a) the PLA coupons were printed at 0° and 90° orientation (Figure 3.22), with 0.2, 0.25, and 0.3 mm layer thickness (LT), 38 and 52 mm/s printing speed (PS), 0.4 mm nozzle diameter (ND), 40% infill density, for a total of 12 test combinations, and with three coupons tested per combination according to ASTM D790. Maximum flexural strength (68 MPa) and modulus were reached at 0° orientation, 38 mm/s PS, and 0.2 mm LT. Although the sample size of 12 coupons is smaller than what recommended for statistical analysis, using the software Minitab® 17 we performed a statistical analysis of the test results, and measured that the effect of layer thickness, orientation, and PS on flexural modulus was about 70, 20, and 10%, respectively. In another study, Jaya Christiyan et al. (2018b) kept the orientation unchanged, and, applying a statistical model and DOE to their experimental data collected per ASTM D790, concluded that the flexural strength (FS) of PLA coupons was influenced by LT, PS, and ND in decreasing order. Particularly, PLA test samples with size 127 × 13 × 3 mm were printed across the following settings: LT 0.2, 0.25, 0.3 mm; PS 30, 40, 50 mm/s; ND 0.4, 0.5, 0.6 mm; printing temperature 180°C; bed temperature 40°C; raster angle: +45/−45°; infill 70% (low value for structural components). FS spanned from 60 to 103 MPa, and peaked at 0.2

TABLE 3.32

Influence of Printing Inputs on Flexural Force of PLA Coupons

Parameter	LT^2	IN	LT×IN	OA	IN×ES	ES	LT
Factor	−14.9	5	5.19	−4	4.73	2.3	1.6
Significant	Yes	Yes	Yes	Yes	Yes	No	No

Source: Luzanin, O., Guduric, V., Ristic, I., Muhic, S., 2017. Investigating impact of five build parameters on the maximum flexural force in FDM specimens – a definitive screening design approach. *Rapid Prototyping Journal* 23 (6): 1088–1098.

Legend: LT layer thickness, IN infill, OA orientation angle, ES extrusion speed.

mm LT, and 0.4 mm ND. The best data fitting model, predicting 90.1% of the variation in FS, was the following:

$$FS = 73.55 - 10.29 \times LT - 9.95 \times PS - 8.91 \times ND + 3.87 \times LT \times ND + 9.78 \times ND^2. \quad (3.10)$$

Luzanin et al. (2017) experimentally measured the influence of LT, deposition angle, infill, extrusion speed and temperature, and their interactions on the maximum flexural force in PLA coupons, employing a new statistical method within DOE, called *definitive screening design* (DSD) allowing to minimize the number of tests while estimating the main effects, some quadratic effects, and some two-factor interactions. The following parameter values were selected: LT 0.1, 0.2, 0.3 mm; orientation angle (OA) 0, 30, 60°; infill (IN) 0.1, 0.2, 0.3, corresponding to 90, 80, and 70% porosity, respectively, and selected because suited to reduce printing time, and matching setting values typically chosen on consumer-grade printers; extrusion speed (ES): 40, 50, 60 mm/s; extrusion temperature (ET): 229, 232, 235°C. Totally 13 combinations were evaluated, and as many coupons were printed out of 1.75 diameter Leaf Green MakerBot PLA filament, and tested in three-point bending. The DOE and analysis of results were conducted using the statistical software JMP® 11. A regression model with seven variables was preferred among other models evaluated, and its factors affecting the flexural force and their relative values are listed in Table 3.32: the factor's influence decreased going from from LT^2 to LT, and the negative and positive signs mean that decreasing a specific factor increases or decreases, respectively, the maximum flexural force. The variables with most impact were LT^2 and IN, those with the least influence ES and LT. Interpretating the influence of LT^2 was difficult, and the authors attributed it to the typical anisotropy intrinsic in the FFF process, and the accentuated anisotropy in their study, caused by low infill levels that resulted in concentrating the bond between adjacent layers in the Z direction rather within each layer. The optimal values to maximize the flexural force were LT 0.22 mm, OA 0°, IN 0.3 mm, and ES 60 mm/s.

3.9.11 Effect of Interlayer and Intralayer Cohesion

The overall strength of printed parts via FFF is critically affected by two types of cohesion (or strength): (a) that between adjacent printed layers (*interlayer*) and (b) that between adjacent filaments or strands within the same printed layer in the XY plane (*intralayer*), and is critical when printed parts undergo loads parallel to the layer plane. Sun et al. (2008) experimentally investigated the mechanisms controlling the interlayer cohesion, and concluded that the bonding determining that cohesion is thermally driven and ultimately determines the integrity and mechanical properties of the printed articles.

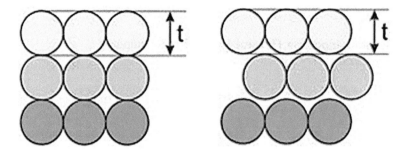

FIGURE 3.42 Layer designs. Left: *top design*. Right: *shifted design*.

Source: Spoerk, M., Arbeiter, F., Cajner, H. et al. 2017. Parametric optimization of intra- and inter-layer strengths in parts produced by extrusion-based additive manufacturing of poly(lactic acid). *J. Appl. Polym. Sci.* 134, 45401. Reproduced with permission from J. Wiley and Sons.

Spoerk et al. (2017) experimented with a commercial PLA (Prirevo, Austria), and demonstrated that maximizing interlayer and intralayer cohesion by optimizing the printing settings resulted in tensile strength values of about 90% of those of test samples compression molded out of the same PLA. The printer settings were kept constant for all specimens, except for die temperature (200, 210, 220, 250°C), layer thickness (0.2. 0.25. 0.3, 0.4 mm), and layer-design (top, shifted) (Figure 3.42). Unidirectional test coupons were printed. The tensile test was performed to measure intralayer cohesion on samples printed with the filament deposited parallel to the sample's width, at 90° to the loading direction. The double cantilever beam (DCB) test was conducted to measure the interlayer adhesion in terms of delamination energy G_{Ic} (Anderson 2017) on samples printed with the filament laid parallel to the sample's length and loading direction. Experimentally, the tensile strength decreased upon increasing layer thickness, and increased with printing temperature in top and shifted designs. Through statistical and regression analysis, the authors derived the analytical models of the tensile strength and G_{Ic} as a function of printing temperature and layer thickness for both layer designs. These models are not applicable to structural components that are typically printed with filaments laid in multiple directions (0, 90, ±45, 30, 60°, etc.) in order to withstand stresses acting in various directions. However, this study came to the following conclusions applicable to any type of printing orientation. Increasing flow rates (mm³/s) produced higher tensile strengths, because higher flow rates led to smaller or negative gaps between adjacent filaments, in turn resulting in smaller void content, fewer discontinuities, and a large volume on which to distribute the load. The increase of tensile strength with raising printing temperature is explained by the following advantages of high temperature: (a) polymer's viscosity is reduced, increasing polymer flow normal to filament direction (*cross-flow*) upon deposition, and hence reducing the air gap between the filaments that weld to each other more strongly; (b) the chains of the semi-molten polymer diffuse more easily across the filament interface, (c) the time span for the diffusion-controlled weld formation is extended.

3.9.12 EFFECT OF TEMPERATURE

This section describes the impact of the environmental and post-printing temperature on physical and mechanical properties of PLA for AM, not the effect of printing temperature that was previously discussed in this chapter.

As to the influence of environmental temperature, the information from material suppliers and technical literature lack values of physical and mechanical properties of PLA for AM measured above and below room temperature, but these values are required to evaluate the suitability of PLA in engineering applications, especially outdoor applications.

TABLE 3.33

Tensile Properties of PLA Filament Extruded from Printer

Temperature	Strength	Modulus	Strain at Break
°C	MPa	MPa	%
30	44.5	1,400	0.25
40	37.0	1,143	0.36
50	23.5	961	0.59
60	>12.6 (no failure)	350	>5.5 (no failure)

Source: Data from Grasso, M., Azzouz, L., Ruiz-Hincapie, P., Zarrelli, M., Ren, G. 2018. Effect of temperature on the mechanical properties of 3D-printed PLA tensile specimens. *Rapid Prototyping Journal* 24 (8): 1337–1346.

TABLE 3.34

Tensile Properties of PLA Test Coupons with 0/90° Infill

Temperature	Strength	Elastic Modulus	Strain at Break
°C	MPa	MPa	%
20	56	2,659	7.9
30	50	2,647	6.8
40	42	2,698	6.5
50	16	1,596	20
60	10	78	20

Source: Data from Grasso, M., Azzouz, L., Ruiz-Hincapie, P., Zarrelli, M., Ren, G. 2018. Effect of temperature on the mechanical properties of 3D-printed PLA tensile specimens. *Rapid Prototyping Journal* 24 (8): 1337–1346. Reproduced with permission from Emerald Publishing Limited.

TABLE 3.35

Tensile Modulus of PLA for AM across Temperature

Temperature	Tensile Modulus
°C	GPa
20	3.50
30	3.07
40	2.56
50	1.65
55	1.23
60	0.51

Source: Data from Grant, A., Ellis, B., Rohani, M. R., Hajiha, R. 2019. Exploring Process-Properties Relationships of 3D-Printed PLA: Towards Process-Informed Simulation and Design. TechConnect Briefs 2019, 143-146. ISBN 978-0-9988782-8-7. TechConnect.org.

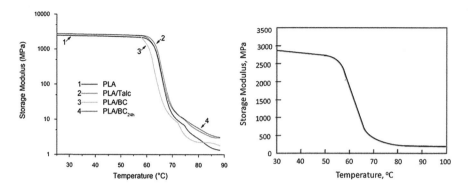

FIGURE 3.43 Left: DMA plot of sheets made of Ingeo™ 4043D PLA: unfilled and as PLA/talc, PLA/plant carbon (PLA/BC), and PLA/plant carbon milled for 24 hours. Right: DMA plot of PLA.

Source: Left: Snowdon, M. R., Wu, F., Mohanty, A. K., Misra, M. 2019. Comparative study of the extrinsic properties of poly(lactic acid)-based biocomposites filled with talc versus sustainable biocarbon. *RSC Adv.* 9, 6752, The Royal Society of Chemistry (RSC) on behalf of the Centre National de la Recherche Scientifique (CNRS) and the RSC. Reproduced under a Creative Commons Attribution 3.0 Unported License. Right: adapted from Silverajah, V. S. G., Ibrahim, N. A., Yunus, W., Hassan, H. A., Woei, C.B. 2012. A Comparative Study on the Mechanical, Thermal and Morphological Characterization of Poly(lactic acid)/ Epoxidized Palm Oil Blend. *Int. J. Mol. Sci.* 13 (5):5878–5898. https://doi.org/10.3390/ijms13055878. Reproduced under the terms and conditions of the Creative Commons Attribution license.

Grasso et al. (2018) measured the tensile properties from 30 to 60°C of a PLA filament in form of (a) wire before and after extrusion from MakerBot Replicator desktop printer, and (b) printed coupons. Table 3.33 illustrates significant and gradual decline of tensile strength and modulus, and increase in ductility (elongation at failure) in the filament. Table 3.34 instead reports test results per ASTM D638 relative to tensile coupons that were all printed with the filament laid at 0/90° but oriented inside each coupon differently in order to be 0°, ±45°, and 30/60° relatively to the loading direction. The values in Table 3.34 are the average values measured at all three testing orientations from 20°C to 60°C. The tensile strength steadily decreased from 20°C to 40°C, and featured larger drops at 50°C and 60°C when approaching T_g (typically around 60°C for PLA). The tensile modulus showed a more gradual decrease across temperatures, with the largest drop from 40 to 50 °C. In conclusion, 40°C resulted the maximum service temperature for engineering applications.

Grant et al. (2019) reported measurements of tensile modulus between 20 and 60°C of an unspecified PLA for AM that are listed in Table 3.35.

Coppola et al. (2018) investigated two PLA grades, Ingeo™ 4032D and 2003D, and extruded them into filaments for FFF, featuring both a T_g of 60°C. Their tensile storage modulus was measured via DMA between 35°C and 55°C: it declined for 4032D from 3.20 GPa to 2.75 GPa, and for 2003D from 2.75 GPa to 2.60 GPa.

Given the scarcity of data on properties of PLA measured not at room temperature, useful information comes from feedstocks for PLA filaments for FFF, although feedstocks' properties are not exactly and necessarily the same as the properties of the resulting filaments and powders for AM. One such feedstock is Ingeo™ 4043D mentioned in Section 3.4. Snowdon et al. (2019) turned it into sheets, and measured their T_g and T_m to be 63°C and 150°C, respectively. They also plotted the storage modulus in tension between 25°C and 90°C: curve 1 in left plot of Figure 3.43 includes unfilled PLA and PLA blended with talc and two forms of carbon extracted from a plant, and is paired - for comparison - to the PLA curve in the right plot of Figure 3.43 relative to an unspecified PLA (Silverajah et al. 2012).

TABLE 3.36

Properties of PLA and Blend PLA/PMMA/BS

Property	Unit	PLA	PLA/PMMA/BS
HDT (at 0.45 MPa)	°C	54	58
T_g	°C	61	62.8
Notched Izod impact strength	kJ/m^2	24	44
Tensile strength	MPa	68	49
Elastic modulus	GPa	3.2	2.5
Elongation at break	%	2.8	116

Source: Bouzouita, A., Notta-Cuvier, D., Delille, R., et al. 2017b. Design of toughened PLA based material for application in structures subjected to severe loading conditions. Part 2. Quasi-static tensile tests and dynamic mechanical analysis at ambient and moderately high temperature. *Polymer Testing* 57:235–244. Reproduced with permission from Elsevier.

TABLE 3.37

Tensile Properties of Blend PLA/PMMA/BS

Test Temperature	Maximum Stress	Modulus	Strain at Break
°C	MPa	GPa	%
Room temperature	38.0	2.46	0.109
50°C	20.1	1.32	>0.5

Source: Adapted from Bouzouita, A., Notta-Cuvier, D., Delille, R., et al. 2017b. Design of toughened PLA based material for application in structures subjected to severe loading conditions. Part 2. Quasi-static tensile tests and dynamic mechanical analysis at ambient and moderately high temperature. *Polymer Testing* 57:235-244. Reproduced with permission from Elsevier.

Bouzouita et al. (2016, 2017b) evaluated the properties above room temperature of a blend of PLA that is described here as a potential feedstock for AM. This blend contained Ingeo™ 4032D PLA, Evonik Plexiglas® 8N poly(methyl methacrylate) (PMMA), and DuPont Biomax® Strong 120 ethylene-acrylate impact modifier (BS) formulated to toughen PLA. The blend was prepared to enhance unfilled PLA in terms of (a) impact resistance and ductility thanks to BS, and (b) temperature resistance thanks to PMMA that features a T_g of 116°C. The composition of 58 wt% PLA, 25 wt% PMMA, and 17 wt% BS attained the best compromise between impact toughness, tensile strength, rigidity, ductility, and HDT. Table 3.36 reports measured properties of the unfilled PLA, and the above blend: in comparison to unfilled PLA, the blend featured 83% higher impact strength, and 41 times elongation at break, but reduced tensile modulus and strength by 22% and 28%, respectively (no free lunch). The improvement in temperature resistance was marginal, since HDT and T_g only rose by 4°C and 1.8°C, respectively. Table 3.37 indicates that maximum tensile stress and modulus of the blend dropped almost by half shifting from room temperature to 50°C.

Zhou et al. (2016) characterized Ingeo™ 3001D, a PLA grade for injection molding, with T_g of 55–60°C, and reported the following changes in response to test temperature raising from 30°C to 55°C: tensile strength from 60 MPa to 14 MPa, yield tensile strength from 50 MPa to 29 MPa, tensile modulus from 2,950 MPa to 1,100 MPa, elongation at break from 41% to 67%.

Temperature can affect PLA performance in form of post-printing heating. After being printed, articles can be kept for some time at a temperature above PLA's T_g, undergoing annealing that consists in initial, gradual heating, followed by a steady temperature, and terminated with slow cooling. Annealing permits the molecular chains to rearrange in a more orderly pattern, and crystallize, and as a result to improve the polymer's mechanical properties.

Pagano et al. (2019) experimentally measured the impact of annealing on tensile strength, modulus, and strain at break of a PLA filament (Plastink) tested according to ISO 527, printed in three orientations, 0°, ±45°, 0/90°, and annealed at 200°C and 230°C. Test data comparing annealed and not annealed coupons indicated that both annealing temperatures improved strength only at 0° and 0/90°, and modulus at all orientations, but did not improve strain. According to Zhou et al. (2016), annealing PLA increases its crystallinity and forms a type of crystals (α) that are more ordered and more densely packed than another type (α'), improving tensile modulus. This is consistent with the fact that mechanical properties of crystalline polymers depend not only on percent crystallinity, but also crystal form and morphology (as mentioned in Section 1.4). Zhou et al. (2016) also reported the influence of annealing on tensile yield strength of Ingeo™ 3001D: across the testing temperature range of 30–60°C, tensile yield strength ranged from 60 MPa to 3 MPa for not annealed 3001D, whereas it increased to 67–70 MPa at 30°C and declined to 12–28 MPa at 60°C for annealed 3001D, with the ranges of values depending on the annealing temperature.

3.9.13 Effect of Filament Color

Plastics are colored by adding compounds (*colorants*) in form of *dyes* and *pigments*: the former ones chemically bond to the plastic to which are added, while the latter ones are completely or nearly insoluble in water. The base polymer, processing conditions, and application guide the choice of type of colorant. Dyes are less numerous than pigments, and compatible only with a limited number of plastics. Very popular pigments are TiO_2 and carbon black, chosen to impart whiteness and blackness. Acid dyes are popular for plastics such as PA in form of textile fibers and not, and PU. Companies like 3DXTECH and FilaCube (both in USA) sell color concentrates for filaments to customers who extrude their own filaments out of PLA and other plastics.

Wittbrodt and Pearce (2015) experimentally studied the effect of color and processing temperature on material properties of a commercial PLA filament from Lulzbot (USA) based on Ingeo™ 4043D, and available in five colors: white, black, blue, gray, and natural. All test coupons were printed at the same temperature of extruder and platform, namely 190°C and 60°C, respectively. As Table 3.38 illustrates, the difference in the measured properties was significant across colors, with tensile strength, yield strength, and strain at break being the following: 51–57 MPa, 46–52 MPa, and 1.9–2.4%, respectively. The crystallinity ranged from 0.93% to 5.05%, depending on the color.

TABLE 3.38
Tensile Properties of PLA Filaments by Lulzbot

Color	Crystallinity	Ultimate Strength	Yield Strength	Strain at Break
	%	MPa	MPa	%
Natural	0.93	57.2	52.5	2.35
Black	2.62	52.8	49.3	2.02
Gray	4.79	50.8	46.1	1.98
Blue	4.85	54.1	50.1	2.13
White	5.05	54.0	50.5	2.22

Source: Data from Wittbrodt, B., Pearce, J. M. 2015. The effects of PLA color on material properties of 3-D printed components. *Additive Manufacturing* 8:110–116.

TABLE 3.39
Tensile Properties of Fil-A-Gehr PLA® Filaments by Gehr Plastics

Color	Ultimate Strength	Modulus
	MPa	GPa
Red	49.2	2.68
Green	48.0	2.93
Blue	47.2	3.00
Black	41.1	2.93
Yellow	39.8	2.66
White	39.2	2.83

Source: https://www.gehrplastics.com/wp-content/uploads//2019/02/FIL-A-GEHR_PLA_EN_datasheet.pdf.

Substantial differences in ultimate tensile strength with filament color are also reported by the company Gehr (Germany, USA) in its FIL-A-GEHR PLA® filament. As indicated in Table 3.39, across six colors the tensile strength and modulus were measured on coupons printed upright in Z direction (Figure 3.15): the former varies from 39.2 MPa to 49.2 MPa (26% change), whereas the latter is less sensitive, spanning from 2.66 GPa to 3 GPa (13% variation). No color outperforms the others in both properties.

3.9.14 CRYSTALLINITY

Crystallinity has been introduced in Section 1.5. A larger amount of crystalline regions in a semi-crystalline polymer improves its tensile properties, and heat resistance, although it reduces its ductility and impact resistance, hence the optimal value of crystallinity is the one that maximizes critical functional properties.

Crystallinity of PLA up to 66% has been documented (Shogren 1997), but in commercial PLA filaments it ranges from about 10% to mid-30%. One efficient way to improve crystallinity of PLA is decreasing the cooling rate of PLA from melt and annealing PLA articles, and the smaller the cooling rate, the higher the crystallinity (Jamshidian et al. 2010). Di Lorenzo (2005) measured crystallization rates of PLA across an ample temperature range, and concluded that the crystallization rate of PLA at temperatures between 100°C and 118°C is very high.

The degree of crystallinity in PLA filaments is affected by its processing conditions. Song et al. (2017) measured via DSC that the crystallinity of IMAKR PLA filament available as (a) as-received, (b) processed through the printer and extruded as standalone filament, (c) printed as coupons, was 8%, 7%, and 38%, respectively. The low values were caused by fast cooling during filament fabrication, and extrusion through printer, whereas the high value derived from slow cooling inside the printed coupon, and on the heated printer platform. However, Pop et al. (2019) published that the crystallinity of Verbatim™ PLA filament was 24% as-received, and 19% as a printed coupon, although the printer platform was kept at 60°C.

Since higher crystallinity in polymers leads to greater shrinkage, articles printed out of more crystalline PLA may end up with not extensive interlayer contact area, and consequently not optimal interlayer cohesion. Wang et al. (2017a) reported that if the PLA is very fluid when melted, the shrinkage can be offset and not impair interlayer cohesion.

TABLE 3.40

Water Uptake (WU) in Biobased and Fossil-Based Feedstocks for AM (except PA 1010)

Material	WU After 800 hrs	Maximum WU	k	n
	% of initial weight	% of initial weight	N/A	N/A
Verbatim™ PLA filament	1.4 (saturation)	N/A	0.933	0.016
Verbatim™ PLA printed	3.3 (no saturation[b])	N/A	0.455	0.113
Verbatim™ ABS filament	1 (saturation)	N/A	0.128	0.392
Verbatim™ ABS printed	5 (no saturation[b])	N/A	0.216	0.145
100% Biobased PA 1010[a]	N/A	1.8	N/A	N/A
100% Biobased PA 11[a]	N/A	1.9	N/A	N/A

Source: Data from Pop, M. A., Croitoru, C., Bedő, T., et al. 2019. Structural changes during 3D printing of bioderived and synthetic thermoplastic materials. *J. Appl. Polym. Sci.* 136, 47382. doi: https://doi.org/10.1002/app.47382.
Notes:
[a]Brehmer B., 2014. Polyamides from Biomass Derived Polymer. In *Bio-Based Plastics*, ed. S. Kabasci, 274–293. Chichester: Wiley, p. 278.
[b]The "no saturation" values appeared very close to saturation.

3.9.15 CHEMICAL COMPOSITION

Cuiffo et al. (2017) analyzed test coupons printed out of an unidentified NatureWorks PLA filament, and reported the following differences between as-received filament and printed filament. Besides C, H, and O, more elements were detected using X-ray fluorescence (XRF), namely P, Ca, Si, Cr, Ti, Cu, and Sn in decreasing order, and present in 2,100 to 13 ppm. The quantity of these elements varied between spooled and printed filament. F.e. P and Si decreased from 2100 and 310 ppm, respectively, in the spooled filament to 1183 and 180 ppm, respectively, in the printed filament. Ca was the second most abundant element (528 ppm and 585 ppm in spooled and printed filament, respectively), and present as $CaCO_3$ added to aid in extrusion and stability of the filament, and its presence is important in biomedical applications of PLA, because carbonate impacts cell growth on printed tissue scaffolds.

3.9.16 THERMAL PROPERTIES BEFORE AND AFTER PRINTING

Cuiffo et al. (2017) analyzed an unspecified filament made from NatureWorks PLA, and via DSC identified shifts in the following temperatures going from as-received to printed filament, namely: T_g from 57.7°C to 53.1°C, cold crystallization temperature from 90.2°C to 110°C, and T_m, from 166.7°C to 155.1°C.

Alin Pop et al. (2019) also observed a similar change in values. They tested Verbatim™ PLA filament that featured nominal T_g and T_m of 58°C and 168°C, respectively, whereas for the as-received filament, and printed coupon T_g was 76°C and 80°C, respectively. Cold crystallization temperature raised from 107°C (as-received) to 123°C (printed), while T_m remained steady at 172°C.

3.9.17 WATER UPTAKE

Alin Pop et al. (2019) assessed the absorption of distilled water in 2.85 mm diameter Verbatim™ PLA filament in form of as-received filament and printed coupons (cylinder 10 mm in diameter and 2 mm thick). In both forms, the rate of weight increase was highest initially, slowed down later, and eventually reached a plateau, approximately following a model formalized by Fick's laws of diffusion. In general, the plateau value of weight increase is the *saturation* value that is reached when polymer and

TABLE 3.41

Properties of Experimental PLA Filament after 7 Days of Exposure in Water and Air

Exposure	Tensile Strength	Tensile Modulus	Elongation at Break	Charpy Unnotched Impact Strength	Charpy Notched Impact Strength	Crystallinity
	MPa	GPa	%	kJ/m^2	kJ/m^2	%
Kept in air	58.0	2.33	1.79	16.5	5.26	41.7
Immersed in tap water	56.5	2.27	1.72	14.6	5.28	45.0

Source: Data from Ecker, J. V., Haider, A., Burzic, I. et al. 2019. Mechanical properties and water absorption behaviour of PLA and PLA/wood composites prepared by 3D printing and injection moulding. *Rapid Prototyping Journal* 25 (4): 672–678.

water are at equilibrium, and remains constant no matter the time of immersion in water and exposure to moisture. The saturation value reached 1.4% in the filament after 50 hrs, while in the coupons it was 3.3% after 800 hrs, close to but not at equilibrium. To place data in perspective, Table 3.40 lists the values of water uptake for PLA along with those for ABS (given its wide utilization in AM), and two biobased PA: PA 11 formulated for AM and PA 1010 not for AM. PA grades are included because PA displays high water uptake among engineering polymers. The "no saturation" value for ABS appeared very close to saturation from the relative plot. At saturation the PLA filament absorbed more water than ABS but less than both PAs. PLA featured larger overall uptake in the printed samples than the filament possibly due to the porosity caused by the interlayer voids in printed samples. The authors modeled the water uptake through equation (3.11), in which Δm_t and Δm_{sat} are the mass (g) variation of the coupon at generic time t and at saturation, respectively, k (min^{-n}) is the water uptake rate, and n (<1) is the power-law diffusion exponent:

$$\frac{\Delta m_t}{\Delta m_{sat}} = kt^n \tag{3.11}$$

Water uptake is important because it penalizes mechanical properties of polymers. However, Ecker et al. (2019) reported that immersing PLA in tap water for one week did not significantly alter their tensile and impact properties. The authors extruded Ingeo™ 3251DPLA as a filament and then printed it in form of test coupons, some of which were stored in air and some immersed in water. The maximum water uptake was 0.75 wt% after 160 hrs, and that did not appear to be the saturation value from the plot of water uptake vs. time. The mechanical test results in Table 3.41 and the standard deviation values indicated no significant difference between air values vs. water values. About impact strength, the drop in unnotched value was countered by a stable notched value, which may lead to conclude that overall the impact resistance was marginally diminished. The recorded increase in crystallinity was possibly attributed to increased mobility of PLA polymeric chains caused by presence of water.

3.9.18 PLASTICIZERS

Selecting plasticizers for PLA depends on the requirements of the specific application. F.e. for food packaging only nontoxic substances approved for food contact are allowed as plasticizers. The plasticizers should be: (a) miscible with PLA, and form a homogeneous blend; (b) not volatile at

TABLE 3.42

Tensile Properties of PLA and PLA with Plasticizers (cPLA1, cPLA2)

Material	Strength	Modulus	Strain at Break	T_g
	MPa	GPa	mm/mm	°C
PLA	67.6	4.3	0.4	61.6
cPLA1	30.7	1.3	1.8	40.1
cPLA2	26.2	0.8	3.2	28.8

Source: Data from Depuydt, D., Balthazar, M., Hendrickx, K., et al. 2018. Production and Characterization of Bamboo and Flax Fiber Reinforced Polylactic Acid Filaments for Fused Deposition Modeling (FDM). *Polym. Compos.* 40:1951–1963.

the processing temperature; (c) disinclined to migrate into the materials in contact with the plasticized PLA (Jamshidian et al. 2010). Typically plasticizers have conflicting effect on physical-mechanical properties of PLA, as they improve ductility, flexibility, and impact strength but reduce tensile modulus, T_g of the amorphous phase and T_m of the crystalline phase (Baiardo et al. 2003). Plasticizers reduce viscosity and hence improve processability.

Depuydt et al. (2018) added plasticizers to a PLA filament for FFF to increase its ductility, and prevent it from breaking during winding, and experimentally noted that the plasticizers did increase elongation at break as intended, but also lowered tensile strength and (severely) modulus, and T_g. Particularly, the plasticizers Proviplast® 2624 (tributyl 2-acetylcitrate, $C_{20}H_{34}O_8$), and Proviplast® 01422 (bis(2-(2-butoxyethoxy)ethyl) adipate, $C_{22}H_{42}O_8$) supplied by Proviron Industries (Belgium) were individually compounded to pellets of an unidentified PLA, resulting in two materials, labeled cPLA1 and cPLA2, respectively. The tensile properties were measured on filaments at 20 mm/min. As illustrated in Table 3.42, the plasticizer in cPLA2 was more effective at increasing the elongation at break (which will aid in processing) but also at reducing tensile modulus and strength, and T_g. It is notable that the PLA studied was quite ductile, since its elongation at break was 40%, far exceeding the typical values of commercial PLA filaments for AM that are below 10%. Overall, the presence of plasticizers limited the application of the resulting filaments, because the reduced values of tensile strength and modulus restricted the range of suitable load-bearing components, and the lower T_g values curbed the maximum service temperature.

Wang et al. (2017a) experimentally assessed that plasticizers impacted the printing quality of PLA parts, and concluded that (a) a threshold value of 10 g/min for MFI is required to achieve an acceptable printing quality (although it is not the only factor affecting it), and (b) measuring MFI can quickly and

TABLE 3.43

Effect of Strain Rate on Tensile Properties of Ingeo™ 3051D PLA

Test Strain Rate	Strength	Yield Strength	Modulus	Strain at Break
1/s	MPa	MPa	GPa	mm/mm
0.003	37–39	44–47	2.25	0.11–0.185
0.006	37–43	48–52	2.32	0.06–0.12
0.01	44–48	52–57	2.50	0.055–0.085

Source: Data from Mirkhalaf, S. M., Fagerström, M. 2019. The mechanical behavior of polylactic acid (PLA) films: fabrication, experiments and modelling. *Mech. Time-Depend. Mater.* https://doi.org/10.1007/s11043-019-09429-w.

economically screen candidate PLA feedstocks for FFF. They also noticed that the concentration of plasticizer in PLA is more influential than its type. Moreover, a high MFI has also the advantage to compensate for shrinkage, and the consequent imperfect interlayer contact and cohesion.

3.9.19 Effect of Speed and Frequency of Applied Load

If PLA is considered for printing functional articles subjected to loads applied not statically but dynamically, the mechanical behavior of PLA under dynamic loading must be analyzed.

Song et al. (2017) tested printed tensile coupons out of IMAKR PLA filament at strain rates of 1.25×10^{-5}/s and 2.5×10^{-4}/s, and recorded a maximum stress of 33 MPa and 53 MPa, respectively. This behavior was expected, since in polymers the higher is the load application rate, and the shorter is the time for the molecular chains to react to it, deform, and fracture, and hence the greater are their strength and stiffness.

Mirkhalaf and Fagerström (2019) measured tensile properties of Ingeo™ 3051D PLA at some strain rates, and recorded the usual polymeric behavior (Table 3.43) by which raising strain rate resulted in raising strength and modulus, and decreasing elongation at break. The variation in values was due to fact that tensile coupons were cut at three angles: 0. 45, and 90° from films not dumbbells. Compared to the PLAs in Table 3.9, 3051D is as strong as the average values, but stands in the lower half of the range for stiffness and upper half of the range for elongation.

The storage moduli in tension and bending differ from the elastic moduli in tension and bending by definition, because the former ones are measured as a response to oscillating deformation or load, while the latter ones are measured upon static and growing deformation. The technical literature on unfilled and blended PLA grades not specific for AM includes plots of storage modulus as a function of the frequency of the applied strain, and all those plots (a) have been originated above room temperature, often beyond T_g of PLA that is above its service conditions, and hence

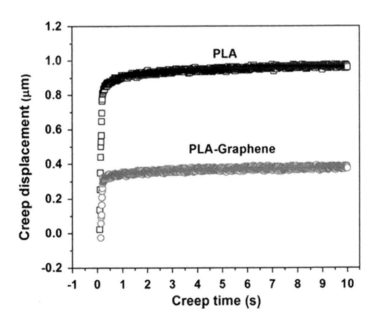

FIGURE 3.44 Creep displacement of commercial PLA and graphene PLA filaments for FFF.

Source: adapted from Bustillos, J., Montero, D., Nautiyal, P. 2018. Integration of graphene in poly(lactic) acid by 3D printing to develop creep and wear-resistant hierarchical nanocomposites. *Polymer Composites* 39:3877–3888. Reproduced with permission from J. Wiley and Sons.

they are not reported here, (b) reported a storage modulus that increased at higher frequencies, as it does the tensile modulus in polymers.

3.9.20 CREEP AND STRESS RELAXATION

Creep is the dimensional change over time of a loaded polymer whose deformation initially is elastic, and later it is not and depends on time. Creep is a manifestation of the viscous nature of polymers, and the fact that their chains keep untangling, rotating, and deforming upon application of a load that is constant over time. From a functional standpoint, creep is important because it relates to the dimensional stability of polymers.

The technical literature on creep in PLA is scarce, probably because most of its current applications do not involve significant long-term loads. The most relevant papers on creep of printed PLA are discussed hereafter.

Bustillos et al. (2018) characterized MakerBot PLA filaments and BlackMagic3D (USA) graphene/PLA. The tensile modulus and nanohardness were 3.9 GPa and 146 MPa for graphene/ PLA respectively, and 3.5 GPa and 125 MPa for PLA, respectively. The authors measured the creep behavior at the microscale on a Universal Surface Tester (Innowep, Germany) that applied static loadings of 10 to 90 mN over a period of 10 s through a 0.8 mm diameter spherical indenter, and recording the continuous deformation under load. Figure 3.44 illustrates the creep displacement under 25 mN load of these two filaments over a very short period of time selected to target orthopedic scaffolds and patient-specific implants, but far shorter than typical durations of a creep test. The response of both filaments is typical for polymers: the displacement was initially large, and then stabilized at a minimum strain rate, indicating a small amount of viscous creep. If the displacement had raised at higher rate after 1 s, that would have revealed that viscous flow was a key contributor to the total creep.

Niaza et al. (2017) studied long-term creep of printed, porous PLA-based scaffolds to replace small trabecular bone defects in implants, and hence their creep measurement extended over hours. The PLA's creep response depended not only on PLA but also scaffold's geometry and porosity. In fact, the scaffolds failed due to accumulated defects such as collapsed pores, micro-cracks, and microdelamination. Pellets of Aldrich PLA (60,000 g/mol MW) were extruded into filaments printed into scaffolds loaded in compression at 10, 12, 14, and 16 MPa for 8 h. Upon load application the strain rose to 1.6–2% depending on the stress, and after 8 h the strain were 2.1, 2.4, and 3.0% at 10, 12, and 14 MPa, respectively, whereas after 4 h at 16 MPa the strain reached 3.5%. Increasing applied stress raised the deformation rate and, hence, the viscous component of the total creep. The smallest amount of creep occurred at 10 MPa, where strain grew from 1.6% to 2.1% after 8 h. A component intended to withstand creep must be designed in a shape that prevents the maximum strain reached in service to exceed the maximum strain tolerated to meet performance requirements.

Stress relaxation is the decrease in stress in a polymer subjected to a deformation that is constant over time. We found no technical literature on stress-relaxation of unfilled PLA, for general use and AM. However, if stresses and deformations are small and they marginally depend on time, the stress-relaxation response can be approximately calculated as the inverse of the creep response, and vice versa, according to the following equation (Nielsen, Landel 1994), in which ε_0, $\varepsilon(t)$, σ_0, and $\sigma(t)$ are the initial strain, the strain after time t, the initial stress, and the stress after time t, respectively:

$$\left(\frac{\sigma_0}{\sigma(t)} \right)_{st.rel.} = \left(\frac{\varepsilon(t)}{\varepsilon_0} \right)_{creep} \tag{3.12}$$

TABLE 3.44

Effect of Recycling on Properties of PLA

PLA	Tensile Yield Strength	Tensile Modulus	Shear Strength	Hardness
	MPa	MPa	MPa	Shore D
ASTM	D638-14	D638-14	D732-10	N/A (handheld durometer)
Virgin	40.4	4,258	33.0	84.8
Recycled 10 times	35.9	4,032	35.3	82.8

Source: Data from Anderson 2017. Mechanical Properties of Specimens 3D Printed with Virgin and Recycled Polylactic Acid. *3D Printing and Additive Manufacturing* 4(2). Mary Ann Liebert Inc. doi:10.1089/3dp.2016.0054.

3.10 PROPERTIES OF RECYCLED PLA

PLA's biodegradability permits to handle wasted PLA through recycling, composting, and dumping in landfills. Recycling is the optimal route in terms of environmental impact, because (a) composting takes 1–3 months in an industrial facility at specific temperatures (Slijkoord 2015), and municipal composting facilities are in limited number, (b) the downside of landfilled waste is obvious, plus decomposing PLA needs oxygen and high temperature that are absent in the landfill.

TABLE 3.45

Mechanical Properties of FFF PLA Virgin and Recycled Once

Material State	Tensile Strength	Tensile Yield Strength	Tensile Modulus	Elongation at Break	Flexural Strength	Flexural Modulus
	MPa	MPa	MPa	%	MPa	MPa
Virgin	30.2	27.7	1,572	2.74	64.5	2,424
Recycled Once	29.5	27.7	1,567	2.45	57.6	2,235

Source: Zhao, P., Rao, C., Gu, F., et al. 2018. Close-looped recycling of polylactic acid used in 3D printing: An experimental investigation and life cycle assessment. *Journal of Cleaner Production* 197:1046–1055. Reproduced with permission from Elsevier.

TABLE 3.46

Properties of FFF PLA Virgin and Recycled Twice

Material State	Number-Average MW	Weight-Average MW	T_g	T_c	T_m
	g/mol	g/mol	°C	°C	°C
Virgin	50,906	106,963	59.9	123.7	165.6
Recycled Once	42,668	89,460	59.5	104.3	168.4
Recycled Twice	15,901	42,037	59.1	98.3	167.6

Source: Zhao, P., Rao, C., Gu, F. et al. 2018. Close-looped recycling of polylactic acid used in 3D printing: An experimental investigation and life cycle assessment. *Journal of Cleaner Production* 197:1046–1055. Reproduced with permission from Elsevier.

Data collected so far point out that the environmental impact of recycling PLA is 50 times and 16 times better than composting and combustion, respectively (Anderson 2017).

Cruz Sanchez et al. (2015) investigated the degradation of a NatureWorks 4043D PLA printing filament subjected for five times to a recycling cycle consisting in processing the filament through a printer and later shredding it. Comparing the tensile properties at cycle 5 to those at cycle 1, stress at break and modulus increased by 5% and 13%, respectively, strain at break contracted by 11%. From cycle 1 to 5, viscosity and MW dropped by 92%, and 56%, respectively, whereas MFI triplicated. The improvement of tensile stress and modulus can be attributed to a balance between MW reduction and increase in crystallinity that can also reduce ductility and lower the strain at break. The authors also summarized some studies on mechanical recyclying of non-AM grades of PLA.

Anderson (2017) measured tensile yield strength and modulus shear strength and hardness of coupons printed with virgin and recycled PLA on a desktop printer, and their values are collected in Table 3.44. The recycled filaments were obtained from coupons previously printed, ground, and extruded again into a printing filament in a cycle repeated up to 10 times. All differences in average property values between virgin and recycled PLA were statistically significant, except for tensile modulus. Some increase in tensile strain (limited) and shear strain (more pronounced) at the end of test revealed a marginal raise in ductility that was consistent with decreased hardness and tensile modulus, and could have also been a sign of improved impact strength. The drop in properties ranged from 2.4% for hardness to 11% for tensile yield strength.

Like Anderson (2017), Zhao Rao et al. (2018) studied the performance of virgin and recycled PLA for FFF. Recycling consisted in shredding, drying, and extruding PLA into a filament. Not only mechanical properties were monitored but also thermal properties and viscosity that is a critical processing parameter for FFF, as we mentioned. The results are summarized in Tables 3.45 and 3.46, and relative to coupons recycled only once and twice, respectively. T_c denotes the crystallization temperature. Notably, after two recycling cycles the viscosity diminished to values unsuitable for further reprocessing and printing. Moreover, break up of molecular chains increased crystallinity and deteriorated thermal stability. Thermomechanical degradation that occurred during melt processing reduced MW.

REFERENCES

Abbas, T. F., Othman, F. M., Ali, H. B. 2017. Influence of layer thickness on compression property of 3D-printed PLA. *Al-Muhandis J. JMISE* 154 (4): 41–48.

Abbas T. F., Othman F. M., Ali H. B. 2018. Influence of layer thickness on impact property of 3d-printed PLA. *Int. Res. J. Eng. Technol. (IRJET)* 5 (2): 1–4.

Afrose, M. F., Masood, S. H., Iovenitti, P., et al. 2016. Effects of part build orientations on fatigue behaviour of FDM-processed PLA material. *Prog. Addit. Manuf.* 1:21–28.

Afrose, M. F., Masood, S. H., Nikzad, M., Iovenitti, P., 2014. Effects of build orientations on tensile properties of PLA material processed by FDM. *Adv. Mater. Res.* ISSN: 1662-8985, 1044-1045:31–34.

Ahmed, S., Jones, F. R. 1990. A review of particulate reinforcement theories for polymer composites. *J. Mater. Sci.* 25:4933–4942.

Ahn, S. H. M., Montero, D., Odell, S., Roundy, P. K. 2002. Wright, Anisotropic material properties of fused deposition modelling ABS. *Rapid Prototyp. J.* 8:248–257.

Ajiro, H., Hsiao, Y.-J., Thi, T.H., Fujiwara, T., Akashi M. 2012. A stereocomplex of poly(lactide)s with chain end modification: simultaneous resistances to melting and thermal decomposition. *Chem. Commun.* 48:8478–8478.

Aldana, D. S., Villa, E. D., Hernández, M. D. D., et al. 2014. Barrier properties of polylactic acid in cellulose based packages using montmorillonite as filler. *Polymers* 6:2386–2403.

Anderson, T. L. 2017. *Fracture Mechanics—Fundamentals and Applications*. Boca Raton: CRC Press Inc. 2006. Anonymous 2006. Making preforms of PLA bottles. *Bioplast. Mag.* 1 (2): 16–18.

Arbeiter, F., Spoerk, M., Wiener, J., Gosch, A., Pinter, G., 2018. Fracture mechanical characterization and lifetime estimation of near-homogeneous components produced by fused filament fabrication. *Polym. Test.* 66:105–113.

ASTM D5045. *Standard Test Methods for Plane-Strain Fracture Toughness and Strain Energy Release Rate of Plastic Materials*. USA: ASTM International. www.astm.org.

ASTM D638. *Standard Test Method for Tensile Properties of Plastics*. USA: ASTM International. www.astm.org.

ASTM D695. *Standard Test Method for Compressive Properties of Rigid Plastics*. USA: ASTM International. www.astm.org.

ASTM D882. *Standard Test Method for Tensile Properties of Thin Plastic Sheeting*. USA: ASTM International. www.astm.org.

ASTM E143. Standard Test Method for Shear Modulus at Room Temperature. USA: ASTM International. www.astm.org.

ASTM E399. *Standard Test Method for Linear-Elastic Plane-Strain Fracture Toughness K_{Ic} of Metallic Materials*. USA: ASTM International. www.astm.org.

ASTM F433. *Standard Practice for Evaluating Thermal Conductivity of Gasket Materials*. USA: ASTM International. www.astm.org.

Auras, R., Harte B., Selke S. 2004. An overview of polylactides as packaging materials. *Macromol. Biosci.* 4:835–864.

Auras R., Harte, B., Selke, S., Hernandez, R. 2003. Mechanical, physical and barrier properties of poly (lactide-films). *J. Plast Film Sheeting* 19:123–135.

Auras, R., Lim, L.-T., Selke, S. E. M., Tsuji, H. 2010. Poly(lactic acid). *Synthesis, Structures, Properties, Processing, and Applications*. Hoboken: John Wiley & Sons.

Azzi, V., Tsai, S., 1965. Anisotropic strength of composites. *Exp. Mech.* 5 (9): 283–288.

Babu, R., O'Connor, K., Seeram, R. 2013. Current progress on bio-based polymers and their future trends. *Prog. Biomater* 2:1–16.

Baghaei, B., Skrifvars, M. 2016. Characterisation of polylactic acid biocomposites made from prepregs composed of woven polylactic acid/hemp–Lyocell hybrid yarn fabrics. *Compos. Part A: Appl. Sci. Manuf.* 81:139–144.

Baghaei, B., Skrifvars, M., Berglin, L. 2013. Manufacture and characterisation of thermoplastic composites made from PLA/hemp co-wrapped hybrid yarn prepregs. *Compos. A: Appl. Sci. Manuf.* 50:93–101.

Bai, H., Huang, C., Xiu, H., Zhang, Q., Fu, Q. 2014. Enhancing mechanical performance of polylactide by tailoring crystal morphology and lamellae orientation with the aid of nucleating agent. *Polymer* 55:6924–6934.

Bai, J., Goodridge, R. D., Hague, R. J. M., Okamoto, M. 2017. Processing and characterization of a polylactic acid/nanoclay composite for laser sintering. *Polym. Compos.* 38:2570–2576.

Baiardo, M., Frisoni, G., Scandola, M., et al. 2003. Thermal and mechanical properties of plasticized poly (L-lactic acid). *J Appl Polym Sci.* 90 (7): 1731–1738.

Bayer, I. S. 2017. Thermomechanical properties of polylactic acid-graphene composites: a state-of-the-art review for biomedical applications. *Materials* 10:748.

Bergsma, J. E., De Bruijn, W. C., Rozema, F. R., Bos, R. R. M., Boering, G. 1995. Late degradation tissue response to poly(l-lactide) bone plates and screws. *Biomaterials* 16:25–31.

Bettini, P., Alitta, G., Sala, G., Di Landro, L. 2017. Fused deposition technique for continuous fiber reinforced thermoplastic. *J. Mater. Eng. Perf.* 26:843–848.

Bogaert, J. C., Coszach, P. 2000. Poly(lactic acids): a potential solution to plastic waste dilemma. *Macromol. Symp.* 153:287–303.

Bouthillier, J. 2016. *NatureWorks Ingeo™ 3D850 PLA Filament Posted on May 13*. http://bootsindustries.com/natureworks-ingeo-3d850-pla-filament/ (Accessed August 31, 2019).

Bouzouita, A., Notta-Cuvier, D., Delille, R., et al. 2017b. Design of toughened PLA based material for application in structures subjected to severe loading conditions. Part 2. Quasi-static tensile tests and dynamic mechanical analysis at ambient and moderately high temperature. *Polym. Testing* 57:235–244.

Bouzouita, A., Notta-Cuvier, D., Raquez, J.-M., et al. 2017a. Poly(lactic acid)-based materials for automotive applications. *Adv. Polym. Sci.* doi:10.1007/12_2017_10.

Bouzouita, A., Samuel, C., Notta-Cuvier, D., et al. 2016. Design of highly tough poly(L-lactide)-based ternary blends for automotive applications. *J. Appl. Polym. Sci.* 133:43402. doi:10.1002/app.43402.

Brischetto, S., Torre, R. 2020. Tensile and compressive behavior in the experimental tests for PLA specimens produced via fused deposition modelling technique. *J. Compos. Sci.* 4:140.

Bustillos, J., Montero, D., Nautiyal, P. 2018. Integration of graphene in poly(lactic) acid by 3D printing to develop creep and wear-resistant hierarchical nanocomposites. *Polym. Compos.* 39:3877–3888.

Caulfield, B., McHugh, P.E., Lohfeld, S. 2007. Dependence of mechanical properties of polyamide components on build parameters in the SLS process. *J. Mater. Process. Technol.* 182 (1–3): 477–488.

Chacón, J. M., Caminero, M. A., García-Plaza, E., Núñez, P. J. 2017. Additive manufacturing of PLA structures using fused deposition modelling: Effect of process parameters on mechanical properties and their optimal selection. *Mater. Des.* 124:143–157.

Chinthapalli, R., Skoczinski, P., Carus, M., et al. 2019. *Bio-based Building Blocks and Polymers – Global Capacities and Trends 2018–2023.* http://bio-based.eu/reports/ (accessed May 31, 2019).

Choren, J. A., Heinrich, S. M., Silver-Thorn, M. B. 2013. Young's modulus and volume porosity relationships for additive manufacturing applications. *J. Mater. Sci.* 48:5103–5112.

Clarinval, A.-M., Halleux, J. 2005. Classification of biodegradable polymers In *Biodegradable Polymers for Industrial Applications*, ed.R. Smith. Cambridge: Woodhead Publishing.

Conn, R. E., Kolstad, J. J., Borzelleca, J. F., et al. 1995. Safety assessment of polylactide (PLA) for use as a food-contact polymer. *Food Chem. Toxycol.* 33:273–283.

Coppola, B., Cappetti, N., Di Maio, L., et al. 2017. *Layered Silicate Reinforced Polylactic Acid Filaments for 3D Printing Of Polymer Nanocomposites.* New York: IEEE.

Coppola, B., Cappetti, N., Di Maio, L., Scarfato, P., Incarnato, L. 2018. 3D printing of PLA/clay nanocomposites: Influence of printing temperature on printed samples properties. *Materials* 11:1947.

Cruz Sanchez, F. A., Lanza, S., Boudaoud, H., Hoppe, S., Camargo, M. 2015. Polymer recycling and additive manufacturing in an open source context: optimization of processes and methods. *Annual International Solid Freeform Fabrication Symposium, ISSF 2015*, Aug 2015, Austin, TX, United States. pp.1591–1600. hal-01523136.

Cui, M., Liu, L., Guo, N., Su, R., Ma, F. 2015. Preparation, cell compatibility and degradability of collagen-modifiedpoly(lactic acid). *Molecules* 20:595–607.

Cuiffo, M. A., Snyder, J., M. Elliott, A. M. et al. 2017. Impact of the fused deposition (FDM) printing process on polylactic acid (PLA) chemistry and structure. *Appl. Sci.* 7(579) doi:10.3390/app7060579.

Cygan, Z. 2009. *Improving Processing and Properties of Polylactic Acid.* http://plasticstrends.com/index.php/last-months-mainmenu-28/218-improving-processing-and-properties-of-polylactic-acid (accessed May 31, 2018).

Cygan, Z. 2013. Biodegradable impact-modified polymer compositions. US Patent US 8,524,832 (accessed August 31, 2019).

Danzer, R., Harrer, W., Supancic, P., et al. 2007. The ball on three balls test – strength and failure analysis of different materials. *J. Eur. Cer. Soc.* 27:1481–1485.

Depuydt, D., Balthazar, M., Hendrickx, K., et al. 2018. Production and characterization of bamboo and flax fiber reinforced polylactic acid filaments for fused deposition modeling (FDM). *Polym. Compos.* 40:1951–1963.

Di Lorenzo, M. L. 2005. Crystallization behavior of poly(l-lactic acid). *Eur. Polym. J.* 41:569–575.

Domingo, M., Puigriol, J. M., Garcia, A. A., et al. 2015. Mechanical property characterization and simulation of fused deposition modeling polycarbonate parts. *Mater. Des.* 83:670–677.

Dorgan, J. R., Lehermeier, H., Mang, M. 2000. Thermal and rheological properties of commercial-grade poly (lactic acid). *J Polym Environ* 8 (1): 1–9.

Dorgan, J. R., Lehermeier, H.J., Palade, L. I., Cicero, J. 2001. Polylactides: properties and prospects of an environmentally benign plastic from renewable resources. *Macromol. Symp.* 175:55–66.

Ecker, J. V., Haider, A., Burzic, I. et al. 2019. Mechanical properties and water absorption behaviour of PLA and PLA/wood composites prepared by 3D printing and injection moulding. *Rapid Prototyp. J.* 25 (4): 672–678.

Edwards, E., Booth, J. C., Roberts, J. K., et al. 2017. Military efforts in nanosensors, 3D printing, and imaging detection. *Proceedings Volume 10167, Nanosensors, Biosensors, Info-Tech Sensors and 3D Systems 1016714. SPIE Smart Structures and Materials + Nondestructive Evaluation and Health Monitoring,* 2017, Portland.

Elsawy, M. A., Kim, K. H., Park, J. W., Deep, A. 2017. Hydrolytic degradation of polylactic acid (PLA) and its composites. *Renew. Sustain. Energy Rev.* 79:1346–1352.

Eppley, B. L. 2005. Use of resorbable plates and screws in pediatric facial fractures. *J. Oral. Maxillofac. Surg.: Off. J. Am. Assoc. Oral. Maxillofac. Surg.* 63 (3): 385–391.

Ezeh, O. H., Susmel, L. 2018. Fatigue behavior of additively manufactured polylactide (PLA). *Proc. Struct. Integr.* 13:728–734.

Fafenrot, S., Grimmelsmann, N., Wortmann, M., Ehrmann, A. 2017. Three-dimensional (3D) printing of polymer-metal hybrid materials by fused deposition modeling. *Materials* 10:1199.

Farah, S., Anderson, D. G., Langer, R. 2016. Physical and mechanical properties of PLA, and their functions in widespread applications – a comprehensive review. *Adv. Drug Deliv. Rev.* 107:367–392.

FDA. n.d. *Inventory of Effective Food Contact Substance (FCS) Notifications.* No.178. https://www.fda.gov/food/packaging-food-contact-substances-fcs/inventory-effective-food-contact-substance-fcs-notifications (accessed January 16, 2021).

Ferreira, R. T. L., Cardoso Amatte, I., Dutra, T. A. 2017. Experimental characterization and micrography of 3D printed PLA and PLA reinforced with short carbon fibers. *Compos. B* 124:88–100.

Flowers, P. F., Reyes, C., Ye, S. et al. 2017. 3D printing electronic components and circuits with conductive thermoplastic filament. *Addit. Manuf.* 18:156–163.

Fu, S.-Y., Feng, X.-Q., Lauke, B., Mai, Y.-W. 2008. Effects of particle size, particle/matrix interface adhesion and particle loading on mechanical properties of particulate–polymer composites. *Compos. B: Eng.* 39 (6): 933–961.

Futerro. 2019. *The PLA.* http://www.futerro.com/products_pla.html (accessed August 31, 2019).

Futerro. n.d. *The Lactide.* http://www.futerro.com/products_lactide.html (accessed January 16, 2021).

Garrison, T.F., Murawski, A., Quirino, R. L. 2016. Bio-based polymers with potential for biodegradability. *Polymers* 8:262.

Gayer, C., Abert, J., Bullemerc, M. 2018. Influence of the material properties of a poly(D,L-lactide)/β-tricalcium phosphate composite on the processability by selective laser sintering. *J. Mech. Behav. Biomed. Mater.* 87:267–278.

Ghiorse, S. 1993. Effect of void content on the mechanical properties of carbon/epoxy laminates. *Sampe Q.* 24:54–59.

Global Market Insights. 2016. https://www.gminsights.com/industry-analysis/lactic-acid-and-polylactic-acid-market (accessed August 28, 2019).

Gomez-Gras, G., Jerez-Mesa, R., Travieso-Rodriguez, J. A., Lluma-Fuentes, J. 2018. Fatigue performance of fused filament fabrication PLA specimens. *Mater. Des.* 140:278–285.

Gong, B., Cui, S., Zhao, Y. et al. 2017. Strain-controlled fatigue behaviors of porous PLA-based scaffolds by 3D-printing technology. *J. Biomater. Sci. Polym. Ed.* 28 (18): 2196–2204.

Government of the Netherlands. 2016. *A Circular Economy in the Netherlands by 2050.* https://www.government.nl › documents › 17037+Circulaire+Economie_EN (accessed May 31, 2019).

Grant, A., Ellis, B., Rohani, M. R., Hajiha, R. 2019. *Exploring Process-Properties Relationships of 3D-Printed PLA: Towards Process-Informed Simulation and Design.* TechConnect Briefs 2019, 143-146. ISBN 978-0-9988782-8-7. TechConnect.org.

Granta, CES 2010 EDUPACK Material and Process Selection Charts page 7.

Grasso, M., Azzouz, L., Ruiz-Hincapie, P., Zarrelli, M., Ren G. 2018. Effect of temperature on the mechanical properties of 3D-printed PLA tensile specimens. *Rapid Prototyping Journal* 24 (8): 1337–1346.

Graupner, N., Herrmann, A. S., Müssig, J. 2009. Natural and man-made cellulose fibre-reinforced poly(lactic acid) (PLA) composites: An overview about mechanical characteristics and application areas. *Composites: Part A* 40:810–821.

Gruber, P. 2005. Polylactides™ NatureWorks PLA. *Biopolym. Online*, pp. 235–239.

Grunewald, S. J. 2015. *Teknor Apex Announces New High-Impact, High-Heat PLA Filament.* https://3dprint.com/59001/teknor-apex-high-impact-pla/ (accessed August 28, 2019).

Gupta, B., Revagade, N., Hilborn, J. 2007. Poly (lactic acid) fiber: an overview. *Prog. Polym. Sci.* 32:455–482.

Hamad, K., Kaseem, M., Yang, H. W., Deri, F., Ko, Y. G. 2015. Properties and medical applications of polylactic acid:A review. *Express Polym. Lett.* 9:435–455.

Harper, C. ed. 1996. Handbook of plastics. *Elastomers & Composites.* New York: McGraw-Hill.

Harris, A. M., Lee, E. C. 2008. Improving mechanical performance of injection molded PLA by controlling crystallinity. *J. Appl. Polym. Sci.* 107:2246–2255.

Hartmann, M., Whiteman, N. 2000. Polylactide, a new thermoplastic for extrusion coating. *Proceedings of the Polymers, Laminations and Coatings Conference*, Chicago, IL, USA, 27–31 August 2000:631–635.

Henton, D. E., Gruber, P., Lunt, J., Randall, J., 2005. In *Natural Fibers, Biopolymers, and Biocomposites*, ed.A. K. Mohanty, M. Mizra, and L. T. Drzal, 527–577. Boca Raton: Taylor and Francis.

Hill, R. 1948. A theory of the yielding and plastic flow of anisotropic metals. *Proc. R. Soc. Lond. A Math. Phys. Sci.* 193:281–297.

Ibrahim, M., Mogan, Y., Jamry, S. N. S., Periyasamy, R. 2016. Resistivity study on conductive composite filament for freeform fabrication on functionality embedded products. *ARPN J. Eng. Appl. Sci.* 11:10.

Jabłoński, A., Kopeć, J., Jatteau S., Ziąbka, S., Rajzer, M. 2018. 3D printed poly L-lactic acid (PLLA) scaffolds for nasal cartilage engineering. *Eng. Biomater.* 21 (144): 15–19.

Jake, R. D. 2021. *3DJake Recycled PLA Filament.* https://www.3djake.com/3djake/rpla (accessed January 16, 2021).

Jamshidi, K., Hyon, S. H., Ikada, Y. 1988. Thermal characterization of polylactides. *Polymer* 29:2229.

Jamshidian, M., Arab-Tehrany, E. A., Imran, M., Jacquot, M., Desobry, S. 2010. Poly-lactic acid: production, applications, nanocomposites, and release studies. *Compr. Rev. Food Sci. Food Saf.* 9:552–571.

Jaya Christiyan, K. G., Chandrasekhar, U., Rajesh, M. N., Venkateswarlu, K. 2018b. Influence of manufacturing parameters on the strength of PLA parts using layered manufacturing technique: a statistical approach. *IOP Conf. Series: Materials Science and Engineering*, 310 012134.

Jaya Christiyan, K. G., Chandrasekhar, U., Venkateswarlu, K. 2018a. Flexural properties of PLA components under various test condition manufactured by 3D Printer. *J. Inst. Eng. (India): C* 99 (3): 363–367.

Jem, K. J., Tan, B. 2020. The development and challenges of poly (lactic acid) and poly (glycolic acid). *Adv. Indus. Eng. Polym. Res.* 3:60–70.

Jones, J. R. 2005. Scaffolds for tissue engineering. In *Biomaterials, Artificial Organs and Tissue Engineering*, ed.L. L. Hench, J. R. Jones, 201–214. Woodhead Publishing Series in Biomaterials.

Kakanuru, P., Pochiraju, K. 2020. Moisture ingress and degradation of additively manufactured PLA, ABS and PLA/SiC composite parts. *Addit. Manuf.* 36:101529.

Koslow, T., de Valensart, G. 2017. *REC 3D Releases Comprehensive Stress Test for 3D Printing Materials.* https://www.filaments.directory/en/blog/2017/01/27/rec-3d-releases-comprehensive-stress-test-for-3d-printing-materials (accessed January 16, 2021).

Kuentz, L., Salem, A., Singh, M., Halbig, M. C., Salem, J. A. 2016. Additive manufacturing and characterization of polylactic acid (PLA) composites containing metal reinforcements. *International Conference and Expo on Advanced Ceramics and Composites.* https://ntrs.nasa.gov/archive/nasa/casi.ntrs.nasa.gov/20160010284.pdf.

Kumar, N., Kulkarni, M., Ravuri, M., et al. 2015. Effects of electroplating on the mechanical properties of FDM-PLA parts. *i-Manager's J. Fut. Eng. Technol.* 10 (3): 29–37.

Lamberti, P., Spinelli, G., Kuzhir, P., et al. 2018. Evaluation of thermal and electrical conductivity of carbon-based PLA nanocomposites for 3D printing. *AIP Conference Proceedings 1981*:020158.

Lasprilla, A. J., Martinez, G. A., Lunelli, B. H., Jardini, A. L., Maciel Filho, R. 2012. Poly-lactic acid synthesis for application in biomedical devices – a review. *Biotechnol. Adv.* 30:321.

Laureto, J.,Tomasi, J., King, J. A., Pearce, J. M. 2017. Thermal properties of 3-D printed polylactic acid-metal composites. *Prog. Addit. Manuf.* 2:57–71.

Letcher, T., Waytashek, M. 2014. Material property testing of 3d-Printed specimen in PLA on an entry-level 3D printer. *Proc. ASME Int. Mech. Eng. Cong. Expos.* Vol. 2a.

Li, L., Wang, T., Sun, J., Yu, Z. 2018. The effect of process parameters in fused deposition modelling on bonding degree and mechanical properties. *Rapid Prototyp. J.* 24 (1): 80–92.

Li, N., Li, Y., Liu, S. 2016. Rapid prototyping of continuous carbon fiber reinforced polylactic acid composites by 3D printing. *J. Mater. Proces. Technol.* 238:218–225.

Lim, L. T., Auras, R., Rubino, M. 2008. Processing technologies for poly(lactic acid). *Prog. Polym. Sci.* 33:820–852.

Liu, H., Zhang, J. 2011. Research progress in toughening modification of poly(lactic acid). *J. Pol. Sci. B.* 49:1051–1083.

Loureiro, N. C., Esteves, J. L. 2019. Green composites in automotive interior parts: A solution using cellulosic fibers. In *Green Composites for Automotive Applications*, ed.G. Koronis, A. Silva, 81–97. Woodhead Publishing Series in Composites Science and Engineering.

Lovett, G., de Bie, F. 2016. *Sustainable Sourcing of Feedstocks for Bioplastics.* https://www.corbion.com/media/550170/corbion_whitepaper_feedstock_sourcing_11.pdf. (accessed August 28, 2019).

Lunt, J. 1998. Large-scale production, properties and commercial applications of polylactic acid polymers. *Polym. Degrad. Stab.* 59:145–152.

Luzanin, O., Guduric, V., Ristic, I., Muhic, S. 2017. Investigating impact of five build parameters on the maximum flexural force in FDM specimens – a definitive screening design approach. *Rapid Prototyp. J.* 23 (6): 1088–1098.

Madhavan Nampoothiri, K., Nair, N.R., John, R.P. 2010. An overview of recent developments in polylactide (PLA) research. *Bioresour, Technol.* 101:8493–8501.

Mainil-Varlet, P. 1998. Effect of in vivo and in vitro degradation on molecular and mechanical properties of various low-molecular-weight polylactides. *J. Biomed. Mater. Res.* 36:360–380.

Mallegni, N., Phuong, T. V., Coltelli, M.-B., Cinelli, P., Lazzeri, A. 2018. Poly(lactic acid) (PLA) based tear resistant and biodegradable flexible films by blown film extrusion. *Materials* 11:148.

Martina, M., Hutmacher, D. W. 2007. Biodegradable polymers applied in tissue engineering research: a review. *Polym. Int.* 56 (2): 145–157.

MatWeb. 2020. *Clariant Natural Color High Impact Polystyrene 3D Printer Filament.* http://www.matweb.com/ search/datasheettext.aspx?matguid=65fe9fe444cd4a14837e466c3702e501 (accessed May 28, 2020).

Meara, D. J., Knoll, M. R., Holmes, J. D., Clark D. M. 2012. Fixation of Le Fort I osteotomies with poly-DL-lactic acid mesh and ultrasonic welding – a new technique. *J. Oral. Maxillofac. Surg.* 70 (5): 1139–1144.

MedlinePlus. n.d. *Lactic Acid Test.* https://medlineplus.gov/ency/article/003507.htm.

Mehdikhani, M., Gorbatikh, L., Verpoest, I., Lomov, S. V. 2019. Voids in fiber-reinforced polymer composites: A review on their formation, characteristics, and effects on mechanical performance. *J. Compos. Mater.* 53 (12): 1579–1669.

Mehta, R., Kumar, V., Bhunia, H., Upadhyay, S. N. 2005. Synthesis of poly(lactic acid): a review. *J. Macromol. Sci. Polym. Rev.* 45:325–349.

Migliaresi, C., de Lollis, A., Fambri, L., Cohn, D. 1991. The effect of thermal history on the crystallinity of different molecular weight PLLA biodegradable polymers. *Clin. Mater.* 8:111–118.

Mirkhalaf, S. M., Fagerström, M. 2019. The mechanical behavior of polylactic acid (PLA) films: fabrication, experiments and modelling. *Mech. Time-Depend. Mater.* doi:10.1007/s11043-019-09429-w.

Miyata, T., Masuku, T. 1998. Crystallization behaviour of poly(l-lactide). *Polymer* 39:5515–5521.

Mohanty, A. K., Misra, M., Drzal, L. T., et al. 2005. *Natural Fibers, Biopolymers, and Biocomposites.* Boca Raton: CRC press.

Mohanty, A.K., Misra, M., Hinrichsen, G. 2000. Biofibres. *Macromol. Mater. Eng.* 276/227:1–24.

Mozafari, H., Dong, P., Hadidi, H., Sealy M. P., Gu L. 2019. Mechanical characterizations of 3D-printed PLLA/steel particle composites. *Materials* 12:1.

Muller, J., González-Martínez, C., Chiralt, A. 2017. Combination of poly(lactic) acid and starch for biodegradablefood packaging. *Materials* 10:952 (CrossRef) (PubMed).

Multi 3D. 2017. *Resources.* https://www.multi3dllc.com/resources/ (accessed May 28, 2020).

Mura006riu, M., Dubois, P. 2016. PLA composites: From production to properties. *Adv. Drug Deliv. Rev.* 107:17–46.

Murariu, M., Laoutid, F., Dubois, P., et al. 2014. Pathways to biodegradable flame retardant polymer (nano) composites. In *Polymer Green Flame Retardants*, ed.C. D., Papaspyrides, P. Kiliaris, 709–773. Amsterdam: Elsevier.

Nadel B. 2019. *The Best Industrial 3D Printers of 2019.* https://www.business.com/categories/best-industrial-3d-printers/ (accessed August 31, 2919).

Nagarajan, V., Mohanty, A. K., Misra, M. 2016. Perspective on polylactic acid (pla) based sustainable materials for durable applications: focus on toughness and heat resistance. *ACS Sustain. Chem. Eng.* 4:2899–2916.

Namiki, M., Ueda, M., Todoroki, A., Hirano, Y. 2014. 3D printing of continuous fiber reinforced plastic. *SAMPE Seattle 2014 International Conference and Exhibition (International SAMPE Technical Conference).*

Nampoothiri, K. M., Nair, N. R., John, R. P. 2010 An overview of the recent developments in polylactide (PLA) research. *Bioresour. Technol.* 101:8493–8501.

NatureWorks LLC. 2007. *Technology Focus Report: Toughened PLA.* https://www.natureworksllc.com/ ~/media/Files/NatureWorks/Technical-Documents/White-Papers/Toughened-PLA-Technology-Focus-pdf.pdf (accessed August 31, 2019).

Niaza, K., Senatov, F., Stepashkin, A., et al. 2017. Long-term creep and impact strength of biocompatible 3D-printed PLA-based scaffolds. *Nano Hybrids Compos.* 13:15–20.

Nielsen L. E. and R. F. Landel 1994. *Mechanical Properties of Polymers and Composites.* New York: Marcel Dekker. 2nd edition.

Osswald, T. A. and G. Menges 2012. *Material Science of Polymers for Engineers*, p. 166. Munich: Hanser. 3rd edition.

Pagano, C., Basile, V., Modica, F., Fassi, I. 2019. Micro-FDM process capability and postprocessing effects on mechanical properties. *Proceedings of PPS-33, AIP Conference Proceedings 2139*, 190002.

Paloheimo, M. n.d. *Tensile or Flexural Strength/Stiffness – Is There Really a Difference?* https://www.plasticprop.com/ articles/tensile-or-flexural-strengthstiffness-there-really-difference/ (accessed August 31, 2019).

Paspali, A., Bao, Y., Gawne, D. T., et al. 2018. The influence of nanostructure on the mechanical properties of 3D printed polylactide/nanoclay composites. *Compos. B: Eng.* 152:160–168.

Paul, B., 1960. Prediction of elastic constants of multiphase materials. *Trans. Metall. Soc. AIME* 218:36.

Pearce, J. M. 2014. *Open-Source Lab How to Build Your Own Hardware and Reduce Research Costs.* Amsterdam: Elsevier. 1st edition.

Pearce, J. M. 2012. Building research equipment with free, open-source hardware. *Science* 337 (6100): 1303–1304.

Perego, G., Cella, G. D., Bastioli, C. 1996. Effect of molecular weight and crystallinity on poly (lactic acid) mechanical properties. *J. Appl. Polym. Sci.* 59:37–43.

Perkowski, C. 2017. Tensile-Compressive Asymmetry and Anisotropy of Fused Deposition Modeling PLA under Monotonic Conditions. Electronic Theses and Dissertations, 2004-2019. 5576. https://stars.library.ucf.edu/etd/5576.

Phattanaphibul, T., Koomsap, P. 2012. Investigation of PLA-based scaffolds fabricated via SVM rapid prototyping. *J. Porous Mater.* 19:481–489.

Phattanaphibul, T., Koomsap, P., Idram, I., Nachaisit, S. 2014. Development of SVM rapid prototyping for scaffold fabrication. *Rapid Prototyp. J.* 20 (2): 90–104.

Phattanaphibul, T., Opaprakasit, P., Koomsap, P., Tangwarodomnukun, V. 2007. Preparing biodegradable PLA for powder-based rapid prototyping. *Proceedings of the 8th Asia Pacific Industrial Engineering and Management Society Conference*, 2007, Taiwan.

Plastics Insights. 2019. *Polylactic Acid Properties, Production, Price, Market and Uses.* https://www.plasticsinsight.com/resin-intelligence/resin-prices/polylactic-acid/ (accessed August 31, 2019).

Pop, M. A., Croitoru, C., Bedő, T., et al. 2019. Structural changes during 3D printing of bioderived and synthetic thermoplastic materials. *J. Appl. Polym. Sci.* 136:47382.

Raquez, J.-M., Habibi, Y., Murariu, M., Dubois, P. 2013. Polylactide (PLA)-based nanocomposites. *Prog. Polym. Sci.* 38:1504–1542.

Rasal, R. M., Janorkar, A. V., Hirta, D. E. 2010. Poly (lactic acid) modifications. *Prog. Polym. Sci.* 35:338–356.

RepRap 2019. Last edited on 22 September. 2019. https://reprap.org/wiki/RepRap (accessed August 28, 2019).

Research and Markets. 2019. *Polylactic Acid Market Report: Trends, Forecast and Competitive Analysis.* https://www.prnewswire.com/news-releases/global-3-7-bn-polylactic-acid-markets-2013-2018--2019-2024---major-players-are-natureworks-total-corbion-pyramid-bioplastic-weforyou-and-zhejiang-hisu-300831460.html (accessed August 28, 2019).

Rosato, Dominick V., Donald V. Rosato and Matthew V. Rosato 2004. *Plastic Product Material and Process Selection Handbook.* Amsterdam: Elsevier Science.

RTP Company. 2011. *Glass Fiber Reinforced Bioplastics.* http://web.rtpcompany.com/info/data/bioplastics/RTP2099X121249A.htm (accessed April 3, 2021).

Saba, N., Jawaid, M., Al-Othman, O. 2017. An overview on polylactic acid, its cellulosic composites and applications. *Curr. Org. Synth.* 14 (2): 156–170.

Safai, L., Cuellar, J. S., Smit, G., Zadpoor, A. A. 2019. A review of the fatigue behavior of 3D printed polymers. *Addit. Manuf.* 28:87–97.

Shogren, R. 1997. Water vapor permeability of biodegradable polymers. *J. Environ. Polym. Degrad.* 5:91–95.

Siakeng, R., Jawaid, M., Ariffin, H., et al. 2019. Natural fiber reinforced polylactic acid composites: a review. *Polym. Compos.* 40:446–463. doi:10.1002/pc.24747.

Siebott, V. 2007. PLA – the future of rigid packaging? *Bioplast. Mag.* 2 (2): 28–29.

Silva, A. L. N., Cipriano, T. F., da Silva, A., et al. 2014. Thermal, rheological and morphological properties of poly (lactic acid) (PLA) and talc composites. *Polímeros Ciência e Tecnol.* 24 (3): 276–282.

Silva, S. P., Sabino, M. A., Fernandes, E. M., et al. 2005. Cork: properties, capabilities and applications. *Int. Mater. Rev.* 50 (6): 345.

Silverajah, V. S. G., Ibrahim, N. A., Yunus, W. M. Z. W., Hassan, H. A., Woei, C. B. 2012. A comparative study on the mechanical, thermal and morphological characterization of poly(lactic acid)/epoxidized palm oil blend. *Int. J. Mol. Sci.* 13:5878–5898.

Sin, L. T., Rahmat, A. R., Rahman, W. A. W. A. 2012. Overview of Poly(Lactic Acid). In *Handbook of Biopolymers and Biodegradable Plastics*, ed. S. Ebnesajjad. Amsterdam: Elsevier.

Siracusa, V., Rosa, M. D., Iordanskii, A. L. 2017. Performance of poly(lactic acid) surface modified films for foodpackaging application. *Materials* 10:850 (CrossRef) (PubMed).

Slapnik, J., Bobovnik, R., Mešl, M., Bolka, S. 2016. Modified polylactide filaments for 3D printing with improved mechanical properties. *Contemp. Mater.* VII (2): 142–150.

Slijkoord, J. W. 2015. *Is Recycling PLA Really Better than Composting?* https://3dprintingindustry.com/news/is-recycling-pla-really-better-than-composting-49679/ (accessed January 16, 2021).

Smart Materials 3D. n. d. https://www.smartmaterials3d.com/en/ (accessed April 3, 2021).

Snowdon, M. R., Wu, F., Mohanty, A. K., Misra, M. 2019. Comparative study of the extrinsic properties of poly(lactic acid)-based biocomposites filled with talc versus sustainable biocarbon. *RSC Adv.* 9:6752.

Sodergard, A., Stolt, M. 2002. Properties of lactic acid based polymers and their correlation with composition. *Prog. Pol. Sci.* 27 (6): 1123–1163.

Song, Y., Li, Y., Song, W., et al. 2017. Measurements of the mechanical response of unidirectional 3D-printed PLA. *Mater. Des.* 123:154–164.

Sood, A. K., Chaturvedi, V., Datta, S., Mahapatra, S. S. 2011. Optimization of process parameters in fused deposition modeling using weighted principal component analysis. *J. Adv. Manuf. Syst.* 2:241–259.

Sood, A. K., Ohdar, R. K., Mahapatra, S. S. 2010. Parametric appraisal of mechanical property of fused deposition modelling processed parts. *Mater. Des.* 31:287–295.

Spoerk, M., Arbeiter, F., Cajner, H., et al. 2017. Parametric optimization of intra- and inter-layer strengths in parts produced by extrusion-based additive manufacturing of poly(lactic acid). *J. Appl. Polym. Sci.* 134:45401.

Sun, Q., Rizvi, G.M., Bellehumeur, C.T., Gu P. 2008. Effect of processing conditions on the bonding quality of FDM polymer filaments. *Rapid Prototyp. J.* 14 (2): 72–80.

Tan, B. H., Muirruri, J. K., Li, Z., He, C. 2016. Recent progress in using stereocomplexation for enhancement of thermal and mechanical property of polylactide. *ACS Sustain. Chem. Eng.* 4:5370–5391.

Tateho. 2020. *Silicon Carbide Whisker.* https://tateho-chemical.com/products/whisker.html (accessed April 26. 2020).

Taubner, V., Shishoo, R. 2001. Influence of processing parameters on the degradation of poly(l-lactide) during extrusion. *J. Appl. Polym. Sci.* 79:2128–2135.

The Guardian. 2013. *Going Dutch: Why the Country Is Leading the Way on Sustainable Business.* https://www.theguardian.com/sustainable-business/blog/dutch-companies-leading-sustainable-business (accessed November 15, 2019).

Tian, X., Liu, T., Wang,Q., et al. 2017 Recycling and remanufacturing of 3D printed continuous carbon fiber reinforced PLA composites. *J. Clean Prod.* 142:1609–1618.

Tian, X., Liu, T., Yang, C., Wang, Q., Li, D. 2016. Interface and performance of 3D printed continuous carbon fiber reinforced PLA composites. *Composites: Part A* 88:198–205.

Torres, I. O., De Luccia, N. 2017. A simulator for training in endovascular aneurysm repair: The use of three dimensional printers. *Eur. J. Vasc. Endovasc. Surg.* 54:247–253.

Torres, J., Cole, M., Owji, A., DeMastry, Z., Gordon, A. P. 2016. An approach for mechanical property optimization of fused deposition modeling with polylactic acid via design of experiments. *Rapid Prototyping J.* 22 (2): 387–404.

Torres, J., Cotelo, J., Karl, J., Gordon, A. P. 2015. Mechanical property optimization of FDM PLA in shear with multiple objectives. *JOM* 67 (5): 1183–1193.

Tres, P. 2014. *Designing Plastic Parts for Assembly.* Munich: Hanser Gardner, p. 51.

Tsouknidas, A., Pantazopoulos, M., Katsoulis, I., et al. 2016. Impact absorption capacity of 3D-printed components fabricated by fused deposition modelling. *Mater. Des.* 102:41–44.

Tsuji, H. 2005. Poly(lactide) stereocomplexes: formation, structure, properties, degradation and applications. *Macromol. Biosci.* 5:569–597.

Tsuji, H., Ikada, Y. 1996. Crystallization from the melt of PLA with different optical purities and their blends. *Macromol. Chem. Phys.* 197:3483–3499.

Tymrak, B. M., Kreiger, M., Pearce, J. M. 2014. Mechanical properties of components fabricated with open-source 3-D printers under realistic environmental conditions. *Materials and Design* 58:242–246.

Urayama, H., Moon, S. I., Kimura, Y. 2003. Microstructure and thermal properties of polylactides with different l- and d-unit sequences: importance of the helical nature of the l-sequenced segments. *Macromol. Mater. Eng.* 288:137–143.

Vadori, R., Mohanty, A. K., Misra, M. 2013. The effect of mold temperature on the performance of injection molded poly (lactic acid)-bioplastic. *Macromol. Mater. Eng.* 298:981–990.

Van de Velde, K., Kiekens, P. 2002. *Polym. Testing* 21 (4): 433–442.

Van den Eynde, M., Van Puyyelde, P. 2018. 3D printing of poly(lactic acid). *Adv. Polym. Sci.* 282:139–158.

Vert, M., Chabot, F., Christel, P. 1981. *Makromol. Chem. Suppl.* 5:30.

Vihaan, Y. 2020. *Electrical Properties of PLA Plastic, 3drific.* Published on June 20, 2020. https://3drific.com/electrical-properties-of-pla-plastic/ (accessed August 28, 2020).

Vink, E. T. H., Rabago, K. R., Glassner, D. A., Gruber, P. R. 2003. Application of life cycle assessment to NatureWorks™ polylactide (PLA) production. *Polym. Degrad. Stab.* 80:403–419.

Wang, L., Gramlich, W. M., Gardner, D. J. 2017b. Improving the impact strength of Poly(lactic acid) (PLA) in fused layer modeling (FLM). *Polymer* 114:242–248.

Wang, S., Capoen, L., D'hooge, D. R., Cardon, L., 2017a. Can the melt flow index be used to predict the success of fused deposition modelling of commercial poly(lactic acid) filaments into 3D printed materials? *Plast. Rubb. Compos.* 47 (3): 1–8.

Weber, C. J., Haugaard, V., Festersen, R., Bertelsen, G. 2002. Production and applications of biobased packaging materials for the food industry. *Food Addit. Contam.* 19:172–177.

Wen, S., Chung, D. D. L. 2006. Self-sensing of flexural damage and strain in carbon fiber reinforced cement and effect of embedded steel reinforcing bars. *Carbon* 44:1496–1502.

Wertz, J. T., Mauldin, T. C., Boday, D. J. 2104. Polylactic acid with improved heat deflection temperatures and self-healing properties for durable goods applications. *ACS Applied Mater. Interface.* 6:18511–18516.

Wittbrodt, B., Pearce, J. M. 2015. The effects of PLA color on material properties of 3-D printed components. *Addit. Manuf.* 8:110–116.

Wu, S. 1990. Chain structure, phase morphology, and toughness relationships in polymers and blends. *Polym. Eng. Sci.* 30 (13): 753–761.

Yamamura, T, Omiya, M, Sakai, T, Viot, P. 2009. Evaluation of compressive properties of PLA/PBAT polymer blends. *Proceedings of the Asian Pacific Conference for Materials and Mechanics 2009 at Yokohama*, Japan.

Yan, Q., Dong, H., Su, J., et al. 2018. A review of 3D printing technology for medical applications. *Engineering* 4 (5): 729–742.

Yang, G. H., Su, J. J., Su, R., et al. 2012. Toughening of poly(L-lactic acid) by annealing: the effect of crystal morphologies and modifications. *J. Macromol. Sci. Part B Phys.* 51:184–196.

Yang, L., Wu, Z., Cao, Y., Yan, Y. 2014. Micromechanical modelling and simulation of unidirectional fibre-reinforced composite under shear loading. *J. Reinf. Plast. Compos.* 34:72–83.

Yang, Y., Zhang, L., Xiong, Z., et al. 2016. Research progress in the heat resistance, toughening and filling modification of PLA. *Sci. China Chem.* 59:1355–1368.

Yao, S. S., Pang, Q. Q., Song, R., et al. 2016. Fracture toughness improvement of poly (lactic acid) with silicon carbide whiskers. *Macromol. Res.* 24 (11): 961–964.

Yao, X., Luan C., Zhang, D., et al. 2017. Evaluation of carbon fiber-embedded 3D printed structures for strengthening and structural-health monitoring. Materials and Design 114:424-432.

Yu, T., Ren, J., Li, S., Yuan, H., Li, Y. 2010. Effect of fiber surface-treatments on the properties of poly (lactic acid)/ramie composites. *Compos. Part A* 41 (4): 499–505.

Zhang, K., Mohanty, A. K., Misra, M. 2012. Fully biodegradable and biorenewable ternary blends from polylactide, poly(3-hydroxybutyrateco-hydroxyvalerate) and poly(butylene succinate) with balanced properties. *ACS Appl. Mater. Interf.* 4 (6): 3091–3101.

Zhao, P., Rao, C., Gu, F., et al. 2018. Close-looped recycling of polylactic acid used in 3D printing: An experimental investigation and life cycle assessment. *J. Cleaner Prod.* 197:1046–1055.

Zhao, Y., Chen, Y., Zhou, Y. 2019. Novel mechanical models of tensile strength and elastic property of FDM AM PLA materials: Experimental and theoretical analyses. *Mater. Des.* 181:108089.

Zhou, C, Guo, H., Li, J., et al. 2016. Temperature dependence of poly(lactic acid) mechanical properties. *RSC Adv.* 6:113762–113772.

Ziemian, C., Sharma M. 2012. Anisotropic mechanical properties of ABS parts fabricated be fused deposition modelling. *Mech. Eng.* 7. doi:10.5772/34233.

4 Polyamide

> *What is the use of a house if you haven't got a tolerable planet to put it on?*
>
> Henry David Thoreau

4.1 OVERVIEW OF SUSTAINABLE AND NON-SUSTAINABLE POLYAMIDES (PAs)

PAs or *nylons* (from the trade name *nylon* given by their manufacturer DuPont in the 1940s) are a family of TP linear highly crystalline polymers whose chemical formula contains repeating amide links –CO–NH– in their molecular chain. A natural version of PAs are proteins and peptides. PAs are widely employed as *commodity plastics* (that is mass-produced) and also as high-performance materials, because they combine affordable cost with high tensile and flexural strengths, elongation at break, functional mechanical properties, and impact resistance from −50 to 200°C, good electrical resistivity and machinability, noise dampening, low coefficient of friction, fatigue resistance, and excellent wear resistance. The last three properties make PAs well-suited for functional parts such as mechanisms, gears, hinges, and similar articles. A well-known downside of PAs is high moisture absorption (*hygroscopy*) from air and water that increases the dimensions and lowers the mechanical performance of the fabricated object. PA can absorb up to 10 times the amount of moisture taken in by other polymers. F.e. at saturation water absorption of PA 11 and PA 12 (two grades of PA for AM) can reach 1.9% and 1.6%, respectively.

PAs are classified as *engineering polymers* (medium physical-mechanical performance) and *high-performance polymers* (top physical-mechanical performance). Chemically, most PAs are polymers or copolymers formed via condensation, by reacting monomers containing equal parts of amine and carboxylic acid, so that amides are formed at both ends of each monomer. PAs are available in several grades identified by the number of carbon atoms present in their ingredient or ingredients employed to produce that specific grade. For example, PA 11 is derived from 11-aminoundecanoic acid, whose formula $H_2N(CH_2)_{10}CO_2H$ contains 11 carbon atoms (Figure 4.1). In PA grades that are identified by two numbers, the first and the second numbers denote the number of carbon atoms in diamine and diacid, respectively. For instance, PA 6,10 (featuring, along with PA 11, the lowest moisture absorption among PAs) derives from two monomers: hexamethylene diamine possessing six carbon atoms, and sebacic acid containing ten carbon atoms. PA 12 is the most popular of all polymers for PBF: it offers similar advantages to those of PA 6,10 and PA 11 but at lower cost, being processed more economically and easily. PAs have many uses, because of their properties and the fact that they can be melt-processed into fibers (swimwear, active wear, innerwear, hosiery, shell fabric, carpets, and tire cords), films, and shapes (car components such as safety bags and engine components, mechanical parts such as machine screws, gears, gaskets).

PA 11 is important in this book because it is formulated from plants as a powder for AM, and hence it is biobased. Non-AM applications of PA 11 and PA 12 are numerous, and are listed here as a benchmark and suggestions for AM applications: bumpers, connectors, consumer sporting goods, electrical, electronic, enclosures, end-use parts manufactured in small and medium volumes, functional prototypes, housings, impellers, living hinges, medical, snap-fit designs, underhood components, vehicle dashboards and grilles, complex thin-walled ductwork for aerospace, motorsports, and unmanned air vehicles.

DOI: 10.1201/9781003221210-4

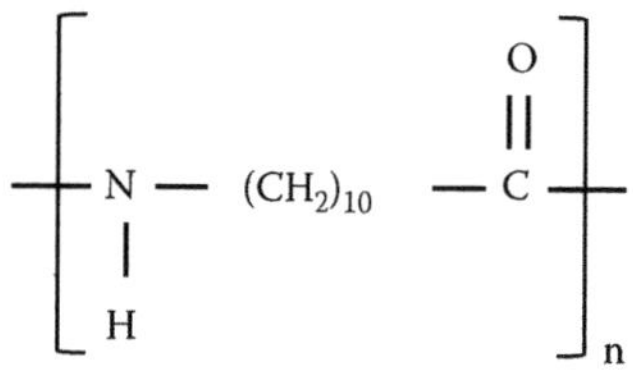

$$\left[\begin{array}{c} N - (CH_2)_{10} - C \\[1ex] \mid \\ H \end{array} \quad \begin{array}{c} O \\ \parallel \\ \\ \end{array} \right]_n$$

FIGURE 4.1 Chemical structure of PA 11.

TABLE 4.1

Properties of Fossil-Based and Sustainable PAs for Conventional Manufacturing

PA Grade	Biobased Content	Suppliers (Product)	Density	Tensile Strength	Yield Tensile Strength	Tensile Modulus	Elongation at Break
	%		g/cm^3	MPa	MPa	MPa	%
6,10. 10,10. 10,12	45–100	Evonik (Vestamid® Terra)	1.03–1.08	N/A	40–61	1,300–2,100	>50
4,10	70	DSM (EcoPAXX® Q210E–H)	1.09	55	95	3,000	30
6. 6,6	0	Various[a]	1.12–1.15	43–94	55–90	690–3,790	15–100
11	100	Arkema (Rilsan®)	1.03–1.25	39–47	40–45	1,000–1,378	>200
11	0	Isoflon, Ashley Polymers	1.04	44–66	N/A	1,550	150–290
12	0	EMS–GRIVORY, RTP Co., GEHR Plastics	1–1.06	41–45	N/A	1,379–2,200	>10–50

Note:

[a]*Handbook of Plastics, Elastomers, and Composites*, 1996, ed. C. A. Harper, Appendix C, 3rd edition, New York: McGraw-Hill.

Table 4.1 lists tensile properties, biobased content, and producers of fossil-based and sustainable PAs for conventional manufacturing processes such as extrusion and injection molding. The amount of biobased content in PAs often depends not only on the biobased diacid but also a sustainable diamine (Winnacker and Rieger 2016). Several commercial sustainable grades of PAs exist for conventional manufacturing (extrusion, injection molding, etc.) (Table 4.2): their biobased content is 45–100%, and they can tapped by AM material formulators as potential feedstocks for AM, with appropriate modifications to be compatible with AM processes. The technology to produce sustainable PAs is owned by a few companies.

In 2020 the global market for PA 11 and PA 12 for AM was valued at about USD 30 million in 2018 and expected to grow at a vigorous growth rate of 27% over the forecast period 2019–2026 (Global Information Inc. 2021), driven by aerospace and defense, automotive, and health care. North America is expected to register high growth, due to a large number of end-use industries adopting AM in USA and Canada. North America and Europe have a large market share, and are expected to continue to dominate the market in the near future.

PA 11 for AM is projected to record high growth because it combines strong mechanical performance (outperforming PA 12 for AM), and sustainability (being 100% biobased). However, PA 12 will dominate the market in the near future, since it has a comparatively established market for PBF, and is available in several grades: impact resistant, fire resistant, glass-, aluminum-, and

TABLE 4.2

Sustainable PAs for Conventional Manufacturing

Biobased PA	Biobased Content (%)	Manufacturer/Product
PA 4,10	100	DSM/EcoPaXX®
PA 10,10	100	EMS-GRIVORY/Grilamid® 1S. Evonik/VESTAMID® Terra DS. DuPont/Zytel® RS LC1000. Arkema/Rilsan® T. Suzhou Hipro Polymers/Hiprolon® 200, Hiprolon® 211.
PA 11	100	Arkema/Rilsan® PA11. Suzhou Hipro Polymers/Hiprolon® 11.
Polyphthalamide (PPA)	>70	Arkema/Rilsan® HT.
PA 6,10	63	BASF/Ultramid® S Balance. EMS-GRIVORY/Grilamid® 2S. Evonik/VESTAMID® Terra HS. Solvay/Technyl® eXten. DuPont/Zytel® RS LC3030. Arkema/Rilsan® S. Suzhou Hipro Polymers/Hiprolon® 70.
Transparent Polyamide	54	Arkema/Rilsan® Clear G830 Rnew
PA 10,T	50	EMS-GRIVORY/Grilamid® HT3. Evonik/VESTAMID® HTplus M3000.
PA 10,12	45	Evonik/VESTAMID® Terra DD. Suzhou Hipro Polymers/Hiprolon® 400.

Source: Jiang, Y., Loos, K. 2016. Enzymatic Synthesis of Biobased Polyesters and Polyamides. *Polymers* 8(243). Freely reproduced under open access Creative Commons Attribution (CC-BY) license, and with kind permission from Dr. K. Loos.

carbon-filled. Major providers of PA 11 and PA 12 are ALM LLC (USA), Arkema (France), CRP Group (Italy), EOS (Germany), Golden Plastics (China), Stratasys and 3D Systems (both in USA). Other grades of PA commercially available for AM are:

- PA 6,6: examples are Novamid® ID1030 and ID1030 CF10, both filaments for FFF made by DSM (The Netherlands), with the latter loaded with carbon fiber to improve structural properties (tensile strength and modulus equal to 110 MPa and 7.6 GPa, respectively) and dimensional stability, and reduce warpage.
- PA 6: an example is Novamid® ID1070, a filament for FFF formulated for applications demanding high temperature resistance in harsh environments: its HDT is 54°C and 104°C at 1.80 MPa and 0.45 MPa, respectively.

Biobased PAs can pair high performance with true sustainability. In fact, unlike several SPs, biobased PAs are among the polymers considered upper performers. The monomers for biobased PAs are partially or completely derived from sustainable feedstocks. Until 2013, all biobased PAs were derived from castor oil (Brehmer 2014).

In the near-future, biobased PAs are expected to increase their presence in a variety of polymer applications, possibly leveraging, f.e., enzymatic polymerization, a route proved itself effective and versatile for producing biobased PAs with several chemical compositions, architectures, and functionalities (Jiang and Loos 2016). R&D activity is generating novel biobased monomers for sustainable PAs, starting from ingredients such as sugar fermented into succinic acid in turn converted into PA (Kabasci and Bretz 2012; Bechthold et al. 2008), and L-arginine, an amino acid produced by our body to help the body build proteins, and extracted from biomass waste streams to formulate PA 4,6 (Könst et al. 2011). Another possible opportunity is blending PAs with biobased polymers containing nitrogen, such as chitin and chitosan, in order to formulate composites with tailored properties (González et al. 2000). PA 10,12 is derived from fossil-based dodecanoic diacid

(DDDA) and decamethylene diamine, but it can already be produced from DDDA in form of lauric acid extracted from palm kernel oil (Brehmer 2014).

4.2   CASTOR OIL (CO)

Some information in this section is derived from a recent and comprehensive review on CO authored by Patel et al. (2016). We recall that PA 11 for AM derives from CO, a vegetable oil, inedible in its native form, obtained by pressing ripe seeds (Figure 4.2) of the castor plant (*Ricinus communis*) without their outer covering (*hull*). The name *castor oil* probably comes from serving as a replacement for *castoreum*, a perfume made from the dried perineal glands of the beaver, *castor* in Latin. CO is a pale yellow or almost colorless transparent viscous liquid with a faint mild odor and nauseating taste. It has been a vital renewable resource for the global specialty chemical industry (Mutlu and Meier 2010), because it is the only commercial source of hydroxylated fatty acid (Severino et al. 2012). Castor plants have the advantages that they feature high yield, generally do not compete with food (because castor beans provide little nutritional value to man and animals), grow in lands of little agricultural value and in semiarid climate (they require very little irrigation and withstand drought) where a few plants can grow, and are naturally pest resistant. Although called *bean*, the castor seed is not a true bean. The seeds contain 40–60% of oil that is rich in triglycerides, and ricin, a water-soluble toxin. Ingesting four to eight seeds can be lethal in adult people. However, if castor oil undergoes a heating process that deactivates ricin, the oil can be ingested safely.

The world production of CO reached 0.8 million tons in 2019, and was predicted to grow to 1.1 million tons in 2025 (Expert Market Research n.d.). The global CO market was valued at USD 1.3B in 2020, and predicted to reach USD 1.5B by 2026 (360 Research Reports 2020) or to USD 1.7B by 2025 (Fior Markets 2020). In 2020, India produced 80% of the world's CO, followed by China and Brazil in decreasing order. End-use markets of CO reflect a broad range of applications: cosmetics, food, industrial, lubrication, paints, pharmaceutical, and soaps. CO is a biodiesel fuel component, and present in the manufacture of adhesives, brake fluids, caulks, dyes, humectants, hydraulic fluids, inks, lacquers, leather treatments, lubricating greases, machining oils, paints, pigments, PU adhesives, refrigeration lubricants, sealants, textiles, washing powders, and waxes (The Chemical Company 2019).

FIGURE 4.2   Castor seeds.

Source: Mubofu, E. B. 2016. Castor oil as a potential renewable resource for the production of functional materials. *Sustain. Chem. Process* 4, 11. https://doi.org/10.1186/s40508-016-0055-8. Reproduced with kind permission from Dr. Mubofu.

The molecular formula and weight of CO are $C_{57}H_{104}O_9$ and 933.45 g/mol, respectively (PubChem n.d.), and its composition includes the following fatty acids: 90% ricinoleic, 4% linoleic, 3% oleic, 1% stearic, and < 1% linolenic. CO's importance for the chemical industry resides in its high content of ricinoleic acid (RA), namely in the presence of the hydroxyl group in RA and the double bond of the ester linkage that together enable to perform a variety of chemical reactions, modifications, and transformations involved in formulating numerous products, such as coatings, inks, lubricants, and paints (Patel et al. 2016). CO features density of 0.959 g/cm³, viscosity of 6–8 poises at 25°C, and boiling point of 313°C at 760 mm Hg (PubChem, n.d.; Kazeem et al. 2014). CO is extracted from seeds by mechanical pressing, solvent extraction, or a combination of both methods (Dasari and Goud 2013). CO can be chemically transformed through a number of methods not only into PA but also polyethers, polyesters, PUs, and interpenetrating polymer networks (Mubofu 2016). Depending on the processing route followed, CO generates two monomers: 11-aminoundecanoic acid to formulate PA 11, or sebacic acid to formulate PA 6,10 and PA 10,10.

4.3   OVERVIEW OF PA 11

The history of PA 11 development started in 1938, when the monomer of PA 11 was formulated from undecanoic acid ($CH_3(CH_3)_9COOH$) obtained from cracking CO. The first industrial monomer plant opened in 1955 in France to supply a polymerization plant. The process developed by the current world leader Arkema for the production of PA 11 from CO starts with RA, whose molecular formula and weight are $C_{18}H_{34}O_3$, and 298.461 g/mol, respectively, and consists of five chemical steps described by Devaux et al. (2011): (1) transesterification of RA with methanol to methyl ricinoleate along with glycerin co-product; (2) cracking of methyl ricinoleate leads to heptaldehyde and methyl undecylenate; (3) hydrolysis to obtain undecylenic acid along with methanol that is recycled to the first step; (4) addition on hydrogen bromide; (5) nucleophilic substitution with ammonia to form 11-aminoundecanoic acid that is the PA 11 monomer.

PA 11 represents a small fraction of the worldwide production of PAs, but it is applied in a broad range of areas such as food packaging, automotive, and offshore applications, thanks to its biocompatibility, good resistance to salt water and oil, excellent piezoelectric and cryogenic properties, and hydrophilicity (Di Lorenzo et al. 2019). The environmental downside of PA 11 is that, although biobased, it is not biodegradable and compostable (Di Lorenzo et al. 2019).

Before delving into PA 11 for AM, we note that parts printed out of PA found their way in (Kauppila 2021):

- Automotive: car components (brackets and interior trim), prototypes of car components, articles for production shops (f.e. tools and fixtures), and pneumatic gripper for the tooling of new parts.
- Medical: customized dental molds and medical prosthetics, gadgets enabling the use of smell in virtual reality.
- Consumer goods: sporting and eyewear items.

4.4   COMMERCIAL PA 11 GRADES FOR AM

4.4.1   OVERVIEW

Table 4.3 lists the material properties reported by the suppliers of commercial PA 11 grades for PBF. Since the materials listed derive from the same feedstock produced by Arkema, their property values are very close. Compared to PLA filaments for FFF, PA 11 powders for PBF are as strong as the strongest PLAs, less stiff, far more ductile in tension, more impact resistant, and have 20–25% lower density, which makes their ratio tensile strength/density higher than that of PLA. In

TABLE 4.3

Properties of Commercial Unfilled PA 11 Powders for PBF

Density After Laser Sintering	Particle Diameter	Tensile Modulus	Tensile Strength	Tensile Stress at Yield	Elongation at Break	Elongation at Yield	Heat Deflection at 0.45 MPa	Heat Deflection at 1.8 MPa	Izod Notched Impact Strength	Charpy Notched Impact Strength	Electrical Volume Resistivity	Dielectric Strength
g/cm³	μm	MPa	MPa	MPa	%	%	°C	°C	J/m	J/m²	Ohm-cm	kV/mm
0.96–1.07	40–95	1,392–1,800	45–54	37–47.5	38–47	5–25	158–188	44–70	69–91	7.4	1.3 x 10¹³	18.5

Sources: Values from data sheets of: Rilsan® Invent Natural, Rilsan® Invent Black, EOS PA 1101, EOS PA 1102 black, ALM P80-ST, ALM PA 850 NAT, ALM PA 860, ALM FR-106, Duraform® EX, ADSINT PA 11, HP HR PA 11, Prodways PA11-SX 1450.

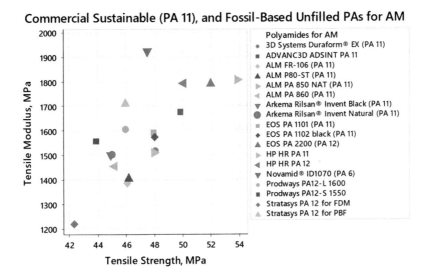

FIGURE 4.3 Tensile properties of unfilled commercial PA powders for AM. Some overlapping data points were offset in order to distinguish them.

Source: Data sheets posted online by suppliers.

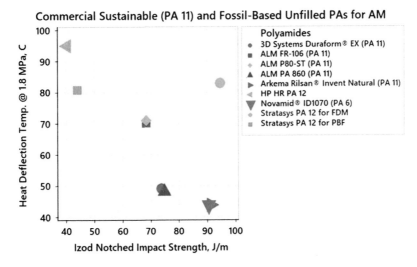

FIGURE 4.4 Impact strength vs. heat deflection temperature of unfilled commercial PA powders for AM. Some overlapping data points were offset in order to distinguish them.

Source: Data sheets posted online by suppliers.

Figure 4.3, values of tensile modulus vs. tensile strength of commercial biobased PA 11 grades listed in Table 4.3 and fossil-based PAs (PA 12, PA 6) for AM are plotted: the PA 11 grades are equivalent to the other PAs in terms of tensile strength and modulus at room temperature. In fact, the tensile strength for PA 11 grades and the other PAs is 45–54 MPa and 42–52 MPa, respectively, while the tensile modulus is 1,392–1,800 MPa and 1,210–1,915 MPa, respectively. Figure 4.4 illustrates the values for HDT and Izod notched impact strength for some sustainable PA 11 and fossil-based PAs plotted in Figure 4.3. Only the plastics whose Izod notched impact

TABLE 4.4

Mechanical Properties of Commercial PLA Filaments and PA Powders, Both for AM. PLA Ranges Include Most Typical Values. Ranges for PAs Are from Table 4.3

Filament	Density	Tensile Strength	Tensile Modulus	Elongation at Break	Izod Notched Impact Strength
	g/cm^3	MPa	MPa	%	J/m
PAs	0.96–1.07	45–54	1,392–1,800	38–47	69–91
PLAs	1.2–1.3	22–48	1,820–4,000	1–8	7–64

strength and HDT were reported by suppliers could be plotted, since some suppliers report Charpy impact strength instead. Referring to the PA 11 and the other PAs in Table 4.3, HDT at 1.8 MPa is 44–70°C and 80–95°C, respectively, while the Izod notched impact strength is 69–91 J/m and 41–94 J/m, respectively.

Finally, in order to place performance of PA powders for AM in perspective, Table 4.4 compares the range of tensile properties and impact strength between PLA filaments for AM and PA powders for AM: PA is more ductile and impact resistant and less stiff.

4.4.2 Rilsan® Invent PA11

Rilsan® Invent PA11 is a very popular powder for PBF available in multiple colors. It derives from Rilsan® PA11, a semi-crystalline TP polymer featuring two phase transitions: glass transition of the amorphous phase around 45°C, and melting of the crystalline phase between 180°C and 189°C, depending on the PA11 grade. For Invent PA11 the water absorption at 50% relative humidity is 1%, and at saturation at 23°C is 2.6%, and, after 25 weeks of immersion in water at 20°C, the length variation is 0.2–0.5% and weight variation is 1.9%.

The tensile properties of Invent PA11 reported in Table 4.5 were most likely measured on an industrial or professional printer equipped with a laser of power ranging from tens to hundreds of watts. Wang et al. (2017) assessed the sensitivity of those properties to a printer's laser power, and measured the tensile properties of ASTM D638-compliant coupons out of Rilsan® Invent Black (comprising Invent PA11 and carbon black particles) built in flatwise and upright positions (Figure 3.15) on a Sintratec (Switzerland) 2 W blue diode (445 nm) desktop laser printer. Tensile strength in flatwise and upright orientations was 49 MPa and 42 MPa, respectively, whereas elongation at break was 65% and 10%, respectively. Since Arkema for the same material reports 45 MPa strength and 45% strain,

TABLE 4.5

Tensile Properties of Printed Coupons out of Rilsan® Invent PA11

Printing Orientation	Strength	Elastic Modulus[a]	Strain at Break	Strength at Yield	Strain at Yield
	MPa	MPa	%	MPa	%
Flat	49.7	1,667	46	42	3.8
Upright	49.0	1,667	38	40	5.0

Source: Data from stress-strain curves published online by Arkema n.d. *3D Printed Marketing Presentation, Technical Polymers.*

Note:

[a] In 0–20 MPa range.

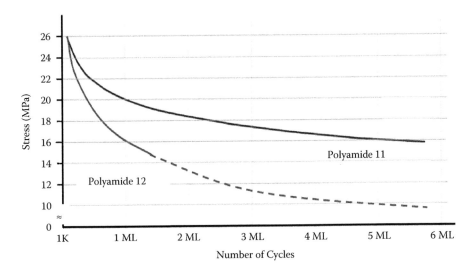

FIGURE 4.5 Tensile fatigue curve of PA 11 (Rilsan® Invent) and PA 12, collected at 5 Hz (ISO 527 1B, notched 0.7 mm).

Source: Modified from Arkema n.d. *3D Printed Marketing Presentation, Technical Polymers.* Reproduced with permission from Arkema S.A.

respectively, in flatwise orientation, the two properties were not penalized by not using an industrial printer. The drop in strength between orientations was due to the fact that in upright orientation interlayer and intralayer defects are perpendicular to the loading direction, and hence they trigger an earlier failure. The increase of strain in upright orientation was possibly caused by an interlayer bonding weaker than the intralayer cohesion among powder particles. Ultimate and yield properties in tension for Rilsan® Invent in Table 4.5 have been derived from stress-strain curves published by Arkema. These values shows that printing orientation has no effect on tensile modulus, marginal effect on tensile strength and yield strength, and more significant effect on elongation at yield and break. Particularly, the upright orientation is associated with the lower strength and elongation at break. Data in Wegner et al. (2015) on tensile properties of PA 11, namely commercial EOS PA 1101, confirmed that anisotropy was present in strain at break, strength, and modulus in decreasing order.

Since Invent PA11 is suited for functional and load-bearing articles, its fatigue data are displayed in Figure 4.5. The curve is an interpolation of the number of cycles at failure, measured on notched coupons shaped according to ISO 527 1B, at specific tensile stress values, 23°C, and load frequency of 5 Hz. Below 16 MPa (about 1/3 of tensile strength value upon static loading), no failures were recorded prior to 5.5 million loading cycles.

The particle size distribution of Rilsan® Invent PA11 and EOS PA 2200 (made of non-biobased PA 12 for PBF) are listed in Table 4.6, with the latter data set added for comparison. The quantity D10 means that, in case of Invent PA11, 10% of the powder sample has a diameter not exceeding 20.6 µm, D50 that 50% of the powder sample has a diameter not above 44.2 µm, and so on. The range of values for Invent PA11 is below 100–150 µm, which is the thickness of the powder layers deposited during PBF. The smaller are the particles, the smoother is the surface finish of the printed part, and the longer is that part's life under fatigue loading. The powder should feature a round shape to adequately flow and generate a dense, continuous layer on the plate heated by the sintering energy source. The denser and more uniform is the layer, the better the physical and mechanical properties of the resulting part are. Figure 4.6 displays a SEM micrograph of Invent PA11 showing rough particles with sharp edges.

One parameter for measuring the density of the deposited layers of powder is the *packing density* that is related to the powder's flowability, and is calculated as the ratio of the density of the

TABLE 4.6

Particle Size Distribution of Rilsan® Invent PA11, and EOS PA 2200 (PA 12)

Material	D_{10}	D_{50}	D_{90}	Material Density	Layer Density	Packing Density
	µm	µm	µm	g/cm³	g/cm³	%
Rilsan® Invent PA11	20.6	44.2	80.1	1.06	0.61	58
EOS PA 2200 (PA 12)	38.8	58.6	88.3	1.03	0.50	49

Source: Verbelen, L., Dadbakhsh, S., Van den Eynde, M., et al, 2016. Characterization of polyamide powders for determination of laser sintering processability. *European Polymer Journal* 75:163–174. Reproduced with permission from Elsevier.

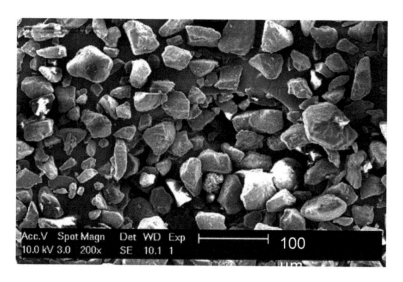

FIGURE 4.6   SEM image of Rilsan® Invent PA11.

Source: Verbelen, L., Dadbakhsh, S., Van den Eynde, M., et al., 2016. Characterization of polyamide powders for determination of laser sintering processability. *European Polymer Journal* 75:163–174. Reproduced with permission from Elsevier.

powder material (*material density*) to the density of one deposited layer (*layer density*). The values of these three properties are reported in Table 4.6 for Invent PA11 and PA 2200 for comparison. Invent PA11 has a packing density greater than 50%, which is on the upper side of the 45–50% range for today's commercially available PBF polymeric powders (Schmid 2018), and is superior to that of PA 2200, although the latter consists of smoother and rounder particles.

Coalescence is the disappearance of the boundary between particles that are in contact and become one, and the more diffused is the coalescence achieved in sintered powders during PBF, the smaller is the porosity, and the greater are the cohesion among particles, and density and mechanical performance of the printed article. Verbelen et al (2016) studied the coalescence of Invent PA11 upon sintering during PBF at 195–205°C: compared to commercial PA powders for AM, Invent PA11 displayed faster coalescence, and generated layers more extensively connected to each other.

The temperature difference between (a) the melting peak during heating and sintering and (b) cold crystallization peak during cooling of the polymer being sintered is related to the polymer's tendency to curl or warp upon sintering during PBF: in *warping* all layers are curved, whereas in *curling* the layers closer to printer's platform are curved but the layers farther from the platform are flat. Verbelen et al. (2016) applied DSC to Invent PA11 to simulate the heating and cooling during sintering, and measured its thermal behavior, including melting and crystallization peaks. Peaks that two were 200°C and about 170°C, respectively, and the resulting 30°C difference was within the range of values for commercial PA powders for AM. The crystallization peak was marginally affected by the maximum heating temperature, possibly due to the presence of an additive acting as nucleating agent promoting crystallization.

Shrinkage upon cooling is also a critical property of a PBF powder, because it can be the main factor preventing dimensional accuracy and part reproducibility, and in some cases leads to curling and warping. In PBF, shrinkage is mainly influenced by crystallization, powder bulk density and characteristics, and thermal contraction. In fact, the total amount of shrinkage β_{total} in PBF upon cooling is a function of shrinkage β_T caused by the change in temperature, shrinkage β_c due to crystallization, and shrinkage β_{powder} resulting from the powder bulk and particle characteristics at bed temperature, according to the following equation (Benedetti et al. 2019):

$$\beta_{total} = \beta_T + \beta_c + \beta_{powder} \qquad (4.1)$$

Shrinkage was measured on Invent PA11 by means of a home-built adapter combined with a TMA running at heating rate of 0.5°C/min, followed by 5 min of equilibration time, and successive cooling at 0.5°C/min. The specific volume (cm^3/g) varied from 0.95 to 1.20 during heating, and withdrew from 1.12 to 0.96 (Verbelen et al. 2016).

4.5 EXPERIMENTAL FILAMENT IN PA 11 FOR FFF

PA 11 is commercially available in form of powder, but some researchers investigated alternative filaments made out of it that would feature the advantage of using FFF printers that are less expensive than PBF printers. A feedstock in form of filament also avoids restricting the choice of fillers to those of size below 20–30 μm fitting the diameter of powder particles, and widens the range of potential fillers for PBF powders.

An experimental PA 11 filament for FFF was evaluated by Herrero et al. (2018), who combined injection molding grade PA 11 Rilsan® Besno with commercial sepiolite nanoclay Pangel S by Tolsa (Spain) in a twin-screw extruder with L/D of 36, temperature profile of 235–250°C and 150 rpm, and obtained composite filaments of 1.7 mm diameter with 0, 1, 3, and 7 wt% sepiolite. Sepiolite is a naturally occurring mineral of sedimentary origin, porous, lightweight and absorbent, and it is a magnesium silicate with chemical formula $Mg_4Si_6O_{15}(OH)_2 \cdot 6H_2O$. Adding sepiolite improved dimensional accuracy of the printed test coupons, tensile strength and modulus, and reduced elongation at break. The values of these tensile properties and failure type (that is ductile vs. brittle) also depended on the printing orientation. A transmission electron microscopy (TEM) photo of PA 11/sepiolite with 93/7 wt% composition displayed a uniform distribution of the nanoclay within PA 11. Tensile coupons were printed in three orientations (Figure 3.15): flatwise, edgewise, and upright. The average tensile test results are reported in Table 4.7. Sepiolite steadily boosted tensile strength and modulus, and steadily decreased strain at break at flatwise and edgewise orientations, whereas at upright orientation the changes in the three properties were insignificant and erratic. The highest and lowest values of strength, modulus, and strain corresponded to flatwise and upright printing orientations, respectively. In comparison to unfilled PA 11 and across orientations, the maximum improvement in strength and modulus was 4–27% and 20–56%, respectively. The highest strength (48.1 MPa) and modulus (1,165 MPa) resulted from 7.3 wt% sepiolite nanoclay. However, referring to the value ranges in Table 4.1, the former value is in the middle for strength, and the latter is below the middle for modulus. The tensile failure was ductile for flatwise and edgewise samples, and brittle for upright coupons. The dimensional accuracy of the printed coupons was expressed by comparing their actual dimensions to their CAD model by means of an ATOS Core 200 3D scanner. Increasing content of sepiolite enhanced the

TABLE 4.7

Test results on Printed Coupons of Invent PA11/Sepiolite

Material	Tensile Strength (MPa)			Tensile Modulus (MPa)			Elongation at Break (%)		
	Flatwise	Edgewise	Upright	Flatwise	Edgewise	Upright	Flatwise	Edgewise	Upright
PA11	38.0	34.9	27.0	749	775	673	38.1	28.8	5.0
PA11/Sepiolite (98.8/1.2 wt%)	41.0	36	29.0	871	815	732	34.9	27.9	5.0
PA11/Sepiolite (97.1/2.9 wt%)	43.1	38.1	28.1	976	907	625	28.2	23.0	5.0
PA11/Sepiolite (92.7/7.3 wt%)	48.1	41	28.1	1,165	990	808	18.9	12.2	4.1

Source: Values from plots in Herrero, M., Peng, F. Núñez Carrero, K., et al. 2018. Renewable Nanocomposites for Additive Manufacturing Using Fused Filament Fabrication. *ACS Sustainable Chem. Eng.* 6:12393–12402.

dimensional accuracy across printing orientations, and the most inaccurate coupons were those printed upright. Compared to unfilled PA, nanoclay may also impart durability, by improving long-term properties like creep and stress relaxation.

Walha et al. (2016) demonstrated that PLA and PA 11 are immiscible and incompatible polymers. However, an experimental filament made of PA 11 and PLA was fabricated by Rasselet et al. (2019), who combined Rilsan® PA11, injection grade Ingeo™ 3251D PLA, and commercially available modified acrylic copolymer Joncryl ADR®-4368, serving as a compatibilizer between PA11 and PLA. PLA/PA11 (80/20 wt%) blends containing 0–3 wt% of Joncryl (accordingly named PLA80-J0 to PLA80-J3) were prepared and tested, and some of them outperformed 3251D in tensile properties: modulus of PLA80-J3 was 3.38 GPa vs. 2.65 GPa for 3251D, strength and elongation at break of PLA80-J2 were 58.5 MPa and 3.25%, respectively vs. 48 MPa and 2.9% for 3251D, respectively. The T_g and T_m of the blends were about the same (60–61°C and 167–170°C, respectively) as those of 3251D (59°C and 170°C, respectively).

4.6 PROCESS OPTIMIZATION FOR LASER PRINTING PA 11

Wegner et al. (2015) experimentally determined the optimal process settings for laser sintering PA 11, namely the effect of laser power, scan speed, hatch distance, layer thickness, and outline energy density on tensile strength, modulus, and strain at break, density, and surface roughness. They identified the combination of process setting values leading to the largest improvement of most properties investigated. The DOE method permitted to cost-effectively analyze 780 test specimens and measure the correlation between the process settings above mentioned and properties of printed coupons. Test coupons were printed out of EOS PA 1101 on a DTM Sinterstation 2500 HS printer. Tensile coupons were printed flatwise and upright. For each set of parameter values, the *volume energy density* (VED) was measured in J/mm³, calculated as laser power divided by the product of scan speed, hatch distance, and layer thickness, and condensing in one variable all the process settings being evaluated. Tensile strength, modulus, and strain at break rapidly raised with the initial increase of VED, and reached and maintained a maximum value of about 50 MPa, 1,600 MPa, and 45%, respectively, corresponding to the middle-upper range of VED values. Tensile strength, modulus, and strain at break also featured anisotropy, with higher values from flatwise coupons, and anisotropy of strength and modulus diminishing as VED increased. Table 4.8 summarizes the lowest and maximum values of the measured properties. The values (J/mm³) of VED corresponding to the peak of each property were: 0.40-0.57 for density, 0.32-0.57 for tensile strength, 0.30-0.60 for tensile modulus, and 0.37-0.57 for elongation at break. *Area density energy*

TABLE 4.8

Properties of Printed Test Coupons in Sustainable EOS PA 1101 as a Function of Volume Energy Density (VED). Ranges of Values Caused by Printing Orientation of Test Coupons

Property Value	Density	Tensile Strength	Tensile Modulus	Elongation at Break
	g/cm^3	MPa	MPa	%
Lowest	0.72–0.94	4–24	300–830	3–11
Maximum	1.02–1.03	47–51	1,600–1,620	23–47

Source: Values from plots in Wegner, A., Harder, R., Witt, G., Drummer, D. 2015. Determination of Optimal Processing Conditions for the Production of Polyamide 11 Parts using the Laser Sintering Process. *International Journal of Recent Contributions from Engineering, Science & IT.* 3, 1. http://dx.doi.org/10.3991/ijes.v3i1.4249.

(J/mm^2) is defined as laser power divided by beam speed and hatch distance (or scan spacing) (Nelson 1993), and represents the amount of energy supplied to the powder particles per unit area of the powder bed surface (Caulfield et al. 2007). The surface roughness was expressed in terms of R_z that reached a minimum in the range 0.30–0.35 of area density energy. Data in Wegner et al. (2015) also illustrated that the anisotropy in strain at break, strength, and modulus was reduced by greater values of VED.

The correlation between density on one hand and tensile strength and modulus on the other hand was measured based the test data graphed by Wegner et al. (2015), and plotted in Figure 4.7. The strength data plotted were calculated from interpolating the average of strength values from flatwise and upright coupons at each VED value, and deriving the best-fitting equation of strength as a function of VED. The same was done with density values at each VED value, and couples of strength and density values at the same VED values were calculated from the best-fitting equations and plotted. This process was repeated with modulus, and all the plotted properties were termed *modeled* to indicate that the values graphed were not actual test values but interpolated and derived from them. The strong linear dependence between tensile properties and density is convenient to optimize the printing process.

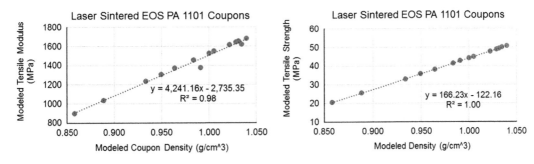

FIGURE 4.7   Tensile strength and modulus of printed coupons of EOS PA 1101 as a function of coupon's density.

Source: Analyzed data taken from plots in Wegner, A., Harder, R., Witt, G., Drummer, D. 2015. Determination of Optimal Processing Conditions for the Production of Polyamide 11 Parts using the Laser Sintering Process. *International Journal of Recent Contributions from Engineering, Science & IT.* 3, 1.

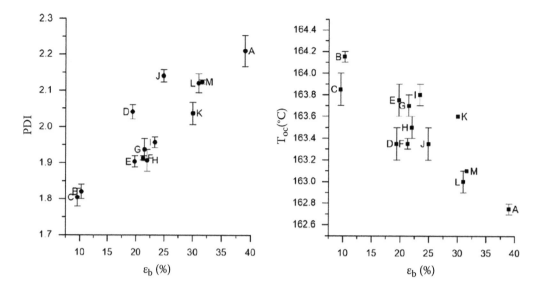

FIGURE 4.8 Correlation between elongation at break (e_b), polydispersity index (PDI), and crystallization onset temperature (T_{oc}) for EOS PA 1101.

Source: Modified from Scherer, B., Kottenstedde, I. L., Bremser, W., Matysik, F.-M. 2020. Analytical characterization of polyamide 11 used in the context of selective laser sintering: Physico-chemical correlations. *Polymer Testing* 91, 106786.

The same process of interpolating test data to derive the influence of density on tensile properties of sintered coupons of PA was applied to experiments on non-biobased sintered PA 12 powder, namely EOS PA2200 (Hofland et al. 2017), and it resulted in some linear dependence on density but not as strong as for PA 1101.

Scherer et al. (2020) measured the effect of heat treating EOS PA 1101 on its elongation at break (ε_b) that is a significant parameter, because the higher ε_b, the more ductile and likely more impact resistant is PA 1101. Particularly, they exposed tensile test coupons to a range of temperatures (153, 162, 171, 180°C) and durations (8, 16, and 24 h), which affected the polydispersity index (PDI) and crystallization onset temperature (T_{co}) of PA 1101. We recall that PDI represents the degree of "non-uniformity" of a distribution of MW values across the polymer chains: the closer PDI is to 1 (its minimum value) the more uniform is the distribution. As illustrated in Figure 4.8, ε_b is positively and negatively correlated to PDI and T_{co}, respectively. Tensile bars with ε_b of 24.9% or greater featured crystallinity below 35.4% and ductile fracture, whereas tensile bars with ε_b not exceeding 10% displayed crystallinity.

4.7 TENSION, FRACTURE, AND FATIGUE PROPERTIES OF PA 11 FOR PBF

Salazar et al. (2014) measured, at different temperatures and after conditioning, tensile and fatigue properties, and fracture toughness of commercial biobased PA 11 and fossil-based PA 12, namely 3D Systems Duraform® EX and Duraform® PA (Stratasys 2021), respectively, and concluded that the former outperformed the latter in all measurements except tensile modulus. The authors fabricated tensile coupons per ASTM D638, fracture coupons featuring *single edge notch bend* (SENB) geometry and initial V-notch (Figure 4.9 left side) per ESIS TC4 protocol (Williams 2001), and fatigue coupons featuring *compact tension geometry* (CT) per ASTM E647 (Figure 4.9 right side). All coupons were tested in the following conditions: (a) at 23°C dry as-processed, (b) at 23°C after being immersed in water at 90°C for 3 months until the weight increase stabilized, and

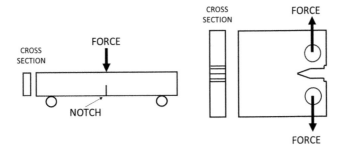

FIGURE 4.9 Left: fracture toughness coupon single edge notch bend. Right: fatigue coupon featuring compact tension geometry.

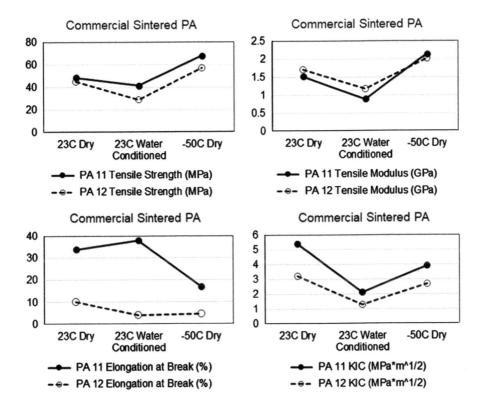

FIGURE 4.10 Variation of properties of sustainable PA 11 and fossil-based PA 12 with test temperature and conditioning.

Source: Data collected from Salazar, A., Rico, A., Rodríguez, J. et al. 2014. Monotonic loading and fatigue response of a bio-based polyamide PA11 and a petrol-based polyamide PA12 manufactured by selective laser sintering. *European Polymer Journal* 59:36–45.

(c) dry as-processed at −50°C after reaching thermal equilibrium. All test results, except fatigue results, are plotted in Figure 4.10, and illustrate the superiority of Duraform® EX over Duraform® PA in strength, elongation, and fracture toughness but not modulus at both test temperature and upon conditioning.

We recall that the fracture parameter K_{Ic} is the critical value of the *stress intensity factor in mode I*, a loading mode characterized by a force that opens a crack and is applied perpendicularly

TABLE 4.9

Parameters Relative to Crack Growth under Fatigue of Commercial PA Grades for PBF: Sustainable PA 11 (Duraform® EX), and Fossil-Based PA 12 (Duraform® PA)

PA Type	Test Temperature. Conditioning	K_{Ic}	ΔK_{th}	$K_{max,c}$	m
	°C	MPa m$^{1/2}$	MPa m$^{1/2}$	MPa m$^{1/2}$	N/A
11	23. Dry	5.4	1.65	5.24	5.4
11	23. Water Conditioned	3.9	2.42	3.88	8.3
11	−50. Dry	2.1	0.62	1.95	4.9
12	23. Dry	3.2	1.71	3.05	13.8
12	23. Water Conditioned	2.7	1.81	2.74	14.5
12	−50. Dry	1.3	0.49	1.22	11.5

Source: Salazar, A., Rico, A., Rodríguez J., et al., 2014. Monotonic loading and fatigue response of a bio-based polyamide PA11 and a petrol-based polyamide PA12 manufactured by selective laser sintering. *European Polymer Journal* 59:36–45. Reproduced with permission from Elsevier.

to the crack's length, as shown in Figure 4.9 right side. K_{Ic} is expressed in stress/√length (MPa/√m), and represents the intrinsic capacity of a material to withstand fracture in the presence of cracks, hence the greater K_{Ic} the more fracture resistant is the material. The resistance to crack growth under fatigue loading was assessed by measuring the following properties (ASTM E1823):

- *Fatigue crack growth threshold K_{th}*, the asymptotic value of the stress intensity factor ΔK at which crack propagation upon fatigue is negligible, being $\Delta K = K_{max} − K_{min}$ the range of K, and K_{max} and K_{min} the maximum and minimum value of K in a loading cycle, respectively.
- The *maximum stress intensity factor* at which the crack under fatigue becomes unstable $K_{max,c}$. The higher $K_{max,c}$ the more crack resistant is the material.
- The parameter m expressing the growth rate of crack length a over the number of loading cycles N, according to the familiar Paris' law equation (4.2) on the rate of growth of a crack in fatigue, in which m and A are constants depending on material, environment, load frequency, temperature, and stress ratio of maximum stress/minimum stress:

$$da/dN = A(\Delta K)^m \qquad (4.2)$$

The results by Salazar et al. (2014) grouped in Table 4.9 demonstrated that Duraform® EX was competitive with Duraform® PA not only in static properties at multiple temperatures, but also fatigue properties. Comparing to the latter, the former featured higher values of $K_{max,c}$ in all test settings, and half average m across test conditions, meaning that Duraform® EX was more resistant to crack upon static loading than Duraform® PA, and, upon fatigue loading, cracks propagated more slowly in it. Across all test conditions the average K_{Ic} of Duraform® EX and Duraform® PA was 3.8 MPa√m and 2.4 MPa√m, respectively. As a benchmark, K_{Ic} of conventional PA 6,6 is 2.5–3 MPa√m (Ku et al. 2005).

4.8 PA 11 NANOCOMPOSITES

The main limiting factor in processing polymer composites via PBF is that obviously only powder composites are suitable for the process, and matrix and filler must be adequately compounded prior to printing (Karevan et al. 2013). Current methods to prepare polymer composite powders for PBF are:

- *Mechanical mixing*: simple and convenient process but often penalized by aggregation of nanofillers preventing their uniform distribution (Yuan et al. 2016).
- *Dissolution-precipitation*: consists in coating the fillers with the base polymer (Zhu et al. 2016), but it requires special solvents for different polymers and consumes many solvents.
- *Two-step route,* comprising *melt blending* and *cryogenic fracture techniques*: it is potentially very effective in large-scale and fast production of composite powders (Goodridge et al. 2011), but melt blending does not yield good filler dispersion and property improvement (Wakabayashi et al. 2010).

Novel PA 11-based composite powders for PBF will be more likely commercially successful if they are made following one of the previously listed processes, instead of spending resources to devise new methods.

We recall that *polymer nanocomposites* are a mixture of two or more materials, in which the matrix is a polymer and the dispersed phase (*filler*) has at least one dimension smaller than 100 nm. Since nanoscale fillers greatly exceed fillers of conventional size in surface area, the interfacial area between matrix and nanofiller is enormous, which (even at marginal amounts of nanofiller uniformly distributed) significantly enhances mechanical strength and stiffness, impact strength, thermal and electrical conductivity, barrier to moisture and gases, flame retardancy, etc. in comparison to conventional composites.

Nanocomposites also broaden the design space, because they enable to produce functional materials with desired properties for targeted applications. Popular examples of nanofillers are CNTs and graphene, nanoclays, halloysite, zeolite, sustainable nanocellulose (nanofibrils and nanocrystals), and nanoparticles of metal alloys.

PA 11 has been investigated as a composite material reinforced by oxides: Al_2O_3, SiO_2, and $BaTiO_3$ in form of nanoparticles. This is also interesting because, through these composites, sustainable PA 11 supports a state-of-the-art technology such as nanotechnology, confirming the role of AM as a leading-edge technology.

Bai et al. (2017) addressed toughening of a commercial sustainable PA 11 by adding multi-walled CNTs, and processed the resulting composite via PBF, building test coupons that featured 54% higher fracture toughness vs. that of unfilled PA 11. The authors combined EOS PA 1101 as matrix, and NanoAmor Materials (USA) multi-walled CNTs featuring 20–30 nm diameter and 10–30 μm length, and followed a phase separation latex method, consisting in coating the surface of the PA 1101 particles with a solvent containing 0.2 wt% CNTs that was evaporated afterwards by drying the PA 1101 powder. The Izod impact strength (ASTM D256) of PA 1101 and PA 1101/CNT was 6.2 kJ/m^2 and 9.5 kJ/m^2, respectively. SEM analysis let to conclude that the presence of CNTs deflected the propagating crack and caused an increment of fracture surface area vs. that of PA 1101, which was associated to increased energy absorption.

Sisti et al. (2018) formulated, via in situ polymerization, PA 11/graphene nanocomposites with varying filler concentrations (0.25, 0.5, 0.75, 1.5, 3 wt%), and investigated the effect of graphene on the molecular, morphological, thermal, and dynamic mechanical properties of the composite. *In situ* polymerization offers the advantage of avoiding (a) melt compounding powders and their possible agglomeration, and (b) casting methods relying on large quantities of solvent. In comparison to unfilled PA 11, adding 0.75 wt% graphene increased the storage modulus consistently from 25°C across the T_g up to 150°C. Graphene behaved as a nucleation agent during crystallization, and raised thermal stability.

Ong et al. (2015) focused on electrical applications of PA 11 nanocomposites for PBF, and formulated and tested Rilsan® PA11 combined with 3 wt% of the multi-walled carbon nanotubes (MWNT) Baytubes® C 150 P by Covestro (Germany). In comparison to PA 11, MWNT increased the electrical conductivity of the nanocomposite but lowered its tensile strength and modulus. The test coupons were printed by varying the printer's area energy density (defined in Section 4.6) in the range 0.026–0.030 J/mm^2. Across that range, the density, tensile strength and modulus of PA

TABLE 4.10

Test Results of PA 11-Based Composites for PBF: Coupons Injection Molded

Material	Tensile Strength	Tensile Modulus	Elongation at Break
	MPa	MPa	%
PA	48.5	1,380	164
PA/FR	41.6	1,870	6.3
PA/FR/K	27–38	1,140–1,630	9.4–40.1
PA/FR/K/NC	34–37	1,920–2,460	3–8
PA/FR/K/NC/MWNT	33	1,460–1,716	17–30

Source: Ortiz, R. 2016. Fire Retardant Polyamide 11 Nanocomposites/Elastomer Blends for Selective Laser Sintering. Master diss., University of Texas at Austin. https://repositories.lib.utexas.edu/handle/2152/39451.

Legend: PA, Rilsan® PA 11; FR, Exolit® OP1312 fire retardant additive; K, Kraton FG1901 G; NC, Cloisite® 30B nanoclay; MWNT, Baytubes C 70 P multi-walled nanotubes.

11/MWNT reached 0.98–1.01 g/cm^3, 35–37 MPa and 9.4–1.05 GPa, respectively, whereas tensile strength and modulus of unfilled Rilsan® PA 11 were 47 MPa and 1.4 GPa, respectively. Electrical conductivity was calculated as the inverse of resistivity that was 10^5 Ohm-cm, far lower than 10^{11} Ohm-cm, an upper limit below which a material can be safely utilized from an electrostatic discharge (ESD) standpoint. The measured resistivity of 10^5 for PA 11/MWNT reveals that the conductivity of the samples was far above the necessary minimum prescribed to be electrically conductive. In comparison, unfilled PA 11 featured resistivity of 10^{14} and higher, and would have not made the ESD cutoff. Details on this PA 11/MWNT composite are in Koo et al. (2017).

Ortiz (2016) prepared formulations of nanocomposites based on Rilsan® PA11 intended for PBF, and tested coupons not printed but injection molded. The other ingredients were: Exolit® OP1312 as a fire retardant additive (abbreviated here as FR), Kraton FG1901 G as a very flexible PS-butylene-ethylene polymer (K), Cloisite® 30B as a nanoclay (NC), and MWNT Baytubes® C 70 P as multi-walled nanotubes. The tested formulations differed in type and amount of ingredients, and they are listed in Table 4.10, along with the tensile test results for all investigated formulations. The nanocomposites trailed PA11 in tensile strength but surpassed it in tensile modulus. Lower strength and higher modulus were accompanied by inferior elongation at break. Since the content of nanoclay (5–7 wt%) and MWNT (2.5–3.5 wt%) and the other ingredients was varied at the same time, it was impossible to single out the effect of nanofillers on tensile and the other properties.

Chung and Das (2008) selected nano-size fumed silica (SiO_2) to reinforce sustainable Rilsan® D80 (Arkema) PA 11, and with the ensuing composite printed via LS demonstration items with functionally graded mechanical properties, and improved tensile and compressive moduli vs. unfilled PA 11. D80 particles had 90–150 μm size, and SiO_2 nanoparticles featured 15 nm particle size, a very large surface area of 395 m^2/g, and density of 0.04 g/cm^3 (Sigma-Aldrich 2021). D80 powder was mixed with 2, 4, 6, and 10 wt% of SiO_2. Employing DOE, printing conditions for LS were optimized to maximize density, dimensionally accuracy, and ease to remove the printed part from its support powder. The optimized printing parameters were:

- Part substrate temperature set point (184°C), roller speed (0.076 m/s), and laser power (4.5 W) for 0, 2, 4, 6, and 10 wt%
- Scan speed: 1.26 m/s for 0 and 2 wt%, 0.90 m/s for 4 and 6 wt%, and 0.64 m/s for 10 wt%.

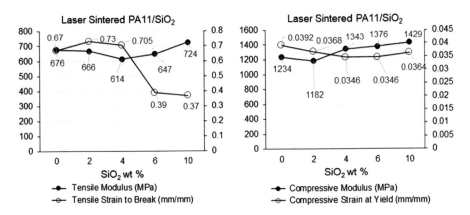

FIGURE 4.11 Properties of experimental PA 11/SiO₂ powder for PBF as a function of filler content.

Source: Modified from Chung, H., Das, S. 2008. Functionally graded Nylon-11/silica nanocomposites produced by selective laser sintering. *Materials Science and Engineering A* 487:251–257. Reproduced with permission from Elsevier.

The SiO_2 particles were uniformly distributed in the composites. Their porosity depended on the composition, and was 0.2–1%, values extremely low and far lower than typical values for experimental polymer composites for AM. Test results from coupons printed with optimized settings are displayed in Figure 4.11. Tensile and compressive moduli increased only at high filler content, and improved at the most by 7% and 16%, respectively. In both cases a rise in stiffness was accompanied by a reduction in ductility, which is typical in polymers. Compressive modulus was about twice the tensile modulus. As a benchmark, the powder PA D80-ST (Advanced Laser Materials, USA) made of PA 11 features tensile modulus and elongation at break equal to 1,392 MPa and 38%, respectively on coupons printed flatwise, outperforming *D80*/SiO₂ in these properties. The potential benefit of incorporating a stiff and strong filler such as SiO_2 in PA 11 includes performance stability over time, and enhancement of long-term properties: creep and stress relaxation. The authors also laser sintered two functionally graded components (Figure 4.12) out of D80 PA 11/SiO₂ by gradually changing composition from 0 to 10 vol% filler across the components' geometry, and accordingly varying the printing settings. One component was a compliant gripper that can serve as a robot hand, and the other was a rotator cuff scaffold; both were harder and stiffer at one end (10 vol% SiO_2), and soft and compliant at the opposite end (0 vol% SiO_2).

Qi et al. (2017) described a new and effective route to prepare a PA 11-based nanocomposite with improved mechanical and electrical properties, and a wider processing window vs. unfilled PA 11. The selected ingredients were:

- Pellets of Rilsan® PA11, chosen for its good piezoelectric properties.
- $BaTiO_3$ particles (Shandong Sinocera Functional Material, China): 500 nm average particle size, aspect ratio of 1, and 6.08 g/cm³ density, and preferred because lead-free and possessing high piezoelectric properties and dielectric permittivity.
- Fumed silica powder (Shanghai Aladdin Bio-Chem Technology, China), with particle size below 10 nm, intended to improve the flowability of the composite powder during sintering, and obtain homogenous deposition on the printer's platform.

The authors devised an original procedure to prepare the composite powder for LS consisting in the following main phases, listed in chronological order:

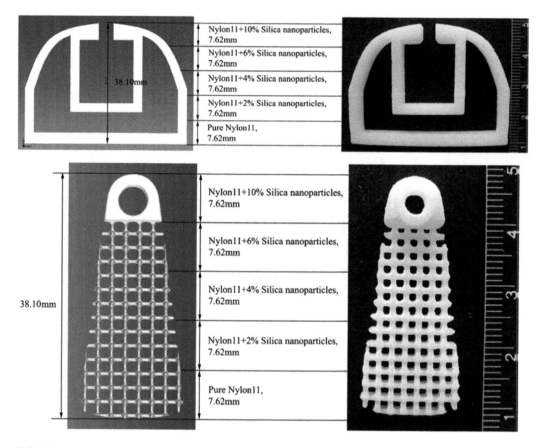

FIGURE 4.12 Two articles functionally graded, printed via PBF out of PA 11/SiO$_2$ powder. Upper pictures, gripper; lower pictures, rotator cuff scaffold. Left picture: model. Right picture: printed article. Note the change in composition across the volume of each item.

Source: Chung, H., Das, S. 2008. Functionally graded Nylon-11/silica nanocomposites produced by selective laser sintering. *Materials Science and Engineering A* 487:251–257. Reproduced with permission from Elsevier.

a. Solid state shear milling: a key operation because it achieved pulverization, dispersion, mixing, mechanical activation, and uniform distribution of the filler in the PA matrix.
b. Blend melting: it was performed in extruder, and produced granulates that underwent the next phase.
c. Cryogenic grinding, drying, and sieving, to obtain the final PA11/BaTiO$_3$ powder.

Test coupons were prepared with the following compositions and relative IDs:

- Unfilled PA11
- BT40: PA11/40 wt% (10 vol%) BaTiO$_3$
- BT60: PA11/60 wt% (20 vol%) BaTiO$_3$
- BT80: PA11/80 wt% (40 vol%) BaTiO$_3$.

Composite coupons incorporating the same amounts of filler as the above composites were also prepared through melt blending and cryogenic grinding but without solid state shear milling, and were labeled: UBT40, UBT60, and UBT80. These coupons served as benchmarks for the BT

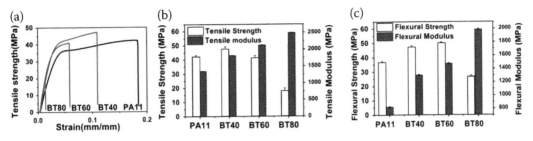

FIGURE 4.13 Mechanical test results of PA 11/BaTiO$_3$ nanocomposite powder for laser sintering.

Source: Qi, F., Chen, N., Wang, Q. 2017. Preparation of PA11/BaTiO$_3$ nanocomposite powders with improved processability, dielectric and piezoelectric properties for use in selective laser sintering. *Materials & Design* 131:135–143. Reproduced with permission from Elsevier.

composites. The measured tensile and flexural properties are summarized in Figure 4.13. These composites outperformed PA11 in tensile and flexural moduli that displayed a steady, almost linear increase from 40 wt% to 80 wt%. The composites surpassed PA11 in tensile strength at 40 wt%, and in flexural strength at 40 wt% and 60 wt%. The stress-strain curves of PA11 and filled PA11 nanocomposite coupons in Figure 4.13 (left) displayed, as expected, that, being BaTiO$_3$ more rigid than PA11, the ductility of the composite material was impaired by BaTiO$_3$, and its elongation at break progressively declined with raising amounts of filler. Solid state shear milling allowed homogeneous dispersion of BaTiO$_3$ and filler–matrix interfacial compatibility that in turn increased the dielectric constant in form of *relative permittivity* (ε_r), *piezoelectric strain coefficient* (d_{33}), and *piezoelectric voltage coefficient* (g_{33}) values that in BT80 reached a maximum of about 5, 50, and 30 times, respectively those in unfilled PA11. The values of ε_r, d_{33}, and g_{33} in any BT composite exceeded those in any UBT composite. The higher d_{33} and g_{33}, the more efficient is the mechanical-to-electrical energy conversion process that is required to meet the demand for self-powering capability of robotic, wireless electronic, and implantable devices (Choi et al. 2013).

Warnakula and Singamneni (2017) formulated and evaluated a composite material for PBF made of an unspecified PA 11 filled with nanoscale alumina (Al$_2$O$_3$). The authors focused on optimizing laser power and scanning speed to laser sinter PA 11/Al$_2$O$_3$ samples, in order to reduce porosity, and increase tensile strength and elongation at break vs. the unfilled PA 11. In sintered coupons of unfilled and composite PA 11 the tensile strength (measured on thin films) ranged from 0.03 MPa to 0.10 MPa, depending on filler content (3, 5, 10 wt%) and values of laser power and scanning speed. This strength range was three orders of magnitude lower than 39–66 MPa, the range for commercial PA 11 grades in Table 4.1. Such low strength was due to the porosity of 23–50% but probably also to absent or poor matrix-filler interfacial adhesion, and agglomeration of Al$_2$O$_3$ particles that impeded efficient load transmission between matrix and filler. The measured porosity was about one order of magnitude higher than the 3–5% measured on laser sintered parts out of PA 12 (Pavan et al. 2018). A positive result was that greater values of laser power and scanning speed noticeably improved strength (due also to lower porosity) that increased across filler content up to almost five times in case of 10 wt% filler. This finding was consistent with results by Wegner et al. (2015).

4.9 EXPERIMENTAL BLENDS OF PA 11

This section describes some experimental, general-purpose blends of PA 11 and a biobased ingredient that may be considered suited to be converted into fully biobased feedstocks for AM.

PA 11 and PLA are immiscible and incompatible, but Walha et al. (2016) combined the epoxy chain extender Joncryl ADR®-4368 with BESNO P40 TL PA 11 and Ingeo™ 2003D PLA, and followed two alternative routes based on the mixing sequence chosen, and made them compatible. In the first approach, all components were mixed in the extruder simultaneously. The resulting blend displayed 1.3% and 12 time increase in tensile modulus and elongation at break, respectively vs. the same blend without Joncryl. The second route consisted in premixing PLA and Joncryl in the extruder until reaching an equilibrium state, and then adding PA 11, followed by further mixing. The subsequent blend featured 8.1% lower tensile modulus but 17 times elongation at break vs. the same blend without Joncryl.

Di Lorenzo et al. (2019) prepared by melt mixing and analyzed PA 11/PBS blends with wt% compositions of 90/10, 80/20, and 60/40 with PBS the lesser ingredient. PBS was selected because it is biobased, biodegradable, and compostable, and its production cost is half of that of PA 11. Rilsan® BESNO TL PA 11, and Bionolle 1001MD PBS were selected. All blend formulations were immiscible (hence they displayed two values of T_g), and PA 11 and PBS were arranged in a particle/matrix morphology featuring roughly spherical PBS particles suspended in the PA 11 matrix, with many small voids surrounding the PBS inclusions. The blends lagged behind unfilled PA 11 in tensile strength and strain (both yield and ultimate) and modulus, and their highest property values were displayed by the 90/10 formulation. Some PA 11/PBS interfacial adhesion, and interaction between amide groups in PA 11 and carbonyl groups in PBS were detected. Adding a compatibilizer to increase interfacial adhesion and further promoting reaction between the functional groups of the two polymers may likely produce a blend with enhanced mechanical performance.

Oliver-Ortega et al. (2016) evaluated composites of Rilsan® PA11 reinforced from 20 wt% to 60 wt% of lignocellulosic fibers from softwood pulp called *stone groundwood* (SGW). About 90% of fibers had lenght of 37-240 μm, a range compatible with the length of AM filaments but not the diameter of all AM powders. SGW successfully reinforced PA11, and all composite formulations outperformed PA11 in tensile strength, with the highest value of 63.9 MPa (vs. 38.3 MPa for PA11) reached at 50 wt% fiber content, but also trailed PA11 in elongation at break, which was 2.8–9.5% vs. 25% for PA11. Advantages of SGW/PA11 were good fiber dispersion, no voids and hence extensive fiber wetting, and good quality interface enabling stress transmission from the matrix to the fibers. The storage modulus of SGW/PA11 markedly exceeded that of PA 11 at all temperatures from −40°C to 120°C, f.e. reaching 3.7 GPa and 1 GPa, respectively at 20°C, and 2.1 GPa and 0.7 GPa, respectively at 50°C.

REFERENCES

360 Research Reports. 2020. *Global Castor Oil Market Research Report 2020.* https://www.360researchreports.com/global-castor-oil-market-15041326 (accessed April 3, 2020).

ASTM D256. 2018. *Standard Test Methods for Determining the Izod Pendulum Impact Resistance of Plastics.* West Conshohocken, PA: ASTM International.

ASTM D638. 2018. *Standard Test Method for Tensile Properties of Plastics.* West Conshohocken, PA: ASTM International.

ASTM E1823 Standard Terminology Relating to Fatigue and Fracture Testing.

ASTM E647 Standard Test Method for Measurement of Fatigue Crack Growth Rates.

Bai, J., Yuan, S., Shen, F. et al. 2017. Toughening of polyamide 11 with carbon nanotubes for additive manufacturing. *Virt. Phys. Prototyp.* 12 (3): 235–240.

Bechthold, I., Bretz, K., Kabasci, S. et al. 2008. Succinic acid: a new platform chemical for biobased polymers from renewable resources. *Chem. Eng. Technol.* 31 (5): 647–654.

Benedetti, L., Brulé, B., Decreamer, N., Evans, K. E., Ghita, O. 2019. Shrinkage behaviour of semi-crystalline polymers in laser sintering: PEKK and PA12. *Mater. Des.* 181:107906.

Brehmer, B. 2014. Polyamides from biomass derived polymer. In *Bio-Based Plastics*, ed. S. Kabasci, 274–293. Chichester: John Wiley and Sons.

Caulfield, B., McHugh, P. E., Lohfeld, S. 2007. Dependence of mechanical properties of polyamide components on build parameters in the SLS process. *J. Mater. Proces. Technol.* 182 (1–3): 477–488.

Choi Y. J., Yoo M.-J., Kang H.-W., et al. 2013. Dielectric and piezoelectric properties of ceramic-polymer composites with 0–3 connectivity type. *J. Electroceram.* 30 (1–2): 30–35.

Chung, H., Das, S. 2008. Functionally graded Nylon-11/silica nanocomposites produced by selective laser sintering. *Mater. Sci. Eng. A* 487:251–257.

Dasari, S. R., Goud, V. V. 2013. Comparative extraction of castor seed oil using polar and non-polar solvents. *IJCET Spec. Issue* 1:121–123.

Devaux, J., Lê, G., Pees, B. 2011. Application of Eco-Profile Methodology to Polyamide 11.

Di Lorenzo, M. L., Longo, A., Androsch, R. 2019. Polyamide 11/Poly(butylene succinate) Bio-based polymer blends. *Materials* 12:2833.

Expert Market Research, n.d. *Global Castor Oil Market.* https://www.expertmarketresearch.com/reports/castor-oil-market (accessed October 22, 2020).

Fior Markets. 2020. *Global Castor Oil Market.* https://www.globenewswire.com/news-release/2020/02/19/1986765/0/en/Global-Castor-Oil-Market-is-Expected-to-Reach-USD-1-72-Billion-by-2025-Fior-Markets.html (accessed October 22, 2020).

Global Information Inc. 2021. *Global 3D PA (Polyamide) Market Size Study.* https://www.giiresearch.com/report/bzc925884-global-3d-pa-polyamide-market-size-study-by-type.html (accessed January 16, 2021).

González, V., Guerrero, C., Ortiz, U. 2000. Chemical structure and compatibility of polyamide–chitin and chitosan blends. *J. Appl. Polym. Sci.* 78 (4): 850–857.

Goodridge, R., Shofner, M., Hague, R. et al. 2011. Processing of a polyamide-12/carbon nanofibre composite by laser sintering. *Polym. Test.* 30 (1): 94–100.

Herrero, M., Peng, F., Nunez Carrero, K., Merino, J. C., Vogt, B. D. 2018. Renewable nanocomposites for additive manufacturing using fused filament fabrication. *ACS Sustain. Chem. Eng.* 6:12393–12402.

Hofland, E. C., Baran, I., Wismeijer, D. A. 2017. Correlation of process parameters with mechanical properties of laser sintered PA12 parts. *Adv. Mater. Sci. Eng.* Article ID 4953173. doi:10.1155/2017/4953173.

ISO 527. *1B Plastics – Determination of tensile properties — Part 1: General principles.* Geneva, Switzerland: ISO.

Jiang, Y., Loos, K. 2016. Enzymatic synthesis of biobased polyesters and polyamides. *Polymers* 8:243.

Kabasci, S., Bretz, I. 2012. Succinic acid: synthesis of biobased polymers from renewable resources. In *Renewable Polymers*, ed. V. Mittal, 355–379. Hoboken: John Wiley & Sons, Scrivener Publishing.

Karevan, M., Eshraghi, S., Gerhardt, R., Das, S., Kalaitzidou, K. 2013. Effect of processing method on the properties of multifunctional exfoliated graphite nanoplatelets/polyamide12 composites. *Carbon* 64:122–131.

Kauppila, I. 2021. *Multi Jet Fusion (MJF) 3D Printing – Simply Explained.* https://all3dp.com/1/multi-jet-fusion-mjf-3d-printing-simply-explained/ (accessed April 26, 2021).

Kazeem, O., Taiwo, O., Kazeem, A. 2014. Determination of some physical properties of castor (Ricinus communis) oil. *Int J Sci Eng Technol.* 3 (12): 1503–1508.

Könst, P. M., Franssen, M. C. R., Scott, E. L., Sanders, J. 2011. Stabilization and immobilization of Trypanosoma bruceiornithine decarboxylase for the biobased production of 1,4-diaminobutane. *Green Chem.* 13:1167–1174.

Koo, J. H., Ortiz, R., Ong, B., Wu, H. 2017. Polymer nanocomposites for laser additive manufacturing. In *Laser Additive Manufacturing – Materials, Design, Technologies, and Applications*, ed. M. Brandt, 205–235. Sawston: Woodhead Publishing.

Ku, H. S., Baddeley, D., Snook, C., Chew, C. S. 2005. Fracture toughness of vinyl ester composites cured by microwave irradiation: preliminary results. *J. Reinf. Plast. Compos.* 24 (11): 1181–1201.

Mubofu, E. B. 2016. Castor oil as a potential renewable resource for the production of functional materials. *Sustain. Chem. Proc.* 4:11.

Mutlu, H., Meier, M. A. R. 2010. Castor oil as a renewable resource for the chemical industry. *Eur. J. Lipid Sci. Technol.* 112 (1): 10–30.

Nelson, J. C. 1993. *Selective Laser Sintering: A Definition of the Process and an Empirical Sintering Model.* PhD diss., University of Texas at Austin.

Oliver-Ortega, H., Oliver-Ortega, L.A., Granda, F. X. et al. 2016. Tensile properties and micromechanical analysis of stone groundwood from softwood reinforced bio-based polyamide11 composites, *Compos. Sci. Technol.* 132:123–130.

Ong, B., Wu, H., Ortiz, R., Yao, E., Koo, J. H. 2015. Electrically conductive polyamide 11 nanocomposites for selective laser sintering: properties characterization. *56th AIAA/ASCE/AHS/ASC Structures, Structural Dynamics, and Materials Conference*, Kissimmee, Florida.

Ortiz, R. 2016. *Fire Retardant Polyamide 11 Nanocomposites/Elastomer Blends for Selective Laser Sintering.* Master diss., University of Texas at Austin. https://repositories.lib.utexas.edu/handle/2152/39451.

Patel, V. R., Dumancas, G. G., Lakshmi, C., Viswanath, K. et al. 2016. Castor oil: properties, uses, and optimization of processing parameters in commercial production. *Lipids Insights* 9:1–12.

Pavan, M., Craeghs, T., Kruth, J.-P., Dewulf, W. 2018. Investigating the influence of X-ray CT parameters on porosity measurement of laser sintered PA12 parts using a design-of-experiment approach. *Polym. Test.* 66:203–212.

PubChem. n.d. *Castor Oil.* https://pubchem.ncbi.nlm.nih.gov/compound/castor_oil (accessed November 15, 2020).

Qi, F., Chen, N., Wang, Q. 2017. Preparation of PA11/BaTiO$_3$ nanocomposite powders with improved processability, dielectric and piezoelectric properties for use in selective laser sintering. *Mater. Des.* 131:135–143.

Rasselet, D., Caro-Bretelle, A.-S., Taguet, A., Lopez-Cuesta, J.-M. 2019. Reactive compatibilization of PLA/PA11 blends and their application in additive manufacturing. *Materials* 12:485.

Salazar, A., Rico A., Rodríguez J., et al. 2014. Monotonic loading and fatigue response of a bio-based polyamide PA11 and a petrol-based polyamide PA12 manufactured by selective laser sintering. *Eur. Polym. J.* 59:36–45.

Scherer, B., Kottenstedde, I. L., Bremser, W., Matysik, F.-M. 2020. Analytical characterization of polyamide 11 used in the context of selective laser sintering: Physico-chemical correlations. *Polymer Testing* 91:106786.

Schmid, M. 2018. *Laser Sintering with Plastics Technology – Processes, and Materials.* Munich: Hanser Verlag.

Severino, L. S., Auld, D. L., Baldanzi, M. 2012. A review on the challenges for increased production of castor. *Agron J.* 104 (4): 853.

Sigma-Aldrich. 2021. *Silica, Fumed.* https://www.sigmaaldrich.com/catalog/product/aldrich/s5130?lang= en&region=US&cm_sp=Insite-_-caSrpResults_srpRecs_srpModel_fumed%20silica-_-srpRecs3-1 (accessed May 28, 2021).

Sisti, L., Totaro, G., Vannini, M. et al. 2018. Bio-based PA11/graphene nanocomposites prepared by in situ polymerization. *J. Nanosci. Nanotechno.* 18 (2): 1169–1175.

Stratasys. 2021. *Materials.* https://www.3dsystems.com/materials (accessed October 22, 2020).

The Chemical Company. 2019. *Castor Oils.* https://thechemco.com/chemical/castor-oils/ (accessed May 31, 2020).

Verbelen, L., Dadbakhsh, S., Van den Eynde, M. et al. 2016. Characterization of polyamide powders for determination of laser sintering processability. *Eur. Polym. J.* 75:163–174.

Wakabayashi, K., Brunner, P. J., Masuda, J. I., Hewlett, S. A., Torkelson, J. M. 2010. Polypropylene-graphite nanocomposites made by solid-state shear pulverization: effects of significantly exfoliated, unmodified graphite content on physical, mechanical and electrical properties. *Polymer* 51 (23): 5525–5531.

Walha, F., Lamnawar, K., Maazouz, A., Jaziri, M. 2016. Rheological, morphological and mechanical studies of sustainably sourced polymer blends based on poly(lactic acid) and polyamide 11. *Polymers* 8 (3): 61.

Wang, Y., DiNapoli, C. M., Tofig, G. A. et al. 2017. Selective laser sintering processing behavior of polyamide powders. *SPE ANTEC® Anaheim* 2017:112–116.

Warnakula, A., Singamneni, S. 2017. Selective laser sintering of nano Al$_2$O$_3$ infused polyamide. *Materials* 10:864.

Wegner, A., Harder, R., Witt, G., Drummer, D. 2015. Determination of optimal processing conditions for the production of polyamide 11 parts using the laser sintering process. *Int. J. Recent Contrib. Eng. Sci. IT* 3 (1): 5–12. doi:10.3991/ijes.v3i1.4249.

Williams, J. G. 2001. K$_C$ and GC at slow speeds for polymers. In *Fracture Mechanics Testing Methods for Polymers, Adhesives and Composites*, ed. D. R. Moore, A. Pavan, J. G. Williams, 11–24. The Netherlands: Elsevier Science Ltd. and ESIS.

Winnacker, M., Rieger, B. 2016. Biobased polyamides: recent advances basic and applied research. *Macromol. Rapid Commun.* 37:1391–1413.

Yuan, S., Bai, J., Chua, C. K., Wei J., Zhou K. 2016. Material evaluation and process optimization of CNT-coated polymer powders for selective laser sintering, *Polymer* 8 (10): 370.

Zhu, W., Yan, C., Shi, Y. et al. 2016. A novel method based on selective laser sintering for preparing high-performance carbon fibres/polyamide12/epoxy ternary composites, *Sci. Rep.* 6. doi:10.1038/srep33780.

5 Polyhydroxyalkanoates

Environment is no one's property to destroy; it's everyone's responsibility to protect.

Mohith Agadi

5.1 OVERVIEW OF POLYHYDROXYALKANOATES (PHAs)

PHAs are linear aliphatic biopolyesters produced by microbes fermenting carbon-rich feedstocks. PHAs form a family of polyesters produced in nature by more than one hundred types of bacteria (Koller et al. 2014) as water-insoluble intracellular inclusions shaped like granules, and serving as a reserve of energy and carbon (Braunegg et al. 1998; Chiulan et al. 2018). Figure 5.1 is a micrograph of *Ralstona eutropha*, a bacterium that under proper conditions generates large amounts of a type of PHA, called poly(3-hydroxybutyrate) (PHB). PHA is produced during fermentation by introducing a culture of specific bacteria in a suitable medium (*substrate*), and feeding them canola oil, cassava, crude palm kernel oil, glucose, methane emissions, plant oil, saccharose, sugar, sugar beets, etc. (Chiulan et al. 2018). Among biobased plastics, PHA is the only biopolymer entirely produced, consumed, and degraded by living organisms (bacteria), hence its ideal life cycle is a closed-loop (Liu et al. 2014) that makes PHA sustainable, because it is produced from biomass and returns to it after being discarded. PHA is insoluble in water, soluble in chloroform and other chlorinated hydrocarbons, relatively resistant to hydrolytic degradation, resistant to ultraviolet light, poorly resistant to acids and bases, biocompatible, and nontoxic thereby suitable for biomedical and food packaging applications (Daminabo et al. 2020), biodegradable in aerobic and anaerobic environments, and compostable (Liu et al. 2014). PHA also feature low permeability towards oxygen and CO_2 (Doherty et al. 2011).

Members of the PHA family exist as homopolymers of hydroxyalkanoic acids (HAs), and copolymers of two or more HAs. PHAs mainly consist of monomeric building blocks of 3-hydroxyalkanoates (3HAs). Until 2014, more than 150 monomers for synthesis of PHA were discovered, but only a few PHAs are commercially available, mainly polyhydroxybutyrate or poly (3-hydroxybutyrate) (PHB or P3HB), poly(3-hydroxybutyrate-co-3-hydroxyvalerate) (PHBV) and poly(3-hydroxybutyrate-co-3-hydroxyhexanoate) (PHBH) that are produced on an industrial scale (Chanprateep 2010). The tensile property values of commercial PHAs and PHBs are listed in Tables 5.1 and 5.2, respectively. PHAs have some material properties similar to those of widespread commodity polymers PE and PP (Holmes 1985). Details on synthesis and structure of PHAs are described by Sudesh et al. (2000).

Physical, tensile, and other properties of commercial PHAs were measured by Corre et al. (2011), and listed in Table 5.3, with tensile values measured on dumbbell-shaped coupons injection molded. Figure 5.2 depicts the chemical structures of PLA, PHB, and polyhydroxyvalerate (PHV).

Typically PHA production comprises the following steps carried out in sequential order: microbial fermentation, separation of microbial biomass (via centrifugation, flocculation, filtration, or sedimentation) from the fermented "broth," biomass drying, and PHA extraction, drying, and packaging (Chen 2010). The process of isolating PHA from the microbial biomass has a critical cost, especially in large-scale productions, and is achieved through three optional routes (Koller et al. 2014). Worldwide suppliers of PHA include Biomer (Germany), Danimer Scientific (USA), Kaneka (Japan), PHB Industrial (Brasil), Shenzhen Ecomann Biotechnology, TianAn Biological

DOI: 10.1201/9781003221210-5

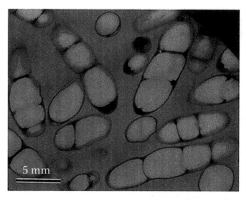

FIGURE 5.1 Micrograph of *Ralstona eutropha*, a bacterium producing a type of PHA, called poly(3-hydroxybutyrate (PHB).

Source: Sudesh, K., Abe, H., Doi, Y. 2000. Synthesis, structure and properties of polyhydroxyalkanoates: biological polyesters. *Prog. Polym. Sci.* 25:1503–1555. Reproduced with permission from Elsevier.

TABLE 5.1
Properties of Commercial PHAs

Property	Tensile Strength	Tensile Modulus	Elongation at Break
Unit	MPa	GPa	%
Value	15–40	1.1–3.5	1–6

Sources: Goonoo, N., Bhaw-Luximon, A., Passanha, P., Esteves, S. R., Jhurry, D. 2017. Third generation poly(hydroxyacid) composite scaffolds for tissue engineering. *J Biomed Mater Res Part B* 105B:1667–1684. Koller M., Salerno A., Braunegg G. 2014. Polyhydroxyalkanoate: Basics, Production and Applications of Microbial Polyesters. In *Bio-Based Plastics – Materials and Applications*, ed. Stephan Kabasci, 137–170. Chichester: John Wiley & Sons. Reproduced with permission from John Wiley & Sons.

TABLE 5.2
Properties of Commercial PHBs

Property	Density	T_m	Tensile Strength	Tensile Modulus	Elongation at Break	Charpy Notched Impact Strength	HDT at 1.8 MPa
Unit	kg/m^3	°C	MPa	MPa	%	kJ/m^2	°C
Value	1.2–1.3	140–174	15–29	800–3,500	6–15	2.4–6.2	50–59

Source: Chiulan I., Frone A. N., Brandabur C. et al., 2018. Recent Advances in 3D Printing of Aliphatic Polyester. *Bioengineering* 5 (1): 2. Data from Biocycle®, Metabolix (now CJ CheilJedang), and Biomer.

Materials, Tianjin GreenBio Materials Yield10 (the last three from China), etc. (Sabapathy et al. 2020). In 2016, the U.S. company Metabolix, a leader in PHA output and technology with several products (fibers for wovens and non-wovens, films, foams), sold its biopolymer intellectual property and PHA business to South Korea's company CJ CheilJedang. Applications of PHA are: (a) agricultural foils and films; (b) medical articles such as bone marrow scaffolds, bone plates, pins, screws, sutures, etc.; (c) single-use packaging for foods, flatware, drug delivery, beverages, consumer products, etc. (Creative Mechanisms 2017). In 2018, the PHA market was about 3% of the global market of biobased polymers equal to 8 million tonnes (Bioplastics Magazine 2020). Another indication about the growing market of PHA is that in 2018 Danimer Scientific announced

TABLE 5.3

Properties of Commercial PHAs

PHA: Name, Producer	Density	MW(p)[a]	MW(im)[b]	Tensile Stress at Break	Tensile Modulus	Elongation at Break	T_g	T_c	T_m
	g/cm^3	g/mol	g/mol	MPa	MPa	%	°C	°C	°C
Enmat Y1000P, Tianan Biologic	1.25	3.4×10^5	2.6×10^5	32.2	2,624	3.5	24.9	115.7	167.1
P226, Biomer	1.25	6.4×10^5	4.8×10^5	15.3	692	27.0	10.5	108.7	162.8

Source: Corre, Y.-M., Bruzaud, S., Audic, J.-L., Grohens, Y. 2011. Morphology and functional properties of commercial polyhydroxyalkanoates: A comprehensive and comparative study. *Polym. Test.* 31:226–235.

Notes:

[a]MW(p) is MW of pellets.

[b]MW(im) is MW after injection molding.

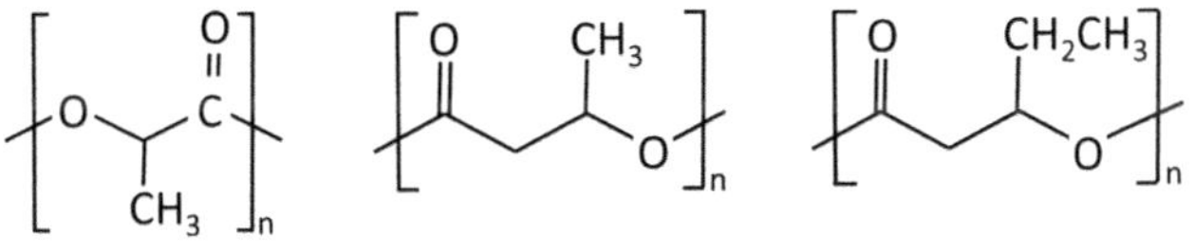

FIGURE 5.2 Chemical structure of PLA (left), PHB (center), and PHV (right).

Source: adapted from Chiulan, I., Frone, A. N., Brandabur, C. et al. 2018. Recent Advances in 3D Printing of Aliphatic Polyester. *Bioengineering* 5:2. Reproduced under open access Creative Common CC BY license.

that it would open in USA the world's first commercial production plant for PHA (Danimer Scientific 2021). The growth of PHA market is expected to benefit from the increasing availability of raw sustainable materials, environmentally friendly procurement policies, and growing demand and use of biodegradable polymers for biomedical, food, and packaging applications.

The chemical composition, main molecular chain, and number of carbon atoms present in the lateral chain of PHAs determines their physical-chemical properties, degradation rate in biological media, and retention of their mechanical strength from short to long periods of time (Goonoo et al. 2017). Depending on the number of carbon atoms present in their lateral chain, PHAs are distinguished in *short chain length* (SCL) *PHAs* (3–5 carbon atoms), and *medium chain length* (MCL) *PHAs* (6 to 14 carbon atoms). The differences in the properties and uses of the two types are shown in Table 5.4: SCL PHA is brittle and rigid (but can also be compliant and very flexible), whereas MCL PHA features low strength and elastomeric behavior. Both types are present in biomedical scaffolds and implants (Chiulan et al. 2018). PHB is well-known among SCL PHAs, but it is difficult to process due to its crystalline nature, hence 3-hydroxyvalerate (HV) molecules are incorporated in PHB, generating the copolymer PHBV.

PHA is produced via biosynthesis from renewable feedstocks such as alcohol, CO_2, edible oils, lactose, lipids, maltose, starch, sucrose, etc. derived in turn from edible feedstocks. Although the route based on natural feedstocks is attractive from a sustainable standpoint, it is not yet economically feasible for industrial products, especially bulk products, primarily due to the raw material cost that absorbs up to half of the production costs. While price of commodity polymers

TABLE 5.4

Properties and Uses of Short Chain Length (SCL) and Medium Chain Length (MCL) PHAs

PHA Type	T_g	T_m	MW	Tensile Strength	Elongation at Break	Degree of Crystallinity	Behavior	Uses[a]
	°C	°C	g/mol	MPa	%			
SCL	−40 to −30	60	$<10^5$	30–40	6,50–1,000[b]	Higher	Rigid and brittle. Elastomeric.	Mostly disposable items, food packaging materials
MCL	0	180	Up to 4×10^6	10	100s	Lower	Low strength. Elastomeric.	High value-added items: biodegradable matrices for drug delivery, surgical implants, and sutures.

Source: Data in Koller, M., Salerno, A., Braunegg, G. 2014. Polyhydroxyalkanoate: Basics, Production and Applications of Microbial Polyesters, chapter in *Bio-Based Plastics – Materials and Applications*, ed. Stephan Kabasci, 137–170. Chichester: John Wiley & Sons. Reproduced with permission from John Wiley & Sons.
Notes:
[a]Chiulan, I., Frone, A. N., Brandabur, C. et al. 2018. Recent Advances in 3D Printing of Aliphatic Polyester. *Bioengineering* 5 (1): 2.
[b]In elastomeric form.

such as PE and PP was around USD 1–2/kg, price of PHAs was estimated to be USD 5–6/kg (Chiulan et al. 2018), and this is obviously a tall obstacle to PHA diffusion. Sources of PHA feedstocks alternative to edible materials are agricultural waste and waste of industries related to agriculture (Solaiman et al. 2006). These sources offer the advantages of sparing food as a feedstock for polymers, lowering production cost, addressing waste disposal problem, and reducing economic dependence from fossil sources, which is beneficial especially in countries lacking the latter ones but possessing agricultural and industrial waste suitable for PHAs. Table 5.5 contains examples of carbon-rich waste streams that can generate raw materials for industrial production of PHA. Particularly, Amaro et al. (2019) described in details the prospects for utilizing whey for production of PHA. Whey is the main by-product of the dairy industry, and is obtained by precipitation and removal of milk casein when cheese is produced. About 120 million tons of whey are produced annually worldwide, of which only 50% go into products present in human and animal feed (Nikodinovic-Runic et al. 2013), and the remaining whey needs to be disposed of. Whey is mainly composed of lactose (39–60 g/L) that can be consumed to produce PHA, with the advantage of dramatically decreasing the cost of PHA production without competing with production of food for people and solving an environmental problem.

PHA is often added to PLA: it reduces its stiffness, increases its flexibility and impact strength, and imparts new properties. F.e. adding 20 wt% of amorphous PHA to PLA reduces PLA's flexural modulus from 3,500 MPa to 2,400 MPa but boosts its Izod notched impact strength from 27 kJ/m to 214 kJ/m (Andrews 2014). Moreover, increasing weight content of PHA from 5% to 35% in blown films amplified their tear strength (N/mm) in machine direction (MD) and transverse direction (TD) from 3.5 (MD) and 4.3 (TD) to 90.7 (MD) and 63.3 (TD).

Estimated at USD 62 million in 2020, the global PHA market size is projected to reach USD 121 million by 2025, and grow at a CAGR of 14.2% (MarketsandMarkets™ 2020).

TABLE 5.5

Waste Sources of Potential Feedstocks for PHA

Carbon Rich Waste Stream	Source
CH_4	Biogas production, landfills, treatment plants, waste water, wetlands
CO_2	Effluent gases
Crude glycerol phase	Biodiesel production
Lignocellulosic	Corn straw, fruit crops, fruit seed, rice straw, sugar bagasse
Molasses	Sugar industry
Starchy waste	Potato processing
Tallow	Rendering and slaughtering industry
Waste oils	Food-related industry and businesses
Whey	Dairy industry

Source: Adapted from Koller, M., Salerno, A., Braunegg, G. 2014. Polyhydroxyalkanoate: Basics, Production and Applications of Microbial Polyesters, in *Bio-Based Plastics – Materials and Applications*, ed. Stephan Kabasci, 137–170. Chichester: John Wiley & Sons. Reproduced with permission from John Wiley & Sons.

TABLE 5.6

Properties of PLA/PHA and PLA Economy Filaments by colorFabb for FFF

Material	Density	T_m	Tensile Strength	Tensile Modulus	Elongation at Break	Flexural Modulus	Flexural Stress at 3.5% Strain	Charpy Notched Impact Strength	Charpy Unnotched Impact Strength
	g/cm³	°C	MPa	MPa	%	MPa	MPa	kJ/m²	kJ/m²
PLA/PHA	1.24	>155	61.5	2,960	5.3	3,295	88.8	2.8	30.8
PLA Economy	1.24	N/A	45	3,400	6	N/A	N/A	7	N/A

Source: Data posted online by colorFabb.

5.2 COMMERCIAL PHAS FOR AM

The company colorFabb sells PLA/PHA, a filament tailored for a FFF, and made from a blend of the two polymers (Table 5.6). This filament is based on Bio-Flex® V 135001, a PLA produced by FkuR, and derived from natural raw materials and completely biodegradable. PLA/PHA contains 88 wt% of PLA, and its PHA contains some HV, and mostly 3-hydroxybutyrate (HB) (Gonzales Ausejo et al. 2018a). The storage modulus of PLA/PHA was measured at 1 Hz on dumbbell-shaped coupons printed upright and flatwise (Figure 3.15), and is plotted in Figure 5.3 (Gonzales Ausejo et al. 2018b): its values at 20°C were 4.20 GPa (flatwise) and 3.47 GPa (upright), and they plunged around the T_g that was 71.6°C and 77.7°C for flatwise and upright coupons, respectively. The storage modulus also indicated a consistent anisotropy in stiffness from below room temperature to 60°C. The greater rigidity of flatwise coupons is attributed to intralayer cohesion being stronger than interlayer bonding, as previously seen in polymeric coupons printed through FFF.

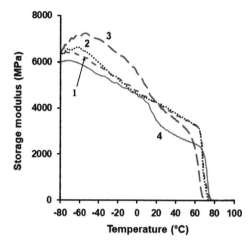

FIGURE 5.3 Storage modulus as a function of temperature for coupons printed in upright and flatwise orientations: line 1 PLA flatwise, line 2 PLA upright, line 3 PLA/PHA flatwise, line 4 PLA/PHA upright.

Source: Gonzales Ausejo, J., Rydz, J., Musiol, M., et al. 2018a. Three-dimensional printing of PLA and PLA/PHA dumbbell-shaped specimens of crisscross and transverse patterns as promising materials in emerging application areas: Prediction study. *Polym. Degrad. Stab.* 156:100–110. Reproduced with permission from Elsevier.

TABLE 5.7

Properties of DURA™ Filament for FFF

Property	Density	Tensile Yield Strength	Tensile Modulus	Elongation at Yield	Toughness	HDT at 0.46 MPa	Melt Flow Index
Unit	g/cm^3	MPa	MPa	%	J	°C	g/10 min
Value	1.29	29	2,754	18	0.68	120	12.5

Source: Data posted online by 3D Printlife.

3D Printlife sells DURA™, a PLA/PHA/PBS (PBS is biodegradable) filament for FFF formulated by ALGIX3D®, a U.S. company specializing in materials from algae feedstocks, and whose properties are reported in Table 5.7. We note that toughness is the energy absorbed by the material before breaking under impact and tension, and is typically measured in J/m^3, not in J. A significant comparison in terms of performance between DURA™ and PLA/PHA is impossible because density is the only property disclosed for both materials, and DURA™ contains PBS, absent in PLA/PHA. 3D Printlife also sells PLAyPHAb™, a FFF filament and a PLA upgrade featuring enhanced elongation at break (2.1%) and impact resistance (9.1 kJ/m^2 and 7.0 kJ/m^2, per ISO 179, depending on the test direction). When comparing properties of PLA/PHA filament with those of filament PLA Economy formulated in 100% PLA by colorFabb, the former is 13% less rigid in tension, and more brittle when notched, but it surpasses PLA Economy in tensile strength by 37%.

colorFabb also sells woodFill (Chapter 6), a filament for FFF based on a PLA/PHA blend and fully sustainable, since it contains 15 wt% of recycled wood from deciduous (shedding leaves) trees. Figure 5.4 illustrates the tensile stress-strain plot of coupons of as-received filaments (not printed test coupons) of woodFill: the strength appears more consistent than elongation at break and (unusual case) modulus, and two tested filaments showed a drop in stress below their strength values which might have been caused by slippage between test fixture grips.

Le Duigou et al. (2016) assessed the effect of printing orientation and immersion in deionized water at room temperature on tensile strength, modulus, and strain at break of coupons made of woodFill Fine filament which is the same material as woodFill. They printed coupons not complying with ISO 527 or

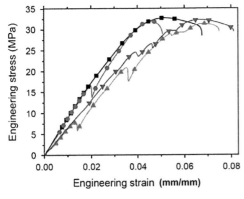

FIGURE 5.4 Tensile stress-strain curve of as-received four colorFabb woodFill filaments.

Source: Adapted from Guessasma, S., Belhabib, S., Hedi Nouri, H. 2019. Microstructure and Mechanical Performance of 3D Printed Wood-PLA/PHA Using Fused Deposition Modelling: Effect of Printing Temperature. *Polymers* 11:1778. Reproduced under open access Creative Common CC BY license.

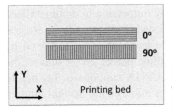

FIGURE 5.5 Schematic top view of printing bed with printed tensile coupons. Black lines denote edges of deposited filament. Drawing not in scale.

ASTM D638 but shaped as 70 × 10 × 3 mm strips. The coupons were printed at 0° and 90° (Figure 5.5), selecting as nozzle and printing bed temperatures 210°C and 70°C, respectively, printing speed 18 mm/s, printing pitch and width 0.322 mm and 0.367 mm, respectively. Table 5.8 shows that the coupons are significantly stiffer, stronger, and more brittle at 0° than at 90°, as expected, since in FFF parts the interlayer cohesion is weaker than the intralayer cohesion, and the number of weak areas increases with the number of printed layers perpendicular to the load direction. The immersion in

TABLE 5.8
Effect of Printing Direction and Water Immersion at Room Temperature on WoodFill Filament

Property	Test Standard	Tensile Modulus	Tensile Strength	Elongation at Break	Water Uptake after 35 Days (WU)	Tensile Modulus after WU	Tensile Strength after WU
Unit		MPa	MPa	%	%	MPa	MPa
Supplier data sheet[a]	ISO 527	3,290	46	5.5	N/A	N/A	N/A
Printed coupons[b], 0°	None	2,222	19.7	0.9	15.4	1,978	14.3
Printed coupons[b], 90°	None	1,667	9.1	2.3	17.3	1,258	7.0
0° value/90° value	None	1.33	2.16	0.39	0.89	1.57	2.04

Sources: [a]colorFabb's online data. [b]Data from plots in Le Duigou, A., Castro, M., Bevan, R., Martin, N. 2016. 3D printing of wood fibre biocomposites: From mechanical to actuation functionality. *Materials and Design* 96:106–114.

deionized water noticeably penalized tensile strength and modulus, and was more damaging in 90° than in 0° coupons, because the latter ones offered the water fewer paths to penetrate inside.

The supplier recommends 195–220°C as a printing temperature for woodFill. Guessasma et al. (2019) attempted at optimizing mechanical performance by varying printing temperature. Test coupons of woodFill were printed at temperatures from 210°C to 250°C, and only marginal improvements in tensile strength and modulus were recorded, with highest strength and modulus resulting from 230°C and 220°C, respectively. Across the printing temperature range, the tensile modulus and strength reached 416–453 MPa and 19.2–20.8 MPa, respectively.

Gonzales Ausejo et al. (2018a, 2018b) concluded that colorFabb PHA/PHA was suitable for applications requiring biocompatibility, after conducting an extensive experimental analysis. Sterilized and dried printed coupons out of PLA/PHA were placed in contact with two well-studied cell cultures: human embryonic kidney cells HEK293, and human lung fibroblasts (cells of animal connective tissue) WI-38. Both cell types grew and proliferated on the surface and inside the lattice of the coupons. The cell appearance showed healthy cultures, indicating that the blend was non-toxic. The hydrolytic degradation of PLA/PHA was also measured by immersion in demineralized water for 70 days at 50°C and 70 days at 70°C. Degradation manifested itself initially as erosion (started at day 42 at 50°C, and day 3 at 70°C, and was detectable as roughness), followed by disintegration. The amorphous phase degraded faster than the crystalline phase. The results of the hydrolytic degradation are illustrated in Figure 5.6 displaying PLA/PHA coupons in (a) before testing, and coupons in (b) after 70 days at 50°C. The greater degradation of the upright coupon (V) vs. that of flatwise coupon (H) was explained by the fact that the interface between adjacent layers acted as inward pathway for the water, and V coupons comprised more layers than H coupons. Yield strength, modulus, and strain at break were recorded in tension (ISO 527) on dumbbell-shaped coupons before immersion, and after immersion for 3 and 7 days at 50°C, and 3 days at 70°C. Compared to before immersion: (a) strength and modulus stayed about unchanged after 7 days at 50°C; after 3 days at 70°C strength decreased, while modulus slightly increased; (b) strain decreased after 7 days at 50°C and 3 days at 70°C, revealing greater brittleness.

Applications of articles printed out of PHAs are prototypes, visual models, casting, possibly implants and biodevices in direct contact with the human body (Mehrpouya et al. 2021), etc. Challenges in printing PHA are significant warpage and low strength. An unusual and ingenious use is biodegradable escape panels and locks for crab, fish, and lobster traps. Typically, each fishing season fishermen lose 10–70% of their total traps that remain active but are never hauled. Animals enter the lost traps, cannot escape, and die since the traps are not recovered. Dead animals

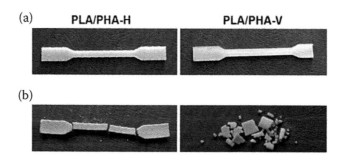

FIGURE 5.6 Coupons printed in flatwise (H) and upright (V) directions undergoing hydrolytic degradation test. (a) before test, (b) after immersed for 70 days at 50°C in demineralized water.

Source: Gonzales Ausejo, J., Rydz, J., Musiol, M. et al. 2018a. Three-dimensional printing of PLA and PLA/PHA dumbbell-shaped specimens of crisscross and transverse patterns as promising materials in emerging application areas: Prediction study. *Polymer Degradation and Stability* 156:100–110. Reproduced with permission from Elsevier.

in the traps attract scavengers looking for food who enter the traps and also become trapped (VIMS n.d.), driving a vicious cycle. A recommended overview of PHAs (PHA, PHB, PHBV, PHBHHx) for AM, their printing methods and applications was published by Mehrpouya et al. (2021). The printing methods evaluated for PHAs are FFF, LS, *computer-aided wet-spinning* (CAWS), and SL. According to the technical literature, the tensile strength and modulus of experimental printed PHAs and blends of them with and without reinforcing fillers are 10–44 MPa and 300–1014 MPa, respectively.

5.3   R&D IN PHAs FOR AM

This section summarizes studies conducted on PHA types in form of feedstocks for AM, mostly FFF.

Novel composite filaments for FFF composed of PHBV and palm fibers (PF), wood fibers (WF), and multi-walled carbon nanotubes (MWCNTs) were fabricated and investigated.

Wu et al. (2017) formulated the PHBV/PF filament by melt mixing PHBV grafted with maleic anhydride (PHBV-g-MA) and silane-treated PF (STPF), and extruding the mix at 130–140 °C. The dispersion of STPF in the PHA-g-MA matrix was highly homogeneous. The treatment had several benefits: strong interface adhesion between PHB and PF without separation between the two because of a greater compatibility of PHA-g-MA with STPF, consistent filament diameter, higher biodegradation rate in soil compared to that of PHBV, and mechanical properties superior to those of PHBV alone and composites of PHBV with untreated fillers. The water resistance of the PHA-g-MA/STPF membranes surpassed that of the PHA/PF membranes, and both materials were nontoxic. The recorded tensile strength of the filament PHBV-g-MA/WF was 6–18 MPa higher than that of untreated composites, and increased with amount of WF content (Wu and Liao 2017a). Adding only 1 wt% of acid-oxidized MWCNTs to PHBV-g-MA generated a filament featuring greatly improved values of tensile modulus, thermal stability, and antibacterial activity. Namely, tensile stress at failure and initial decomposition temperature increased by 16 MPa and 72 °C, respectively (Wu and Liao 2017b).

Investigations were also conducted on PHB in form of powder for the LS version of PBF. Oliveira et al. (2007) selected PHB to print porous structures via LS about 2.5 mm thick, with 1 mm diameter holes, mimicking scaffolds for tissue engineering. The authors solved the following problems: nonuniform powder layer caused by dust dragged by the printer's roller, and warpage of the printed object after depositing several layers. Pereira et al. (2012) successfully applied SLS to Biocycle®1000 PHB (PHB Industrial, Brazil), and printed porous 3D cubes about 10 mm in side, containing internal channels 0.8 mm in diameter and perpendicular to each other, and featuring actual dimensions and geometry close to those of the 3D model. The printing parameters were: laser power 16 W, laser scan speed 2 m/s, laser beam spot 450 μm, scan spacing 0.15 mm, and bed temperature 100°C. DSC and NMR testing proved that the PHB powder heated for 32 h but not fused during printing did not degrade, and could be reused without exhibiting any significant change in T_g, T_m, T_c, and chemical composition.

PHBV combined with nano-sized osteoconductive (that is promoting bone growth) calcium phosphate (Ca-P) has been proven suited to print scaffolds for bone TE. One application of AM is to print synthetic bone grafts for bone grafting that consists in, transplanting bone tissue for fixing problems with bones and joints by leading to bone regeneration. 3D scaffolds are required in scaffold-based bone regeneration, because they mimic the structure and function of extracellular matrix (ECM), which is the 3D network of extracellular macromolecules (such as collagen, enzymes, and proteins) that provides biochemical and structural support to the surrounding cells. Ca-P/PHBV composite is a strong candidate material for synthetic scaffolds, because PHBV is biodegradable and biocompatible with blood and tissue, and nano-size Ca-P is osteoconductive, with high surface-area-to-volume ratio, and, when combined with biopolymers, mimics the structure of natural bone (Duan et al. 2010). Selecting Ca-P/PHBV nanocomposite microspheres of 10–30 nm diameter, Duan et al.

TABLE 5.9

Compressive Properties of Printed Scaffolds

Property	Strength, Dry	Strength, Wet[a]	Modulus, Dry	Modulus, Wet[a]
Unit	MPa	MPa	MPa	MPa
PHBV	0.48	0.22	5.03	3.16
Ca-P/PHBV	0.55	0.20	6.40	3.59

Source: Values from plots in Duan, B., Wang, M., Zhou, W. Y. et al. 2010a. Three-dimensional nanocomposite scaffolds fabricated via selective laser sintering for bone tissue engineering. *Acta Biomater.* 6:4495–4505.
Note:
[a]After 21 days immersed in phosphate-buffered saline at 37°C.

(2010) printed via LS tetragonal scaffolds exhibiting about 60% porosity, and the following advantages vs. unfilled PHBV: (a) the nanoscale of Ca-P provided a better cell response and osteoconductivity, (b) the nanocomposite microspheres led to better dispersion of Ca-P, (c) improved cell proliferation, and (d) LS provided a geometrically controlled microstructure with totally interconnected pores. The microspheres were made dispersing 15 wt% Ca-P in a chloroform solution with PHBV 12 mol% from ICI (UK). Scaffolds in unfilled PHBV were also printed as a baseline, shaped as square-based prisms, 9 × 9 × 16.5 mm in size, with cubic pores 0.8 mm in side, and 0.5 mm wall thickness. Porosity of PHBV and Ca-P/PHBV scaffolds was 64.6% and 62.6%, respectively. Table 5.9 reports values of compressive strength and modulus measured on scaffolds as printed, and after immersion for 21 days at 37°C in a *phosphate-buffered saline*, an aqueous solution simulating body fluids. Cell proliferation on the scaffolds was measured at 4, 7, and 14 days, and proved to be successful on PHBV and Ca-P/PHBV.

It was also demonstrated that Ca-P/PHBV nanocomposite microspheres could be printed in complex shapes (Duan and Wang 2010b), and shaped as femoral condyle (femur's lower extremity), whose surface incorporated gelatin and heparin (a blood thinner) to control bone growth (Duan and Wang 2010c). Scaffolds composed of Ca-P/PHBV nanocomposite microspheres could also feature an additional biological activity by releasing therapeutics previously included in the formulation of Ca-P/PHBV microspheres (Duan and Wang 2010a).

Traditional methods to fabricate scaffolds for bone TE are particle leaching, freeze-drying, and hard templating, but they have limited control of the pore structure and mechanical properties (Zhao et al. 2014). An AM process called *bioextrusion* (section 2.3.7) was selected to overcome these limitations, and print bioactive scaffolds for bone TE composed of a blend of mesoporous (that is with pore diameters between 2 and 50 nm) bioactive glass (MBG) and poly(3-hydroxybutyrate-co-3-hydroxyhexanoate) (PHBHHx). MBGs stimulate more bone generation than other bioactive ceramics by bonding with bone faster than other bioactive ceramics (Jones 2013) due to their greater surface area. The feedstocks were formulated as a paste with PHBHHx:MBG mass ratios 1:3, 1:5, and 1:7, and extruded by EnvisionTEC 3D Bioplotter printer in form of cylindrical scaffolds with 6 mm diameter and 3 mm height, featuring controlled architecture, and compressive strength of 5–12 MPa that depended on the ratio PHBHHx:MBG and fell in the range of that of human trabecular bone (2–12 MPa). The presence of biocompatible PHBHHx polymer accelerated bone growth, and promoted adhesion and proliferation of marrow-derived stem cells, and the blend was effective in repairing skull defects in rats (Zhao et al. 2014).

Thaxton (2016) experimentally devised a process to extrude a filament made from Metabolix PHA pellets using an EX2 Filabot extruder with a speed-controlled drill. The author described in details all the steps to optimize the result and overcome difficulties, including temperature differences between extruder's nozzle and the LCD display that measured the extruder's nozzle temperature, calibrating the temperature and drill speed (in order to give the PHA pellets more time to melt in the hopper, and prevent chunky filaments), and so on. Eventually, the extruder temperature and drill speed were set at 170°C and 15 rpm, and a system was built that comprised, in its process sequence, extruder, cooling water bath, filament's diameter gage, spring tensiometer to keep the filament taut, and spool winding apparatus. The system was equipped with feedbacks based on Arduino circuits to control the rotation speed of the spooler and extruder's temperature, depending on the filament's diameter measured by the automated gage.

Wimmer et al. (2015) formulated, extruded, and experimentally evaluated a filament for FFF composed of combinations of a PLA/PHA blend filled with sustainable powders (cellulose, lignin, talcum, and wood flour). The powder size was critical, because even particle size <250 µm would frequently block the printer's nozzle. The best performing combinations were: (a) PLA Ingeo™ 4043D with 15 wt% PHA and 15 wt% cellulose pulp, because of its optimal combination of printability and mechanical performance, and (b) PLA/PHA/nanocellulose, displaying improved mechanical properties over PLA/PHA/cellulose pulp.

Because PHB and PLA have close melting temperatures, they can be melted and blended together into a fully biodegradable feedstock combining advantages of both ingredients. Studying the influence of MFI, plasticizer type, and crystallinity on printing quality of PLA filaments for FFF, Wang et al. (2017) reported that adding Biomer® P304 PHB to commercial PLAs produced a PHB/PLA filament with the following benefits vs. PLA: (a) PHB possessed a T_m lower than that of PLA, leading to a greater and more suitable MFI for printing; (b) PHB also enabled PLA to reach higher crystallinity during the cooling phase of the printing process without the need for post-printing annealing (with consequent time and cost saving), leading to superior heat resistance and mechanical strength; (c) PLA accelerated crystallization of PHB, and increased stiffness of PHB.

Printed 3D scaffolds for bone TE out of PHBHHx and PCL were manufactured by means of an AM version of *wet spinning*, a process fabricating fibers that consists of extruding a polymeric solution into a coagulation batch with tetrahydrofuran, where the extruded filament precipitates, and forms a continuous polymeric fiber (Mota et al. 2017; Puppi et al. 2017). By optimizing the printing parameters, Puppi et al. (2017) printed scaffolds with several pore sizes and internal architectures, and achieved good control of the fiber alignment and 3D fully interconnected porous networks, with 79–88% porosity, 47–76 µm fiber diameter, and 123–789 µm pore size, depending on the PHBHHx/PCL composition. PHBHHx featuring 300,000 g/mol MW was supplied by Tsinghua University (China), and purified before use. Scaffolds were fabricated layer upon layer on the printer Roland MDX-40A 3D Milling Machine, in-house customized, by replacing the milling head with a syringe pump and a glass syringe fitted with a 0.41 mm internal diameter stainless steel needle. The PHBHHx/PCL blend composition considerably influenced the scaffold morphological, thermal, and mechanical properties. The printed scaffolds displayed various lattice structures, with fibers oriented in 0/90°, and 0/90/±45° patterns, and distance in XY between two adjacent fibers of 200, 500, and 1000 µm (Figure 5.7). The pore size in the XY plane was 116, 123, 357, and 789 µm. The compressive properties measured on the scaffolds varied based on pore size and fiber orientation, and fell in the following ranges: yield stress 0.39–0.49 MPa, modulus 0.71–1.40 MPa, and yield strain 37–39%. These properties exceeded the to compressive properties of PHA scaffolds fabricated by means of other techniques reported in previous studies. Moreover, PHBHHx proved to be biocompatible, by supporting in rodents the proliferation of cells that are precursors of bone cells. Wet spinning can potentially become an effective process to develop customized PHBHHx-based blend scaffolds for TE.

Batchelor (2016) developed a laboratory procedure for fabricating a filament out of PHA that may be of interest to formulators of PHAs for FFF. He choose PHA pellets manufactured by Metabolix, and consisting in a mixture of PHB and the copolymer poly(3HB-co-4HB). Employing a Filabot EX2 extruder equipped with speed control, a water bath to cool the filament exiting from

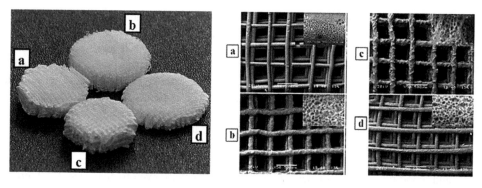

FIGURE 5.7 Scaffolds printed using computer-aided wet spinning made of: (a) PHBHHx; (b) PHBHHx/ PCL 3:1; (c) PHBHHx/PCL 2:1; (d) PHBHHx/PCL 1:1. Left: full size. Center and right: SEM micrographs of top view.

Source: Adapted from Puppi, D., Morelli, A., Chiellini, F. 2017. Additive Manufacturing of Poly(3-hydroxybutyrate-co-3-hydroxyhexanoate)/poly(e-caprolactone) Blend Scaffolds for Tissue Engineering. *Bioengineering* 4:49. Reproduced under open access Creative Common CC BY license.

the extruder, and controlling the force pulling the filament from the extruder, he fabricated a filament of 100% PHA with diameter within tolerance wound on a spool fit for FFF printers.

The work by Menčik et al. (2018) is interesting because they formulated and tested fully sustainable and degradable FFF filaments made of PHB, PLA, and commercial plasticizers Citroflex® (Vertellus LLC) composed of esters of natural citric acid. Citric acid is a weak acid that occurs naturally present in citrus fruits, and in USA is most often derived from fermentation of corn. Esters of citric acid have been long utilized in food and beverage flavoring, and medical and pharmaceutical applications (ChemPoint 2020). The plasticizers served to reduce brittleness and increase ductility of the PHA/PLA blend. Four filaments were extruded, each with composition PHB/PLA/plasticizer (60/25/15 wt%): PHB and PLA were Biomer® P3HB (MW of 410,000 g/mol), and Ingeo™ 4060D (MW of 180 000 g/mol), and the plasticizer was one of four commercial grades of Citroflex®: C-4, A-4, B-6, and A-6. The effect of plasticizers on tensile (ISO 527-1) and thermal properties, and degradation was experimentally measured. Test results in Table 5.10 point out that adding any plasticizer penalized all reported properties except, as anticipated, ductility (elongation at break) vs. those of PHB/PLA. Tensile strength and modulus, and T_g were especially penalized, and the lower the T_g, the lower the maximum service temperature. This was expected because introducing low MW plasticizer into the PHB/PLA blends caused greater mobility of macromolecular chains, and lowered T_c, T_m, and T_g. Not surprisingly, there was a direct correlation between MW of plasticizers and filament's properties: in almost all cases the greater the MW, the higher the tensile strength and modulus, T_m, T_g, and the lower the elongation at break. A-6 and B-6 plasticizers resulted in the best tensile strength and modulus, because they featured the combination of properties with highest values (although the lowest elongation at break). One downsize of plasticizers is that they can migrate from the bulk material toward the surface, and, in doing so, they reduce material flexibility and ductility. The migration of plasticizers was assessed by collecting the weight loss upon continuous exposure in a drying oven at 110°C, a temperature not reached in most common applications, given the T_g of the tested blends (Table 5.10), and probably chosen to accelerate the test. The weight loss after 15 days at 110°C ranged from 32.5% to 95%. The authors also measured degradation per ISO 20200 in simulated composting thermophilic conditions by keeping thin films (6.5 × 10 × 0.08 cm) of PHB/ PLA/A-4 in an air-circulation oven at constant temperature of 58°C. The films completely disintegrated under visual observation after 65 days (Figure 5.8).

TABLE 5.10
Effect of Plasticizers on Properties of PHB/PLA Filament for FFF

Material (PHB/ PLA/ Plasticizer)	Chemical Name	Plasticizer MW	Tensile Modulus	Tensile Strength	Elongation at Break	T_m	T_g	Crystallinity
		g/mol	GPa	MPa	%	°C	°C	%
PHB/PLA[a]	N/A	N/A	2.41	39.1	5	172	58.4	63
PHB/PLA/C-4	Tributyl Citrate	360	0.66	19.8	155	136	23.2	55
PHB/PLA/A-4	Acetyl Ttributyl Citrate	403	0.69	20.3	187	164	25.5	59
PHB/PLA/A-6	Acetyl Trihexyl Citrate	486	1.16	22.1	21	167	49.8	66
PHB/PLA/B-6	n-Butyryl tri-n-hexyl Citrate	514	1.46	23.5	11	168	45.9	60

Source: Menčik, P., Prikryl, R., Stehnová, I., et al. 2018. Effect of Selected Commercial Plasticizers on Mechanical, Thermal, and Morphological Properties of Poly(3-hydroxybutyrate)/Poly(lactic acid)/Plasticizer Biodegradable Blends for Three-Dimensional (3D) Print. *Materials* 11:1893. Reproduced under open access Creative Common CC BY license.
Note:
[a]7 days after printing.

Kontarova et al. (2020) continued the work conducted by Menčik and et al. (2018), and studied the effect of the following plasticizers on the printability and mechanical and thermal properties of PHB/PLA blends: acetyl tris(2-ethylhexyl) citrate, tris(2-ethylhexyl) citrate, and poly(ethylene glycol)bis(2-ethylhexanoate). After assessing that plasticizers were miscible, PHB/PLA/plasticizer

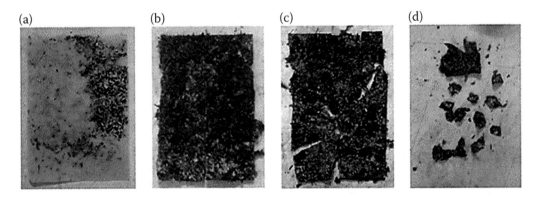

FIGURE 5.8 Results of disintegration test on film of PHB/PLA/A-4 showed in chronological order: (a) day 4, (b) day 29, (c) day 42, (d) day 55.

Source: Adapted from Menčik, P., Prikryl, R., Stehnová, I. et al. 2018. Effect of Selected Commercial Plasticizers on Mechanical, Thermal, and Morphological Properties of Poly(3-hydroxybutyrate)/Poly(lactic acid)/Plasticizer Biodegradable Blends for Three-Dimensional (3D) Print. *Materials* 11:1893. Reproduced under open access Creative Common CC BY license.

FIGURE 5.9   Printed temperature tower.

Source: Kontarova, S., Prikryl, R., Melcová, V. et al. 2020. Printability, mechanical and thermal properties of poly(3-hydroxybutyrate)-poly(lactic acid)-plasticizer blends for three-dimensional (3D) printing. *Materials* 13, 4736. doi: 10.3390/ma13214736. Reproduced with kind permission from Dr. S. Kontarova.

blends of 60/25/15 wt% were prepared using a corotating meshing twin-screw extruder that produced pellets that were in turn converted in a single-screw extruder into a filament for FFF. Plasticizers were added to improve not only the mechanical properties but also the printability of the blends. Several properties were measured on printed coupons. Plasticizers boosted elongation at break from about 25% to a maximum of about 150% and impact strength, quantified on un-notched, notched and tensile (ISO 8256) coupons, but lessened tensile strength and modulus. The values of HDT at 0.45 MPa and 1.8 MPa in the blends lagged those in PHB/PLA. Viscosity was evaluated on an oscillating rheometer, and decreased from 1,000 Pa·s in PHB/PLA to 350–600 MPa·s in the blends, not surprisingly. The printing temperature was varied between 180°C and 205°C, and affected the amount of warping in the printed articles, although in different ways depending on the plasticizer. This paper is notable also because it describes a new method for a fast and qualitative evaluation of printing quality. This method consists in printing for each material being evaluated a *temperature tower* (Figure 5.9) in which each floor: (a) is printed at a temperature gradually decreasing from bottom to top floor, and (b) contains varying geometric elements: circular, rectangular, and triangular holes, diagonals, colonnades, stringings, and bridges. After the tower is printed, each geometric element is visually graded from 1 (poorest quality) to 5 (best quality) and an average grade is calculated.

FURTHER READINGS

Koller, M. ed. 2017. *Advances in Polyhydroxyalkanoate (PHA) Production*. Special issue published in Bioengineering. Basel: MDPI. 1st edition. http://www.mdpi.com/journal/bioengineering/special_issues/PHA.

Koller M., Salerno A., Braunegg G. 2014. Polyhydroxyalkanoate: basics, production and applications of microbial polyesters. In *Bio-Based Plastics – Materials and Applications*, ed. Stephan Kabasci, 137–170. Chichester: John Wiley & Sons.

REFERENCES

Amaro, T., Rosa, D., Comi, G., Iacumin, L. 2019. Prospects for the use of whey for polyhydroxyalkanoate (PHA). *Prod. Front. Microbiol.* 10:992.

Andrews, M. 2014. Mirel™ PHA polymeric modifiers & additives. *Add. Com. 2014*, Barcelona, October 21–22.

Batchelor, W. 2016. *PHA Biopolymer Filament for 3D Printing*. Dissertation for Bachelor of Science degree in Physics, College of William and Mary (accessed April 3, 2020).

Bioplastics Magazine. 2020. *The Global Bio-Based Polymer Market in 2019 – A Revised View*. (https://www.bioplasticsmagazine.com/en/news/meldungen/20200127-The-global-bio-based-polymer-market-in-2019-A-revised-view.php (accessed April 3, 2020).

Braunegg, G., Lefebvre, G., Genser, K. F. 1998. Polyhydroxyalkanoates, biopolyesters from renewable sources: physiological and engineering aspects. *J. Biotechnol.* 65 (2–3): 127–161.

Chanprateep, S. 2010. Current trends in biodegradable polyhydroxyalkanoates. *J. Biosci. Bioeng.* 110 (6): 621–632.

ChemPoint. 2020. Citroflex® B-6. https://www.chempoint.com/products/vertellus-specialties/citroflex-citric-acid-esters/citroflex-citric-acid-esters/citroflex-b-6 (accessed October 22, 2020).

Chen, G.-Q. 2010. Industrial production of PHA. In *Plastics from Bacteria: Natural Functions and Applications*, ed.G.-Q. Chen, Microbiology Monographs, Vol. 14. Berlin: Springer. doi:10.1007/978-3-642-03287-5_6.

Chiulan, I., Frone, A. N., Brandabur, C. et al. 2018. Recent advances in 3D printing of aliphatic polyester. *Bioengineering* 5:2.

Corre, Y.-M., Bruzaud, S., Audic, J.-L., Grohens, Y. 2011. Morphology and functional properties of commercial polyhydroxyalkanoates: A comprehensive and comparative study. *Polym. Test.* 31:226–235.

Creative Mechanisms. 2017. *Everything You Need to Know About PHA*. https://www.creativemechanisms.com/blog/everything-you-need-to-know-about-pha-polyhydroxyalkanoates (accessed April 3, 2020).

Daminabo, S. C., Goel, S., Grammatikos, S. A. et al. 2020. Fused deposition modeling-based additive manufacturing (3D printing): techniques for polymer material systems. *Mater. Today Chem.* 16:100248.

Danimer Scientific. 2021. *Danimer Scientific Opens World's First Commercial PHA Production Facility in Winchester, KY*. https://danimerscientific.com/commercial-manufacturing-new-plant/ (Accessed January 16, 2021).

Doherty, W. O. S., Mousavioun, P., Fellows, C. M. 2011. Value-adding to cellulosic ethanol: Lignin polymers. *Ind. Crops Prod.* 33:259–276.

Duan, B., Wang, M. 2010a. Encapsulation and release of biomolecules from Ca-P/PHBV nanocomposite microspheres and three-dimensional scaffolds fabricated by selective laser sintering. *Polym. Degrad. Stabil.* 95:1655–1664.

Duan, B., Wang, M. 2010b. Customized Ca–P/PHBV nanocomposite scaffolds for bone tissue engineering: design, fabrication, surface modification and sustained release of growth factor. *J. R. Soc. Interface* 7:S615–S629.

Duan, B., Wang, M., Zhou, W. Y. et al. 2010. Three-dimensional nanocomposite scaffolds fabricated via selective laser sintering for bone tissue engineering. *Acta Biomater.* 6:4495–4505.

Gonzales Ausejo, J., Rydz, J., Musiol, M. et al. 2018a. A comparative study of three-dimensional printing directions: The degradation and toxicological profile of a PLA/PHA blend. *Polym. Degrad. Stab.* 152:191–207.

Gonzales Ausejo, J., Rydz, J., Musiol, M. et al. 2018b. Three-dimensional printing of PLA and PLA/PHA dumbbell-shaped specimens of crisscross and transverse patterns as promising materials in emerging application areas: Prediction study. *Polym. Degrad. Stab.* 156:100–110.

Goonoo, N., Bhaw-Luximon, A., Passanha, P., Esteves, S. R., Jhurry, D. 2017. Third generation poly(hydroxyacid) composite scaffolds for tissue engineering. *J. Biomed. Mater. Res. Part B* 105B:1667–1684.

Guessasma, S., Belhabib, S., Hedi Nouri, H. 2019. Microstructure and mechanical performance of 3D printed wood-PLA/PHA using fused deposition modelling: effect of printing temperature. *Polymers* 11:1778.

Holmes, P. A. 1985. Applications of PHB – A microbially produced biodegradable thermoplastic *Phys. Technol.* 16 (1): 32–36.

Jones, J. R. 2013. Review of bioactive glass: From Hench to hybrids. *Acta Biomater.* 9 (1): 4457–4486.

Koller, M., Salerno, A., Braunegg, G. 2014. Polyhydroxyalkanoate: basics, production and applications of microbial polyesters. In *Bio-Based Plastics – Materials and Applications*, ed.S. Kabasci, 137–170. Chichester: John Wiley & Sons.

Kontarova, S., Prikryl, R., Melcová, V. et al. 2020. Printability, mechanical and thermal properties of poly(3-hydroxybutyrate)-poly(lactic acid)-plasticizer blends for three-dimensional (3D) printing. *Materials* 13:4736.

Le Duigou, A., Castro, M., Bevan, R., Martin, N. 2016. 3D printing of wood fibre biocomposites: From mechanical to actuation functionality. *Mater. Des.* 96:106–114.

Liu, Q., Zhang, H., Deng, B., Zhao, X. 2014. Poly(3-hydroxybutyrate and Poly(3-hydroxybutyrate-co-3-hydroxyvalerate) (PHBV): Structure, Property and Fiber. *Int. J. Polym. Sci.* article ID 374368.

MarketsandMarkets™. 2020. *Polyhydroxyalkanoate (PHA) Market.* https://www.marketsandmarkets.com/Market-Reports/pha-market-395.html (accessed October 22, 2021).

Mehrpouya, M., Vahabi, H., Barletta, M., Laheurte, P., Langlois, V. 2021. Additive manufacturing of polyhydroxyalkanoates (PHAs) biopolymers: Materials, printing techniques, and applications. *Materials Science and Engineering: C,* 217:112216.

Menčik, P., Prikryl, R., Stehnová, I. et al. 2018. Effect of selected commercial plasticizers on mechanical, thermal, and morphological properties of poly(3-hydroxybutyrate)/poly(lactic acid)/plasticizer biodegradable blends for three-dimensional (3D) print. *Materials* 11:1893.

Mota, C., Wang, S.-Y., Puppi, D. et al. 2017. Additive manufacturing of poly[(R)-3-hydroxybutyrate-co-(R)-3-hydroxyhexanoate] scaffolds for engineered bone development. *J. Tissue Eng. Regen. Med.* 11:175–186.

Nikodinovic-Runic, J., Guzik, M., Kenny, S. T., et al. 2013. Carbon-rich wastes as feedstocks for biodegradable polymer (polyhydroxyalkanoate) production using bacteria. *Adv. Appl. Microbiol.* 84:139–200.

Oliveira, M. F., Maia, I. A., Noritomi, P. Y. et al. 2007. Construção de Scaffolds para engenharia tecidual utilizando prototipagem rápida. *Matérial* 12 (2): 373–382.

Pereira, T. F., Oliveira, M. F., Maia, I. A. et al. 2012. 3D printing of poly(3-hydroxybutyrate) porous structures using selective laser sintering. *Macromol. Symp.* 319:64–73.

Puppi, D., Morelli, A., Chiellini, F. 2017. Additive manufacturing of poly(3-hydroxybutyrate-co-3-hydroxyhexanoate)/poly(ε-caprolactone) blend scaffolds for tissue engineering. *Bioengineering* 4:49.

Sabapathy, P. C., Devaraj, S., Meixner, K., et al. 2020. Recent developments in polyhydroxyalkanoates (PHAs) production – a review. *Bioresour. Technol.* 306:123132.

Solaiman, D. K. Y., Ashby, R. D., Foglia, T. A., Marmer, W. N. 2006. Conversion of agricultural feedstock and co-products into poly(hydroxyalkanoates). *Appl. Microbiol. Biotechnol.* 71:783–789.

Sudesh, K., Abe, H., Doi, Y. 2000. Synthesis, structure and properties of polyhydroxyalkanoates: biological polyesters. *Prog. Polym. Sci.* 25:1503–1555.

Thaxton, R. D. 2016. *Extrusion of Polyhydroxyalkanoate Filament for use in 3D Printers.* BS degree in Physics diss., The College of William and Mary.

VIMS. *Fact Sheet: Polyhydroxyalkanoate (PHA) Biodegradable Escape Panel (Biopanel) for Crab, Lobster, and Fish Traps.* https://www.vims.edu/ccrm/_docs/marine_debris/biodegradablepanel_factsheet.pdf (accessed April 3, 2020).

Wang, S., Capoen, L., D'hooge, D. R., Cardon, L. 2017. Can the melt flow index be used to predict the success of fused deposition modelling of commercial poly(lactic acid) filaments into 3D printed materials? *Plast. Rubb. Compos. Macromol. Eng.* 47 (1): 9–16.

Wimmer, R., Steyrer, B., Woess, J. et al. 2015. 3D printing and wood. *Pro Ligno* 11 (4): 144–149.

Wu, C.-S., Liao, H., Cai, Y.-X. 2017. Characterisation, biodegradability and application of palm fibre-reinforced polyhydroxyalkanoate composites. *Polym. Degrad. Stabil.* 140:55–63.

Wu, C.-S., Liao, H. 2017a. Fabrication, characterization, and application of polyester/wood flour composites. *J. Polym. Eng.* 37:689–698.

Wu, C.-S., Liao, H. 2017b. Interface design of environmentally friendly carbon nanotube-filled polyester composites: Fabrication, characterization, functionality and application. *Express Polym. Lett.* 11:187–198.

Zhao, S., Zhu, M., Zhang, J. et al. 2014. Three dimensionally printed mesoporous bioactive glass and poly(3-hydroxybutyrate-co-3-hydroxyhexanoate) composite scaffolds for bone regeneration. *J. Mater. Chem. B* 2:6106–6118.

6 Wood-Filled Feedstocks

Earth provides enough to satisfy every man's needs, but not every man's greed.

Mahatma Gandhi

6.1 WOOD OVERVIEW

Wood is the sustainable material *par excellence*. It is the oldest structural and engineering material, because it is ubiquitous, readily available, not expensive, durable, tough, lightweight, and, due to its efficient structure and anatomy, possesses values of strength/density (*specific strength*) and tensile modulus/density (*specific stiffness*) competing with those of polymers, metals, and alloys. In fact, along the grain, balsa, oak and pine feature 20–30 MNm/kg and 100 MNm/kg of specific stiffness and strength, respectively, whereas polymers exhibit specific stiffness of 0.1–10 MNm/kg, and specific strength of 6–70 MNm/kg (University of Cambridge 2002). Some consider wood a carbon neutral material, arguing that growing trees absorb carbon as CO_2 from the atmosphere, and release the same amount of CO_2 when burned, but not everybody agrees (Crawford 2008; Johnson 2009).

Wood is made of natural polymers, and is less expensive, stiffer, and stronger than many commodity synthetic polymers, making it an attractive ingredient to be added to them in order to reduce the material cost. Typically, wood is classified in two groups, defined based on their anatomical and botanical features instead of hardness, as the group name may incorrectly imply:

- *Softwoods*: cedar, cypress, Douglas-fir, fir, larch, pine, spruce
- *Hardwoods*: ash, aspen, balsa, bamboo, birch, boxwood, cherry, ebony, elm, hickory, mahogany, maple, oak, walnut.

Softwood features strong interfacial adhesion to polymers and low strength, whereas hardwood possesses poor interfacial adhesion and high strength (Kumar et al. 2011; Gardner et al. 2015). Wood is porous, fibrous, and anisotropic. It is mainly composed of hollow, elongated, spindle-shaped cells (*fibers*) arranged parallel to each other along the tree's trunk. These fibers are strongly bonded together, and constitute the structural component of the wood tissue. Their center (*lumen*) is hollow, and partially or completely filled with deposits, such as resins, gums, or growths from near cells. Fibers' length greatly varies, and averages about 1 mm in hardwoods and 3–8 mm in softwoods. Fiber diameters are typically 15–45 μm (Miller 1999; Clemons 2008). An example of wood anatomy is illustrated in Figure 6.1 showing 3D views of side (left) and top (right) cross sections of loblolly pine.

Wood contains two major chemical components, both polymers: carbohydrate (cellulose and hemicellulose) and lignin, respectively 65–75% and 18–35 wt%. Minor amounts of other materials, mostly in the form of *extractives* (organic compounds soluble in solvents, usually 4–20 wt%) and ash (inorganic residue usually around 1 wt%) complete wood's composition. Overall, wood's elemental composition in wt% is about 50 carbon, 44 oxygen, and 6 hydrogen (Pettersen 1984). Cellulose, hemicellulose, and lignin containing hydroxyl are distributed throughout the cell wall. Wood's chemical composition cannot be defined precisely for a given tree species or even for a given tree, because it depends on the tree part (root, stem, or branch), type of wood (i.e. normal,

DOI: 10.1201/9781003221210-6

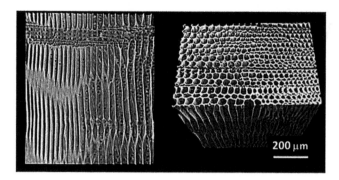

FIGURE 6.1 Tridimensional views of side (left) and top (right) cross sections of loblolly pine.

Source: Modified from Mayo, S. C., Chen, F., Evans, R. 2010. Micron-scale 3D imaging of wood and plant microstructure using high-resolution X-ray phase-contrast microtomography. *Journal of Structural Biology* 171 (2): 182–188. Reproduced with permission from Elsevier.

tension, or compression wood) (Shigo 1986), geographic location, climate, and soil conditions. As a an indicative example, the chemical composition in wt% of incense cedar, red maple, white oak, southern red oak, loblolly pine, and ponderosa pine is: cellulose 37–47, hemicellulose 19–30, lignin 21–34, extractives 2–5, and ask 0.2–0.5 (Pettersen 1984).

Cellulose is a highly (typically 60–90 wt%) crystalline, linear polymer composed of about 9,000–15,000 sugar molecules (*anhydroglucose units*), and it is the component providing most of wood's strength and structural stability. Hemicellulose is a family of branched polymers also structurally contributing to wood, although composed of several 5- and 6-carbon sugars, whose MW values are well below those of cellulose. Lignin is an amorphous, crosslinked polymer network consisting of an irregular array of variously bonded hydroxy- and methoxy-substituted phenylpropane units (Pettersen 1984). Lignin is more extensively described in Chapter 9.

Wood is less expensive than commodity TPs. F.e., in 2015 the price in USD/kg of one ton of sawdust and one ton of polypropylene (PP) were about 100 (Agricultural Marketing Resource Center 2018) and 1000 (Plastics Insight 2016), respectively, whereas currently one ton of sawdust and one ton of PLA cost USD 150 and USD 1,200–2,000, respectively. Wood is commonly used in the TP composite industry due to significant advantages: high tensile modulus, low price, good machinability, problem-free disposal, and being abundant and renewable (Ayrilmis 2018). Some examples of producers of wood products, such as wood flour and sawdust (that are present in feedstocks for AM and non-AM plastics), are listed below:

- American Wood Fibers (USA): wood flour (https://awf.com/index.php/wood-fuel-pellets/)
- Eden Products Ltd. (UK): wood products (flour and flakes) from 53 μm to 4 mm size (https://www.edenproductsltd.co.uk/edenproducts/products/woodflour.php)
- J. Rettenmaier & Soehne (Germany): wood fibers and lignocellulose (https://www.jrs.eu/jrs_en/wood-fiber/lignocel.php)
- PJ Murphy Forest Products (USA): wood flour and sawdust (https://pjmurphy.net/services/wood-flour/)
- P.W.I. Industries (Canada): hardwood and softwood sawdust and wood flour (https://www.pwi-industries.com/en/products).

6.2 ADVANCES IN FEEDSTOCKS AND PROCESSES FOR WOOD AM

Wood can be an AM feedstock especially in countries rich in forests, where it represents a very affordable sustainable resource. Considerable research has been conducted to turn wood and forest products into AM feedstocks (Li et al. 2016). These efforts have spanned from large-scale building manufacturing to small-scale and specialized value-added bioprinting. One example is the material *microtimber* developed by the University of Sidney (Australia): it combines microground macadamia nutshells and ABS mixed with a binding agent to extrude filaments for FFF, and is described in Chapter 10 (Girdis et al. 2017). Wood flour and pulp, along with cellulose and lignin, have been forest products converted into feedstocks for AM processes: bioprinting, hot extrusion, power binding, and slurry extrusion/spray. Obviously, among possible products of wood AM are replacements for wood and wood-looking products that are not only cost-competitive, but also possess features absent in other wood products such as customized features that enable customers to personalize the dimensions and style of their furniture.

New AM processes specific for forest-derived feedstocks have been investigated. A Canadian-U.S. start-up firm, 4 AXYZ, has introduced *stratified additive manufacturing*, a new technology to print industry-grade, small, uniformly cut wood pieces that are secured in layers to form larger objects and fabricate customizable furniture more affordable for consumers. Additionally, 4 AXYZ planned to develop Smart Wood that labels wood-based components printed with embedded electronics, sensors and conductive metals suitable, f.e., to print a window linked to a thermostat that opens and closes depending on changes in weather (Top for 3D Printing 2014).

Since China is one of the leading countries in forested area (2 million sq. km), it is not surprising that Chinese researchers (Xu et al. 2016) developed a brass nozzle featuring a variable diameter of 2−5 mm, and mounted on a printer built to process wood powder and adhesive (volume ratio 2:1, respectively), mixed together by the rotating blades of the mixer included in the printer. The diameter was controlled by a magnetostrictive material called Terfenol-D made of $Tb_xDy_1Fe_2$ ($x{\approx}0.3$) that is capable to reach a strain of 0.002 m/m in response to a magnetic field. Terfenol-D outperforms other piezoelectric ceramics: it is more deformable, it withstands greater stress values, and it has a shorter response time to changes in the magnetic field. Other printer's key components are the system to convey the mix of powder and adhesive to the nozzle (consisting in pump, suction pipe, plunger, and pipe valve), and extrusion screw.

In the next sections, physical and mechanical properties of commercial and experimental wood-filled polymers for AM are illustrated, but these properties and cost should not be the only criteria to choose AM over other fabrication technologies, as explained in Chapter 2. Other criteria include f.e. producing designs with shapes too costly or impossible to achieve with conventional processes, reducing part count, fabricating an article having multiple functions, reproducing an item originally fabricated via a conventional method whose blueprint is lost, and so on.

6.3 COMMERCIAL WOOD/POLYMERS FOR AM

Wood is combined with polymers to produce materials for FFF and LS, because of any combination of the following reasons: lowering the cost and weight of polymers, imparting a characteristic finish, and being wood inexpensive, renewable and biodegradable. Several wood/polymer filaments for FFF are commercially available, but those listed below, along with their ingredients, have material properties reported by their suppliers (Table 6.1):

- Fibrolon® 3D Wood by FKuR (Germany): PLA, copolyester, wood fiber, additives
- EasyWood™ by Formfutura (Netherlands): PLA, 40 wt% grinded cedar wood particles
- Wood by MG Chemicals (Canada): PLA, wood
- woodFill by colorFabb (Netherlands): PLA, polyesters, 10–20 wt% wood fibers, additives

TABLE 6.1
Properties of Commercial Wood/Polymer Filaments for FFF

Property	Unit	Fibrolon® 3D Wood (FkuR)	woodFill (colorFabb)	Timberfill (Fillamentum)	ISO Standards for Woodfill, Fibrolon®, Timberfill	EasyWood™ (Formfutura)	Wood (MG Chemicals)	ASTM Standards for Wood, EasyWood™
Density	g/cm³	1.19	1.15	1.28	1183	1.2	1.25	D1505
Melting temperature	°C	>155	>155	145–160	3146C	145	130–210	N/A
Melt flow rate (190°C/2.16 kg)	g/10 min	2.5–4.5	4.0–7.5	25.6	1133	4.5	N/A	N/A
Tensile Strength	MPa	47	46	33	527	71[b]	37[b]	D882
Tensile Stress at Break	MPa	38	42	N/A	527	N/A	N/A	N/A
Elongation at Break	%	6.5	5.5	2.9	527	171[b]	6[b]	D882
Tensile Modulus	MPa	2,900	3,290	2,800	527	1,930[b]	3,100[b]	D882
Flexural Stress/ Strength	MPa	64[a]	70[a]	N/A	178	59.6	N/A	N/A
Flexural Modulus	MPa	2,950	3,930	N/A	178	2,584	N/A	N/A
Flexural Strain	%	N/A (no break)	5	N/A	178	N/A	N/A	N/A
Charpy Notched Impact Strength	kJ/m²	4.4	4.2	2.8	179-1/1 eA	7.1	N/A	N/A
Charpy Unmotched Impact Strength	kJ/m²	21	19	15.1	179-1/1 eU	N/A	N/A	N/A
Glass Transition Temperature	°C	N/A	N/A	55–60	N/A	N/A	N/A	N/A

Source: Online data posted by suppliers.
Notes:
[a] At 3.5% strain.
[b] In machine direction.

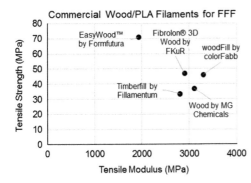

Commercial Wood/PLA Filaments for FFF

FIGURE 6.2   Plot of tensile strength vs. tensile modulus of commercial wood/PLA filaments for FFF.

Source: Data sheets posted online by suppliers.

- Timberfill by Fillamentum® (Czech Republic): PLA, 15 wt% spruce wood fibers, starch, polyalkanoates, proteins
- Laywoo-D3™ by CC-Products (Germany): blend of co-polyesters, 40 wt% recycled wood from various tree species, water, plasticizers (PEG). It smells like wood and can be cut, ground, and painted like wood. Its tensile strength and density are 24 MPa and 0.94 g/cm^3, respectively.

We must emphasize that the tensile properties in Table 6.1 are collected following two standards, ASTM D882 and ISO 527: the former requires test coupons shaped like a strip of constant cross section, the latter calls for test coupons shaped like a dumbbell. The different geometries generate differing values of the same properties on the same material, and, therefore, comparing the five materials to each other in tension from the values in Table 6.1 is only indicative. The tensile strength and modulus values in Table 6.1 are also plotted in Figure 6.2. The tensile properties of filaments in Table 6.1 are comparable to those of the commercial PLA filaments listed in Table 3.11 that span the following ranges: strength 22–71 MPa, modulus 1.9–3.4 GPa, and elongation to break 1–47%, although most values of PLA's elongation stay below 10%. The printing temperature for the filaments in Table 6.1 ranges from 170°C to 240°C depending on the material.

Since wood/polymer composites for AM contain particulate fillers, they may compete against particulate wood polymer composites (PWPCs). Particulate fillers are powdered substances made of particles typically smaller than 100 μm. PWPCs are composed of TPs or TS plastics filled with powder generally of beech, cedar, fir, larch, mango, maple, oak, pine, poplar, sal, and spruce, and are fabricated through injection molding, thermoforming, extrusion, compression molding, and hand lay-up (Chand and Fahim 2008; Gardner et al. 2015; Schwarzkopf and Burnard 2016). In 2015, applications of PWPCs in the United States were decking, molding/trim, fencing, building/ construction, and others, and the demand of PWPCs was expected to reach USD 5.9 billion in 2020 in the United States (Khan et al. 2020). Examples of articles made of PWPCs are car interior components, compostable cups, park benches, prefab houses, furniture elements, toys, siding and cladding, trim and molding, and vanity cases. Tables 6.2 and 6.3 report benchmark values of tensile and flexural properties of non-AM PWPCs based on TS matrices (epoxy, phenolic, and polyester), and TP matrices (PE, PP, and PVC), respectively. Since commercial wood-filled composites for AM are filaments based on PLA which is a TP, comparing them to non-AM PWPCs has to take in account the difference in the matrix, and the critical influence that matrix has on composite's properties. Another popular family of wood/polymer composite is *particleboard* that typically comprises three layers: two faces consisting of fine wood particles, and the core made of coarser material. Specific gravity and major mechanical properties of industrial particleboard products are listed in Table 6.4, as a baseline for AM wood-materials, but particleboard are flat, and typically selected for flooring, stair tread underlayment, case goods, and furniture cores,

TABLE 6.2

Mechanical Properties of Particulate Wood/TS Composites

TS Composite	Tensile Strength	Tensile Modulus	Flexural Strength	Flexural Modulus
	MPa	GPa	MPa	GPa
Hardwood	5.77–45.6	0.96–5.77	4.2–78.2	1.81–4.60
Softwood	18.2–34.5	0.024–4.47	20.0–61.8	2.60–4.48

Source: Data from Khan M. Z. R., Srivastava S. K., Gupta M. K. 2020. A state-of-the-art review on particulate wood polymer composites: Processing, properties and applications. *Polym. Test.* 89:106721. Reproduced with permission from Elsevier.

whereas an efficient application of AM is for articles with complex geometry not achievable with particleboard.

Printing articles not completely solid but with infill below 100% decreases weight and material cost but penalizes mechanical properties vs. those relative to 100% infill. Martikka et al. (2018) measured the decline of mechanical properties of test coupons printed out of woodFill, Laywoo-D3, and PrintPlus PLA (3Dfactories, Germany), with PrintPlus PLA chosen as a benchmark because unfilled: some coupons were printed without voids, and others with inside voids. Infill values of 23% and 55% produced tensile coupons with 42.4% and 66.4% filled by volume, respectively. Table 6.5 lists the retained tensile strength, that is the ratio of strength of coupons printed with

TABLE 6.3

Mechanical Properties of Particulates Wood/TP Composites

TP Composite	Tensile Strength	Tensile Modulus	Flexural Strength	Flexural Modulus
	MPa	GPa	MPa	GPa
Hardwood	22.8–37.0	1.40–3.12	28.9–49.8	1.44–3.34
Softwood	20.5–37.5	2.28–2.70	N/A	N/A

Source: Data from Khan M. Z. R., Srivastava S. K., Gupta M. K. 2020. A state-of-the-art review on particulate wood polymer composites: Processing, properties and applications. *Polym. Test.* 89:106721. Reproduced with permission from Elsevier.

TABLE 6.4

Mechanical Properties of Industrial Particleboard Products

Specific Gravity	Tensile Strength	Tensile Modulus	Flexural Strength	Flexural Modulus
N/A	MPa	GPa	MPa	GPa
0.60–0.77	5.6–10.9	1.6–3.7	15.2–22.8	2.8–4.0

Source: Data from Mc Natt J.D. 1973. Basic engineering properties of particle board. Res. Pap. FPL–206. Madison, WI: U.S. Department of Agriculture, Forest Service, Forest Products Laboratory. Publication available at https://www.fs.usda.gov/treesearch/pubs/37440.

TABLE 6.5

Properties of Wood/PLA and PLA Coupons with Infill of 42.4% (A) and 66.4% (B)

Material	Retained Tensile Strength[a]		Tensile Strength/Weight[b]	Tensile Modulus/Weight[b]
	%		MPa/g	GPa/g
	A	B	A, B	A, B
PrintPlus PLA	25	41	3.39–3.44	0.23–0.24
woodFill	18	27	1.38–1.48	0.14–0.16
Laywoo-D3	15	20	0.98–1	0.06–0.07

Source: Martikka O., Kärki T., Wu Q. 2018. Mechanical Properties of 3D-Printed Wood-Plastic Composites. Key. Eng. Mater. 777:499–507

Notes:

[a]Ratio of strength of coupons with voids (57.6 or 33.6%) over the strength of coupons without voids.

[b]Values from coupons with filled volume of 42.4% (A) or 66.4% (B).

voids (57.6% and 33.6 vol%) to the strength of coupons without voids. Below 100% infill, the wood/PLA filaments trailed unfilled PLA in retained strength and in both strength and modulus divided by weight, and resulted structurally less weight-efficient.

An experimental study conducted by Yang (2018) revealed the effect of printing temperature on density, surface color, tensile and flexural strengths and moduli, and compressive strength of wood fiber/PLA. The author chose the commercial filament EasyWood™, mentioned earlier in this chapter. Test samples were printed at 200°C, 210°C, 220°C, and 230°C, and designated as WP200, WP210, WP220, and WP230, respectively. For all conditions, the platform temperature and printing speed were kept constant at 50°C and 30 mm/s. Figure 6.3 shows no significant improvement but some significant degradation. In fact, with increasing printing temperature, compressive strength and density increased, while tensile and flexural strengths and moduli, water absorption (after 24 h immersion in water) and moisture content declined. The loss in the tensile and flexural properties can result from two phenomena. The first is the formation of acidic products from hemicellulose degradation while the wood/PLA filament advanced through the heater. These acids cause depolymerization and shortening of the cellulose polymeric chain, and splitting of C–C and C–O chemical

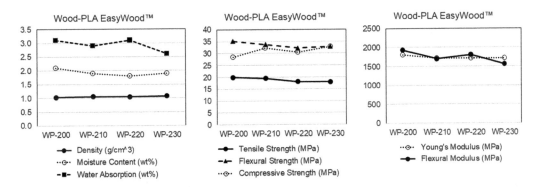

FIGURE 6.3 Effect of printing temperature on properties of commercial wood/PLA filament for AM.

Source: Data from Yang, T.-C. 2018. Effect of Extrusion Temperature on the Physico-Mechanical Properties of Unidirectional Wood Fiber-Reinforced Polylactic Acid Composite (WFRPC) Components Using Fused Deposition Modeling. *Polymers* 10:976.

bonds at the intrapolymer level in wood fibers (Garrote et al. 2001; Çolak et al. 2007). Other authors (Kubojima et al. 2000; Yildiz et al. 2006; Ayrilmis et al. 2011a, 2011b; Aydemir et al. 2015) have confirmed the reduction of flexural properties of wood/PLA composites at high temperature. The second phenomenon is the deterioration of the bonding between wood and PLA at increasing temperature, resulting in a less efficient load transfer between the two components. Although numerically different, moisture content and water absorption values were statistically equal across the printing temperatures, which may be explained by similar wettability of the PLA matrix on the surfaces of wood fibers in all samples. Density slightly increased with greater printing temperatures; since PLA's viscosity drops with increasing printing temperature (Hwang et al. 2015), lower viscosity may have resulted in lower void content, in turn causing higher density.

6.4 WOOD FOR AM

Wood has been studied, especially in form of flour (WF), as a filler for AM feedstocks available as powder and (mostly) filaments. WF is finely pulverized wood that has a consistency fairly equal to sand and sawdust, and is composed of particles not larger than 850 μm average size, corresponding to 20 U.S. standard mesh (Clemons 2010). The best quality of WF is made from hardwood because of its durability and strength. Currently, WF is a filler for non-AM TP and TS plastics, for example in exterior building products, especially railing and decking. Other applications of plastics filled with WF and natural fibers are consumer products, doors, furniture, fencing, flooring, panels, roofing, siding, and industrial infrastructure (Morton et al. 2003). WF and natural fibers are combined with PE, PVC, and PP. Because WF starts degrading slightly above 200°C, it is typically present in plastics that are processed below 200°C. WF is derived from scrap wood, and is processed commercially often from post-industrial materials such as chips, planer shavings, and sawdust. Several WF grades are available depending upon wood species and particle size (USDA 2010). The most common WF for plastic composites in USA derive from pine, oak, and maple. Most commercial WFs selected as fillers in TPs have a particle size range of 180–425 μm. As an example of price, a 50-lb bag of WF is sold at USD 33 (Douglas & Sturgess n.d.), and the larger the purchased quantity, the lower the unit price. Clemons (2010) contains more details on WF.

The bulk density of WF typically is about 0.19–0.22 g/cm³, depending on species, particle size, and moisture content, but density of the wood cell wall is about 1.44–1.50 g/cm³. As a filler, WF is compressible. Hence, if high pressure is applied to fabricate wood-plastic composites, the pressure can collapse the hollow fibers in WF or fill them with polymer, and WF's density will approach wood cell wall's density (Clemons 2010). However, even WF's high density values are well below the density of inorganic fillers, providing an advantage in applications targeting low weight. In fact, glass and carbon fibers have density of 2.5 g/cm³ and 1.8 g/cm³, respectively. Wood expands on heating, and contracts on cooling. Values of the CTE of completely dry wood are positive in all directions, and depend on the measuring direction: along, radial, and tangential to grain. The average CTE is about 70×10^{-6}/K (Clemons 2010). According to Kellogg (1981), dry wood's thermal conductivity k increases linearly with its density ρ (g/cm³) according to the equation below:

$$k = 0.200\rho + 0.024 \tag{6.1}$$

Assuming that density of compressed WF in a composite is 1.5 g/cm³, k of WF is 0.324 W/(mK).

6.5 EXPERIMENTAL WOOD-FILLED COMPOSITES FOR AM

6.5.1 INTRODUCTION

In this section recent and selected papers on experimental wood-filled composites for AM have been summarized because they: (a) address major aspects of these composites, such as different wood-binder combinations, enhancement of performance, and interfacial compatibility, and (b) provide information on material properties additional to that disclosed by suppliers of commercial AM wood composites.

6.5.2 Wood Content. Matrix Materials

Obviously, crucial in wood/polymer composites for AM is selecting the amount of wood that maximizes the composite's mechanical properties. Kariz et al. (2018) studied the effect of wood content on tensile properties of wood/PLA filaments for FFF, and reported that the tensile strength and modulus of a wood/PLA composite exceeded those of unfilled PLA, and peaked at 10 wt% (strength) and 20 wt% (modulus) wood content, with tensile strength increasing from 55 MPa to 57 MPa, and tensile modulus from 3.27 GPa to 3.94 GPa, and decreased at higher wood content (Figure 6.4). However, from the reported values of standard deviation, 57 and 55 MPa did not appear statistically different. Wood also adversely affected the surface finish and void content of printed parts that were respectively rougher and more porous compared to parts printed from unfilled PLA. Involved in this study were beech wood powder (WP) with average particle size below 0.237 mm, and general purpose, extrusion grade Ingeo™ 2003D PLA in granulated form. WP and 2003D were combined in five mixtures containing 10, 20, 30, 40, and 50 wt% WP pelletized, and extruded to produce five filaments with 1.45–1.75 mm diameter. 2003D was also extruded without WP and served as a control. In comparing the maximum strength (57 MPa) and modulus (3.94 GPa) measured by Kariz et al. (2018) with the values in Table 6.1, (a) 57 MPa exceeded the range 33-47 MPa measured on dumbbells, and was in the middle of the range 37-71 MPa recorded on films , and (b) 3.94 GPa exceeded the highest value from dumbbells and films. After peaking, the tensile strength in the composite filament steadily dropped, until it reached a minimum value of 30 MPa, equal to 55% the value of 2003D. The elastic modulus of WP/PLA behaved differently: it increased with 10–40 wt% WP content surpassing 2003D's value, remained practically stable from 20 to 40 wt%, and at 50 wt% fell below 2003D's value. The decline in properties was explained by the agglomeration of wood particles at higher amounts (even visible on the surface of printed coupons), and the deriving ineffective load transfer between wood and

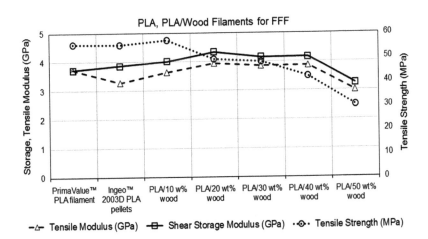

FIGURE 6.4 Test results for PLA, commercial PLA filament, and experimental PLA/wood powder filament for FFF.

Source: Data from Kariz, M., Sernek, M., Obućina, M., Kitek Kuzman, M., 2018. Effect of wood content in FDM filament on properties of 3D printed parts. *Materials Today Communications* 14:135–140.

PLA. Another explanation of lack of property improvement in WP/PLA is that the added wood particles featured a relatively small aspect ratio, and did not contribute to the mechanical properties as much as fibers or "long" particles would have. However, in general larger particles pose the risk of clogging the nozzle of the printer's extruder. The T_g and hence the service temperature of 2003D and WP/PLA was about the same. 65–66°C. The storage modulus in torsion of 2003D was 0.59 GPa, while for WP/PLA was 0.26–0.39 GPa depending on the composition: maximum values were reached at 20 wt% (0.39 GPa) and 10 wt% (0.38 GPa).

PLA is not the only material considered as a matrix for wood-filled AM materials. In fact, Gardan et al. (2016) combined WF and modified starch to formulate a fully sustainable and biodegradable filament for FFF, where WF was the reinforcement and the starch was the binder. *Starch* (or *amylum*) is a polymeric carbohydrate consisting of a large number of glucose (sugar) units, produced by most green plants as energy storage, and contained in large amounts in staple foods such as corn, potatoes, rice, wheat, and cassava. Pure starch is a white, tasteless, and odorless powder, and its chemical formula is $(C_6H_{10}O_5)_n(H_2O)$. It comprises two types of molecules: the linear, water soluble, and helical *amylose* (around 20 wt%), and the branched, insoluble in water *amylopectin* (about 80 wt%). Starch molecules arrange themselves in the plant in semi-crystalline granules, whose size depends on the plant, being f.e. about 2 μm in rice, and up to 100 μm in potatoes. The ingredients of the studied filament were:

- Beech flour LIGNOCEL® HB 120 by J. Rettenmaier & Soehne, comprising fibers 40–120 μm long
- Hydroxypropyl starch E1440: a starch typically added to food, modified with propylene oxide to become more resistant to thermal degradation and bacterial attack, and more stable against acid, alkaline, and starch-degrading enzymes
- Demineralized water.

One 1.2 mm diameter filament was fabricated by means of a printing head installed on a CNC machine, and cold extruded at 20°C, 60% relative humidity, and 1.3 m/min speed. The percent content of wood and starch was not reported. The filament WF/starch showed good adhesion between starch and wood upon microscopic analysis, and its tensile properties were: strength 5.45 ± 0.36 MPa, modulus 600 ± 96 MPa, and strain at break 1.25 ± 0.36%. Strength and modulus are about 1/10 and 1/5, respectively, of the same properties of wood /PLA reported by Kariz at al. (2018), and well below the ranges of values in Table 6.1, making this wood/starch composite not competitive with commercial wood/PLA filaments for load-bearing applications. Another downside is the large scattering in test results. Four coupons were tested: two displayed brittle failure, and two ductile break although within 1.5% strain at failure. A starch's feature is that its granules attach to each other and form larger granules. In the fabricated filament the starch wrapped and agglomerated the wood fibers during cooling, and formed a composite material. The interfacial adhesion between matrix and reinforcement in composites is critical to maximize the composite's mechanical properties, and microscopic analysis revealed that this adhesion was present in the tested filament. One shortcoming of this material was the humidity uptake by the starch and hemicellulose contained in the WF. The authors did not report the mechanical properties of hydroxypropyl starch E1440, but several types of starch exist featuring values of tensile strength and modulus likely to result in a wood/starch filament with higher tensile strength and modulus than those in the one fabricated. F.e., Rodriguez et al. (2006) tested a potato starch film characterized by tensile strength and elongation to break of 44.1 MPa and 5.9%, respectively, while Chang et al. (2006) studied a tapioca starch film displaying tensile strength and modulus of 28 MPa and 2.6 GPa, respectively.

Probably as a way to improve adhesion between wood particles and matrix, Kariz et al. (2016) experimentally investigated wood/adhesive composites in form of a FFF filament, and prepared two composites, by mixing ground beech WP featuring particle size up to 0.237 mm with two adhesives, used one at the time:

TABLE 6.6
Flexural Properties of Wood Powder (WP)/Adhesive

Materials	Flexural Modulus	Flexural Strength
	MPa	MPa
PVAc-17.5% WP	13	3
PVAc-20% WP	45	5
UF-15% WP	1,930	19
UF-17.5% WP	2,002	18

Source: Data from Kariz M., Sernek M., Kitek Kuzman M. 2016. Use of wood powder and adhesive as a mixture for 3D printing. *Eur. J. Wood Prod.* 74:123–126.

- TP polyvinyl acetate (PVAc). PVAc is also known as wood glue, school glue, and, in USA, Elmer's glue. PVAc is a polyvinyl ester, and its formula and density are $(C_4H_6O_2)_n$ and 1.19 g/cm^3.
- Urea-formaldehyde (UF) is a TS resin, with chemical formula and density $[(O)CNHCH_2NH]_n$ and 1.5 g/cm^3. UF's advantages are high tensile strength, flexural modulus, hardness, and HDT, and low water absorption. Over 70% of UF produced is employed as adhesives to bond particleboard, medium density fiberboard, hardwood plywood, etc.

The wood content in the filaments ranged from 12.5 wt% to 25 wt%. First, the filament composition was optimized based on the measurement of the force required to extrude the wood mixture. The higher was the wood content, the higher the viscosity and extruding force were for both composites, with the latter increasing exponentially. Four formulations were selected because they could be extruded with equipment driven by compressed air. The formulations and relative test results are listed in Table 6.6. The values are averages of five coupons. The extent of the standard deviation values suggests that the strength and modulus values across wood content, except for modulus of PVAc mixtures, are statistically the same, leading to conclude that the change in the amount of filler was too small to be influential. The authors reported no mechanical property of unfilled PVAc and UF, preventing the assessment of WP's contribution to the composites' bending performance. However, they concluded that all tested mixtures were inadequate for structural applications. Table 6.7 lists average values of tensile strength and modulus for grades of PVAc and UF not utilized by Kariz et al. (2016) but mentioned here as a benchmark. Notably, the researchers developed a new process called *liquid/paste deposition modeling* and based on a home-built printer, equipped with an extruder derived from a modified grease gun, an interchangeable nozzle with different diameters, and a container for the WP-adhesive mixture that was pushed out by a piston driven by compressed air manually regulated according to the viscosity of the different mixtures.

Pitt et al. (2017) have also investigated a WF/UF filament for FFF, and prepared two versions of it: WF/UF, and WF/UF/glass fibers (GF). Both versions featured mechanical properties comparable to those of commercial wood-based composites, such as particleboards. The following ingredients were combined:

- WF: EPWF 110 (Eden Products, UK), a wood waste feedstock from European white softwood, with 75 μm median particle size, and a moisture content of 14% acceptable for use with wood adhesives

TABLE 6.7

Mechanical Properties of Polyvinyl Acetate (PVAc) and Urea-Formaldehyde (UF)

Material	Tensile Modulus	Tensile Strength	Flexural Strength at Yield
	MPa	MPa	MPa
PVAc film[a]	320	10.2	NA
UF[b,c]	9,000	30, 65	100

Sources: [a]Gardan et al. 2016. Poly(vinyl alcohol) – poly(vinyl acetate) composite films from water systems: formation, strengthdeformation characteristics, fracture. *IOP Conference Series: Mater. Sci. Eng.* 111, 012009. [b]MakeItFrom 2020. Urea Formaldehyde (UF). https://www.makeitfrom.com/material-properties/Urea-Formaldehyde-UF (accessed April 26, 2020). [c]SubsTech 2012. Thermoset Urea Formaldehyde (UF) www.substech.com/dokuwiki/doku.php?id=thermoset_urea_formaldehyde_uf (accessed April 26, 2020).

- UF: Cascorit 1205 (Casco Products, Sweden) adhesive for wood and hardener 2545. Pot life of the UF/hardener system is 8 h, sufficient time for formulation, preparation and printing of the pastes.
- GF: Vitrostrand (Owens Corning, USA) milled fibers, about 100 µm long, made from Advantex® E-Glass.

The authors mixed WF and UF (plus hardener) in 13 wt% and 87 wt%, respectively, and WF, GF, and UF (plus hardener) in 9, 10, 81 wt%, respectively, and fabricated test coupons for tensile and flexural testing: half coupons were molded by pouring the two formulations in rectangular molds, and the remainder were printed by extruding the two formulations through a 1.6 mm diameter nozzle into a FFF filament. As the pastes were extruded, the fibers progressively aligned themselves along the extruding direction. The mixed pastes required a viscosity sufficiently low to be easily extruded through the nozzle, but sufficiently high to enable the extruded material to keep its shape until curing, and the layers to support the adjacent upper layers without widening. All printed and molded coupons were cured in oven according to a multi-step cycle from 50°C to 110°C. Results of tensile (ISO 527) and three-point flexural (ISO 178) testing are shown in Table 6.8 as average and standard deviation values, and ratio of values from printed coupons to values from molded coupons. Printed coupons outperformed molded coupons in all measured properties by 19–66%, which could have stemmed from: (a) in molding the coupons "the pastes were simply placed into rectangular-shaped molds," whereas printed coupons were extruded under 20 psi of pressure, which yielded higher density and largely lower void content; (b) aligned GF in printed samples vs. randomly oriented GF in molded samples. Adding GF to printed coupons raised tensile strength and flexural modulus by 29% and 28%, respectively, but left flexural strength basically unchanged. Since the lower half of bending coupons is subjected to tension, and their upper half to compression, we infer that WF/UF/GF formulation did not effectively perform in compression. In fabricating particleboards, pressure is applied to wood to compress the particles, and control the thickness, hence it would be insightful to compare the coupons printed by Pitt to coupons of the same material molded under pressure. However, the stiffness of printed coupons is comparable to similar commercial materials. In fact, the flexural modulus of printed coupons is 6.8–8.7 GPa, whereas that of wood panel products (such as particleboards, fiberboards, and plywood) is 2.8–8.6 GPa, and wood/plastic is 1.5–4.2 GPa (Cai and Ross 2010).

TABLE 6.8

Test Results of Printed and Molded Wood/Composites

Wood/Composite	Weight Fraction	Void Content	Tensile Strength (TS)	Flexural Strength (FS)	Flexural Modulus (FM)	TS Printed/TS Molded	FS Printed/FS Molded	FM Printed/FM Molded
	%	%	MPa	MPa	GPa			
WF/UF[a] Molded	13/87	12.3 ± 0.88	9.7 ± 1	44 ± 3	5.4 ± 0.4	N/A	N/A	N/A
WF/UF[a] Printed	13/87	7.3 ± 1.21	17 ± 1	57 ± 5	6.8 ± 0.6	N/A	N/A	N/A
WF/UF/GF[b] Molded	9/81/10	12.3 ± 1.05	16 ± 1	38 ± 5	5.9 ± 0.7	1.66	1.23	1.19
WF/UF/GF[b] Printed	9/81/10	4.9 ± 0.46	22 ± 2	58 ± 2	8.7 ± 0.2	1.27	1.41	1.36

Source: Modified from Pitt, K., Lopez-Botello, O., Lafferty, A. D., 2017. Investigation into the material properties of wooden composite structures with in-situ fibre reinforcement using additive manufacturing. *Compos. Sci. Technol.* 138:32–39. Reproduced under the terms of the Creative Commons CC-BY license.

Notes:

[a] Wood flour/urea-formaldehyde.

[b] Wood flour/urea-formaldehyde/glass fibers.

FIGURE 6.5 Cross section of spruce chips (SC) filled samples. Left: SC/cement/water. Right: SC/gypsum/water.

Source: Adapted from Henke, K., Treml, S. 2013. Wood based bulk material in 3D printing processes for applications in construction. *Eur. J. Wood Prod.* 71:139–141. Reproduced with permission from Springer-Verlag.

6.5.3 Wood/Concrete for AM

In AM materials, wood has not only been combined to polymers, but also to gypsum and cement, in order to be utilized in construction, which is fitting, being wood one of the oldest building materials, with evidence indicating that homes built over 10,000 years ago were made of timber (Woods 2016). Composite materials comprising wood as an aggregate and cement as a binder have been selected for construction components (like cement-bound particle boards) since the 1920s (Filipaj 2010). Wood has been a successful construction material over very long time, because it combines strength, stiffness, low cost, lightweight, and low thermal conductivity.

Henke and Treml (2013) employed wood as agglomerate and various materials as binders in a non-FFF, layer-based deposition process to fabricate buildings. They experimented with spruce chips (SC) featuring bulk density of 192 kg/m^3, and particle size of 0.8–2 mm, combined with gypsum, methylcellulose, sodium silicate, and differing types of cement. The composite materials providing the best results were (Figure 6.5):

- SC/gypsum/water, with SC/gypsum weight ratio of 0.33, and water/gypsum weight ratio of 0.75. This formulation yielded good contour accuracy and chip consolidation.
- SC/cement/water, featuring SC/cement weight ratio of 0.15 and water/cement weight ratio of 0.80. The resulting material had bulk density of 0.7–0.8 g/cm^3, and bending strength of 0.5–0.95 MPa, comparable to wood wool (wood slivers cut from logs) lightweight building boards.

Test coupons were printed according to the layer-based process schematically pictured in Figure 6.6. In step 1, a layer B of wood and binder was laid on the platform P. In step 2, water aerosol was sprayed onto B by actuator A and activated the binder only in the area of B exposed by mask M1. Step 2 was repeated in step 3 until step 4, using various amounts of water and masks (M2, etc., MX) with different openings. In Step 5 the part was completed.

Henke et al. (2016, 2017) formulated a novel concrete, made of wood-filled cement with performance comparable to that of concrete with density of 800–2,000 kg/cm^3 (*lightweight concrete*), and developed an extrusion-based AM process to shape this material into buildings. The main ingredients of the new concrete were Portland-limestone CEM II A-ll 32.5 R cement (1,000 g), water (610 g), and LIGNOCEL® 9 softwood chips (160 g), with a cement:wood volume ratio of 1:1. Also incorporated were ingredients imparting the concrete certain characteristics, such as accelerators to speed up concrete's hardening, pigments to change concrete's color, corrosion inhibitors to minimize the corrosion of steel and steel bars in concrete, and bonding agents to form

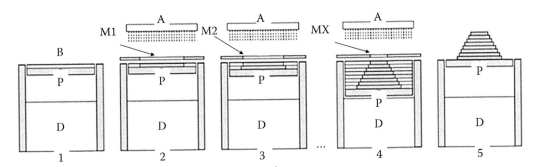

FIGURE 6.6 Schematic of process of the selective activation of thin layers consisting of mixtures of bulk-material and binding agents. B mixture of bulk and binder, P platform, M1-MX masks, A activator, D drive and control.

Source: Adapted from Henke, K., Treml, S. 2013. Wood based bulk material in 3D printing processes for applications in construction. *Eur. J. Wood Prod.* 71:139–141. Reproduced with permission from Springer-Verlag.

a bond between previously and newly laid concrete. The final wood/cement possessed density of 995 kg/m^3, and thermal conductivity of 0.25 W/(mK). Test coupons were poured relying on a numerically controlled printing system, consisting in a custom designed and custom-built extruder mounted on an industrial KUKA KR 150 L110 6-axis robot. The extruder included a 20 mm diameter nozzle expelling a strand with oval cross section measuring 25 × 10 mm. Compression and bending testing were carried out on coupons according to DIN EN 12390–3:2009-07, and DIN EN 196–1:2005-05, respectively. The test results (Figure 6.7) indicated that the extrusion-based printing produced a material with mechanical performance similar to that of conventional concrete pouring. Large-scale components were successfully printed to test scalability of material and process. A hollow insulation wall structure measuring 500 × 1500 × 930 mm was printed in 7 h without including time breaks. Moreover, the wood/cement's workability also was assessed: the new concrete responded to sawing, milling, drilling, and sanding like a metal. In conclusion, the

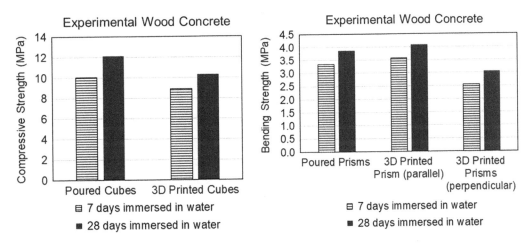

FIGURE 6.7 Test results comparing printed vs. poured wood/concrete.

Source: Data from Henke, K, Talke, D., Winter S. 2016. Additive Manufacturing of Building Elements by Extrusion of Wood Concrete. *World Conference on Timber Engineering* (WCTE 2016).

novel concrete featured the following advantages: load-bearing capacity, lightweight, thermal insulation, and containing a sustainable material.

6.5.4 WOOD/PLA COMPATIBILITY AND INTERFACIAL ADHESION

Since wood is hydrophilic and PLA hydrophobic, they feature poor interfacial compatibility, resulting in poor interfacial adhesion and modest mechanical properties (Wang et al. 2007; Wang et al. 2008). Several investigators have studied ways to enhance this compatibility, including Guo et al. (2018), who experimentally demonstrated that adding TPU to WF/PLA improved strength of WF/PLA, and grafting PLA with glycidyl methacrylate (GMA) enhanced interfacial compatibility, and increased strength of PLA/WF/TPU. They prepared a filament composed of Ingeo™ 4032D PLA, and nominal 100 mesh poplar WF, and selected the following materials as tougheners: (a) TPU by BASF with density of 1.19 g/cm^3, and Shore hardness 85 A, (b) Capa™ 6500 PCL with MW 50,000 g/mol, and MFI 6–8 g/10 min, (c) Engage® 8150 polyethylene-octene copolymer (POE) with density of 0.87 g/cm^3 and MFI 0.5 g/10 min. TPU is recyclable (Wang et al. 2018). PCL is not formulated from renewable sources but is biodegradable and nontoxic (Lang et al. 1999), and widely chosen to toughen PLA. In order to improve WF/PLA interfacial compatibility, graft copolymers (GC) containing GMA were added, applying an efficient and environmentally friendly method. Finally, PE wax was added as a lubricant to improve the processability during extrusion. GMA was used alone and in presence of dicumyl peroxide (DCP) acting as initiator. Figure 6.8 illustrates that adding WF to PLA distinctly lowered PLA's impact, tensile and flexural strengths, and that, among PCL, POE, and TPU individually added to PLA/WF, TPU was the only additive to improve all three properties: in fact, impact, tensile, and flexural strength of WF/PLA/TPU exceeded those of PLA/WF raised by 51%, 34%, and 10%, respectively. SEM images revealed that TPU increased compatibility between PLA and WF, and FTIR spectra confirmed that PLA, WF and the additives were compounded successfully, and no chemical reactions occurred during the melt blending. Grafting GMA increased the interfacial compatibility, because GMA contains two functional groups: the acrylic group that bonds to PLA's polymeric chain, and the

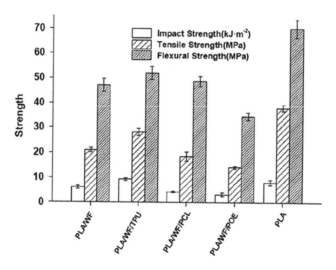

FIGURE 6.8 Mechanical properties of PLA and PLA composites for FFF, comprising wood flour (WF), and tougheners (TPU, PCL and POE).

Source: Guo, R., Ren, Z., Bi, H. et al. 2018. Effect of toughening agents on the properties of poplar wood flour/poly(lactic acid) composites fabricated with Fused Deposition Modeling. *Eur. Polym. J.* 107:34–45. Reproduced with permission from Elsevier.

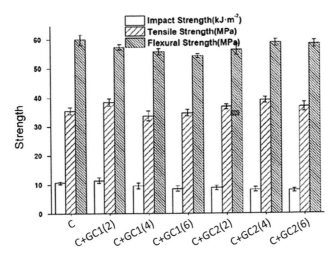

FIGURE 6.9 Mechanical properties of WF/PLA composites for FFF. Legend: C control = PLA/WF/TPU/ PE. GC grafted copolymer. GC1(X) = PLA/WF/GMA/DCP, GC2(X) = PLA/GMA/DCP, X = 2, 4, 6 wt% of GC added to C.

Source: Guo, R., Ren, Z., Bi, H. et al., 2018. Effect of toughening agents on the properties of poplar wood flour/poly(lactic acid) composites fabricated with Fused Deposition Modeling. *European Polymer Journal* 107:34–45. Reproduced with permission from Elsevier.

epoxy group that links to the hydroxyl groups in WF (Xu et al. 2012). GMA and DCP were added to PLA/WF/TPU/PE (control C) in 2, 4, and 6 wt%, and relative samples of C were tested to measure impact, tensile, and flexural strengths (Figure 6.9). Only the formulation C+GC1(2) with 2 wt% of GMA/DCP outperformed C, by raising impact and tensile strengths by 7.8% and 8.4%, respectively, although it trailed in flexural strength. SEM micrographs of tested coupons confirmed improvement in WF/PLA interfacial adhesion in the presence of GMA, because when WF particles were pulled out of PLA, the interface was clear in coupons without GMA, but was blurred and without voids in coupons with GMA. By feeding a PLA/WF/TPU composite filament developed in this study to a FFF printer, a table and some chairs were fabricated.

It was shown earlier that wood/PLA interfacial compatibility is inadequate, and must be enhanced in order to drive up the mechanical performance of wood/PLA composites for FFF. The paper by Tao et al. (2017) confirmed that wood/PLA compatibility is poor, and described how it reduces mechanical properties of wood/PLA composites. The authors combined 4032D PLA pellets with 5 wt% of QM0.4L aspen wood WF (Chishun LLC, China) from dried wood sawdust with 14 μm particle size, and extruded a 1.75 mm diameter filament. Employing the same PLA but unfilled, they also extruded a 1.75 filament. Test coupons for tension, TGA, TMA, and DSC were printed out of PLA and wood/PLA via FFF at 210°C through a 0.4 diameter nozzle. By adding WF to PLA, tensile strength and T_g dropped from 27 MPa and 67°C for PLA to 22 MPa and 60°C for WF/PLA. SEM analysis confirmed poor interfacial bonding between wood and PLA. Figure 6.10 displays SEM micrographs of the fracture surface of tensile coupons: the PLA's surface on the left is smoother than wood/PLA's surface on the right, and some gaps (black arrows) appear at the interface of wood (white arrows) and PLA, revealing poor interfacial bonding between the two components.

Coupling agents (more) and plasticizers (less) have been studied as additives to improve compatibility between WF and PLA (Xie et al. 2017). For example, Dow DuPont Fusabond® M603 a coupling agent made of ethylene copolymer for wood/polymer composites that not only strengthens the composite but also improves wood fiber dispersion, wood/polymer bonding, and reduces water absorption. Common coupling agents are silanes and titanates. Lee et al. (2008)

FIGURE 6.10 SEM micrographs of printed samples of PLA and wood/PLA. White and black arrows indicate woof flour and gaps, respectively.

Source: Modified from Tao, Y., Wang, H., Li, Z. et al. 2018. Development and Application of Wood Flour-Filled Polylactic Acid Composite Filament for 3D Printing. *Materials* 10:339. Reproduced under the terms and conditions of the Creative Commons Attribution (CC BY) license.

extruded polymer blends not specific for AM by combining 1 and 3 wt% silane with WF, PLA and talc, and measured that, vs. plain PLA, silanes enhanced tensile properties: strength raised from 60 MPa to 85 MPa, and modulus from 1,300 MPa to 2,043 MPa. *Plasticizer* are monomeric liquids and low-MW materials that improve melt flow properties, reduce brittleness, and enhance flexibility. Basically, they are small molecules that push the molecules of the polymer to which they are added further apart, and thus weaken the forces between the polymer's molecules, and render the polymer softer and more flexible, acting almost as "lubricant" between segments of the polymer chains. Examples of plasticizers are phthalate esters. Plasticizers are present in PVC cables and (mostly) films, rubbers adhesives, and coatings.

Xie et al. (2017) evaluated the effect of three plasticizer formulations on the properties and processability of a FFF filament made of WF/PLA. They combined 270 g of 140–160 mesh size poplar WF and 630 g of unspecified PLA, and, incorporating glycerol and tributyl citrate (TBC), prepared the following formulations (with percentages by weight) which were extruded into FFF filaments:

- WF/PLA/4% glycerol (WP4G)
- WF/PLA/2% glycerol/2% TBC (WP22)
- WF/PLA/4% TBC (WP4T).

Test coupons of the above formulations were printed, and their test results are reported in Table 6.9. The 4% TBC yielded the highest interfacial compatibility and tensile properties, letting WF penetrate the crystalline regions of PLA, whereas 4% glycerol imparted the highest flowability during filament extrusion. Tensile stress concentrated at the WF/PLA interface, and better interface compatibility enabled the composite to withstand greater stress values, and made WP4T the strongest and most ductile tested formulation. SEM images confirmed the degree of compatibility, since cross sections unveiled a decreasing amount of void and smoother surface going from WP4G to WP22 to WP4T. The optimal values of MFI for filament extrusion must fall within a suitable range: if MFI is very low, the pressure required to extrude becomes very high, if MFI is very high, the filament will keep its shape once laid under the weight of above layers . All MFI values measured fell in the proper range, and WP4G yielded the highest MFI, corresponding to the lowest pressure required for extrusion. TGA ascertained the material's processing and maximum service temperature, and pointed out that the mass loss of all formulations was gradual and in 5–10% range until 250°C, with WP4T

TABLE 6.9

Test Results for Wood Flour (WF)/PLA/Plasticizer Composites

Formulations	Tensile Strength	Elongation to Break	MFI	Crystallinity Degree	Water Absorption
	MPa	%	g/min	%	%
WP4G[a]	18.6	1.15	14.8	40.3	9.6
WP22[b]	21.4	1.56	9.7	39.1	7.4
WP4T[c]	24.5	1.78	7.1	38.4	3.1

Source: Data from Xie, G., Zgang Y., Lin, W. 2017. Plasticizer combinations and performance of wood flour-poly(lactic acid) 3D printing filaments. *BioResources* 12 (3): 6736–6748.

Notes:

[a]WF/PLA/4 wt% glycerol.

[b]WF/PLA/2 wt% glycerol/2 wt% tributyl citrate (TBC).

[c]WF/PLA/4 wt% TBC.

and WP4G degrading the least and the most, respectively. Between 250°C and 400°C the degradation rate plunged, and all materials lost almost their entire mass.

Among coupling agents improving wood/PLA interfacial adhesion, methylenediphenyl-diisocyanate (MDI) was proven effective by Petinakis et al. (2009). As reported in Table 6.10, when added to a WF/PLA composite not specifically formulated for AM, 1 wt% of MDI increased tensile strength and modulus by 11% and 134%, respectively vs. unfilled PLA, and by 11% and 20%, respectively vs. PLA/WF (60/40 wt%).

Optimizing interface adhesion in wood/PLA composites not specific for AM was pursued in the study by Faludi et al. (2013) that is mentioned here because it may be insightful for wood/PLA composites formulated for AM. The authors disclosed that reactive coupling agents N,N-(1,3-phenylene dimaleimide) (BMI) and 1,1-(methylenedi-4,1-phenylene) bismaleimide (DBMI), separately used, distinctly raised the tensile strength and modulus in a generic wood/PLA composite. In fact, the average values between BMI and DBMI of tensile strength and modulus of wood/PLA were higher up to 124% and 44%, respectively than those of wood/PLA without BMI and DBMI. The authors combined 4032D PLA, Filtracel EFC 1000 wood fiber (Rettenmaier and Söhne)

TABLE 6.10

Test Results of Wood Flour (WF)/PLA Composite Not Specific for AM

Material	Tensile Strength	Tensile Modulus	Elongation at Break
	MPa	GPa	%
PLA	60.0	3.79	7.3
PLA/WF (60/40 wt%)	60.2	7.40	1.1
PLA/WF/MDI[a] (70/30/1 wt%)	66.7	6.54	2.0
PLA/WF/MDI[a] (60/40/1 wt%)	65.7	8.87	1.2

Source: Data from Petinakis, E., Yu, L., Edward, G., et al., 2009. Effect of Matrix–Particle Interfacial Adhesion on the Mechanical Properties of Poly(lactic acid)/Wood-Flour Micro-Composites. *J. Polym. Environ.* 17:83–94. Reproduced with permission from Springer-Verlag.

Note:

[a]MDI methylenediphenyl-diisocyanate.

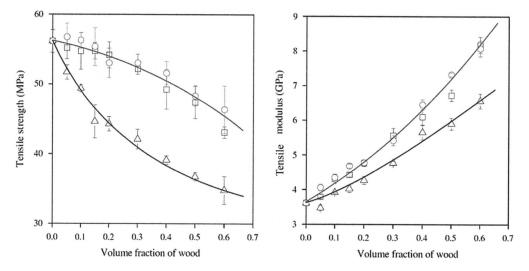

FIGURE 6.11 Tensile strength (left) and modulus (right) of wood/PLA composites with and without coupling agents (BMI, DBMI) not formulated specifically for AM. Symbols: □ BMI, ○ DBMI, and △ no coupling agent.

Source: Faludi, G., Dora, G., Renner, K., et al. 2013. Improving interfacial adhesion in PLA/wood bio-composites. *Composites Science and Technology* 89:77–82. Reproduced with permission from Elsevier Ltd.

featuring 210 µm average particle size and 6.8 aspect ratio, and 5 wt% of BMI and DBMI separately added. Plates 1 mm thick were compression molded with different fiber volume fraction (V_f) values. Figure 6.11 points out that tensile strength and modulus decreased and increased, respectively, with greater V_f (indicating weak but not absent interfacial adhesion), and that both properties improved by adding BMI and DBMI in comparison to wood/PLA without coupling agents. The authors tentatively attributed the positive effect of BMI and DBMI on modulus to the possibility that high amount of debonding occurred between wood and PLA without coupling agents, causing the formation of voids that penalized the composite's stiffness. Similarly, lower strength without BMI or DBMI might have been caused by wood fibers debonding from PLA upon loading at an earlier time than it occurred in presence of coupling agents. Indeed, acoustic emission testing revealed that the cumulative number of acoustic signals corresponding to local deformation processes (such as debonding and fiber fracture) doubled in wood/PLA without BMI or DBMI vs. wood/PLA with them.

6.5.5 Surface Properties, Self-Shaping Design, New Process

In some applications, printed parts may be the substrate for thin overlays, liquid surface coatings, and finishes such as paint and varnish. In those cases, the wettability and surface roughness of the printed substrate are factors affecting bonding adhesion between printed layer and adjacent layer, and bond performance during use. The wettability of a solid surface by a liquid is, in fact, linked directly to the contact angle θ between a droplet of that liquid and that solid surface (Nguyen et al. 2021), illustrated in Figure 6.12: the lower θ, the more the liquid spreads over the surface, and the

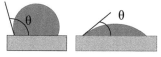

FIGURE 6.12 Schematic representation of contact angle θ. Left: large angle θ, low wettability, hydrophobic surface. Right: small θ, high wettability, hydrophilic surface.

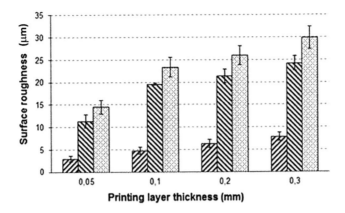

FIGURE 6.13   Surface roughness values parallel to printing direction of wood/PLA printed specimens.

Source: Modified from Ayrilmis, N. 2018. Effect of layer thickness on surface properties of 3D printed materials produced from wood flour/PLA filament. *Polymer Testing* 71:163–166. Reproduced with permission from Elsevier.

higher the surface's wettability by the liquid, and the stronger the adhesion of it to the surface. More exactly, wettability is good when θ is below 90°, and is poor when θ exceeds 90°. Ayrilmis (2018) experimentally studied the effect of the thickness of deposited layers on wettability and surface roughness of printed specimens using a commercial unspecified wood/PLA filament (30/70 wt%) featuring 1.75 mm diameter and 1 wt% moisture content. Increasing layer thickness raised wettability, and surface roughness, which in turn strengthened the adhesive bonding at the interface between printed substrate on one hand, and varnish and paint on the other. The surface roughness was measured through three parameters (ISO 4287): mean arithmetic deviation of profile (R_a), mean peak-to-valley height (R_z), and maximum peak-to-valley height (R_y). The values of R_a, R_z, and R_y collected in direction parallel and perpendicular to the specimen's printed direction were plotted in Figures 6.13 and 6.14, respectively, as a function of the layer's thickness. R_a, R_z, and R_y

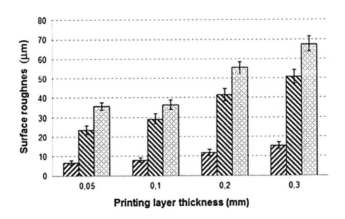

FIGURE 6.14   Surface roughness values perpendicular to printing direction of wood/PLA 3D printed specimens.

Source: Modified from Ayrilmis, N. 2018. Effect of layer thickness on surface properties of 3D printed materials produced from wood flour/PLA filament. *Polymer Testing* 71:163–166. Reproduced with permission from Elsevier.

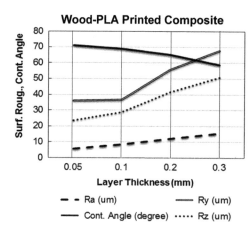

FIGURE 6.15 Effect of layer thickness on surface roughness (R_a, R_y, R_z) of printed wood/PLA measured in mm perpendicularly to the printing direction, and contact angle.

Source: Data from Ayrilmis, N. 2018. Effect of layer thickness on surface properties of 3D printed materials produced from wood flour/PLA filament. *Polymer Testing* 71:163–166.

steadily increased with layer thickness, and the printing direction notably affected the roughness of the specimens. In this study and AM precesses in general, as the layer thickness grows, the geometrical accuracy of the printed objects decreases but so does its printing time. Inversely, with smaller layer thickness, the surface roughness diminishes, but printing time and overall manufacturing cost increases. In load-bearing applications surface roughness affects fatigue performance, because roughness may act as microscopic defects becoming cracks. In this study a rise in the layer thickness multiplied the numbers of minute pores on the surface of the samples, causing the absorption of water, and producing higher wettability. The contact angle θ was measured at 5, 10, 20, and 30 s after water drops were applied, because the longer the time interval the lower θ. The averages of θ values at different time values were plotted in Figure 6.15, and indicate that wettability and surface roughness steadily increased in response to raising layer thickness.

The presence of natural fibers (wood, flax, hemp, etc.) reinforcing polymeric matrices induces a relatively high amount of moisture and water absorbed in the resulting composites, which degrades their mechanical properties and durability (Azwa et al. 2013). Inspired by nature, Le Duigou et al. (2016) exploited the anisotropic properties of parts printed through FFF, and the moisture absorption by wood fibers, and designed and fabricated an ingenious self-shaping component that represents an example of *four-dimensional (4D) printing* (Tibbits 2014; Pei 2014; Ding et al. 2018; Wu et al. 2018), or self-assembly. 4D printing is the process of 3D printing objects that change their shape (f.e folding) once printed without utilizing any electromechanical device but in response to external stimuli such as moisture, light, and heat (Section 2.11). In fact, Le Duigou et al. (2016) experimented in one area of 4D printing termed *biomimetic 4D printing* (Sydney Gladman 2016; Studart 2018) that fabricates nature-inspired designs, and mimicked the mechanism employed by pine cones to release their ripe seeds in the environment by opening their scales in response to decreasing moisture content. Both upper and lower side of the cone's scales (Figure 6.16) contain stiff cellulose fibers that are embedded in a flexible matrix, and oriented in two directions with respect to the scale's long axis, namely parallel to it in the upper side, and perpendicular to it in the lower side. When the moisture decreases, the scale shrinks, but, because shrinking is greater perpendicularly to the fibers' direction, the scale's lower side shrinks more than its upper side, causing the scale to bend downward, open, and release its seeds. The authors exploited the freedom allowed by FFF in orienting the deposited filament, and printed a strip featuring a similar actuation mechanism. They choose the filament woodFill Fine, featuring 2.85 mm diameter, and composed of 15 wt% recycled wood mixed to PLA/PHA, and mimicked the cellulose fiber orientation, by depositing the filament (Figure 6.17 (a)) along the strip length (0° layer) in the strip's upper half and perpendicularly to the strip length (90° layer) in the strip's lower half. Based on equations developed to model the bending of bimetal strips uniformly heated (Timoshenko 1925), the authors designed a self-shaping strip that bent upon immersion in

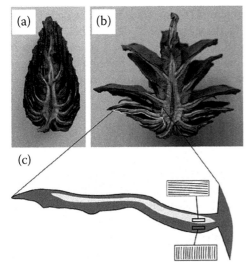

FIGURE 6.16 Sections of pine cone: (a) closed; (b) open; (c) cone's scale open, with lines representing orientations of cellulose fibrils in the upper side (white band) and the lower side (dark portion) of the scale.

Source: Burgert, I., Fratzl, P. 2009. Actuation systems in plants as prototypes for bioinspired devices. *Phil. Trans. R. Soc. A* 367:1541–1557. Reproduced with permission from The Royal Society.

deionized water at room temperature (Figure 6.17 (b), (c)), and also controlled and maximized the strip curvature, by exploiting the FFF capabilities, namely by varying: (a) the ratio of 0° layer's thickness over 90° layer's thickness, (b) the printing parameters, and hence the tensile modulus of the 0° layer's thickness and 90° layer's thickness, and consequently the water uptake in terms of rate and maximum amount.

Rosenthal et al. (2018) employed the *liquid deposition modeling* (LDM) process already applied by Kariz et al. (2016) (Section 6.5.2) with the goal to significantly exceed the wood content reached by the latter authors. They printed test coupons out of a paste-like suspensions composed of ground beech sawdust and methylcellulose (MC) dissolved in water, with wood content up to 89 wt% in dry mass. MC was selected as a binding agent and lubricant, because it is renewable, and already present as a binder (and thickener) in adhesives, cosmetics, food, paints, and pharmaceuticals. The mixtures comprised fractions of the ground sawdust with two particle size

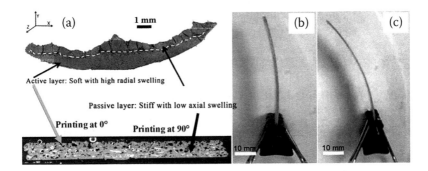

FIGURE 6.17 (a) Upper: principle of hygromorphic design, that is responding to changes in environmental humidity. (a) Lower: strip printed out of wood/PLA with filament deposited at 0° (black) and 90° (grey) with respect to strip's length in upper and lower half of strip, respectively. Example of hygromorphic biocomposite made via FFF before (b) and after immersion (c). Scale in (b) and (c) is 10 mm.

Source: Modified from Le Duigou, A., Castro, M., Bevan, R., Martin, N., 2016. 3D printing of wood fibre biocomposites: From mechanical to actuation functionality. *Materials and Design* 96:106–114. Reproduced with permission from Elsevier.

distributions: fraction A contained particles with size up to 0.25 mm; fraction B included fraction A particles mixed at a ratio of 1:1 with particles having a size above 0.25 mm and below 0.4 mm. MC powder (Carl Roth, Germany) featured viscosity of 3,660 mPa s, and was added to water in ratio of 1:20 and 1:30. After dissolving and swelling, the gel-like mass was mixed with the ground sawdust according to the following mixtures (mass-related values), and turned into paste:

- M1: MC:water 1:30, MC:wood fraction A 11:89,
- M2: MC:water 1:20, MC:wood fraction B 15.5:84.5,
- M3: MC:water 1:20, MC:wood fraction A 14.5:85.5.

The pastes were processed at 1 mm/min on a standard printer equipped with a self-made extruder consisting of: (a) a cylindrical plastic cartridge featuring internal diameter of 27 mm, and a nozzle 8 mm in diameter and 51 mm in length; (b) a NEMA 17 linear stepper motor driving a piston towards the nozzle. Pretesting led to the following lessons learned: (a) a ratio MC/water too small resulted in phase separation, with water oozing out of the nozzle and wood remaining in the cartridge; (b) a wood content too high prevented the stepper motor to extrude the mixture; (c) since water is removed during drying, the higher is the water content, the larger is the volumetric shrinkage. The average test results are collected in Table 6.11: density was 0.33–0.48 g/cm^3, and bending strength and modulus were 2.3–7.4 MPa and 285–733 MPa, respectively. Binder/water ratio and wood particle size affected the physical properties: density, flexural strength and modulus improved with raising viscosity of dissolved MC and decreased with growing particle size. Volumetric shrinkage after drying was 17.3–20.0%.

Zhang et al. (2016) investigated wood/plastic composites (WPC) made of powders of bamboo, microcrystalline cellulose, cornstalk, eucalyptus, pine, and rice husk as fillers, and PP and copolyester hot-melt adhesive (cPET) as a matrix, plus coupling agents and viscosity reducers. The powders were obtained by smashing agroforestry waste into particles of diameter up to 100 µm. Particles 60 µm in size produced printed samples with better quality than particles 80 µm in size. Since the WPs were chemically polar, whereas PP and cPET were nonpolar, the WPs were alkali-treated and dried to improve their interfacial bonding to PP and cPET, with cPET resulting more suitable for LS than PP. The WPCs were formulated for LS, and LS parameters were optimized based on the feedstocks. Porosity decreases mechanical properties. Laser sintered WPC articles contained porosity up to 50%, except cellulose WPC whose porosity was 28–31%. A method to reduce porosity is infiltrating printed articles with wax or resin, with the latter leading to higher

TABLE 6.11

Test Results of Coupons Printed out of Wood, MC, and Water Using LDM

Mixture	Density	Height After Printing	Height After Drying	Height Reduction	Flexural Strength	Flexural Modulus
	g/cm^3	mm	mm	%	MPa	MPa
M1	0.33	43.5	36.0	17.2	2.3	285
M2	0.40	44.0	36.1	18.0	5.1	547
M3	0.48	43.8	35.1	19.9	7.4	733

Source: Adapted from Rosenthal, M., Henneberger, C., Gutkes, A., Bues, C. T. 2018. Liquid Deposition Modeling: a promising approach for 3D printing of wood. *European Journal of Wood and Wood Products* 76:797–799. Reproduced with permission from Springer-Verlag.

mechanical properties than with wax. Among all WPCs, WPC with cellulose exhibited the highest cost and mechanical strength: after epoxy infiltration, tension, flexure, and impact strengths were 11 MPa, 19 MPa, and 5.51 kJ/m^2, respectively. Tensile and flexural strengths were lower than in unfilled PLA for AM. The optimal content of cellulose was 25 wt%. Applications of these WPCs are investment castings, models, prototypes, and artistic articles.

6.5.6 RECYCLED WOOD FURNITURE WASTE

Since the furniture industry produces a large amount of wood-based waste (that only in Michigan exceeds 150 ton/day), Pringle et al. (2018) devised a process to recycle furniture waste into a feedstock for a WPC filament for FFF, and print furniture components. Their process started with (a) board scraps made of light-density fill (LDF) and medium-density fill (MDF) and (b) melamine-particleboard-paper impregnated with phenolic resins, and consisted in four steps: (a) pulverizing the board scraps into a 80 μm particle size powder; (b) mixing the powder with Ingeo™ 4043D PLA serving as matrix for the resulting composite filament; (c) extruding the mixture into a WPC 1.65 mm diameter filament; (d) processing the WPC filament on a printer with the following settings: nozzle diameter 0.5 mm, layer height 0.15 mm, printing temperature 185°C, print speed 62.5 mm/s. No flow disruption was detected during printing due to moisture inside the filament, but, since wood and PLA are hygroscopic, the filament should be stored in a controlled low-humidity environment. Phenolic and melamine resins had a negligible effect on the filament extrusion, because both are TS polymers and do not liquefy at the extrusion temperature, but they behave as solid particulates and affect the filament's material properties, namely higher amounts of resin may increase mechanical properties, water, and heat resistance. Some test articles were successfully printed. However, in comparison to unfilled PLA filaments, there were more instances of nozzle clogging and general filament blockages, most likely caused by inconsistent particle size. The effect of printing settings on the printed items, and some aspects in their process that needed optimization in order to maximize the results were elucidated. Discarded printed articles can be recycled into WPC filaments, although the degradation of polymers undergoing heating upon recycling should be assessed, in order to evaluate a possible ensuing decrease of the filament's performance. Future investigation should focus on: (a) wettability of wood-waste powder to possibly increase the filler amount while avoiding agglomeration, and (b) evaluation of other polymeric matrices.

FURTHER READINGS

Clemons C. 2008. Raw materials for wood-polymer composites. In *Wood-polymer* Composites, ed. K. Oksman Niska, M. Sain, 1–22. Boca Raton: CRC Press.

USDA (United States Department of Agriculture). 2010. *Wood handbook – Wood as an Engineering Material*. General Technical Report FPL-GTR-190. Madison, WI, Centennial Edition. https://www.fpl.fs.fed.us/documnts/fplgtr/fpl_gtr190.pdf (accessed April 26, 2020).

REFERENCES

Agricultural Marketing Resource Center. 2018. *Sawdust*. https://www.agmrc.org/commodities-products/biomass/sawdust (accessed April 26, 2020).

Aydemir, D., Kiziltas, A., Erbas Kiziltas, E., Gardner, D. J., Gunduz, G. 2015. Heat treated wood-nylon 6 composites. *Compos. Part B Eng.* 68:414–423.

Ayrilmis, N. 2018. Effect of layer thickness on surface properties of 3D printed materials produced from wood flour/PLA filament. *Polym. Test.* 71:163–166.

Ayrilmis, N., Jarusombuti, S., Fueangvivat, V., Bauchongkol P. 2011a. Effect of thermal-treatment of wood fibers on properties of flat-pressed wood plastic composites. *Polym. Degrad. Stab.* 96:818–822.

Ayrilmis, N., Jarusombuti, S., Fuengvivat, V., Bauchongkol P. 2011b. Effects of thermal treatment of rubber wood fibres on physical and mechanical properties of medium density fibreboard. *J. Trop. Forest Sci.* 23:10–16.

Azwa, Z. N., Yousif, B. F., Manalo, A. C., Karunasena, W. 2013. A review on the degradability of polymeric composites based on natural fibres. *Mater. Des.* 47:424–442. doi:10.1016/j.matdes.2012.11.025.

Cai, Z., Ross, R.J. 2010. Mechanical Properties of wood-based composite materials. In *Wood handbook, Wood as an Engineering Material*, Chapter 12. Madison, WI: Forest Products Laboratory, USDA, Forest Service.

Chand, N., Fahim, M. 2008. Wood reinforced polymer composites. In *Tribology of Natural Fiber Polymer Composites*, ed. N. Chand and M. Fahim, 180–196. Cambridge: Woodhead Publishing Limited.

Chang, Y. P., Abd Karim, A., Seow, C. C. 2006. Interactive plasticizing–antiplasticizing effects of water and glycerol on the tensile properties of tapioca starch films. *Food Hydrocoll.* 20 (1): 1–8.

Clemons, C. M. 2010. Wood flour. In *Functional Filler for Plastics*, ed.M. Xanthos, 269–290. Wiley-VCH.

Clemons, C. 2008. Raw materials for wood-polymer composites. In *Wood-polymer Composites*, ed. K. Oksman Niska and M. Sain, 269–290. Boca Raton: CRC Press.

Çolak, S., Çolakoglu, G., Aydin, I., Kalaycioglu, H. 2007. Effects of steaming process on some properties of eucalyptus particleboard bonded with UF and MUF adhesives. *Build. Environ.* 42:304–309.

Crawford, M. 2008. *Is Burning Wood Really A Long-Term Energy Descent Strategy?* https://www.transitionculture.org/2008/05/19/is-burning-wood-really-a-long-term-energy-descent-strategy/ (accessed April26, 2020).

Ding, Z., Yuan, C., Dunn, M., Qi, H. 2018. Direct 4D printing by using multimaterial additive manufacturing. *Bull. Am. Phys. Soc.* abstract id.V41.003.

Douglas & Sturgess. n.d. www.douglasandsturgess.com/product/FM-1174.html (accessed April 26, 2020).

Faludi, G., Dora, G., Renner, K. et al. 2013. Improving interfacial adhesion in pla/wood biocomposites. *Compos. Sci. Technol.* 89:77–82.

Filipaj, P. 2010. *Architektonisches Potential von Dämmbeton*. vdf Hochschulverlag AG an der ETH Zürich.

Gardan, J., Nguyen, D. C., Roucoules, L., Montay, G. 2016. Characterization of wood filament in additive deposition to study the mechanical behavior of reconstituted wood products. *J. Eng. Fibers Fabr.* 11 (4): 56–63.

Gardner, D.J., Han, Y., Wang, L. 2015. Wood-plastic composite technology. *Curr. Forestory Rep.* 1:139–150.

Garrote, G., Dominiguez, H., Parajó, J. C. 2001. Study on the deactylation of hemicelluloses during the hydrothermal processing of Eucalyptus wood. *Holz. Roh. Werkst.* 59:53–59.

Girdis J., Gaudion L., Proust N., Loschke S., Dong A. 2017. Rethinking timber: investigation into the use of waste macadamia nut shells for additive manufacturing. *JOM* 69 (3): 575–579.

Guo, R., Ren, Z., Bi, H. et al. 2018. Effect of toughening agents on the properties of poplar wood flour/poly (lactic acid) composites fabricated with Fused Deposition Modeling. *Eur. Polym. J.* 107:34–45.

Henke, K, Talke, D., Winter, S. 2016. Additive manufacturing of building elements by extrusion of wood concrete. *World Conference on Timber Engineering (WCTE 2016)*.

Henke, K., Talke, D., Winter, S. 2017. Additive manufacturing of building elements by extrusion of wood concrete. *World Conference on Timber Engineering*.

Henke, K., Treml, S. 2013. Wood based bulk material in 3D printing processes for applications in construction. *Eur. J. Wood Prod.* 71:139–141.

Hwang, S., Reyes, E.I., Moom, K. S., Rumpf, R. C., Kim, N. S. 2015. Thermal-mechanical characterization of metal/polymer composite filaments and printing parameter study for fused deposition modeling in the 3D printing process. *J. Electron. Mater.* 4:771–777.

ISO 4287 1997. Amendment 1. 2009. *Geometrical Product Specifications (GPS). Surface Texture, Profile Method. Terms. Definitions and Surface Texture Parameters*. Geneva, Switzerland: International Organization for Standardization. https://www.iso.org/standard/10132.html.

Johnson, E. 2009. Goodbye to carbon neutral: Getting biomass footprints right. *Environ. Impact Assess. Rev.* 29 (3): 165–168.

Kariz, M., Sernek, M., Kitek Kuzman, M. 2016. Use of wood powder and adhesive as a mixture for 3D printing. *Eur. J. Wood Prod.* 74:123–126.

Kariz, M., Sernek, M., Obućina M., Kitek Kuzman M. 2018. Effect of wood content in FDM filament on properties of 3D printed parts. *Mater. Today Commun.* 14:135–140.

Kellogg, R. M. 1981. Physical properties of wood. In *Wood: Its Structure and Properties*, ed. F.F. Wangaard, 191–223. College Park: Pennsylvania State University.

Khan, M. Z. R., Srivastava, S. K., Gupta, M. K. 2020. A state-of-the-art review on particulate wood polymer composites: Processing, properties and applications. *Polym. Test.* 89:106721.

Kubojima, Y., Okano, T., Ohta, M. 2000. Bending strength and toughness of heat-treated wood. *J. Wood Sci.* 46:8–15.

Kumar, V., Tayagi L., Sinha S. 2011. Wood flour–reinforced plastic composites: a review. *Rev. Chem. Eng.* 27 (5-6): 253–264.

Lang, M., Bei, J., Wang, S. 1999. Synthesis and characterization of polycaprolactone/poly(ethylene oxide)/polylactide tri-component copolymers, *J. Biomater. Sci. Polym. Ed.* 10 (4): 501–512.

Le Duigou, A., Castro, M., Bevan, R., Martin, N. 2016. 3D printing of wood fibre biocomposites: From mechanical to actuation functionality. *Mater. Des.* 96:106–114.

Lee, S.-Y., Kang, I.-A., Doh, G.-H. et al. 2008. Thermal and mechanical properties of wood flour/talc-filled polylactic acid composites: effect of filler content and coupling treatment. *J. Thermoplas. Compos. Mater.* 21:209–223.

Li, T., Aspler G., Kingsland, A. et al. 2016. 3D printing – a review of technologies, markets, and opportunities for the forest industry *J. Sci. Technol. Forest Prod. Process.* 5 (2): 30–31.

Martikka O., Kärki T., Wu Q. 2018. Mechanical properties of 3D-printed wood-plastic composites. *Key Eng. Mater.* 777:499–507.

Miller, R. B. 1999. Structure of wood. In *The Wood Handbook: Wood as an Engineering Material.* USDA Forest Service, Forest Products Laboratory General Technical Report FPL-GTR-113. Madison, WI. https://www.fpl.fs.fed.us/documnts/fplgtr/fplgtr113/ch02.pdf (accessed April 26, 2020).

Morton, J., Quarmley, J., Rossi, L. 2003. Current and emerging applications for natural and wood fiber-plastic composites. *Proceedings for the Seventh International Conference on Wood Fiber-Plastic Composites*, 3–6. Forest Products Society, Madison, WI.

Nguyen, H. N. G., Zhao, C.-F., Millet, O. et al. 2021. Effects of surface roughness on liquid bridge capillarity and droplet wetting. *Powder Technol.* 378:487–496.

Pei, E. 2014. 4D printing: dawn of an emerging technology cycle. *Assem. Autom.* 34 (4): 310–314.

Petinakis, E., Yu, L., Edward, G. et al. 2009. Effect of matrix–particle interfacial adhesion on the mechanical properties of poly(lactic acid)/wood-flour micro-composites. *J. Polym. Environ.* 17:83–94.

Pettersen, R. C. 1984. *The Chemical Composition of Wood.* Chapter 2, pp. 57–126. Washington: American Chemical Society.

Pitt, K., Lopez-Botello, O., Lafferty, A. D. et al. 2017. Investigation into the material properties of wooden composite structures with in-situ fibre reinforcement using additive manufacturing. *Compos. Sci. Technol.* 138:32–39.

Plastics Insight. 2016. *United Sates Polypropylene Price Forecast.* https://www.plasticsinsight.com/united-states-polypropylene-price-forecast-polymer-grade-propylene-pgp/ (accessed April 26, 2020).

Pringle, A. M., Rudnicki, M., Pearce, J. M. 2018. Wood furniture waste–based recycled 3-D printing filament. *Forest Prod. J.* 68 (1): 86–95.

Rodriguez, M., Osés, J., Ziani, K., Maté, J. I. 2006. Combined effect of plasticizers and surfactants on the physical properties of starch based edible films. *Food Res. Int.* 39 (8): 840–846.

Rosenthal, M., Henneberger, C., Gutkes, A., Bues, C. T. 2018. Liquid deposition modeling: a promising approach for 3D printing of wood. *Eur. J. Wood.* 76 (2): 797–799.

Schwarzkopf, M. J., Burnard, M. B. 2016. Wood-plastic composites—performance and environmental impacts. In *Environmental Impacts of Traditional and Innovative Forest-based Bioproducts*, ed. A. Kutnar, and S. S. Muthu, 19–43. Singapore: Springer.

Shigo, A. 1986. *A New Tree Biology: Facts, Photos, and Philosophies on Trees and their Problems and Proper Care.* Durham: Shigo and Trees, Associates.

Studart, A. 2018. 4D printing of morphing soft materials inspired by nature. *Bull. Am. Phys. Soc.* 16(1). http://meetings.aps.org/Meeting/MAR18/Session/V41.4 (accessed April 26, 2020).

Sydney Gladman, A., Matsumoto, E. A. et al. 2016. Biomimetic 4D printing. *Nat. Mater.* doi:10.1038/NMAT4544, MacMillan.

Tao, Y., Wang, H., Li, Z., Li, P., Shi, S. Q. 2017. Development and application of wood flour-filled polylactic acid composite filament for 3D printing. *Materials* 10:339.

Tibbits, S. 2014. 4D printing: multi-material shape change. *Architect. Des.* 84 (1): 116–121.

Timoshenko, S. 1925. Analysis of bi-metal thermostats. *J. Opt. Soc. Am.* 11:233–255.

Top for 3D Printing. 2014. *4 AXYZ Hits Indiegogo with Groundbreaking Wood 3D Print Project.* http://top43dprinting.com/4-axyz-hits-indiegogo-with-groundbreaking-wood-3d-print-project/ (accessed April 26, 2020).

University of Cambridge. 2002. *Specific Stiffness – Specific Strength*. http://www-materials.eng.cam.ac.uk/mpsite/interactive_charts/spec-spec/basic.html (accessed April 26, 2020).

USDA. 2010. *Wood Handbook. Wood as an engineering materials*. Forest Products Laboratory. United States Department of Agriculture Forest Service. General Technical Report FPL-GTR-190, Madison, WI.

Wang, H. L., Yu, J. T., Fang, H. G. et al. 2018. Largely improved mechanical properties of a biodegradable polyurethane elastomer via polylactide stereocomplexation. *Polymer* 137:1–12.

Wang, N., Yu, J. G., Chang, P. R., Ma, X. F. 2008. Influence of formamide and water on the properties of thermoplastic starch/poly(lactic acid) blends. *Carbohydr. Polym.* 71 (1): 109–118.

Wang, N., Yu, J. G., Ma, X. F. 2007. Preparation and characterization of thermoplastic starch/PLA blends by one-step reactive extrusion. *Polym. Int.* 56 (11): 1440–1447.

Woods, S. 2016. A history of wood from the stone age to the 21st century. *Architect Magazine*. https://www.architectmagazine.com/technology/products/a-history-of-wood-from-the-stone-age-to-the-21st-century_o (accessed April 26, 2020).

Wu, J.-J., Huang, L.-M., Zhao, Q., Xie, T. 2018. 4D printing: history and recent progress. *Chin. J. Polym. Sci.* 36:563–575.

Xie, G., Zhang, Y., Lin, W. 2017. Plasticizer combinations and performance of wood flour-poly(lactic acid) 3D printing filaments. *BioResources* 12 (3): 6736–6748.

Xu, P., Hu, M., Na, L., Gao, X. 2016. Magnetic controlled variable diameter nozzle for wood powder 3D printer. *Rev. Tec. Ing. Univ. Zulia* 39 (8): 77–84.

Xu, T., Tang, Z., Zhu, J. 2012. Synthesis of polylactide-graft-glycidyl methacrylate graft copolymer and its application as a coupling agent in polylactide/bamboo flour biocomposites. *J. Appl. Polym. Sci.* 125 (S2): E622–E627.

Yang, T.-C. 2018. Effect of extrusion temperature on the physico-mechanical properties of unidirectional wood fiber-reinforced polylactic acid composite (WFRPC) components using fused deposition modeling. *Polymers* 10:976.

Yildiz, S., Gezer, E. D., Yildiz, U. C. 2006. Mechanical and chemical behavior of spruce wood modified by heat. *Build. Environ.* 41:1762–1766.

Zhang H., Guo Y., Jiang K. et al. 2016. A review of selective laser sintering of wood-plastic composites. *Proceedings of the 26th Annual International Solid Freeform Fabrication 2016*.

7 Cellulose

Sooner or later, we will have to recognize that the Earth has rights, too, to live without pollution. What mankind must know is that human beings cannot live without Mother Earth, but the planet can live without humans.

Evo Morales

7.1 OVERVIEW OF CELLULOSE

Cellulose is the most abundant sustainable material. Although cellulose is available in any plant, wood pulp supplies the main part of cellulose employed today, f.e. for boards, paper, and textiles. Historically, cellulose has been utilized for as long as man has existed, first as fuel and shelter, and later for clothing and writing material such as papyrus. Cotton fibers are the purest form of cellulose, since nearly 90% of the cotton fibers is cellulose. Cotton is still the main source of cellulose for textile, but the demand for wood-based raw material is increasing due to the environmental drawbacks associated with cotton cultivation and processing, such as high usage of water and pesticide. Both cotton- and wood-based cellulose can also be utilized as the basis for cellulose derivatives, such as cellulose ethers and esters. These are widely present f.e. in pharmaceuticals, food, construction materials, and paint. As an AM feedstock, cellulose appears in multiple forms, roles, applications, and AM processes, thanks to its benefits: low cost and widespread availability. Its range of utilization in AM spans from mature areas, such as construction, to cutting edge technologies, such as TE, electronics for Internet, pharmacology, and medicine, which may be surprising and fascinating, considering that the source of cellulose is mostly plants. Cellulose is also attracting growing interest for energy storage devices, where it serves as a gel binder in flexible electrodes, a reinforcing agent in polymer electrolytes, a separator in Li-ion batteries, and a nanoporous template in supercapacitors (Shao et al. 2015). Cellulose is present in only one commercial AM material, Formi 3D, but it has been the focus of an intense activity of research leveraging AM processes to devise new applications (Olsson and Westman 2013).

Cellulose combines excellent mechanical properties with the advantages of being a natural, fully renewable, and biodegradable material (Adusumali et al. 2006). Cellulose is the polymer most abundantly produced in nature, being present in plants, animals, algae, fungi, minerals, and bacteria, and contributes for about 40% to the carbon fraction in plants (Heinze 2016). In theory, cellulose is an inexhaustible resource, and yields 1.5×10^{12} tons of biomass every year (Olsson and Westman 2013), which is comparable to the planetary reserves of main mineral and fossil sources (Klemm et al. 2005). The main function of cellulose in plants is to provide them with structural properties, and most cellulose resides in the secondary walls of higher plants.

Cellulose can be obtained from natural and synthetic sources. Wood contains 40–50 wt% of cellulose. Vegetable (or natural) fibers are also a source of cellulose, present in them in varying amounts, as detailed in Table 7.1. Interestingly from a sustainable perspective, a process to extract cellulose from agricultural waste has been recently reported (Mnuruddin et al. 2011). Natural fibers possess the advantages of low density, high values of strength/density and stiffness/density, low cost, and biodegradability (Li et al. 2007).

Chemically, cellulose (Figure 7.1) is a high-MW homopolymer with chemical formula $(C_6H_{12}O_5)_n$ (Kamide 2005) consisting of *glucose* (D-glucopyranose $C_6H_{12}O_6$) ring-shaped units joined together by β-1,4 bonds, that is bonds established between the carbon atom in location 1 on

DOI: 10.1201/9781003221210-7

TABLE 7.1
Vegetable Fibers Containing Cellulose

Fiber	Bagasse	Bamboo	Barley	Coconut	Cotton	Flax	Hemp	Jute	Kapok	Kenaf	Ramie	Rice	Straw	Wheat
Cellulose content (wt%)	32–48	26–55	31–45	54–65	85 to >90	43–80	57–77	41–65	70–75	37–57	70–91	28–48	40–50	29–51

Source: Heinze, 2016. Han, J. S., Rowell, J. S., 1997. Chemical Composition of Fibers. In *Paper and Composites from Agro-based Resources*, ed. R. M. Rowell, R. A. Young, J. K. Rowell, 83–134. 1997 USDA, Forest Products Laboratory, https://www.fpl.fs.fed.us/documnts/pdf1997/han97a.pdf. Wang, Q., Sarkar, J. 2018. Pyrolysis behaviors of waste coconut shell and husk biomasses. *Int. J. of Energy Prod. & Mgmt.* 3, 1.

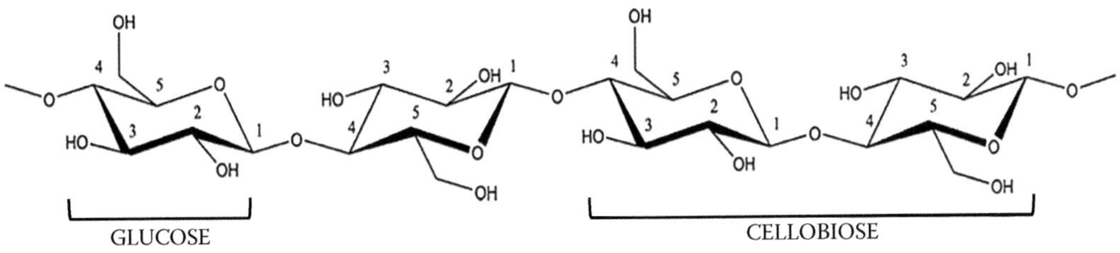

FIGURE 7.1 Chemical structure of cellulose. Each of the four units is glucose. Cellobiose comprises two glucose units. Numbers 1 to 6 identify locations of the carbon atoms on the ring.

Source: Modified from Heinze, T. 2016. Cellulose: Structure and Properties. *Adv Polym Sci* 271:1–52. doi: 10.1007/12_2015_319. Reproduced with permission from Springer International.

one unit and the carbon atom in location 4 on the adjacent unit, and featuring the OH group above the ring's plane. The bonds result in alternating turning of the cellulose chain axis by 180°, as pictured in Figure 7.1 showing schematically four units of the cellulose structure. The repeating unit of cellulose is called *cellobiose*, formed by two molecules, and 1.3 nm long. Native cellulose has molecular chains long 500–15,000 nm, depending on its origin (Heinze 2016).

Cellulose is a semi-crystalline polysaccharide, and its molecules are strongly connected through intermolecular and intramolecular hydrogen bonding and van der Waals forces, which results in the formation of small crystals called *microfibrils* that are embedded in a disordered, amorphous matrix, and in turn form larger fibers (Swift 1977). The length of the microfibrils is of the order of micrometers, and their species-specific diameter is 2–25 nm (O'Sullivan 1997). Figure 7.2 is a schematic view of microstructure of plants, showing microfibrils, cellulose, and hemicellulose that is a branched polymer crosslinked to microfibrils and composed of sugars.

Because of the strong hydrogen bonding interactions, cellulose exhibits relatively high tensile modulus, up to 800 MPa (Bledzki and Gassan 1999; Mohanty et al. 2000) vs. other polysaccharides. Cellulose is tough, fibrous, hydrophilic, and insoluble in water and most common organic solvents (Habibi et al. 2010; Raquez et al. 2013).

The mechanical properties of cellulose are typically reported not for isolated cellulose but for blends and compounds including cellulose as one of their ingredients. The value of the elastic modulus of pure crystalline cellulose is not yet well-defined (Denoyelle 2011). Sakurada et al. (1962) employing X-ray diffraction estimated an approximate value of 140 GPa that was confirmed by later molecular simulation techniques (Tanaka and Iwata 2006). Tensile properties were measured on four regenerated cellulose fibers, namely viscose, modal, lyocell, and rayon for tire cord, and compared to flax and glass, with flax and glass considered because they are widely employed as natural and technical fibers, respectively (Adusumali et al. 2006). *Regenerated cellulose* is a family of materials obtained by converting natural cellulose into a soluble cellulosic derivative, and from it forming fibers (f.e. rayon) and films (f.e. cellophane) (Niaounakis 2017). The test results are collected in Table 7.2, and include tensile strength and modulus, elongation at break, and energy to fracture (or toughness). Among the regenerated fibers, rayon features the best combination of values, and, with lyocell, leads in tensile strength and modulus. Rayon also displayed the highest toughness, and viscose the most ductility (expressed as elongation at break). The four cellulose fibers underperformed flax and glass in tensile strength and modulus, but were comparable in toughness, exceeding 33 MJ/m^3 (Agnarsson et al. 2010), the value for Kevlar®, a synthetic fiber well-known for its outstanding mechanical performance.

The worldwide production of cellulose increased from 295 million tonnes in 2003 to 390 million tonnes in 2013 (Italia 2015). The top exporters of cellulose are USA (USD 1.28 billion), Germany (USD 1.05 billion), and China (USD 520 million). The top importers are China (USD 467 million), USA (USD 418 million), and India (USD 319 million) (Observatory of Economic

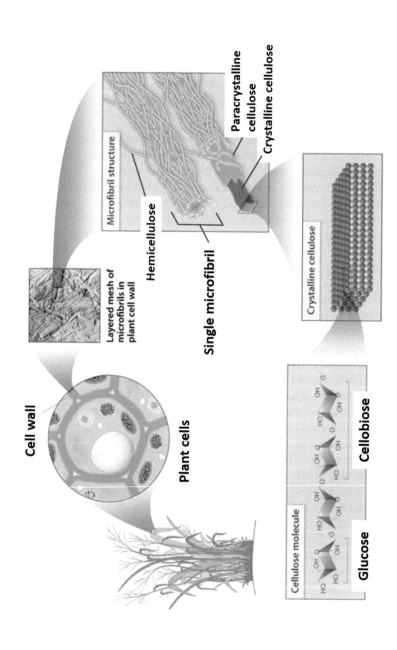

FIGURE 7.2 Schematic view of microstructure of plants showing microfibrils, cellulose, and hemicellulose.

Source: Adapted from US DOE. 2005. Genomics:GTL Roadmap, DOE/SC-0090, U.S. Department of Energy Office of Science. (p. 204) (website). Office of Biological and Environmental Research of the U.S. Department of Energy Office of Science. science.energy.gov/ber/. Freely reproduced.

TABLE 7.2

Selected Properties of Fibers Made of Regenerated Cellulose, Flax, and Glass

Property	Regenerated Cellulose	Density	Tensile Modulus	Tensile Strength	Elongation at Break	Toughness
Unit		g/cm^3	GPa	MPa	%	MJ/m^3
Viscose	yes	1.5	10.8	340	15.4	32.7
Modal	yes	1.5	13.2	437	10.4	37.2
Lyocell	yes	1.5	23.4	556	8.7	34.5
Rayon	yes	1.5	22.2	778	10.7	40.8
Flax	no	1.5	40	904	1.4	6.7
Glass	no	2.5	70	3,000	4.3	54.3

Source: Adusumali, R.-B., Reifferscheid, M., Weber, H., et al. 2006. Mechanical Properties of Regenerated Cellulose Fibres for Composites. *Macromol. Symp.* 244:119–125. Reproduced with permission from John Wiley and Sons.

Complexity n.d.). Some of the largest producers of chemical cellulose include Akzo Nobel (The Netherlands), Ashland, Celanese, Eastman, FMC (all in USA), and Rhodia Acetow (Germany).

7.2 COMMERCIAL AM MATERIALS CONTAINING CELLULOSE

As anticipated in Section 7.1, the only commercial material for AM containing cellulose is Formi 3D, a cellulose/PLA composite, containing 15–50 wt% cellulose, sold as pellets (Figure 7.3) by UPM Biocomposites (Finland), and as filament by Octofiber. In other words, Formi 3D can be extruded into filaments for FFF or utilized as pellets for an AM process called *fused granular fabrication* (FGF). Formi 3D has yellowish color, and mild wood odor. It is available in three grades whose physical and mechanical properties are detailed in Table 7.3. According to UPM, Formi 3D is suited for FGF and its large (2–8 mm) diameter nozzles, thanks to its dimensional stability, self-supporting molten stage properties, shear thinning melt flow, fast cooling, very low shrinkage, and lack of internal stresses due to cellulose. We recall that a FFF filament shows shear thinning when it efficiently flows through a small diameter deposition nozzle, and after deposition quickly behaves like a solid and tmaintains its shape even under the weight of layers deposited on it (Compton and Lewis 2014; Studart 2016). UPM Formi 3D 40 can be blended with PLA or wood/

FIGURE 7.3 Left: granules of Formi 3D, cellulose/PLA composite for additive manufacturing. Right: leaf Bridge printed with filament made by Octofiber with UPM Formi 3D.

Source: UPM Formi 3D, Customer Presentation 2020. Reproduced with permission from UPM Formi.

TABLE 7.3

Properties of Formi 3D Cellulose/PLA Material, and Benchmark AM Materials

Property	Unit	Test Method	3D 19/20[a]	3D 40[a]	Fillamentum® Timberfill (wood/PLA)	Futerro PLA General Purpose
AM Process	N/A	N/A	FGF[a]	FFF	FFF	FFF
Cellulose Content	wt%	N/A	20	40	15[c]	0
Density	g/cm^3	ISO 1183	1.2	1.2	1.28	1.24
Tensile Strength	MPa	ISO 527	39	48	33	55
Tensile Modulus	MPa	ISO 527	3,600	5,400	2,800	3,500
Elongation at Break	%	ISO 527	4	2	2.9	6
Unnotched Charpy Impact Strength	kJ/m^2	ISO 179/1eU	20	14	15	3.5[d]
T_m	°C	ISO 11357	140–180	135–180	145–160	145–175
T_g	°C	ISO 11357	65	60	55–60	52–60
MFI[b]	g/10 min	ISO 1133	16	7	25.6	2–4

Source: Data sheet posted online by UPM Bioplastics, Fillamentum®, and Futerro.
Notes:
[a]FGF fused granular fabrication.
[b]190°C/10 kg for Formi 3D.
[c]Wood content.
[d]Notched Izod Impact Strength (kJ/m²), ISO 180.

plastic composites, and with PLA its recommended amount is below 25 wt%. Formi 3D is paintable (f.e. with acrylic), soft to sand, and can be glued to itself with PVAc adhesive, also known in USA as carpenter's glue and Elmer's glue.

The tensile property data for UPM Formi 3D grades indicate that doubling the cellulose content is effective in improving significantly the tensile strength and modulus. In fact, tensile strength of 3D 40 exceeds that of 3D 20 and 3D 19/20 by 71% and 23%, respectively, and tensile modulus of 3D 40 surpasses that of 3D 20 and 3D 19/20 by 108% and 50%, respectively. On the other hand, as occurred with other polymers (Bagotia et al. 2015), higher tensile modulus is accompanied by lower ductility and impact strength. Furthermore, as expected, higher cellulose content increases the composite's viscosity, which is inversely related to the MFI. Table 7.3 also shows that, in terms of physical and mechanical properties, Formi 3D performs similarly to two commercial FFF filaments, reported as benchmarks: wood/PLA Timberfill by Fillamentum®, and unfilled PLA by Futerro. Particularly, tensile modulus of 3D 40 exceeds that of Timberfill and Futerro PLA, but tensile strength of Futerro PLA surpasses that of all Formi 3D grades.

Commercial FFF filament of Formi 3D are sold by Octofiber in 1.75 mm and 2.85 mm grades, and are compatible with more than 12 printer brands. The printer settings for OctofiberFormi 3D filament are: bed and nozzle temperature 60°C and 210–225°C, respectively, and printing speed 10–30 mm/s.

Formi 3D 2.8 mm filament made by Octofiber was selected to build the Leaf Bridge, a small bridge 2 m × 80 cm × 80 cm in size (Figure 7.3) that shows the capabilities of AM in an unusual use: a components that is large, for outdoors, and combines structural and aesthetic functions. The bridge weighed 93 kg, and was printed in 17 parts on a conveyor belt-type printer whose head was connected to an ABB robot arm. The bridge's beam was printed in 10 sections, assembled with glue and secured with steel bars and nuts. The leaves serving as railings were printed as hollow sections, filled with PU foam to prevent buckling, and then glued together. The railing and the beam were joined through glued tongue-and-groove joint. The bridge parts were painted with tinted alkyd primer paint before assembly, and after assembly a second layer of primer was applied. To prevent slipping, the footway on the bridge was painted with the primer mixed with additional non-slip granules. Lastly, layers of acrylic spray were applied to impart the final colors.

7.3 EXPERIMENTAL CELLULOSE FOR AM

7.3.1 INTRODUCTION

Cellulose combine low cost of production with several benefits such as being sustainable, biodegradable, widely available, hydrophilic but not dissolvable in water, lightweight, strong and stiff, and hence it has been frequently investigated as an AM feedstock for uses ranging from serving as simple paper substrate to being a building block in some of current leading innovations in material science and fabrication technologies, spanning from biomedical sensing to energy production and storage (microbatteries), to Internet of Things (capacitors and inductors for flexible miniaturized circuits), biomimetic design, TE, and artificial organs. Depending on the cost/benefit ratio, parts printed out of cellulose can also serve more mature markets oil and gas, building and construction, automotive, aerospace, marine, and coatings.

Cellulose is involved in a recent and growing application area for AM that is bone tissue engineering. After injury, bone heals itself by self-regeneration if the fractures are smaller than a certain (*critical*) size. Bone fractures at or above critical size are treated with metal and ceramic implants or bone grafting that is a surgical procedure consisting in replacing missing bone with bone harvested from the same patient, donors, cadavers, and species other than human. However, grafting is limited by donor site morbidity (any complication at the site as it heals), donor bone supply shortage, infection, corrosion, stress shielding, and secondary surgery. For large bone fractures and voids, scaffolds are an alternative to grafting, and serve as a 3D template to guide, stimulate, and support the growth of bone tissue on its surfaces. Candidate materials for bone scaffolds must satisfy several requirements, detailed in Section 1.15, such as biocompatibility, biodegradability and bioresorbability (Vert et al. 1992), load-bearing mechanical properties, porosity, be sterile and bioactive, and controlled delivery of bioactive molecules or drugs (Hutmacher 2000; Brown et al. 2009; Porter et al. 2009; Ghassemi et al. 2018). AM is well suited to fabricate bone scaffolds, because it produces geometries customized to the size of specific bone defects and controls scaffold's mesh resolution, surface finish, and mechanical performance.

Very recent and comprehensive reviews of applications of cellulose in AM were published by Wang et al. (2018a) and Dai et al. (2019). The types of cellulose employed in AM can be classified as cellulose, and derivatives of cellulose, in turn grouped in cellulose ethers and esters, microcellulose, and nanocellulose (Figure 7.4). Overall, the following compounds will be discussed: *cellulose acetate* (CA), *carboxymethyl cellulose* (CMC), *ethyl cellulose* (EC), *hydroxyethyl cellulose* (HEC), *hydroxypropyl cellulose* (HPC), *hydroxypropyl methylcellulose* (HPMC), and *methyl cellulose* (MC).

Many experimental applications of cellulose processed via AM illustrated hereafter targeted pills in form of tablets and capsules. AM has proved itself to be effective in precisely controlling fine-tuned release rate and released quantity of multiple medications in one pill, but AM is a process slower than conventional fabrication processes. The commercial success of printing pills depends on its cost vs. the cost of conventionally fabricated (although less advanced in "performance" and design) pills.

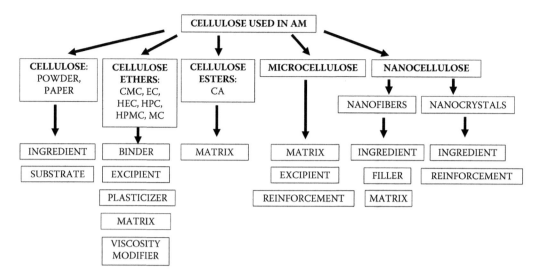

FIGURE 7.4 Cellulose types employed in AM, and their function. Legend: cellulose acetate (CA), carboxymethyl cellulose (CMC), ethyl cellulose (EC), hydroxyethyl cellulose (HEC), hydroxypropyl cellulose (HPC), hydroxypropyl methylcellulose (HPMC), and methyl cellulose (MC).

Source: Prepared from data in Dai, L., Cheng, T., Duan, C., et al. 2019. 3D printing using plant-derived cellulose and its derivatives: A review. *Carbohydrate Polymers* 203:71–86. https://doi.org/10.1016/j.carbpol. 2018.09.027.

7.3.2 Cellulose Powder and Paper

7.3.2.1 Cellulose as Substrate

We start with two advanced applications of the most conventional and oldest form of cellulose, plain paper, but these applications leverage the specific advantages of AM. Paper substrates possess high porosity and large specific surface area, and swell in aqueous environments, which increases their sensitivity as sensor materials for chemical detection of molecules of a chemical constituent (Gu and Huang 2013). Paper-based *microfluidic analytical devices* (µPADs) are basically the most advanced version of test paper strips. More exactly, µPADs are inexpensive point-of-care diagnostic devices that detect biochemical changes in the body, and convey to the user the results in a simple format (Martinez et al. 2007; Cate et al. 2014; He et al. 2015; Tang et al. 2016). Paper utilizes its natural capillary action to transport a liquid sample instead of pumps, and is also lightweight, biocompatible, easily disposed of, easy to use, store, transport, and modify, compatible with biological assays and colorimetric tests whose results can be detected by the naked eye or a cellular phone (Chong et al. 2013). He et al. (2016) exploited the above advantages of cellulose in the form of cellulose powder, and the ability of AM to fabricate complex, small, and accurate geometries, and fabricated a novel µPAD out of PLA on a desktop FFF printer. Cellulose powder (α-cellulose), a polysaccharide composed of long chains of linked D-glucose units, is an organic compound with formula $(C_6H_{10}O_5)_n$ and particle size of 74–125 µm. The µPAD was also recyclable, and designed for specific analytical sequences. The fabrication process was the following. First, a 5 × 5 cm PLA hydrophobic substrate with open hydrophilic microchannels on its surface was fabricated from 1.75 mm filament by Alkht Co. (China) on a FFF desktop printer. The resolution of the printer exceeded the µPAD fabrication requirement of 500 µm. Next, the substrate's surface was coated with a thin layer of Dow Corning polydimethylsiloxane (PDMS) to seal the microscopic gaps induced by FFF. Finally, the microchannels were filled with a mixture of deionized water and cellulose powder (Sigma-Aldrich, China) ready for use after 30-min oven-drying. Figure 7.5 is a top view photo of the µPAD with microchannels filled for a pH test. Since

FIGURE 7.5   Top view of µPAD with microchannels filled for a pH test.

Source: modified from He, Y., Gao, Q., Wu, W.-B., et al. 2016. 3D Printed Paper-Based Microfluidic Analytical Devices. *Micromachines* 7, 108; doi: 10.3390/mi7070108, MDPI (Switzerland). Reproduced under the terms and conditions of the Creative Commons Attribution (CC-BY) license.

channels with different depths can be printed, different values of the capillary flow speed through cellulose powder in the microchannels can be programmed, and specific analytical sequences can be designed. One more advantage of a printed µPAD is that, being FFF desktop printers in-expensive, this device can be fabricated at the point-of-care or point-of-test, or as a lightweight platform for analytical chemistry.

The *Internet of Things* (IoT) is one of the latest technologies, and defines a system of inter-connected computing devices, mechanical, electrical and digital machines, objects, people, and even animals, all possessing unique identifiers (UIDs) and capable to autonomously and wirelessly sense, collect, and transfer data over a network without requiring human-to-human or human-to-computer interaction. Mariotti et al. (2017) have linked cellulose and IoT, by combining a cellulose-based paper and AM to fabricate high-performing planar inductors and capacitors on a mechanically flexible, and environmentally friendly substrate, and also reduce the production cost and time, and chemical waste with respect to traditional technologies. Particularly, standard Mitsubishi photographic paper featuring thickness of 230 µm, dielectric constant (ε_r) of 2.9, and loss tangent (tanδ) of 0.06 at 1 GHz served as a substrate to print capacitors and inductors ex-hibiting a maximum inductance and capacitance of 1.4 nH/mm^2 and 6.5 pF/mm^2, respectively, with the latter value one order of magnitude higher (i.e. better) than those obtained with state-of-the-art technologies on less lossy (i.e. less inclined to dissipate electrical or electromagnetic en-ergy) flexible substrates. Inductors and capacitors were fabricated using two novel AM processes: *copper adhesive laminate* (Alimenti et al. 2012; Mariotti et al. 2013), and *multilayer inkjet printing* (Rida et al. 2007; Cook et al. 2014; Roselli et al. 2014). The usability of the capacitors and inductors printed for practical radio-frequency (RF) components and circuits was experimentally demonstrated by combining them in a printed RF mixer, a device changing the frequency of an electromagnetic signal while keeping unaffected every other characteristic (such as phase and amplitude) of the initial signals. A 5 × 5 mm passive RF mixer was laminated with copper adhesive on cellulose paper, and exhibited an experimental value of conversion loss (CL) below 10 dB, falling in the range of 8–12 dB measured on similar mixer topologies using traditional substrates and technologies. CL is a most important parameter for passive mixers, and the lower CL the better the mixer.

Switching now to an older and less advanced application, Jo et al. (2014) employed FFF to improve variable tactile patterns such as those used in elevators, paving blocks, plastic credit cards, and Braille system for visually impaired individuals. Employing AM instead of mechanical Braille devices such as embossers, it is easier to generate Braille letters of diverse and desired size or shape (for example larger size for newly blind people or blind children learning Braille) and produce tactile maps with multilayered thicknesses. In fact, the authors successfully employed a FFF printer to control size, thickness, and shape of the printed dots on the cellulose paper substrate, specifically width and thickness of letters, and diameter, height, and spacing of Braille dots. After the polymeric tactile patterns were formed, a thermal reflow process was conducted on a hot plate as a post-processing step that markedly improved surface smoothness due to the surface tension effect, and enhanced interfacial adhesion strength of the printed pattern on paper substrate. Contrary to punched patterns, printed dots maintained their original shape and proved to be durable after tribology test. The authors employed the FFF desktop printer ROKIT (3DISON, South Korea), PLA filament, and cellulose paper A4 sheets with density of 80 g/m^2. Dots and letters were printed layer by layer, with a layer thickness of 0.3–0.4 mm. Speed and cost were not compared between conventional fabrication and printing of Braille characters.

7.3.2.2   Cellulose as Ingredient

At elevated temperatures cellulose does not melt but decomposes, hence to obtain liquid cellulose and combine it with other ingredients, cellulose has to be dissolved or chemically modified. However, cellulose is insoluble and only partly soluble in water and most common solvents due to the preferential formation of intra- and intermolecular hydrogen bonds (Olsson and Westman 2013), and to date only a few solvent systems can dissolve cellulose. Current aqueous and non-aqueous cellulose solvents suffer from insufficient solvation power and high toxicity (Olsson and Westman 2013). Hereafter two dissolution processes successfully employed for cellulose for AM are mentioned: one based on N-methylmorpholine-N-oxide (NMMO) (Olsson and Westman 2013), the other on *ionic liquids* (ILs), a family of liquids that consist entirely of ions and have a T_g or T_m below 100°C (Mäki-Arvela et al. 2010; Hauru et al. 2012; Gunasekera et al. 2016). The NMMO process offers the benefit of having been successfully implemented on industrial scale, while ILs easily dissolve cellulose and are less hazardous than other solvents.

Employing cellulose dissolved in NMMO, Li et al. (2018) claimed to be the first ones to print complex structures with controlled, ordered, and interconnected pores out of cellulose without any binding agent, obtaining a material with tensile and compressive modulus of 161 MPa and 12.9 MPa, respectively. Their process was the following:

- Preparing a cellulose/NMMO solution, by completely dissolving 5g of pulp in 100 ml of NMMO (50 wt% H$_2$O). The pulp consisted in bleached wood pulp or cotton linters (fine, silky fibers that adhere to the seeds of the cotton plant after separating cotton fibers from their seeds) featuring cellulose content above 90 wt%.
- Printing the solution through a 3D-Bioplotter® (EnvisionTEC, Germany) printer at 70°C through a 22-gage needle (0.413 mm inner diameter), and extruding a gel that solidified quickly.
- Immersing the printed object in Milli-Q® Ultrapure Water (Millipore Corporation, USA), to remove NMMO, and also turning the object into cellulose hydrogel (HG).
- Freezing the object at −70°C for 12 hours
- Freeze-drying the object to final pure cellulose product (Figure 7.6)

This process was instrumental to develop a final article possessing not only a structure with more controlled and finer complexity (smaller pore size, thinner fibers and interconnected pores on the

FIGURE 7.6   Final cellulose printed object from cellulose dissolved in N-methylmorpholine-N-oxide (NMMO).

Source: Modified from Li, L., Zhu, Y., Yang, J. 2018. 3D bioprinting of cellulose with controlled porous structures from NMMO. *Mater. Lett.* 210:136–138. Reproduced with permission from Elsevier Ltd.

TABLE 7.4
Properties of Articles Printed from Cellulose Dissolved in NMMO

Cellulose	Tensile Modulus	Tensile Strength	Compressive Modulus	Compressive Strength
	MPa	MPa	MPa	MPa
Final Form	160.6	2.6	12.9	5.7
Gel Form	73.2	1.7	7.6	4.1

Source: Values from Li, L., Zhu, Y., Yang, J. 2018. 3D bioprinting of cellulose with controlled porous structures from NMMO. *Mater. Lett.* 210:136–138.

side wall) than that at gel state, but also crystalline cellulose instead of amorphous cellulose, and hence higher mechanical properties, reported in Table 7.4. This process shows that, if the cost of material and process are competitive, printed cellulose has potential for industrial applications, especially those that would need complicated 3D structures.

Markstedt et al. (2014) demonstrated the fabrication via FFF of spatially tailored gel structures made from cellulose dissolved in IL. These structures have numerous possible applications, such as, f.e., packaging, medicine tablets, scaffolds for TE, and sensors (Kim et al. 2006; Pacquit et al. 2007). The authors tested three types of cellulose (Table 7.5), featuring differing degrees of polymerization (DP), all dissolved in an IL: 97% pure 1-ethyl-3-methylimidazolium acetate (EmimAc) (BASF). For each cellulose/IL solution three concentrations were tested: 1, 2, and 4 wt %. DP is a critical structural property of cellulose, because it expresses the cellulose's chain length that affects the mechanical properties of paper products and composite materials containing

TABLE 7.5
Types of Cellulose Dissolved in 1-Ethyl-3-Methylimidazolium Acetate

Cellulose	Supplier	Degree of Polymerization
Avicel-PH 101	Fluka-Sigma Aldrich	150–300
Dissolving pulp	Domsjö AB	750
Bacterial nanocellulose	N/A	2,000–8,000

Source: Markstedt, K., Escalante, A., Toriz, G., Gatenholm, P. 2017. Biomimetic Inks Based on Cellulose Nanofibrils and Cross-Linkable Xylans for 3D Printing. *ACS Appl. Mater. Interfaces* 9:40878–40886.

cellulose, and the solubility of cellulose in a given solvent (Liu et al. 2014). Fabrication was carried out on a commercial printer MakerBot Replicator™ 2, customized by replacing the heating element and filament head with a 5 ml syringe holder, and vertical syringe actuator. The syringe contained the IL cellulose solution, whose flow rate was 10 µl/min, and printing speed 10 mm/s. The best printing solution contained 4 wt% of dissolving pulp: it exhibited shear thinning behavior, allowing to use low pressure in the syringe pump. The best method to achieve coagulation of the printed layers (i.e. changing from liquid to solid state) was dispensing the solution on an agar gel, which induced coagulation almost immediately, and adhesion between adjacent printed layers. After complete removal of the IL, the printed structures were made of amorphous cellulose gels with interconnected porous structure.

Gunasekera et al. (2016) selected cellulose in liquid form as a feedstock for inkjet printing. Namely, they took two ILs, EmimAc and 1-butyl-3-methylimidazolium acetate, and dissolved Sigma Aldrich microgranular cellulose in each IL. Since both ILs resulted in solutions with high viscosity (above 80 mPa s) not suitable for inkjet printing, two more solvents were added: 1-butanol, and dimethyl sulfoxide (DMSO). The former did not work and caused precipitation, whereas DMSO was successful, because it kept cellulose soluble and obtained a viscosity suitable for printing. The working solutions had the following wt% compositions: cellulose 1–3.2, 1-butyl-3-methylimidazolium acetate 52–66, DMSO 32–47.

7.3.3   Cellulose Esters

7.3.3.1   Introduction

Cellulose esters and ethers are the most important cellulose derivatives and have been commercially produced for many years. *Cellulose esters* have the chemical structure pictured in Figure 7.7, and are commonly derived from natural cellulose by reaction with organic acids, anhydrides, or acid chlorides (Edgar 2004). Common cellulose esters include *cellulose acetate* (CA), *cellulose acetate propionate* (CAP), and *cellulose acetate butyrate* (CAB). Since cellulose esters can be chemically modified and feature a good range of properties, they are versatile, and utilized in form of coatings, fibers, films, plastics, automotive finishes, eyewear, industrial coatings, LCD displays, separation media, tool handles, etc. A main producer of cellulose esters is Eastman (USA). More recently, cellulose esters have been studied for the development of new advanced materials and composites (Ciolacu et al. 2013). Cellulose esters have been often involved in AM and have performed different functions, such as modifying the viscosity of AM inks, and imparting them *shear thinning*.

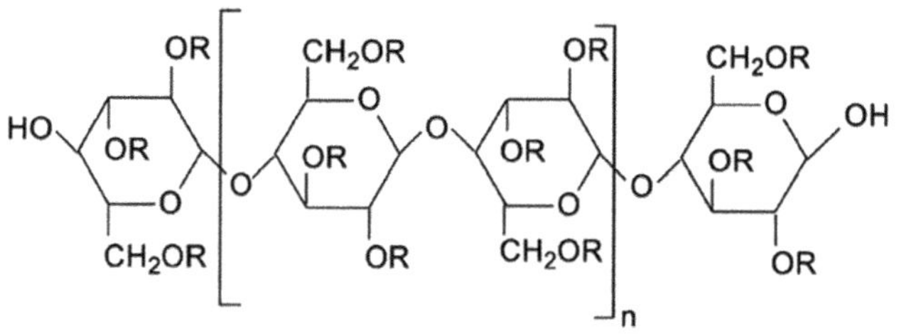

FIGURE 7.7 Structure of cellulose ester. The R group changes depending on the type of cellulose ester: H for cellulose, acetyl for cellulose acetate, acetyl and propionyl for cellulose acetate propionate, and butyryl for cellulose acetate butyrate.

Source: Mohanty, A. K., Wibowo, A., Misra, M., Drzal, L. T. 2004. Effect of process engineering on the performance of natural fiber reinforced cellulose acetate biocomposites, *Composites Part A: Applied Science and Manufacturing*, 35 (3): 363–370. Reproduced with permission from Elsevier.

TABLE 7.6

Properties of CA

Density	MW	T_m	Tensile Strength	Tensile Modulus	Hardness	Water Absorption in 24 h	HDT at 1.8 MPa	Max Service Temperature	CTE	Thermal Conductivity at 23°C
g/cm³	g/mol	°C	MPa	GPa	Rockwell	%	°C	°C	10^{-6}/K	W/mK
1.27–1.34	264.23	260	12–110	1.0–4.0	34–125	1.9–7.0	48–86	53–67	80–180	0.16–0.36

Source: http://www.goodfellow.com/A/Cellulose-Acetate-Polymer.html, https://pubchem.ncbi.nlm.nih.gov/compound/Cellulose-acetate#section=Chemical-and-Physical-Properties (accessed January 16, 2021). https://www.azom.com/properties.aspx?ArticleID=1461 (accessed January 16, 2021).

7.3.3.2 Cellulose Acetate (CA)

CA is the acetate ester of cellulose, and its molecular formula is $C_{10}H_{16}O_8$. CA's major properties are listed in Table 7.6. CA is the most important organic ester because of its broad application in fibers and plastics. Thanks to its typical properties that include good toughness, deep gloss, high transparency, and high absorbency, and formability and low cost, CA features multiple applications: pharmaceuticals (tablet excipients), photography as film base, X-ray films, adhesives, eyeglasses frames (especially to achieve special graphic effects such as imitation tortoise shell), cigarette filters, diapers, surgical products, tool handles, and apparel textile. Commercial CA is made from wood pulp that is processed using acetic anhydride to form acetate flake from which film, fibers, etc. are made. In 2018 CA was the second most produced biobased polymer in the world with 0.93 million tonnes per year and its produced volume is predicted to decline in 2024 (Table 1.14). Some CA producers are Celanese and Eastman (both headquartered in USA). The price of CA in bulk is $3–7/kg, based on supplier, type (pellets, flakes, etc.), and quantity (Alibaba n.d.; Made-in-china n.d.; Zauba n.d.).

Pattinson and Hart (2017) combined CA, as a feedstock, and a customized filament-based printing process to fabricate parts with the following advantages, demonstrating that AM enables to design a material with tailored properties and multiple functionality:

- High tensile modulus and strength, superior to those of PLA and ABS
- Isotropic tensile strength and modulus, vs. anisotropy of FFF TPs
- High toughness, outperforming PLA and ABS
- Antimicrobial property.

The process started by dissolving CA (Sigma-Aldrich 180955 and 419028 with average MW ~30,000 g/mol and ~50,000 g/mol, respectively) in acetone, chosen because it is an inexpensive solvent that evaporates rapidly, and can be recycled if collected through condensation. The resulting mixture's viscosity increased exponentially with CA's concentration. The optimal concentration was 25–35 wt%, because it provided a balance between flowability through the nozzle and shape retention of the extruded line. The mixture was loaded into a syringe holder attached to

TABLE 7.7

Mechanical Properties of Test Coupons Printed in CA

Loading Direction vs. Printing Direction	Parallel			Perpendicular			N/A	N/A	N/A
	Tensile Strength	Tensile Modulus	Toughness	Tensile Strength	Tensile Modulus	Toughness	Tensile Strength	Tensile Modulus	Toughness
	MPa	GPa	MJ/m³	MPa	GPa	MJ/m³	MPa	GPa	MJ/m³
CA As-Printed	45	2.2	3.0	44.7	2.2	1.7	N/A	N/A	N/A
CA Converted to Cellulose	N/A	N/A	N/A	N/A	N/A	N/A	55	2.5	4.4

Source: Data from Pattinson, S. W., Hart, A. J. 2017. Additive Manufacturing of Cellulosic Materials with Robust Mechanics and Antimicrobial Functionality. *Adv. Mater. Technol.* 2, 1600084.

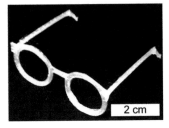

FIGURE 7.8   Miniature eyeglass frames printed out of CA.

Source: Pattinson, S. W., Hart, A. J., 2017. Additive Manufacturing of Cellulosic Materials with Robust Mechanics and Antimicrobial Functionality. *Adv. Mater. Technol.*, 2, 1600084. Reproduced with permission from John Wiley and Sons.

an air pump mounted on Printrbot Simple Metal desktop printer, and extruded through a 0.2 mm diameter nozzle. The mixture CA/acetone was very adhesive, and the printed layer bonded to each other, and hardened as the acetone evaporated. Table 7.7 lists the mechanical properties of printed test coupons in CA. Figure 7.8 is a photograph of a miniature eyeglass frames printed out of CA. As anticipated, the tensile strength and modulus are isotropic, except toughness, because it was calculated as the area under the tensile stress-strain curve, and hence it was influenced by being the elongation at break in parallel direction almost twice that in perpendicular direction. In order to increase the mechanical properties of the printed part, CA was converted into cellulose by immersing the part in a solution of NaOH, which, based on FTIR spectra, reduced the amount of acetate groups (Khatri et al. 2012) while increasing that of hydrogen bonding (Liu and Hsieh 2002), and considerably boosted tensile strength and modulus, and toughness (Table 7.7), whose values were competitive with, and even exceeded, those of ABS, PLA, and PA for AM, reported in Table 7.8. Moreover, low standard deviation for all the values in Table 7.7 pointed out that the AM process and feedstock generated a material with consistent performance. Finally, the authors demonstrated that antimicrobial functionality could be incorporated in the printed part by adding antimicrobial agents to the CA/acetone mixture and printed discs. A solution containing Escherichia coli bacteria was poured on the discs and control samples, and, after 20 h of exposure to fluorescent light, the number of bacteria in the solution dispensed on the discs contained 95% less bacteria than the solution poured on the control samples. This result indicates that this material and process have potential for manufacturing sterilized medical and surgical tools in locations that are remote or have difficult access to such supplies.

7.3.4   Cellulose Ethers

7.3.4.1   Introduction

Cellulose ethers are sustainable, plant-derived, abundant, economical, biodegradable, water soluble, and chemically stable and physiologically safe compounds (Dai et al. 2019). They are produced by the chemical modification of cellulose, and serve as stabilizers, thickeners, and viscosity modifiers in many industries, including food, construction, oil field chemicals, personal care products, pharmaceuticals, paper, adhesives, and textiles. They include CMC, EC, HEC, HPMC, HPC and MC. CMC is the cellulose ether mostly consumed worldwide, accounting for half of the total consumption volume in 2018 (IHS Markit 2019). HPMC, HPC, and HEC are safe to ingest, biocompatible, reasonably priced, and available in several grades, and suitable as materials for shells and coatings enclosing drugs that progressively swell and erode, and enabling to control the drug's release rate (*erodible delivery systems*) (Gazzaniga et al. 2008).

7.3.4.2   Carboxymethyl Cellulose (CMC)

CMC, or *cellulose gum*, is characterized by the carboxymethyl group (–CH$_2$–COOH). It is often utilized as its sodium salt, called *sodium carboxymethyl cellulose* and containing the functional group –CH$_2$COONa. CMC's molecular structure is shown in Figure 7.9. As a benchmark, tensile strength and elongation at break of CMC films are 27.1 MPa and 17.7%, respectively (Ebrahimi

TABLE 7.8
Property Values of ABS, PLA, and PA for AM

Material, Fabrication Process	ABS, Extrusion AM			PLA, Extrusion AM			PA, Laser Sintering AM		
Property	Tensile Strength	Tensile Modulus	Toughness	Tensile Strength	Tensile Modulus	Toughness	Tensile Strength	Tensile Modulus	Toughness
Unit	MPa	GPa	MJ/m^3	MPa	GPa	MJ/m^3	MPa	GPa	MJ/m^3
Value	22–34	1.6	2.2	39	1.5	0.8	50–58	0.6–2.6	2.2–6.3

Source: Data from Pattinson, S. W., Hart, A. J. 2017. Additive Manufacturing of Cellulosic Materials with Robust Mechanics and Antimicrobial Functionality. *Adv. Mater. Technol. 2,* 1600084.

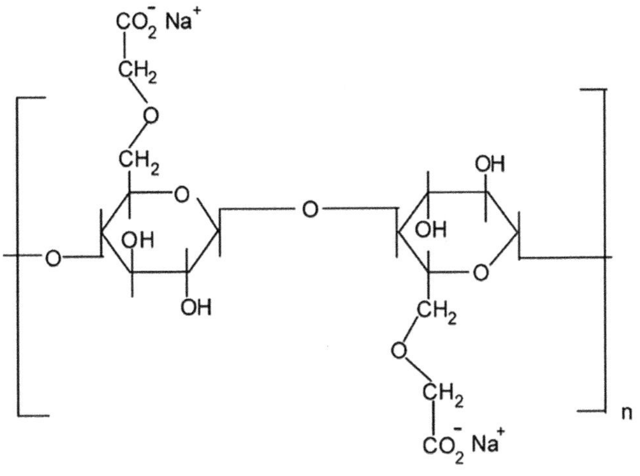

FIGURE 7.9 Molecular structure of carboxymethyl cellulose (CMC).

Source: Biswal, D. R., Singh, R. P. 2004. Characterisation of carboxymethyl cellulose and polyacrylamide graft copolymer. *Carbohydrate Polymers* 57:379–387. Reproduced with permission from Elsevier.

et al. 2018). According to CP Kelco (USA), producer of CMC with trade names CEKOL® and Finnfix®, CMC's functions are rheology modification, water-binding, film forming, protein protection, and particle stabilization, while CMC's applications are beverages, papermaking, mining, household products, and health and personal care (CP Kelco 2020). Other CMC producers are AEP Colloids (USA), Anqiu Eagle Cellulose (China), Ashland (trade name Aqualon™, USA), and Dow Wolff Cellulosics (Germany).

CMC sodium salt was mixed as a binder to glass frit and DI water and printed via extrusion into experimental foamy glass articles (Klein et al. 2015). More details are in Section 7.3.4.4.

CMC is widely present in composite materials for AM due to its viscosity thickening capability and thixotropic rheology.

Miniaturized, low-cost, and high-throughput devices made of biological molecules play a critical role in: biosensing, drug screening, environmental monitoring, forensic investigation, military defense, etc. (Pignataro 2009). One process to deposit biomolecules is inkjet printing (Section 2.7), consisting in ejecting picoliter-size droplets of fluid at a frequency up to tens of kHz (Cibis and Kruger 2008). Brindha et al. (2016) developed formulations of bioinks for inkjet printing based on a solution of phosphate buffered saline (PBS), and bovine serum albumin (BSA). The former is an aqueous solution consisting of a mixture of a weak acid and its conjugate base, the latter is a protein derived from cows. These bioinks featured values of density, surface tension, and viscosity very similar to those of commercial inkjet printing inks. Typical fluids for inkjet printing have a density close to 1,000 kg/m^3 (Derby 2012), viscosity of 0.003–0.020 kg/(m s) and surface tension 20–70 mN/m (Gonzalez-Macia et al. 2010). The authors employed CMC as a viscosity modifier, and studied the effect of adding 1% of CMC to BSA-based bioinks, and assessed that this addition raised viscosity and surface tension of bioinks, resulting in ink drops with smaller diameter, but did not affect the activity of the biological molecules, which is a requirement of bioink formulation. Moreover, the binding ability of bioinks containing CMC was improved by means of glutaraldehyde crosslinking on glass surface. Glutaraldehyde is an organic compound employed as disinfectant, medication, preservative, and fixative. Simulation results showed that inks with greater viscosities required higher jetting velocities.

Park et al. (2017) leveraged the capability of FFF to manufacture parts with custom-designed materials, and fabricated a three-layer lithium miniature battery using CMC-based electrically conductive composite pastes, specifically formulated to reach a very low percolation threshold. This threshold is the lowest concentration of a filler at which an insulating material is converted to conductive material, and, hence, an electrical pathway is formed throughout a sample (Wypych 2016). Specifically, they added Sigma Aldrich CMC powder (MW ~250,000 g/mol) serving as a matrix to a DI water solution of electrically conductive silver nanowire (AgNW), with 100–150 nm diameter, ~20 μm length, and 133–200 aspect ratio, serving as a filler. AgNW was selected because it possesses high electrical conductivity, and, when added to a matrix, it produces a composite with low percolation thresholds. CMC was chosen because of its viscosity thickening property, and shear thinning behavior. AgNW/CMC nanocomposite pastes with differing AgNW concentrations (0.3, 0.7, 1.1, 1.5, and 1.9 vol%) were prepared, in order to conduct computational simulation and experimental evaluation, and the AgNW concentration relative to the optimal percolation threshold was 0.7 vol%. The composite pastes contained the following ingredients:

- Anode: CMC, $Li_4Ti_5O_{14}$ (<200 nm particle size, Sigma Aldrich), AgNW
- Cathode: CMC, $LiFePO_4$ (<5 μm particle size, Sigma Aldrich), AgNW
- Separator electrolyte: poly(ethylene oxide) (PEO, MW~900,000 g/mol, Sigma Aldrich), $LiClO_4$, TiO_2.

The printing apparatus integrated the commercial printer Orion Delta™ (SeeMeCNC, USA) and the paste extrusion system Discov3ry (Structur3D Printing, Canada), with the Orion's printing head modified to hold a paste extrusion syringe. The syringe pump of the original Discov3ry paste extruder was replaced with a hydraulic pumping mechanism equipped with two syringes: a 60 mL pumping syringe receiving primary pressure from the Discov3ry motor, and a 5 mL extrusion syringe located at the printing head for material extrusion. The printing parameters for each layer varied depending on the layer printed (anode, cathode, and electrolyte). The deposition of each layer was followed by drying at 60–70°C, depending on the layer, and is pictured in Figure 7.10. The battery area was about 15 × 15 mm. The optimal electrical percolation threshold was obtained with 0.7 vol% of AgNW. A reliable conductivity of 1.19×10^2 S/cm was obtained from the paste

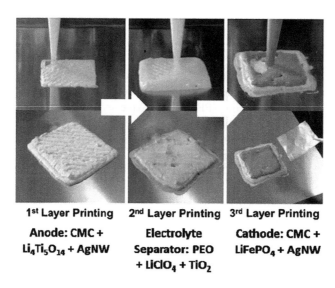

1st Layer Printing **2nd Layer Printing** **3rd Layer Printing**

Anode: CMC + **Electrolyte** **Cathode: CMC +**
$Li_4Ti_5O_{14}$ + AgNW **Separator: PEO** **$LiFePO_4$ + AgNW**
 + $LiClO_4$ + TiO_2

FIGURE 7.10 Images of printing in sequence anode, electrolyte and cathode to form a three-layered battery.

Source: Modified from Park, J. S., Kim, T., Kim, W. S. 2017. Conductive cellulose composites with low percolation threshold for 3D printed electronics. *Scientific Reports* 7:3246. Reproduced under a Creative Commons Attribution 4.0 International License.

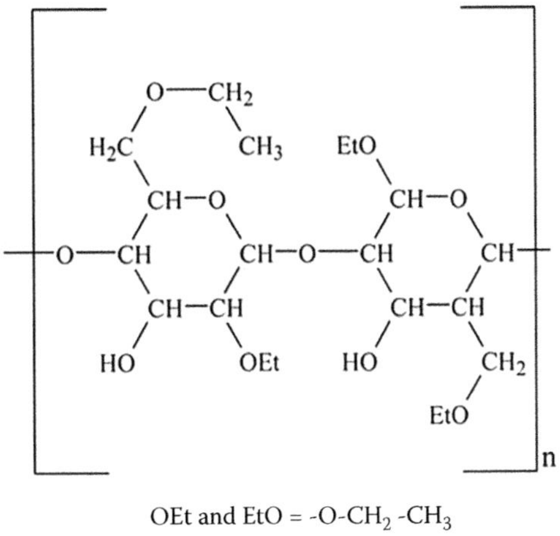

FIGURE 7.11 Molecular structure of ethyl cellulose (EC).

Source: McKeen, L. W. 2012. Renewable Resource and Biodegradable Polymers. In *Film Properties of Plastics and Elastomers*, ed. L. W. McKeen, 353–378, 3rd edition. William Andrew. Reproduced with permission from Elsevier.

with 1.9 vol% of AgNW and 20 vol% of CMC. The voltage of the printed battery was 1.8 V, falling in the range of 1.5–1.8 V characteristic of lithium iron phosphate/lithium titanate batteries.

Tanwilaisiri et al. (2016) combined CMC (as a binder), water/ethanol, and activated carbon (a form of carbon featuring small pores that increase the surface area available for adsorption and chemical reaction) to print electrodes for an electrochemical supercapacitor functioning as an energy storage device with high power densities and long life cycles that are characteristic of fitting portable electronic devices, electric vehicles, and emergency power supplies. A paste of CMC, water/ethanol, and activated carbon was prepared, and extruded from a syringe mounted on a commercial FFF desktop printer, chosen for being affordable to buy and run. The supercapacitor had a size of 3 × 5 cm, required no assembly, and was printed in one session, by extruding: PLA filaments to build the frames, paste of conductive paint, and electrode paste. Since the specific capacitance of the supercapacitor was about 0.20 F/g, the printing process for the supercapacitor proved to be a satisfactory fabrication method.

7.3.4.3 Ethyl Cellulose (EC)

EC (Figure 7.11) derives from cellulose by converting the hydroxyl –OH groups on the repeating glucose units into the ethoxy groups $-OCH_2CH_3$. Rekhi and Jambhekar (1995) and McKeen (2012) published comprehensive reviews on EC. EC is insoluble in water and glycerol, and soluble in esters, aromatic hydrocarbons, alcohols and ketones. Its color is white to slightly yellow. Selected properties of EC are listed in Table 7.9. EC features the following (Chemical Book 2017, Food-Info 2017):

TABLE 7.9

Properties of EC

Density	Molecular Weight	Viscosity[a]	Particle Size Mean[a]	Particle Size Max[a]
g/cm^3	g/mol	mPa s	μm	μm
1.07–1.18	454.5	3–110	3–60	100–150

Source: Dow, ETHOCEL™ Ethyl cellulose A Technical Review, Technical Bulletin, 02–2016.
Note:
[a]Relative to Ethocel™ and depending on its grade.

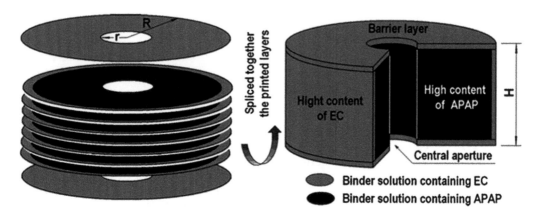

FIGURE 7.12   Schematic drawing of the novel printed drug delivery devices (DDD).

Source: Yu, D.-G., Branford-White, C., Ma, Z.-H., et al. 2009. Novel drug delivery devices for providing linear release profiles fabricated by 3DP. *International Journal of Pharmaceutics*, 370:160–166. Reproduced with permission from Elsevier.

- Functions: binder, film former, filler, coating, emulsifier, dietary fiber, anti-clumping agent, and thickener.
- Applications: food, suntan gels, creams, lotions, coatings, hot-melt adhesives, transfer inks, vitamin and mineral tablets, industrial processes.

Commercial EC is available in a range of viscosity that increases as the length of the polymer molecules grows. Examples of commercial grades of EC are ETHOCEL™, and Aquacoat® ECD, manufactured by Dow® for food and pharmaceutical applications and FMC, respectively. EC suppliers are located in USA, China, Europe, Japan, and India, selling at a price depending on packaging (from less than 1 kg to 25 kg) and viscosity (4 to 300 cP). F.e. Sigma Aldrich sells 1 kg package for USD 373. ETHOCEL™ is a colorless, odorless, and physiologically inert polymer for pharmaceutical uses, ranging in viscosity (cP 3–110) and MW. Powder size is: mean 5–60 µm, max 100–150 µm. It is soluble in ethanol, acetone, isopropanol, methanol, and combinations of all (ETHOCEL™ 2016). In terms of mechanical properties, films comprising 15 g of EC and 100 cc of one of several solvents exhibited tensile yield strength of 41.2–43.2 MPa, elongation at yield 6–12%, and specific gravity of 0.85–1.24 (Mahnaj et al. 2013).

The work by Yu et al. (2009) is notable because it leveraged EC's insolubility in water and AM's suitability for complex design and material formulations at very small scale. 3DP™ (Section 2.6.1), a version of BJ, and EC were combined to manufacture a drug tablet with a novel doughnut-shaped design (Figure 7.12) that provided drug release with linear rate over the entire drug dissolution time, and met adequate values of mechanical properties (hardness, friability). Test tablets were printed out of hydrophilic HPMC as matrix, hydrophobic EC as release-retardant material, and two binders, and consisted of three sections: drug-free top and bottom barrier layers, and in-between drug-loaded layers, schematically pictured in Figure 7.12 left. The drug was acetaminophen (APAP), a medication for treating pain and fever, present in Tylenol®. Because the top and bottom layers were inert and impermeable to water penetration and drug diffusion, the drug was released only from the outer and inner cylindrical surface. Over time, as the outer surface decreased, the inner surface increased synchronously in the dissolution process, and the tablet's total eroded surface area remained stable over time even initially, contrary to the fact that typically drug release rates at the beginning are faster than later on. Moreover, tablets printed via 3DP™ outperform tablets conventionally made in terms of adhesion between the barrier layers and the drug-contained regions. The top and bottom layers were formed by depositing droplets of a binder containing 2 w/v% of EC in 90 vol% of ethanol in water onto spread HPMC powder. The 0.2 mm thick outer cylindrical shell was made by depositing the above mentioned binder on HPMC powder containing APAP. The regions with drug inside the shell were laid by dispensing a binder containing 10 w/v% of APAP in 90 vol% ethanol in water onto powder comprising APAP, HPMC, EC, polyvinylpyrrolidone K30 and colloidal SiO_2 in the weight ratio of 60/20/10/9.5/0.5, respectively. The tablet's center corresponding to its hole was made of HPMC powder with no binder. Tablets with various values of the drug dosage were tested, and the dosage could be adjusted while slightly affecting the release profiles, which could lead to customization of medications. Since the drug release rate was independent of the tablet's height, the dosage could be easily varied by increasing or decreasing the number of printed layers containing the drug. EC came from Shandong Ruitai Chemical (China), and HPMC was Dow® METHOCEL™ E50 Premium. The spacing of droplets along the raster direction was 40 µm, the line-to-line spacing 100 µm, and layer thickness 200 µm.

The paper by Kempin et al. (2017) also dealt with a pharmaceutical application of EC, but this time EC served as feedstock of a drug-impregnated filament utilized to manufacture drug-loaded implants through FFF, which presents the advantage of customizing implants' shape and size to individual patients. Particularly, ETHOCEL™ Standard 45 Premium EC was evaluated with respect to its suitability for fabricating drug-loaded implants via FFF, and the implants' drug release behavior. Quinine, a medication for lupus and arthritis, was chosen as a drug, and incorporated into EC by a solvent casting technique. First, EC was dissolved in acetone, to which an ethanolic solution of quinine was added. This mixture was homogenized upon stirring for a few minutes, poured onto a glass plate as a thin layer, and placed in a drying oven to let the volatile solvent evaporate. The resulting film was cut into pieces, and fed into an extruder, along with triacetin, a colorless oily liquid triglyceride found in fruit, added as a plasticizer. The extruder, set at nozzle temperature of 59°C, around the T_g of the blend EC and triacetin, generated a 2.6 mm diameter filament. Implant test samples, shaped as hollow cylinders with 5 mm outer diameter, 3 mm inner diameter, and 3 mm height, were printed via FFF at nozzle temperature of 180°C, below EC's degradation temperature. A hollow cylinder geometry was chosen because such implant could be inserted, f.e., after surgery in the area of the paranasal sinuses (a group of four paired air-filled spaces around the nose) and simultaneously release medication, and allow drainage of secretions through its hole. After 100 days of in vitro testing, the released quinine was very low and amounted to 4.5 wt%. The implant's flexibility induced by triacetin may allow to insert the implant by first transporting it in compressed form with an endoscope to pass constrictions, and then let it spread out at the application site.

7.3.4.4 Hydroxyethyl Cellulose (HEC)

HEC is obtained from cellulose by replacing $-OH$ groups with $-CH_2CH_2OH$. Its chemical formula is $C_{36}H_{70}O_{19}$, and its chemical structure is illustrated in Figure 7.13. Some of its properties are (PubChem® n.d.): MW 806.9 g/mol, specific gravity 1.003 g/cm^3, Brookfield viscosity 75 cP at 5% solids (low MW), 5,000 cP at 1 wt% solids (high MW). HEC is non-ionic, and exhibits gel thickening, emulsifying, bubble-forming, and water-retaining and stabilizing properties. It is soluble in water and glycerol but generally not in organic solvents. A commercial HEC is Natrosol™ (Ashland), available as a white powder and possessing low to medium MW. Natrosol™ is added as a binder, bond strengthener, cement extender, coating and optical brightener aid, coating polymer, filtration control additive, green strength enhancer, protective colloid, rheology controller and modifier, lubricity and workability enhancer, suspension and stabilization agent, shape retention enhancer and thickener. Natrosol™ is applied in adhesives and sealants, advanced ceramics, building and construction, ceramics, pottery and porcelain, oil and gas technologies, metal castings and foundry, paint and coatings, personal care, pharmaceutical, ophthalmic, pulp and paper (Ashland 2020).

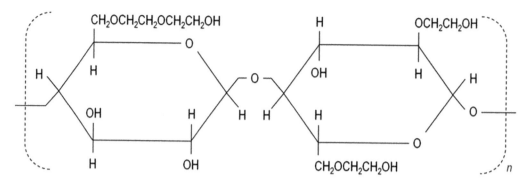

FIGURE 7.13 Molecular structure of hydroxyethyl cellulose (HEC).

Source: Teixeira, M. A., Amorim, M. T. P., Felgueiras, H. P. 2020, Poly(Vinyl Alcohol)-Based Nanofibrous Electrospun Scaffolds for Tissue Engineering Applications. *Polymers* 12, 7. Reproduced from open access article under the terms and conditions of the Creative Commons Attribution (CC BY).

Sun et al. (2013) employed HEC and HPC as viscosity modifiers in viscoelastic printing inks formulated to fabricate with μm resolution a microbattery (Figure 7.14) featuring interdigitated (interlock imitating the fingers of two clasped hands) architecture, designed to fit micro-electromechanical systems (MEMS), biomedical sensors, wireless sensors and actuators, and capable to provide twice the energy density than that in larger batteries. The following aqueous solutions of nanoparticles, DI water, ethylene glycol, and glycerol made up the inks to print the battery's high aspect ratio (thin-walled) elements:

- Anode: 9 wt% HPC, 1 wt% HEC, $Li_4Ti_5O_{12}$ (LTO) powder of 50 nm mean diameter
- Cathode: 8 wt% HPC, 2 wt% HEC, $LiFePO_4$ (LFP) powder of 180 nm mean diameter.

Inks were formulated in a multistep process that comprised particle dispersion, centrifugation, and homogenization. The printing equipment included 3-axis micropositioning stage, syringes with 3 ml barrel, and 30 μm diameter nozzle, and printing speed was 250 μm/s. The printed articles were annealed at 600°C for 2 h. The inks can be deposited over areas ranging from 100's of $μm^2$ to 1 mm^2 with minimum feature sizes as small as 1 μm. In terms of areal power density (mW/cm^2) and areal energy density (J/cm^2), the printed battery outperformed most of its rechargeable counterparts.

Typically, glass objects are fabricated from glass in a melt state, from which a glob is taken, blown, and let solidify and cool in its final shape. However, printing glass directly from a melt, layer

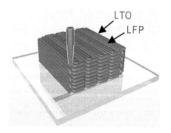

FIGURE 7.14 Left: drawing of interdigitated $Li_4Ti_5O_{12}$–$LiFePO_4$ (LTO-LFP) battery. Right: SEM images of printed and annealed version of the battery.

Source: Modified from Sun, K., Wei, T.-S., Ahn, B. Y., et al. 2013. 3D Printing of Interdigitated Li-Ion Microbattery. *Adv. Mater.* 25:4539–4543. Reproduced with permission from John Wiley and Sons.

by layer is impossible, because glass is very sensitive to the cooling process and temperature gradients between layers. To achieve a glass state, the melt glass has to be cooled rapidly to avoid crystallization of the material, and then annealed to remove cooling-induced stress. In order to print glass articles, Klein et al (2015) followed a different glass-making process, called *kiln glass method*, in which ground glass or *frit* is poured into a mold to be shaped, and fused together in an oven (*kiln*). In this technique, the smaller the frit size the finer the detail on the finished piece, but the less transparent the piece is. The authors studied a glass feedstock suited for a lab method consisting in: (a) preparing a paste out of glass frit and binder to be printed at room temperature via an extrusion method, (b) drying, (c) burning the binder off, and (d) melting and fusing together the glass grains. The investigators prepared and tested three mixtures, all based on woodland brown 0203-08 glass frit: two mixtures also contained one binder, and DI water. HEC and CMC sodium salt were chosen as binders because they performed well in pastes processed via for FFF. The firing schedule consisted in steps designed to dry the frit, cool, and quench the melt into a glass state, and slowly cool to anneal the glass samples. Samples with HEC contained fewer bubbles (although with a higher bubble concentration in the center) than samples with CMC sodium salt that formed foams with an even bubble distribution The material properties of the final objects were crucially dependent on the frit size of the glass powder selected during shaping, the chemical formula of the binder, and the firing procedure. Frit sizes below 250 μm resulted in a constant volume of pores below 5%, but decreasing frit size led to an increase in the number of pores which in turn increased opacity.

Avery et al. (2014) studied the rheology of glass pastes for AM, and compared HEC to xanthan gum as binders in glass frit for AM. The wt% composition and tensile modulus of paste with HEC were glass/water/HEC = 1/1/0.05, and 69 GPa average, respectively. The authors assessed that HEC increased tensile yield stress vs. xanthan gum, and resulted in porosity similar to that of frit-only samples. The value of the yield stress was found to depend on the choice of binder and particle size, and lower yield stresses occurred in presence of larger particle sizes and xanthan gum binder.

7.3.4.5 Hydroxypropyl Cellulose (HPC)

HPC is an ether and polysaccharide derived from cellulose in which hydroxyl groups on the cellulose backbone have been replaced with hydroxypropyl groups. Its molecular formula is $C_{36}H_{70}O_{19}$, and its molecular structure is shown in Figure 7.15. HPC has a combination of

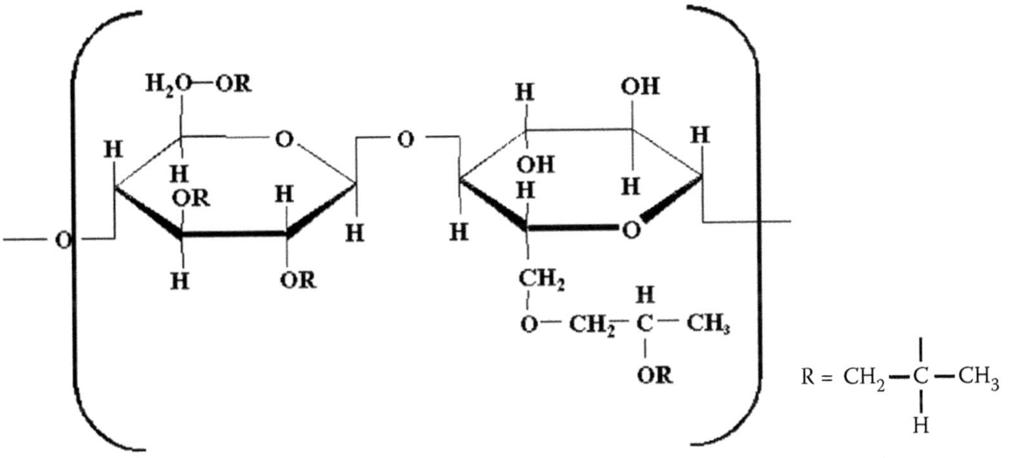

FIGURE 7.15 Molecular structure of hydroxypropyl cellulose (HPC).

Source: Modified from Mezdour, S., Cuvelier, G., Cash, M. J., Michon, C. 2007. Surface rheological properties of hydroxypropyl cellulose at air–water interface. *Food Hydrocolloids* 21:776–781. Reproduced with permission from Elsevier.

TABLE 7.10

Properties of HPC Films for Pharmaceutical Use

Molecular Weight	Viscosity	Tensile Strength	Elastic Modulus	Elongation at Break
g/mol	cP	MPa	MPa	%
40,000–140,000	2–10	16.3–25.6	523–603	2.7–7.3

Source: Takeuchi, Y., Umemura, K., Tahara, K., Takeuchi, H. 2018. Formulation design of hydroxypropyl cellulose films for use as orally disintegrating dosage forms. *J. Drug Deliv. Sci. Technol.* 46:93–100.

hydrophobic and hydrophilic groups, hence below 45°C (its lower critical solution temperature) is readily soluble in water, whereas above the 45°C, it is not soluble.

Table 7.10 lists, as baseline data, the measured tensile properties of films (Takeuchi et al. 2018) composed of 25 wt% Nippon Soda/Nisso HPC, 37.5 wt% water, and 37.5 wt% ethanol, and possessing a range of MW and viscosity values. Other HPC films were cast starting from a water solution of 5 wt% HPC powder with average MW of 1,195,000 g/mol. The films exhibited values of tensile strength and modulus of 16 MPa and 703 MPa, respectively, similar to those in Table 7.10 (Yanagida and Matsuo 1992).

Outside AM, HPC serves as lubricant, thickener, viscosity modifier, coating, film matrix, drug tablet, binder, and foam and suspension stabilizer. Commercial HPC grades not formulated for AM should be considered and evaluated for AM. Examples of those grades are Klucel™ (Ashland), and CELNY™ (Nippon Soda Co./Nisso). The markets and application of HPC are pharmaceutical, food, and personal care. As an example, the Nisso HPC powder size is mesh 40–330, and D_{50} is 20–190 μm, with D_{50} is the diameter at which 50% of a sample's mass comprises particles featuring smaller diameter. Klucel™ is a non-ionic TP water-soluble cellulose ether with viscosity of 75–150 at 25°C, and 5 wt% concentration in water. It combines organic solvent solubility, surface activity, thickening and stabilizing properties of other water-soluble cellulose polymers. Its particle size is mostly 595–841 μm, its MW 40,000–1,150,000 g/mol depending on the grade, and its viscosity 75–6500 cP depending on grade and concentration in water.

Melocchi et al. (2015) effectively combined HPC and FFF to prototype and manufacture capsules for the time-controlled release of oral drugs in the digestive system. FFF accelerated the screening of capsule formulation and design, and the HPC capsules printed (Figure 7.16) performed satisfactorily, and in a manner comparable to analogous HPC injection molded capsule shells branded as Chronocap™ (Gazzaniga 2010), namely the drug release profiles (amount and time lag) of printed capsules and molded capsules were almost identical. Since no FFF filament in HPC was available, the first step was extruding a 1.75 mm diameter filament in laboratory from previously oven-dried at 40°C Klucel™ LF HPC mixed with 2–10 wt% of PEG 1500 (Clariant Masterbatches, Italy) as a plasticizer. Trials were run to optimize HPC/PEG composition, extrusion temperature, screw speed, and pulling/calibrating device speed, and the resulting filament was processed on a desktop printer at 210°C. Design and fabrication trials using FFF were conducted, varying tip diameter (0.25, 0.3, 0.2 mm) and printer operating settings. Eventually, capsules were printed that were very similar in geometry, and weight to molded capsules. FFF also provided adequate resolution and dimensional control to achieve the design feature to overlap and join the two halves of the capsule.

7.3.4.6 Hydroxypropyl Methylcellulose (HPMC)

HPMC, or *hypromellose*, is a semi-synthetic, inert, viscoelastic polymer present in a variety of commercial products: eye drops, tile adhesives, cement renders, gypsum products, pharmaceutical, paints, coatings, food, cosmetics, detergents, cleaners, etc. It derives from cellulose by partially converting cellulose's hydroxyl groups into hydroxypropyl and methoxyl groups. HPMC is an

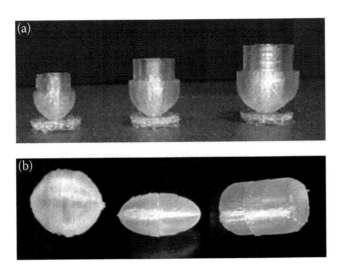

FIGURE 7.16 Printed hollow PLA structures before (a) before raft (support base) removal, and (b) assembled capsular devices with different size and shape.

Source: Melocchi, F., Parietti, G., Loreti, A. et al. 2015. 3D printing by fused deposition modeling (FDM) of a swellable/erodible capsular device for oral pulsatile release of drugs. *J. Drug. Deliv. Sci. Technol.* Part B 30:360–367. Reproduced with permission from Elsevier.

odorless, flavorless, transparent, stable, oil-resistant, nontoxic, edible, nonionic polymer with a linear structure of glucose molecules, in which its matrix is stabilized using hydrogen bonds (Ghadermazi et al. 2019). Its chemical formula is $C_{56}H_{108}O_{30}$ and its chemical structure is depicted in Figure 7.17. HPC is added to food as an emulsifier, thickening and suspending agent, and as an alternative to animal gelatin. HPMC is a solid, and slightly off-white to beige powder, and may be formed into granules. Its compound forms colloids when dissolved in water.

Smay et al. (2002) leveraged Methocel™ F4M, a commercial grade of HPMC, to increase viscosity of colloidal inks printed in self-supporting articles for: sensors, photonics, composites, TE, and advanced ceramics. Using inks containing a concentrated silica dispersions and poly-ethylenimine, self-supporting, 0–90° lattice samples featuring spacing of 250–750 µm were

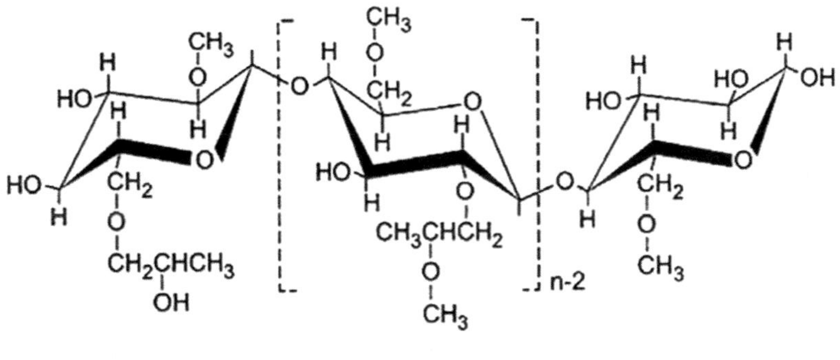

FIGURE 7.17 Molecular structure of hydroxypropyl methylcellulose (HPMC).

Source: Modified from Ghadermazi, R., Hamdipour, S., Sadeghi, K., Ghadermazi, R., Khosrowshahi, A. 2019. A. Effect of various additives on the properties of the films and coatings derived from hydroxypropyl methylcellulose - A review. *Food Sci. Nutr.* 7: 3363–3377. Reproduced under the terms of the Creative Commons Attribution License from Wiley Periodicals.

printed, through an AM process for ceramics called *robocasting* developed by Robocasting Enterprises (Robocasting 2018).

The study by Khaled et al. (2014) is notable because it leveraged HPMC and AM for personalized medicine, and demonstrated that even a desktop printer costing less than USD 1,000 can produce bilayer pharmaceutical tablets with multiple active ingredients for immediate release followed by sustained release. More in general, AM can potentially produce medicines with accurate and personalized dosages in a single tablet, involving even multiple active ingredients present as a single blend or as individual layers, which may be challenging for conventional processes that manufacture pharmaceutical tablets by pressing the active ingredients and excipients together. The drug fabrication could even take place on a desktop printer at the point of care, i.e. hospital or patient's home. The on-demand drug preparation enabled by AM will also address another issue: the tendency of some drugs (such as nitroglycerin to treat angina pectoris) to degrade upon storage. The physical and mechanical properties of the printed tablets, such as hardness, thickness, weight variation, and friability were measured and compared to the commercial tablets, and their values were within the acceptable ranges established by the international standards. The success of printing medications also depends on being cost-competitive with conventional production of medications.

Khaled et al. (2015) extruded on a regenHU printer a *polypill* (Figure 7.18), a single drug tablet featuring three compartments for as many drugs to be released at independent rates via two release mechanisms: diffusion through gel layers, and osmotic release through a controlled porosity shell. The drugs were physically separated in the multi-compartment tablet to avoid incompatibility issues, and to allow greatest flexibility in manipulating each drug. HPMC type 2910 (29 wt% methoxyl groups, 10 wt% hydroxypropyl groups) was an ingredient of the paste printed to form the polypill. HPMC hydro-alcoholic gel served as a binder to form a smooth and highly consistent paste, extruded through a 400 μm printing tip. The printed tablets incorporated an osmotic pump with the drug captopril and sustained release compartments with the drugs nifedipine and glipizide. This combination of medicines could possibly treat diabetics suffering from hypertension.

For medications, a release profile (relationship between release percentage and time) depending on no other factors (Fu and Kao 2010) but time (zero-order) is preferred. Wang et al. (2006) selected HPMC and 3DP™ to print a near zero-order controlled-release formulations for a water-soluble drug. Namely, they printed a core plus shell tablet with HPMC serving as a powder matrix to be consolidated via 3DP™ (using an aqueous binder containing 50 wt% of pseudoephedrine hydrochloride) and providing pathways for the drug to diffuse outward at a controlled

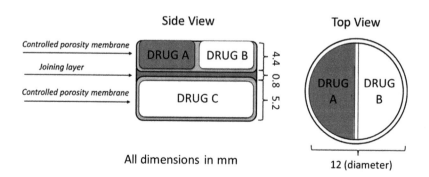

FIGURE 7.18 Cross sections of multi-compartment printed pill, or *polypill*.

Source: Modified from Khaled, S. A., Burley, J. C., Alexander, M. R. et al. 2015. 3D printing of tablets containing multiple drugs with defined release profiles. *International Journal of Pharmaceutics* 494:643–650. Reproduced with permission from Elsevier.

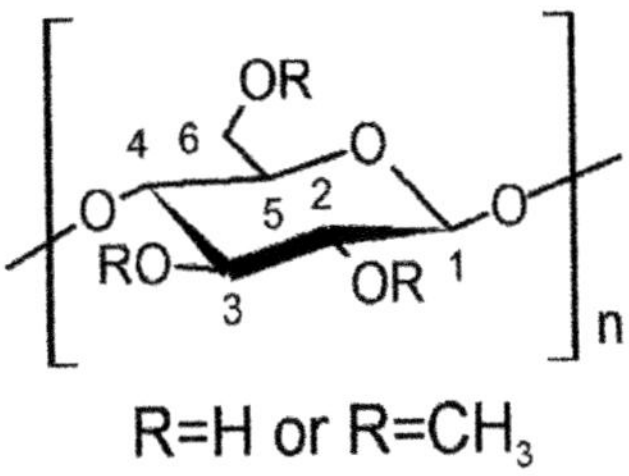

FIGURE 7.19 Molecular structure of methylcellulose (MC).

Source: Nasatto, P., Pignon, F., Silveira, J., et al. 2015. Methylcellulose, a Cellulose Derivative with Original Physical Properties and Extended Applications. *Polymers* 7:777–803. Reproduced under the terms and conditions of the Creative Commons Attribution license.

rate. HPMC appeared to be the key drug release-regulating material, because the overall drug release rates increased with the HPMC content in the powder blend in a near-linear relationship.

Boetker et al. (2016) leveraged the customization capability of AM, and combined HPMC (food grade, water-soluble Shin-Etsu Metolose® 60SH) and PLA to print via FFF oral drug tablets enabling flexible dosing and precision medication, and demonstrated that AM can fabricate custom-made, precision products with tailored drug release characteristics. First a printing mixture of HPMC, PLA, and nitrofurantoin (an antimicrobial drug to treat urinary tract infections) as an active pharmaceutical ingredient was extruded as a filament for FFF, then disk-shaped samples 10 mm diameter × 2 mm height were printed. The release of nitrofurantoin from the disks depended on the amount of HPMC, and higher content of HPMC resulted in higher drug release.

7.3.4.7 Methylcellulose (MC)

Methylcellulose (MC) is one of the most important commercial cellulose ethers, present in many industrial applications. MC is the simplest cellulose derivative, in which (Figure 7.19) methyl groups ($-CH_3$) substitute the hydroxyls at C-2, C-3 and/or C-6 positions of anhydro-d-glucose units (Nasatto et al. 2015). MC has chemical formula $C_{29}H_{54}O_{16}$, dissolves in water and organic solvents, and displays both hydrophilic and lipophilic properties. It is not digestible, non-allergenic, and nontoxic. It is present in food and cosmetic products as an emulsifier and thickener, in construction materials as a performance additive, in the production of papers and textiles as a sizing, and as a mild glue and binder in wallpaper pastes, pastel crayons, and medications. Examples of manufacturers of MC are Ashland, Dow Chemicals, JRS, and Shin-Etsu.

Henke and Treml (2013) tested MC as a binder for sprue chips, 0.8–2 mm in size, processed through a laboratory powder-based AM process, involving aerosolized water as activator, and a mask limiting the area of powder layer to be wetted, and hence to be consolidated in a solid object. However, results were not satisfactory, since the printed items exhibited poor mechanical strength.

Thakkar et al. (2017) selected MC as a plasticizer to promote the adhesion between the binder (bentonite clay) and adsorbent particles (aminosilica) for successfully printing aminosilica-based adsorbents monolithic articles (such as porous cylinders 1.5 mm in height × 2 mm in diameter) for capturing CO_2. The aminosilica materials retained their structural, physical, and chemical properties in the printed monoliths.

7.3.5 Cellulose/PLA

Cellulose is a promising additive for PLA, because, besides being biobased and biodegradable, it increases crystallinity, and improves mechanical and thermal properties of PLA (Oksman et al. 2003). A main challenge in combining cellulose and PLA is the difficulty to achieve uniform dispersion of cellulose in PLA, being cellulose hydrophilic, and PLA hydrophobic and tending to aggregate (Raquez et al. 2013).

One route to improve dispersion of cellulose in PLA, and simultaneously enhance the bonding between the two materials has been reported by Murphy and Collins (2018), and basically consisted in using *microcrystalline cellulose* (MCC) surface-modified with Lica® 38, a titanium-derived thermally stable compound, that chemically bridges PLA and cellulose, strengthening the bonding between the two, and enhancing the mechanical properties of the resulting composite material (Elshereksi et al. 2017).

Wang et al. (2017) reinforced PLA with micro-nano cellulose, and printed it in form of composite wire rods. They introduced a process to downscale cellulose fiber to the micro- and nanolevel that resulted in a mixture of microcellulose (majority) and nanocellulose fibers, or micro/nano cellulose fibers (MNCs). MNCs showed properties differing from those of unmixed microcellulose fibers and nanocellulose fibers, and offered several advantages: biodegradability, low cost, low density, easy separation, renewability, sequestration of CO_2, chemical and physical modification, etc. MNCs can also reinforce numerous polymers. The authors formulated MNC/PLA filament for FFF with maximum filler quantity (to minimize PLA cost), featuring interface compatibility and printability, while achieving mechanical properties comparable to those of unfilled PLA. Silane as PLA/cellulose coupling agent, and PEG as plasticizer were selected, besides MNC and PLA,. A MNC/PLA composite filament composed of 30 wt% cellulose, 65 wt% PLA, 5 wt% PEG was prepared, and exhibited mechanical properties similar to those of unfilled PLA, melt flow rates suited for FFF, no weight gain after being soaked 24 h in water, and successfully underwent wood-like processing, such as planing, sawing, and polishing.

Kearns (2017) explored the use of cellulose derived from recycled cotton as a filler for polymeric FFF filaments. Cotton has some advantageous properties as a filler: (a) it is a good conductor of heat, favoring uniform heating and melting throughout the filament; (b) it gains strength when wet and hence is suited in presence of moisture; (c) its crystallinity does not decrease during processing; (d) its aspect ratio exceeds that of other cellulosic materials and is more effective than the latter ones at improving strength and stiffness in composites. Filaments were extruded from pellets made of different amounts of recycled cotton, Ingeo™ 4043D PLA, and 1 wt% lubricant. Filaments' tensile properties were: strength 43–61 MPa, modulus 116–121 MPa, and strain at break 4.6–6.9%, depending on the cotton content. Tensile properties of dumbbell coupons from the above pellets depended on the cotton content, and were: strength 82–122 MPa, modulus 366–372 MPa, and strain at break 3.9–4.8%. Increasing cotton from 10 to 15 to 20 wt% raised strenght and strain but left basically unchanged the modulus.

7.3.6 MICROCRYSTALLINE CELLULOSE (MCC)

7.3.6.1 Introduction

MCC is the crystalline portion of cellulose fibrils that also contain an amorphous region. MCC is obtained from partially depolymerized and purified cellulose (Xiang et al. 2016; Kambli et al. 2017), using enzymes or chemical treatments such as steam explosion or acid hydrolysis to remove the amorphous regions in the cellulose microstructure. MCC is a renewable and biodegradable, white or almost white, odorless powder, with particle size varying within the range of tens of μm depending on the sources and preparation methods. MCC is insoluble in water, ethanol, and diluted mineral acids, and is added in products ranging from cosmetics and pharmaceuticals to foods (Lavanya et al. 2011; Xu et al. 2018). The tensile modulus of a MCC was measured by Eichhorn and Young (2001) to be equal to 25 GPa. Recent and comprehensive reviews of MMC were published by Trache et al. (2016) and Trache (2017).

MCC is present as a filler in the following industries (Hindi 2017):

- Polymer composites: it imparts high strength and stiffness, and provides high surface area for bonding. However, it suffers from moisture absorption, poor wettability, incompatibility with most polymeric matrices and limitation in processing temperature (Trache 2017)
- Pharmaceutical: as a binder, excipient, and disintegrant for making oral tablets (Thoorens et al. 2014; Li et al. 2019), and in vitamin supplements
- Food: as an anticaking, bulking agent, emulsifier, fat substitute, texturizer, and thickener
- Cosmetics (Chauhan et al. 2009; Ohwoavworhua and Adelakun 2010).

In considering MCC as an AM feedstock, it is worth noting that MCC is combined with many polymers (some of which biodegradable) to form composites. Some of these polymers may be candidates for MCC-filled composites for AM, such as:

- PBS: PBS is a crystalline polyester with formula $(C_8H_{12}O_4)_n$ and a melting temperature exceeding 100°C, and suited for high temperatures. A 100% biobased and compostable version of PBS was formulated by Succinity from succinic acid based on renewable feedstocks, such as glucose, sucrose, and biobased glycerol, and, in the future, also from second generation sustainable feedstocks (Succinity n.d.).
- Starch: WillowFlex FX1504 (BioInspiration) is the first flexible printing filament made from compostable raw materials, and its main ingredient is non-GMO corn starch.
- PE: biobased PE (also known as *renewable polyethylene*) is made from ethanol derived from sugar cane, sugar beet, and wheat grain, etc. that becomes ethylene after a dehydration process. Biobased PE is produced by Braskem (Brazil). Moreover, HDPE is the most common recycled plastic. We found only two HDPE filament for FFF made by Noxcel Science (Centipark 2021) and in Canada (Filaments.ca n.d.). Kreiger et al. (2013) derived FFF filaments from trashed milk jugs made of HDPE by means of RecycleBot, an open source waste plastic extruder that turns waste plastic and natural polymers into printing filaments.
- PP: biobased PP is produced by Braskem, Dow Chemical, Exxon Mobil, FKuR Kunststoff, LyondellBasell, Solvay, Washington Penn Plastic, etc. (Eurowatchers 2018). PP filaments for FFF are already available from FormFutura (Centaur PP™), Ultimaker, and Verbatim.
- PVA: it is a biodegradable but not biobased polymer. However, PVA filaments are sold as a support filament for FFF (GreenCycles, MatterHackers) that is water-soluble and eliminated after printing the part.

7.3.6.2 Commercial MCC

MCC is manufactured globally by more than ten suppliers (Thoorens et al. 2014). DuPont manufactures Avicel®, the first commercialized MCC, available in several pharmaceutical grades, all labeled with PH prefix. They feature particle size of 20, 50, 100, 150 and 180 µm, and reciprocally differ in terms of compactibility (ability to be compressed), flow, density, and moisture content (DuPont 2020). The most common source of pharmaceutical MCC is wood. Because MCC has high strength, low density, biodegradability, and favorable mechanical properties, its potential utilization for producing nanocellulose and biocomposites has generated attention and interest in academia and industry. MCC's properties affect quality processing and mechanical behavior of articles made out of it. Among all the material properties of MCC, those affecting blending, drying, flow, mechanical performance, stability, and wetting are particle size distribution and surface area. On the other hand, bulk and true density, elastic modulus, and brittleness also affect MCC's mechanical performance (Thoorens et al. 2014).

7.3.6.3 Experimental Formulations of MCC for AM

Murphy and Collins (2018) combined injection molding grade Ingeo™ 3001D PLA with 1, 3, and 5 wt% of surface-modified MCC (mMCC), and, separately, unmodified MCC, preparing seven formulations, whose thermal and mechanical properties were experimentally assessed. Manufacturing the PLA/MCC and PLA/mMCC filaments consisted of the following steps:

- Modifying MCC's surface: dissolving Kenrich LICA 38 in tetrahydrofuran (THF) and mixing it with MCC, stirring, filtering, washing repeatedly with THF.
- Casting films of PLA/MCC and PLA/mMCC, then crashing them into granules.
- Extruding 1.55 mm filaments from the above granules.

TABLE 7.11

Properties of PLA and PLA Blended with MCC and Modified MMC (mMCC)

Material	Tensile Storage Modulus at 35°C	T_g (DMA)
	MPa	°C
PLA	2,300	73.4
PLA/1MCC	2,354	73.2
PLA/3MCC	2,466	74.1
PLA/5MCC	2,354	76.9
PLA/1mMCC	2,357	72.5
PLA/3mMCC	2,500	75.5
PLA/5mMCC	2,357	77.6

Source: Data from plots in Murphy, C. A., Collins, M. N. 2018. Microcrystalline Cellulose Reinforced Polylactic Acid Biocomposite Filaments for 3D Printing. *Polym. Compos.* 1311–1320.

MCC and mMCC raised tensile storage modulus and, marginally, T_g (Table 7.11), with mMCC being slightly more effective than MCC. In fact, mMCC raised the modulus from 2,300 to 2,500 MPa, whereas MCC increased storage modulus from 2,300 to 2,466 MPa. The amount of MCC and mMCC yielding the largest increase in storage modulus was 3 wt%. The stiffening action by cellulose was attributed to the fact that it favored mobility in the PLA molecular chains, and hence its crystallinity (Herrera et al. 2015). The increase in crystallinity was confirmed by the DSC plots. The effect of cellulose on tensile strength and elongation at break was omitted. However, Herrera et al. (2015), experimenting with non-AM PLA filled with cellulose nanofibers (CNF) and glycerol triacetate (GTA) serving as plasticizer, recorded that elongation at break and toughness surged from 2 to 31% and from 1 to 8 MJ/m^3, respectively, vs. unfilled PLA. Therefore, a similar positive effect by MCC on elongation at break and toughness of AM PLA may be expected. About including LICA 38 in a commercial filament, the supplementary cost of buying and adding it to the filament has to be weighed against the associated property improvement.

Katstra et al. (2000) choose Avicel® PH-301 MCC powder (Figure 7.20) as a feedstock to additively manufacture two types of tablets for drugs orally taken and released after lag time: tablets exhibiting erosion mechanism, and tablets featuring diffusion mechanisms. The 3DP™ process was evaluated because regarded a process superior to the current fabrication methods of

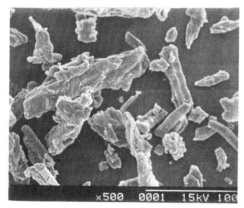

FIGURE 7.20 SEM photograph of granules of Avicel® PH 301 microcrystalline cellulose.

Source: Wu, J.-S., Ho, H.-O, Sheu, M.-T. 2001. Influence of wet granulation and lubrication on the powder and tableting properties of codried product of microcrystalline cellulose with β-cyclodextrin. *European Journal of Pharmaceutics and Biopharmaceutics* 51:63–69. Reproduced with permission from Elsevier.

pharmaceutical devices in terms of dosage accuracy, and spatial positioning. PH-301 served as the material for the tablet and the matrix for the binders in 3DP™. PH-301 has 0.34–0.45 g/cm³ bulk density, 50 μm particle size, and 5.5–7.0 pH. The following polymeric binder solutions, displaying different dissolution properties, were tested:

- 80/20 wt% ethanol/Eudragit® E-100 (Rohm Pharma), a methacrylic ester copolymer with 150,000 MW. Binder content was 8.9–17.9 vol% of MCC
- 80/20 wt% acetone/Eudragit® RLPO (Rohm Pharma), an ammonio-methacrylic acid copolymer with 150,000 MW. Binder amount was 9–16.7 vol% of MCC.

Tablets with E-100 showed erosion followed by dissolution, and full release of drug within 50 to 110 minutes, whereas tablets with RLPO exhibited swelling without erosion and full release of drug within 6 to 9 h. As the polymer content increased in both binders, the overall drug release time stretched, with lag time going up, and release peak rate declining. The AM parameters can be set to deliver specific quantities of binder, which in turn controls the timing of the drug release from the tablets. Experiments revealed that: (a) the measured amount of drug coincided with the quantity of drug set through the printer settings, and validated the accuracy of dosage control using 3DP™; (b) 3DP™ supplied the exact quantity of drug in the center of the tablet, demonstrating dosage control at the microgram level, and accurate drug placement. 3DP™ possessed greater accuracy than conventional mixing and pressing technique for drug packaging. Moreover, the printed tables underwent hardness and friability tests, and their test values were comparable to those of tablets of commercial drug brands.

Another pharmaceutical application of MCC described by Rowe et al. (2000) is very interesting, because it demonstrates how AM, particularly 3DP™, enables to efficiently fabricate articles featuring complicated designs and multiple functionalities that conventional manufacturing cannot achieve at all, or can but expensively. Employing 3DP™, Avicel® PH-301, and different binders allowed to fabricate several tablet types, with each type combining materials featuring diverse properties, and hence functionalities, namely, four types of tablets that delivered oral drugs at different time rate and, due to different solubility, selectively in several organs of the human body. The tablet designs (Figure 7.21) described by Rowe et al. (2000) can serve as a model for other articles requiring time-controlled release of specific quantities of some material in response to some environmental stimulus.

MMC Pharmacel® 102 was selected by Khaled et al. (2014) as a disintegrant in the layer for immediate release of drug in a bilayer drug tablet printed on a desktop extrusion-based printer.

Switching field of application, MCC was evaluated as a filler (60 and 80 μm particle size) for a copolyester hot-melt adhesive (PES) to form a wood/plastic composite (WPC) for LS by Zhang

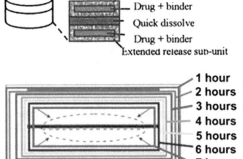

FIGURE 7.21 Two tablet designs printed using 3DP™ featuring multiple release mechanisms.

Source: Rowe, C. W., Katstra, W. E., Palazzolo, R.D. et al. Multimechanism oral dosage forms fabricated by three dimensional printing™ 2000. *Journal of Controlled Release* 66:11–17. Reproduced with permission from Elsevier Ltd.

et al. (2016). Test coupons were printed out of MCC/PES with 11–14 W laser power, and in-filtrated with epoxy, which not only decreased porosity and increased mechanical properties, but also reduced surface roughness. The MCC/PES with 60 μm particles showed better surface quality and mechanical properties than MCC/PES with 80 μm particles. The optimal version of infiltrated MCC/PES (25/75 wt%) exhibited tensile, flexural, and impact strengths (with impact not specified whether notched or unnotched) equal to 11 MPa, 19 MPa, and 5.5 kJ/m^2, respectively. PA is the most popular polymer for laser sintering. As a comparison, the tensile and flexural strengths of the commercial PA for LS called EOS PA 2201 are 48 MPa and 58 MPa, respectively. The authors suggested that suitable applications for MCC/PES would be those benefiting from high precision, low price and wood-like appearance, such as investment casting, prototyping, exhibit fabrication, and artwork, but they did not mention functional parts.

7.3.7 Nanocellulose (NC)

7.3.7.1 Introduction

NC is highly crystalline, and possesses outstanding characteristics including excellent mechanical properties (stiffness ~150 GPa), dimensional stability, low thermal expansion coefficient, remarkable reinforcing capability, anisotropic shape, tailorable surface chemistry, and interesting optical prop-erties (Kim et al. 2015; Abitbol et al. 2016). Thanks to its properties and being sustainable, NC attracts growing interest for applications in materials science and biomedical engineering, and is being studied in photonics, films and foams, surface modifications, nanocomposites, and medical devices.

NC exists in two forms: *cellulose nanofiber* (CNF) and *cellulose nanocrystal* (CNC). CNF comprises the ordered and disordered regions of the cellulose chains, has lengths in the microscale and diameters in the nanoscale, and is also known as *nanofibrillated cellulose* (NFC), *micro-fibrillated cellulose* (MFC), and *cellulose nanofibrils*. CNCs are the ordered regions in the cellulose chains, feature diameters in the range of nanometer, and are also called *nanocrystalline cellulose*, *cellulose (nano)whiskers*, and *rod-like cellulose microcrystals* (Sultan et al. 2017).

NC is extracted from natural resources according to chemical methods and mechanical methods. An example of the former ones is acid hydrolysis for extraction of CNC (Siró and Plackett 2010), and an example of the latter ones is a recent environmentally-friendly method (*aqueous counter collision*) to extract CNF that relies on no chemical modifications but only two water jets to convert feedstocks into nano-elements (Kondo et al. 2014). The extracted NC can be aligned, processed, and modified for films, long fibers, powders and suspensions, and serve as sustainable material. The density, tensile strength and modulus, and thermal expansion coefficient of CNC are 1.6 g/cm^3, 7.5 GPa, 145 GPa, and 3–22 ppm/K, respectively (Moon et al. 2011). CNFs with high crystallinity possess elevated mechanical strength and stiffness, but they are too short to work as fibers, and hence are converted to long fiber by spinning, stretching, and drying (Kafy et al. 2017). CNF is expected to be a very effective reinforcing ingredient in composite materials (Abitbol et al. 2016).

NC is suited for AM as a substrate to impart printing functionality, and an ingredient in gel bioinks to form 3D structures (Rees et al. 2015).

Metal-organic frameworks (MOFs) are crystalline materials that contain a type of covalent bond between metal cations and anionic organic linkers. MOFs can be fabricated as films, discs, sheets and monoliths for many advanced applications comprising nanomedicine, quantum dot semi-conductors, biosensing, biochemical motors, molecular rotors, etc. (Sultan et al. 2017). The study by Sultan et al. (2019) is notable because it reported the in situ growth of MOFs onto a type of CNFs (TOCNFs) for printing advanced materials, using CNF as a carrier phase for MOFs. The TOCNF was selected due to its unique properties: high aspect ratio, satisfactory mechanical properties, capability to form gel, and print biomedical scaffold for successful cartilage and bone regeneration.

7.3.7.2 Cellulose Nanofibers (CNFs)

Håkansson et al. (2016) for the first time printed and dried 3D structures featuring a controlled architecture from a hydrogel (HG) consisting of 2 wt% CNF and 98 wt% water. They printed a wet HG (that is adequate for biomedical applications), and converted it by solidification into a dry artifact that did not collapse, potentially opening up new applications. The authors printed sample items and evaluated four drying procedures: (a) air drying, (b) air drying with surfactants, (c) solvent exchange before drying, and (d) freeze-drying in liquid nitrogen. The last method was successful, and did not alter the external dimensions of the printed items. The air-dried (case (a)) item collapsed, mostly in z-direction. In case (b), all directions shrank equally, and the structure was kept at ~30% of the wet size. In case (c) the printed structure shrank but retained its 3D features. The solidification methods greatly affected the mechanical properties of the printed items. The surface texture of the dry object could be modified with the time of freezing and the amount of added surfactant, and it was also possible to design pores on the millimeter scale. The CNF HG could be functionalized to meet various performance requirements, f.e. by adding conductive carbon nanotubes.

Rees et al. (2015) evaluated NC from Monterey pine as an AM bioink for modifying film surfaces. The NC underwent some chemical pretreatments leading to different morphology and surface chemistry: treatment A was (TEMPO)-mediated oxidation, and the subsequent material was employed to cast substrate films; treatment B was a combination of carboxymethylation and periodate oxidation, and produced a homogeneous material with short nanofibrils, having widths and lengths below 20 nm and 200 nm, respectively that had rheological properties suited to an AM bioink compatible with an EnvisionTEC 3D Bioplotter. Two items were printed, shaped as a cylinder with a grid structure, and 25 mm in diameter and 6 mm in height: one deposited on the cast film, and the other displaying a 3D grid comprising nine layers to mimic porous scaffolds. The printed constructs were then freeze-dried. The open porosity of the printed articles can possibly carry and release antimicrobial components, and the NC evaluated did not support bacterial growth. Both features are advantageous for wound dressing applications.

Lille et al (2018) studied the suitability of CNF and AM for food. Particularly, they assessed the applicability of AM and CNF to food pastes made of protein, starch and fiber-rich materials, as a starting point in the development of healthy and customized snack products. CNF was selected in order to provide shape stability after printing. The printability of food containing CNF, starch, milk powder, oat and fava bean protein, and their mixtures was evaluated in terms of ease and uniformity of extrusion, and precision and stability of the printed pattern. CNF was prepared from once dried bleached birch Kraft pulp from Finland. The dry pulp composition in wt% was 73 cellulose, 26 hemicellulose and 1 lignin. The pulp suspension (2 wt% of dry matter) was pre-refined by grinding. Mixtures of CNF/starch and CNF/milk powder were prepared by adding starch and milk powder into the CNF gel and mixing with a spoon, without adding water. The printing device consisted in pumps to feed mixtures with varying viscosities, and a CAD-based XYZ-motion system to control the position of the syringes dispensing the mixtures. The best printing precision and shape stability was obtained with a semi-skimmed milk powder-based paste. The yield stress of the paste affected the shape stability after printing. Post-processing consisted in oven drying and freeze-drying, and the former was most successful at high initial contents of solids in the printed samples. Extrusion-based AM is a promising method for producing foods that are healthy, and have targeted nutritional compositions, but it requires additional research to optimize the mechanical properties of the printed materials. Printing food is illustrated in Chapter 14.

Torres-Rendon et al. (2016) studied CNF for TE, and processed HGs of CNF into complex shapes that were selected as a sacrificial template to prepare freestanding tubular cell constructs that were highly biocompatible, and enabled the growth of mouse fibroblasts into merging cell layers. The authors fabricated hollow fibers using extrusion-based AM through a circular die into a coagulation bath. The hollow fibers had the nanofibrillar porosity of the CNF HG, and were 200 mm long by 2.5 mm in outer diameter, with submillimeter wall thickness. Their dimensions were tunable. The printed CNF tubes were deposited as a coil at the bottom of an ethanol bath, dried at room conditions, and then rehydrated when needed. Rehydration was followed by cell culturing, and cytocompatibile degradation of the sacrificial CNF templates, leaving behind fibroblast tubes. The tensile strength and stiffness of the CNF tubes were 5.3–36.5 kPa, and 55–304 kPa, respectively, depending on their composition (glutaraldehyde solution and $CaCl_2$). Their stiffness could be tuned within a range attractive for biomaterials, and their mechanical robustness exceeded most typical natural biopolymers such as collagen, fibrin, gelatin, and hyaluronic acid. AM enables HG geometries more complex than tubes, resulting in more complex cell cultures characterized by hierarchical and communicating structures.

Torres-Rendon et al. (2015) selected CNFs to print HGs that worked as sacrificial templates to form scaffolds with controlled porosity and topographically complex structure geometry for bone tissue regeneration. The templates were printed on a EnvisionTEC medical Mini Multi Lens printer based on digital laser processing (Section 2.5.1.2), and formed scaffolds shaped as 15 mm edge cubes with pore diameter below 1 mm. Not only human mesenchymal stem cells adhered to the resulting scaffolds, but a collagen-mimetic coating with calcium phosphate was added to the scaffolds to act as a chemical cue and induce differentiation in osteogenic tissues. The method developed can be extended to soft tissues as well.

Markstedt et al. (2015) selected NFC and alginate to formulate a bioink that married the outstanding shear thinning properties and non-toxicity of NFC with the fast crosslinking of alginate, and was suited for bioprinting soft tissue with living cells. Printability was evaluated in terms of printer parameters and shape fidelity. A bioprinter 3DDiscovery by RegenHU (Switzerland) equipped with a 300 μm diameter nozzle was chosen. The following NFC/alginate formulations were printed and evaluated, with values in wt%: 90/10, 80/20, 70/30, and 60/40. Not only 2D grid-like structures and a 3D disc were printed, but also anatomically shaped cartilage structures, such as a sheep meniscus and a 27 × 18 mm human ear based on MRI and CT images as blueprints. Human chondrocytes bioprinted in the bioink exhibited a cell viability of 73% and 86% after 1 and 7 days of culture, respectively. The results pointed out that the NFC-based bioink was a suitable HG for bioprinting with living cells, and demonstrated the potential use of NC for bioprinting living tissues and organs.

Shao et al. (2015) investigated HGs for AM composed of MFC and lignosulfonate (LiS) as sustainable precursors of carbon items, such as electrodes, and measured the HGs' rheological properties to assess their suitability for AM. Two series of test cuboids (L × W × H = 2 × 2 × 1 cm) composed of 2 wt% MFC suspensions and LiS contents of 0 to 50 wt% were printed using a 0.96 mm syringe needle and a printing speed of 35 mm/s on a commercial desktop printer Fab@Home Model 3 (Seraph Robotics). To acquire their final, solid, and permanent shape, the first series was initially dried at room temperature for 5 days and then in oven at 110°C for 48 h, whereas the second series was frozen in a refrigerator, and then freeze-dried. All dried samples were carbonized in oven up to 800°C. MFC/LiS systems displayed a complex rheological behavior that was affected by the LiS concentration: 10 wt% was the maximum LiS concentration that could be chosen to print cuboids of good quality (minimal shape variations), very low apparent density (63 kg/m^3), and high electrical conductivity (5.5–55 S/m).

Gladman et al. (2016) combined NFC, a monomer, photoinitiator, enzyme/glucose, and deionized water to formulate a HG composite ink composed of stiff cellulose fibrils embedded in a soft acrylamide matrix that imitated the composition of plant cell walls. That ink allowed to print articles that changed their shape in response to external stimuli, similarly to plants, whose leaves, flowers, etc. react to humidity, light, and so on. NFC was unbleached and processed directly from soft wood pulp. The maximum NFC concentration that allowed for smooth, clog-free print behavior was 0.8 wt%. Unidirectional filled and unfilled HG test samples were built on an Aerotech

ABG 10000 printer, and featured composite architectures with localized, anisotropic swelling behavior controlled by the alignment of cellulose fibrils along planned four-dimensional printing pathways. Based on mathematical models, the alignment patterns were designed to achieve specific 3D target shapes.

NFC can also promote cartilage regeneration. In fact Nguyen et al. (2017) included NFC in two bioink formulations to treat cartilage lesions that can progress into secondary osteoarthritis and cause severe and frequent clinical problems. NFC provided structural and mechanical support, and, in the case of cartilage, simulated the bulk collagen matrix. One formulation contained NFC and alginate (NFC/A), the other comprised NFC and hyaluronic acid (NFC/HA), both mixed with human-derived induced pluripotent stem cells (iPSCs). The bioinks were co-printed with irradiated human chondrocytes on a RegenHu 3DDiscovery bioprinter with a 300 μm nozzle. The NFC/HA was not successful because it led to a decrease of pluripotency. Instead, pluripotency was initially maintained in items printed out of NFC/A 60/40 wt% dry, and after five weeks, cartilaginous tissue with collagen and tumor-free was observed in them, along with a significant increase in cell number within the cartilaginous tissue.

CNFs have resulted in limited improvement of PLA's mechanical properties most likely because hydrophilic NC tends to aggregate within the hydrophobic PLA. A solution to this problem is chemical modification of NC, because chemically modified NC usually terminates with hydrophobic functional groups that are similar to PLA in polarity, and lead to a more uniform distribution of NC within PLA. Dong et al. (2017) followed this approach when employing CNFs (Daicel FineChem, Japan) as a filler to enhance mechanical properties of PLA. L-lactide monomers were grafted onto CNFs via ring-opening polymerization, and formed PLA grafted cellulose nanofibers (PLA-g-CNFs). PLA-g-CNFs and unfilled PLA were then blended in chloroform, dried, melt extruded together in PLA-g-CNFs/PLA composite filaments for AM, successively thermally annealed, and finally underwent mechanical, thermal, and morphological characterization. PLA was successfully grafted on the surface of CNFs, in concentration of 33 wt%, and produced highly crystallized grafted PLA. The extruded composite filaments were suitable for AM. In comparison to unfilled PLA, incorporating PLA-g-CNFs in PLA improved storage modulus at room and high temperature: at 90°C this property jumped from 616 MPa for PLA to 1702 MPa for the composite. Post-extrusion annealing provided 63% and 25% increase for tensile strength and modulus of the filaments, respectively. T_g was left almost unchanged in not annealed and annealed filaments.

Markstedt et al. (2017) leveraged CNFs' shear thinning properties and added it to all-wood-based gel inks suitable for printing furniture, electronics, health care products, and so on. CNFs (Stora Enso, Sweden) was mixed with xylan, a polysaccharide hemicellulose extracted from spruce that contributed with its crosslinking properties to impart final stability to the printed ink. Xylan became crosslinkable after being functionalized with tyramine (a naturally occurring amine) at different degrees. Multiple ink compositions were evaluated through rheology measurements and printing tests on 3DDiscovery bioprinter. The tested compositions (wt% values) were made of water (86.8–92.3), CNFs (2–3.3), and xylan-tyramine (5.8–10.6). The maximum storage modulus after crosslinking was 25 kPa reached by the formulation with 86.8/2.6/10.6 wt% of water/CNFs/xylan-tyramine. The degree of tyramine substitution and the ratio of CNFs to xylan-tyramine in the prepared inks influenced the printability and crosslinking density. In conclusion, xylan-tyramine enabled to crosslink the printed items and form freestanding gels, and at the same time did not impair the excellent printing properties of CNFs.

PP is one of the most sold TPs but shrinks and warps during printing. Since CNFs have a low thermal expansion coefficient (0.1 ppm/K), they can potentially contain the shrinkage of PP, making it fit for FFF, besides enhancing PP's mechanical properties. Because CNFs are hydrophilic, research has focused on solvent mixing as the strategy for manufacturing CNF composites rather than melt mixing, although melt compounding fibers and TPs is the typical industrial method to make TP composites. Wang et al. (2018b) studied a novel PP/CNF composite by experimentally comparing the properties of the following formulations: PP, PP/MAPP, PP/CNF, PP/CNF/MAPP, with MAPP being maleic anhydride PP. Dried CNFs had a non-circular cross section but the diameter of the equivalent area was 9.6 µm, and aspect ratio 1.25, making the shape of the dried CNFs more spherical than fibril-like. Adding 3 wt% of CNFs to PP did not impact the flexural properties of pure PP, irrespective of the presence of MAPP, but 10 wt% of CNFs improved flexural strength and modulus of the composite by 5.9% and 26.8%, respectively compared to those of PP, and limited strain creep, making the composite more dimensionally stable than PP under constant stress. Adding MAPP raised the flexural strength of PP by 12.9% vs. unfilled PP. The improved stiffness of PP composites with CNFs is attributed to CNFs' own stiffness, whereas the increased strength in short fiber-filled polymer composites is due to enhanced stress transfer at fiber-matrix interface, lower stress concentrations at fiber ends, and crack deflection.

7.3.7.3 Cellulose Nanocrystals (CNCs)

Cellulose microfibrils are divided into two distinct regions: highly ordered (crystalline) and less ordered (non-crystalline or amorphous) regions. When wood pulp is mechanically or chemically treated, and then subjected to acid hydrolysis to remove the amorphous regions, it leaves behind individual crystallites that are are anisotropic particles and known as CNCs. Depending on the cellulose source and processing conditions, the average dimensions of CNCs (from plant source) are 5–70 nm in diameter and 100–250 nm in length (Klemm et al. 2011). CNCs have intricate intra- and inter-cellulose chain hydrogen bonding that give rise to remarkable mechanical properties, such as axial elastic modulus of 110–220 GPa (Li et al. 2017), far higher than those of microfibrils. High stiffness and strength, low weight and density, sustainability, biocompatibility, biodegradability, recyclability, and large availability of cellulose make CNCs attractive engineering materials. Another advantage of CNCs is that they are oriented by applying electric and magnetic fields, and by mechanical shearing, electrospinning, and dry spinning (Siqueira et al. 2017). CNCs have been explored as reinforcing elements in composites (Dufresne 2012) and as a nanofiller for food packaging (Dhar et al. 2014). CNCs are prepared by sulfuric acid and by hydrochloric acid hydrolysis, and their chemical and physical properties vary with the hydrolysis conditions (e.g. acid type, acid concentration, reaction temperature, and reaction time) and the cellulose source chosen (Kargarzadeh et al. 2012).

Direct ink writing (DIW) is an AM process enabling to extrude a broad variety of materials at the microscale level, but it is challenging to formulate inks for DIW exhibiting shear thinningCompton 2014Studart 2016. Siqueira et al. (2017) prepared DIW inks composed of 10 wt % and 20 wt% CNCs dispersed in a solution containing 2-hydroxyethyl methacrylate (HEMA) monomer, Dymax polyether urethane acrylate (PUA) oligomer, and a photoinitiator cured after printing by exposure to UV light. HEMA and PPUA were preferred because of their good compatibility with unmodified CNC. CNC was prepared via sulfuric acid hydrolysis of eucalyptus pulp at the USDA Forest Service, and measured an average length and diameter of 120 nm and 6.5 nm, respectively, and aspect ratio of 18.5. CNC was employed unmodified and chemically modified on its surface using methacrylic anhydride, in order to further improve its dispersibility and interfacial bonding between particles and matrix after polymerization. CNC reinforced two matrices: the stiff and brittle HEMA/PUA-BR571, and the soft and rubbery HEMA/PUA-BR3741. The latter enhanced flexibility and impact resistance, while PUA-BR571 imparted hardness and toughness. Using these aqueous inks, cellulose-based structures with a high degree of CNC particle alignment along the printing direction were printed via DIW. Table 7.12 contains the test results of tensile strength and modulus for inks printed via DIW as films with and without CNCs. Unmodified CNC improved the modulus considerably but not the strength (if we consider the average strength in longitudinal and transversal directions) of the stiff matrix. However, modified CNCs raised both modulus and strength of the soft matrix. Since CNCs are expensive, mixing CNFs and CNFs would reduce cost, and balance quality and cost. Following this versatile method, items can be printed out of sustainable materials that have reinforcements located along specific directions and exhibit tailored responses to mechanical loads.

TABLE 7.12

Properties of Inks Printed Via DIW as Films with and without CNCs

Material	Tensile Strength	Tensile Modulus
	MPa	MPa
HEMA[a]/PUA[b]-BR751	45.7	938
HEMA[a]/PUA[b]-BR751/modified CNC 10 wt% in longit. dir.	44.8	1,250
HEMA[a]/PUA[b]-BR751/modified CNC 20 wt% in longit. dir.	49.1	1,688
HEMA[a]/PUA[b]-BR751/modified CNC 20 wt% in transv. dir.	40.7	1,406
HEMA[a]/PUA[b]-BR3741	5.4	4
HEMA[a]/PUA[b]-BR3741/unmodified CNC 10 wt% in longit. dir.	7.3	36
HEMA[a]/PUA[b]-BR3741/unmodified CNC 10 wt% in transv. dir.	6.2	40

Source: Data from plots in Siqueira, G., Kokkinis, D., Libanori, R., et al. 2017. Cellulose nanocrystal inks for 3D printing of textured cellular architectures. *Adv. Funct. Mater.* 27, 1604619.

Notes:

[a]2-hydroxyethyl methacrylate.

[b]polyether urethane acrylate.

Aerogel is a gel comprised of a microporous solid in which the dispersed phase is a gas (IUPAC 2014). Aerogels are solids with very high porosity and sound absorption, and extremely low density and thermal conductivity. They are employed in oil and gas, building and construction, automotive, aerospace, marine, performance coatings (Grand View Research 2020), and biomedical (Maleki et al. 2016). Manufacturers of aerogels are Aerogel Technologies, Aspen Aerogels, BASF SE, Cabot, and Dow Corning, all headquartered in USA except BASF (Germany). Li et al. (2017) employed DIW and CNC to fabricate pure CNC aerogels while controlling 3D structures and inner pore architecture. Freeze-drying printed articles permitted to fabricate CNC aerogel structures with minimal structural shrinkage and damage, differently from traditional cellulosic aerogel processing, and the resulting artifacts can likely be applied in tissue scaffold templates, drug delivery, packaging, etc., due to their sustainability, biocompatibility, and biodegradability. CNC aerogels at 11.8, 15, 20, and 30 wt% in water were prepared. Trials were run with nozzle tip of 200, 400, and 500 μm diameter, and, as expected, the smallest nozzle yielded the smoothest surface. After freeze-drying, aerogels with 11.8–30 wt% of CNC had 127–399 mg/cm^3 density and 92-75% porosity. The compressive modulus varied from 7 MPa (as printed) to 9 MPa (cross-linked). The quality of printed items (Figure 7.22) can be improved by increasing CNC amount and printing resolution. Successfully printed items included articles without support material, and dual pore scaffolds whose 3D structure and inner pore architecture met specific cell integration requirements. Since the global aerogel market size is expected to total USD 3.3 billion by 2025, with 22.6% CAGR between 2018 and 2025 (Grand View Research 2020), the business opportunities for printing aerogels are attractive. A scaffold with dual pore structure (Figure 7.23) incorporates structural pores and random pores, and it is advantageous over traditional fully random porous structures for facilitating cell growth: the larger structural pores enhance nutrient and oxygen transport, and promote homogenous cell proliferation throughout the entire scaffold, whereas the smaller random pores provide the large surface areas required for high-density cell growth.

Cataldi et al. (2018) aimed at producing fully biodegradable nanocomposite filaments made of CNCs and PVOH that would be compatible with commercial printers for FFF and outperform PVOH in stiffness, dimensional stability, and thermal properties, ultimately in order to expand PVOH applications. PVOH is a synthetic polymer soluble in water, effective in film forming and emulsifying, and possessing an adhesive quality. It has no odor and is nontoxic, resistant to grease,

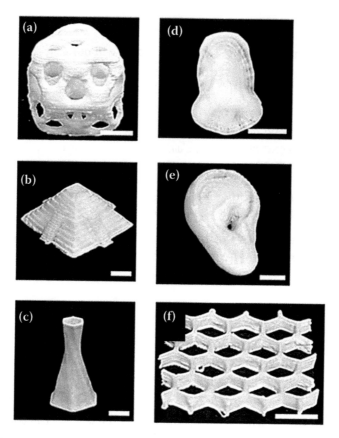

FIGURE 7.22 DIW 3D printed articles from 20 wt% CNC aerogel using 500 μm nozzle tip: (a) octet cube, (b) pyramid, (c) hexagonally twisting vase, (d) nose model, (e) ear model, and (F) honeycomb. Displayed scale bars are 1 cm.

Source: Li, V., Dunn, C., Zhang, Z., Deng, Y., Qi, H. J. 2017. Direct Ink Write (DIW) 3D Printed Cellulose Nanocrystal Aerogel Structures. *Scientific Reports* 7:8018. Reproduced under a Creative Commons Attribution 4.0 International License (http://creativecommons.org/licenses/by/4.0/).

oils, and solvents, ductile but strong and flexible, and functions as a strong oxygen and aroma barrier. The CNC/PVOH filaments contained 2–20 wt% of CNCs that were rod-like shaped (200–500 nm in length, 5–10 nm in diameter), and produced by solution mixing, grinding, and extrusion. CNCs were well dispersed in PVOH, both in filament and printed articles. In filaments and printed coupons, CNCs progressively enhanced thermal stability by raising T_g. Raising filler content proportionally increased dynamic storage and loss moduli, and reduced creep compliance. Ultimate mechanical properties of the filaments also improved by adding CNCs up to 10 wt%, with maximum reinforcement effect observed at 2 wt% of CNCs resulting in 81% and 45% increase of tensile energy to break and stress at break, respectively. Tensile strength improved by 73% at 5 wt % of CNCs. Storage modulus of printed coupons at 23°C jumped by 290% with 10 wt% of CNCs. Enhancement in thermal stability was evidenced by higher T_g of the printed parts (from 34°C to 48°C), while T_m remained practically unaffected. Decline in creep compliance signaled an improvement in dimensional stability, and was measured upon increasing filler content. The good adhesion between CNCs and PVOH improved toughness of the printed coupons.

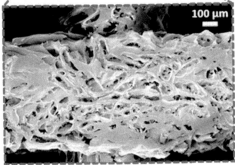

FIGURE 7.23 SEM images of dual-pore scaffold aerogel printed via DIW in a 0–90° configuration. Upper image shows structural pores, lower image random pores. A 20 wt% CNC gel and 500 μm nozzle tip were used.

Source: Li, V., Dunn, C., Zhang, Z., Deng, Y., Qi, H. J. 2017. Direct Ink Write (DIW) 3D Printed Cellulose Nanocrystal Aerogel Structures. *Scientific Reports* 7:8018. Reproduced under a Creative Commons Attribution 4.0 International License (http://creativecommons.org/licenses/by/4.0/).

Hong (2015) formulated a printable nanocomposite for bone TE comprising PCL combined with several amounts of surface oxidized CNCs (SO–CNCs). PCL is an aliphatic polyester, and a synthetic bioresorbable biopolymer. Bioresorbable polymers are suited for biomedical applications, because scaffolds made of them do not require a secondary surgical procedure to remove the implants after healing which would be painful and expensive. PCL is a hydrophobic and semi-crystalline linear polymer, with T_g and T_m of −60°C, and 60°C, respectively. PCL has been approved by USDA for use in drug delivery devices, sutures, long-term implants, and adhesion barriers, and hence is the most popular aliphatic polyester in medical applications, specifically in drug delivery systems and medical devices (Chandra and Rustgi 1998; Okada 2002; Nair and Laurencin 2007). Moreover, its good processability favors it over other polymers that are candidates for biomedical applications. The recent reviews on printed PCL for medical applications by Asa'ad et al. (2016), Chia and Wu (2015), and Shirazi et al. (2015) are suggested for details. Hong (2015) derived CNCs from bleached milled softwood pulp (Temalfa 93A by Tembec, Canada) through acid hydrolysis, and prepared SO–CNC/PCL composite by applying the following route: (a) solvent exchange of aqueous SO–CNC suspension, (b) physical mixing, (c) solvent casting, and (d) melt compounding extrusion. This route was chosen in order to overcome the challenge of homogeneously dispersing hydrophilic SO–CNCs in hydrophobic PCL with minimal use of harmful organic solvents. Chemical and physical properties and biocompatibility of SO–CNC/PCL were analyzed, and scaffolds out of it were printed via FFF on a commercial desktop printer, and tested in creep. The test results were positive, and proved that SO–CNCs met various requirements to be a viable reinforcing nanofiller in scaffold materials for bone TE applications. Specifically, the carboxyl groups on the surface of SO–CNCs significantly increased calcium ion binding ability which could improve the biomineralization, by inducing mineral formation required for bone TE uses. The addition of SO–CNCs increased compressive modulus and creep resistance of the PCL scaffolds: the former from 3.9 MPa for PCL to 31.5 MPa for SO–CNC/PCL, and the creep shear modulus from 1.2 MPa for PCL to 6.4 MPa for SO–CNC/PCL. The maximum tensile strength and modulus of 18.2 MPa and 493 MPa, respectively were attained at 10 wt% SO–CNC, and constituted an increase of 75% and 153%, respectively over unfilled PCL. Since tensile strength and modulus of human cancellous bone (the spongy tissue of mature adult bone typically present inside spine vertebral bones and the ends of the long bones) are 2.54 MPa and 483 MPa, respectively (Røhl et al. 1991), SO–CNC/PCL at 10/90 wt% is adequately strong and stiff for bone scaffolds. Raising the filler amount led to steadily stronger, stiffer, and less ductile materials. Adding 10 wt% of SO–CNCs changed T_g and T_m to −64.5°C and 56°C , and slightly reduced crystallinity from 42.9% to 39.7%. SO–CNC/PCL and the AM process employed were not toxic to the cells that are precursors of cells synthetizing bone. SO–CNCs accelerated calcium and mineral deposits on the surface of the scaffolds comparing to PCL scaffolds, which is favorable to bone growth. In order to assess hydrophilicity, the contact angle was measured in DI water and simulated blood fluid (SBF), a liquid containing ion concentrations similar to those of the inorganic constituents of human blood plasma. Since the lower the contact angle between a liquid and a surface the more wetted that surface is, the decrease of contact angle in DI water from 87° to 82.6°, and in SBF from 88.8° to 82.6° was a positive result, because indicated that the initial surface hydrophobicity of PCL was slightly reduced by adding the hydrophilic SO–CNCs.

CNC has been studied as a filler for polymers compatible with the AM process of stereolithography (SL) (Section 2.5.1.1). Kumar et al. (2012) sought to reduce brittleness and raise maximum service temperature of a resin for stereolithography (SLR), and were the first to select CNC for that purpose because CNC had previously proved itself effective as a filler at increasing stiffness in epoxies. CNC was isolated via sulfuric acid hydrolysis of cotton cellulose pulp. The composition of SLR was a mixture of monomers based on epoxy and acrylate chemistry, among others. The cured SLR had a density of 1.2 g/cm^3. The SLR/CNC nanocomposites were formulated with 0.5, 0.5, 1, 2, and 5 wt% of CNC, and featured an interpenetrating network structure. The presence of CNC impacted some properties of the nanocomposites: tensile strength raised from 69 MPa (SLR) to 82 MPa (2 wt% CNC), tensile modulus increased from 3.1 MPa (SLR) to 4.1 GPa (5 wt% CNC), strain at break dropped from 3.8% (SLR) to 1.6% (5 wt% CNC), and T_g improved from 95°C (SLR) to 107°C (5 wt% CNCs). The storage modulus below and above T_g steadily grew with higher CNC content. Sample parts were printed to prove that CNC-filled resins could be processed with standard printers for SL.

Feng et al. (2017a) formulated a mixture of lignin-coated cellulose nanocrystals (L-CNC) and methacrylate (MA) resin to be printed as a novel nanocomposite, potentially suited for TE and electronic components. Lignin was added to NC, because the presence of covalent bonds between NC and lignin increased the thermal stability of the latter, and lignin's hydrophobicity made NC less polar, improving NC dispersion in a non-polar matrix such as MA. L-CNC by American Process (USA) contained 3–6 wt% lignin. Samples with 0, 0.1, 0.5, and 1 wt% of L-CNC mixed with MA were made on a Formlabs printer. Regardless of the L-CNC content, postcuring about doubled tensile strength and modulus, and lowered elongation at break to about ¼ vs. no post-curing. Table 7.13 shows the highest values and relative filler content of tensile properties of L-CNC/MA composites before and after postcure. The onset degradation temperature in L-CNC/MA was marginally higher than in MA (289°C): the largest increment before postcuring was at 0.5 wt% (293°C), and after post-curing at 0.1 wt% (304°C). T_g only increased from 96°C in MA to a maximum of 100°C at 0.5 wt% after postcuring.

In another application, L-CNC was added to ABS, and formed L-CNC/ABS nanocomposites also printed through SL (Feng et al. 2017b). Amounts of L-CNC were 2, 4, 6, 8 and 10 wt%. Test results pointed out that adding L-CNC foamed ABS, and reduced the density of the printed nanocomposite samples from 0.94 g/cm^3 to 0.83 g/cm^3, due to two phenomena: (a) the high temperature during extrusion and printing caused the decomposition of cellulosic fibers, and the gases generated during these processes and trapped inside the material led to the formation of pores which remained after the printing process; (b) the cellulose nanocrystals acted as foaming nuclei and formed microcellular foams. The ratio tensile strength/density decreased at all filler contents, but the ratio tensile modulus/density (GPa/g/cm^3) increased from 1.09 in ABS to 1.13

TABLE 7.13

Highest Values of Tensile Properties of L-CNC/MA Composites Printed Via SL

Material	Tensile Strength	Filler content	Tensile Modulus	Filler content	Elongation at Break	Filler content
	MPa	wt%	GPa	wt%	%	wt%
Not postcured	36.2	0.1	0.67	0.1	10.3	0
Postcured	68.8	0.5	1.19	0.1	3.01	0.1

Source: Feng, X., Yang, Z., Chmely, S., et al. 2017. Lignin-coated cellulose nanocrystal filled methacrylate composites prepared via 3D stereolithography printing: Mechanical reinforcement and thermal stabilization. *Carbohydr. Polym.* 169:272–281.

in L-CNC(6 wt%)/ABS. L-CNC narrowly raised the maximum degradation temperature by 2–7°C (depending on L-CNC content) in the composites. As in the previous case of L-CNC and MA, L-CNC dispersed well, aided by the hydrophobicity of lignin that made NC less polar, and more compatible with non-polar ABS. Interfacial adhesion between L-CNC and ABS was also satisfactory.

REFERENCES

Abitbol, T. A., Cao, Y. et al. 2016. Nanocellulose, a tiny fiber with huge applications. *Curr. Opin. Biotechnol.* 39:76–88.

Adusumali, R.-B., Reifferscheid, M., Weber, H. et al. 2006. Mechanical Properties of Regenerated Cellulose Fibres for Composites. *Macromol. Symp.* 244:119–125.

Agnarsson, I., Kuntner, M., Blackledge, T. A. 2010. Bioprospecting finds the toughest biological material: extraordinary silk from a giant riverine orb spider. *PLoS ONE* 5 (9): e11234. doi:10.1371/journal.pone.0011234.

Alibaba. n.d. www.alibaba.com (accessed April 26, 2019).

Alimenti, F., Mezzanotte, P., Dionigi, M., Virili, M., Roselli, L. 2012. Microwave circuits in paper substrates exploiting conductive adhesive tapes. *IEEE Microw. Wireless Compon. Lett.* 22 (12): 660–662.

Asa'ad, F., Pagni, G., Pilipchuk, S. P., et al. 2016. 3D-printed scaffolds and biomaterials: review of alveolar bone augmentation and periodontal regeneration applications. *Int. J. Dentis.* Article ID 1239842. 10.1155/2016/1239842.

Ashland. 2020. Natrosol™ HEC. https://www.ashland.com/industries/pharmaceutical/oral-solid-dose/natrosol-hydroxyethylcellulose (accessed May 28, 2020).

Avery, M. P., Klein, S., Richardson, R., et al. 2014. *The Rheology of Dense Colloidal Pastes Used in 3D-Printing.* Hewlett-Packard: HP Laboratories HPL-2014-29.

Bagotia, N., Singh, B., Choudhary, V., Sharma, D. K. 2015. Excellent impact strength of ethylene-methyl acrylate copolymer toughened polycarbonate. *RSC Adv.* 5:87589–87597.

Bledzki, A. K., Gassan, J. 1999. Composites reinforced with cellulose based fibres. *Prog. Polym. Sci.* 24 (2): 221–274.

Boetker, J., Water, J. J., Aho, J. et al. 2016. Modifying release characteristics from 3D printed drug-eluting products. *Eur. J. Pharmaceut. Sci.* 90:47–52.

Brindha, J., Privita Edwina, R. A. G., Rajeshb, P. K., Rani P. 2016. Influence of rheological properties of protein bio-inks on printability: a simulation and validation study. *Mater. Today: Proc.* 3:3285–3295.

Brown, B. N., Valentin, J. E., Stewart-Akers, A. M., McCabe, G. P., Badylak, S. F. 2009. Macrophage phenotype and remodeling outcomes in response to biologic scaffolds with and without a cellular component. *Biomaterials* 30 (8): 1482–1491.

Cataldi, A., Rigotti, D., Nguyen, V. D. H., Pegoretti, A. 2018. Polyvinyl alcohol reinforced with crystalline nanocellulose for 3D printing application. *Mater. Today Commun.* 15:236–244.

Cate, D. M., Adkins, J. A., Mettakoonpitak, J., Henry, C. S. 2014. Recent developments in paper-based microfluidic devices. *Anal. Chem.* 87:19–41.

Centipark. 2021. *3D Printer Filament (HDPE)*. https://www.centipark.com/store/index.php?route=product/product&product_id=173 (accessed January 16, 2021).

Chandra, R., Rustgi, R. 1998. Biodegradable polymers. *Prog. Polym. Sci.* 23 (7): 1273–1335.

Chauhan, Y. P., Sapkal, R. S., Sapkal, V. S., Zamre, G. S. 2009. Microcrystalline cellulose from cotton rags (waste from garment and hosiery industries). *Int. J. Chem. Sci.* 7 (2): 681–688.

Chemical Book. 2017. *Ethyl Cellulose*. https://www.chemicalbook.com/ChemicalProductProperty_EN_CB6165620.htm (accessed May 28, 2020).

Chia, H. N., Wu, B. M. 2015. Recent advances in 3D printing of biomaterials. *J. Biol. Eng.* 9:4.

Chong, H., Koo, Y., Collins, B. et al. 2013. Paper-based microfluidic point-of-care diagnostic devices for monitoring drug metabolism. *J. Nanomed. Biother. Discov.* 3:1.

Cibis, D. and Kruger, K., 2008. Influencing parameters in droplet formation for DoD printing of conductive inks. *Proc 4th IMAPSIACerS Int. Conf. and Exhib. on Ceramic Interconnect and Ceramic Microsystems Technologies (CICMT)*, 417–423, Munich, Germany.

Ciolacu, D., Olaru, L., Suflet, D., Olaru, N. 2013. Cellulose esters – from traditional chemistry to modern approaches and applications. In *Pulp Production and Processing: From Papermaking to High-Tech Products*, ed. Valentin I. Popa, 253–298. Smithers Rapra.

Compton, B. G., Lewis, J. A. 2014. 3D-printing of lightweight cellular composites. *Adv. Mater.* 26 (34): 5930–5935.

Cook, B. S., Mariotti, C., Cooper, J. R. et al. 2014. Inkjet-printed, vertically-integrated, high-performance inductors and transformers on flexible LCP substrate. In *IEEE MTT-S Int. Microw. Symp. Dig.*, 1–4.

CP Kelco. 2020. *CP Kelco Cellulose Gum/Carboxymethyl Cellulose*. https://www.cpkelco.com/products/cellulose-gum-cmc/ (accessed May 28, 2020).

Dai, L., Cheng, T., Duan, C. et al. 2019. 3D printing using plant-derived cellulose and its derivatives: a review. *Carbohydr. Polym.* 203:71–86.

Denoyelle, T. 2011. *Mechanical Properties of Materials Made of Nano-Cellulose*. https://www.diva-portal.org/smash/get/diva2:396162/FULLTEXT01.pdf.

Derby, B. 2012. Printing and prototyping of tissues and scaffolds. *Science* 338:921–926.

Dhar, P., Bhardwaj, U., Kumar, A., Katiyar, V. 2014. Cellulose nanocrystals: a potential nanofiller for food packaging applications. In *Food Additives and Packaging*, Chapter 17, *ACS Symposium Series*, vol. 1162, 197–239.

Dong, J., Li, M., Zhou, L. et al. 2017. The influence of grafted cellulose nanofibers and postextrusion annealing treatment on selected properties of poly(lactic acid) filaments for 3D printing. *J. Polym. Sci., Part B: Polym. Phys.* 55:847–855.

Dufresne, A. 2012. *Nanocellulose: From Nature to High Performance Tailored Materials*. Berlin: de Gruyter GmbH.

DuPont. 2020. *Nutrition & Biosciences*. https://www.pharma.dupont.com/pharmaceutical-products/avicelr-for-solid-dose-forms.html (accessed May 28, 2020).

Ebrahimi, B., Mohammadi, R., Rouhi, M. et al. 2018. Survival of probiotic bacteria in carboxymethyl cellulose-based edible film and assessment of quality parameters. *LWT – Food Sci. Technol.* 87:54–60.

Edgar, K. J. 2004. Cellulose Esters, Organic. In *Encyclopedia of Polymer Science and Technology*, vol. 9, 129–158, New York: John Wiley & Sons.

Eichhorn, S. J., Young, R. J. 2001. The Young's modulus of a microcrystalline cellulose. *Cellulose* 8:197–207.

Elshereksi, N. W., Ghazali, M., Muchtar, A., Azhari, C. H. 2017. Review of titanate coupling agents and their application for dental composite fabrication. *Dental Mater. J.* 36 (5): 539–552.

Ethocel™. 2016. *Ethylcellulose, A Technical Review*. Midland: The Dow Chemical Company, February 2016.

Eurowatchers. 2018. *Global Bio-Based Polypropylene Market Automotive*. https://www.polyestertime.com/global-biobased-polypropylene-market-automotive/ (accessed May 28, 2020).

Feng, X., Yang, Z., Chmely, S. et al. 2017a. Lignin-coated cellulose nanocrystal filled methacrylate composites prepared via 3D stereolithography printing: Mechanical reinforcement and thermal stabilization. *Carbohyd. Polym.* 169:272–281.

Feng, X., Yang, Z., Rostom, S. S. H. 2017b. Structural, mechanical, and thermal properties of 3D printed L-CNC/acrylonitrile butadiene styrene nanocomposites. *J. Appl. Polym. Sci.* 34:45082.

Filaments.ca. n.d. *HDPE Filament – Natural – 1.75mm*. https://filaments.ca/products/hdpe-filament-natural-1–75mm (accessed May 31, 2020).

Food-Info. 2017. *462 Ethylcellulose.* http://www.food-info.net/uk/e/e462.htm (accessed May 28, 2020).

Fu, Y., Kao, W. J. 2010. Drug release kinetics and transport mechanisms of non-degradable and degradable polymeric delivery systems. *Expert Opin. Drug Deliv.* 7 (4): 429–444.

Gazzaniga, A. 2010. ChronoCap: a novel pulsatile delivery system. *BioPharmaNet.* http://www.tefarco.it/bpn/chronocap%20_gazzaniga_mi2010.pdf.

Gazzaniga, A., Palugan, L., Foppoli, A., Sangalli, M. E. 2008. Oral pulsatile delivery systems based on swellable hydrophilic polymers. *Eur. J. Pharm. Biopharm.* 68:11–18.

Ghadermazi, R., Hamdipour, S., Sadeghi, K., Ghadermazi, R., Khosrowshahi, A. 2019. A. Effect of various additives on the properties of the films and coatings derived from hydroxypropyl methylcellulose – a review. *Food Sci Nutr.* 7:3363–3377.

Ghassemi, T., Shahroodi, A., Ebrahimzadeh, M. H. et al. 2018. Current concepts in scaffolding for bone tissue engineering. *Arch. Bone Jt. Surg.* 6 (2): 90–99.

Gladman, A. S., Matsumoto, E. A., Nuzzo, R. G., Mahadevan, L., Lewis, J. A. 2016. Biomimetic 4D printing. *Nat. Mater.* 15:413–419.

Gonzalez-Macia, L., Morrin, A., Smyth, M. R., Killard A. J. 2010. Advanced printing and deposition methodologies for the fabrication of biosensors and biodevices. *Analyst* 135 (5): 845–867.

Grand View Research. 2020. *Aerogel Market.* https://www.grandviewresearch.com/press-release/global-aerogel-market (accessed May 28, 2019).

Gu, Y., Huang, J. 2013. Colorimetric detection of gaseous ammonia by polyaniline nanocoating of natural cellulose substances. *Colloids Surf. A: Physicochem. Eng. Aspects* 433:166–172.

Gunasekera, D., Kuek, S., Hasanaj, D. et al. 2016. Three dimensional ink-jet printing of biomaterials using ionic liquids and co-solvents. *Faraday Discuss.* 190:509.

Habibi, Y., Lucia, L. A., Rojas, O. J. 2010. Cellulose nanocrystals: chemistry, self-assembly, and applications. *Chem. Rev.* 110 (6): 3479–3500.

Håkansson, K., Henriksson, I. C., de la Peña Vázquez, C., et al. 2016. Solidification of 3D printed nanofibril hydrogels into functional 3D cellulose structures. *Adv. Mater. Technol.* 1:1600096.

Hauru, L. K. J., Hummel, M., King, A. W. T., Kilpeläinen, I., Sixta, H. 2012. Role of solvent parameters in the regeneration of cellulose from ionic liquid solutions. *Biomacromolecules* 13:2896–2905.

He, Y., Gao, Q., Wu, W.-B., et al. 2016. 3D printed paper-based microfluidic analytical devices. *Micromachines* 7:108.

He, Y., Wu, Y., Fu, J. Z., Wu, W. B. 2015. Fabrication of paper-based microfluidic analysis devices: A review. *RSC Adv.* 5:78109–78127.

Heinze, T., 2016. Cellulose: structure and properties. *Adv. Polym. Sci.* 271:1–52.

Henke, K., Treml, S. 2013. Wood based bulk material in 3d printing processes for applications in construction. *Eur. J. Wood Prod.* 71 (1): 139–141.

Herrera, N., Mathew, A. P., Oksman, K. 2015. Plasticized polylactic acid/cellulose nanocomposites prepared using melt-extrusion and liquid feeding: mechanical, thermal and optical properties. *Compos. Sci. Technol.* 106:149.

Hindi, S. S. Z. 2017. Microcrystalline cellulose: the inexhaustible treasure for pharmaceutical industry. *Nanosc. Nanotechnol. Res.* 4 (1): 17–24.

Hong, J. K. 2015. *Bioactive Cellulose Nanocrystal Reinforced 3D Printable Poly(ε-caprolactone) Nanocomposite for Bone Tissue Engineering.* Ph. D. diss., Virginia Polytechnic Institute and State University.

Hutmacher, D. W. 2000. Scaffolds in tissue engineering bone and cartilage. *Biomaterials* 21 (24): 2529–2543.

IHS Markit. 2019. *Cellulose Esters.* https://ihsmarkit.com/products/cellulose-ethers-chemical-economics-handbook.html (accessed May 28, 2020).

Italia, C. 2015. The international cellulose market. *Paper Industry World.* https://www.paperindustryworld.com/the-international-cellulose-market/ (accessed March 6, 2021).

IUPAC. 2014. *Compendium of Chemical Terminology – Gold Book.*

Jo, W., Kim, D. H., Lee, J. S. et al. 2014. 3D printed tactile pattern formation on paper with thermal reflow method. *RCS Adv.* 4:31764–31770.

Kafy, A., Kim, H. C., Zhai, L. et al. 2017. Cellulose long fibers fabricated from cellulose nanofibers and its strong and tough characteristics. *Scientific Rep.* 7:17683.

Kambli, N. D., Mageshwaran, V., Patil, P. G., Saxena, S., Deshmukh, R. R. 2017. Synthesis and characterization of microcrystalline cellulose powder from corn husk fibres using bio-chemical route. *Cellulose* 24 (12): 5355–5369.

Kamide, K. 2005. *The Cellulose and Cellulose Derivatives,*p. 4. The Netherlands: Elsevier.

Kargarzadeh, H., Ahmad, I., Abdullah, I. et al. 2012. Effects of hydrolysis conditions on the morphology, crystallinity, and thermal stability of cellulose nanocrystals extracted from kenaf bast fibers. *Cellulose* 19 (3): 855–866.

Katstra, W. E., Palazzolo, R. D., Rowe, C. W. et al. 2000. Oral dosage forms fabricated by three dimensional printing™. *J. Controll. Release* 66:1–9.

Kearns, A. J. 2017. *Cotton Cellulose Fibers for 3D Print Material.* Master diss., North Carolina State University.

Kempin, W., Franz, C., Koster, L. C., Schneider F., et al. 2017. Assessment of different polymers and drug loads for fused deposition modeling of drug loaded implants. *Eur. J. Pharma. Biopharmaceut.* 115:84–93.

Khaled, S., Burley, J. C., Morgan, R., Alexander, M., Roberts, C. 2014. Desktop 3D printing of controlled release pharmaceutical bilayer tablets. *Int. J. Pharmaceut.* 461:105–111.

Khaled, S. A., Burley, J. C., Alexander, M. R. et al. 2015. 3D printing of tablets containing multiple drugs with defined release profiles. *Int. J. Pharmaceut.* 494:643–650.

Khatri, Z., Wei, K., Kim, B. S., Kim, I. S. 2012. Effect of deacetylation on wicking behavior of co-electrospun cellulose acetate/polyvinyl alcohol nanofibers blend. *Carbohydr. Polym.* 87 (3): 2183–2188.

Kim, J., Yun, S., Ounaies, Z. 2006. Discovery of cellulose as a smart material. *Macromolecules* 39:4202–4206.

Kim, J. H., Shim, B. S., Kim, H. S. et al. 2015. Review of Nanocellulose for Sustainable Future Materials. *Int. J. Precis. Eng. Manufactur. – Green Technol.* 2 (2): 197–213.

Klein, S., Avery, M. P., Richardson, R. et al. 2015. 3D printed glass: surface finish and bulk properties as a function of the printing process. *Proc. of SPIE 9398*, 93980R.

Klemm, D., Heublein, B., Fink, H.-P. et al. 2005. Cellulose: fascinating biopolymer and sustainable raw material. *Angew. Chem. Int. Ed.* 44:3358–3393.

Klemm, D., Kramer, F., Moritz, S. et al., 2011 Nanocelluloses: a new family of nature-based materials. *Angew. Chem. Int. Ed.* 50 (24): 5438–5466.

Kondo, T., Kose, R., Naito, H., Kasai, W. 2014. Aqueous counter collision using paired water jets as a novel means of preparing bio-nanofibers. *Carbohydr. Polym.* 112:284–290.

Kreiger, M., Anzalone, G. C., Mulder, M. L., Glover, A., Pearce, J. M. 2013. Distributed recycling of post-consumer plastic waste in rural areas. *MRS Online Proceedings* Library, 1492. mrsf12–1492-g04–06.

Kumar, S., Hofmann, M., Steinmann, B. et al. 2012. Reinforcement of stereolithographic rapid prototyping with cellulose nanocrystals. *ACS Appl. Mater. Interfaces* 4:5399–5407.

Lavanya, D., Kulkarni, P. K., Dixit, M. et al. 2011. Sources of cellulose and their applications – A review. *IJDFR* 2 (6): 19-38..

Li, L., Zhu, Y., Yang, J. 2018. 3D bioprinting of cellulose with controlled porous structures from NMMO. *Mater. Lett.* 210:136–138.

Li, V., Dunn, C., Zhang, Z., Deng, Y., Qi, H. J. 2017. Direct ink write (DIW) 3D printed cellulose nanocrystal aerogel structures. *Scientific Rep.* 7:8018.

Li, X., Tabil, L. G., Panigrahi, S. 2007. Chemical treatments of natural fiber for use in natural fiber-reinforced composites: a review. *J. Polym. Environ.* 15 (1): 25–33.

Li V., C.-F., Dunn C. K., Zhang Z., et al. 2017 Direct ink write (DIW) 3D printed cellulose nanocrystal aerogel structures. *Scientific Rep.* 7:8018.

Li, M., He, B., Zhao, L. 2019. Isolation and characterization of microcrystalline cellulose from cotton stalk waste. *BioRes.* 14 (2): 3231–3246.

Lille, M., Nurmela, A., Nordlund, E. et al. 2018. Applicability of protein and fiber-rich food materials in extrusion-based 3D printing. *J. Food Eng.* 220:20–27.

Liu, H., Hsieh, Y. L. 2002. Ultrafine fibrous cellulose membranes from electrospinning of cellulose acetate. *J. Polym. Sci. Part B Polym. Phys.* 40:2119–2129.

Liu, L., Chang, H., Jameel, H., Park, S. 2014. Determination of cellulose degree of polymerization in lignocellulosic biomass. *AIChE Annual Meeting.* https://www.aiche.org/academy/videos/conference-presentations/determination-cellulose-degree-polymerization-lignocellulosic-biomass (accessed April 26, 2019).

Made-in-china. n.d. www.made-in-china.com (accessed April 26, 2019).

Mahnaj, T., Ahmed, S. U., Plakogiannis, F. M. 2013. Characterization of ethyl cellulose polymer. *Pharm Dev Technol.* 18 (5): 982–989.

Mäki-Arvela, P., Anugwom, I., Virtanen, P., Sjöholm, R., Mikkola, J. P. 2010. Dissolution of lignocellulosic materials and its constituents using ionic liquids – a review. *Indust. Crops Prod.* 32:175–201.

Maleki, H. et al. 2016. Synthesis and biomedical applications of aerogels: Possibilities and challenges. *Adv. Colloid Interf. Sci.* 236:1–27.

Mariotti, C., Alimenti, F., Roselli, L., Tentzeris, M. M. 2017. High-performance RF devices and components on flexible cellulose substrate by vertically integrated additive manufacturing technologies. *IEEE Trans. Microw. Theory Tech.* 65 (1): 62–71.

Mariotti, C., Alimenti, F., Mezzanotte, P. et al. 2013. Modeling and characterization of copper tape microstrips on paper substrate and application to 24 GHz branch-line couplers. *Proceedings of the 43rd European Microwave Conference*, 794–797. Nuremberg, Germany.

Markstedt, K., Escalante, A., Toriz, G., Gatenholm, P. 2017. Biomimetic inks based on cellulose nanofibrils and cross-linkable xylans for 3D printing. *ACS Appl. Mater. Interf.* 9:40878–40886.

Markstedt, K., Mantas, A., Tournier, I. et al. 2015. 3D bioprinting human chondrocytes with nanocellulose–alginate bioink for cartilage tissue engineering applications. *Biomacromolecules* 16:1489–1496.

Markstedt, K., Sundberg, J., Gatenholm, P. 2014. 3D bioprinting of cellulose structures from an ionic liquid. *3D Print. Addit. Manuf.* 1 (3): 115–121.

Martinez, A. W., Phillips, S. T., Butte, M. J., Whitesides, G. M. 2007. Patterned paper as a platform for inexpensive, low-volume, portable bioassays. *Angew. Chem. Int. Ed.* 46:1318–1320.

McKeen, L. W. 2012. Renewable resource and biodegradable polymers. In *Film Properties of Plastics and Elastomers*, 3rd edition, ed. L. W. McKeen, 353–378. New York: William Andrew.

Melocchi, F., Parietti G., Loreti A., et al. 2015. 3D printing by fused deposition modeling (FDM) of a swellable/erodible capsular device for oral pulsatile release of drugs. *J. Drug Deliv. Sci. Technol.* 30:360–367.

Mnuruddin, M., Chowdhury, A., Haque, S. A. et al. 2011. Extraction and characterization of cellulose micro fibrils from agricultural wastes in an integrated biorefinery initiative. *Cellulose Chem. Technol.* 45 (5-6): 347–354.

Mohanty, A. K., Wibowo, A., Misra, M., Drzal, L. T. 2004. Effect of process engineering on the performance of natural fiber reinforced cellulose acetate biocomposites. *Compos. A: Appl. Sci. Manuf.* 35 (3): 363–370.

Mohanty, A. K., Misra, M., Hinrichsen, G. 2000. Biofibres, biodegradable polymers and biocomposites: an overview. *Macromol. Mater. Eng.* 276–277 (1): 1–24.

Moon, R. J., Martini, A., Nairn, J., Simonsen, J., Youngblood, J. 2011. Cellulose nanomaterials review: structure, properties and nanocomposites. *Chem. Soc. Rev.* 40 (7): 3941–3994.

Murphy, C. A., Collins, M. N. 2018. Microcrystalline cellulose reinforced polylactic acid biocomposite filaments for 3D printing. *Polym. Compos.* 39:1311–1320.

Nair, L. S., Laurencin, C. T. 2007. Biodegradable polymer as biomaterials. *Prog. Polym. Sci.* 32 (8-9): 762–798.

Nasatto, P., Pignon, F., Silveira, J. et al. 2015. Methylcellulose, a cellulose derivative with original physical properties and extended applications. *Polymers* 7:777–803.

Nguyen, D., Hägg, D. A., Forsman, A. et al. 2017. Cartilage tissue engineering by the 3D bioprinting of iPS cells in a nanocellulose/alginate bioink. Scientific Reports 7:658.

Niaounakis, M. 2017. *Management of Marine Plastic Debris*. Oxford, Cambridge: Elsevier.

O'Sullivan, A., 1997. Cellulose: the structure slowly unravels. *Cellulose* 4:173–207.

Observatory of Economic Complexity. n.d. https://atlas.media.mit.edu/en/resources/about/.

Ohwoavworhua, F. O., Adelakun, T. A. 2010. Non-wood fibre production of microcrystalline cellulose from *Sorghum caudatum*: Characterisation and tableting properties. *Ind. J. Pharmaceut. Sci.* 72 (3): 295–301.

Okada, M. 2002. Chemical syntheses of biodegradable polymers. *Polym. Prog. Polym. Sci.* 27 (1): 87–133.

Oksman, K., Skrifvars, M., Selin, J. F. 2003. Natural fibres as reinforcement in polylactic acid (PLA) composites. *Compos. Sci. Technol.* 63:1317.

Olsson, C., Westman, G. 2013. Direct dissolution of cellulose: background, means and applications, *Cellulose. Fundamental Aspects*, ed. T. G. M. Van De Ven 143–178. IntechOpen. https://www.intechopen.com/books/cellulose-fundamental-aspects/direct-dissolution-of-cellulose-background-means-and-applications.

Pacquit, A., Frisby, J., Diamond, D. et al. 2007. Development of a smart packaging for the monitoring of fish spoilage. *Food Chem.* 102:466–470.

Park, J. S., Kim, T., Kim, W. S. 2017. Conductive cellulose composites with low percolation threshold for 3D printed electronics. *Scientific Rep.* 7:3246.

Pattinson, S. W., Hart, A. J. 2017. Additive manufacturing of cellulosic materials with robust mechanics and antimicrobial functionality. *Adv. Mater. Technol.* 2:1600084.

Pignataro, B. 2009. Nanostructured molecular surfaces: Advances in investigation and patterning tools. *J. Mater. Chem.* 19:3338–3350.

Porter, J. R., Ruckh, T. T., Popat, K. C. 2009. Bone tissue engineering: a review in bone biomimetics and drug delivery strategies. *Biotechnol. Prog.* 25 (6): 1539–1560.

PubChem®. n.d. *Hydroxyethylcellulose.* https://pubchem.ncbi.nlm.nih.gov/compound/4327536 (accessed May 31, 2020).

Raquez, J. M., Habibi, Y., Murariu, M., Dubois, P. 2013. Polylactide (PLA)-based nanocomposites. *Prog. Polym. Sci.* 38:1504.

Rees, A., Powell, L. C., Chinga-Carrasco, G. et al. 2015. 3D bioprinting of carboxymethylated-periodate oxidized nanocellulose constructs for wound dressing applications. *BioMed Res. Int.* doi:10.1155/2015/925757.

Rekhi, G. S., Jambhekar, S. S. 1995. Ethylcellulose – a polymer review. Drug *Develop. Indust. Pharm.* 21 (1): 61–77.

Rida, A., Yang, L., Tentzeris, M. M. 2007. Design and characterization of novel paper-based inkjet-printed UHF antennas for RFID and sensing applications. *Proc. IEEE Int. Antennas Propag. Symp.* 2749–2752, Honolulu, USA.

Robocasting. 2018. https://robocasting.com/ (accessed May 28, 2020).

Røhl, L., Larsen, E., Linde, F. et al. 1991. Tensile and compressive properties of cancellous bone. *J. Biomech.* 24 (12): 1143–1149.

Roselli, L., Carvalho, N. B., Alimenti, F. et al. 2014 Smart surfaces: Large area electronics systems for Internet of Things enabled by energy harvesting. *Proc. IEEE* 102(11):1723–1746.

Rowe, C. W., Katstra, W. E., Palazzolo, R. D. et al. 2000. Multimechanism oral dosage forms fabricated by three dimensional printing™. *J. Controll. Release* 66:11–17.

Sakurada, I., Nukushina, Y., Ito, T. 1962. Experimental determination of the elastic modulus of crystalline regions in oriented polymers. *J. Polym. Sci.* 52:651–660.

Shao, Y., Chaussy, D., Grosseau, P., Beneventi, D. 2015. Use of microfibrillated cellulose/lignosulfonate blends as carbon precursors: impact of hydrogel rheology on 3D printing. *Ind. Eng. Chem. Res.* 54:10575–10582.

Shirazi, S. F. S., Gharehkhani, S., Mehrali, M. et al. 2015. A review on powder-based additive manufacturing for tissue engineering: selective laser sintering and inkjet 3D printing. *Sci. Technol. Adv. Mater.* 16:033502.

Siqueira, G., Kokkinis, D., Libanori, R. et al. 2017. Cellulose nanocrystal inks for 3D printing of textured cellular architectures. *Adv. Funct. Mater.* 27:1604619.

Siró, I., Plackett, D. 2010. Microfibrillated cellulose and new nanocomposite materials: a review. *Cellulose* 17 (3): 459–494.

Smay, J. E., Gratson, G. M., Shepherd, R. F., Cesarano, J., Lewis, J. A. 2002. Directed colloidal assembly of 3D periodic structures. *Adv. Mater.* 14:1279–1283.

Studart, A. R. 2016. Additive manufacturing of biologically-inspired materials. *Chem. Soc. Rev.* 45:359–376.

Succinity. n.d. *Biobased Polybutylene Succinate (PBS).* http://www.succinity.com/images/succinity_broschure.pdf (accessed May 28, 2020).

Sultan, S., Abdelhamid, H., Zou, X., Mathew, A. P. 2019. CelloMOF: nanocellulose enabled 3D printing of metal–organic frameworks. Adv. Funct. Mater. 29 (1805372): 1–12.

Sultan S., Siqueira G., Zimmermann T., Mathew A. P. 2017. 3D printing of nano-cellulosic biomaterials for medical applications. *Current Opinion in Biomedical Engineering* 2:29–34.

Sun, K., Wei, T.-S., Ahn, B. Y. et al. 2013. 3D printing of interdigitated Li-ion microbattery. *Adv. Mater.* 25:4539–4543.

Swift, M. J. 1977. The ecology of wood decomposition. *Sci Progr. Oxf.* 64:175–199.

Takeuchi, Y., Umemura, K., Tahara, K., Takeuchi, H. 2018. Formulation design of hydroxypropyl cellulose films for use as orally disintegrating dosage forms. *J. Drug Deliv. Sci. Technol.* 46:93–100.

Tanaka, F., Iwata, T. 2006. Estimation of the elastic modulus of cellulose crystal by molecular mechanics simulation. *Cellulose* 13:509–517.

Tang, M., Wang, G., Kong, S. K., Ho, H. P. 2016. A review of biomedical centrifugal microfluidic platforms. *Micromachines* 7:26.

Tanwilaisiri, A., Zhang, R., Xu, Y., Harrison, D., Fyson, J. 2016. A manufacturing process for an energy storage device using 3D printing. *IEEE International Conference on Industrial Technology (ICIT)*, 888–891, Taipei, Taiwan.

Thakkar, H., Eastman, S., Al-Mamoori, A. et al. 2017. Formulation of aminosilica adsorbents into 3D-printed monoliths and evaluation of their CO2 capture performance. *ACS Appl. Mater. Interf.* 9:7489–7498.

Thoorens, G., Krier, F., Leclercq, B., Carlin, B., Evrard, B. 2014. Microcrystalline cellulose, a direct compression binder in a quality by design environment: a review. *Int. J. Pharmaceut.* 473 (1–2): 64–72.

Torres-Rendon, J. G., Femmer, T., DeLaporte, L. et al. 2015. Bioactive gyroid scaffolds formed by sacrificial templating of nanocellulose and nanochitin hydrogels as instructive platforms for biomimetic tissue engineering. *Adv. Mater.* 27:2989–2995.

Torres-Rendon, J. G., Köpf, M., Gehlen, D. et al. 2016. Cellulose nanofibril hydrogel tubes as sacrificial templates for freestanding tubular cell constructs. *Biomacromolecules* 17:905–913.

Trache, D. 2017. Microcrystalline cellulose and related polymer composites: synthesis, characterization and properties. In *Handbook of Composites from Renewable Materials*, ed. Vijay Kumar Thakur, Manju Kumari Thakur, Michael R. Kessler, 61–92. New York: Scrivener Publishing.

Trache, D., Hazwan Hussin, M., Chuin, C. T. H. et al. 2016 Microcrystalline cellulose: isolation, characterization and bio-composites application – a review. *Int. J. Biol. Macromol.* 93:789–804.

Vert, M., Li, M., Spenlehauer, G., Guerin, P. 1992. Bioresorbability and biocompatibility of aliphatic polyesters. *J. Mater. Sci.-Mater. Med.* 3 (6): 432–446.

Wang, C.-C., Tejwani, M. R., Roach, W. J. et al. 2006. Development of near zero-order release dosage forms using three-dimensional printing (3-DP™) technology. *Drug Develop. Indus. Pharm.* 32:367–376.

Wang, L., Gardner, D. J., Bousfield, D. W. 2018b. Cellulose nanofibril-reinforced polypropylene composites for material extrusion: Rheological properties. *Polym. Eng. Sci.* 58 (5): 793–801.

Wang, Q., Sun, J., Yao, Q. et al. 2018a. 3D printing with cellulose materials. *Cellulose.* 25:4275–4301.

Wang, Z., Xu, J., Lu, Y. et al. 2017. Preparation of 3D printable micro/nanocellulose-polylactic acid (MNC/PLA) composite wire rods with high MNC constitution. *Indus. Crops Prod.* 109:889–896.

Wypych, G. 2016. Physical properties of fillers and filled materials. In *Handbook of Fillers*, ed. G. Wypych, 303–371. Scarborough:ChemTec Publishing.

Xiang, L. Y., Mohammed, M. A. P., Baharuddin, A. S. 2016. Characterisation of microcrystalline cellulose from oil palm fibres for food applications. *Carbohydr. Polym.* 148:11–20.

Xu, W., Wang, X., Sandler, N. et al. 2018. Three-dimensional printing of wood-derived biopolymers: a review focused on biomedical applications. *ACS Sustainable Chem. Eng.* 6:5663–5680.

Yanagida, N., Matsuo, M. 1992. Morphology and mechanical properties of hydroxypropyl cellulose cast films crosslinked in solution. *Polymer* 33 (5): 996–1005.

Yu, D.-G., Branford-White, C., Ma, Z.-H. et al. 2009. Novel drug delivery devices for providing linear release profiles fabricated by 3DP. *Int. J. Pharmaceut.* 370:160–166.

Zauba. n.d. www.zauba.com (accessed May 28, 2019).

Zhang, H., Guo, Y., Jiang, K. et al. 2016. A review of selective laser sintering of wood-plastic composites. *Proceedings of the 27th Annual International Solid Freeform Fabrication Symposium.*

8 Bamboo

The only way forward, if we are going to improve the quality of the environment, is to get everybody involved.

Richard Rogers

8.1 OVERVIEW OF BAMBOO

Bamboo is a family of tall, woody perennial evergreen plants that comprise 1718 species (Bamboo Genera List 2020), and grows in tropical and subtropical to mild temperate regions, with the most concentration and species in East and Southeast Asia and on islands of the Indian and Pacific oceans. Some species are also present in the southern USA. Bamboo has a hollow stem called *culm* (Figure 8.1) with cellulose fibers aligned along its length carrying nutrients between the leaves and roots (Mwaikambo 2006). Bamboo is an anisotropic material, featuring mechanical properties varying in the longitudinal, transverse, and radial directions. Longitudinal fibers are aligned within a lignin matrix, divided by nodes located along the culm's length. The thickness of the culm wall tapers from its base to its top (Sharma et al. 2015). The microstructure of bamboo plant is schematically illustrated in Figure 8.1 that details cross sections of its elements, from culm to elementary fibers to microfibrils.

Bamboo is the world's fastest-growing woody plant, developing 1/3 faster than the fastest-growing tree (Rathod and Kolhatkar 2013). It grows approximately 7.5–40 cm a day, with the world record being 1.2 m (4 feet) in 24 hours in Japan (Kumar et al. 2017). It also excels in biomass production, yielding 40 tons/year or more per hectare in managed stands, and accounts for around 1/4 and 1/5 of biomass produced in tropical and subtropical regions, respectively (Kumar et al. 2017). Bamboo possesses other environmental advantages. It has successfully been used to rehabilitate soil ravaged by brick making, and sites of abandoned tin mines. It shelters top soil from the onslaught of tropical downpours, preserves many exposed areas, providing microclimate for forest regeneration and watershed protection. It is frequently introduced into the banks, streams, or in other vulnerable areas for rapid control of soil erosion; in fact, the roots of one bamboo can bind up to 6 m^3 of soil (Kumar et al. 2017). Bamboo attains maturity in three years, while wood takes 20 years and, after maturity, the tensile strength of a bundle of bamboo fibers is 300–400 MPa (Yu et al. 2011), comparable to the tensile strength of AISI 1018 mild steel, which is 440 MPa. Bamboo absorbs CO_2 and releases over 30% more oxygen into the atmosphere compared to an equivalent mass of trees. It is naturally anti-bacterial: it eliminates and prevents over 70% of bacteria that attempt to grow on it, whether bamboo is in its natural or fabric form. Because of its abundance, low cost, attractive physical and mechanical properties, and versatility as a raw material (powder, strip, shorter and longer fiber), bamboo and its related industries provide income, food, and housing to over 2 billion people worldwide (Rathod and Kolhatkar 2013). Therefore, bamboo is twice as valuable, because it is not only sustainable but also significantly contributes to well-being and economy of local populations. In countries producing bamboo, such as China and India (the number one and two world producers, respectively), bamboo heavily contributes to the national economy, because it is largely utilized and exported. In developing countries, it is one of the main building materials, surpassed only by wood (Mwaikambo 2006).

Bamboo was one of the early man-made composites: some examples of bows made of it combined with pine resin from as early as 1,200 AD (Jones 1998) are recorded. Some products

DOI: 10.1201/9781003221210-8

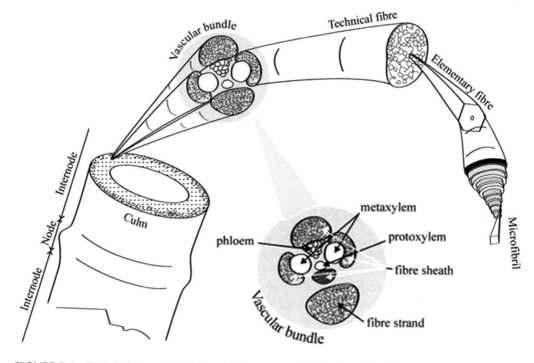

FIGURE 8.1 Exploded view of the hierarchical structure of a bamboo plant, from culm to microfibril.

Source: adapted from Depuydt, D. E., Sweygers, N., Appels, L., Ivens, J., van Vuure, A. W. 2019. Bamboo fibres sourced from three global locations: A microstructural, mechanical and chemical composition study. *J. Reinf. Plast. Compos.* 38 (9): 397–412. Reproduced under the STM Permissions Guidelines from SAGE Journals.

derived from bamboo are: boards, flooring, furniture, woven products, textiles, handicrafts, food, chemical extracts, charcoal, surfboards, cutlery, tableware, and vinegar. At least until the early 1980s the construction industry utilized it for foundations, frames, flooring, roofing, walls, partitions, ceiling, doors, windows, pipes, reinforcement of concrete, and scaffolding (McClure 1981). In rural communities of developing countries, bamboo is popular for building domestic housing. The global market of bamboo is estimated to exceed USD 60 billion, and is rapidly growing (Knight et al. 2017).

8.2 BAMBOO PROPERTIES

Bamboo outperforms most types of wood and wood products in tensile strength and modulus (Knight et al. 2017), and is lighter than major types of construction wood: oak, maple, and hickory. Engineering materials comprising bamboo utilize its fibers because of their mechanical properties. In fact, bamboo fibers BFs impart the plant its strong mechanical performance (Amada and Untao 2001; Lo et al. 2004). The chemical composition of BFs depends on many factors (species, culm age, location along the height of the culm and within the culm wall, geographic location, climate, soil condition, and so on), however its main components are always *cellulose* and *hemicellulose*, two carbohydrates, and *lignin*, a polymer that bonds cellulose and hemicelluloses within cell walls, and bonds cells together. Examples of bamboo compositions, along with fiber density and diameter, are reported in Table 8.1, and refer to three species growing in South America, Asia, and Europe. Table 8.2 lists values of properties of BFs and glass fibers that can serve as a benchmark in formulating BF-filled plastics for AM. The tensile properties possess widely ranging values

TABLE 8.1
Chemical Composition, Fiber Diameter, and Density of Selected Bamboo Species

Bamboo Species	Fiber Diameter	Fiber Density	Cellulose	Hemicellulose	Lignin	Extractives	Ash
	μm	g/cm^3	wt%	wt%	wt%	g/100 g	g/100 g
Guadua angustifolio Kunth (South America)	144–185	1.41–1.44	53.7–54.6	14.0–15.2	31.0–31.4	0.7	0.6
Dendrocalamus membranaceus Munro (Asia)	134–228	1.41–1.42	50.4–58.5	15.3–16.8	26.7–32.8	0.0–9.7	0.1–0.3
Phyllostachys nigra Boryana (Europe)	167–177	1.38–1.42	49.2–52.8	16.5–17.9	29.3–34.2	4.7–7.9	0.2–0.5

Source: Depuydt, D. E., Sweygers, N., Appels, L., Ivens, J., van Vuure, A.W. 2019. Bamboo fibres sourced from three global locations: A microstructural, mechanical and chemical composition study. *J. Reinf. Plast. Compos* 38 (9): 397–412. Reproduced under the STM Permissions Guidelines from SAGE Journals.

TABLE 8.2
Properties of Bamboo and Glass Fibers

Material	Specific Gravity	Diameter	Length	Tensile Strength	Tensile Modulus	Elongation at Break
		μm	mm	MPa	GPa	%
Bamboo fibers	0.6–1.1	10–400	2.7	140–1,910	11–89	1.4–6.5
E-Glass fibers	2.5	13	7	2,000–3,500	70	2.5

Sources: Han, J. S., Rowell, J. S. 1997. Chemical composition of fibers. In *Paper and composites from agro-based resources*, ed. R. Rowell, R. Young, J. Rowell, 83–134. New York: CRC Lewis Publisher. Kozlowski, R., Wladyka-Przybylak, M. 2004. Use of Natural Fiber Reinforced Plastics. In *Natural Fibers, Plastics and Composites*, ed. F. T. Wallenberger, N.E. Weston, 249–274. New York: Springer Science Business Media. Célino, A., Fréour, S., Jacquemin, F., Casari, P. 2014. The hygroscopic behavior of plant fibers: a review. *Frontiers in Chemistry* 1, 43. Rao, K. M. M., Rao, K. M. 2007. Extraction and tensile properties of natural fibers: Vakka, date and bamboo. *Composite Structures* 77 (3): 288–295. Osorio, L., Trujillo, E., Van Vuure, A.W., Verpoest, I. 2011. Morphological aspects and mechanical properties of single bamboo fibers and flexural characterization of bamboo/epoxy composites. *J. Reinf. Plast. Compos.* 30:396–408. Nurul Fazita, M. R., Jayaraman, K., Bhattacharyya, D. et al. 2016. Green Composites Made of Bamboo Fabric and Poly (Lactic) Acid for Packaging Applications—A Review. *Materials* 9 (6): 435.

because they depend not only on the above mentioned factors affecting chemical composition (Awalluddin et al. 2017), but also on the method of extracting fibers from the plant (Osorio et al. 2011), which are chemical, mechanical, and chemical-mechanical (Rao and Rao 2007; Liu et al. 2012). Therefore, in selecting the suitable BFs as a filler, it is useful to know that mechanical extraction increases tensile strength and modulus in comparison to chemical extraction. Figure 8.2 includes images of a bamboo fiber bundle after mechanical extraction. BFs cannot be used by themselves but have to be embedded in a plastic matrix, and hence their strength and stiffness is weakened by those of the matrix. However, their values of tensile strength and modulus divided by specific gravity exceed those of mild steel, as reported in Table 8.3. A reason to add bamboo to polymers is also to reduce the material cost of the derived composite material.

BFs can outperform wood fibers in tensile properties. In fact, Yu et al. (2014) reported that the tensile strength and modulus of eleven Chinese bamboo species were 1.21–1.91 GPa and 25–46 GPa,

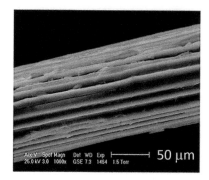

FIGURE 8.2   Bamboo (*Guadua angustifolia*) fibers after mechanical extraction: fiber bundle (left), and SEM images of fiber bundle (right). The fiber bundle is composed of several elementary fibers.

Source: Osorio, L., Trujillo, E., Van Vuure, A. W., Verpoest I. 2011. Morphological aspects and mechanical properties of single bamboo fibers and flexural characterization of bamboo/epoxy composites. *J. Reinf. Plast. Compos.* 30 (5): 396–340. Reproduced under the STM Permissions Guidelines from SAGE Journals.

TABLE 8.3

Comparison of Absolute and Specific Properties of Bamboo Fibers and Mild Steel

Material	Specific Gravity (SG)	Tensile Modulus (E)	Tensile Strength (TS)	E/SG	TS/SG
	N/A	GPa	MPa	GPa	MPa
Bamboo Fibers[a]	0.8–1.1	48–89	391–1,000	44–111	355–1,250
AISI 1018 Mild Steel[b]	7.85	205	440	26.1	56.1

Notes:

[a]Nurul Fazita, M. R., Jayaraman, K., Bhattacharyya, D. et al. 2016. Green Composites Made of Bamboo Fabric and Poly (Lactic) Acid for Packaging Applications—A Review. *Materials* 9 (6): 435.
[b]https://www.azom.com/article.aspx?ArticleID=6115 (accessed January 16, 2020).

respectively, and exceeded tensile strength and modulus of two softwood trees, Chinese Fir and Chinese Red Pine, that were 0.77–0.81 GPa and 17.6–18.2 GPa, respectively.

BFs feature an almost linear tensile stress-strain curve up to failure, as plotted in Figure 8.3: BFs extracted using a chemical method (C in Figure 8.3), or a mechanical route (M in Figure 8.3), and their performance is compared to fibers derived from date and royal palm (*Roystonea regia*) (Rao and Rao 2007).

Before being chosen for AM, bamboo had been a reinforcement in engineered composite materials that represented a sustainable, biodegradable, and less expensive alternative to glass-filled and carbon-filled composite materials. Beside their cost and environmental benefits (replacing non-sustainable fibers such as glass and carbon with BFs increases the sustainable portion of the resulting composite material), BFs offer the following advantages - compared to synthetic fibers - that make BFs valuable for composite materials (Liu et al. 2012):

- They derive from abundant source.
- Their cellulose constituent has low density, producing attractive specific mechanical properties.

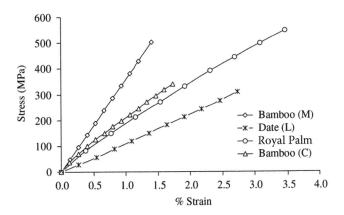

FIGURE 8.3 Tensile stress-strain curve of three natural fibers: bamboo, date, and royal palm. (M) and (C) indicate fiber extracted using mechanical and chemical method, respectively, (L) means fiber from dry leaves.

Source: modified from Rao, K. M. M., Rao, K. M. 2007. Extraction and tensile properties of natural fibers: Vakka, date and bamboo, *Compos. Struct.* 77 (3): 288–295. Reproduced with permission from Elsevier.

- They are flexible and not brittle such as carbon and glass, and do not break around sharp curvatures, leaving intact their aspect ratio and hence their mechanical enhancement.
- They are not toxic, and cause no health problems such as skin irritation and respiratory problems, and are antibacterial.
- Their surface can be modified.
- Their production require low amount of energy. The energy for manufacturing common engineered polymers, such as PP, HDPE, PU, and PA is 6–25 times the amount to produce bamboo fibers, therefore, switching from unfilled PP, HDPE, PU, and PA to the same polymer filled with bamboo reduces the energy to produce the end-use material (Knight et al. 2017).
- They benefit the economy of rural areas not farming food.

One additional reason to print with bamboo-filled polymers is economic: in countries where it grows in abundance, bamboo is less expensive than other reinforcing materials, such as carbon and glass, and hence companies operating in countries rich with bamboo have an economic incentive to use it. On the other side, it is challenging to extract BFs of good quality, and, as with other natural fibers, the scattering in diameter, physical and mechanical properties is large even within the same batch, as explained before.

The following species of bamboo are preferred for their mechanical performance and are recommended as a source of fibers for AM feedstocks:

- Moso (*Phyllostachys edulis*), native to China, Korea, and Taiwan
- Guadua angustifolia, native to South America, and, with Moso, the most common species employed in building applications, and the most important American bamboo
- Rubro (*Phyllostachys rubromarginata*), native to Central China.

The superior mechanical performance of bamboo derives from its fiber being its main constituent, and the unidirectional arrangement of its fibers (Wai et al. 1985). Tensile strength and modulus are proportional to its cellulose content, and the elongation at break depends on the orientation angle of the microfibrils along the stalk (2–10°). Tensile strength, modulus, and strain at break are maximum in a single fiber, drop in a fiber bundle, and drop further in a *regenerated* fiber, obtained when bamboo cellulose is dissolved and wet- or melt-spun into fiber (Table 8.4).

TABLE 8.4

Tensile Properties of Bamboo Fibers in Various Form

Fiber Form	Strength	Modulus	Strain at Break
	MPa	GPa	%
Single fiber	916	13.6	12.6
Fiber bundle	387	2.7	16.7
Regenerated fiber	290	1.8	71.2

Source: Liu, D., Song, J., Anderson, D. P., Chang, P. R., Hua, Y. 2012. Bamboo fiber and its reinforced composites: structures and properties. *Cellulose* 19:1449–1480. Reproduced with permission from Springer Science Business Media.

A supplier of BFs is Bambooder®, a Dutch company that, besides manufacturing powder and short fibers, has developed a patented technology to extract "long" BFs, and to make an endless bamboo thread for high-performance composite applications, such as lower weight vehicles. Particularly, Bambooder® extracts fibers without chemicals, and supplies long fibers in form of threads, net, and woven and unidirectional fabrics, and short fiber preform and prepreg mats (Bambooder® 2020). Shanghai Tenbro® (Tenbro® 2013) is a supplier of BFs and textiles made of them. Its fibers are produced without any chemical added, and are guaranteed antibacterial. Their length is 38–86 mm, strength (or *tenacity*) is 2.1–2.4 cN/dtex, and elongation at break 18–26%. cN means centinewtons, and dtex measures linear density, expressed in grams per 10 km of filament. Examples of other suppliers of bamboo yarns and fabrics are Lakshmi Mills in India (Lakshmi Mills n.d.), and BambroTex® (BambroTex® 2007).

According to Khalil et al. (2012) andLiu et al. (2012), bamboo has been selected as a filler in the following non-AM composites:

- Oil-derived polymers: polyester, epoxy, phenolic, PP, HDPE, PP/PLA, PP/PE, PVC, PA, and PS.
- Sustainable polymers: polysaccharide, soy protein, PHAs, PBS, and PLA.

Most of the above polymers are also feedstocks for AM, and hence can potentially be formulated in a version compatible with AM. Bamboo-filled composite materials not for AM are available as laminates instead of filament or powder, but some lessons learned from their development and optimization may apply to bamboo-filled AM feedstocks.

8.3 COMMERCIAL BAMBOO FILAMENTS FOR AM

Examples of commercial feedstocks for AM containing bamboo are the following filaments for FFF:

- eBamboo 3D filament by eSUN (China): 1.75 mm diameter, bamboo powder-filled PLA, featuring a matte, wood, frosted texture, and wood aroma (eSUN 2007)
- Bamboo PRI-MAT 3D Filament: 2.85 mm diameter, fully biodegradable, PLA filled with 20 wt% BFs (Global 3D 2020)
- PopBit® Wood Filament for 3D Printer: 1.75 mm and 3 mm diameter, 0.02 mm BF-filled PLA, features real wood texture, and is suited for a variety of arts and crafts (PopBit® n.d.)
- bambooFill by colorFabb: PLA/PHA blend (90/10 wt%) plus 20–30 wt% BFs, and additives. Discontinued in 2017 but mentioned here, because its properties can represent a benchmark for next bamboo-based AM materials.

The range of values of physical and mechanical properties of eBamboo and bambooFill are listed in Table 8.5. The performance of these bamboo filaments is shown in perspective, by plotting in

TABLE 8.5

Properties of Commercial Bamboo-Filled Filaments for FFF: BambooFill (colorFabb) and eBamboo (eSUN)

Density	Melt Flow Rate (190°C/ 2.16 kg)	Tensile Strength	Tensile Modulus	Elongation at Break	Flexural Strength	Flexural Stress at 3.5% Strain	Flexural Modulus	Unnotched Charpy Impact Strength	Notched Charpy Impact Strength	Izod Impact Strength	Heat Distortion Temperature (0.45 MPa)
g/cm^3	g/10 min	MPa	GPa	%	MPa	MPa	GPa	kJ/m^2	kJ/m^2	J/m	°C
1.2	6–14	28–46	2.36–3.30	3.7–8	35–55	70	2.26–3.60	22	3	43	48

Source: Datasheets posted online by the suppliers. Landes, S., Letcher, T. 2020. Mechanical Strength of Bamboo Filled PLA Composite Material in Fused Filament Fabrication. *J. Compos. Sci.* 4, 159. Reproduced from an open access article distributed under the terms and conditions of the Creative Commons Attribution (CC BY) license (http://creativecommons.org/licenses/by/4.0/).

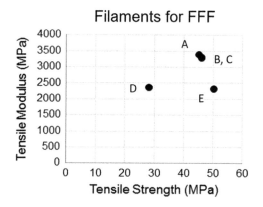

FIGURE 8.4 Tensile properties of commercial and experimental filaments for FFF: (A) PLA (PLA Economy), (B) bamboo/PLA/PHA (bambooFill), (C) wood/PLA (woodFill), (D) bamboo/PLA (eBamboo), (E) lignin/PLA (experimental).

Figure 8.4 the tensile strength and modulus of them, and other AM filaments: a wood-filled PLA filament, a lignin/PLA filament, and unfilled PLA (PLA Economy by colorFabb). PLA Economy was selected because: (a) it is likely made of the same PLA as bambooFill, being supplied by the same producer; (b) it possesses upper values in strength and stiffness among PLA filaments. Lignin/PLA was experimental (Gkartzou et al. 2017). The two bamboo-filled filaments do not outperform neat PLA in modulus, and bambooFill has practically the same strength, hence their benefits vs. PLA do not consist in offering a superior mechanical performance (including also ductility, impact resistance, and thermal resistance), but a different appearance that may lead to applications in furniture, appliances, and architecture, and likely a lower cost to produce. In fact, the Oak Ridge National Lab (ORNL), Resource Fiber LLC (USA), and Techmer PM (USA) teamed up and developed a BF/PLA filament, and built out of it sections, later assembled together, to form an outdoor large pavilion and seats displayed at the entrance of the design fair 2016 Miami Design (Scott 2016).

Landes and Letcher (2020) conduced a comprehensive characterization of eBamboo through physical and mechanical characterization of ASTM-compliant test coupons printed at 100% infill in the following raster orientations: 0°, 90°, 0/90°, and −45/45°, with the angles measured between filament direction and coupon's longitudinal axis. The measured properties at 0° orientation are collected in Table 8.6, and failure in all test types was ductile. The notched Izod impact strength at 0° was 42.9 J/m. Notably, in Table 8.6 the shear strength (measured according to ASTM D5379-

TABLE 8.6

Tensile Properties of Coupons of eBamboo Tested along 0° Printing Orientation

Load Mode	Ultimate Strength	Elastic Modulus	Yield Strength	Strain at Break
	MPa	GPa	MPa	%
Tension	29.2	2.36	27.6	2.4
Flexure	54.6	2.26	46.3	4.6
Compression	48.1	1.95	N/A	N/A
Shear	29.1	1.13	N/A	N/A

Source: Landes, S., Letcher, T. 2020. Mechanical Strength of Bamboo Filled PLA Composite Material in Fused Filament Fabrication. *J. Compos. Sci.* 4, 159. Reproduced from an open access article distributed under the terms and conditions of the Creative Commons Attribution (CC BY) license (http://creativecommons.org/licenses/by/4.0/).

TABLE 8.7

Mechanical Properties of Coupons of eBamboo

Printing Orientation	Tension		Flexure		Shear		Compression		Notched Izod Impact Strength
	Strength	Modulus	Strength	Modulus	Strength	Modulus	Strength	Modulus	Strength
Degree	MPa	GPa	MPa	GPa	MPa	GPa	MPa	GPa	J/m
0	29.2	2.36	54.6	2.26	29.1	1.13	48.4	1.95	42.9
90	31.1	2.49	49.4	2.15	N/A	N/A	N/A	N/A	13.9
0/90	30.0	2.49	52.4	2.22	N/A	N/A	N/A	N/A	N/A
−45/45	32.5	2.63	56.8	2.40	25.1	1.32	48.6	1.89	21.0

Source: Landes, S., Letcher, T. 2020. Mechanical Strength of Bamboo Filled PLA Composite Material in Fused Filament Fabrication. *J. Compos. Sci.* 4, 159. Reproduced from an open access article distributed under the terms and conditions of the Creative Commons Attribution (CC BY) license (http://creativecommons.org/licenses/by/4.0/).

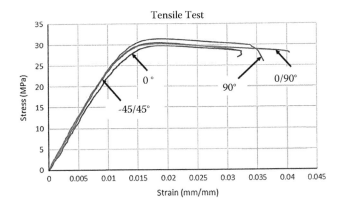

FIGURE 8.5 Representative tensile stress-strain curves of test coupons printed at different orientations out of eBamboo (eSUN) filament.

Source: modified from Landes, S., Letcher, T. 2020. Mechanical Strength of Bamboo Filled PLA Composite Material in Fused Filament Fabrication. *J. Compos. Sci.* 4, 159; doi:10.3390/jcs4040159. Reproduced from an open access article distributed under the terms and conditions of the Creative Commons Attribution (CC BY) license (http://creativecommons.org/licenses/by/4.0/).

19) is about equal to the tensile stregth, instead of being lower as tipically in unfilled polymers (Section 1.4) and filled polymers. In comparison, the tensile strength of PP loaded with 30 wt% BFs was 16.5 MPa (Thwe and Liao 2002). Strength and modulus measured at all orientations are grouped in Table 8.7. Representative tensile stress-strain curves of test coupons printed at varying orientations are plotted in Figure 8.5. The variation in property values with orientation depended on the stress type: it was the smallest for tension and flexure (10–15%), and the maximum for impact (threefold). In tension and flexure eBamboo is almost isotropic, which is an advantage in designing load-bearing components, A magnified view of the eBamboo's cross section displayed a uniform distribution of the bamboo powder (Figure 8.6).

Although bambooFill was discontinued, its properties (filament cross section, surface finish, thermal properties, and water uptake) measured by Pop et al. (2019) may be instructive for formulating bamboo-filled filaments for FFF, but also for sustainable and non-sustainable filled polymeric filaments for FFF. bambooFill filament displays a circular cross section (Figure 8.7),

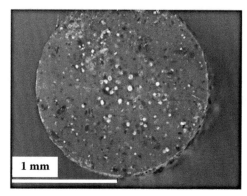

FIGURE 8.6 Magnified view of the cross section of eBamboo filament.

Source: modified from Landes, S., Letcher, T. 2020. Mechanical Strength of Bamboo Filled PLA Composite Material in Fused Filament Fabrication. *J. Compos. Sci.* 4, 159; doi:10.3390/jcs4040159. Reproduced from an open access article distributed under the terms and conditions of the Creative Commons Attribution (CC BY) license (http://creativecommons.org/licenses/by/4.0/).

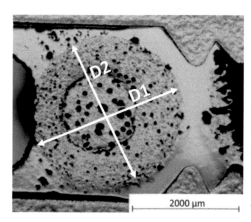

FIGURE 8.7 Optical micrograph of cross section of bambooFill filament.

Source: Pop, M. A., Croitoru, C., Bedő, T., et al. 2019. Structural changes during 3D printing of bioderived and synthetic thermoplastic materials. *J. Appl. Polym. Sci.*, 136, 47382. Reproduced with permission from John Wiley and Sons.

with an average diameter of 2.75 mm (from 2.83 and 2.67) versus nominal 2.85 mm. The BF particles are displayed as black dots: they have an average diameter of 49 μm, and are fairly uniformly distributed. The presence of BFs affected the polymers layer distribution and adhesion, and caused defects (air bubbles, etc.) during extrusion from the printer. The thermal properties of bambooFill measured via DSC were: T_g 62.7°C, T_m 152.5°C, degradation temperature 333.8°C, crystallinity 6.4%, and cold crystallization temperature 101.8°C. Compared to commercial PLAs in Chapter 3, bambooFill has higher T_g and lower T_m, which is advantageous because it translates in greater maximum service temperature and lower energy to soften it during printing extrusion. The fact that bambooFill degrades at a lower temperature than PLA is due to the lower degradation temperature of bamboo's ingredients (such as hemicellulose) than PLA's, and possibly to bambooFill's crystallinity trailing that of unfilled PLA's.

The control of surface finish on printed parts is important not only for aesthetic reasons but also functionality, since surface finish of parts impacts their friction, wear, fatigue, corrosion, and so on. An example of AM parts where surface finish is critical are inserts for microinjection molding (Davoudinejad et al. 2017), models for wind tunnel tests, and orthopedic implants. Surface finish of bambooFill optically measured in terms of average roughness R_a was 57.9 μm, while, f.e., R_a for Verbatim® PLA was 63.4 μm.

Applications of AM polymers are generally intended for non-wet service conditions, and typically water uptake is not reported by their suppliers, hence, the plot of distilled water uptake over time of filaments and printed discs made of bambooFill, PLA, and ABS in Figure 8.8 is instructive. In it each data point plotted is the average of five samples, and the test results are labeled as follows: bambooFill filament (1) and relative printed coupons (2), ABS filament (3) and ABS

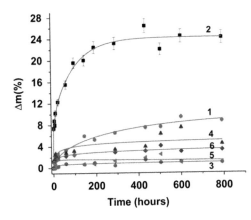

FIGURE 8.8 Water uptake rate expressed as % mass increase of (1) bambooFill filament and (2) relative printed coupons, (3) ABS filament, (4) ABS printed coupons, (5) PLA filament, and (6) PLA printed coupons.

Source: Pop, M. A., Croitoru, C., Bedő, T., et al. 2019. Structural changes during 3D printing of bioderived and synthetic thermoplastic materials. *J. Appl. Polym. Sci.,* 136, 47382. Reproduced with permission from John Wiley and Sons.

printed coupon (4), PLA filament (5) and PLA printed coupon (6). The data points are curve fitted using equation (8.1), where the mass uptake of distilled water is expressed as ratio of water uptake Δm_t at a specific time t over the water uptake at equilibrium Δm_{eq}, n (0.454 for bambooFill filament) is the water uptake rate at which the maximum value is reached at equilibrium, whereas k (0.156 for bambooFill filament) is the diffusion coefficient and affects the shape of the uptake curve:

$$\Delta m_t / \Delta m_{eq} = kt^n \tag{8.1}$$

In Figure 8.8, after 800 hours (about 33 days) of immersion, bambooFill filament reached about 9% weight increase (and apparently not at equilibrium), about eight times more than that of PLA filament, likely due to the hydrophilic nature of the bamboo particles.

Information on the molecular components and structures of bambooFill was acquired from FTIR spectra. The filament spectra displayed the absorption features of a polyester blend, as expected being PLA and PHA both polyesters. The spectra also revealed that: (a) the crystalline phase of PHA increased in the printed sample by about five times vs. the filament, implying an increased order in the PHA phase taken place during the melt-extrusion printing process; (b) crystallinity in PLA declined from filament to printed sample.

8.4 EXPERIMENTAL BAMBOO-FILLED PLA FILAMENTS FOR AM

Depuydt et al. (2018) developed and tested filaments for FFF out of PLA filled with 15 wt% BFs from Bambooder® featuring different length/diameter (*l/d*) ratio. They started with PLA pellets and extruded three filaments: PLA alone, PLA mixed with two plasticizers (cPLA1, cPLA2) from Proviron Industries (Belgium), and PLA mixed with cPLA1, cPLA2, and BFs. The tests results (Table 8.8) were compared among themselves and to eBamboo and bambooFill filaments. cPLA1 and cPLA2 were added in order to increase ductility and prevent the PLA filament from breaking during winding on its spool, and did raise the elongation at break to about four and eight time, respectively, vs. 100% PLA, but severely penalized tensile strength and modulus. Similarly, adding the bamboo fibers cut tensile strength and modulus of PLA down to values similar to those of PLA/plasticizers: depending on the plasticizer and fiber *l/d*, and comparing to PLA, the tensile strength and modulus of bamboo/PLA/plasticizer filaments dropped to 32–42% and 42–65%, respectively. The *l/d* value in the composites was 4–5, not adequate to carry more load than PLA, and additionally fibers acted as defects and introduced stress concentrations: these phenomena reduced strength and strain at break. T_g of the composite filaments was also distinctly reduced, dropping to

TABLE 8.8

Properties of Experimental and Commercial Bamboo (B)/PLA Filaments for FFF. B1–B4 Are Bamboo Fibers with Different Length/Diameter Values, cPLA1, cPLA2 Are Plasticizers

Filament Composition	Commercial Name	Ultimate Tensile Strength	Tensile Modulus	Elongation at Break	T_g
		MPa	GPa	%	°C
PLA/PHA/B	bambooFill	25.8	2.4		59.2
PLA/B	eBamboo	28–29.2	2.36	2.4–7	N/A
PLA	N/A	68.1	4.3	40	61.6
PLA/cPLA1	N/A	30.7	1.3	180	40.1
PLA/cPLA2	N/A	26.2	0.8	320	28.8
B1cPLA1	N/A	23.2	1.6	12	39.8
B2cPLA1	N/A	24.3	1.8	13	39.2
B3cPLA1	N/A	28.4	2.4	8.4	41.9
B3cPLA2	N/A	23.2	2.6	5.5	27.1
B4cPLA1	N/A	25.1	2.8	2.9	37.6
B4cPLA2	N/A	22.4	2.7	4.9	27.6

Source: Depuydt, D., Balthazar, M., Hendrickx, K., et al. 2018. Production and characterization of bamboo and flax fiber reinforced polylactic acid filaments for fused deposition modeling (FDM). *Polym. Compos.* 40:1951–1963. Reproduced with permission from John Wiley and Sons.

an impractical range of 27.1–41.9°C from 61.6°C for PLA. Compared to PLA/plasticizers, BFs greatly improved the tensile modulus and left unchanged the tensile strength.

Shin et al. (2018) formulated and evaluated three versions of a composite PLA filament for FFF, each with bamboo flour from one of the following plants native to Korea: Moso, Timber, and Henon. The investigated unfilled PLA featured: tensile strength and modulus of 6.85 MPa and 4.08 GPa, respectively, and T_m and T_g of 170–230°C and 50°C, respectively. In comparison, the ranges of tensile strength and modulus of most commercial PLA filaments for FFF are 22–71 MPa and 1.8–4 GPa, respectively. Bamboo was crushed using ball milling, and sieved using a 100 mesh screen. Then, the surface of the sieved powder was immersed and stirred in a solution of 3 wt% of (3-aminopropyl)triethoxysilane and 99% ethanol, and dried. The solution acted as a coupling agent between bamboo and PLA. Bamboo powder and pellets were loaded into an extruder, processed at temperature of 120–185°C, and converted into a 1.75 mm diameter filament for FFF. Tensile specimens were fabricated by injection molding the mixture collected from the extruder. The bamboo/PLA mixture ratios (wt%) tested were 10/90, 20/80, and 30/70. All nine filament formulations trailed PLA in tensile strength and modulus, except Timber/PLA 10/90, whose strength was 7.12 MPa. The tensile strength and modulus of the nine filaments were 5.16–7.12 MPa, and 2.09–2.91 MPa, respectively. It was speculated that bamboo/PLA underperformed vs. PLA because: (a) bamboo powder possessed a very low *l/d* that made it ineffective at reinforcing and stiffening; (b) SEM analysis of fracture surfaces of the filaments revealed uniform distribution of bamboo particles but numerous voids in all composites that greatly compromised strength and stiffness. As the content of bamboo powder increases, the color of the filaments became darker, and their surface rougher.

Bamboo fillers must offer good interfacial interaction with a polymer matrix in terms of physical and, chemical absorption, electrostatic interaction, and mechanical connection, if they have to effectively contribute to the mechanical performance of the ensuing composite material. Interfacial adhesion is enhanced by acting on the fiber and adding a coupling agent. The former method includes changing fiber surface morphology, and removing compounds impairing interfacial bonding strength. Coupling agents are chemicals acting like a bridge between filler (or fiber) and matrix: they typically react with the filler surface and possess at least one side group that reacts and bonds with the polymer matrix or at least is compatible with it. Coupling agents are also called *compatibilizers*, although the latter ones are chemicals formulated to promote interfacial adhesion between two polymers that are otherwise immiscible.

Long et al. (2019) aimed at developing a BF/PLA composite filament for FFF with superior performance by enhancing the compatibility between BFs and PLA. The authors applied chemical modifications (based on alkali, silane, and diisocyanates) of BFs to improve their compatibility with PLA, and studied the influence of alkali-treated BFs (ABF) content, compatibilizer (maleated PP [MAPP]) content, and PP/PLA proportion on the mechanical and thermal properties of the subsequent composites. The BFs featured from 60- to 80-mesh and 10–20 aspect ratio. The PLA filament exhibited tensile strength of 49.7 MPa and strain at break of 3.5%. Testing injection molded coupons revealed that adding MAPP up to 5 wt% raised tensile strength and strain at break (Figure 8.9), and flexural modulus and strength (Figure 8.10). The peak tensile strength obtained was 33.7 MPa, below 65.6 MPa of unfilled PLA injection molded. MAPP also boosted notched impact strength from 2.55 k/m^2 (no MAPP) to a maximum of 3.15 kJ/m^2 (5 wt% MAPP). The enhancement of mechanical properties was attributed to the fact that irregular grooves and cracks induced by modifying BF facilitated the infiltration of PLA and MAPP into the fibers.

Alkaline treatment (such as water solution of NaOH) can improve the properties of bamboo fibers and interfacial adhesion between fibers and matrix (Shah et al. 2016).

Kang and Kim (2011) chemically treated the surface of 5 mm long BFs mixed with Ingeo™ 4032D PLA in order to improve the mechanical and thermal properties of the derived composite. They investigated an alkali pretreatment (immersion in a 0.7% NaClO$_2$ solution) to remove lignin (that is less structurally effective than cellulose), followed by immersion in an ethanol/water solution of a silane coupling agent: 3-aminopropyl triethoxysilane (or vinyl trimethoxysilane), although the latter silane contains vinyl functional groups unsuitable for BF treatment. Even though the composite was not formulated for AM, the research findings might apply to an AM version of

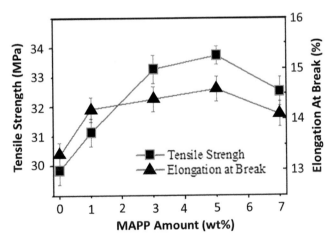

FIGURE 8.9 Effect of MAPP amount on tensile properties of BF/PP/PLA composites.

Source: modified from Long, H., Wu, Z., Dong, Q., et al. 2019. Mechanical and thermal properties of bamboo fiber reinforced polypropylene/polylactic acid composites for 3D printing. *Polym Eng Sci*, 59:E247-E260. Reproduced with permission from Society of Plastics Engineers and John Wiley and Sons.

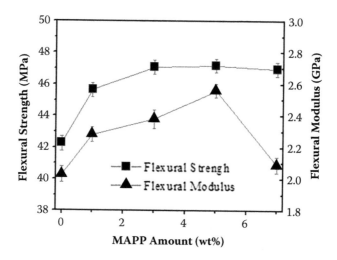

FIGURE 8.10   Effect of MAPP amount on flexural properties of BF/PP/PLA composites.

Source: modified from Long, H., Wu, Z., Dong, et al. 2019. Mechanical and thermal properties of bamboo fiber reinforced polypropylene/polylactic acid composites for 3D printing. *Polym Eng Sci*, 59:E247-E260. Reproduced with permission from Society of Plastics Engineers and John Wiley and Sons.

that composite. The test results are summarized in Table 8.9, with the ranges depending on the silane type and fiber content (10, 20, 30 wt%): compared to unfilled PLA, both silanes increased tensile and flexural moduli in the composite but significantly penalized tensile and flexural strengths.

Xiao et al. (2019) addressed the brittleness of bamboo/PLA for AM by investigating two toughening agents: (a) PBAT combined with ethylene-co-methyl acrylate glycidyl methacrylate acting as an interfacial compatibilizer; (b) a commercial core/shell impact modifier. Filaments for FFF were extruded compounding together bamboo powder, a commercial grade PLA (20/80 wt% bamboo/PLA) and one toughening agent at the time. The PBAT-toughened feedstock displayed higher impact strength and ductility, filament quality, processability, and smoother surface than the core/shell toughened feedstock.

Wang et al. (2014) targeted the poor compatibility between PLA and bamboo flour by adding to them PLA-g-glycidyl methacrylate (GMA) as a compatibilizer. PLA-g-GMA was synthesized by grafting GMA onto PLA. Composites were prepared by blending PLA (85 wt%) and bamboo flour (15 wt%) with concentrations of PLA-g-GMA from 5% to 25% of the total bamboo/PLA amount.

TABLE 8.9

Mechanical Properties of Non-AM Composite Made of Silane-Treated BF/PLA

Material	Tensile Strength	Tensile Modulus	Flexural Strength	Flexural Modulus
	MPa	GPa	MPa	GPa
PLA	53	1.85	18.7	1.50
BF/PLA	12–40	1.80–1.90	7–17.3	1.75–2.85

Source: Data from Kang, J. T., Kim, S. H. 2011. Improvement in the Mechanical Properties of Polylactide and Bamboo Fiber Biocomposites by Fiber Surface Modification. *Macromol. Res.* 19 (8): 789–796.

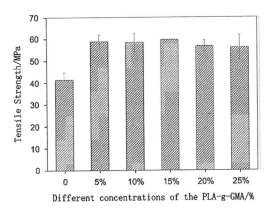 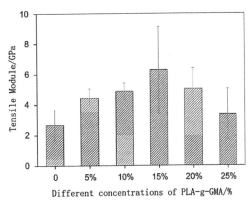

FIGURE 8.11 Properties of bamboo flour/PLA/PLA-g-GMA for AM. The bars indicate standard deviation.

Source: Wang, Y., Weng, Y., Wang, L. 2014. Characterization of interfacial compatibility of polylactic acid and bamboo flour (PLA/BF) in biocomposites. *Polymer Testing* 36:119–125. Reproduced with permission from Elsevier.

PLA-g-GMA was very effective: bamboo/PLA/PLA-g-GMA outperformed bamboo/PLA in tensile strength and modulus (Figure 8.11), impact strength (from 10 kJ/m^2 to a maximum of 14 kJ/m^2), and thermal decomposition temperature (from 295°C to 306°C).

Some insight about the selection of ingredients, composition, and manufacturing process aimed at producing a bamboo powder/PLA filament featuring satisfactory performance was shared by Zhao et al. (2015) who experimented with differing amounts of Ingeo™ 4032D PLA, bamboo powder (0.27 mm size), polyethylene glycol (PEG600) as a lubricant, and a plant-derived ester as a plasticizer. The optimal printing performance was characterized by continuous and smooth flow at intermediate to low pressure, intermediate viscosity, no nozzle clogging, printed articles with fine resolution, and was achieved by the composition comprising 50 g PLA, 20 g bamboo, 1.5 wt% PEG, and 1.5 wt% ester. Printer nozzle was 0.4 mm in diameter, and optimal printing parameters were: temperature 180°C, speed 30 mm/s, and layer thickness 0.20–0.40 mm. The authors dried PLA pellets and bamboo powder before mixing them in a twin-screw extruder at a heating temperature range above the PLA's T_m (155–170°C) and below the bamboo's onset degradation temperature (230°C). Drying the ingredients, and extruding below bamboo's degradation temperature are obviously steps required when mixing bamboo with every polymer, and the latter step restricts the choice of polymer, because polymers melting above bamboo degradation cannot be selected.

A factor to account for in enhancing mechanical properties of BF/PLA composites for AM is the route followed to obtain the added BF. Tokoro et al. (2008) tested injection molded coupons made of PLA combined with BF derived through three alternative processes: (a) short BF bundles consisting of filaments mechanically extracted from bamboo chips by rolling and cutting machines; (b) alkali-treated filaments derived from long BF bundles; (c) steam-exploded filaments. In BF/PLA the highest bending strength was recorded from steam-exploded filaments, but impact strength was not greatly improved by any fillers.

The mechanical performance of composites made of bamboo/sustainable polymers for AM can be improved for the following reasons: the R&D activity conducted in this area has a short history, but promising results have been achieved, and bamboo's high ratio mechanical performance/cost can drive its utilization in load-bearing applications. Possible ways to succeed are reducing voids, increasing the actual *l/d* of BF inside the filament, improving the compatibility between BF and the polymers, and applying any relevant lesson learnt from the R&D work conducted to develop conventional bamboo/polymer composites.

REFERENCES

Amada, S., Untao, S. 2001. Fracture properties of bamboo. *Compos. B: Eng.* 32:451–459.

Awalluddin, D., Ariffin, M. A. M., Osman, M. H. et al. 2017. Mechanical properties of different bamboo species. *MATEC Web of Conferences 138, 01024. EACEF 2017.* doi:10.1051/matecconf/201713801024.

Bamboo Genera List. 2020. *Guadua Bamboo.* https://www.guaduabamboo.com/bamboo-genera (accessed May 31, 2020).

Bambooder®. 2020. *Our Lightweight Products.* https://www.bambooder.nl/products (accessed May 31, 2020).

BambroTex®. 2007. http://www.bambrotex.com/ (accessed May 31, 2020).

Davoudinejad, A., Charalambis, A., Bue Pedersen, D., Tosello, G. 2017. Evaluation of surface roughness and geometrical characteristic of additive manufacturing inserts for precision injection moulding. *33rd Conference of the Polymer Processing Society*, Cancun, Mexico. https://www.researchgate.net/publication/322068906.

Depuydt, D., Balthazar, M., Hendrickx, K. et al. 2018. Production and characterization of bamboo and flax fiber reinforced polylactic acid filaments for fused deposition modeling (FDM). *Polym. Compos.* 40:1951–1963.

eSUN. 2007. *eSUN eBamboo 3D Filament.* http://www.esun3d.net/products/208.html (accessed May 31, 2020).

Gkartzou, E., Koumoulos, E. P., Charitidis, C. A. 2017. Production and 3D printing processing of bio-based thermoplastic filament. *Manufacturing Rev.* 4:1.

Global 3D. 2020. *Bamboo PRI-MAT 3D Filament.* https://global3d.pl/en/filaments/79-bamboo-pri-mat-3d-filament-for-3d-printers.html (accessed May 31, 2020).

Jones, R. M. 1998. *Mechanics of Composite Materials.* Boca Raton: CRC Press.

Kang, J. T., Kim, S. H. 2011. Improvement in the mechanical properties of polylactide and bamboo fiber biocomposites by fiber surface modification. *Macromol. Res.* 19 (8): 789–796.

Khalil, H. P. S. A., Bhat, I. U. H., Jawaid, M. et al. 2012. Bamboo fibre reinforced biocomposites: a review. *Mater. Des.* 42:353–368.

Knight, D., Slaven, L., Vaidya, U. 2017. *Biobased Bamboo Composite Development.* Oak Ridge National Laboratory, Report ORNL/TM-2017/316.

Kumar, N., Mathur, U., Phulwari, B., Choudhary, A. 2017. Bamboo as a construction material. *IJARIIE-ISSN(O)* 3(1):2395–4396. http://ijariie.com/AdminUploadPdf/BAMBOO_AS_A_CONSTRUCTION_MATERIAL_ijariie3659.pdf (accessed May 31, 2021).

Lakshmi Mills. n.d. http://www.lakshmimills.com/yarns/ (accessed May 31, 2020).

Landes, S., Letcher, T. 2020. Mechanical strength of bamboo filled PLA composite material in fused filament fabrication. *J. Compos. Sci.* 4:159.

Liu, D., Song, J., Anderson, D. P., Chang, P. R., Hua, Y. 2012. Bamboo fiber and its reinforced composites: structures and properties. *Cellulose* 19:1449–1480.

Lo, T. Y., Cui, H. Z., Leung, H. C. 2004. The effect of fiber density on strength capacity of bamboo. *Mater. Lett.* 58:2595–2598.

Long, H., Wu, Z., Dong, Q. et al. 2019. Mechanical and thermal properties of bamboo fiber reinforced polypropylene/polylactic acid composites for 3D printing. *Polym. Eng. Sci.* 59: E247–E260.

McClure, F. A. 1981. *Bamboo as a Building Material.* Peace Corps, Washington D.C.: Appropriate Technologies for Development, US Dept. of Agriculture. https://files.eric.ed.gov/fulltext/ED242878.pdf (accessed May 31, 2020).

Mwaikambo. L. Y. 2006. Review of the history, properties, and applications of plant fibres. *Af. J. Sci. Technol. (AJST) Sci. Eng.* 7 (2): 120–133.

Osorio, L., Trujillo, E., Van Vuure, A. W., Verpoest, I. 2011. Morphological aspects and mechanical properties of single bamboo fibers and flexural characterization of bamboo/ epoxy composites. *J. Reinf. Plast. Compos.* 30:396–408.

Pop, M. A., Croitoru, C., Bedő, T. et al. 2019. Structural changes during 3D printing of bioderived and synthetic thermoplastic materials. *J. Appl. Polym. Sci.* 136:47382.

PopBit®. n. d. *Wood Filament for 3D Printer.* https://www.popbit3d.com/wood-filament/3d-wood-filament/wood-filament-for-3d-printer.html (accessed April 26, 2020).

Rao, K. M. M., Rao, K. M. 2007. Extraction and tensile properties of natural fibers: Vakka, date and bamboo. *Compos. Struct.* 77 (3): 288–295.

Rathod, A., Kolhatkar, A. 2013. *Bamboo: An Alternative Source for Production of Textiles.* http://www.woodema.org/conferences/2013_Gdansk_presentations/205_Rathod_Kolhatkar.pdf (accessed March 6, 2021).

Scott, C. 2016. *SHoP Architects and Partners Create Giant 3D Printed Bamboo Pavilion for Design Miami.* https://3dprint.com/154720/shop-architects-3d-printed-bamboo/ (accessed May 31, 2020).

Shah, A., Sultan, M., Jawaid, M. , et al. 2016. A review of the tensile properties of bamboo fiber reinforced polymer composites. Bio Resources 11 (4): 10654–10676.

Sharma, B., Gatóo, A., Bock, M., Ramage, M. 2015. Engineered bamboo for structural applications. *Constr. Building Mater.* 81:66–73.

Shin, Y. J., Yun, H. J., Lee, E. J., Chung, W. Y. 2018. A study on the development of bamboo/PLA biocomposites for 3D printer filament. *J. Korean Wood Sci. Technol.* 46 (1): 107–113.

Tenbro®. 2013. http://www.tenbro.com/ (accessed May 31, 2020).

Thwe, M. M. Liao, K. 2002. Effects of environmental aging on the mechanical properties of bamboo-glass fiber reinforced polymer matrix hybrid composites. Composites - Part A: Applied Science and Manufacturing 43 (1): 33:52.

Tokoro, R., Vu, D. M., Okubo, K. et al. 2008. How to improve mechanical properties of polylactic acid with bamboo fibers. *J. Mater. Sci.* 43:775–787.

Wai, N. N., Nanko, H., Murakami, K. 1985. A morphological study on the behavior of bamboo pulp fibers in the beating process. *Wood Sci Technol* 19:211–222.

Wang, Y., Weng, Y., Wang, L. 2014. Characterization of interfacial compatibility of polylactic acid and bamboo flour (PLA/BF) in biocomposites. *Polym. Testing* 36:119–125.

Xiao, X., Chevali, V., Wang, H. 2019. *Toughening of Polylactide/Bamboo Powder Biocomposite for 3D Printing. ICCM22 2019.* Melbourne, VIC: Engineers Australia, 4927–4934.

Yu, Y., Tian, G., Wang, H., Fei, B., Wang, G. 2011. Mechanical characterization of single bamboo fibers with nanoindentation and microtensile technique. *Holzforschung* 65 (1): 113–119.

Yu, Y., Wang, H. K., Lu, F., Tian, G. L., Lin, J. G. 2014. Bamboo fibers for composite applications: a mechanical and morphological investigation. *J. Mater. Sci.* 49:2559–2566.

Zhao, D., Cai, X., Shou, G. et al. 2015. Study on the preparation of bamboo plastic composite intend for additive manufacturing. *Key Eng. Mater.* ISSN: 1662-9795, 667:250–258.

9 Lignin

I'd put my money on the sun and solar energy. What a source of power! I hope we don't have to wait until oil and coal run out before we tackle that. I wish I had more years left.

Thomas A. Edison

9.1 OVERVIEW OF LIGNIN

Lignin is the most abundant sustainable material after cellulose, being the second constituent (18–35 wt%) of wood (Pettersen 1984), the main sustainable source of the high-performance aromatic polymers (García Calvo-Flores and Dobado 2010), and a CO_2 neutral alternative feedstock to petrochemical polymers for formulating chemicals and plastic (Laurichesse and Luc Avérous 2014). Lignin is located in the cell walls of wood, and provides mechanical strength and rigidity by binding cellulose and hemicelluloses (two major ingredients of wood) within cell walls, and binding cells together. Lignin in wood is usually a pale-yellow substance. The term *lignin* comes from the Latin word *lignum*, meaning wood. Lignin also resides in plant stems, as a critical ingredient of their tissues conducting water and minerals (Sarkanen and Ludwig 1971).

Chemically, lignin is a natural amorphous polymer, with density of 1.3–1.4 g/cm^3, comprising crosslinked phenolic heteropolymers that are connected to each other through several linkages in a 3D network, and form an extremely complex structure (Figure 9.1) that, f.e., in the case of spruce lignin, comprises 16 aromatic units. Commercial lignins are mostly brown powders. Lignin does not melt upon heating, but softens at its T_g, and changes from a glassy solid to a rubbery plastic (Holtzapple 2003). In nature lignin is synthetized from three aromatic monomers (*monolignols*) that constitute its building blocks: p-coumaryl alcohol, coniferyl alcohol, and sinapyl alcohol (Upton and Kasko 2016). The composition of lignin depends on its plant family: grass typically contains all three monolignols, hardwood lignins contain coniferyl and sinapyl alcohol, and softwood lignins contain mostly coniferyl alcohol (Dorrestijn et al. 2000).

Some types of lignin, termed *technical lignins* or *industrial lignins* (Berlin and Balakshin 2014), are industrially produced as the by-product of the wood pulp and paper industry, whose main goal is to isolate the cellulosic fibers present in wood and plants for further processing. The properties of industrial lignin greatly differ from those of lignin found in plants. Four industrial processes are followed to extract lignin from woodUpton 2016: *sulfite*, *kraft*, *soda*, and *organosolv*. The sulfite process extracts lignin from wood chips immersed in aqueous solution of various salts, such as sulfites (SO_3^{2-}) or bisulfites (HSO^{3-}), at about 175°C. It is by far the most common route to produce lignin, generating about 1000 ton/year, and, compared to the kraft method, it produces lower yields and weaker fibers, but it removes a greater percentage of lignin (Holtzapple 2003). In the kraft process the feedstock is added to a mixture of NaOH and Na_2S, heated at 150–180°C (Upton and Kasko 2016). The kraft lignin is mostly utilized as fuel in pulping mills instead of feedstock for producing chemicals and materials. In the soda process the wood fiber feedstock is heated at 140–170°C in a pressurized reactor in presence of an aqueous solution of 13–16% of NaOH, where lignin gets separated from cellulose (Doherty and Rainey 2006). The soda process is usually applied to non-wood-based sources, such as flax and sugar cane (Upton and Kasko 2016). The organosolv process consists in immersing the feedstock in an aqueous organic solvent, such as acetone, at 140–220°C, simultaneously isolating individual streams of hemicellulose, cellulose,

DOI: 10.1201/9781003221210-9

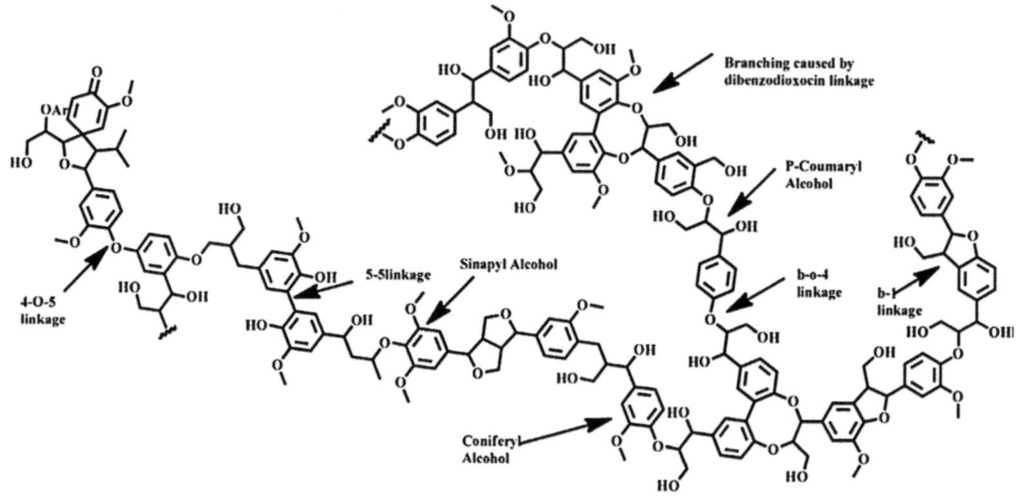

FIGURE 9.1 Detailed chemical structure of lignin. Each hexagon is an aromatic unit.

Source: Kumar, A., Anushree, Kumar, A. J., Bhaskar, T. 2020. Utilization of lignin: A sustainable and eco-friendly approach. *Journal of the Energy Institute* 93 (1): 235–271. Reproduced with permission from Elsevier.

SODA

KRAFT

LIGNOSULFONATE

ORGANOSOLV

FIGURE 9.2 Pictures of typical technical lignins and their relative extraction process.

Source: Bruijnincx, P., Weckhuysen, B., Gruter, G.-J., Westenbroek, A., Engelen-Smeets, E. 2016. Lignin valorization. Utrecht University, Platform Agro-Paper-Chemistry. https://www.worldcat.org/title/lignin-valorisation-the-importance-of-a-full-value-chain-approach/oclc/951662206 (accessed January 16, 2021). Reproduced with kind permission from Prof. P. Bruijnincx.

and lignin. Figure 9.2 shows pictures of typical technical lignins produced through the previous processes. Figure 9.3 is a SEM micrograph of kraft lignin powder sold by Sigma Aldrich.

Table 9.1 lists technical lignins and their extraction process, feedstocks, and features. The source (hardwood, softwood, grass) and method of extraction of lignin affect its structure (Lora and Glasser 2002), MW, amount of each functional groups, and T_g, as illustrated in Table 9.2. Particularly, depending on the tree source, T_g of lignin ranges from 89.9°C in aspen to 141°C in tulip poplar. In turn, MW, share of functional groups, and T_g of lignin determine its general physicochemical properties (Lora and Glasser 2002; He et al. 2013) and applications. Lignin's T_g

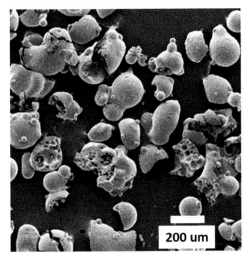

FIGURE 9.3 SEM micrographs of Sigma Aldrich kraft lignin powder.

Source: Modified from Rao, N. R., Rao, T. V., Reddy, S. V. S. R., Rao, B. S. 2015. The effect of gamma irradiation on physical, thermal and antioxidant properties of kraft lignin. *Journal of Radiation Research and Applied Sciences* 8 (4): 621–629. Reproduced under the CC BY-NC-ND license (http://creativecommons.org/licenses/by-nc-nd/4.0/).

TABLE 9.1
Typical Technical Lignins and Their Extraction Process, Feedstocks, and Features

Lignin	Extraction Process	Feedstocks	Features
Lignosulfonates	Sulfite	Hardwoods, softwoods	Water-soluble. High density of functional groups. Used as stabilizers, dispersing agents, surfactants and adhesives. Relatively high content of ash and carbohydrates (low purity). MW 15,000-50,000 mol/g. T_g 130°C. It contains sulfur. Price 180-500 USD/ton.
Kraft lignin	Kraft	Hardwoods, softwoods	Highly condensed. Moderately pure, with impurities in form of sulfur species. MW 1,000-3,000 mol/g. T_g 140-150°C. Price 260-500 USD/ton.
Soda lignin	Soda	Annual crops, hardwoods	Sulfur-free. At best moderately pure, with small amounts of ash, carbohydrates and possibly vinyl ester. MW 800-3,000 mol/g. Tg 140°C. Price 200-300 USD/ton.
Organosolv lignin	Organosolv	Hardwoods, softwoods, annual crops	Sulfur-free. Highly pure. MW 500-5,000 mol/g. T_g 90-110°C. Price 280-520 USD/ton.

Source: Bruijnincx, P., Weckhuysen, B., Gruter, G.-J., Westenbroek, A., Engelen-Smeets, E. 2016. Lignin valorization. Utrecht University, Platform Agro-Paper-Chemistry. https://www.worldcat.org/title/lignin-valorisation-the-importance-of-a-full-value-chain-approach/oclc/951662206. Ebers, L.-S., Arya, A., Bowland, C. C., et al. 2021. 3D printing of lignin: Challenges, opportunities and roads onward. Biopolymers 112 (6): e23431

also depends on its MW, and it spans from about 127°C in low-MW (4,300 mol/g) lignin to 193°C in high-MW (85,000 mol/g) lignin (Holtzapple 2003).

The book chapter by Stark et al. (2016) illustrates techniques for characterizing lignin, and its properties, and includes MW, chemical structure, thermal and mechanical properties. The DMA plot of kraft lignin from ponderosa pine in Figure 9.4 quantifies the variation of storage modulus (stiffness) and tanδ (the energy dissipated by the material when stressed or strained) between −30°C and 250°C. These DMA data are important because they were collected on lignin alone, whereas typically DMA is performed on lignin blended with a polymeric matrix. However, in the

TABLE 9.2

Number-Average Molecular Weight (MWn) and Functional Groups of Technical Lignins

Lignin Type	MW$_n$	COOH	OH Phenolic	O–CH$_3$
	g/mol	%	%	%
Soda (bagasse)	2,160	13.6	5.1	10.
Soda (wheat straw)	1,700	7.2	2.6	16
Kraft (softwood)	3,000	4.1	2.6	14
Organosolv (hardwood)	800	3.6	3.7	19
Organosolv (bagasse)	2,000	7.7	3.4	15

Source: Doherty, W. O. S., Mousavioun, P., Fellows, C. M. 2011. Value-adding to cellulosic ethanol: Lignin polymers, *Industrial Crops and Products* 33 (2): 259–276. Reproduced with permission from Elsevier.

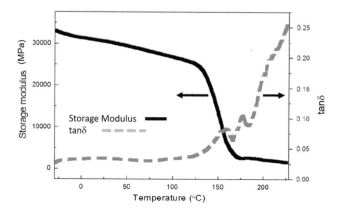

FIGURE 9.4　DMA plot of powdered kraft lignin.

Source: Modified from Stark, N. M., Yelle, D. J., Agarwal, U. P. 2016. Techniques for characterizing lignin. In *Lignin in Polymer Composites*, ed. O. Faruk, M. Sain, 49–66. Oxford: Elsevier. Reproduced with permission from Elsevier Books.

case of Figure 9.4, because DMA was carried out on powder instead of solid coupons, the calculation of storage modulus was not very accurate, because it did not account for the unknown surface area of the powdered coupon. The storage modulus decreased steadily and slowly from about 30 GPa at room temperature to about 25 GPa at 125°C, where it dropped at much higher rate down to 2.5 GPa at about 170°C. This is a high value for polymers at 170°C, considering that widely consumed polymers such as PC and PS offer no stiffness at the same temperature (Dunson n.d.). The value of 2.5 GPa at about 170°C also exceeds the storage modulus of temperature-resistant polymers PET (Dunson n.d.) and PEEK (de Jesus Silva et al. 2016), displaying at 175°C a storage modulus of 0.25 GPa and 0.62 GPa, respectively. The tanδ curve in Figure 9.4 shows transitions at around 158°C and 177°C. The DMA test was conducted in single cantilever mode, at 0.03% strain and 1 Hz frequency, employing a specific test fixture that consisted in a stainless tray in which the powder was capped by an upper tray that kept the powder inside. Figure 9.5 displays the TMA plot collected on kraft lignin and acetylated (reacted with acetic acid) kraft lignin, and illustrates the lignin's thermal degradation manifested by its weight loss.

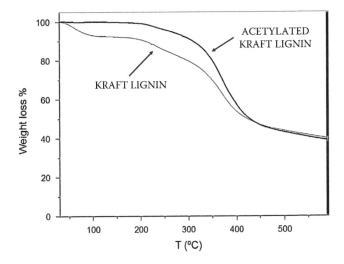

FIGURE 9.5   TMA curve of kraft lignin and acetylated kraft lignin.

Source: Gordobil, O., Delucis, R., Egüés, I., Labidi, J. 2015. Kraft lignin as filler in PLA to improve ductility and thermal properties. *Industrial Crops and Products* 72:46–53. Reproduced with permission from Elsevier.

9.2   MARKET AND APPLICATIONS OF LIGNIN

The global production of lignin was about 100 million tonnes/year valued at USD 732.7 million in 2015 (Bajwa et al. 2019). Lignin is produced by producers of chemical pulp (Graça et al. 2012; Rozite et al. 2013), and 98% of it is burned to generate energy, and only 2% is used for other purposes (Yang et al. 2020). The revenues of worldwide lignin market in 2019 exceeded USD 730 million (Global Market Insights 2020a) or USD 950 million (Grand View Research 2020) depending on the data source, and are forecast to surpass USD 1 billion by 2026 (Global Market Insights 2020b). Lignosulphonates constitute the largest product segment, accounting for over 88% of the global share in 2015, followed by kraft lignin, organosolv lignin, and others (Grand View Research 2020). The production of lignosulphonates was reported in 2014 to be around 1.2 million tons per year (Berlin and Balakshin 2014). Key industry players include Borregaard LignoTech (Norway), Tembec (France), Domtar and MeadWestvaco (both in USA), GreenValue (Switzerland), West Fraser (Canada), Domsjö Fabriker (Sweden), and Stora Enso (Finland). In 2018, Storaenso launched a new biobased lignin called Lineo™ capable to replace fossil-based phenolic resins. The price of lignin is reported in Table 9.1. Bruijnincx 2016 In comparison, laboratory quantities of lignin are far pricier. F.e., 500 g of ligninis are sold for USD 201 by Millipore Sigma. Sigma Aldrich 2020.

Lignin is not only fully sustainable and biodegradable, but also inexpensive, thermally stable, hydrophobic, antioxidant, flame retardant, antimicrobial, and resistant to weather, temperature, and chemicals. Moreover, the presence of the aromatic ring confers good structural properties. Due to the above features, its high content of carbon, reactive functional groups, high production volumes, and good rheological and viscoelastic properties, lignin has a wide range of applications, including binders in animal feed, soil stabilizer in agriculture, liquid fertilizers, micronutrient carriers, concrete mixtures, leather tanning, ceramics, gypsum-boards, pigments, dyes, low-cost precursor for carbon fibers, and so on (García Calvo-Flores and Dobado 2010). Lignin also appears in feedstocks for AM.

A most sound environmental use of lignin is as a feedstock to formulate polymers (Saito et al. 2012) possessing high thermal stability and good flow resistance (Holmberg et al. 2016a, 2016b), adhesives (Wang et al. 1992; Danielson and Simonson 1998a, 1998b; Bonini et al. 2005; Stewart 2008; Zhang et al. 2013; Ghaffar and Fan 2014), chemicals, and carbon fibers (Kadla et al. 2002a,

2002b). Upton and Kasko (2016) published a recent and wide-ranging review on methods to synthetize polymers derived from lignin, monolignols and lignin-derived chemicals, and argued that lignin can play a major role in reducing the consumption of fossil feedstocks. In fact, through degradation methods and chemical processing, lignin has been converted into vanillin, dimethyl sulfide, and dimethyl sulfoxide, and has served as a macromonomer for the synthesis of PUs (Mahmood et al. 2016), PETs, epoxide resins, hydrogels, vinyl-based graft copolymers, and phenolic resins. Finally, lignin is often chemically modified (*functionalized*) to decrease its brittleness, increase solubility in organic solvents, improve processability and reactivity. Lignin is also being studied for advanced materials and applications, such as nanomaterials (Duval and Lawoko 2014; Tran et al. 2016; Roopan 2017), biomedicine (Figueiredo et al. 2018), and biotechnologies (Roopan 2017).

The near future of lignin depends, among other factors, on (García Calvo-Flores and Dobado 2010; Upton and Kasko 2016): (a) enhancing the technology for separating lignin from hemicellulose and cellulose more cleanly, efficiently, and economically, in order to produce a raw feedstock more homogeneous and with more consistent properties; (b) establishing methods to chemically functionalize lignin into products without expensive reagents and/or complicated synthetic routes; (c) developing new depolymerization and upgrading techniques.

9.3   COMMERCIAL LIGNIN AM FILAMENTS

TwoBEears (Germany) sells bioFila®, a filament for FFF containing lignin, and available in two forms: Silk and Linen. bioFila® Silk is a blend of biopolymers with a silky look, whereas bioFila® Linen features a line texture and a natural look (Aderholt 2014). Table 9.3 is released by twoBEars, and lists physical and mechanical properties of both filaments: compared to unspecified PLA and ABS, the bioFila® filaments have similar yield tensile stress, lower modulus, and far higher impact strength. However, as a further comparison, the range of tensile (ultimate not yield) strength and modulus of the commercial PLA filaments described in Chapter 2 is 22–71 MPa and 1.8–4 GPa, respectively.

9.4   EXPERIMENTAL LIGNIN BLENDS

This section deals with blends of lignin and selected sustainable and non-sustainable polymers that are not specific for AM but have been selected because they are also available as grades for commercial AM materials (Tables 1.2 and 1.3).

Blending lignin with polymers requires miscibility between the components and complementary functionality, that is functionality enhancing the qualities of each component. However, that is not straightforward, because lignin molecules are polar, and strongly interact with each other. Lignin

TABLE 9.3

Properties of BioFila® Silk and BioFila® Linen Filaments, PLA, and ABS

Material	Density	Tensile Stress at Yield	Tensile Modulus	Charpy Impact Test	Softening Temperature	T_m
	°C	MPa	GPa	kJ/m^2	°C	°C
Silk	1.25	53	2.50	60	57	165
Linen	1.40	51	2.70	58	58	153
PLA	1.28	55	5.10	8	60	180
ABS	1.07	48	2.2	55	60	180

Source: Data posted online by the suppliers.

functionality is improved by attaching to lignin reactive functional groups, such as acrylates, amines, epoxies, phenols, vinyl, etc. In order to form blends of lignin and polymers that are not completely miscible, lignin is modified chemically or by plasticization to improve its dispersion in the polymer, or a compatibilizer is added to increase lignin/polymer interfacial adhesion (Kun and Pukánszky 2017).

Various formulations of lignin with and without chemical modification are blended with polymers (Gordobil et al. 2014; Naseem et al. 2016; Kun and Pukánszky 2017). Lignin can also react with other polymers and produce copolymers, thanks to its compatibility with numerous chemical reactions (García Calvo-Flores and Dobado 2010).

Li et al. (2003) investigated the mechanical and thermal behavior of fully sustainable and biodegradable blends of PLLA and lignin manually mixed. Test results disclosed the existence of intermolecular hydrogen-bonding interaction between PLLA and lignin. The authors added 10-40 wt% of lignin which diminished tensile strength, modulus, and strain at break that reached their highest values at 10 wt% lignin (46.8 MPa, 1,888 MPa, and 3.3%, respectively), and all steadily decreased upon increasing lignin content. In comparison, tensile strength, modulus, and strain at break for PLLA were 68 MPa, 1,894 MPa, and 5.1%, respectively. Furthermore, lignin would accelerate the thermal degradation of PLLA when the content of lignin surpassed 20 wt%.

PHB is the simplest compound of PHA, a family of sustainable polyesters obtained by bacterial fermentation (Section 1.13) and commercialized as a FFF filament by colorFabb. Ghosh et al. (2000) reported that adding up to 20 wt% lignin to PHB decreased tensile strength, modulus, and strain at break, greatly retarded crystallization, and lowered T_m.

Kadla and Kubo (2004) studied blends of lignin and synthetic polymers, and reported that lignin was miscible with PEO and PET, but not with PVA and PP, all non-sustainable polymers. Blending lignin with isocyanate prepolymer increased T_g, water resistance, tensile strength, elastic modulus (over 35 times), and lowered elongation to break (Chauhan et al. 2014). Adding lignin to isotactic PP (Canetti et al. 2006) and biobased polymer PHB (Bertini et al. 2012) raised the thermal resistance of both blends. Gordobil et al. (2015) reviewed the technical literature on composites containing lignin and the following TPs: ABS (Akato et al. 2015), PP (Rozman et al. 2000; Toriz et al. 2002), PET, PVC, PS, LDPE, PU (Saito et al. 2013), and linear LDPE. Lignin has also been combined with TS polymers, such as epoxy resins (Wood et al. 2011). Doherty 2011

Pucciariello et al. (2004) blended lignin from straw with HDPE and PS, but lignin greatly lowered maximum tensile stress (values in MPa): in HDPE from 34.5 to 18.8 (20 wt% lignin), in Aldrich PS from 29.6 to 12 (10 %wt lignin), and in Dow PS from 28.8 to 8.5 (20 wt% lignin). However, the upside was that lignin acted as a stabilizer against the UV radiation for PS.

A broad review of lignin application is authored by Doherty et al. (2011).

9.5   EXPERIMENTAL LIGNIN-FILLED FEEDSTOCKS FOR AM

Lignin is being studied as a suitable filler in polymers for AM, because it is abundant, inexpensive, sustainable, biodegradable, nontoxic, possesses thermal stability and good mechanical properties at and above room temperature. It has however some drawbacks: its properties depend on the identity of the plant source and method of isolation and hence they lack consistency, it is brittle, the stress transfer at matrix-lignin interphase is poor, and high content of lignin mixed in polymers leads to agglomeration and increases the composite's brittleness and melt viscosity during extrusion.

The effects of lignin on properties of polymeric matrices with which it was blended to form filaments for FFF, and the properties and performance of those filaments were elucidated by Nguyen et al. (2018a, 2018b) and Gkartzou et al. (2017).

Gkartzou et al. (2017) selected Ingeo™ 2003D as PLA and Indulin AT, both without chemical treatment, and a purified kraft pine lignin. They formulated and extruded a lignin/PLA filament for FFF, and experimentally measured the effect of lignin content and printing parameters on the tensile and thermal properties of coupons printed from the filament. Four lignin/PLA blends, with content of unmodified kraft lignin of 5, 10, 15, and 20 wt% were studied, and, compared to unfilled PLA, lignin

TABLE 9.4

Properties of Lignin, ABS, NBR41, and Their Composites for FFF

Material	Tensile Modulus	Ultimate Tensile Strength	Elongation at Break	T_g	Heat Capacity	Temperature T(5)[a]
	GPa	MPa	%	°C	J/g°C	°C
Lignin	N/A	N/A	N/A	86.2	0.490	252
ABS	1.91	54.09	8.30	104.1	0.281	336
NBR41	N/A	N/A	N/A	−12.8	0.548	383
AL64	1.82	20.50	1.21	95.8	0.205	305
ANL514	1.41	39.79	3.48	90.5	0.201	314
ANL613	1.19	31.16	7.19	93.1	0.211	323
ANL712	1.36	35.59	10.52	94.6	0.246	324
ANLC4141	2.44	50.76	2.84	90.1	0.175	311
ANLC5131	2.64	64.68	3.85	90.9	0.279	318
ANLC6121	2.31	53.70	3.89	92.3	0.218	329

Source: Nguyen, N. A., Bowland, C. C., Naskar, A. K. 2018. Mechanical, thermal, morphological, and rheological characteristics of high performance 3D-printing lignin-based composites for additive manufacturing applications. *Data in Brief* 19:936–950. Reproduced with permission from Elsevier.

Note:

[a]Temperature corresponding to 5 wt% loss.

Legend: Materials: A=ABS, L=lignin, N=NBR41, C=carbon fibers. Numbers indicate the weight percentage of each component divided by 10, and are reported in the same order as the components.

decreased the tensile strength (from 55.9 MPa to 41.3 MPa) and strain at break (from 4.6% to 1.9%), and left about unchanged the tensile modulus (from 2.31 GPa to 2.41 GPa), T_g and T_m of the filaments and printed coupons. The higher the lignin content, the greater the reduction of strength and strain. All blends formed heterogeneous systems, due to the poor compatibility between PLA matrix and un-modified lignin, which had been previously reported in technical literature. Lignin formed aggregates of large size. Interface separation and sliding between aggregates and PLA occurred, and revealed very weak secondary interactions between lignin and PLA.

Nguyen et al. (2018a, 2018b) developed a method to combine lignin, a TP matrix, and other ingredients, in order to prepare a filament for FFF composed of partially renewable material featuring tensile and bending properties higher than those of the matrix, with strong interlayer adhesion in the printed objects, and, contrary to wood-based materials, suited for the high-speed melt extrusion experienced during FFF. The materials involved were hardwood lignin, acrylonitrile butadiene rubber (NBR41) with 41 mol% nitrile content, ABS, and unsized polyacrylonitrile (PAN) 1/8-inch carbon fibers (CFs). The presence of CFs enhanced the tensile and bending properties of the resulting composite, and bridged adjacent deposited layers, improving the interlayer adhesion. Combining lignin and nitrile rubber had been already studied by Bova et al. (2016). The tensile and thermal test results are grouped in Table 9.4 and Figure 9.6. Adding lignin alone to ABS decreased slightly modulus, more than halved strength, and cut elongation at break to 1/7, compared to unfilled ABS; in other words, lignin resulted in a far weaker and more brittle blend. The addition of rubber to ABS and lignin raised strength and deformability but decreased stiffness, as expected, due to NBR41's low tensile modulus. The blends that outperformed ABS and all other materials in tensile strength and modulus were the ABS/NBR41/lignin/CF blends, since PAN CFs have tensile modulus, strength, and strain at failure equal to 227–392 GPa (Singer 1989), 2.42–5.03 GPa, and strain at a failure of 0.6–1.95%, respectively. The blend ANLC5131, with a composition ABS/NBR41/lignin/CF equal to 50/10/30/10 wt%, reached the highest tensile

TABLE 9.5

Composition (wt%) of Experimental Composites with Best Performance for FFF

Lignin	Cellulose	Water	Acetone
46.52	4.65	30.23	18.60
43.96	5.49	21.98	28.57
32	8.00	36	24
58.26	4.85	9.71	27.18
61.90	4.84	15.47	17.79

Source: Data from Liebrand, T. R. H. 2018. 3D printed fiber reinforced lignin. M.Sc. thesis diss., Delft University of Technology.

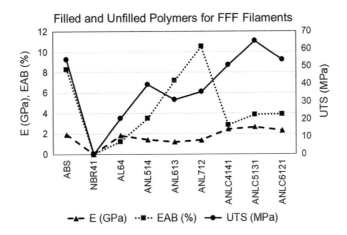

FIGURE 9.6 Tensile strength (UTS), modulus (E) and elongation at break (EAB) of ingredients and relative composites for FFF filaments. Materials: A=ABS, L=lignin, N=NBR41, C=carbon fibers. Numbers indicate the weight percentage of each component divided by 10, and are reported in the same order as the components.

Source: Data from Nguyen, N. A., Bowland, C. C., Naskar, A. K. 2018. Mechanical, thermal, morphological, and rheological characteristics of high performance 3D-printing lignin-based composites for additive manufacturing applications. *Applied Materials Today* 12:138–152.

modulus and strength. Adding lignin to ABS reduced T_g, being T_g of lignin almost 20°C lower than that of ABS, and the more lignin, the lower the blend's T_g. The improvement in tensile modulus with CFs was confirmed by DMA test data: the bending storage modulus of ABS, ABS/NBR41/lignin-514 (50/10/40 wt%), and ABS/NBR41/lignin/CF-4141 (40/10/40/10 wt%) were 1.72 GPa, 1.97 GPa, and 3.90 GPa, respectively. The thermal stability of the materials was measured via TGA, and revealed that all materials tested lost only about 5% of their weight up to about 300°C, and around 400°C suffered a significant weight loss of 50–90% of their initial weight at a very rapid rate. Lignin represented an exception and featured a weight loss of approximately 10% at 300°C, and a more gradual weight reduction than that of all the other materials. Table 9.4 lists the values of T(5) that is the temperature corresponding to 5 wt% loss, and in ABS/NBR41/lignin/CF blends obviously increasing the content of ABS raised T(5). Heat capacity was included in Table 9.4 because it quantifies the amount of energy necessary to melt and extrude the polymer as a

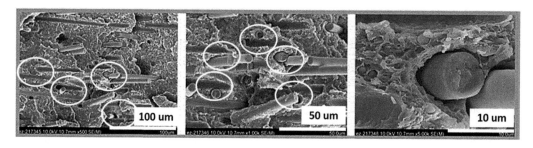

FIGURE 9.7   SEM micrographs of ABS/NBR41/lignin/carbon fibers-5131 (50/10/30/10). The white ovals indicate the breaking of physical crosslinks between carbon fibers and matrix (fiber percolation).

Source: Modified from Nguyen, N. A., Bowland, C. C., Naskar, A. K. 2018. Mechanical, thermal, morphological, and rheological characteristics of high performance 3D-printing lignin-based composites for additive manufacturing applications. *Data in Brief* 19:936–950. Reproduced with permission from Elsevier.

filament, and it is related to the material's processability as a feedstock for FFF. Tear testing was carried out on ABS, ABS/NBR41/lignin-514 (50/10/40 wt%), and ABS/NBR41/lignin/CF-4141 to measure the energy required to tear the weld line between two adjacent deposited layers, and quantify the interlayer mechanical adhesion. The tearing force for the CF composite surpassed 4000 N/mm, more than twice that for the other two materials tested. Figure 9.7 includes SEM photographs of ABS/NBR41/lignin/carbon fibers-5131: CFs acted as a rigid physical network, stiffening and reinforcing the whole composite material, but they were also not always fully bonded to the matrix, as in the locations indicated by the white ovals.

Liebrand (2018) formulated 20 experimental compositions of a lignin/cellulose/acetone/water feedstock for FFF, and, rather than focusing on physical and mechanical properties, qualitatively assessed their processability in terms of homogeneity, viscosity, and interlayer adhesion. Lignin and cellulose were extracted via the kraft process from softwood, and were respectively Indulin AT powder, and bleached Skogcell 90 Z fibers sifted through a 10-mesh sieve. All ingredients were mixed together, and, after liquid evaporation, test-extruded. The compositions providing the best results overall are listed in Table 9.5, with 32–61 wt% lignin content in them.

Vaidya et al. (2019) blended unmodified lignin from biorefinery and PHB, and melt extruded them into composite filaments, ultimately converted into films subjected to chemical, microscopic, and thermal characterization. Test results pointed out that the filament was a particulate composite with a PHB-rich surface and lignin particles at the filament's core. Lignin did not noticeably affect the decomposition, T_m, and T_c of PHB. On the other hand, lignin/PHB composite manifested a shear thinning profile which improved layer adhesion during printing. In fact, 20 wt% of lignin changed the composite's melt viscosity to a value favorable to printing. Finally, a printing test was run on lignin/PHB (20/80 wt%) to assess printing behavior and quality of printed items: lignin/PHB featured better quality, and 34–78% less warpage vs. unfilled PHB.

Dominguez-Robles et al. (2019) proved that the multiple benefits of lignin can translate into multifunctional articles, such as mesh for wound dressing combining mechanical protection, antioxidant and antimicrobial properties. A new filament for FFF was devised via hot melt extrusion from PLA pellets coated with lignin powder, and biobased castor oil. The lignin was BioPiva 100, a softwood kraft type, and its amount in PLA was 0.5–3 wt%. PLA was Ingeo™ PLA 3D850, possessing processability and printability specifically developed for FFF. First, the PLA pellets were coated with castor oil, and subsequently lignin powder was added and fed to the 3devo Next 1.0 filament extruder, running at 170–190°C and 5 rpm speed, and producing a 2.85 mm diameter filament. Lignin lowered the filament breaking load but increased its wettability. Particularly, all formulations of lignin/PLA filament trailed PLA in the breaking load and attained a maximum value of 104 N at 3 wt% lignin, while PLA filament broke at 114 N. Since a targeted application of lignin/PLA was health care, and particularly wound dressing, the following characteristics of the composite filament were assessed:

- Stability in the human body: it was quantified by immersing printed discs in a phosphate-buffered saline (PBS) solution with pH 7.4 simulating the human body environment: the discs immersed for 30 days at 37°C did not lose weight.
- Antioxidant activity performed by scavenging radicals present in lignin, which is critical in would healing: 2,2-diphenyl-1-picrylhydrazyl, a free radical damaging the living cells, was added to lignin/PLA, and in 5 h its concentration was reduced to 20% of its initial value.
- Antimicrobial properties: curcumin, a natural anti-inflammatory and wound healing compound, was applied on the surface of meshes printed out of lignin/PLA that released it outward into the PBS solution at a rate increasing with the size of the mesh holes.

Wasti et al. (2021) assessed the effect of two plasticizers on lignin/PLA filaments for FFF, namely PEG 2000 (a high-quality research grade PEG) and Struktol® TR451, added separately and in varying concentrations (0.25-5 wt%) to a 80/20 wt% lignin/PLA blend. PLA and lignin were Ingeo™ 2003D and Attis Innovations hardwood organosolv type, respectively. Figure 9.8 illustrates the tensile test results on filaments out of PLA, and lignin/PLA with and without plasticizers. The naming convention was: L for lignin, P for PEG, and S for Struktol®, and the numbers after L, P, and S indicated wt%. As shown in Table 9.6, the composite material formulations trailed 2003D in strength and strain at maximum load, but PLA_L20 outperformed PLA in modulus. The plasticizers in the formulations in Figure 9.8 improved the strength vs. PLA_L20, with the highest value (51 MPa) reached by PLA_L20_P2 but still below PLA's strength. A 2 wt% of PEG

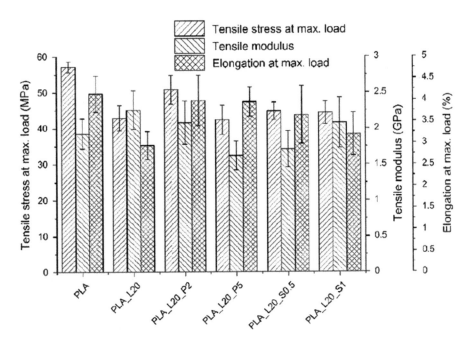

FIGURE 9.8 Tensile properties of experimental FFF filaments made of PLA, lignin/PLA, and lignin/PLA/ plasticizers denoted by P and S. L stands for lignin.

Source: Wasti, S., Triggs, E., Farag, R., et al. 2021. Influence of plasticizers on thermal and mechanical properties of biocomposite filaments made from lignin and polylactic acid for 3D printing. *Composites Part B* 205, 108483. Reproduced with permission from Elsevier.

TABLE 9.6

Tensile Properties of FFF Filaments Made of PLA, Lignin/PLA, and Lignin/PLA/Plasticizers

Materials	Strength	Modulus	Strain at Maximum Load
	MPa	GPa	%
PLA	57	1.9	4.1
PLA_L20	42.5	2.25	2.9
Maximum value in composite, and composite name code in parentheses	52 (PLA_L20_P2)	2.25 (PLA_L20)	4 (PLA_L20_P2)
Minimum value in composites, and composite name code in parentheses	42 (PLA_L20_P5)	1.6 (PLA_L20_P5)	3 (PLA_L20)

Source: Wasti, S., Triggs, E., Farag, R., et al. 2021. Influence of plasticizers on thermal and mechanical properties of biocomposite filaments made from lignin and polylactic acid for 3D printing. *Composites Part B* 205, 108483. Reproduced with permission from Elsevier.

Legend: L for lignin, P for PEG, and S for Struktol®, and numbers after L and P indicate wt%.

improved tensile strength and elongation at maximum load of PLA_L20 by 19% and 35%, respectively, while Struktol® enhanced elongation at a maximum load by 24%.

PS is a fossil-based polymer, and a feedstocks for some commercial filaments, such as high impact PS (HIPS) sold by Matter Hackers, and iMakr, and CFs/expanded PS (Table 1.2). Wasti et al. (2020) formulated FFF filaments out of HIPS filled with 5, 10, 15 and 20 wt% of lignin. Up to 10 wt% of lignin resulted in mechanical property values similar to those of unfilled PLA, but tensile strength and elongation at break were degraded in presence of 15 wt% and 20 wt% of lignin, respectively.

Employing lignin for AM is not limited to FFF. Ebers et al. (2021) reported experimental feedstocks containing lignin for VP, and also briefly described lignin types and their characteristics and producers.

Zhang et al. (2019) incorporated different amounts (0.2, 0.4, 0.5, 0.8, 1.0 wt%) of softwood kraft lignin into a commercial methacrylate resin (Formlabs RS-F2-GPCL-04), and successfully formulated composites for SL with improved tensile properties vs. those of the unfilled resin: namely, lignin raised the tensile strength by 46–64% and modulus by 14–37% in the postcured printed composites, depending on lignin amount. SEM analysis of the tested composites revealed a fracture surface that was rougher than that of the samples without lignin, and probably dissipated the stress better than a smoother surface, and hence might have enhanced mechanical properties. The authors also discovered that adding unmodified lignin hindered the curing process during printing: in fact, the addition of 1 wt% lignin to the resin resulted in curing less than half of the resin and lowering tensile strength and modulus.

Ibrahim et al. (2019) combined an urethane acrylate-based resin (PU) (Wanhao Precision Casting, China) for digital light processing (DLP), unmodified organosolv lignin extracted from oil palm empty fruit bunch fibers (procured from Szetech Engineering, Malaysia) as a filler, and graphene nanoplatelets (GNP) by Sigma Aldrich (Germany) as a reinforcement. Via DLP four set of coupons were printed: (a) PU, (b) PU and 0.2, 0.6, 3 wt% of lignin, (c) PU-GNP, and (d) PU, plus 0.2, 0.6, 3 wt% of lignin, and GNP. The extracted lignin contained syringyl, guaiacyl, and hydroxyl molecules that are highly compatible with the photo-curable PU. Moreover, organosolv lignin acted as a compatibilizer for GNP. The hardness (measured via nanoindentation), tensile strength and modulus of the PU printed coupons were (all in MPa) 27.3, 21.2, and 9.77, respectively. The formulation with the largest property improvement was PU/0.6 lignin/GNP featuring 238%, 29%, and 30% increase in hardness, tensile strength and modulus, respectively. The

TABLE 9.7

Tensile Properties of Unfilled (PR48) and Lignin-Filled Resins (LR) for SL

Resin	Lignin Content	Tensile Strength	Tensile Modulus	Elongation at Break
	wt%	MPa	GPa	%
PR48	0	11.0	0.65	1.87
LR5	5	11	0.64	3.0
LR10	10	18	0.48	6
LR15	15	15	0.37	7.6

Source: Sutton, J. T., Rajan, K., Harper, D. P., Chmely, S. C. 2018. Lignin-Containing Photoactive Resins for 3D Printing by Stereolithography, *ACS Applied Materials & Interfaces* 10 (42): 36456–36463. Reprinted with permission from American Chemical Society.

formulation with the second largest property improvement was PU/0.6 lignin displaying 43%, 16%, and 19% raise in hardness, tensile strength and modulus, respectively. In PU/0.6 lignin/GNP the GNP were uniformly distributed thanks to lignin. There are synergistic interactions in lignin-GNP, lignin-PU, and graphene-PU, and lignin-graphene. Particularly, lignin-GNP in PU facilitated distribution, significantly, increasing the mechanical properties of the printed composites.

Sutton et al. (2018) proved that modified lignin could be incorporated in resins for VP in larger amounts than unmodified lignin. They took organosolv lignin modified through acylation with methacrylic anhydride, and added 5, 10, and 15 wt% of it to commercial acrylate resins, and prepared new photoactive feedstocks for SL labelled LR5, LR10, and LR15, from lignin/resin and the wt% amount of legnin. Lignin had 86% purity, 2973 mol/g MW, and was extracted from pulp-grade wood chips of hybrid poplar. The acrylate resins were ethoxylated pentaerythritol tetra-acrylate (Sartomer SR494), and aliphatic urethane acrylate (Allnex Ebecryl 8210). A resin photoinitiator and a reactive diluent were also added. The commercial resin PR48 resin (Colorado Photopolymer Solutions, USA) was selected as a control. Table 9.7 contains the tensile properties of PR48 and experimental LRs measured on coupons fabricated on a Formlabs Form 1+ printer: all LRs outperformed PR48 in ductility (elongation at break), and generated uniformly fused, and high-resolution prints that featured improved material toughness upon a lower UV dosage. No LR exceeded PR48 in modulus, but LR10 and LR15 surpassed PR48 in strength.

On one hand lignin is suited for ink-based DW (Section 2.10.2), because this AM technology processes feedstocks at room temperature, below lignin's degradation temperature, but on the other hand lignin's rheological behavior is unfavorable to ink-based DW. Only a few studies investigated the applicability of lignin to ink-based DW, and they focused on blends of lignin and polymers forming gels and pastes that were ultimately crosslinked for final shape retention. Following are some recent examples of such studies.

Zhang et al. (2020) selected colloidal lignin particles produced from softwood kraft lignin powder, crosslinked with Ca^{2+} ions, and combined them with cellulose nanofibers and alginate, and printed the resulting blend for soft TE. The blend resulted compatible with HepG2 cells that are human liver cells selected for studies of drug metabolism and toxicity.

Jiang et al. (2020) added up to 83 wt% soda lignin to Pluronic® F-127, a soft triblock copolymer, containing acrylate and acting as a crosslinking agent. After printing, parts were freeze-dried and oven-cured at 120°C to produce crosslinked, water-insoluble articles, featuring 5 GPa tensile modulus and UV-blocking properties.

Gleuwitz et al. (2020) formulated fully biobased, liquid crystalline, shear thinning inks for ink-based DW, starting from lyotropic gel-forming hydroxypropyl cellulose (HPC) with up to 25 wt%

beech organosolv lignin in acetic acid aqueous solution. Shear casting of the blend solutions was followed by chemical crosslinking with citric acid-based crosslinkers and a dimerized fatty acid, which produced water-insoluble, anisotropic films predisposed to swelling in water. Along the shear direction, the films reached 80 MPa, 3.5 GPa, and 5 MJ/m^3 in tensile strength and modulus, and toughness, respectively. Interestingly, HPC and lignin alone were not compatible with DIW, but the blend of them was compatible, because HPC and lignin provided shear thinning for extrusion and shape retention at rest, respectively.

9.6 NEAR FUTURE OF LIGNIN FOR AM

The near future of lignin for AM is promising because of lignin's favorable properties above mentioned, but it also hinges on controlling its widely variable properties, improving the understanding of the chemical and physical aspects of its molecule, and enhancing its compatibility with polymer matrices, through surface modification or graft copolymerization (Thakur et al. 2014), and hence, ultimately, its efficiency as a strengthening and stiffening filler. The studies described above point out that work is needed to understand how to stiffen and strengthen polymers by adding lignin. Ebers et al. (2021) mentioned the following challenges and opportunities for effectively employing lignin in AM. The difference in properties among technical lignins can be exploited to formulate appropriate feedstocks for specific AM processes: f.e. modifying lignin for solubility and reactivity with organic monomers is beneficial for SL, whereas heat-induced flowability is advantageous in FFF. The performance of printed parts containing lignin also depends on controlling the thermodynamic miscibility, in the form of polymer blend compatibility in feedstocks for FFF, or solubility in a liquid reactive monomer in resin feedstocks for VP. Phase separation of lignin-based polymer blends during FFF impedes lignin contribution to load-bearing properties, and must be avoided. Furthermore, the structure/processing/morphology/performance relationships for every pair of technical lignin and AM process should be elucidated in reference to specific applications.

REFERENCES

Aderholt, M. 2014. *TwoBears Unveils New Bio-degradable, Renewable Filament, bioFila®*. https://3dprint.com/1873/twobears-renewable-filament-biofila/ (accessed January 16, 2020).

Akato, K., Tran, C. D., Chen, J., Naskar, A. K. 2015. Poly (ethylene oxide)-assisted macromolecular self-assembly of lignin in ABS matrix for sustainable composite applications. *ACS Sustain. Chem. Eng.* 3 (12): 3070–3076.

Bajwa, D., Pourhashem, G., Ullah, A. H., Bajwa, S., 2019. A concise review of current lignin production, applications, products and their environment impact. Industrial Crops and Products139: 111526.

Berlin, A., Balakshin, M. 2014. Industrial lignins: Analysis, properties, and applications. In *Bioenergy Research: Advances and Applications*, ed.V. Gupta, M. Tuohy, C. Kubicek, J. Saddler, F. Xu, 315–336. Amsterdam: Elsevier.

Bertini, F., Canetti, M., Cacciamani, A. et al. 2012. Effect of ligno-derivatives on thermal properties and degradation behaviour of poly(3-hydroxybutyrate)-based biocomposites. *Polym. Degrad. Stab.* 97:1979–1987.

Bonini, C., D'Auria, M., Emanuele, L. et al. 2005. Polyurethanes and polyesters from lignin. *J. Appl. Polym. Sci.* 98:1451–1456.

Bova, T., Tran, C. D., Balakshin, M. Y. et al. 2016. An approach towards tailoring interfacial structures and properties of multiphase renewable thermoplastics from lignin–nitrile rubber. *Green Chem.* 18 (20): 5423–5437.

Bruijnincx, P., Weckhuysen, B., Gruter, G.-J., Westenbroek, A., Engelen-Smeets, E. 2016. *Lignin Valorization*. Utrecht: Utrecht University, Platform Agro-Paper-Chemistry.

Canetti, M., Bertini, F., De Chirico, A., Audisio, G. 2006. Thermal degradation behaviour of isotactic polypropylene blended with lignin. *Polym. Degrad. Stab.* 91:494–498.

Chauhan, M., Gupta, M., Singh, B., Singh, A., Gupta, V. 2014. Effect of functionalized lignin on the properties of lignin–isocyanate prepolymer blends and composites. *Eur. Polym. J.* 52:32–43.

Danielson, B., Simonson, R. 1998a. Kraft Lignin in phenolformaldehyde resin. Part 1. Partial replacement of phenol by kraft lignin in phenol-formaldehyde adhesives for plywood. *J. Adhes. Sci. Technol.* 12:923–939.

Danielson, B., Simonson, R. 1998b. Kraft lignin in phenol-formaldehyde resin. Part 2. Evaluation of an industrial trial. *J. Adhes. Sci. Technol.* 12:941–946.

de Jesus Silva, A. J., Berry, N. G., Ferreira da Costa, M. 2016. Structural and thermo-mechanical evaluation of two engineering thermoplastic polymers in contact with ethanol fuel from sugarcane. *Mat. Res.* 19(1)). doi:10.1590/1980-5373-MR-2015-0480.

Doherty, W. O. S., Mousavioun, P., Fellows, C. M. 2011. Value-adding to cellulosic ethanol: Lignin polymers. *Ind. Crops Prod.* 33:259–276.

Doherty, W. O. S., Rainey, T. J. 2006. Bagasse fractionation by the soda process. *Proceedings of theAustralian Society of Sugar Cane Technologists*, ed. D. Hogarth, 2–5. Queensland: Mckay.

Dominguez-Robles, J., Martin, N. K., Fong, M. L. et al. 2019. Antioxidant PLA composites containing lignin for 3D printing applications: a potential material for healthcare applications. *Pharmaceutics* 11:165.

Dorrestijn, E., Laarhoven, L. J. J., Arends, I. W. C. E., Mulder, P. 2000. The occurrence and reactivity of phenoxyl linkages in lignin and low rank coal. *J. Anal. Appl. Pyrolysis* 54:153–192.

Dunson, D. n.d. *Characterization of Polymers using Dynamic Mechanical Analysis (DMA)*. https://www.eag.com/resources/whitepapers/characterization-polymers-using-dynamic-mechanical-analysis-dma/ (accessed May 31, 2020).

Duval, A., Lawoko, M. 2014. A review on lignin-based polymeric, micro- and nano-structured materials. *React. Funct. Polym.* 85:78–96.

Ebers, L.-S., Arya, A., Bowland, C. C., et al. 2021. 3D printing of lignin: Challenges, opportunities and roads onward. Biopolymers112 (6): e23431 https://doi.org/10.1002/bip.23431.

Figueiredo, P., Lintinen, K., Hirvonen, J. T. et al. 2018. Properties and chemical modifications of lignin: Towards lignin-based nanomaterials for biomedical applications. *Prog. Mater. Sci.* 93:233–269.

García Calvo-Flores, F., Dobado, J. A. 2010. Lignin as renewable raw material. *Chem. Sus. Chem.* 3 (11): 1227–1235.

Ghaffar, S. H., Fan, M. 2014. Lignin in straw and its applications as an adhesive. *Int. J. Adhes. Adhesiv.* 48:92–101.

Ghosh, I., Jain, R. K., Glasser, W. G. 2000. Blends of biodegradable thermoplastics with lignin esters. In *Lignin: Historical, Biological, and Materials Perspectives*, Glasser et al., ACS Symposium Series. Washington: American Chemical Society.

Gkartzou, E., Koumoulos, E. P., Charitidis, C. A. 2017. Production and 3D printing processing of bio-based thermoplastic filament. *Manufact. Rev.* 4:1.

Gleuwitz, F. R., Sivasankarapillai, G., Siqueira, G., et al. 2020. Lignin in bio-based liquid crystalline network material with potential for direct ink writing. ACS Applied Bio Mater. 3: 6049.

Global Market Insights. 2020a. *Lignin Market Size*. https://www.gminsights.com/industry-analysis/lignin-market (accessed January 16, 2020).

Global Market Insights. 2020b. *Lignin market Size to Exceed $1 Billion by 2026*. https://www.gminsights.com/pressrelease/lignin-market-size (accessed January 16, 2020).

Gordobil, O., Delucis, R., Egüés, I., Labidi, J. 2015. Kraft lignin as filler in PLA to improve ductility and thermal properties. *Indus. Crops Prod.* 72:46–53.

Gordobil, O., Egüés, I., Llano-Ponte, R., Labidi, J. 2014. Physicochemical properties of PLA lignin blends. *Polym. Degrad. Stab.* 108:330–338.

Graça, F. M. P., Rudnitskaya, A., Faria, F. A. 2012. Electrochemical impedance study of the lignin-derived conducting polymer. *Electrochim. Acta* 76:69–76.

Grand View Research. 2020. *Lignin Market Size, Share & Trends Report*. https://www.grandviewresearch.com/industry-analysis/lignin-market (accessed January 16, 2020).

He, Z.-W., Yang, J., Lu, Q.-F., Lin, Q. 2013. Effect of structure on the electrochemical performance of nitrogen- and oxygen-containing carbon micro/nanospheres prepared from lignin-based composites. *ACS Sustain. Chem. Eng.* 1:334–340.

Holmberg, A. L., Nguyen, N. A., Karavolias, M. G. et al. 2016a. Softwood lignin-based methacrylate polymers with tunable thermal and viscoelastic properties. *Macromolecules* 49 (4): 1286–1295.

Holmberg, A. L., Reno, K. H., Nguyen, N. A. et al. 2016b. Syringyl methacrylate, a hardwood lignin-based monomer for high-T_g polymeric materials. *ACS Macro Lett.* 5 (5): 574–578.

Holtzapple, M. T. 2003. Lignin. *Encyclopedia of Food Sciences and Nutrition*. Cambridge: Academic Press. 2nd edition.

Ibrahim, F., Mohan, D., Sajab, M., et al. 2019. Evaluation of the compatibility of organosolv lignin-graphene nanoplatelets with photo-curable polyurethane in stereolithography 3D printing. Polymers 11 (10): 1544.

Jiang, B., Yao, Y., Liang, Z., et al. 2020. Lignin-Based Direct Ink Printed Structural Scaffolds. Small 16: 31.

Kadla, J. F., Kubo, S. 2004. Lignin-based polymer blends: analysis of intermolecular interactions in lignin-synthetic polymer blends. *Compos. A* 35:395–400.

Kadla, J. F. Kubo, S., Gilbert, R., Venditti, R. 2002a Lignin-based carbon fibers. In *Chemical Modification, Properties, and Usage of Lignin,* ed.Thomas Q. Hu, 121–137. Kluwer Academic/Plenum Publishers.

Kadla, J. F., Kubo, S., Venditti, R. A., Gilbert, R. D., Compere, A. L. 2002b. Lignin-based carbon fibers for composite fiber applications. *Carbon.* 40 (15): 2913–2920.

Kun, D., Pukánszky, B. 2017. Polymer/lignin blends: Interactions, properties, applications. *Eur. Polym. J.* 93:618–641.

Laurichesse, S., Luc Avérous, L. 2014. Chemical modification of lignins: Towards biobased polymers. *Prog. Polym. Sc.* 39:1266–1290.

Li, J., He, Y., Inoue, Y. 2003. Thermal and mechanical properties of biodegradable blends of poly(l-lactic acid) and lignin. *Polym. Int.* 52:949.

Liebrand, T. R. H. 2018. *3D Printed Fiber Reinforced Lignin.* M. Sc diss., Delft University of Technology. https://repository.tudelft.nl/islandora/object/uuid%3A2856a86c-d862-48b1–924e-1f3ce74647b3 (accessed January 16, 2020).

Lora, J. H., Glasser, W. G. 2002. Recent industrial applications of lignin: A sustainable alternative to non-renewable materials. *J. Polym. Environ.* 10:39–48.

Mahmood, N., Zhongshun, Y., Schmidt, J., Xu, C. 2016. Depolymerization of lignins and their applications for the preparation of polyols and rigid polyurethane foams: A review. *Renew. Sustain. Energy Rev.* 60:317–329.

Naseem, A., Tabasum, S., Zia, K. M. et al. 2016. Lignin-derivatives based polymers, blends and composites: a review. *Int. J. Biol. Macromol.* 93 (Part A): 296–313.

Nguyen, N. A., Bowland, C., Naskar, A. K. 2018a. A general method to improve 3D-printability and inter-layer adhesion in lignin-based composites. *Appl. Mater. Today* 12:138–152.

Nguyen, N. A., Bowland, C., Naskar, A. K. 2018b. Mechanical, thermal, morphological, and rheological characteristics of high performance 3D-printing lignin-based composites for additive manufacturing applications. *Data in Brief* 19:936–950.

Pettersen, R. C. 1984. The chemical composition of wood. In *The Chemistry of Solid Wood*, ed. R. M. Rowell, 57–126, Washington DC: American Chemistry Society.

Pucciariello, R., Villani, V., Bonini, C. et al. 2004. Physical properties of straw lignin-based polymer blends. *Polymer* 45 (12): 4159–4169.

Roopan, S. M. 2017. An overview of natural renewable bio-polymer lignin towards nano and biotechnological applications. *Int. J. Biol. Macromol.* 103:508–514.

Rozite, L., Varna, J., Joffe, R., Pupurs, A. 2013. Nonlinear behavior of PLA and lignin-based flax composites subjected to tensile loading. *J. Thermoplast. Compos. Mater.* 26:476–496.

Rozman, H., Tan, K., Kumar, R. et al. 2000. The effect of lignin as a compatibilizer on the physical properties of coconut fiber–polypropylene composites. *Eur. Polym. J.* 36 (7): 1483–1494.

Saito, T., Brown, R. H., Hunt, M. A. et al. 2012. Turning renewable resources into value-added polymer: development of lignin-based thermoplastic. *Green Chem.* 14 (12): 3295–3303.

Saito, T., Perkins, J. H., Jackson, D. C. et al. 2013. Development of lignin-based polyurethane thermoplastics. *RSC Adv.* 3 (44): 21832–21840.

Sarkanen K. V. and C. H. Ludwig ed. 1971. *Lignins: Occurrence, Formation, Structure, and Reactions.* New York: J Wiley and Sons.

Sigma Aldrich. 2020. *Lignin.* https://www.sigmaaldrich.com/catalog/product/aldrich/370959?lang=en& region=US. (accessed September 6, 2018).

Singer, L. S. 1989. *Carbon Fibers.* In *Concise encyclopedia of composite materials*, ed.A. Kelly, 47–55. Oxford: Pergamon Press.

Stark, N. M., Yelle, D. J., Agarwal, U. P. 2016. Techniques for characterizing lignin. In *Lignin in Polymer Composites*, ed. O. Faruk and M. Sain, 49–66. Oxford: Elsevier, 1st edition.

Stewart, D. 2008. Lignin as a base material for materials applications: chemistry, application, and economics. *Indus. Crops Prod.* 27:202–207.

Sutton, J. T., Rajan, K., Harper, D. P., Chmely, S. C. 2018. Lignin-containing photoactive resins for 3D printing by stereolithography. *ACS Appl. Mater. Interf.* 10 (42): 36456–36463.

Thakur, V. K., Thakur, M. K., Raghavan, P., Kessler, M. P. 2014. Progress in green polymer composites from lignin for multifunctional applications: a review. *ACS Sustain. Chem. Eng.* 2:1072–1092.

Toriz, G., Denes, F., Young, R. 2002. Lignin-polypropylene composites. Part 1: Composites from unmodified lignin and polypropylene. *Polym. Compos.* 23 (5): 806–813.

Tran, C. D., Chen, J., Keum, J. K., Naskar, A. K. 2016. A new class of renewable thermoplastics with extraordinary performance from nanostructured lignin-elastomers. *Adv. Funct. Mater.* 26 (16): 2677–2685.

Upton, B. M., Kasko, A. M. 2016. Strategies for the Conversion of Lignin to High-Value Polymeric Materials: Review and Perspective. *Chem. Rev.* 116:2275–2306.

Vaidya, A. A., Collet, C., Gaugler, M., Lloyd-Jones, G. 2019. Integrating softwood biorefinery lignin into polyhydroxybutyrate composites and application in 3D printing. *Materials Today Communications* 19:286–296.

Wang, J., Banu, D., Feidman, D. 1992. Epoxy-lignin polyblends: effects of various components on adhesive properties. *J. Adhes. Sci. Technol.* 6:587–598.

Wasti, S. 2020. *Production of Bio-Composite Filament using Lignin, Polylactic Acid and High Impact Polystyrene for Additive Manufacturing (3D Printing).* MSc diss., Auburn University.

Wasti, S., Triggs, E., Farag, R. et al. 2021. Influence of plasticizers on thermal and mechanical properties of biocomposite filaments made from lignin and polylactic acid for 3D printing. *Compos. B: Eng.* 205:108483.

Wood, B. M., Coles, S. R., Maggs, S., Meredith, J., Kirwan, K. 2011. Use of lignin as a compatibiliser in hemp/epoxy composites. *Compos. Sci. Technol.* 71 (16): 1804–1810.

Yang, J., An, X., Liu, L. et al. 2020. Cellulose, hemicellulose, lignin, and their derivatives as multi-components of bio-based feedstocks for 3D printing. *Carbohydr. Polym.* 250:116881.

Zhang, S., Li, M., Hao, N., Ragauskas, A. J. 2019. Stereolithography 3D printing of lignin-reinforced composites with enhanced mechanical properties. *ACS Omega* 4 (23): 20197–20204.

Zhang, W., Ma, Y., Xu, Y., Wang, C., Chu, F. 2013. Lignocellulosic ethanol residue-based lignin-phenol-formaldehyde resin adhesive. *Int. J. Adhes. Adhesiv.* 40:11–18.

Zhang, X., Morits, M., Jonkergouw, C., et al. 2020., Three-dimensional printed cell culture model based on spherical colloidal lignin particles and cellulose nanofibril-alginate hydrogel. Biomacromolecules 21 (1875): 1875–1885.

10 Trees and Natural Fibers

Why use up the forests which were centuries in the making and the mines which required ages to lay down, if we get the equivalent of forest and mineral products in the annual growth of the hemp fields?

Henry Ford

10.1 FEEDSTOCKS FROM TREES AND NATURAL FIBERS

This chapter describes sustainable feedstocks mostly in form of filaments for AM made of sustainable and non-sustainable polymers combined with fillers from plants (flax, harakeke, hemp, sisal, etc.) and trees (coconut, cork, resins) that are commercially available, or have been experimentally developed. Bamboo and wood are also natural fibers and AM feedstocks, but they are discussed in Chapters 8 and 6, respectively. The feedstocks illustrated in this chapter should be evaluated not only in terms of physical and mechanical properties, but also considering the following advantages that they have over their synthetic counterparts: they are locally sourced; have minimal cost of transportation among points of source, production, and use; support local economies; and reduce imports. It is also remarkable that, although they are of some of the oldest utilized natural fibers, they are still part of the latest engineering feedstocks and fabrication technologies, and participate in the AM "revolution."

Natural materials fall in four families (Ashby et al. 1995):

- *Polymers* and *polymer composites*: they comprise cellulose, chitin, collagen, cuticle, keratin, silk, and tendon, and outperform engineering polymers (except aramid fiber) in tensile strength and modulus.
- *Elastomers*: they are artery, cartilage, elastin, muscle, skin, etc., and feature moduli and densities similar to those of engineering elastomers.
- *Ceramic* and *ceramic composites*: they include antler, coral, bone, dentine, enamel, and shell, and are made up of ceramic particles (calcite, hydroxyapatite, etc.) in a matrix of collagen. Compared to engineering ceramics, they possess lower elastic moduli but similar tensile strengths and superior toughness by a factor of about 10.
- *Cellular materials*, such as cork, cancellous (lightweight, sponge-like, porous) bone, palm, and wood: they are characterized by low densities due to their large volume of voids, and are almost always anisotropic because of the shape and orientation of their cells and fibers.

The tensile strength and elastic modulus of natural materials are plotted in Figure 10.1, and the ranges of their values are 0.5–2,000 MPa and 0.001–200 GPa, respectively, whereas tensile strength and elastic modulus of engineering materials are 2–2,000 MPa and 0.001–1,000 GPa, respectively (Ashby et al. 1995). However, natural materials are lighter than metals and ceramics, and hence the property values of natural materials divided by their density compete with those of man-made engineering materials. Interestingly, Ashby et al. (1995) pointed out that natural materials perform remarkably well as structural components. In fact, the value of $E^{1/2}/\rho$ (that measures beam stiffness in $GPa^{1/2}/(Mg/m^3)$, being E and ρ tensile modulus and density, respectively, is about 10 for wood and 3.1 for aluminum alloys. The index $\sigma^2/(E\rho)$, with σ tensile strength, quantifies the ability to store elastic energy at minimum weight, and is higher for silk than for the

DOI: 10.1201/9781003221210-10

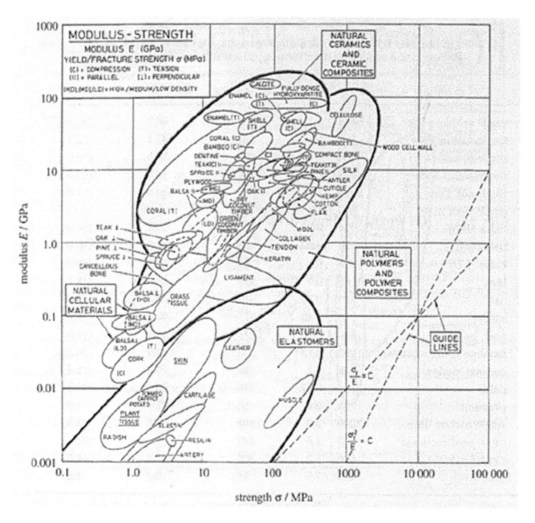

FIGURE 10.1 Chart of tensile strength and elastic modulus of natural materials.

Source: Ashby, M. F., Gibson, L. J., Wegst, U., Olive, R. 1995. The mechanical properties of natural materials. I. Material property charts. *Proc. R. Soc. Lond. A* 450:123–140. Reproduced with permission from The Royal Society.

best spring steel. The index $(EJ_c)^{1/2}$, being J_c (kJ/m^2) the fracture energy per unit area, measures the resistance to fracture, and is greater for shell and enamel than any man-made ceramic.

10.2 CORK

10.2.1 OVERVIEW OF CORK

Cork is a biological tissue located in the outer bark of cork oak tree (*Quercus suber*), and serves as a protective layer. Cork is a material sustainable, biodegradable, buoyant, scarcely permeable to fluids, viscoelastic, effectively insulating against temperature, sound and vibrations, impact resistant, fire retardant, recyclable, and nontoxic. It is composed of anisotropic closed cells, whose walls have a chemical composition depending on several factors and comprising, in wt%, suberin (33–50), lignin (13–29), polysaccharides (6–25), and extractives (8.5–24) (Fernandes et al. 2015).

TABLE 10.1

Properties of Cork

Property	Density	Tensile Yield Strength	Tensile Modulus	Compressive Modulus[a]	Fracture Toughness K_{IC}	Maximum Service Temperature	Thermal Conductivity	CTE
Unit	kg/dm^3	MPa	MPa	MPa	MPa(m)$^{1/2}$	°C	W/(mK)	10^{-6}/K
Values	0.1–0.2	0.3–2	10–50	8–20	0.05–0.1	120–130	0.04–0.5	150–200

Source: Ashby, M. F., Gibson, L. J., Wegst, U., Olive, R. 1995. The mechanical properties of natural materials. I. Material property charts. *Proc. R. Soc. Lond. A*, 450:123–140.

Note:

[a]Silva, S. P., Sabino, M. A., Fernandes, E. M., et al. 2005. Cork: properties, capabilities and applications. *International Materials Reviews* 50 (6): 345–365.

Suberin is an aliphatic polymer composed of glycerol and long-chain fatty acids that provides structural integrity, lignin is an aromatic polymer, and extractives are dichloromethane, ethanol, and water. A detailed chemical analysis of cork is published by Pereira (2013). Cork forests cover over 2.2 million hectares worldwide, and are found in Algeria, France, Morocco, Italy, Portugal, Spain, Tunisia, and Turkey (Cork Information Bureau 2010). Spain and Portugal produce each about 1/3 of cork in the world. Cork trees live on average 150 years, and sometimes exceed 250 years (Cooke 1951). Cork production is considered sustainable, because bark is harvested by stripping it (without damaging the trunk) instead of cutting the tree down. The first harvest occurs when the tree is 20–30 years old, and is repeated between 8 to 11 times during the tree's life span. Cork is used not only as a wine stopper but also in aerospace, artefacts, buildings, constructions, environment (collecting spilled oil), gaskets, industry, sport, etc. and is turned into many items (Janevski 2020). Relevant to AM feedstocks is that cork is employed as granulates and agglomerates in many products, where it is combined with aluminum sheets, fibers, glue, natural and synthetic resins, polymers, rubber, etc. (Mestre 2014).

The main properties of cork are listed in Table 10.1. Figure 10.2 illustrates the cellular structure of cork, and the variation of its cells' orientation across axial, radial, and tangential directions. Since cork's mechanical properties are anisotropic, their values depends on the measurement direction. Thermal decomposition of cork starts around 200°C, and increases with raising temperature until cork turns into ash at about 485°C (Sen et al. 2014). Pereira (2015) describes in detail how structure and chemistry determine cork's peculiar properties.

Cork is added to polymers to form non-AM composites utilized in aviation, buildings, furniture, naval construction, and transport (Fernandes et al. 2014). It is also mixed with biodegradable polymers, such as PCL and sustainable (biobased and biodegradable) polymers such as PLLA, and PHBV (Vilela et al. 2013; Fernandes et al. 2015).

Its properties and sustainability make cork a prime candidate material meeting requirements of design for sustainability (Janevski 2020), which consists in factoring environmental issues in the design process and material selection. For recent reviews on cork and cork composites, Gil (2009) and Gil (2015) are suggested.

10.2.2 Commercial Cork-Based Filaments for AM

To our knowledge, the only commercial feedstocks containing cork and PLA for AM are two filaments for FFF: corkFill by colorFabb, and EasyCork™ by Formfutura. Their properties are listed in Table 10.2, along with properties of PLA filaments for FFF sold by the same suppliers, and included here as a benchmark, assuming that each supplier uses its same own PLA for unfilled and

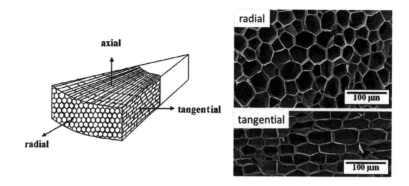

FIGURE 10.2 Sections of cork, illustrating the different cell orientations across axial, radial, and tangential directions. Left: drawing of cellular disposition. Right: SEM micrographs of radial and tangential sections after boiling.

Source: Adapted from Silva, S. P., Sabino, M. A., Fernandes, E. M., et al. 2005. Cork: properties, capabilities and applications. *Int. Mater. Rev.*, 50 (6): 345–365. Reproduced with permission from Taylor and Francis.

filled PLA. In comparing cork-filled and unfilled PLA (PLA Economy) by colorFabb, the former trails the latter in tensile strength, modulus, and notched impact strength, and leads in elongation at break. Since cork is impact resistant, the fact that corkFill is less impact resistant than its matrix may indicate, among other characteristics, a weak filler-matrix interface mechanical bonding. Comparing cork-filled and unfilled PLA (EasyFil™) by Formfutura in tensile properties is not possible, because these properties were measured on dumbbell coupons of EasyCork™, and on film coupons of EasyFil™, with the film displaying values of tensile strength and elongation at break far (and oddly) greater than those typical for PLAs.

10.2.3 EXPERIMENTAL CORK FEEDSTOCKS FOR AM

Daver et al. (2018) formulated a composite filament for FFF made of cork/PLA, with cork content of 5–50 wt% (Figure 10.3). They selected Ingeo™ 4032D PLA and Amorim cork powder with particle size distribution values such that 10%, 50%, and 90% of the particles had size below 272, 446, and 733 μm, respectively. PLA and cork were manually and melt mixed, and extruded into 1.75 mm diameter filament. Test coupons were printed on a MakerBot Replicator 2 with 100% infill. Comprehensive testing indicated that the cork did not strengthen and stiffen PLA: all composites lagged unfilled PLA in tensile yield strength, modulus and strain at break, and impact strength (Figure 10.4). Increasing amount of cork had the following effects: it lowered tensile yield strength and modulus, and raised elongation at break; it decreased impact strength, storage and loss modulus; it left unchanged T_g, and thermal degradation. These effects and the fact that all property variations were not linear with the cork content revealed that the mechanical bonding between cork and PLA was poor, and adding an interface adhesion promoter would be beneficial. The composite formulation with 5% cork possessed the highest tensile yield strength, modulus, and strain at break: 38.3 MPa, 2,816 MPa, and 0.8% respectively, still significantly inferior to PLA's corresponding values of 60 MPa, 3,345 MPa, and 1.5%. To improve the ductility of cork/PLA, and facilitate its extrusion during printing, the authors added 1–15 wt% of tributyl citrate (TBC) plasticizer to cork/PLA (5/95 wt%) granule pre-mix. Compared to PLA, all formulations with TBC displayed inferior tensile yield strength, modulus, and strain at break, except that with 10 wt% TBC formulation that displayed higher strain at break than PLA's. Since cork/PLA trailed PLA in mechanical

TABLE 10.2
Commercial Filaments for FFF Made of Cork-Filled PLA and Unfilled PLA

Material	Cork Weight Fraction	Density	Tensile Strength	Tensile Modulus	Elongation at Break	Flexural Strength	Flexural Modulus	Charpy Unnotched Impact Strength	Charpy Notched Impact Strength	T_m
	%	kg/m^3	MPa	MPa	%	MPa	MPa	kJ/m^2	kJ/m^2	°C
PLA-Cork (colorFabb)	10–20	1.18	40	2,475	11	58[a]	2,490	36	4	>155
EasyCork™ (Formfutura)	<20	1.03	19.4	1,050	15	N/A	N/A	N/A	5–6	160
PLA Economy (colorFabb)	0	1.2–1.3	45	3,400	6	N/A	N/A	N/A	7	N/A
EasyFil™ (Formfutura)	0	1.24	110[b]	3,310[b]	160[b]	N/A	N/A	N/A	7.5	210

Sources: Data posted online by colorFabb, Formfutura.
Notes:
[a]Stress at 3.5% strain.
[b]Values measured on films per ASTM D882.

FIGURE 10.3 SEM micrograph of cork powder from cork/PLA filament for FFF. Notice the miniscule cells within each powder particle.

Source: Daver, F., Lee, K. P. M., Brandt, M., Shanks, R. 2018. Cork-PLA composite filaments for fused deposition modeling. *Compos. Sci. Technol.* 168:230–237. Reproduced with permission from Elsevier.

performance, functional benefits of preferring the former over the latter would be leveraging cork's thermal and acoustic insulation, and vibration damping.

Fernandes et al (2015) described fabrication, and tensile and thermal properties of two cork-based composites for AM: cork/PLLA and cork/PHBV composites. The authors selected Amorim cork granules with particle size 0.5–1 mm, and specific weight of 166 kg/m^3, Cargill Dow PLLA with 99.6% L-lactide content and 69,000 g/mol MW, and PHBV by PHB Industrial with 12% 3-hydroxyvalerate (HV) content and 426,000 g/mol MW. The main properties, measured on injection molded coupons, are collected in Table 10.3. In comparison to unfilled polymers, both composites featured lower density and tensile properties, and a marginal change in T_g and T_m. Cork/PLLA featured 19% higher tensile strength but 16% lower tensile modulus than cork/PHBV, which is consistent with the fact that PLLA leads and trails PHBV in tensile strength and modulus, respectively. Going from PLA to 30/70 wt% cork/PLA, the drop in tensile yield strength and modulus was 74% and 65%, respectively on printed coupons (Daver et al. 2018), whereas it was 51% and 32%, respectively on injection molded coupons (Fernandes et al. 2015). This can be explained by the fact that typically, thanks to high pressure, injection molding produces lower void content than extrusion.

Brites et al. (2017) combined standard HDPE with differing amount of cork powders and coupling agent, in order to test and select the formulation providing the best combination of morphological, physical, and mechanical properties in test coupons printed via FFF. Two cork powders were investigated: virgin cork (Amorim Cork Composites, Portugal) with average particle size of 0.5–1 mm, and cork waste from floor covering manufacturing and contaminated with varnish, cork, wood fiber, PVC, PU, etc. HDPE was Dow KS 10100 with 4 g/10 min MFI, and 955 kg/m^3 specific weight. The coupling agent was DuPont™ Fusabond® E265. HDPE was modified with maleic anhydride, and featured 12 g/10 min MFI, and 950 kg/m^3 specific weight. Table 10.4 illustrates the composition and properties of printed HDPE and cork/HDPE. All tested composites trailed HDPE in tensile strength and modulus, with cork waste composites reaching higher values in both properties than virgin cork. Differently than what reported in Daver et al. (2018), the more cork, the higher tensile strength. Not the same was for modulus: increasing virgin cork led to lower modulus, whereas waste cork resulted in composite's modulus higher at 30 wt% cork than at 15 wt% but lower at 50 wt% than at 30 wt%. Although the same weight fractions (15% and 30%) of cork waste and virgin cork were utilized, higher density of cork waste translated in lower volume fraction of it, and hence higher volume fraction of HDPE, which contributed to higher property values, according to the authors. It is also possible that PVC and PU present in cork waste resulted in better cork/HDPE interface bonding, and hence higher properties than those in virgin cork/HDPE.

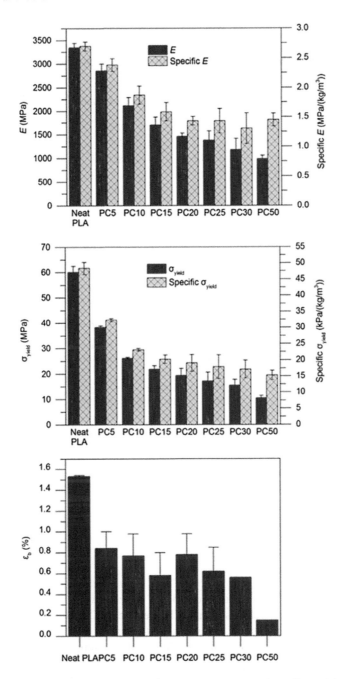

FIGURE 10.4 Tensile properties of PLA and cork/PLA for FFF: absolute and specific modulus (top), absolute and specific yield strength (middle), and strain at break (bottom). Bars on top of columns indicate standard deviation.

Source: Daver, F., Lee, K. P. M., Brandt, M., Shanks, R. 2018. Cork-PLA composite filaments for fused deposition modeling. *Compos. Sci. Technol.* 168:230–237. Reproduced with permission from Elsevier.

10.3 NATURAL FIBERS AND THEIR POLYMER COMPOSITES

Figure 10.5 illustrates a detailed classification of *natural* and *synthetic* fibers of technological use. Natural fibers are the most common SPs available.

TABLE 10.3

Properties of Injection Molded Coupons of Experimental Unfilled and Cork-Filled Composites

Material	Cork Amount	Density	Tensile Strength	Tensile Modulus	Elongation at Break	T_g	T_m
	wt%	g/cm^3	MPa	MPa	%	°C	°C
PLLA	0	1.24	52.2	1,257	5.5	57.3	155.3
Cork/PLA	30	1.20	25.7	851	3.4	58.4	154.2
PHBV	0	1.23	30.3	1,490	2.9	−0.6	160.6
Cork/PHBV	30	1.21	21.6	1,012	3.8	4.6	153.7

Source: Data from Fernandes, E. M., Correlo, V. M., Mano, J. F., Reis, R. L. 2015. Cork-polymer biocomposites: mechanical, structural and thermal properties. *Mater. Des.* 82:282–289. Reproduced with permission from Elsevier.

TABLE 10.4

Composition and Properties of Cork/HDPE Composites for AM

Material	HDPE	Cork Waste	Virgin Cork	Coupling Agent	Density	Tensile Strength	Tensile Modulus	Elongation at Break	T_m
Unit	wt%	wt%	wt%	wt%	g/cm^3	MPa	MPa	%	°C
Test Method	N/A	N/A	N/A	N/A	ISO 1183	ISO 527	ISO 527	ISO 527	DSC
HDPE	100	0	0	0	0.96	24.3	645	N/A	133
CPC1	80	0	15	5	0.73	16.8	626	31.7	137
CPC2	65	0	30	5	0.66	17.3	569	21.2	138
CPC3	80	15	0	5	0.70	18.6	665	52.7	136
CPC4	65	30	0	5	0.61	19.4	952	10.7	137
CPC5	45	50	0	0	0.83	21.0	852	3.7	136

Source: Data from Brites, F., Malça, C., Gaspar, F. et al. 2017. Cork Plastic Composite Optimization for 3D Printing Applications. *Procedia Manuf.* 12:156–165. Reproduced with permission from Elsevier.

This chapter focuses on AM feedstocks composed of *cellulose/lignocellulose* (or plant-derived) fibers: flax, hemp, sisal, etc. Cellulose/lignocellulose fibers (CLFs) have been leveraged as engineering materials since prehistoric times, as in the case of flax and hemp. CLFs are lignocellulosic because they consist of cellulose microfibrils running along the length of CLFs, and embedded in an amorphous matrix of lignin and hemicellulose. Each fibril features a complex layered structure consisting in a thin outer wall encircling a thicker inner layer that comprises three layers, the middle one being the thickest and most affecting mechanical properties (Prasad 1989). CLFs are potential alternatives to synthetic fibers for the composite industry, thanks to their environment-friendly nature, flexibility of use, low cost, local availability, high strength/density and stiffness/density ratios, high impact strength (Prasad 1989), and renewability (Stevens and Mussig 2010). Selected properties of CLFs are listed in Table 10.5. However, fiber quality and properties are affected by several factors at various stages of fiber production: plant growth, harvesting, fiber extraction, and supply, with each factor in turn depending on other variables, as shown in Table 10.6. The chemical composition of natural fibers is reported by Mochane et al. (2021).

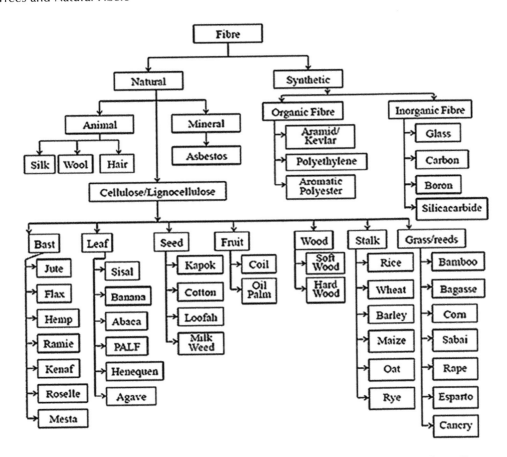

FIGURE 10.5   Classification of natural and synthetic fibers. PALF stands for pineapple leave fiber.

Source: Jawaid, M., Abdul Khalil, H. P. S. 2011. Cellulosic/synthetic fiber reinforced polymer hybrid composites: A review. *Carbohydr. Polym.* 86 (1): 1–18. Reproduced with permission from Elsevier.

Combining CLFs and SPs forms fully sustainable composites serving as AM feedstocks. PLA is the most widespread SP for AM, and adding CLFs to PLA filaments for FFF would result in a composite with superior mechanical properties to those of PLA, but effectively blending the two constituents is hindered by incompatibility between hydrophobic PLA and hydrophilic fibers, and fiber agglomeration. The two issues are tackled by modifying the fiber surface with physical and chemical methods such as acetylation, cyanoethylation, esterification, coupling agents (Section 8.4) (including silanes, isocyanates, zirconates, titanates, and chitosan), functional polyolefins (f.e. maleated PP), bi-functional molecules, and grafting monomers (Petinakis et al. 2013).

The surface of fibers for AM composites can be modified by selecting a method among the routes followed to modify the surface of fibers in general: physical (mechanical, solvent extraction, electrical discharge), chemical (peroxide, coupling agent, bleaching, enzyme, alkali), and physico-chemical (Mochane et al. 2021).

Mechanical properties in CLF-reinforced polymer composites are governed by several factors, including, among others: content, aspect ratio (length/diameter) and orientation of fibers; moisture content; porosity, and voids between matrix and fibers and among strands; fiber wettability; matrix-fiber interfacial bonding; and uniform fiber distribution with the matrix. The fiber-matrix interface in turn depends on the surface topography of the fiber, and the chemical compatibility

TABLE 10.5
Properties of Natural Fibers

Material	Density (D)	Tensile Strength (TS)	Tensile Modulus (E)	Elongation at Break	Specific Strength (TS/D)	Specific Modulus (E/D)
	N/A	MPa	GPa	%	MPa	GPa
Bamboo	0.6–1.1	140–230	11–17	N/A	127–283	10–28
Coconut	1.15	131–175	4–6	15–40	114–152	3.5–5.2
Flax	1.54	345–2,000	27.5–85	1–4	224–1,299	18–55
Harakeke[a]	1.27	778	32	4.5	612	25
Hemp	1.47	368–800	17–70	1.6	250–544	12–48
Sisal	1.45–1.5	350–700	2–22	2–7	233–483	1.3–15
E-glass	2.5	2,000–3,500	70	2.5	800–1,400	28
Carbon	1.4	4,000	230–240	1.4–1.8	2,857	164–171

Source: Célino, A., Fréour, S., Jacquemin, F., Casari, P. 2014. The hygroscopic behavior of plant fibers: a review. *Front. Chem.* 1, 43. Reproduced being an open-access article distributed under the terms of the Creative Commons Attribution License (CC BY).
Note:
[a]Le, T. M., Pickering, K. L. 2015. The potential of harakeke fibre as reinforcement in polymer matrix composites including modelling of long harakeke fibre composite strength. *Composites: Part A* 76:44–53.

TABLE 10.6
Factors Affecting Fiber Quality during Natural Fiber Production

Stage	Factors Affecting Fiber Quality
Plant growth	Species of plant
	Crop cultivation
	Crop location
	Fiber location in plant
	Local climate
Harvesting	Fiber ripeness, affecting: cell wall thickness, fiber coarseness, adherence between fiber and surrounding structure
Fiber extraction	Decortication process
	Type of retting method
Supply	Transportation conditions
	Storage conditions
	Age of fiber

Source: Thakur, V. K., Thakur, M. K. 2014. Processing and characterization of natural cellulose fibers/thermoset polymer composites, *Carbohydr. Polym.* 109:102–117. Reproduced with permission from Elsevier.

between fiber surface and resin matrix. CLFs have quite an irregular surface, which should the-oretically enhance the fiber-matrix interfacial bond (Prasad 1989). Porosity can be reduced by lowering matrix viscosity, which in turn can be accomplished by adding appropriate plasticizers, f.e. a citrate (a salt or ester of citric acid) (Arrieta et al. 2014; Wan et al. 2015). All the above factors should be accounted for in developing such composites for AM. Moreover, the role of the

complex morphology of natural fibers in their reinforcing mechanisms needs to be further elucidated (Mazzanti et al. 2019).

Several plant-derived fillers are combined with polymers in AM feedstocks, thanks to creativity
and the low cost of developing and producing a new filament for FFF, but fabricating a reliable
filament suited for functional and load-bearing components in a range of service conditions is a
more costly, broad, and selective process.

A recent and extensive review of CLF-filled PLA composites in general and not specific for AM
was published by Siakeng et al. (2018).

10.4   SISAL

Sisal fiber is extracted from the leaves of the sisal plant (*Agave sisalana*) that is cultivated in
Brazil, East Africa, Haiti, India, Indonesia, Mexico, etc. The plant has 200–250 leaves, and each
leaf contains 1,000–1,200 fiber bundles, of which only 4% is used as fibers when the plant reaches
maturity (Puttegowda et al. 2018; Senthilkumar et al. 2018), but it can be cultivated in large
quantities thanks to its short plantation time. Sisal fiber contains by wt% 65–68 cellulose, 10–22
hemicellulose, 10–14 lignin, and 10–22 moisture (Senthilkumar et al. 2018). Sisal fiber is moderately crystalline (Chand and Fahim 2008). Markets for sisal are industries of paper, ropes and
twine, spin and carpet, etc., and some products are buffing cloths, carpets, dartboards, filters,
geotextiles, handicrafts, strings, bath exfoliates, paint brushes, and specialized paper. Sisal fiber
features low cost, good impact strength, and high strength/density and stiffness/density, and, added
to TSs, TPs, and biopolymer, it improves their mechanical strength. It is also present in plaster, and
other cement materials, also due to its thermal and acoustic insulation properties, and excellent
toughness (Senthilkumar et al. 2018).

The factors affecting mechanical properties of sisal fiber-reinforced polymer composites are
relevant when the composites with same components are formulated for AM: fiber architecture,
loading, orientation and chemical treatment, critical fiber length, interfacial adhesion, morphological changes on fiber surfaces, type of matrix, fillers, and additives (Senthilkumar et al. 2018).

Since technical information on sustainable feedstocks with sisal for AM is scarce, the papers by
Rajesh et al. (2015) and Dangtungee et al. (2015) on sisal/PLA and sisal/PHBV non-AM composites, respectively, provide a benchmark in terms of material properties, effect of sisal, sisal
treatment, and sisal-matrix interaction for developing AM feedstocks with the same ingredients.
However, the properties of parts are also governed by the manufacturing process. The samples by
Rajesh and Dangtungee were fabricated via injection molding and hot compression, respectively,
whereas parts printed via FFF were penalized by porosity, relatively weak interlayer adhesion,
residual stresses during cooling, etc., and hence the values of mechanical properties measured by
Rajesh and Dangtungee should be considered a lower bound.

Rajesh et al. (2015) tested coupons with 5, 10, 15, 20, and 25 wt% of PLA and untreated and
alkali (10%NaOH and H_2O_2)-treated sisal 3 mm fiber. Tensile and flexural strength increased with
sisal content up to 20 wt%, and dropped after that, tensile and flexural moduli instead improved up
to 25 wt%, whereas elongation at break dropped at 5 wt% and hovered around that value up to 25
wt%. Impact strength of untreated sisal/PLA improved up to 25 wt%, while that of treated sisal/
PLA improved at 5 wt% but steadily decreased at higher content. Water absorption in untreated
and treated fiber steadily increased with sisal content, and reached a maximum at 25 wt%, and a
maximum in all formulations after 28 h: untreated sisal/PLA absorbed up to 9 wt%, while treated
sisal/PLA absorbed up to 6 wt%. Property values at 20 wt% are collected in Table 10.7: treated
sisal/PLA led untreated sisal/PLA in all properties except impact strength.

Dangtungee et al. (2015) experimented with sisal of 0.25 and 5 mm length and 200–400 μm
diameter, and content of 5, 10, 20, and 30 wt%. Test results are illustrated in Table 10.8 and
Figures 10.6, 10.7, and 10.8 relative to tensile strength, tensile modulus, and impact strength,
respectively. Adding sisal to PHBV was beneficial to tensile modulus, and (especially) impact

TABLE 10.7

Mechanical Properties of Sisal/PLA Injection Molded Composites

Material	Fiber Content	Tensile Strength	Tensile Modulus	Elongation at Break	Flexural Strength	Flexural Modulus	Impact Strength
	wt%	MPa	GPa	%	MPa	GPa	kJ/m^2
PLA	0	52	2.08	4.4	96	2.80	2.6
Untreated Sisal/PLA	20	62	2.40	2.6	111	6.25	5.4
Treated Sisal/PLA	20	67	2.50	2.7	112	8.30	3.2

Source: Data from Rajesh, G., Ratna Prasad, A. V., Gupta, A. 2015. Mechanical and degradation properties of successive alkali treated completely biodegradable sisal fiber reinforced poly lactic acid composites. *J. of Reinf. Plastics and Composites* 34 (12): 951–961.

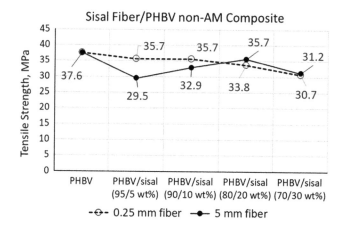

FIGURE 10.6 Tensile strength values for injection molded coupons of sisal fiber/PHBV non-AM composite.

Source: Data from Dangtungee, R., Tengsuthiwat, J., Boonyasopon, P. Siengchin, S. 2015. Sisal natural fiber/clay-reinforced poly(hydroxybutyrate-cohydroxyvalerate) hybrid composites. *J. Thermoplast. Compos. Mater.* 28 (6): 879–895.

strength but detrimental to tensile strength. The surprising result that greater modulus enhancement came from the shorter fiber was explained by the fact that the test coupons were cut from panels thinner than 5 mm, and the 5 mm fiber were randomly aligned during molding while overall the 0.25 mm fibers were more uniformly oriented. In conclusion, the benefits of adding sisal to PHBV are reducing composite cost, and raising stiffness and impact resistance. Since silane may improve the interfacial interaction in natural fiber polymer composites (Li et al. 2007), the authors immersed 5 mm sisal fibers for 24 h in solution of 3 vol% of a silane coupling agent (namely bis (triethoxysilylpropyl)tetrasulfide) and acetone, followed by drying the fibers in oven. Test results on PHBV/sisal 80/20 wt% untreated and treated with silane revealed that silane boasted the tensile modulus and strength by 10% and 5%, respectively.

Coelho et al. (2019) demonstrated that natural fibers are feedstocks suited for an AM process different than FFF, namely BJ (Section 2.6) that consists in selectively depositing a liquid bonding agent to join regions of powder spread on a bed. By means of BJ they produced a composite

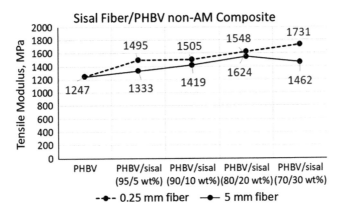

FIGURE 10.7 Tensile modulus values for injection molded coupons of sisal fiber/PHBV non-AM composite.

Source: Data from Dangtungee, R., Tengsuthiwat, J., Boonyasopon, P., Siengchin, S. 2015. Sisal natural fiber/clay-reinforced poly(hydroxybutyrate-cohydroxyvalerate) hybrid composites. *J. Thermoplast. Compos. Mater.* 28 (6): 879–895.

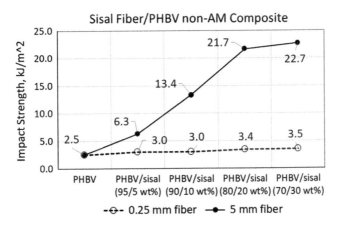

FIGURE 10.8 Impact strength values for injection molded coupons of sisal fiber/PHBV non-AM composite.

Source: Data from Dangtungee, R., Tengsuthiwat, J., Boonyasopon, P. Siengchin, S. 2015. Sisal natural fiber/clay-reinforced poly(hydroxybutyrate-cohydroxyvalerate) hybrid composites. *J. Thermoplast. Compos. Mater.* 28 (6): 879–895.

comprising a gypsum powder matrix (3D Systems VisiJet PXL Core made of 80–90% of $CaSO_4 \cdot 0.5H_2O$), and 1–2 mm long commercial sisal fibers that were wetted by a binder (3D Systems VisiJet PXL Clear) and the infiltrant ethil-2-cyanoacrylate adhesive with 24 h curing time, viscosity of 10–30 cps. Test coupons were printed on a Projet460 Plus BJ printer, and subjected to thermal analysis, flexural testing, and porosity measurements. DOE was employed to analyze efficiently all combinations of printing orientation, and presence and absence of sisal, and measure the effect of each variable. Sisal improved mechanical strength: the average flexural strength of infiltrated (full strength) coupons with sisal and infiltrated coupons without it were 17.1 MPa and 14.8 MPa, respectively. Apparent porosity increased with sisal, which seemed to improve infiltration, and resulted in larger infiltrant-gypsum-sisal bonding area, and hence higher flexural strength.

TABLE 10.8

Properties of Sisal/PHBV Composite Made Via Hot Compression

Material	Fiber Content	Tensile Strength	Tensile Modulus	Impact Strength
	wt%	MPa	GPa	kJ/m^2
PHBV	0	37.6	1,247	2.5
0.25 mm Sisal/PHBV	20	33.8	1,624	3.4
5 mm Sisal/PHBV	20	35.7	1,548	21.7

Source: Dangtungee, R., Tengsuthiwat, J., Boonyasopon, P. Siengchin, S. 2015. Sisal natural fiber/clay-reinforced poly (hydroxybutyrate-cohydroxyvalerate) hybrid composites. *J. Thermoplast. Compos. Mater.* 28 (6): 879–895.

10.5 FLAX

10.5.1 Overview of Flax

Flax, or *linseed*, is an annual food and fiber plant, and a most widely utilized natural fiber. Historically, flax is also one of the first fibers to be extracted, spun, and woven into textiles, dating back to 30,000 BC (Kvavadze et al. 2009). Flax is grown in cooler regions of the world in two primary varieties: one is farmed for its oil and the other (*linseed*) for its fiber. Canada is the largest producer of flaxseed in the world (40%) with China, USA, and India together accounting for 40% of world production (SaskFlax 2020).

Flax fibers are taken from the "inner bark" (*bast*) surrounding the stem. Flax fiber is a cellulose polymer like cotton, but, thanks to its more crystalline structure, it is stronger and stiffer. Flax fibers are composed (wt% values) of cellulose (62–75), hemi-cellulose (11–20.6), pectin (1.8–2.3), wax (1.5–1.7), and moisture (7.9–12) (Yan et al. 2014). Density, tensile strength, modulus, and strain at break of flax, other natural fibers, carbon, and E-glass are collected in Table 10.5, and they vary according to their reporting source. Average tensile strength and modulus of flax are lower than those of E-glass, but can exceed them in terms of specific values, since density is lower in flax than in E-glass. Like bamboo, hemp, jute, kenaf, and sisal, flax is a natural fiber that replaces glass fibers as a reinforcement in engineering composites (Yan et al. 2012)

TABLE 10.9

Properties of Commercial Flax/PLA Composite Filaments for FFF

Material	Density	Tensile Strength	Tensile Modulus	Elongation at Break	Flexural Strength	Flexural Modulus	Hardness	T_m	T_g	MFR
	g/cm^3	MPa	GPa	%	MPa	GPa	Shore D	°C	°C	g/10 min
Extrudr FLAX	1.45	43	N/A	22.3	30	3.4	N/A	160–200	N/A	15
PLA Flax	1.00	39[a]	3.4	2.0	N/A	2.3	77	N/A	54	17

Source: Data posted online by the suppliers.

Note:

[a]https://www.makershop3d.com/expert-filament/743-pla-fibre-flax.html?search_query=flax&results=1.

TABLE 10.10

Properties of Non-AM Flax/PLA Composites with Short Fibers Randomly Oriented

Composite Fabrication	Flax Content	Tensile Strength	Tensile Modulus	Sources
	wt%	MPa	GPa	
Compression molding	30	53	8.3	(1)
Injection molding	30	54	6.3	(1)
Injection molding	10	43	3.9	(1)
Injection molding	20	49	5.1	(2)
Injection molding	30	54	6.3	(2)

Sources: (1) Gopalakrishna, K., Reddy, N., Zhao, Y. 2020. Biocomposites from Biofibers and Biopolymers. In *Biofibers and Biopolymers for Biocomposites*, ed. A. Khan, S. Mavinkere Rangappa, S. Siengchin, A. Asiri, 91–110. Cham: Springer Nature. (2) Bax, B., Mussig, J. 2008. Impact and tensile properties of PLA/Cordenka and PLA/flax composites. *Compos. Sci. Technol.* 68:1601–1607.

10.5.2 COMMERCIAL FLAX-BASED FEEDSTOCKS FOR AM

The only commercial flax-based materials for AM we identified are two filaments for FFF: Extrudr FLAX (Extrudr, Austria) and PLA Flax (Nanovia, France) (Nanovia 2020). Their properties reported by their suppliers are in Table 10.9. PLA Flax is available in 1.75 and 2.85 mm diameter, and contains flax fibers featuring 215 µm length and 10 µm diameter. The Extrudr FLAX is food-safe, FDA-approved, and biodegradable. Neither supplier discloses the fiber content.

Table 10.10 includes tensile strength and modulus of non-AM flax/PLA composites to provide some baseline values.

10.5.3 EXPERIMENTAL FLAX-BASED FEEDSTOCKS FOR AM

Depuydt et al. (2018) and Balthazar (2016) developed a flax/PLA filament for FFF, and assessed the effect of *l/d* on the filament's tensile and thermal properties, with *l* and *d* the length and diameter of the flax fiber, respectively. In their comprehensive and detailed investigations the authors started from the micromechanical models for fiber composites, analyzed the factors critical to maximizing the filament's quality and performance, such as fiber wettability by the matrix, filament porosity, fiber's *l/d* ratio, fiber orientation, extruder speed, etc., and concluded explaining how to control and measure all the above factors. They selected pellets of virgin PLA (Proviron Industries), tape of unidirectional flax fibers called FlaxTape™ (Lineo), cut in 2 mm (F1), and 5 mm (F2) length, and extruded two filaments with PLA, flax, and a plasticizer P1 (Proviron Industries) with the following composition: PLA/flax/P1 equal to nominal 74.8/15/10.2 wt%. The plasticizer was added not only to reduce PLA's brittleness and wind the filament on spools without breaking it, but also to decrease porosity and improve interlayer adhesion and layer finish (Stoof et al. 2017). Before compounding, *l/d* was approximately 12 and 40 for F1 and F2, respectively, but after compounding flax fibers into a PLA matrix, the fiber lengths were significantly reduced, with *l/d* for F1 dropping to 8. This fiber shortening has to be accounted for in producing the filament, because it radically changes the actual *l/d* from the nominal one (Berzin et al. 2014). Tensile properties were measured on single extruded filaments not dumbbell coupons, and are shared in Table 10.11, along with their T_g. The plasticizer notably penalized strength, modulus, and T_g of PLA, and adding flax to cPLA1 (PLA/plasticizer) lowered strength (slightly) and strain but almost doubled modulus. The low T_g basically prevents outdoor use of this material, and one way to boost T_g is to diminish the amount of P1. The following details are conducive to optimizing filament fabrication and performance. A twin-screw extruder is preferred to a single-screw extruder for extruding composites, and was employed by the authors, because the latter one has lower investment cost and

TABLE 10.11

Properties of Experimental Flax/PLA Filaments for FFF

Material	Composition	Nominal Fiber Content	Tensile Strength	Tensile Modulus	Elongation at Break	T_g
		wt%	MPa	GPa	%	°C
F1cPLA1	PLA + 2 mm flax	15	29.9	2.4	0.74	38.3
F2cPLA1	PLA + 5 mm flax	15	29.2	2.3	0.98	N/A
cPLA1	PLA + plasticizer	0	31.0	1.3	1.8	41.1
PLA	PLA	0	68.1	4.3	0.4	61.6

Source: Data from Depuydt, D., Balthazar, M., Hendrickx, K. et al. 2018. Production and Characterization of Bamboo and Flax Fiber Reinforced Polylactic Acid Filaments for Fused Deposition Modeling (FDM). *Polym. Compos.* doi:10.1002/pc.24971.

maintenance and is simpler to operate, but it requires several additives to achieve good mixing, whereas the twin-screw model minimizes material losses, provides consistent products, and mixes well with fewer additives (Balthazar 2016). Filament porosity penalizes mechanical properties, and since porosity is mainly caused by moisture, drying fibers before compounding reduces porosity (Rothon 2003). The flax-PLA interfacial adhesive strength is satisfactory, but not the wetting of flax by molten PLA, which could be improved by increasing the acid groups in plasticizer or PLA (Balthazar 2016).

The paper by Badouard et al. (2019) is notable because the authors fabricated and analyzed a composite filament extruded for FFF comprising 1 mm Alize flax fibers andIngeo™ PLLA 7001D, and compared tensile properties of flax/PLLA printed coupons to those of flax/PLLA injection molded coupons. The tensile modulus, strength and elongation at break of the elementary scutched flax fibers were 53.8 GPa, 1,215 MPa, and 2.2%, respectively. Tensile test samples were fabricated through the following steps: fiber drying, extruding flax and PLLA through twin screw extruder into pellets at 20 rpm and 190°C, pellet drying, injection molding pellets into test coupons, and extruding pellets into a 2.85 mm diameter filament for FFF, processed on a Prusa i3 Rework printer at 190°C. In the composite, flax fibers showed a trimodal distribution of length values with peak values equal to 20 μm, and (mostly) 125 μm and 650 μm. Table 10.12 shows that flax/PLLA for FFF trailed PLLA in strength and elongation and sligthly led it in modulus, and was surpassed by injection molded flax/ PLLA in strength and modulus, due to the porosity present in the printed filament (because of the

TABLE 10.12

Properties of Experimental Flax/PLA Injection Molded (IM) and as Filaments for FFF

Material	Flax Amount	Tensile Strength		Tensile Modulus		Elongation at Break	
	wt%	MPa		MPa		%	
		Avg[a]	SD[b]	Avg[a]	SD[b]	Avg[a]	SD[b]
Flax fibers	100	1,215	500	53,800	14,300	2.2	0.6
PLLA, IM	0	55.2	0.9	3,760	56	2.8	0.4
Flax/PLLA, FFF	10	34.2	2.6	3,968	245	1.2	0.1
Flax/PLLA, IM	10	53.2	1.8	4,765	166	1.9	0.2

Source: Adapted from Badouard, C., Traon, F., Denoual, C. et al. 2019. Exploring mechanical properties of fully compostable flax reinforced composite filaments for 3D printing applications. *Ind. Crops & Products* 135:246–250.
Notes:
[a]Average.
[b]Standard deviation.

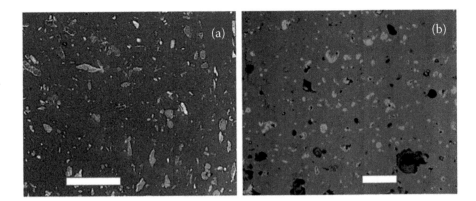

FIGURE 10.9 SEM photographs of cross section of composites of flax (10 wt%) and PLLA: injection molded (a) and printed (b). The scale bar is 500 μm.

Source: Adapted from Badouard, C., Traon, F., Denoual, C., et al. 2019. Exploring mechanical properties of fully compostable flax reinforced composite filaments for 3D printing applications. *Ind. Crops & Products* 135:246–250.

relative low pressure applied to extrude the filaments) but absent in the injection molded samples (Figure 10.9). Table 10.12 also indicates that flax caused higher scattering in tensile strength and modulus vs. unfilled PLLA, with scattering equal to standard deviation/average.

10.6 HEMP

10.6.1 Overview of Hemp

Hemp (*Cannabis sativa L.*) is an annual fast-growing plant, and one of the oldest plants farmed, going back more than 12,000 years to the beginnings of pottery (Shahzad 2012). Hemp grows preferably in the temperate zones between the 25th and 55th parallels on either side of the equator and is widespread in the Northern Hemisphere. It is present in many commercial items, including animal feed, biodegradable plastics, biofuel, clothing, food (seeds), insulation, paper, paints, shoes, textiles, ropes, and sails. The main constituents of hemp fibers are cellulose (55–78), which is the mechanically strongest constituent, followed by hemicellulose (2–22), pectin (0.8–18), lignin (3–13), and others (0.8–7), with values in wt% and depending on the reporting source (Shahzad 2012).

Table 10.13 lists some properties of hemp fibers that possess relatively high mechanical properties (Pickering et al. 2016). The values in Table 10.13 span a wide range because they derive from values published by several authors. More property values of hemp fibers are in Table 10.5 reported by a different source.

TABLE 10.13
Properties of Hemp Fibers

Specific Gravity	Length	Diameter	Tensile Strength	Tensile Modulus	Elongation at Break
N/A	mm	μm	MPa	GPa	%
1.50	5–55	12–32	270–1,235	20–70	1.0–4.2

Source: Adapted from Shahzad, A. 2012. Hemp fiber and its composites – a review. *J. Compos. Mater.* 46 (8): 973–986.

TABLE 10.14

Power Consumption and Harmful Emissions in Producing 1 Kg of Hemp and Glass Fiber

Parameter	Unit	Hemp Fiber	Glass Fiber
Power consumption	MJ	3.4	48.3
CO_2 emissions	kg	0.64	20.4
NO_x emissions	g	0.95	2.9
SO_x emissions	g	1.2	8.8

Source: Adapted from Shahzad, A. 2012. Hemp fiber and its composites – a review. *J. Compos. Mater.* 46 (8): 973–986. https://doi.org/10.1177/0021998311413623.

Hemp fiber is biodegradable and does not require any herbicides or pesticides to grow. It is superior to its synthetic counterpart glass fiber from an environmental standpoint, if we consider the amount of energy required to produce the fibers, and harmful emissions such as CO_2, NO_x, and SO_x associated with their production, as detailed in Table 10.14. Particularly, CO_2 emissions and power consumption for glass are 32 times and 14 times those for hemp, respectively. Another advantage of hemp is that it can replace wood as an engineering filler and reduce deforestation. An overview of industrial hemp fibers was recently authored by Manaia et al. (2019) who mentioned the following industries as users of industrial hemp: textiles, automotive, composites, fiberboard, heat-insulating materials, concrete, sound insulation and/or sound absorption, sporting goods, musical instruments, and brake pads. Other applications of hemp are for construction (Schwarzova et al. 2017), and medical preparations treating chronic diseases, multiple sclerosis, neuropathic pain, and epilepsy (Fiorini et al. 2020). Improving hemp fiber properties and hence widening its application range requires additional R&D, and addressing some issues, such as producing composites with consistent properties.

10.6.2 COMMERCIAL AND EXPERIMENTAL HEMP/POLYMER COMPOSITES

A review of hemp-filled composites was published by Shahzad (2012). Hemp was mixed with TPs such as PU (Czlonka et al. 2020), TS (epoxy), and biodegradable matrices and formed composites featuring a good mechanical performance that improves when hemp fiber surface treatments are applied to enhance the fiber-matrix interfacial bonding. However, since weathering conditions have detrimental effects on hemp (Manaia et al. 2019), outdoors applications have to be carefully designed. Commercial and experimental versions of hemp/polymer composites exist.

Table 10.15 contains property values of available hemp/polymer composites not for AM that can be considered as a benchmark in developing hemp/polymer composites for AM. The polymers listed include PLA (sustainable), PBS (biodegradable and available in biobased and non-biobased form), and PP (fossil-based). Sankhla et al. (2020) reported mechanical, thermal, and morphological properties of experimental hemp fiber composites.

10.6.3 COMMERCIAL HEMP FILAMENT FOR AM

Commercial hemp/PLA filaments for FFF are Entwined Hemp by 3D-Fuel™ and Kanèsis Hemp by Kanèsis (Italy), with the latter one containing hemp wastes from industrial processing. Table 10.16 lists their tensile properties along with those of unfilled PLA filament produced by 3D-Fuel™ in order to compare the effect of hemp on the composite's properties, assuming that 3D Fuel™ employs the same PLA for for its unfilled PLA filament and hemp/PLA filament.

TABLE 10.15

Properties of Injection Molded (IM) and Compression Molded (CM) Hemp/Polymer Composites

Composite	Hemp Amount	Fiber Orientation	Density	Tensile Strength	Tensile Modulus	Elongation at Break	Izod Notched Impact Strength	DTUL at 1.8 MPa
	wt%	N/A	g/cm^3	MPa	GPa	%	kJ/m^2	°C
Hemp/ PLA, IM[a]	30	Random	N/A	75	7.9	N/A	N/A	N/A
Hemp/ PLA, CM[a]	35	Multi-directional	N/A	85	12.7	N/A	N/A	N/A
Hemp/ PLA, CM[a]	40	Multi-directional	N/A	45	7.4	N/A	N/A	N/A
Hemp/ PBS, IM[b]	25	N/A	N/A	N/A	2.65	N/A	7.2	78
Hemp/PP, NAFILean PF2 555, IM[b]	20	N/A	0.98	N/A	2.65	N/A	7.5	72
Hemp/PP, REFINE® PF3 434, IM[b]	30	Random	1.02	44	4.0	4	5.6	52

Sources:[a] Siakeng, R., Jawaid, M., Ariffin, H., 2018. Natural fibre reinforced polylactic acid composites: A review. *Polymer Composites* January 2018. doi: 10.1002/pc.24747. [b]Automotive Performance Materials. https://www.apm-planet.com/pdf/APM-TDS-REFINE-PF3-434.pdf (accessed April 3, 2021).

TABLE 10.16

Properties and Print Temperature of Commercial Hemp and PLA Filaments for FFF

Material	Hemp Content	Tensile Strength	Tensile Yield Strength	Tensile Modulus	Elongation at Break	DTUL	Notched Izod Impact Strength	Print Temperature
	wt%	MPa	MPa	MPa	%	°C	KJ/m^2	°C
Entwined™ Hemp	≤2	38	16	N/A	4.6	N/A	N/A	190–220
Standard PLA	0	41	37	3,200	1.8	N/A	N/A	180–205
Kanesis Hemp	17.5	N/A	41.8	4,420	3.5	50.8	2.4	165–185

Source: Data posted online by the suppliers.

10.6.4 Experimental Hemp-Based Filaments for AM

One advantage of hemp is that it is less expensive than PLA (Jaksic et al. 2017), and hence, added to PLA, it reduces the cost of the resulting composite which, however, has to be traded off against performance. Since the information on the properties of commercial hemp/PLA for AM is scarce,

FIGURE 10.10   Digested fibers: hemp (left) and harakeke (right).

Source: Stoof, D., Pickering, K., Zhang, Y. 2017. Fused Deposition Modelling of Natural Fibre/Polylactic Acid Composites. *J. Compos. Sci.* 1, 8. Reproduced under the terms and conditions of the Creative Commons Attribution (CC BY) license (http://creativecommons.org/licenses/by/4.0/).

the study by Jaksic et al. (2017) is valuable, because it documented the experimental characterization in terms of tensile and thermal properties of a laboratory-developed hemp/PLA filament, and compared its properties to those of wood/PLA. The study did not disclose values but stress-strain curves, and TGA, DSC, and FTIR plots. The tested hemp is industrial, and the study omitted the hemp's content in the filament, possible use of coupling agent, source of PLA, and details on the filament fabrication, except that it was extruded. From the tensile stress-curves of two dumbbell coupons, we calculated for hemp/PLA the following values of tensile strength, modulus, and strain at break: 58.6 ± 0.9 MPa, 3.84 ± 0.42 MPa, and 2.8 ± 0.2%, respectively. The curves displayed that hemp/PLA behaved almost linearly up to failure. The TGA plot showed no weight loss going from 0°C to about 275°C, and hence no thermal degradation in hemp/PLA filament before and after printing.

Stoof et al. (2017) illustrated the experimental process to extrude a hemp/PLA filament for FFF in detail, and the effect of hemp content on the filament's tensile strength and modulus. Hemp

TABLE 10.17

Tensile Properties of Experimental Hemp/PLA and Harakeke/PLA Filaments for FFF

Material	PLA Amount	Tensile Strength	Tensile Modulus
	wt%	MPa	GPa
PLA	0	35.5	2.50
Hemp/PLA	10	37.7	3.42
Hemp/PLA	20	28.9	3.42
Hemp/PLA	30	23.9	N/A
Harakeke/PLA	10	33.9	2.66
Harakeke/PLA	20	37.2	4.28
Harakeke/PLA	30	28.3	4.08

Source: Values derived from plots in Stoof, D., Pickering, K., Zhang, Y. 2017. Fused Deposition Modelling of Natural Fibre/Polylactic Acid Composites. *J. Compos. Sci.* 1(8). doi:10.3390/jcs1010008.

fibers by Hemp Farm (NZ) were sifted through an 8-mm sieve, immersed in alkali/water solution (5% NaOH), transferred to a digester, where they were heated from 20°C to 160°C during a 90-min interval and kept for 30 min at 160°C, washed in tap water, and dried at 103°C for 48 h, with a final diameter of 28.3 ± 8.3 μm (Figure 10.10 left). The PLA chosen was Ingeo™ 3052D powder for injection molding, featuring MFR of 14 g/10 min, T_g of 55–60°C, tensile yield strength of 62 MPa, flexural modulus of 3,600 MPa, and elongation at break of 3.5%. PLA powder was dried in the oven at 85°C for 48 h prior to mixing with hemp. The hemp/PLA filament was prepared combining PLA powder and 10, 20, and 30 wt% hemp in a mixer at 180°C and 16 rpm, sifting through an 8-mm sieve, and finally extruding the resulting granules in a twin-screw extruder into a 3-mm diameter filament. Dumbbell-shaped tensile coupons were printed and tested. Table 10.17 indicates that 10 wt% hemp improved tensile strength in PLA, but greater amounts of hemp impaired the strength, which was possibly caused by higher amount of porosity, uneven volume distribution, and agglomeration. The tensile modulus at 10 wt% and 20 wt% hemp was higher than that in unfilled PLA which may confirm good interfacial bonding between matrix and fibers at least at low strain. The tensile modulus at 30 wt% hemp was not reported.

10.7 HARAKEKE

Harakeke is the Maori name for *Phormium tenax*, an evergreen perennial plant native to Australia and New Zealand from which fibers are extracted to make textiles, ropes, sails, sacks, and crafts. Tensile properties of harakeke are listed in Table 10.5. Since harakeke fiber has similar properties to those of sisal fiber, and the latter one is chosen to reinforce polymer composites, harakeke may be an effective ingredient in composites as well.

Stoof et al. (2017) developed a process to extrude a harakeke/PLA filament for FFF (similar to the process for hemp/PLA) and reported the effect of fiber amount on its tensile strength and modulus. Harakeke fibers by Templeton Flax Mill (NZ) were sifted through an 8-mm sieve, immersed in alkali solution (5% NaOH plus 2% Na_2SO_4), transferred to a digester to be heated from 20°C to 160°C during a 90-min interval and kept for 40 min at 170°C, washed in tap water, and dried at 103°C for 48 h (Figure 10.10 right). The fibers' final diameter was 13 μm. PLA was Ingeo™ 3052D. Harakeke filament was fabricated going through the same process conditions and equipment for hemp/PLA: mixing PLA powder and 10, 20, and 30 wt% harakeke, and extruding a 3-mm diameter filament. Dumbbell-shaped tensile coupons were printed and tested. Table 10.17 points out that harakeke raised strength and modulus of unfilled PLA: particularly 20 wt% improved both, whereas 10 wt% and 30 wt% increased only modulus. The modulus is by definition measured at low stress (exactly, linearly elastic stress), hence the fact that it improved at all fiber contents means that there was an adequate transfer of stress from PLA to harakeke at low stress level. The fact that tensile strength at 10 wt% and 30 wt% trailed PLA's strength may be explained by a required minimum content of harakeke (20 wt%), and that above that value the void amount and/or fiber uneven distribution might have penalized the composite's strength.

10.8 NUTSHELL AND NUT SKIN

10.8.1 COCONUT SHELL

Tran et al. (2017a) extruded and tested composite filaments for FFF comprising cocoa bean shell waste (CBSW), a by-product of the chocolate industry, and PCL, a synthetic, semi-crystalline, biocompatible, and biodegradable polymer, and a feedstock for commercial FFF filaments. PCL is easy to process, has a low melting temperature (60°C), and high decomposition point (350°C). As a feedstock, CBSW is sustainable and inexpensive, and already successfully utilized as a filler in silicone matrices (Tran et al. 2017b). PCL powder and CBSW were supplied by Polysciences (USA) and Ferrero (Italy), respectively. PCL was used as is, whereas CBSW was reduced into particles of

TABLE 10.18

Properties of Coconut Bean Shell Waste/PCL (CBSW/PCL) Composite for FFF

Material	Filler Content	Tensile Modulus	Elongation at Break
	wt%	MPa	%
PCL	0	304 ± 20	702 ± 6
CBSW/PCL	10	334 ± 8	697 ± 6
CBSW/PCL	20	329 ± 12	538 ± 18
CBSW/PCL	30	356 ± 9	495 ± 30
CBSW/PCL	40	338 ± 22	129 ± 56
CBSW/PCL	50	312 ± 12	23 ± 8

Source: Tran, T. N., Bayer, I. S., Heredia-Guerrero, J. A. et al. 2017. Cocoa Shell Waste Biofilaments for 3D Printing Applications *Macromol. Mater. Eng.* 302:1700219.

50 μm average diameter, blended with PCL at different ratios (10, 20, 30, 40, 50 wt% of CBSW), dried, and fed into an extruder where it was homogeneously dispersed in PCL, and formed into a 1.75 mm diameter filament that displayed featureless, smooth, compact, and nonporous surface, with CBSW particles well dispersed within PCL without clusters and clumps. The test results are included in Table 10.18: CBSW increased the tensile modulus of PCL from 304 MPa up to a maximum of 356 MPa (30 wt% CBSW). In printed objects the filament displayed good interlayer adhesion and fine resolution. Through extensive testing, the authors detected: (a) no chemical interactions between PCL and CBSW; (b) no significant alteration of PCL's crystalline structure from melt extruding PCL and CBSW, and hence no degradation in PCL's mechanical properties; (c) marginal reduction of thermal stability in composite versus PCL, expressed by the degradation temperature declining from 400°C in PCL filament to 387°C in CBSW/PCL 50/50 wt% filament. Potential products of CBSW/PCL are biomedical and household items. The paper by Tran et al. (2017a) is also recommended as a guide to comprehensive laboratory analysis of superficial and internal morphology, crystallinity, thermal degradation, etc. of newly developed filaments in general.

10.8.2 MACADAMIA NUTSHELL

Wood/polymer composites (WPCs) consist of recycled polymers, and, mostly, wasted wood flour. In the last two decades WPCs have represented a sustainable and affordable alternative to natural wood products, especially in the areas of buildings, construction, and furniture, thanks to their mechanical properties and durability (Oksman Niska and Sain 2008). However, since the demand for WPCs is growing, it is questionable whether wood supply could meet future demand of WPCs. A sustainable and affordable material without wood that is alternative to WPCs is polymer matrix composites filled with macadamia nutshell waste (MNW), with macadamia nutshell a material similar to wood in chemical composition but not in density and structure. Girdis et al. (2017) developed a FFF filament that was made of ABS, MNW, and maleic anhydride binder, and offered the double environmental advantage of sparing wood and reducing waste. The authors selected macadamia nutshell fiber with average particle diameter of 94 μm, and, before filament extrusion, average *l/d* of 0.6. ABS, fibers, and binder were mixed and extruded into a 1.75 mm diameter filament out of which compression and tensile test samples were built on a commercial Leapfrog Creatrprinter with 30% and 100% infill, respectively. The authors measured tensile and compressive properties divided by density, and the test results are listed in Table 10.19, along with properties of woodFill filament by colorFabb added as a baseline. In comparing the test results the difference in wood content across the composites has to be considered. The 19 wt% MNW/ABS trails woodFill in compressive properties, but outperforms

TABLE 10.19

Mechanical Properties of Macadamia Nutshell Waste (MNW)/ABS Composite for FFF

Material	Filler	Filler Content	Specific Tensile Strength	Specific Tensile Modulus	Elongation at Break	Specific Compression Strength	Compression Strain
		wt%	MPa/ (g/cm^3)	GPa/(g/cm^3)	%	MPa/(g/cm^3)	%
ABS	None	0	N/A	N/A	N/A	28.8	12.5
MNW/ABS	MNW	19	23.3	2.04	2.5	14.7	16.7
MNW/ABS	MNW	29	N/A	N/A	N/A	12.9	23.0
woodFill (wood/ PLA)	Wood	10–20[a]	17.8	1.62	1.6	18.5	20.5

Source: Girdis, J., Gaudion, L., Proust, G. et al. 2017. Rethinking Timber: Investigation into the Use of Macadamia Nut Shells for Additive Manufacturing. *JOM* 69 (3): 575–579. doi: 10.1007/s11837-016-2213-6. Published by The Minerals, Metals & Materials Society (TMS).

Note:

[a] Referred to content of wood and additives together according to colorFabb's MSDS posted online.

TABLE 10.20

Test Results of PLA/Almond Skin Powder (AS) Filament for FFF

Composition	Filament diameter	Tensile Strength	Elongation at Break	Modulus of Toughness
	mm	MPa	%	MPa
PLA	1.69–1.99	42.8–47.8	7–13	1.01–2.89
PLA/2.5 wt% AS	1.94–2.03	44.6–53.4	5–6	1.01–1.34
PLA/5 wt% AS	1.97–2.11	21.1–30.4	9–20	1.23–2.07

Source: Data from Singh, R., Kumar, R., Preet, P., Singh, M., Singh, J. 2019. On mechanical, thermal and morphological investigations of almond skin powder-reinforced polylactic acid feedstock filament. J. Thermoplast. Compos. Mater. 1–19. https://doi.org/10.1177/0892705719886010.

it in tensile properties, aided by the fact that the measured density of ABS and PLA was 1.06 and 1.28 g/cm^3, respectively. In analyzing tensile and compressive properties one must consider that a different infill was used between the two sets of coupons: in compressive loading, localized buckling of the outside layers was the dominant failure mode across all samples, and it is possible that the woodFill samples were more resistant to buckling than MNW/ABS.

10.8.3 ALMOND SKIN

Almond skins (AS) are a waste by-product of blanched almond production. Singh and her team combined almond skin powder as filler and PLA as matrix in a new filament to print TE scaffolds via FFF. PLA3052D (Nature Tech, India) and natural almond skin obtained as waste from the almond oil industry were selected. The skin was dried in oven, then crushed inside a cryogenic ball mill grinder to a particulate having average size of 50 μm. AS in amount of 2.5% and 5 wt% was added to PLA, and extruded into filaments of unfilled PLA, and AS/PLA featuring diameter of

1.7–2.1 mm. Extrusion was experimented at varying values of torque (0.1, 0.2, 0.3 Nm) and load (10, 12.5, 15 kg) and nine combinations of AS content and extrusion parameters were tested. According to Table 10.20 (Singh et al. 2019) 2.5 wt% AS/PLA resulted in the highest range of tensile strength, namely the formulation extruded at 0.3 Nm and 10 kg featured tensile strength, elongation at break, and modulus of toughness of 53.4 MPa, 6%, and 1.44 MPa, respectively. The modulus of toughness measures the material's capability to withstand plastic stress (*ductility*). The presence of AS marginally affected T_g: upon shifting from 0 to 2.5 to 5 wt%, T_g varied from 58.3 to 57.5 to 59.7°C, respectively. One value reported of surface roughness R_a was 173 nm for 2.5 wt% AS/PLA. The authors developed an equation enabling to predict compressive strength, modulus, and strain based on values of composition, torque, and load.

On the same 2.5 wt% AS/PLA formulation, Singh et al. (2020) also recorded the compressive strength and modulus, hardness, and surface roughness. A total of nine combinations of infill value (60, 80, 100%), infill angle (45, 60, 90°), and printing speed (50, 70, 90 mm/s) were tested, following a Taguchi-based DOE. Not surprisingly, the three combinations with 100% infill density provided the largest set of values in compressive strength (about 37 MPa) and modulus (275–388 MPa), and among those combinations the highest modulus (388 MPa) and modulus of toughness (2.24 MPa) came from 90° and 70 mm/s, and from 45° and 90 mm/s, respectively. The surface roughness R_a for 100% infill density combinations was 6.5–11.8 nm.

10.9 PLANT-BASED WASTE

Sustainable materials in form of waste materials from plants and seashells were studied by Zeidler et al. (2018) as a prospective feedstock to print custom-shaped, low-cost packaging for sensitive components, such as temperature-sensitive electronic components. Their work is interesting also because the requirements for packaging are different than those for engineering applications under significant loads, and include low-cost, resistance to minor loads, lightweight, and being disposable. Same requirements hold for architectural models, movie set and stage props, and visual aids. An environmentally friendly packaging material would biodegrade quickly. Zeidler et al. (2018) focused on the AM process BJ. The advantage of BJ is its capability to produce customized shapes at a cost and lead time inferior to those of conventional fabrication processes, but, on the other hand, the question is whether the printing speed of 20 mm/h of build height in state-of-the-art printers allows to make a cost-competitive product. The answer may be larger and faster printers. The investigators considered several feedstock candidates, including rice husk, and experimented with: (a) powder made of miscanthus (or silvergrass), apricot stone, sea shells, and wood (larch, maple); (b) lignin sulfonate, sodium silicate, and PVA as binders; (c) acrylic resin, epoxy, stearin from suet and tallow, as coatings and infiltrating materials to raise the mechanical strength of and impart water resistance to the printed parts. Sample parts were printed in specific shapes out of apricot stone, larch and maple wood flour, miscanthus, and seashell to prove it was possible to achieve accurate and complicated shapes. Combining AM and miscanthus was already researched in a project funded by the BMW company (Klemm et al. 2015).

10.10 ALGAE

10.10.1 Overview of Algae

Algae can be broadly described as organisms that carry out oxygen-producing photosynthesis, are not "higher plants" and embryophytes (land plants), and include bacterial and eukaryotic organisms (Raven and Giordano 2014). However, since algae look like plants, for convenience, AM feedstocks comprising algae were included in this chapter. Alga means *seaweed* in Latin. There are thousands of algae species that reciprocally differ in structure and composition (Graham et al. 2009; Dodge 2012). Approximately they can be split in three types: green, brown, and red algae. In terms of structure, algae are similar to land plants and consist of stiff and strong cellulose-based fibrils that are

TABLE 10.21

Diameter and Tensile Properties of Dried Filaments of Green Alga (*Cladophora Glomerata*)

Filament	Diameter	Tensile Strength	Tensile Modulus	Elongation at Yield
	μm	MPa	MPa	%
Early season	78 ± 20	40.9	50 ± 61	1.3
Mid-season	99 ± 26	19.4	21 ± 18	1.4

Source: Johnson, M., Shivkumar, S., Berlowitz-Tarrant, L. 1996. Structure and properties of filamentous green algae. *Mater. Sci. Eng. B* 38:103–108.

embedded in a soft amorphous matrix, but, contrary to plant fibers, cellulose content is below 10 wt% of their dry mass, and the amount of all their constituents largely differ within the species (Bulota and Budtova 2015). Algae use photosynthesis to transform CO_2 and sunlight into energy so efficiently that they can double their weight several times a day (Prosina 2020). During the photosynthesis process algae produce oil and can generate 15 times more oil per acre than other plants harvested for biofuels, such as corn and switchgrass (University of Virginia 2008). Added to biodegradable plastics, such as PLA, algae increase their biodegradability. Algae are commercially and industrially cultivated to produce bioplastics, chemical feedstock, fertilizer, food, food ingredients such as omega-3 fatty acids and natural food colorants and dyes, fuel (bioethanol), and pharmaceuticals.

Johnson et al. (1996) studied structure and properties of the green alga (*Cladophora glomerata*) from freshwater rivers in California, and their findings are summarized here as a benchmark. First, the microstructure and tensile properties of the algae may depend on the period during which the algae are harvested. The algal filaments are composed of cylindrical cells 40–100 μm in diameter, and 100–400 μm in length. The cell walls consist of crystalline cellulosic microfibrils and are about 10% of the cell diameter. The tensile properties of dry filaments exceed those of wet and soaked filaments. Since algae present in composites are in dry form, Table 10.21 includes values of tensile properties of dried filaments, showing how harvest time distinctly affects the values, and the large scattering in modulus, which is also due to the intrinsic variations typical in natural, not man-made filaments. Tensile strength and modulus of green alga are orders of magnitude lower than those of the natural fibers in Table 10.5, and hence green alga is distinctly less suited than those fibers for load-bearing applications.

10.10.2 Algae/Polymer Composites

Algae have been present in non-AM composites in two forms (Bulota and Budtova 2015): (a) untreated, milled, and pulverized fillers, in order to reduce composite's cost and carbon footprint, and consume algae waste; (b) treated reinforcing fibers. R&D activity performed on algae/polymer composites not specific for AM provides valuable information in developing the same type of feedstock for AM, and is described in the references (including also biobased polymers) cited in Bulota and Budtova (2015).

Bulota and Budtova (2015) investigated non-AM composites of Ingeo™ 3051D PLA filled with different algae (red, brown, and green), and prepared via melt mixing, and experimented with various lengths (200–400 μm and below 50 μm) and concentration (2, 20, 40 wt%) of algae. Algae were dried, sifted, and shaped as flakes. Figure 10.11 shows test results of injection molded coupons, with Young's modulus synonym of tensile modulus. Overall the presence of algae is detrimental to the tensile properties, and the only improvement is in modulus when the algae content reached 40 wt%. The decrease in elongation at break in Figure 10.11(c) and (d) may signal

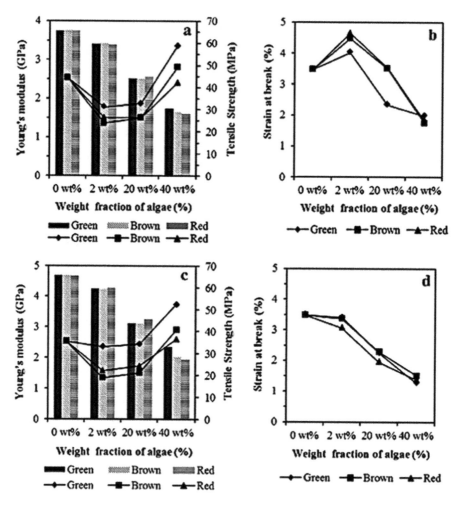

FIGURE 10.11 Test results for non-AM algae/PLA composites with the initial size of particles below 50 mm (a, b), and between 200 and 400 mm (c, d). Columns represent tensile strength, points represent tensile modulus and strain at break.

Source: Bulota, M., Budtova, T. 2015. PLA/algae composites: Morphology and mechanical properties. *Composites: Part A* 73:109–115. Reproduced with permission from Elsevier.

a decline in impact strength. Adhesion between algae flakes and PLA can be improved by removing salts from algae surface through a chemical treatment.

10.10.3 COMMERCIAL ALGAE-BASED AM FEEDSTOCKS

3D Printlife (US) commercializes Alga™, a 100% biodegradable filaments for FFF incorporating algae and PLA formulated by ALGIX3D®. Alga™ is offered in 1.75 and 2.85 mm diameter and various colors, is printed at 195–210°C, and features the properties in Table 10.22. According to 3D Printlife, the target for Alga™ is prototyping, education and design and artistic projects.

Cellink (Sweden) sells Alginate Kit, a bioink composed of sodium alginate for applications ranging from 3D cell culture to drug delivery to bioink components. The kit is in form of powder to be dissolved in phosphate-buffered saline solution, and achieve the ideal concentration for a

TABLE 10.22

Properties of Commercial ALGA™ Filament by 3D Printlife

Material	Density	Tensile Yield Strength	Tensile Modulus	Elongation at Yield	Toughness	Failure Mode	Melting Temp.	DTUL at 1.8 MPa	MFI at 195°C
	g/cm^3	MPa	MPa	%	J	N/A	°C	°C	g/10 min
Alga™	1.25	41	2,860	3.06	0.15	Brittle	160	55	27

Source: Data posted online by 3D Printlife.

specific application. The printed item is then coated with a crosslinking agent to form a stable network.

10.10.4 EXPERIMENTAL ALGAE-BASED AM FEEDSTOCKS

Prosina (2020) proposed algae as an ingenious approach for establishing long-term habitation on the Moon and Mars, by employing a printed mixture of regolith-filled algae to build habitable structures on those celestial bodies without resorting to mining and sifting. *Regolith* is a layer of loose, heterogeneous superficial deposits (dust, soil, broken rock, and likes) that covers solid rock on Earth, Moon, and Mars. The algae will be grown on-site in laboratory, and regolith will be harvested on-site. Algae present the following advantages: they lend themselves to be processed into TPs, can be converted into many everyday consumables including clothing, are rich in proteins, and can be transformed into food (dietary supplements and "superfoods"). The author discussed benefits, utilization, processing of algae, required infrastructure, and associated challenges.

10.10.5 EXPERIMENTAL ALGINATE-BASED AM FEEDSTOCKS

Alginic acid, or *algin*, is a polysaccharide that is present in the cell walls of brown algae and imparts structural properties. Algin is hydrophilic and forms a viscous gum when hydrated. Salts of algin with metals such as sodium and calcium are called *alginates*. Alginate is an attractive biomaterial for TE because it is biocompatible, slightly toxic, relatively inexpensive, and produces a mild and rapid formation of hydrogel (HG) (Lee and Mooney 2012), with the latter characteristic leading to stable geometry after printing. Brown algae contain an average of 40% alginate. Alginates are utilized by agricultural, biomedical (drug delivery, TE, wound healing), cosmetic, food, medical, paper, pharmaceutical (ingredient), and textile printing industries, with biomedical, medical, and pharmaceutical making up 20% of the global market. Cosmetic, food, and pharmaceutical industries primarily utilize alginates as thickeners and stabilizers. The price of alginates varies depending on its purity and applications: alimentary grade sodium alginate is USD 6.5–11/kg, while pharmaceutical grade is USD 13–15.5/kg. Global alginate market was USD 624 million in 2016, and is expected to grow at 2–3% per year. Pereira and João Cotas (2020) published a wide-ranging overview of alginates, from which some of above information were derived.

Alginate has been largely employed as HG in EBB because of its low cost, biocompatibility, good bioprintability, fast gelation, despite its low adhesion to living cells and spreading without HG modification (Ozbolat and Hospodiuk 2016). Several researchers have printed bioinks out of alginate HG for TE, and drug release. Chung et al. (2013) printed biological cells from bioinks of alginate, and bioinks of alginate/gelatin blends for muscle tissue regeneration, and compared in printability both bioinks to pre-crosslinked alginate and alginate solution. Ahn et al. (2012) printed 3D cell-laden alginate scaffolds highly porous and with uniformly designed pore size and shape, also for tissue regeneration. Fabricating 3D thick tissues is still a major challenge, because it requires a suitable vascular network that is even more important in thick tissues and functional organs conducting high metabolic activities, such as heart and liver. Hence, another AM application in TE is printing vascular networks integrated in printed organs. In fact, Zhang et al. (2013) selected alginate and chitosan HGs to print microfluidic channels serving as a vascular network incorporated in printed organs. Zhang et al. (2015) introduced a novel method to print vessel-like microfluidic channel in form of hollow tubes from alginate and chitosan HGs, and studied the effect of the biomaterial and its flow rheology on geometric properties of the microfluidic channels. The channels were designed to serve as vascular conduits.

Nishiyama et al. (2009) developed a new inkjet-based bioprinter, and built 3D biological gel structures composed of living cells and HG in vitro. The authors selected an alginate HG obtained through the reaction of a sodium alginate solution with a calcium chloride solution, and accurately printed structures shaped as lines, planes, laminated structures, and tubes characterized by good cell survivability.

Davoodi et al. (2015) printed microparticles for controlled released of drugs, namely 15–50 μm microspheres made of alginate shell around a core of PLGA, a type of PLA in which D- and L-lactic acid forms are present in equal ratio. According to the experimental results, the microspheres could be a promising device for controlled drug release. Microcapsules containing biological cells have been developed in medicine to cure various diseases, including cancer and diabetes: the capsules feed the cells, allowing blood carrying oxygen and nutrients inside them, and protect the cells from attack by the immune systems.

Islet cells are specific stem cells. Ma et al. (2013) printed, via co-axial electro-jetting, a new type of alginate-based HG microcapsules with core-shell structures for islet cells that confined the islets in the core region of the capsules, and protected the islets by preventing them from protruding outside capsules and receiving inadequate immuno-protection.

Consistently with its use in food, alginate has also been evaluated in AM of food. Namely, Dankar et al. (2020) investigated the effects of alginate (and other ingredients) on the mechanical and rheological properties of potato puree prepared via microwave heating and boiling, and ultimately the effect of alginate on the suitability of the potato puree to be printed.

10.10.6 Experimental Agarose-Based AM Feedstocks

Agarose is a family of high-MW (120,000 g/mol) polysaccharides based on galactose (a monosaccharide sugar about as sweet as glucose), and extracted from seaweed found mostly in the Pacific and Indian Oceans. Agarose is insoluble in cold water, but readily dissolves in boiling water. Upon cooling, agarose chains form side-by-side aggregates that condense into a 3D, interlocking network held together by non-covalent hydrogen bonds. Because each individual gel fiber contains many agarose chains, agarose gels possess relatively high mechanical strength (Mayer and Fiechter 2013) even in diluted concentrations. A few grades of agarose with different T_m are marketed (Ozbolat and Hospodiuk 2016). Agarose is chosen for its thickening and gelling properties in food, industrial, and pharmaceutical applications.

Although agarose is fragile, features low adhesion to living cells, and requires heated printing equipment, it is selected as a bioink HG for EBB because of its low cost, compatibility with living cells, low melting and gelling temperatures, high mechanical properties, stability, and capability to produce an immuno response. Examples of items bioprinted in agarose are scaffolds for human adult stem cells (Awad et al. 2004), and 3D structures with human stem cells encapsulated into agarose HGs (Duarte Campos et al. 2013).

TABLE 10.23

Mechanical Properties of Selected Varieties of Rice Straws

Rice Straw Variety	Tensile Modulus	Tensile Strength	Elastic Modulus in Bending	Bending Strength	Shear Strength
	MPa	MPa	MPa	MPa	MPa
Alikazemi (Iran)[a]	N/A	N/A	543	7.46	8.59
Hashemi (Iran)[a]	N/A	N/A	687	8.93	13.08
MTU2077, crushed (India)[b]	2,427	69.7	N/A	N/A	N/A
MTU2077, uncrushed (India)[b]	3,323	74.6	N/A	N/A	N/A

Notes:
[a]Tavakoli, M., Tavakoli, H., Azizi, M. H., Haghayegh, G. H. 2010. Comparison of Mechanical Properties between Two Varieties of Rice Straw. *Adv. J. Food. Sci. Technol.* 2 (1): 50–54.
[b]Ratna Prasad, A. V., Murali Mohan Rao, K., Moran Rao, K., Gupta, A. V. 2007. Tensile and impact behavior of rice straw-polyester composites. *Indian Journal of Fibre & Textile Research* 32:399–403.

10.11   RICE STRAW

Rice straw (RS) or *paddy straw* is the vegetative part of the rice plant (*Oryza sativa*), and is cut during grain harvest or later. RS "consists of jointed hollow culms, leaf blades, and threshed-out panicles. The main rice culm, also called *rice stem*, is made up of a series of round, hollow internodes separated by solid nodes" (Bassyouni et al. 2014). The length of RS varies with the rice species and cultivation method, and typically reaches 1–2 m. RS has the following chemical composition in wt%: cellulose 38.3, hemicellulose 28.0, lignin 14.9, and ash 18.8 (Zhang and Cai 2008). Structurally, RS is a reinforced composite of aligned cellulose fibers forming microfibrils 3–4 nm in diameter. The microfibrils are bound together by hemicelluloses and lignin, and form macrofibrils having diameter of 10–25 nm (Bassyouni et al. 2014). Table 10.23 includes tensile, bending, and shear properties of three varieties of RS from technical literature. Like for other natural fibers, and differently from synthetic fibers, mechanical properties of rice straw feature large scattering, because they depend on several factors: variety, harvest time, ground composition, chemical treatment if applied, etc.

RS is an abundant and underutilized residue of agricultural production that may be burned (and hence wasted, polluting and contributing to climate change), left on the field before the next ploughing, ploughed down as a soil improver, or utilized, after physical or chemical treatment, as a feed for livestock, as a major forage in rice-producing areas (Kadam et al. 2000). There are options to make a functional use of RS, thus reducing the need to burn the excess waste: RS can be a source of fuel and a natural filler in wood/plastic composites (Grozdanov et al. 2006), and composites with matrix made of PVA, PS (Ismail et al. 2011), and PP (Majid et al. 2016). The benefits of using RS as a natural fiber are low cost, biodegradability, high strength-to-weight ratio, but its shortcomings are high rate of moisture absorption and inconsistent fiber sizes. Since natural fibers are inexpensive, they are combined with low-cost polymers to form composite materials, and hence TP polymers such as HDPE, LDPE, PP, PVC, and ABS are commonly selected.

In 2014, a press article announced that a plastic filament for FFF containing straw was commercially available (Hobson 2014). The Chinese company Jinghe formulated a material by grinding up various dried crops (wheat, rice, and cotton) that were typically disposed of by burning. The resulting sawdust was mixed with additives (PP, a silane coupling agent, and ethylene bis(stearamide), and extruded into pellets to be melt extruded into a filament (Hobson 2014). However, when writing this book, we did not find online any commercial feedstock for AM containing straw.

TABLE 10.24

Properties and Cost of Rice Straw/ABS Filament for FFF

Rice Straw Amount	Tensile Strength	Tensile Modulus	Flexural Strength	Flexural Modulus	Water Absorption	Total Material Cost
wt%	MPa	MPa	MPa	MPa	wt%	USD/kg
0	39	540	52	1,750	1.47	1.057
5	30.5	521	41.2	1,687	2.19	1.021
10	26.5	500	32.3	1,425	2.95	0.958
15	29	558	51.5	2,312	4.13	0.949

Source: Osman, M. A., Atia, M. R.A. 2016. Investigation of ABS-rice straw composite feedstock filament for FDM. *Rapid Prototyp. J.* 24/6:1067–1075.

In developing a RS/plastic composite filament for FFF, the following aspects should be considered in order to maximize the filament's performance (Bassyouni et al. 2014):

- *Thermal stability*: RS should be processed below 190°C, because otherwise it decomposes, and produces volatile organic compounds that form pores lowering the mechanical properties of resulting parts.
- *Moisture amount*: since RS contains cellulose that is hydrophilic, it does not combine well with hydrophobic polymers. RS can absorb 7–10 wt% moisture, which lowers the mechanical properties in composites. A method to reduce the moisture uptake in RS is grafting RS with acrylic or vinyl monomers.
- *Particle size*: RS can be employed as refined fiber, strands, and ground pellets, and typically is employed in ground form having size range of 100–1000 μm for non-AM composites. The particle size controls the composite's mechanical and thermal properties. At constant amount of RS, a larger particle size increases the composite's tensile strength, because greater particles transfer stress more effectively at the filler-matrix interface than smaller particles.

Osman and Atia (2018) found a lack of technical literature covering the use of RS-reinforced plastic composites for FFF, and investigated RS combined with ABS, measuring various properties. Namely, virgin ABS pellets were ground down to powdery form, sieved, and particles 0.71-1.70 mm in size were collected. Washed and dried RS was obtained from local farmlands, ground to produce flour, then sieved, and particles 0.105-0.15 mm in size were selected. RS was dried in oven until its mass stabilized. ABS powder was combined with 0, 5, 10, 15, and 20 wt% of RS flour in a mixer, and fed to Filastruder, a low-cost, benchtop single-screw extruder that, running at 190°C, produced a 1.75 mm diameter RS/ABS filament. Test coupons were printed with a 0.2 mm layer height, 100% infill density, 250°C nozzle temperature, and 100°C bed temperature on a Printrbot Simple Metal printer for FFF, fitted with a 1 mm nozzle. Table 10.24 lists the test results. Composites with all filler contents trailed unfilled ABS in tensile and flexural strengths and moduli, except for the moduli at 15 wt%. Increasing content of RS led to greater water absorption due to the hydrophilic nature of RS, and lower cost, since the cost in USD/kg of ABS and RS selected for the experiments was 1 and 0.28, respectively. Since RS was uniformly distributed across the ABS matrix, the authors attributed the drop in strength upon adding RS to porosity and weak interfacial bond between ABS and RS.

Majid et al. (2016) investigated RS fibers/PP composites in which one composite contained untreated fibers, and the other RS fibers immersed in a water solution of 5 wt% of NaOH for 24 h and dried. A significant increase (approximately from 10% to 70% depending on type of RS fiber and its amount) in tensile strength of treated RS/PP compared to untreated RS/PP was recorded. Possibly, this treatment can be effective also for RS present in polymeric filaments for FFF.

However, chemical surface treatment of natural fibers might alter their crystallinity and ultimately penalize the mechanical performance of the resulting composite material (Bassyouni et al. 2014).

10.12 BEER, COFFEE, AND WINE WASTE

Since the AM feedstocks described in this chapter derive from plants and trees, two commercial FFF filaments made by 3D-Fuel® out of beer and coffee are mentioned here. Wound Up™ is a coffee/PLA filament, made of <10 wt% coffee by-products and Ingeo™ 3D850 PLA, and featuring the following tensile properties: strength 34 MPa, yield strength 16 MPa, and strain at break 4.1%. Parts printed exhibit a rich brown color and a noticeable natural grain. 3D-Fuel Buzzed™ is made of PLA and <8 wt% of waste by-products from the beer-making process. It generates items with a rich golden color and natural grain. No material properties are disclosed for Buzzed™, but we expect that, like Wound Up™, it is intended for cups, glasses, and mugs, and other decorative items.

Andrew Kudless, a San Francisco-based designer and architect, took grape skins discarded during the champagne-making process in France, upcycled them into a feedstock for AM, and printed ice buckets employing BJ (Section 2.6). Others followed his lead and printed wine goblets with grape scraps. In a similar vein, Marina Ceccolini, an Italian-based designer, formulated a printing feedstock out of food found in landfills: coffee grounds, orange peels, peanut shells, potato starch, and tomato skins (Krueger 2017). We realize that the items by Kudless and Ciccolini are niche products, even if sold in "large" numbers.

Recycling agro-waste to fabricate plastics is not new. Bayer et al. (Fondazione Istituto Italiano di Tecnologia 2015) devised a new process that employed an organic acid to transform cellulosic inedible agro-waste (cocoa pod husks, oat hulls, orange peel, parsley stems, rice husks, spinach stems, etc.) into amorphous plastics displaying a wide range of mechanical properties: from weak and stretchable to strong and stiff plastics. A further advantage is that the acid can be recycled in a closed system of production.

10.13 GRAINS

Dried distillers' grain with solubles (DDGS) is a dried cereal by-product of the distillation process of alcohol from wheat grain to produce ethanol, and contains 10–12 wt% moisture. After fermentation of grain to make ethanol, the alcohol is removed, and the residue is dried to yield DDGS that is usable as a source of energy and protein in animal diets (fish, pigs, poultry) due to its high content of protein, water-soluble vitamins, and minerals (Chatzifragkou et al. 2018). The online price of DDGS, here reported only as indicative, widely changes varying from USD 50 to USD 450 per metric ton, with a minimum order of 10 and more metric tons (alibaba.com 2020a), whereas PLA pellets are sold for USD 3000–4000 per metric ton (alibaba.com 2020b).

Tisserat et al. (2015) formulated and tested a fully sustainable DDGS/PLA filament for FFF with the intent to produce a filament less expensive than commercial filaments in unfilled PLA, and PLA filled with bamboo or wood. DDGS were chemically modified by extracting oils with hexane, then milled, ground, and sieved. Particles smaller than 63 μm in diameter were mixed with Ingeo™ 2003D PLA and extruded into pellets, in turn extruded into a 1.75 mm diameter filament for FFF, according to the formulations in Table 10.25 that comprised PLA, DDGS, Paulownia wood (PW), and Osage orange wood (OOW). A version (termed EDDGS) of the above filaments was prepared by encasing the composite filaments entirely in one build layer of unfilled PLA, and hence increasing the PLA content of the filament. Test coupons were printed on a MakerBot Replicator 2X with infill 100%, extrusion temperature and speed of 220°C and 90 mm/s, respectively, and build platform at 70°C. Measured values of tensile strength, modulus, and strain at break are in Table 10.26: no formulation outperformed PLA in tensile strength, only PLA/EDDGS/PW exceeded PLA in modulus, and three formulations (PLA/DDGS, PLA/EDDGS, PLA/EDDGS/OOW) resulted more ductile than PLA, exceeding it in elongation at break. The authors recognized the need for an effective interfacial binding between PLA and fillers in order to improve load transfer, and ultimately enhance strength and stiffness of the composite filaments. The

TABLE 10.25

Composition of Experimental Grain/PLA Filaments for FFF

Material	Composition			
	PLA	DDGS[a]	PW[b]	OOW[c]
	wt%	wt%	wt%	wt%
PLA	100	0	0	0
PLA/DDGS[a]	75	25	0	0
PLA/DDGS[a]/PW[b]	75	12.5	12.5	0
PLA/DDGS[a]/OOW[c]	75	12.5	0	12.5

Source: Tisserat, B., Liu, Z., Finkenstadt, V. et al. 2015. 3D printing biocomposites. Society of Plastics Engineers (SPE) 10.2417/spepro.005690.

Notes:

[a]Dried distillers' grain with solubles.

[b]Paulownia wood.

[c]Osage orange wood.

TABLE 10.26

Properties of Experimental Grain/PLA Filaments for FFF

Material	Tensile Strength	Tensile Modulus	Elongation at Break
	MPa	MPa	%
PLA	44.7	334	16.5
PLA/DDGS[a]	22.5	261	17.1
PLA/EDDGS[a]	41.9	322	21.3
PLA/DDGS[a]/PW[b]	24.6	311	12.5
PLA/EDDGS[a]/PW[b]	38.3	346	16.1
PLA/DDGS[a]/OOW[c]	18.9	248	14.0
PLA/EDDGS[a]/OOW[c]	40.7	291	20.8

Source: Tisserat, B., Liu, Z., Finkenstadt, V. et al. 2015. 3D printing biocomposites. Society of Plastics Engineers (SPE) 10.2417/spepro.005690.

Notes:

[a]Dried distillers' grain with solubles.

[b]Paulownia wood.

[c]Osage orange wood.

decline in tensile strength and modulus could have also been caused by voids or uneven dispersion of filler, but the composite filaments' cross section was not analyzed to verify this hypothesis.

10.14 RESINS FROM TREES

Resins from trees are complex mixtures of various compounds including acids, alcohols, waxes, etc. produced by broadleaved trees and conifers, with conifers generating far larger amounts. Tree resins are converted into products such as phenolic-formaldehyde adhesives, widely employed for bonding wood panels in exterior applications.

TABLE 10.27

Properties of Experimental Filaments for FFF Made of Resins from Trees

Tree	Density	Tensile Strength	Compressive Strength	Extrusion Temp.	Printing Temp.	T_m
	g/cm^3	MPa	MPa	°C	°C	°C
Frankincense	1.25	0.16	0.32	65–78	80–90	78
Benzoin	1.31	0.26	0.54	70–95	80–90	95
Myrrh	1.02	0.17	0.29	70–90	80–100	89

Source: Horst, D. J., Tebcherani, S. M., Kubaski, E. T. 2016.Thermo-Mechanical Evaluation of Novel Plant Resin Filaments Intended for 3D Printing. *Int. J. Eng. Res.* 5 (10): 582–586.

Horst et al. (2016) developed filaments for FFF made of sustainable semi-crystalline TP resins extracted from three trees: (a) *Stirax benzoin* or benzoin, from India, South-East Asia, Malaysia, and Indonesia; (b) *Commiphora myrrha* or myrrh, native to the Arabian peninsula and Africa; (c) *Boswellia papyrifera* or frankincense, growing in Ethiopia, Nigeria, Sudan, and other African countries. The purchased resins, without undergoing any treatment such as purification and distillation, were converted through a Filastruder extruder from powder into a 1.75 mm diameter filament that was processed on a FFF desktop printer at 80–100°C to print tensile and compressive coupons. Table 10.27 lists selected properties and processing settings. The tensile and compressive strength were two orders of magnitude lower than same properties of PLA and ABS, which are the most popular sustainable and non-sustainable filaments for home printing, respectively. Another major drawback was that the filaments were too brittle to be wound on a reel, and were manufactured as 1 m × 1.75 mm sticks. These two shortcomings make these filaments inadequate for load-bearing applications.

FURTHER READINGS

Pereira, Helena ed. 2007. *Cork: Biology, Production and Uses*. Amsterdam: Elsevier.

Senthilkumar, K., Saba, N., Rajini, N. et al. 2018. Mechanical properties evaluation of sisal fibre reinforced polymer composites: a review. *Constr. Build. Mater.* 174 (2018): 713–729.

Siakeng, R., Jawaid, M., Ariffin, H. 2018. Natural fibre reinforced polylactic acid composites: a review. *Polym. Compos.* doi:10.1002/pc.24747.

Zia, K. M., M. Zuber, and M. Ali ed. 2017. *Algae Based Polymers, Blends, and Composites: Chemistry, Biotechnology and Material Sciences*. Amsterdam: Elsevier.

REFERENCES

Ahn, S., Lee, H., Puetzer, J., Bonassar, L. J., Kim, G. 2012. Fabrication of cell-laden three dimensional alginate-scaffolds with an aerosol cross-linking process. *J. Mater. Chem.* 22:18735.

alibaba.com. 2020a. https://www.alibaba.com/showroom/dried-distiller-grains-with-soluble.html (accessed April 3, 2020).

alibaba.com. 2020b. https://www.alibaba.com/showroom/dried-distiller-grains-with-soluble.html (accessed April 3, 2020).

Arrieta, M. P., Samper, M. D., López, J., Jiménez, A. 2014. Combined effect of poly(hydroxybutyrate) and plasticizers on polylactic acid properties for film intended for food packaging. *J. Polym. Environ.* 22:460–470.

Ashby, M. F., Gibson, L. J., Wegst, U., Olive, R. 1995. The mechanical properties of natural materials. I. Material property charts. *Proc. R. Soc. Lond.* A 450:123–140.

Awad, H. A., Wickham, M. Q., Leddy, H. et al. 2004. Chondrogenic differentiation of adipose-derived adult stem cells in agarose, alginate, and gelatin scaffolds. *Biomaterials* 25:3211–3222.

Badouard, C., Traon, F., Denoual, C. et al. 2019. Exploring mechanical properties of fully compostable flax reinforced composite filaments for 3D printing applications. *Indus. Crops Prod.* 135:246–250.

Balthazar, M. 2016. *Production and Mechanical Properties of Bamboo and Flax Fibre Reinforced PLA Compounds for Fused Deposition Modeling*. Master Thesis, Katholieke Universiteit Leuven.

Bassyouni, M., Waheed, Ul, Hasan, S. 2014. The use of rice straw and husk fibers as reinforcements in composites. In *Biofiber Reinforcements in Composite Materials*, ed.O. Faruk and M. Sain, 385–422. Amsterdam: Elsevier.

Berzin, F., Vergnes, B., Beaugrand, J. 2014. Evolution of lignocellulosic fibre length along the screw profile during twin screw compound with polycaprolactone. *Compos. A* 59:30–36.

Brites, F., Malça, C., Gaspar, F. et al. 2017. Cork plastic composite optimization for 3D printing applications. *Proc. Manuf.* 12:156–165.

Bulota, M., Budtova, T. 2015. PLA/algae composites: Morphology and mechanical properties. *Composites: Part A* 73:109–115.

Chand N. and M. Fahim 2008. Sisal reinforced polymer composites. In *Tribology of Natural Fiber Polymer Composites*, 84–107. Sawston: Woodhead Publishing.

Chatzifragkou, A., Charalampopoulos, D. 2018. Distiller's dried grains with solubles (DDGS) and intermediate products as starting materials in biorefinery strategies. In *Sustainable Recovery and Reutilization of Cereal Processing By-Products*, ed. Charis M. Galanakis, 63–86. Sawston: Woodhead Publishing.

Chung, J. H. Y., Naficy, S., Yue, Z. et al. 2013. Bio-ink properties and printability for extrusion printing living cells. *J. Biomater. Sci. Polym. Ed.* 1 (7): 763–773.

Coelho, A., Thiré, R., Araujo, A. 2019. Manufacturing of gypsum–sisal fiber composites using binder jetting. *Addit. Manuf.* 29:100789.

Cooke, G. B. 1951. Cork, bark of the exotic quercus suber. *Scientif. Month.* 72(3):169–179. http://www.jstor.org/stable/20224 (accessed April 3, 2020).

Cork Information Bureau. 2010. *Cork – Environmental Importance*. https://www.apcor.pt/wp-content/uploads/2015/07/Environmental_Importance_of_Cork.pdf (accessed April 3, 2020).

Czlonka, S., Strąkowska, A., Kairytė, A. 2020. The impact of hemp shives impregnated with selected plant oils on mechanical, thermal, and insulating properties of polyurethane composite foams. *Materials* 13:4709.

Dangtungee, R., Tengsuthiwat, J., Boonyasopon, P., Siengchin, S. 2015. Sisal natural fiber/clay-reinforced poly(hydroxybutyrate-cohydroxyvalerate) hybrid composites. *J. Thermoplast. Compos. Mater.* 28 (6): 879–895.

Dankar I., Haddarah A., Sepulcre F., Pujolà M. 2020. Assessing mechanical and rheological properties of potato puree: effect of different ingredient combinations and cooking methods on the feasibility of 3D printing. *Foods* 9:21. doi:10.3390/foods9010021.

Daver, F., Lee, K. P. M., Brandt, M., Shanks, R. 2018. Cork-PLA composite filaments for fused deposition modeling. *Compos. Sci. Technol.* 168:230–237.

Davoodi, P., Feng, F., Xu, Q. et al. 2015. Coaxial electrohydrodynamic atomization: microparticles for drug delivery applications. *J. Control. Release* 205:70–82.

Depuydt, D., Balthazar, M., Hendrickx, K. et al. 2018. Production and characterization of bamboo and flax fiber reinforced polylactic acid filaments for fused deposition modeling (FDM). *Polym. Compos.* 40 (5) : 1951–1963.

Dodge, J. D. 2012. *The Fine Structure of Algal Cells*. Cambridge: Academic Press doi:10.1088/1757-899X/810/1/012070.

Duarte Campos, D. F., Blaeser, A., Weber, M. et al. 2013.Three-dimensional printing of stem cell-laden hydrogels submerged in a hydrophobic high-density fluid. *Biofabrication* 5 (1): 015003.

Fernandes, E. M., Correlo, V. M., Mano, J. F., Reis, R. L. 2014. Polypropylene-based cork–polymer composites: processing parameters and properties. *Compos. Part B Eng.* 66:210–223.

Fernandes, E. M., Correlo, V. M., Mano, J. F., Reis, R. L. 2015. Cork-polymer biocomposites: mechanical, structural and thermal properties. *Mater. Des.* 82:282–289.

Fiorini, D., Scortichini, S., Bonacucina, G. et al. 2020. Cannabidiol-enriched hemp essential oil obtained by an optimized microwave-assisted extraction using a central composite design. *Indus. Crops Prod.* 154:112688.

Fondazione Istituto Italiano di Tecnologia. 2015. *Bioplastics and Bioelastomers*. https://iit.it/technology-transfer-docs/505-new-technology-teaser-bioplastics-and-bioelastomers/file (accessed April 3, 2019).

Gil L. 2009. Cork composites: a review. *Materials* 2:776–789.

Gil L. 2015. New cork-based materials and applications. *Materials* 8:625–637.

Girdis, J., Gaudion, L., Proust, G. et al. 2017. Rethinking timber: investigation into the use of macadamia nut shells for additive manufacturing. *JOM J. Miner. Met. Mater. Soc.* 69 (3): 575–579.

Graham, L. E., Graham, J. M., Wilcox, L. W. 2009. *Algae*. San Francisco: Benjamin-Cummings Publishing Company.

Grozdanov, A., Buzarovska, A., Bogoeva-Gaceva, B. et al. 2006. Rice straw as an alternative reinforcement in polypropylene composites. *Agronom. Sustain. Develop.* 26 (4): 251–255.

Hobson, J. 2014 *Straw Based Filament?* https://hackaday.com/2014/05/01/straw-based-filament/ (accessed April 3, 2020).

Horst, D. J., Tebcherani, S. M., Kubaski, E. T. 2016.Thermo-mechanical evaluation of novel plant resin filaments intended for 3D printing. *Int. J. Eng. Res. Technol. (IJERT)* 5 (10): 582–586.

Ismail, M. R., Yassen, A. A. M., Afify, M. S. 2011. Mechanical properties of rice straw fiber-reinforced polymer composites. *Fibers Polym.* 12 (5): 648–656.

Jaksic, N., Druelinger, M., Mendicello, L. 2017. Industrial hemp fibers as reinforcing agents in 3D-printing filament composites. *1stInstitute of Cannabis Conference, Colorado State University*, Pueblo, Colorado.

Janevski, S. 2020. *Design for Sustainability as a Key Driver for Exploring the Potential of Cork Material.* Ottawa, Canada: Carleton University.

Johnson, M., Shivkumar, S., Berlowitz-Tarrant, L. 1996. Structure and properties of filamentous green algae. *Mater. Sci. Eng. B* 38:103–108.

Kadam, K. L., Forrest, L. H., Jacobson, W. A. 2000. Rice straw as a lignocellulosic resource: collection, processing, transportation, and environmental aspects. *Biomass Bioenergy* 18:369–389.

Klemm, D., Meyer, W., Glowa, G., Zeidler, H. 2015. *Use of 3D Printing for the Production of Miscanthus Straw Packaging.* Germany: Technische Universität Dresden. https://tud.qucosa.de/landing-page/?tx_dlf[id]=https%3A%2F%2Ftud.qucosa.de%2Fapi%2Fqucosa%253A28619%2Fmets (accessed April 3, 2020).

Krueger, A. 2017. *Why 3D Printing with Wine Grape Skins Is the Future.* May 25, 2017 https://www.foodandwine.com/wine/3d-printing-wine-grape-skins (accessed April 3, 2020).

Kvavadze, E., Bar-Yosef, O., Belfer-Cohen, A. et al. 2009. 30,000-year-old wild flax fibres. *Science* 325 (5946): 1359.

Lee, K. Y., Mooney, D. J. 2012. Alginate: properties and biomedical applications. *Prog. Polym. Sci.* 37:106–126.

Li, X., Tabil, L. G., Panigrahi, S. 2007. Chemical treatments of natural fiber for use in natural fiber-reinforced composites: a review. *J. Polym. Environ.* 15:25–33.

Ma, M., Chiu, A., Sahay, G. et al. 2013. Core–shell hydrogel microcapsules for improved islets encapsulation. *Adv. Healthcare Mater.* 2:667–672.

Majid, M. A., Attahirah, M., Hishammudin, M. et al. 2016. Tensile properties of treated and untreated paddy straw fiber using sodium hydroxide strengthened with polypropylene resin. *MATEC Web of Conf.* 47:01019.

Manaia, J. P., Manaia, A. T., Rodriges, L. 2019. Industrial hemp fibers: an overview. *Fibers* 7 (12): 106.

Mayer, H. K., Fiechter, G. 2013. *Comprehensive Analytical Chemistry.* https://www.sciencedirect.com/topics/chemistry/agarose (accessed April 3, 2020).

Mazzanti, V., Pariante, R., Bonanno, A. et al. 2019. Reinforcing mechanisms of natural fibers in green composites: Role of fibers morphology in a PLA/hemp model system. *Compos. Sci. Technol.* 180:51–59.

Mestre, A. 2014. *Cork Design – A Design Action Intervention Approach towards Sustainable Product Innovation.* PhD diss. https://repository.tudelft.nl (accessed April 3, 2020).

Mochane, M. J., Magagula, S. I., Sefadi, J. S., Mokhena, T. C. 2021. A review of green composites based on natural fiber-reinforced polybutylene succinate (PBS). Polymers 13, 1200.

Nanovia. 2020. *PLA Flax.* https://nanovia.tech/en/ref/pla-flax-flax-fibre-reinforced/ (accessed April 3, 2020).

Nishiyama, Y., Nakamura, M., Henmi, C. et al. 2009. Development of a three-dimensional bioprinter: construction of cell supporting structures using hydrogel and state-of-the-art inkjet technology. *J. Biomech. Eng.* 131 (3): 035001.

Oksman Niska, K., Sain, M. ed. 2008. *Wood-Polymer Composites.* XV. Cambridge: Woodhead.

Osman, M. A., Atia, M. R. A. 2018. Investigation of ABS-rice straw composite feedstock filament for FDM. *Rapid Prototyp. J.* 24 (6): 1067–1075.

Ozbolat, I., Hospodiuk, M. 2016. Current advances and future perspectives in extrusion-based bioprinting. *Biomaterials* 76:321–343.

Pereira, H. 2013. Variability of chemical composition of cork. *BioResources* 8 (2): 2246–2256.

Pereira, H. 2015. The rationale behind cork properties: a review of structure and chemistry. *BioResources* 10 (3): 6207–6229.

Pereira, L., João Cotas, J. 2020. Introductory chapter – alginates –a general overview. In *Alginates – Recent Uses of This Natural Polymer*, ed. P. L. Pereira. London: IntechOpen. doi:10.5772/intechopen.77849.

Petinakis, E., Yu, L., Simon, G., Dean, K. 2013. In *Fiber Reinforced Polymers – The Technology Applied for Concrete Repair*, ed.M. A. Masuelli. London: IntechOpen.

Pickering, K. L., Efendy, M. A., Le, T. M. 2016. A review of recent developments in natural fibre composites and their mechanical performance. *Compos. B Part A* 83:98–112.

Prasad, S. V. 1989. Natural-fiber based composites. In *Concise Encyclopedia of Composite Materials*, ed.A. Kelly. Oxford: Pergamon.

Prosina, A. 2020. *Algae-Based Printer Ink As the Way to Foster In-Situ Resource Utilization in Habitation Structures.* https://www.researchgate.net/publication/340077158.

Puttegowda, M., Mavinakere Rangappa, S., Jawaid, M. et al. 2018. Potential of natural/synthetic hybrid composites for aerospace applications. In *Sustainable Composites for Aerospace Applications*, ed.M. Jamwaid, M. Thariq, 315–351. Sawston: Woodhead.

Rajesh, G., Ratna Prasad, A. V., Gupta, A. 2015. Mechanical and degradation properties of successive alkali treated completely biodegradable sisal fiber reinforced poly lactic acid composites. *J. Reinf. Plast. Compos.* 34 (12): 951–961.

Raven, J. A., Giordano, M. 2014. Algae. *Curr. Biol.* 24 (13): R591–R595.

Rothon, R. ed. 2003. *Particulate-filled Polymer Composites.* Shawbury: iSmithers Rapra Publishing. 2nd edition.

Sankhla, D., Sancheti, M., Ramachandran, M. 2020. Review on mechanical, thermal and morphological characterization of hemp fiber composite. *IOP Conf. Series: Materials Science and Engineering,* 810 012070. Bristol: IOP Publishing.

SaskFlax. 2020. *What Is Flax?* https://www.saskflax.com/industry/index.php (accessed April 3, 2020).

Schwarzova, I., Stevulova, N., Melicha, T. 2017. Hemp fibre reinforced composites. *Environmental Engineering,10th Int. Conf.*, Lithuania.

Sen, A., Van den Bulcke, J. N., Van Hacker, J., Pereira, H. 2014. Thermal behavior of cork and cork components. *Thermochim. Acta* 528:94–100.

Senthilkumar, K., Saba, N., Rajini, N. et al. 2018. Mechanical properties evaluation of sisal fibre reinforced polymer composites: a review. *Constr. Build. Mater.* 174:713–729.

Shahzad A. 2012. Hemp fiber and its composites – a review. *J. Compos. Mater.* 46 (8): 973–986.

Siakeng, R., Jawaid, M., Ariffin, H. 2018. Natural fibre reinforced polylactic acid composites: a review. *Polym. Compos.* doi:10.1002/pc.24747.

Singh, R., Kumar, R., Preet, P., Singh, M., Singh, J. 2019. On mechanical, thermal and morphological investigations of almond skin powder-reinforced polylactic acid feedstock filament. *J. Thermoplas. Compos. Mater.* doi:10.1177/0892705719886010.

Singh, R., Kumar, R., Singh, M., Preet, P. 2020. On compressive and morphological features of 3D printed almond skin powder reinforced PLA matrix. *Mater. Res. Express* 7:025311.

Stevens, C., Mussig, J. 2010. *Industrial Applications of Natural Fibres: Structure, Properties and Technical Applications.* Hoboken, NJ: John Wiley & Sons.

Stoof, D., Pickering, K., Zhang, Y. 2017. Fused deposition modelling of natural fibre/polylactic acid. *Compos. J. Compos. Sci.* 1 (1): 8.

Tisserat, B., Liu, Z., Finkenstadt, V. et al. 2015. 3D printing biocomposites. *2015 Society of Plastics Engineers (SPE)*.doi:10.2417/spepro.005690.

Tran, T. N., Bayer, I. S., Heredia-Guerrero, J. A. et al. 2017a. Cocoa shell waste biofilaments for 3D printing applications. *Macromol. Mater. Eng.* 302:1700219.

Tran, T. N., Heredia-Guerrero, J. A., Mai, B. T. et al. 2017b. Bioelastomers Based on Cocoa Shell Waste with Antioxidant Ability. Adv. Sustain. Syst.1 (7): 1700002.

University of Virginia. 2008. *Algae: Biofuel of The Future? ScienceDaily.* www.sciencedaily.com/releases/2008/08/080818184434.htm (accessed April 3, 2020).

Vilela, C., Sousa, A. F., Freire, C. S. R. et al. 2013. Novel sustainable composites prepared from cork residues and biopolymers. *Biomass Bioener.* 55:148–155.

Wan, T., Yang, G., Du, T., Zhang, J. 2015. Tri-(butanediol-monobutyrate) citrate plasticizing poly(lactic acid): Synthesis, crystallization, thermal, and mechanical properties. *Polym. Eng. Sci.* 55:205–213.

Yan, L., Chouw, N., Jayaraman, K. 2014. Flax fibre and its composites – a review. *Composites: B* 56:296–317.

Yan, L. B., Chouw, N., Yuan, X. W. 2012. Improving the mechanical properties of natural fibre fabric reinforced epoxy composites by alkali treatment. *J. Reinf. Plast. Comp.* 31 (6): 425–437.

Zeidler, H., Klemm, D., Böttger-Hiller, F. et al. 2018. 3D printing of biodegradable parts using renewable biobased materials. *15th Global Conference on Sustainable Manufacturing Procedia Manufacturing*, vol. 21, 117–124.

Zhang, Y., Yu, Y., Akkouch, A. et al. 2015. In vitro study of directly bioprinted perfusable vasculature conduits. *Biomater. Sci.* 3:134–143.

Zhang, Q., Cai, W. 2008. Enzymatic hydrolysis of alkali-pretreated rice straw by Trichoderma reesei ZM4-F3. *Biomass Bioener.* 32 (12): 1130–1135.

Zhang, Y., Yu, Y., Ozbolat, I. T. 2013. Direct bioprinting of vessel-like tubular microfluidic channels. *J. Nanotechnol. Eng. Med.* 4:21001.

11 Carbohydrates

If all mankind were to disappear, the world would regenerate back to the rich state of equilibrium that existed ten thousand years ago. If insects were to vanish, the environment would collapse into chaos.

E. O. Wilson

11.1 INTRODUCTION

Carbohydrate, or *saccharide*, is a biomolecule consisting of carbon, hydrogen, and oxygen atoms, with a hydrogen:oxygen atom ratio often of 2:1. Carbohydrates abound in nature, are multifunctional, and involved in every form of biological processes. They include sugars, starch, and cellulose, and are divided into four chemical groups: *monosaccharides* (contain one sugar unit, such as glucose, galactose, fructose, etc.), *disaccharides* (two sugar units, such as maltose, lactose, and sucrose), *oligosaccharides* (three to ten sugar units), and *polysaccharides* (many sugar units, such as starch, glycogen, and cellulose). *Sucrose* is table sugar, granulated sugar, cane sugar, and beet sugar. Carbohydrates perform numerous functions in living organisms. F.e. polysaccharides store energy (as starch and glycogen do), and provide structural integrity in plants (as cellulose does), crustaceans and insects (as chitin, a derivative of glucose, does). Carbohydrate polymers from natural sources have been studied and exploited in the pharmaceutical and biotechnological industries for many years (Tai et al. 2019).

11.2 STARCH

11.2.1 Introduction

Starch is a polysaccharide that is composed of a large number of glucose units connected by glycosidic bonds, and is produced by most green plants as energy storage. It is the most common carbohydrate in human diet, and abundant in cereals (corn, rice, wheat), potatoes, and processed food based on cereal flour, such as bread, pasta, and pizza. It provides 70–80% of the calories consumed by people worldwide (Thomas and Atwell 1999). Starch is the second most abundant biomass material in nature, and found in crop seeds, plant roots, and stalks. It represents a significant mass fraction in many agricultural commodities such as cereals (30–70 wt%), tubers (60–90 wt%), and vegetables (25–50 wt%) (Gardan and Roucoules 2011). Fruits are also rich in starch. It is inexpensive to produce, compostable, totally biodegradable, and industrially mostly derived from corn, potato, rice, tapioca, and wheat. Native or raw starch occurs in the form of granules, whose size, shape and molecular arrangement inside them depend on the plant species, and the genetic-environment interactions (Horstmann et al. 2017). Starch is made up of two polysaccharides: the mostly linear *amylose* and the branched *amylopectin*, featuring MW of 0.2–2 $\times 10^6$ and 100–400 $\times 10^6$ g/mol, respectively. The ratio of the two polysaccharides depends on the botanical origin of the starch: f.e., potato starch contains 20 wt% amylose and 80 wt% amylopectin (Le 2020). Since amylose and amylopectin have unique physical and chemical properties, their relative proportions influence the overall properties of starches. F.e. rice grains rich in amylose feature large volume expansion and high degree of flakiness upon cooking, cook dry, are less tender, and become hard upon cooling, whereas rice grains with intermediate amylose content cook moist and tender, and do not harden upon cooling (Panesar and Kaur 2016). The main use of starch

DOI: 10.1201/9781003221210-11

is for food applications (60% of the market), split in confectionery and drinks (31%) and processed foods (29%); non-food and feed represent the rest of the market (Avérous and Halley 2014). Starch is added to food to promote gelling, thickening, adhesion, moisture retention, stabilizing, film forming, texturizing, and antistaling (Horstmann et al. 2017). Starch granules do not dissolve in water, but form a suspension. Currently, corn is the main source for producing starch polymers. Major factors governing the mechanical properties of starch are amylose content, crystallinity, T_g, types and contents of plasticizers, and sources of starch (Gupta et al. 2012). In 2021 ReportLinker (France) estimated that the global starch market in 2020 reached 120 million tonnes.

In presence of a plasticizer, and under specific values of pressure, shear, water content, and time, the crystalline structure of starch is converted into the amorphous form of *thermoplastic starch* (TPS) (Van Le 2020) that is preferred in industrial applications, because granules of TPS can be easily processed by traditional plastics processing technologies (Aldas et al. 2020). TPS composites can exceed 50% in starch content.

Starch is present as a biodegradable filler in traditional plastics with the goal of facilitating and accelerating the biodegradation that is triggered by microbial consumption of starch. In fact, buried carrier bags out of PE and 6% starch disintegrated in 2–3 years instead of hundred years for PE only (Maddever and Chapman 1987). Another use of starch is as a food additive to improve some food's properties: water holding capacity, heat resistant behavior, binding, thickening, and so on (Abbas et al. 2010). Examples of companies utilizing starch to produce plastics are BIOP Biopolymer Technologies (Germany), Novamont (Italy), and Rodenburg Biopolymers (The Netherlands). An interesting example of starch-based plastic is Novamont Mater-Bi®, a family of biodegradable and compostable plastics derived from cornstarch and other feedstocks, and applied in agriculture (compostable clips, mulching films, pheromone dispensers for insect mating disruption), carrier bags, food service (products approved for food contact: flatware, straws, hot and cold cups, and reusable dishes), food and non-food packaging (bags, butter wrappers, cardboard trays, tubular netting, wrapping paper), and organic waste collection (Novamont 2015). The above applications are high-volume and inexpensive, hence currently not suited for AM. Mater-Bi® is offered in different grades, all containing starch but combined with several ingredients: CA, PVA, PCL, and PBAT. In 2019, Novamont announced that it would raise its production capacity of Mater-Bi® to 150,000 ton per year. Table 11.1 lists major physical-mechanical properties of three grades of Mater-Bi® and PS and LDPE, two petrochemical plastics with similar property values. The capability of tuning properties of Mater-Bi® from rigid to stretchable should be investigated in order to replicate them in starch feedstocks for AM, targeting applications with different performance requirements. Due to its in-termolecular and intramolecular hydrogen bond network (Lv et al. 2017), starch is hydrophilic, and that has to be accounted for when formulating AM feedstocks for load-bearing articles. Chang et al. (2006) measured that the tensile strength, modulus, and strain at break of dry films from tapioca starch made not for AM (Kapal ABC brand) imported from Thailand were 27.5 MPa, 2.6 GPa, and 1%, respectively, indicating a rigid and brittle material.

In order to make biodegradable films, starch has also been blended with SPs, such as chitosan (Bourtoom and Chinnan 2008), water-soluble CMC (Suvorova et al. 2000), and flax cellulose nanocrystals whose presence resulted in a tensile strength about 50 times that of 100% starch films (Cao et al. 2008). Polymers for films feature high elongation at break and impact strength, but are not as strong and stiff as it is required in typical load-bearing parts and their functional prototypes, hence the blends mentioned previously should have their tensile strength and modulus adequately enhanced for AM applications. Conversely, AM applications can be investigated that meet the unmodified properties of starch-based materials.

Starch may be profitably printed for medical applications (Lu et al. 2009), if the latter ones combine "solutions" customized to the patients who will be likely willing to pay a premium over a non-customized solution. Compared to other medical polymers, starch-based polymers have the following advantages (Sinha and Kumria 2001; Mendes et al. 2001; Marques et al. 2002; Mendes 2001Azevedo et al. 2003Sinha 2001): (a) good biocompatibility, (b) biodegradability, (c)

TABLE 11.1

Properties of Sustainable Novamont Mater-Bi® Starch-Based Plastic and Petrochemical Commodity Plastics

Properties	Unit	Mater-Bi® DI01A[a]	Mater-Bi® YI10U[b]	Mater-Bi® ZIO1U[b]	LDPE[b]	PS[b]
Density	g/cm³	1.20	1.35	N/A	N/A	1.04–1.09
Melt Flow Index	g/10 min	35	10–15	1.5	0.1–22	8–12
Melting Temperature	°C	N/A	N/A	N/A	N/A	N/A
Tensile Strength	MPa	48	25–30	28	8–10	35–64
Tensile Modulus	MPa	2,700	2,100–2,500	180	100–200	2,800–3,500
Elongation at Break	%	22	2.6	780	150–600	1–2.5
Elongation at Yield	%	2.5	N/A	N/A	N/A	N/A
Izod Unnotched Impact Strength	kJ/m²	46	N/A	N/A	N/A	N/A
Izod Notched Impact Strength	kJ/m²	4	N/A	N/A	N/A	N/A

Notes:

[a] www.materialdatacenter.com.

[b] Bastioli, C. 1998. Properties and applications of Mater-Bi starch-based materials. *Polym. Degrad. Stab* 59:263–212.

degradation products are non-toxic, (d) adequate mechanical properties. An example of medical uses of starch is biodegradable bone cements that provide immediate structural support, and are combined with bioactive particles that promote bone growth at the cement-bone interface and in the volume opened up by polymer degradation (Boesel et al. 2004). An additional AM application of starch may be pharmaceutical, following suit of non-AM pharmaceutical use of starch.

PLA is the most popular SP for AM, and blending it with starch will improve its ductility by increasing its elongation at break and impact strength. In fact, starch filaments are less brittle than PLA filaments, and have higher T_g. However, mechanical properties of starch/PLA blends obtained through conventional processes are poor because of the poor interfacial interaction between hydrophilic starch and hydrophobic PLA (Wang et al. 2008).

Methods to improve starch/PLA compatibility are gelatinization of starch, and adding one of the following ingredients: compatibilizer, glycerol, formamide, and water, with the last three alone or combined (Lu et al. 2009).

Recommended reviews of starch and its polymer technology were published by Bastioli (2004) and Bastioli et al. (2012).

11.2.2 COMMERCIAL STARCH FEEDSTOCKS FOR AM

Starch is present in commercial feedstocks for these AM processes: FFF, BJ, and bioplotting, described in Sections 2.3, 2.6, and 2.3.7, respectively.

Biome3D (3D-Fuel™) supplies a biodegradable odor-free flexible TP filament for FFF made from plant starches, and featuring properties listed in Table 11.2. Its recommended printing settings are: print temperature 190–210°C, heated bed 50°C, and print speed of 50 to 100 mm/s depending on part size.

In BJ starch is bound with aqueous inks and turned into solid parts more suited for models, prototyping, tooling, and visual aids than for functional parts. BJ printers by Z Corp are based on 5-color printheads (cyan, magenta, yellow, black, and clear) to deposit small droplets of colored

TABLE 11.2

Properties of Commercial Starch-Based Filament Biome3D for FFF

Tensile Strength	Tensile Stress at Yield	Tensile Modulus	Elongation at Break	Notched Izod Impact
MPa	MPa	GPa	%	J/m
35	26	2.0	6.2	29

Source: Data posted online by 3D-Fuel™.

binders that solidify areas of a powder layer deposited on a bed. The powder material VisiJet PXL, often referred to as *starch, plaster,* or *gypsum powder,* has a granular finish and is brittle coming out of the printer. After printing, the parts are cleaned of any powder, coated in super glue to enhance cohesion, strength, and stiffness, and sprayed with a clear lacquer to protect and smooth the surface.

In the U.S. patent 7,807,077 B2 by Voxeljet on BJ (Hochsmann and Ederer 2010) starch is included as a binder (in form of pregelatinized, acid-modified, or hydrolyzed starch), and as a filler (f.e. maltodextrin).

Herr et al. (2016, 2018) included starch (from corn, microcrystalline cellulose, pea, potato, rice, wheat) in the formulation of edible filaments for printers described in their patents filings, and comprising, besides starch, an oil as an active ingredient imparting taste, odor or medicinal benefit, polyvinylpyrrolidone, and a strong disintegrant. The formulation is converted into a powderized water soluble polymer mixed with excipient ingredients, including plasticizer, colored/dyed Arabic gum, gelling agent, fillers, flour, binding or thickening agent, lubricant, and preservative, and it is hot melt extruded into 1.75 and 3 mm diameter filaments that are printed at 70–90°C.

11.2.3 Experimental Starch Feedstocks for AM

Not surprisingly, starch has been investigated as a feedstock for food prepared via AM. Very recently, Theagarajan et al. (2020) investigated the printability of rice starch in products for the food industry, which has important practical interest given the fact that rice starch in the form of rice flour is widespread as a key food ingredient. The authors studied the rheological behavior of rice starch, and the effect of nozzle diameter (1.2, 1.5, 1.7 mm), print speed (800, 1500, 2200 mm/min), and motor speed (120, 180, 240 rpm) on the printability of rice starch, evaluated uniformity and ease of extrusion, and examined the printed items in terms of quality, binding property, finishing, texture, layer definition, shape, dimensional stability, and appearance of the deposited thread. The physical and chemical properties of rice starch principally depend on its amylose and amylopectin content that in the rice studied were 20% and 80%, respectively. Larger content of amylose led to more stable material, more pronounced shear thinning behavior, and greater water absorption. Rice starch was mixed thoroughly with an equal amount of water to prepare a dough-like rice paste that represented the feedstock, and behaved as a non-Newtonian fluid, with viscosity decreasing upon increasing applied shear. An in-house developed extrusion-based printer was employed, supported by a controlled air pressure system operating up to 4 bars, and equipped with a 30 cm^3 syringe barrel filled with the rice paste. Best printability was obtained at higher motor speeds (180, 240 rpm) and lower printing speeds (800, 1500 mm/min). The highest printing rate was reached with 1.7 mm diameter, 240 rpm, and 800 mm/min.

Chen et al. (2019) assessed that rice, corn, and potato starch were suited for printing food via hot extrusion. The authors prepared homogeneous suspensions at 5, 10, 15, 20, 25, and 30 wt% concentrations for each individual starch, and measured the relationship between rheological properties (flow stress, yield strength, storage modulus) and printability. The amylose content for potato, corn, and rice starch was 34.5, 24.1, and 26.5 wt%, respectively. Each starch sample proved suited for AM because it showed a shear-thinning behavior and self-support, revealed by storage modulus decreasing at higher strains and recovering at lower strains. The starch suspensions with concentrations of 15–25 wt% heated to 70–85°C were optimal, because they resulted in values of flow stress of 140–722 Pa, yield stress of 32–455 Pa, and shear modulus of 1.15–6.91 MPa that translated in excellent extrusion processability, and adequate mechanical integrity to achieve high geometric resolution (0.80–1.02 mm line width).

A Chinese patent application (WO2018/171630A1) was filed about methods of preparing and processing a feedstock containing a composite starch for printing food in form of dough, but also non-food items. The feedstock is a mixture of cornstarch and both potato starch and cassava starch or one of those starches, and is prepared as follows: starch and protein are mixed together, water is added for gelatinization, and, when gelatinized, butter is poured for kneading the mix into a dough that is printed into objects that are suitable for successive processing such as baking, boiling, deep frying, pan frying, and steaming. The feedstock is also compatible with plasticine printers, and can generate models, props, and sand tables (terrain models for military planning and wargaming) of any shape.

Yang et al. (2018) combined potato starch and lemon juice gel into a new printed food, and investigated the effect of various amounts of potato starch on the rheological properties and mechanical properties of lemon juice gels. The mechanical properties and rheological behavior of the composition 15 g/100 g of lemon juice gel/potato starch was suitable for printing objects of different shapes and geometries.

Zheng et al. (2019) investigated food-related characteristics (color, texture, morphology, etc.) of printed samples made from corn, potato, and wheat starch gels. Their paper shares a comprehensive analysis of feedstocks for food printing that can serve as a guide. Each starch was mixed with distilled water at a weight ratio of 1:4, and heated in water bath. The gelatinization of potato, wheat, and cornstarch occurred at 67, 70, and 72°C, respectively. All starch gels were successfully printed on a Shinnove S2 (China) food printer, equipped with 1.2 mm diameter plastic nozzle, extrusion column at 45°C, 0.7 mm layer height, and 30 mm/s nozzle speed. Several characteristics describing the performance as an AM edible feedstock were assessed: hardness, springiness, and cohesiveness, peak viscosity, color, chemical changes, etc. All three starches were printed with good accuracy and reproducibility, and without chemical changes in gelatinized and printed samples. Wheat starch samples featured the highest shape fidelity (while corn and potato revealed significant shrinkage), lowest viscosity, and best extrudability and storage properties, and most regular microstructure.

Starch has been considered for AM also in form of HG. In fact, Maniglia et al. (2020) demonstrated that regular native cassava starch, kept in dry conditions at 130°C for 2 and 4 h (*dry heating treatment*, DHT) and converted into HG, acquired new properties that expanded the use of cassava starch in AM. Varying the DHT duration (2 h vs. 4 h), the maximum gelation temperature (65, 75, 85, and 95°C), and gel refrigerated storage time (1 d vs. 7 d), the gels' strength was assessed through the gels' firmness measured on a texture analyzer from the energy required to penetrate the material. The gels after 4 h of DHT were firmer than gels after 2 h of DHT, and both were stronger than those produced without DHT. Stronger gels are typically preferred in AM applications, such as food printing. Gel printability was evaluated, in order to determine whether food could be printed, maintain dimensional stability, and support its own weight without spreading. All gels formulated were successfully extruded by a syringe, and produced a straight line. The 4 h DHT gel produced thinner lines than 1 h DHT gel and gel without DHT, did spread less on the surface (consistently with its highest firmness), formed a test shape with the highest resolution among all gels, and proved to be the best candidate for food printing.

Gardan et al. (2016) formulated a feedstock for FFF by combining modified starch, beech flour, and water. The modified starch was the food additive hydroxypropyl starch E 1440, and was included to increase viscosity through crosslinking and impart binding properties, and remained completely biodegradable. Beech flour was hardwood LIGNOCEL® HB 120 (Rettenmaier & Soehne). The optimal composition from a printing and mechanical performance standpoint was the following: starch 40 g, 24.6 g of beech flour with 40 μm fiber length, and 246 mil of water. The tensile strength, modulus, and elongation at break of a filament extruded out of the above blend were 5.45 MPa, 600 MPa, and 1.25%, respectively, indicating brittle failure. The filament's microscopic structure seemed that of a nonwoven composite with fibers oriented in the extrusion direction. The adhesion between beech fibers and matrix appeared adequate, and similar to that of a nonwoven composite. This is not surprising because starch's granules bond to each other and form larger granules. During cooling, starch wrapped around the beech flour and agglomerated the fibers. Chemically modified starches such as E 1440 develop viscosity and binding properties at low temperature and "at normal heat temperatures in a water-limited environment" (Gardan et al. 2016).

Kuo et al. (2016) converted starch into TPS by debranching and plasticization, blended TPS, ABS, compatibilizers, impact modifiers, and pigments, and extruded them into white and black 1.75 mm diameter filaments for FFF. The lab-made TPS/ABS filaments favorably compared to a commercial ABS filament in terms of printability, mechanical and thermal properties, and emissions of volatile organic compounds (VOCs). The TPS/ABS blend was satisfactorily processed. The specific ingredients were: glycerol and water as starch plasticizers, enzyme α-isoamylase to debranch starch, ABS U200B from XYZprinting Co., styrene-maleic anhydride (SMA) copolymer as compatibilizer between TPS and ABS, methyl-methacrylate butadiene styrene (MBS) as impact modifier, TiO_2 and carbon black (CB) as pigments. Test values of HDT and impact strength in Table 11.3 lacked further details in the paper by Kuo et al. (2016) on how they were measured, since HDT can be measured at 0.46 MPa and 1.8 MPa, and impact strength on notched and unnotched coupons. In Table 11.3, we added the widespread commercial ABS filament by Stratasys as a significant benchmark. SMA was effective in improving tensile strength, and flexural strength and modulus that reached the maximum values (76.8 MPa, 113.8 MPa, and 2528 MPa, respectively) in the formulation TPS/ABS/SMA 30/70/1 wt%. The formulations with largest amount of MBS exhibited the highest impact strength values, whereas CB resulted in the maximum HDT (103°C). Since starch proved effective at modifying ABS, it should be evaluated in combination with SPs to possibly develop fully sustainable AM feedstocks.

Starch has been processed via 3DP™ (Section 2.6.3), a type of AM processes that consists in selectively depositing a liquid bonding agent (*binder*) to join regions of a powder bed, and falls under the family of BJ.

Leong et al. (2003) utilized 3DP™ and printed scaffolds out of hydroxyapatite, cellulose, starch powders, and a water-based polymeric binder. By varying printing speed, flow rate and location of liquid binder drops, the scaffold's microstructure can be tailored, and complex and consistent structures with channels and controlled structural anisotropy, and high degree of pore interconnectivity can be fabricated, without need for supporting material.

Lam et al. (2002) developed and investigated a sustainable and biocompatible feedstock for 3DP™ composed of a starch-based blend of polymer powders and distilled water as a binder, and demonstrated that starch was suitable for printing porous 3D scaffolds for biomedical applications. The blend contained cornstarch/dextran/gelatin (50/30/20 wt%), all biomaterials. Cylindrical scaffolds of five different designs were printed, post-processed to enhance the mechanical and chemical properties, and analyzed in terms of physical and mechanical properties. Four types of porous cylindrical scaffolds (12.5 mm diameter × 12.5 mm height) featuring either cylindrical or rectangular pores, and a solid cylinder serving as a control were printed on the Z Corp Z402 printer out of natural polymers and plaster of Paris, combined with a water-based ink as a binder. In order to maintain the integrity of the scaffolds and increase their strength, the printed scaffolds were heated at 100°C for 1 h, infiltrated with different amount of a copolymer solution comprising 75% PLLA and 25% PLC in dichloromethane (CH_2Cl_2), and left to evaporate at room temperature. Infiltration improved resistance to water absorption, and, significantly, compressive modulus and

TABLE 11.3

Composition and Properties of TPS/ABS Blends (A, B, C, D, E, F) and Commercial ABS for FFF

Blend	TPS	ABS	SMA	MBS	TiO$_2$	CB	Tensile Strength	Flexural Strength	Flexural Modulus	Impact Strength	HDT	MFI	VOCs
	wt%	wt%	wt%	wt%	wt%	wt%	MPa	MPa	MPa	J/m	°C	g/10 min	mg/(m^{2}hr)
A	30	70	0	0	0	0	61.3	90.9	2,107	8.26	92	34.8	N/A
B	30	70	1	0	0	0	76.8	113.8	2,528	12.35	99	48.3	N/A
C	30	70	1	1	0	0	74.2	109.9	2,411	14.83	97	47.6	N/A
D	30	70	1	2	0	0	73.2	109.0	2,372	18.57	96	47.1	N/A
E	30	70	1	2	5	0	73.1	108.3	2,332	18.23	101	46.7	0.05
F	30	70	1	2	0	5	73.4	108.7	2,362	18.35	103	46.8	0.05
ABS U200B	0	100	N/A	N/A	N/A	N/A	68.6	101.6	2,254	15.00	100	39.2	.09
ABS M30[a]	0	100	N/A	N/A	N/A	N/A	28–32	48–60	1,760–2,060	128[b]	82[c]	28.3	N/A

Source: Adapted from Kuo, C.-C., Liu, L.-C., Teng, W.-F., et al. 2016. Preparation of starch/acrylonitrile-butadiene-styrene copolymers (ABS) biomass alloys and their feasible evaluation for 3D printing applications. *Compos. B. Eng.* 86:36–39. Reproduced with permission from Elsevier.

Notes:

[a] Stratasys. Ranges depend on variation of values with printing direction.

[b] Notched Izod.

[c] At 0.46 MPa.

Abbreviations: TPS thermoplastic starch, SMA styrene-maleic anhydride, MBS methyl-methacrylate butadiene styrene, CB carbon black, VOCs volatile organic compounds.

TABLE 11.4

Compressive Properties of Starch-Based Scaffolds and Solid Cylinders

Sample	Treatment	Yield Strength	Modulus
		MPa	MPa
Scaffolds (porous)	Un-infiltrated	0.18–1.12	3.07–12.4
Scaffolds (porous)	Infiltrated	0.60–1.77	11.2–55.2
Scaffolds (porous)	Infiltrated and soaked[a]	N/A	0.08–0.10
Solid cylinders (non-porous)	Un-infiltrated	1.73	33.5
Solid cylinders (non-porous)	Infiltrated	1.68	59.7
Solid cylinders (non-porous)	Infiltrated and soaked[a]	N/A	0.27

Source: Lam, C. X. F., Mo, X. M., Teoh, S. H., Hutmacher, D. W. 2002. Scaffold development using 3D printing with a starch-based polymer. *Mater. Sci. Eng. C* 20 (1–2): 49–56.

Note:

[a] 4 h at 37°C in phosphate-buffered saline solution to simulate in vivo conditions.

yield strength, as illustrated in Table 11.4 where the values ranges depended on the scaffolds' geometry. The porosity values achieved in the printed scaffolds corresponded to the design values, with some minor differences due to fabrication side effects such as shrinkage. Microporosity decreased after infiltration, since the copolymer had infiltrated the micropores and voids.

Suwanprateeb (2006) also targeted starch for BJ, and studied whether applying post-printing infiltration twice instead of once to improve strength would be beneficial. Aiming at biomedical applications, the author employed the following mixture of sustainable and biocompatible powders with particle sizes of 20–150 µm: starch/cellulose fiber/sucrose/maltodextrin (from vegetable starch) 40/15/25/20 wt%. Test samples were built on the Z Corp Z400 printer: some were infiltrated with a water-based binder once and some twice, and all were cured at 105°C for 30 min, then post-cured at 120°C for 30 min. Flexural strength, modulus, and strain were measured upon single infiltration (4 GPa, 27.5 MPa and 1.9%, respectively) and double infiltration, and the change between the two treatments was insignificant for dry coupons but beneficial for coupons tested after being immersed in water for 24 h and dried at 100°C for 1 h between two infiltrations. We suppose that the second infiltration resulted in a barrier opposing penetration of water.

Fabricating filament for FFF out of starch-based TP compounds through melt processing is not easy, and has been attempted by compounding starch with SPs (cellulose, natural rubber, PLA, PHB, etc.) and synthetic polymers, such as PVA (Zia et al. 2017). Dimonie et al. (2019) investigated in detail how to improve the quality of FFF filaments made up of cornstarch physically modified with PVA. Broad experimental characterization led to the conclusion that cornstarch and PVA could be converted into TP compounds for high-quality FFF filaments via melt modification. In fact, filaments for FFF featuring silky and smooth surface, few defects, natural color, and acceptable ovality and diameter tolerance were produced.

Butler et al. (2020) studied printability and biocompatibility of potato starch and chitosan blends for EBB scaffolds for neural cell applications, because chitosan increased biocompatibility of bioink, while starch improved its printability. They tested the following starch/chitosan formulations, with values in wt%: 100/0, 75/25, 50/50, 25/75, 0/100. The optimal bioink composition in terms of printability and biocompatibility with neural cells was starch/chitosan 50/50 wt%. The bioinks were formulated using glycerol and gelatin to increase their flexibility and MFI, and acetic acid and water as solvents. The resultant solution was then stirred for 6 h at room temperature, and then in a hot plate at 100°C for 5 min to achieve polymerization, and become a homogenous mixture. An open source printer (Velleman K8200, Belgium) was modified to be run as a bioprinter. The bioink was loaded into a custom-made temperature-controlled syringe mounted on the bioprinter, and scaffolds with an orthogonal grid, and pore sizes of 1 × 1 mm were built. Mammalian cell-based assays with mouse neural cells (Neuro-2a ATCC® CCL-131 cells by Neurodyn Life Sciences, Canada) were chosen to test the biocompatibility. The cells were grown at 37°C in a synthetic cell culture medium with 10% fetal bovine serum and 1% penicillin-streptomycin solution. The authors evaluated printability through equation (11.), in which L and A are the perimeter and the area of the pore, respectively, Pr is printability which characterizes the degree of the gelatinization of the bioink, and $F = (a - b)/b$ is the *flattening factor*, with a and b width and height of the filament cross section, respectively:

$$Pr = \frac{L^2}{16A} \times (1 - F) \tag{11.1}$$

Equation (11.1) is derived from the equation by Ouyang et al. (2016) on printability by adding the factor $(1 - F)$. Setting the acceptable printability range at $Pr = 0.95$–1.05, the only bioink formulation that met the printability requirement was starch/chitosan 100/0 wt%, but this formulation was unsuited for bioprinting neural cells because it degraded rapidly, which hindered the adhesion and attachment of neural cells, and ultimately resulted in minimal cell growth in media. Increasing content of chitosan decreased the printability of the bioink. Starch/chitosan formulations of 50/50 wt% featured higher rates of cell growth vs. the control, but had the downside that they were prepared in pH 3.5 since pure chitosan only gels in acidic environments. Compatibility with human body requires a pH >3.5 during bioink preparation, and leads to precipitation of chitosan which can be avoided by neutralizing the bioink after printing and before cell seeding.

Lv et al. (2017) formulated a blend with cornstarch, PLA, wood flour (WF) and glycerol as plasticizer in which 3, 9, 15, and 21 wt% of WF enhanced tensile and flexural strength compared to those of starch/PLA/glycerol 30/70/12 wt%. Their maximum values (>45 MPa and >75 MPa for tensile and flexural strength, respectively) were reached by PLA/WF/starch/glycerol 70/9/21/12 wt % and are suitable for load-bearing applications. Hence, their blends should be investigated in a version compatible with AM.

11.3 SUGAR

11.3.1 INTRODUCTION

Sugars are carbohydrates or *saccharides* found in the tissues of most plants that feature a sweet taste. *Common sugar*, or *table sugar* or *granulated sugar*, typically refers to *sucrose*, a crystalline molecule ($C_{12}H_{22}O_{11}$) composed of two monosaccharides, glucose and fructose, joined by a covalent chemical bond (Figure 11.1). Honey and fruit are ample natural sources of a family of sugars called *monosaccharides* (Section 11.1). There are four types of sugar: *fructose* (fruit sugar), *glucose* (grape sugar, blood sugar), *lactose* (dairy sugar), and *sucrose* (table sugar). Sugarcane and

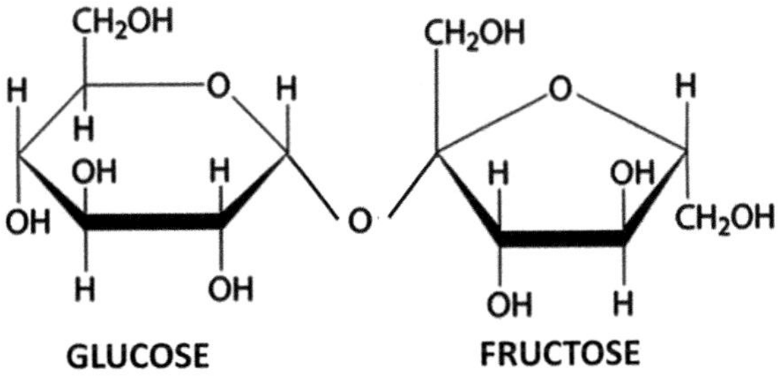

FIGURE 11.1 Chemical structure of sucrose.

sugar beet are rich in sucrose that is commercially extracted to be converted into refined sugar. Lactose is only found in milk, and is the only sugar that cannot be extracted from plants. Fructose, glucose, and maltose (another sugar) are derived from cornstarch. In 2019, the combined world production of sugarcane reached 1.95 billion tonnes (Shahbandeh 2021).

11.3.2 EXPERIMENTAL SUGAR FEEDSTOCKS FOR AM

Braskem (Brazil) produces a biobased, sustainable version of PE called BioPE™ derived from sugar contained in sugarcane and sugar beet. Filgueira et al. (2018) selected BioPE™ to develop for the first time a fully sustainable fiber-filled PE filament for FFF that met requirements related to surface finish, void content, warping, bending, interlayer adhesion, and water uptake, and over-came the incompatibility between natural hydrophilic fibers and hydrophobic matrix. The authors selected two BioPE™ grades: one with MFI 20 g/10 min (labeled BioPE1) and the other with MFI 4.5 g/10 min (labeled BioPE2), and thermomechanical pulp (TMP) fiber from Norway spruce chips with initial length and diameter of 1.5 mm and 33 μm, respectively. Since PE is hydrophobic whereas TMP fibers are hydrophilic, fiber surface was modified, by grafting one of two hydro-phobic compounds and antioxidant food additives, octyl gallate (OG) or lauryl gallate (LG), with the aid of laccase, an environmentally friendly enzyme. The maleic anhydride PE Licocene MA 4351 by Clariant was also added to improve the compatibility between TMP fiber and PE. Blends of 14 different compositions of the above ingredients (TMP fiber content was 10 and 20 wt%) were extruded into 2 mm diameter filaments for FFF at 150–160°C and 155–165°C using higher and lower MFI, respectively. Some filaments were extruded twice to improve the filament's quality. Under SEM, the filaments extruded twice showed fibers uniformly distributed and absence of large voids, and fibers modified with LG and OG were more uniformly distributed than fibers un-modified. Compared to filaments made of BioPE1, those made of BioPE2 were smoother, featured lower thickness variation, and showed better fiber blending. Importantly, fiber-matrix interfacial adhesion might have depended on the degree of hydrophobicity acquired by the TMP fibers, and chemical structure of PE. Porosity was very large (Figure 11.2), and likely to markedly penalize the mechanical properties (not measured). Namely, in filaments made of unfilled BioPE1 and BioPE2 void content was 0.1%, but in fiber-BioPE1 and fiber-BioPE2 it jumped to 8–47% and 11–27% respectively: porosity at 10 wt% fiber content was lower than at 20 wt%, and lower at smaller MFI than at greater MFI. Higher MFI and fiber content were associated to higher water uptake than lower MFI and fiber content. BioPE2 outperformed BioPE1 in printability that was assessed in terms of swelling, shrinkage, geometrical fidelity, and interlayer adhesion. Surface finish was evaluated visually not instrumentally on printed samples of BioPE2, and appeared not significantly different between unmodified and modified fibers. Samples printed with LG-TMP fibers showed a similar smoothness without warping or curling for both 10 wt% and 20 wt% fiber content, while items with OG-TMP fibers where smoother at 20 wt% than at 10 wt%.

Hamidi et al. (2017) focused on support materials, i.e. materials that are necessary to print specific geometric features such as hollow features and overhung surfaces and are removed from the final printed articles. Support materials can represent a profitable area in AM, because they do not have to meet all the performance requirements of the building materials that remain in the printed articles, and these performance requirements are even more demanding in the case of functional and load-bearing components. The authors experimented with melted sugar as a material supporting parts printed out of silicone elastomer and PLA fiber. Sugar has the advantages of being sustainable, low-cost (its price is 3-5% that of conventional polymeric supporting materials), environmentally friendly, soluble in water, and considerably faster to remove than other com-mercially available support materials, leaving no trace, and hence requiring no post-printing fin-ishing step. Complex structures serving as components for soft robots were fabricated on an inexpensive desktop printer. The results pointed out that further development of this combination sugar-AM would lead to printing complex parts combining high quality and low cost.

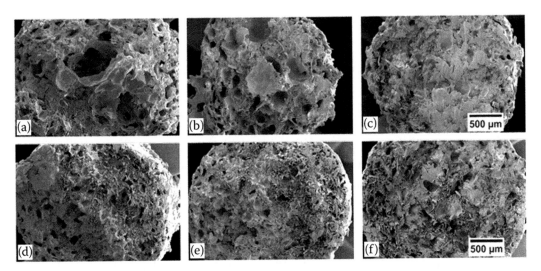

FIGURE 11.2 SEM images of the biocomposite filaments. BioPE1 containing 20% of untreated TMP fibers (a); 20% LG-treated TMP fibers (b); 20% OG-treated TMP fibers (c); BioPE2 containing 20% of untreated TMP fibers (d); 20% LG-treated TMP fibers (e); and 20% OG-treated TMP fibers (f). All the images were acquired at 50X magnification. Abbreviations: bioPE biobased PE, TMP thermomechanical pulp, OG octyl gallate, and LG lauryl gallate.

Source: Filgueira, D., Holmen, S., Melbø, J. K. et al. 2018. 3D Printable Filaments Made of Biobased Polyethylene Biocomposites. *Polymers* 10, 314. Reproduced with permission under the terms and conditions of the Creative Commons Attribution (CC BY) license (http://creativecommons.org/licenses/by/4.0/).

Pollet et al. (2020) investigated sugar as a feedstock to print models mimicking the human blood vascular system (the network of arteries, capillaries, and veins), in order to learn more about its functioning. Sugar glass is a brittle transparent form of sugar that looks like glass, and is chosen to simulate shattering glass in movies, plays, and photographs. The authors demonstrated that AM using sugar glass is a promising route to duplicate the vascular system's geometry, and reproduce elements of the microvasculature, i.e. channels featuring a diameter ranging from 1 mm down to 5 µm. On a custom-made, dedicated, pressure-based extrusion printer, the authors fabricated circular cross-sectional channels, 120–569 µm in diameter, reciprocally connected in a network by intersecting, merging and bifurcating, and able to handle pressurized fluid. Networks of fibers intersecting each other made of sugar glass (sucrose, glucose, water) and carbohydrate glass (sugar glass, dextran, water) were printed with control over their diameter on a glass plate. Next, the printed networks were embedded in polydimethylsiloxane (PDMS) and cured, and then dissolved when the PDMS cast was immersed in water for 24 h, leaving behind a network of empty channels. Measurements of T_g, viscosity, and storage modulus were instrumental to set the printing pressure and temperature. Employing particle image velocimetry, the flow behavior inside the empty channels was analyzed, and the quality of the printing process was validated. Fiber quality and stringing properties depend on the rheological properties of the feedstock, and printing parameters, including temperature and nozzle positioning speed. The ideal feedstock would feature: high T_g, consistent formation of small diameter fiber, and a clean start/stop of printing. Carbohydrate glass displayed robust fiber formation at small diameters, but poor control over start/stop properties, whereas sugar glass showed clean start/stop properties, but less control over fiber diameter. Eventually, sugar glass was preferred, and printed at 83°C, 0.15 mm nozzle, and 50 mm/min.

Polysaccharides can be classified according to their electric charge: *cationic* (chitin, chitosan), *anionic* (alginate, hyaluronic acid, etc.), and *nonionic* (cellulose, dextran, starch). In presence of divalent cations, such as calcium, anionic polysaccharides form gels acting as a feedstock for AM.

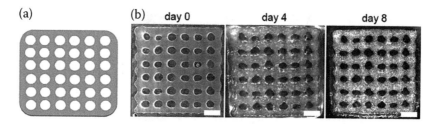

FIGURE 11.3 Lattice structures printed with bioink and their dimensional change in 4 and 8 days. (a) Digital model of lattice structure. (b) Lattice structures at days 0, 4, and 8. White scale bar = 2 mm.

Source: Jia, J., Richards, D. J., Pollard, S., et al. 2014. Engineering alginate as bioink for bioprinting. *Acta Biomaterialia* 10 (10): 4323–4331. Reproduced with permission from Elsevier.

Tai et al. (2019) published an overview of the main natural anionic polysaccharides from algae, bacteria, and plants, and focused on the recent developments of bioprinting processes using alginate as polysaccharide. Alginate (Chapter 10) is a suitable sustainable feedstock for TE applications, because it is biocompatible, has low toxicity and relatively low cost, and quickly forms hydrogel by adding Ca^{2+} (Lee and Mooney 2012). Alginate bioinks are commercially available, f.e. by CELLINK, Sweden.

Jia et al. (2014) researched bioinks, and followed a process that can be duplicated for bioink formulation: 30 alginate HGs with various oxidation percentages and concentrations were prepared and investigated, and a set of bioinks applicable for several TE applications was developed. Two key material properties of alginate solutions are viscosity and density, and their effect on the printability of alginate solutions was systematically investigated, in order to detect a range of material properties of alginates suitable for bioprinting. Four alginate solutions with varied biodegradability containing human adipose-derived stem cells (hADSCs) into lattice-structured, cell-laden HGs were printed with high accuracy (Figure 11.3). These alginate-based bioinks modulated proliferation and spreading of hADSCs without affecting the structure integrity of the lattice structures after eight days in a cell culture.

Voisin et al. (2018) selected polysaccharides as sustainable lightweight materials with high surface area for a wide range of applications, such as energy storage, packaging, and thermal insulation, and assessed AM as a technology to produce such materials, since AM can generate articles with tailored performance and geometry. The rheological and mechanical properties of polysaccharide-based composite foams made of mixtures of methylcellulose (MC), cellulose nanofibrils (CNF), montmorillonite (MMT), glyoxal (an organic compound added as a crosslinker), and tannic acid were investigated and optimized. MC was added for its surface activity and potential interaction with CNF and MMT, while MMT was selected as a rheological additive and strengthening filler in the dry foams. The foams were water-resilient, and featured high porosity, and values of compressive modulus and yield strength that outperformed other biobased foams and commercially available expanded PS. The foams could be surface modified, lending themselves to applications for air purification and thermal insulation.

He et al. (2015) investigated sugar as a feedstock to print a type of cutting-edge device, called *microfluidic chip* or *lab-on-a-chip* that is utilized to conduct R&D in biology, chemistry, and biomedicine. A microfluidic chip consists in a network of microchannels molded or etched into a material (f.e. ceramics, semiconductors, metals, and polymers), and connected together in order to perform a number of functions (mix, pump, sort, etc.) on a fluid that is exchanged with the outside environment through input and output (Casquillas et al. 2018). AM can be a successful alternative method to photolithography and laser writing to manufacture microfluidic devices, because it meets the requirements of cost and efficiency. The authors utilized AM and sugar as a sacrificial material, and demonstrated fast and effective printing of 2D and 3D microfluidic chips on a low-cost sugar printer that comprised: redesigned sugar extruder, PDMS extruder, 3D motion stage, temperature control device, and pneumatic control device. Candidate sugars were screened in terms of: (a) T_m suited for the printing temperature; (b) stable printing quality and high repeatability; and (c) low-cost fabrication. Maltitol (USD 52/kg) is a sugar-derived alcohol employed as sugar substitute. Maltitol was preferred over glucose, fructose, maltose, and sucrose (their price is USD 13–129/kg), because it did not easily oxidize when melt, and possessed appropriate surface tension and high water solubility. The fabrication process was straightforward: (a) maltitol was loaded in the extruder, melt at 150°C, and pushed out by air pressure to form 2D and 3D sugar tracks on a supporting layer; (b) PDMS was poured on the sugar structures and cured in oven for 25 min at 85°C; (c) the printed artifacts were immersed in boiling water for 10 min to dissolve the maltitol tracks, and leave behind PDMS with hollow microchannels. This method did not require further sealing as with conventional methods, which added to the other advantages: low cost and minimal time spent from design to fabrication (1 h or less). About the cost, the printer selected for this study cost about USD 800, and components such as temperature controller, pneumatic control device, electronic accessories and PTFE hoses are commercially available, whereas the sugar extruder was custom made. In comparison, printers fabricating fluidic devices range from about to USD 2,000 (MiiCraft, Germany) to USD 330,000 (Objet Connex 350, Israel, USA).

Recent advances in gravity-fed AM of molten glass have resulted in large-scale extrusion of molten material. Since equipment and techniques for printing molten glass are expensive and complex because of process temperatures exceeding 1000°C, Leung (2017) investigated a mixture of sucrose sugar and corn syrup that substituted molten glass utilized for conducting research in molten material-fed AM. Molten sugar is optically transparent, and it features similar temperature-viscosity relationship and solidification properties as molten glass, but at a far lower temperature (100–150°C). In fact, soda lime glass exhibits a T_g around 520°C, and is printable around 1000°C, while a sucrose and fructose mixture features T_g of 57–69°C, and T_m around 172°C. The low working temperature of sugar (100–150°C) resulted in a less complex and costly molten material reservoir, and reduction of heating energy required, and working hazards associated with high-temperature materials. A further advantage of employing sugar-based materials is their potential to build large and inexpensive objects. In fact, in 2016 retail sugar cost around USD 1.44/kg, while retail PLA cost around USD 53/kg. The authors designed a low-cost, desktop size printer with a temperature-controlled sugar reservoir made from easily available parts, and successfully printed in molten sugar articles having similar complexity to glass prints. The feedstock comprised 1000 g of Domino® white table sugar (the main printing material due to its low cost and high availability), mixed with 100 g of water, and then with 500 g of Karo® corn syrup. The mixture was heated at 154°C boiling water off, poured into the printer reservoir, and let cool down to the correct printing temperature. The water and corn syrup lowered the T_m and T_g of the mixture, and increased the temperature tolerance before crystallization occurred. Leung et al. (2017) described in detail the design and components of the printer (having a total cost of USD 2,000) so that others can duplicate it. The build rate was 40 mm³/s using 10 mm diameter nozzle, and maximum build volume was 240 × 300 × 90 mm. The techniques described can be easily employed for investigating AM of other molten materials, and its aspects, such as: design features (minimum radii, draft angle, etc.), optical properties, and process parameters (layer height, printing speed, nozzle design, multicolor material feed, and toolpath strategy).

REFERENCES

Abbas, K. A., Khail, S. K., Hussin, A. S. M. 2010. Modified starches and their usages in selected food products: a review. *Study J. Agric. Sci.* 2 (2): 90–100.

Aldas, M., Rayón, E., López-Martínez, J., Arrieta, M. P. 2020. A deeper microscopic study of the interaction between gum rosin derivatives and a mater-bi type bioplastic. *Polymers* 12:226.

Avérous, L. R., Halley, P. J. 2014. Starch polymers: from the field to industrial products. In *Starch Polymers: From Genetic Engineering to Green Applications*, ed. P. J. Halley, L. R. Avérous, 3–12. New York: Elsevier.

Azevedo, H. S., Gama, F. M., Reis, R. L. 2003. In vitro assessment of the enzymatic degradation of several starch based biomaterials. *Biomacromolecules* 4:1703–1712.

Bastioli, C. 2004. Starch. In *Encyclopedia of Polymer Science and Technology*. doi:10.1002/0471440264.pst348.

Bastioli, C., Magistrali, P., Gesti Garcia, S. 2012. Starch in polymers technology. In *Degradable Polymers and Materials: Principles and Practice*, ed. Khemani, Scholz, 87–112. ACS Symposium Series, 2nd edition.

Boesel, L. F., Mano, J. F., Reis, R. L. 2004. Optimization of the formulation and mechanical properties of starch based partially degradable bone cements. *J. Mater. Sci. Mater. Med.* 15:73–83.

Bourtoom, T., Chinnan, M. S. 2008. Preparation and properties of rice starch-chitosan blend biodegradable film. *LWT-Food Sci. Technol.* 41:1633–1641.

Butler, H. M., Naseri, E., MacDonald, D. S. et al. 2020. Optimization of starch- and chitosan-based bio-inks for 3D bioprinting of scaffolds for neural cell growth. *Materialia* 12:100737.

Cao, X., Chen, Y., Chang, P. R., Muir, A. D., Falk, G. 2008. Starch-based nanocomposites reinforced with flax cellulose nanocrystals. *Express Polymer Lett.* 2:502–510.

Casquillas, G. V., Houssin, T., Durieux, L. 2018. Microfluidics and microfluidic devices: a review. *Elveflow Microfluidic Instruments RSS*. https://www.elveflow.com/microfluidic-reviews/general-microfluidics/microfluidics-and-microfluidic-device-a-review/ (accessed November 15, 2020).

Chang, Y. P., Abd Karima, A., Seow, C. C. 2006. Interactive plasticizing–antiplasticizing effects of water and glycerol on the tensile properties of tapioca starch films. *Food Hydrocolloids* 20:1–8.

Chen, H., Xie, F., Chen, L., Zheng, B. 2019. Effect of rheological properties of potato, rice and corn starches on their hot extrusion 3D printing behaviors. *J. Food Eng.* 244:150–158.

Dimonie, D., Damian, C., Trusca, R., Rapa, M. 2019. Some aspects conditioning the achieving of filaments for 3d printing from physical modified corn starch. *Mater. Plast.* 56 (2): 351–359.

Filgueira, D., Holmen, S., Melbø, J. K. et al. 2018. 3D printable filaments made of biobased polyethylene biocomposites. *Polymers* 10:314.

Gardan, J., Nguyen, D. C., Roucoules, L., Montay, G. 2016. Characterization of wood filament in additive deposition to study the mechanical behavior of reconstituted wood products. *J. Eng. Fibers Fabr.* 11 (4): 56–63.

Gardan, J., Roucoules, L. 2011. Characterization of beech wood pulp towards sustainable rapid prototyping. IDMME – Virtual Concept 2010, France, 2010-10 – Research in Interactive Design – 2011.

Gupta, M., Brennan, C., Tiwari, B. K. 2012. Starch-based edible films and coatings. In *Starch-Based Polymeric Materials and Nanocomposites: Chemistry, Processing, and Applications*, ed. J. Ahmed, B. K. Tiwari, S. H. Imam, M. A. Rao, 239–254. Boca Raton: CRC Press.

Hamidi, A., Jain, S., Tadesse, Y. 2017. 3D printing PLA and silicone elastomer structures with sugar solution support material. *Proc. SPIE 10163, Electroactive Polymer Actuators and Devices (EAPAD)* (17 April 2017). doi:10.1117/12.2258689.

He, Y., Qiu, J., Fu, J. et al. 2015. Printing 3D microfluidic chips with a 3D sugar printer. *Microfluid Nanofluid* 19:447–456.

Herr, Ashley G., Cohen, Paige E. 2016. Edible 3D printer filament. Publ. No. 2016/0066601 A1, filed September 6, 2015, and published March 10, 2016.

Herr, Ashley G., Cohen, Paige E. 2018. Edible 3D printer filament. Publ. No. 2018/0310601 A1, filed April 20, 2018, and published November 1, 2018.

Hochsmann, R., Ederer, I. 2010. Methods and systems for the manufacture of layered three-dimensional forms. Pat. No.: US 7,807,077 B2, filed Jun. 11, 2004, and issued Oct. 5, 2010.

Horstmann, S. W., Lynch, K. M., Arendt, E. K. 2017. Starch characteristics linked to gluten-free products. *Foods* 6:29.

Jia, J., Richards, D. J., Pollard, S. et al. 2014. Engineering alginate as bioink for bioprinting. *Acta Biomater.* 10 (10): 4323–4331.

Kuo, C.-C., Liu, L.-C., Teng, W.-F. et al. 2016. Preparation of starch/acrylonitrile-butadiene-styrene copolymers (ABS) biomass alloys and their feasible evaluation for 3D printing applications. *Compos. B: Eng.* 86:36–39.

Lam, C. X. F., Mo, X. M., Teoh, S. H., Hutmacher, D. W. 2002. Scaffold development using 3D printing with a starch-based polymer. *Mater. Sci. Eng. C* 20 (1–2): 49–56.

Le, V. 2020. *Bio-Plastic Production from Starch*. Degree diss., Centria University of Applied Sciences, Finland.

Lee, K. Y., Mooney, D. J. 2012. Alginate: Properties and biomedical applications. *Prog. Polym. Sci.* 37:106–126.

Leong, K. F., Cheah, C. M., Chua, C. K. 2003 Solid freeform fabrication of three-dimensional scaffolds for engineering replacement tissues and organs. *Biomaterials* 24:2363–2378.

Leung, P. Y. V. 2017. Sugar 3D printing: additive manufacturing with molten sugar for investigating molten material fed printing. *3D Print. Addit. Manufact.* 4 (1): 13–18.

Lu, D. R., Xiao, C. M., Xu, S. J. 2009. Starch-based completely biodegradable polymer materials. *eXPRESS Polymer Lett.* 3 (6): 366–375.

Lv, S., Gu, J., Tan, H., Zhang, Y. 2017. The morphology, rheological, and mechanical properties of wood/flour/starch/poly(lactic acid) blends. *J. Appl. Polym. Sci.* doi:10.1002/APP.44743.

Maddever, W. J., Chapman, G. M. 1987. Making plastics biodegradable using modified starch additions. *Proceedings of the SPI Symposium on Degradable Plastics*, Washington DC.

Maniglia, B. C., Lima, D. C., Matta Junior, M. D. et al. 2020. Preparation of cassava starch hydrogels for application in 3D printing using dry heating treatment (DHT): A prospective study on the effects of DHT and gelatinization conditions. *Food Res. Int.* 128:108803.

Marques, A. P., Reis, R. L., Hunt, J. A. 2002. The biocompatibility of novel starch-based polymers and composites: In vitro studies. *Biomaterials* 23:1471–1478.

Mendes, S. C., Reis, R. L., Bovell, Y. P. et al. 2001. Biocompatibility testing of novel starch-based materials with potential application in orthopaedic surgery: a preliminary study. *Biomaterials* 22:2057–2064.

Novamont. 2015. MATER-BI®. http://materbi.com/en/ (accessed November 15, 2020).

Ouyang, L., Yao, R., Zhao, Y., Sun, W. 2016. Effect of bioink properties on printability and cell viability for 3D bioplotting of embryonic stem cells. *Biofabrication* 8 (3): 035020.

Panesar, P. S., Kaur, S. 2016. Rice: types and composition. In *Encyclopedia of Food and Health*, ed.B. Caballero, P. M. Finglas, F. Toldrá, 646–652. Academic Press.

Pollet, A. M. A. O., Homburg, E. F. G. A., Cardinaels, R., den Toonder, J. M. J. 2020. 3D sugar printing of networks mimicking the vasculature. *Micromachines* 11(43). doi:10.3390/mi11010043.

Shahbandeh M. 2021. *World Sugar Cane Production from 1965 to 2019*. https://www.statista.com/statistics/249604/sugar-cane-production-worldwide/ (accessed November 15, 2021).

Sinha, V. R., Kumria, R. 2001. Polysaccharides in colon-specific drug delivery. *Int. J. Pharmaceut.* 224:19–38.

Suwanprateeb J. 2006. Improvement in mechanical properties of three-dimensional printing parts made from natural polymers reinforced by acrylate resin for biomedical applications: a double infiltration approach. *Polym. Int.* 55:57–62.

Suvorova, A. I., Tyukova, I. S., Trufanova, E. I. 2000. Biodegradable starch-based polymeric materials. *Russ. Chem. Rev.* 69:451–459.

Tai, C., Bouissil, S., Gantumur, E. et al. 2019. Use of anionic polysaccharides in the development of 3D bioprinting technology. *Appl. Sci.* 9:2596.

Theagarajan, R., Moses, J. A., Anandharamakrishnan, C. 2020. 3D extrusion printability of rice starch and optimization of process variables. *Food Bioprocess Technol.* 13:1048–1062.

Thomas, D., and W. Atwell 1999.*Starches*. St. Paul: Eagan Press Handbook, American Association of Cereal Chemists.

Voisin, H. P., Gordeyeva, K., Siqueira, G. et al. 2018. 3D Printing of strong lightweight cellular structures using polysaccharide-based composite foams. *ACS Sustain. Chem. Eng.* 6:17160–17167.

Wang, N., Yu, J. G., Chang, P. R., Ma, X. 2008. Influence of formamide and water on the properties of thermoplastic starch/poly(lactic acid) blends. *Carbohydr. Polym.* 71:109–118.

WO2018/171630A1. 2018. Composite starch 3D printing material preparation and process patent application. https://patents.google.com/patent/WO2018171630A1/en (accessed November 15, 2020).

Yang, F., Zhang, M., Bhandari, B., Liu, Y. 2018. Investigation on lemon juice gel as food material for 3D printing and optimization of printing parameters. *LWT – Food Sci. Technol.* 87:67–76.

Zheng, L., Yu, Y., Tong, Z., Zou, Q., Han, S., Jiang, H. 2019.The characteristics of starch gels molded by 3D printing. *J. Food Process. Preserv.* 43:13993. doi:10.1111/jfpp.13993.

Zia, U. D., Xiong, H. G., Fei, P. 2017. Physical and chemical modification of compounds: a review. *Food Nutr. Sci.* 57:2691–2705.

12 Hydrogels

What we are doing to the forests of the world is but a mirror reflection of what we are doing to ourselves and to one another.

Chris Maser

12.1 INTRODUCTION

Hydrogel (HG) is a water-swollen gel comprising a crosslinked 3D polymeric network. *Gel* is a soft, solid, or solid-like material consisting in two or more components one of which is a liquid present in substantial quantity (Almdal et al. 1993). A well-known example of gel is the food Jell-O™ made by heating gelatin in water, and cooling the mixture. The large, strand-like protein molecules of gelatin wiggle around in hot water, and, as the mixture cools down, they possess less energy, and are less mobile, until eventually they bond together at points along the strands, forming pockets trapping the surrounding liquid and a 3D structure, or matrix, that gives Jell-O™ its structural integrity (Scientific American 1999). Polymers forming hydrogels (HGs) comprise biobased and synthetic polymers, polymer blends, nanocomposites, functional polymers, and cell-laden systems (Li et al. 2020).

HGs represent the largest share of ingredients for bioinks. A *bioink* is a mixture of biomaterial and living cells for *bioprinting* (Section 2.3.7), a family of AM processes consisting in printing structures and organs from viable biological molecules, biomaterials, and cells. Bioprinting is a version of *biofabrication*, consisting in producing complex living and non-living biological products from raw materials such as living cells, molecules, extracellular macromolecules, and biomaterials (Mironov et al. 2009). HG-based bioinks introduce us to *tissue engineering* (TE), a fascinating, relatively new, and advanced field of medicine concerned with replacing and regenerating human cells, tissues, and organs, in order to reestablish normal functionality (Van Cauwenberghe 2019) from degeneration, disease, and trauma. TE combines principles of life sciences and engineering, and is similar to regenerative medicine, because both share the goal of restoring, maintaining, and improving functions of damaged tissues and organs, but the former grows cells and tissue in laboratory setting (Celirix 2018), while the latter stimulates the body to generate its own cells using stem cell therapy and treatments. TE is key in medicine because it represents a way to overcome problems associated with transplants, such as shortage of donors, diseases, infections, and rejection of the organ and tissue by the host. Artificially developed tissues are also valuable in research, especially in drug development.

Besides TE, biomedical applications of printed HGs encompass cancer research, in vitro disease modeling, high-throughput drug screening, surgical preparation, soft robotics, and flexible wearable electronics (Li et al. 2020).

HGs are prepared from biobased polymers such as collagen, gelatin, and silk, and synthetic polymers such as poly(acrylic acid), poly(methacrylic acid), and PVA. Table 12.1 lists additional examples of biobased and synthetic polymers for HGs, and their relative advantages and limitations. HGs are utilized as membranes for biosensors, in contact lenses, artificial heart and skin, and drug delivery, and are the most widely consumed biopolymers in soft and hard (bone, cartilage, and vascular tissues) TE, thanks to their mechanical properties, compatibility with biological tissues, and their highly swollen 3D environment that is very similar to soft tissues, and permits diffusion of

TABLE 12.1

Biobased and Synthetic Polymers for Hydrogels: Adavantages and Limitations

	Biobased Polymers	Synthetic Polymers
Examples	Agarose.Alginate. Chitosan. Collagen. Gelatin. Fibrin. Hyaluronan.	Poly(acrylic acid) (PAA). Poly(methacrylic acid) (PMMA). Poly(vinyl alcohol) (PVA). Poly(ethylene glycol). (PEG). Poly (hydroxyethyl methacrylate) (pHEMA).
Advantages and Limitations	Almost always biocompatible. Biodegradable. Closely resemble the tissues they replace. Difficult to isolate from biological tissues. Limited versatility. Expensive. High batch-to-batch variation. Possible chronic immunogenic response. Structural modifications often limited.	Reliably produced. Versatile control polymer architecture and structure. May use toxic reagents. Not always biocompatible and biodegradable.

Source: Information from Lee, K. Y., Mooney, D. J. 2001. Hydrogels for Tissue Engineering. *Chem. Rev.* 101 (7): 1869–1880.

nutrients, growth factors, and cellular waste and regeneration of damaged tissues. Non-biomedical applications of HGs relate to agriculture, food, water purification, and so on (Dutta et al. 2019).

Bioprinted items exploit some features of AM: fine dimensional resolution, custom geometry, tunable properties, and small articles. Bioprinted items are not low-price and mass-produced articles but are part of the healthcare market, where the price is not a factor as critical for commercial success as in other markets, and hence this market may represent an attractive business opportunity for AM.

HGs for biomedical applications must meet a specific set of property requirements. For starters, HGs obviously have to be biocompatible. Their mechanical properties (controlled by the cross-linking degree) should allow to maintain the physical integrity of the item made of HG until the carried living cells are released at a set rate for a set time (Dutta et al. 2019). Equally critical to the applicability of HGs is possessing adequate permeability (or transport properties), and allow inward diffusion of nutrients, oxygen, and biomolecules, and outward release of biomolecules and waste (Gasperini et al. 2014), which controls cell's survival, expressed by *cell viability* that is the ratio of the number of living cells over the total population of cells. Transport properties depend on chemical composition, structure, and crosslinking density of HGs. Other properties and features that are critical for effective HGs are: adhesion to tissue, chemical composition, contraction and swelling, crosslinking density, protein adsorption, protective shell, rheological properties, stimuli responsiveness, surface topography, and water content.

It is challenging to accurately control the geometry of soft HGs (because of their limited mechanical properties), and impart to them an ordered structure enabling to perform their functions effectively as required in regenerative medicine and TE.

The methods to prepare HGs by forming their network can be broadly classified into two categories (Ali and Ahmed 2018):

- *Physical* methods, based on physical interactions among polymer chains: heating, cooling, ionic interaction, hydrogen bonding, etc. They lead to reversible networks.
- *Chemical* methods, based on forming new covalent bonds: grafting, and crosslinking techniques. They form irreversible networks.

Physical and chemical networks can also be combined. Crosslinking of polymer chains is the main mechanism behind formation of HGs, and can take place by means of an external crosslinking agent, chemical modification, or exposure to high-energy radiation. The three common gelling mechanisms are (Gasperini et al. 2014):

- *Thermal gelation*: in response to a temperature change, the polymer molecules reorganize from random coil to helix, and the helices assemble in clusters, joined together by the untwined regions. This type of gelation can be reversible.
- *Ionic crosslinking*: the sections of the polymer backbone carrying the charge bind with ions of opposite charge.
- *Chemical crosslinking* (f.e. photocrosslinking): when exposed to UV light, the photoinitiator molecules in solution form radicals, and crosslink the polymer chains.

Bioinks selected to bioprint articles (especially for TE) must exhibit the following characteristics: being soft to minimize damages to the surrounding tissues, possessing features similar to those of the natural extracellular matrix (ECM) material in human body, providing a highly hydrated environment for proliferation of living cells, physically protecting cells and fragile drugs, and being able to be modified with ligands to improve adhesion to cells (Min et al. 2015; Li et al 2018a).

HGs suited for bioinks must also meet specific requirements related to printability. According to Li et al (2018a) the behavior and suitability of HGs as bioinks depend on these characteristics:

- *Viscosity*: this property is largely determined by concentration and MW of the HG polymer, and printing temperature, and must be adequate for the HG to maintain the intended shape once printed.
- *Shear stress*. Upon being extruded from the printer's nozzle or syringe, the HG and the cells embedded in it undergo shear stress, whose value depends on printing pressure, nozzle's diameter, and bioink's viscosity. The higher the stress the lower the cell viability (Blaeser et al. 2016), but printing resolution is increased with narrower nozzles that produce higher shear stresses.
- *Shear rate*. Since HG-based bioinks are non-Newtonian fluids, their viscosity depends on the shear rate, and the latter changes as the bioink advances in the printer through diminishing cross sections.
- *Shear thinning*: this property refers to the non-Newtonian behavior of a polymer in which the viscosity decreases with increasing shear rate. During extrusion HG undergoes shear stress, which results in the breaking of the physical crosslinks and alignment of the polymer chains, leading to fewer entanglements and lower viscosity, which in turns facilitates the flow of bioink from nozzle and syringe. A HG that, once extruded, very quickly regains its higher viscosity in order to maintain the layer's shape is preferred for bioprinting, because of the short time existing between depositions of successive layers, and allows low values of dispensing pressure.
- *Thixotropic behavior*: it takes place when the bioink transitions from gel to liquid and vice versa due to presence or absence of shear stress, respectively, and it takes a measurable time for the viscosity to change accordingly.
- *Interfacial bonding*: it refers to interlayer adhesion that should minimize interfacial defects that cause poor stackability and mechanically weak 3D constructs. This characteristic can be quantified through a lap shear test (Li et al. 2018b).
- *Shape fidelity*: it is the degree of geometric correspondence between actual dimensions of the printed item and the dimensions of its digital 3D model. Shape fidelity is controlled by the diameter of printer's nozzle or syringe, printing speed, and feedstock dispensing pressure, among other parameters, and is enhanced by crosslinking, high viscosity at no shear stress, and high printing resolution. Webb and Doyle (2017) experimented with a HG blend of 7 wt%

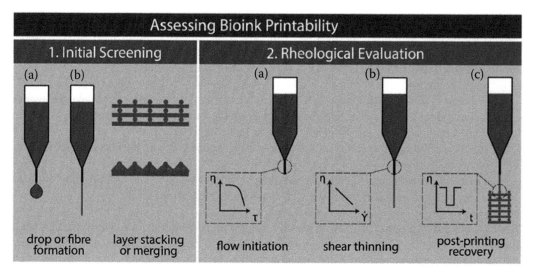

FIGURE 12.1 Two-step screening process for formulating acceptable bioinks at research level.

Source: Paxton, N., Smolan, W., Böck, T., et al. 2017. Proposal to assess printability of bioinks for extrusion-based bioprinting and evaluation of rheological properties governing bioprintability. *Biofabrication* 9, 044107. Reproduced under the terms of the Creative Commons Attribution 3.0 license.

alginate and 8 wt% gelatin, and disclosed a simple method of determining optimum printing parameters for EBP maximizing printing accuracy and cell viability. Shape fidelity is critical for the extruded filament of HG in order to maintain its shape and also support the subsequently printed structure without collapsing.

Paxton et al. (2017) argued that a fast and simple method for screening HGs for bioprinting was needed, and proposed a two-step screening process for formulating new acceptable bioinks (Figure 12.1) that consisted in:

- Initial screening to assess: (a) filament formation versus droplet formation, (b) stability of stacked layers without merging between layers.
- Rheological measurements to characterize: (a) flow initiation properties and yield stress, (b) shear thinning to predict extrusion process and cell survival, and (c) bioink's recovery behavior (that is increased viscosity) after printing.

Wang et al. (2018) applied a set of specific equations to characterize and screen candidate HGs for bioprinting in terms of printability, combination of fluid properties, and print parameters.

He et al. (2016) systemically studied the printability of alginate/gelatin HGs by starting with printing lines before printing 3D structures. These one-dimensional printing experiments allowed the authors to find optimized printing parameters, such as pressure and feed rate.

The reviews of HGs for AM by Malda et al. (2013) and Li et al. (2018a) are recommended. The former one included AM processes, concentrations, gelation methods, printing quality, and cytocompatibility of the mentioned HGs. The latter one mentioned names and makers of commercial bioprinters based on robotic dispensing and inkjet processes.

12.2 SUSTAINABLE HGs FOR AM

12.2.1 INTRODUCTION

Biobased HGs (alginate, chitosan, collage, gelatin, and so on) are frequently utilized in TE because of their following advantages: biodegradability, biocompatibility (they inherently contain bioactive regions imparting them good compatibility with living cells), mechanical properties comparable to those of living cells, a monodispersed MW, and a more defined structure than that of synthetic HGs. However, biobased HGs trail synthetic HGs in strength and stiffness, and hence, when bioprinting a construct layer by layer in biobased HGs, the layer may spread, deform, and even collapse under the weight of its upper layers. Despite a potential release of toxic substances during degradation, and low biocompatibility, synthetic HGs are still preferred for bioprinting, since they enable fine tuning of chemical and physical properties while degrading at a controlled rate (Raphael et al. 2017).

A way to improve mechanical properties of biobased HGs is adding a suitable biocrosslinker or plasticizer (Ali and Ahmed 2018). However, biobased HGs are less stable than synthetic HGs, feature properties varying among species, tissues, and batch, and have immunogenicity issues. *Immunogenicity* and *antigenicity* need to be defined. "Despite being used interchangeably, the terms immunogenicity and antigenicity have distinct meaning. The term immunogenicity refers to the ability of a substance to induce cellular and humoral [in extracellular fluids] immune response, while antigenicity is the ability to be specifically recognized by the antibodies generated as a result of the immune response to the given substance. While all immunogenic substances are antigenic, not all antigenic substances are immunogenic" (Ilinskaya and Dobrovolskaia 2016). Synthetic HGs are often poorly biocompatible and generate non-natural degradation products (Li et al. 2018a).

Another source of biobased HGs are edible polymers, mainly consisting of polysaccharides, proteins, and lipids. These feedstocks have the benefits of reducing environmental contamination, and increasing sustainability and recyclability. Application of edible polymeric HGs span many areas including food industry, agriculture, drug delivery, and TE (Ali and Ahmed 2018).

In developing HGs for AM, it is important to know that their mechanical characteristics are among their principal limitations. Table 12.2 reports, as benchmark values for sustainable HGs, tensile properties of two synthetic HGs prepared from N-vinyl caprolactam/ethylene glycol dimethacrylate (PVCL) and hydroxypropylcellulose (HPC)/divinyl sulfone: their tensile strength, modulus, and strain at break are 58–130 kPa, 30–765 kPa, and 15–239%, respectively. Being quite soft, some methods to measure physical and mechanical properties of HGs have been cleverly

TABLE 12.2

Properties of PVCL and HPC Synthetic Hydrogels at 22°C

Hydrogel	Nominal Polymer Concentration at Synthesis	Polymer Concentration at 22°C	Tensile Modulus	Ultimate Tensile Strength	Elongation at Break
	% (w/v)	% (w/v)	kPa	kPa	%
PVCL[a]	47	12.2	30.1	98	239
HPC1[b]	10	9.9	67	58	51
HPC2[b]	20	20.8	765	130	15

Source: Hinkley, J. A., Morgret, L. D., Gehrke, S. H. 2004. Tensile properties of two responsive hydrogels. *Polymer* 45:8837–8843. Reproduced with permission from Elsevier.

Notes:

[a]PVCL N-vinyl caprolactam/ethylene glycol dimethacrylate.

[b]HPC hydroxypropylcellulose/divinyl sulfone.

customized. Ahearne et al. (2008) published a concise review (rich in references) of mechanical characterization of HGs, from which the summary below was extracted. Tensile properties are measured by stressing a strip or a ring. Compressive properties are assessed by compressing the HG between two plates. The *indentation test* consists in indenting a HG at a single point to a preset displacement depth, and measuring the reaction force required to make the indentation; since the indenter is connected to a transducer recording the force, the HG's elastic modulus and stress relaxation data are derived from the force-displacement curve. Specific tests for HGs are *bulge test*, *spherical ball inclusion* (Lin et al. 2005), and *micropipette aspiration* (Boudou et al. 2006). The *bulge test* (or *inflation test*) involves inflating the HG through a window opened in its substrate, and measuring the resulting displacement as a function of the applied pressure by means of a CCD camera or a laser (Tsakalakos 1981). A finite element model analyzes the data and derives the mechanical properties.

The following techniques differ from the previous ones because they enable non-destructive monitoring of HG's mechanical properties under cells culture conditions. One technique, termed *long focal microscopy-based indentation*, applies to thin HGs: the sample is clamped around its outer edge and subjected to a spherical load; its resulting time-dependent deformation is measured in situ by means of a long focal distance microscope connected to a CCD camera, and its mechanical and viscoelastic properties are derived through a theoretical model from the deformation displacement. Another technique for non-destructive monitoring is called *optical coherence tomography-based spherical microindentation*, and is suited for thicker HGs. It consists in measuring the time-dependent depth of indentation caused by a spherical load on top of the HG, followed by applying Hertz contact theory to calculate the HG's mechanical properties.

Besides physical, mechanical, rheological, printability tests, and so on, additional measurements are required to assess whether specific HGs are suited for bioprinting, and these measurements comprise chemical and biological assays and evaluations: biocompatibility, degradability, degradation rate, cell viability, cell adhesion, crosslink density, swelling, and so on. Even atomic force microscopy (AFM) is useful because nano- to micro-scale structural and mechanical cues are associated with biological responses in both native ECM materials and synthetic constructs (Raphael et al. 2017).

Billiet et al. (2012) provided an extensive list of HGs (described in this chapter) made up of sustainable and non-sustainable polymers, and their combinations, and the AM processes compatible with them. The HGs were grouped according to AM processes, categorized in laser-based methods, nozzle-based methods, 3DP™, and MJ.

12.2.2 AGAROSE

Agarose is a natural linear polysaccharide obtained from the cell walls of red algae (Gasperini et al. 2014) with chemical formula $C_{24}H_{38}O_{19}$ and structure in Figure 12.2. It is water soluble above 65°C, and gradually gels between 17°C and 40°C, depending on its MW, and chemical

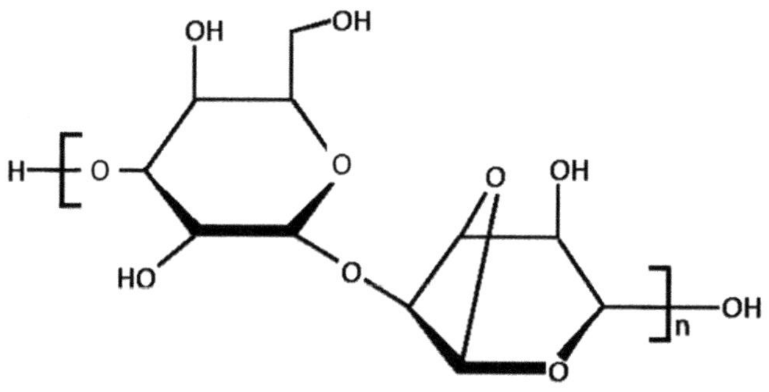

FIGURE 12.2 Chemical structure of agarose.

TABLE 12.3

Properties of Low-Viscosity Agarose Gels

Agarose Concentration		Tensile Strength	Tensile Modulus	Elongation at Break	Compressive Strength	Compressive Modulus	Compresive Strain
wt%	10^4 mol/ liter	kPa	MPa	mm/mm	kPa	kPa	mm/mm
2.5–10	4.6–17.6	78–453	0.4–3.7	0.11–0.14	104–606	1.5–2580	0.33–0.38

Source: Values from Normand, V., Lootens, D., Amici, E., et al. 2000. New Insight into Agarose Gel Mechanical Properties. *Biomacromolecules* 1 (4): 730–738.

modifications. Such low gelation temperatures complicate its use as a bioprinting ink. Agarose gels through the formation of intermolecular hydrogen bonds upon cooling, leading to the aggregation of double helices. Once gelled, it is stable and does not swell and liquefy, until heated to 65°C (Li et al. 2020). Low cell adhesion and low degradation rate are major drawbacks of agarose HGs (Zhang et al. 2012). The elastic moduli of agarose HGs can be varied from about 1 kPa to a few thousand kPa, values within the stiffness range of natural tissues (Normand et al. 2000). The strength and elastic moduli in tension and compression of low viscosity agarose gels are listed in Table 12.3. Agarose HG has been selected at very low concentrations for robotic dispensing systems, and as a sacrificial material to build a mold and pattern (Suntornnond et al. 2017), rather than printing agarose with cells directly.

Bertassoni et al. (2014) printed agarose strands as microchannels for vascularized TE constructs using a micromolding technique. Namely, they embedded functional and perfusable microchannels inside methacrylated gelatin (GelMA), star poly(ethylene glycol-co-lactide) acrylate (SPELA), PEG dimethacrylate (PEGDMA), and PEG diacrylate (PEGDA) HGs at different concentrations. Particularly, it was demonstrated that printed vascular networks in GelMA HGs can improve mass transport, cellular viability and differentiation within the cell-laden tissue constructs.

Agarose HG is also combined with other HGs, such as chitosan, to enhance thermoresponsive properties for soft TE applications (Miguel et al. 2014).

Agarose HG has been present in: regeneration of cartilage, cornea, enamel, and nerve; delivery of drugs, genes, and growth factors; and immunoengineering (Li 2020). It also shows potential for regeneration of bone (Fedorovich et al. 2008).

Agarose HG is compatible with pneumatic- and mechanical-extrusion-based bioprinters at low concentration of 1–5 w/v% (Kucukgul et al. 2015).

Agarose was one of the ingredients of Celleron®, a polymeric biodegradable compound for bioprinting formulated by D. Thomas at Swansea University (UK), and containing PLLA/PLGA, phospholipids, ibuprofen, graphene, collagen, and antibiotics for TE (Williams 2015). The formulation was processed on a 3Dynamic Systems Alpha custom-designed bioprinter. Once printed and in presence of a biological activator, Celleron® ferments, becomes microporous with increase in surface area and mechanical strength, and provides a path deep into the structure for the migration of cells. Protein growth factors are then saturated into the porous scaffold to turn it into a biologically attractive composite. Thomas and his team printed a prototype of human ear out of Celleron®, and experimented with mesenchymal stem cells possessing the advantage that they can be differentiated into various tissues, provide trophic support, and also modulate immune response.

12.2.3 ALGINATE

Alginate is a water-soluble, linear polysaccharide copolymer derived from bacteria and brown seaweed. Its polymeric chain contains a sequence of three blocks linked to each other through covalent bonds: a block composed of consecutive mannuronic acid residues (M-block), a block comprising consecutive guluronic acid residues (G-block), and a block of alternating G- and M-blocks (GM-block), as pictured in Figure 12.3. Its gelation can occur through ionic and covalent crosslinking (Li et al. 2018a). The crosslinking density of alginate HGs is a function of the G-blocks and MW (Li et al. 2020). The viscosity of alginate solution and the stiffness of its HG depend upon the polymer's concentration and MW (Kong et al. 2002). Alginate HGs degrade slowly. Favorable properties such as biocompatibility, low cost, and ease of gelation have made alginate HGs attractive in drug stabilization and delivery, TE, and wound healing applications (Lee and Mooney 2012). Alginate HG has been selected for bioprinting due to its easy use, and capacity to be manipulated. It is rated safe and biocompatible by the U.S. Food and Drug Administration (FDA). A prevalent form of alginate in applications is *sodium alginate*, employed by food, pharmaceutical, and textile printing industries. However, alginate HGs suffer from some limitations affecting printing, starting with their mechanical weakness that hinders shape fidelity. Li et al. (2016) increased the concentration of alginate, and obtained finer dimensional resolution. Shape fidelity is also enhanced by adding a physical crosslinking agent such as $CaCl_2$ (Atabak Ghanizadeh et al. 2015). Another limitation is that alginate inherently does not adhere to mammalian cells, and requires to be accordingly modified (Lee and Mooney 2012).

Most mechanical properties of alginate HGs are measured in compression and shear, but Drury et al. (2004) measured tensile properties of alginate HGs as a function of alginate type, formulation, gelling conditions, incubation, and strain rate, and published the following values: ultimate strength 3–36 kPa, elastic modulus 5–53 kPa, and strain at break 70–85% at 0.10/s strain rate. Generally, the initial tensile behavior and properties of alginate HGs were highly dependent on the choice of the alginate polymer and how it was processed: high guluronic acid containing alginate polymers yielded stronger and more ductile HGs than high mannuronic acid containing

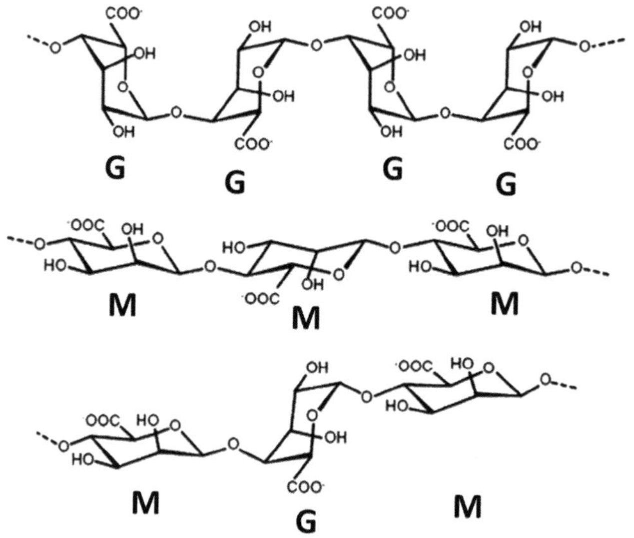

FIGURE 12.3 Chemical structure of alginate: G-block, M-block, and GM-block.

Source: Li, H., Tan, C., Li, L., 2018. Review of 3D printable hydrogels and constructs *Mater. Des.* 159:20–38. Reproduced with permission from Elsevier.

alginates. The tensile strength, strain at break, and modulus diminished by increased phosphate concentrations, among other factors.

Kong et al. (2002) reported that raising the concentration of high-MW alginate from 2 to 5 wt% enhanced the shear modulus of the resulting HG from 25 to 50 kPa. Alginate have been tested in the combinations alginate/gelatin, alginate/gelatin/fibrinogen, and alginate/gelatin/chitosan (Billiet et al. 2012).

Alginate HGs have been utilized for: cancer therapy (ovarian cancer, tumor models); delivery of drugs, and proteins; immunoengineering; regeneration of bone, cartilage, myocardial, nerve, and cell encapsulation and transplantation.

AM methods processing alginate HGs are inkjet printing, pneumatic extrusion, positive displacement extrusion, and laser-based processes.

12.2.4 Carrageenan

Carrageenans are a family of water-soluble, linear, sulfated polysaccharides extracted from the red seaweeds of the class *Rhodophyceae*, common in the Atlantic Ocean near British Isles, Europe, and North America. Carrageenan is present in food production (under the EU additive E-number E407 or E407a) as a thickening, gelling, and stabilizing agent, and in non-food industries, as excipients in pharmaceutical pills and tablets. Native carrageenan have MW of 1.5×10^6 to 2×10^7 g/mol, while food-grade carrageenan features MW of 1×10^5 to 8×10^5 g/mol (Necas and Bartosikova 2013). Products including carrageenan are: cosmetics, dietetic formulations, infant formula, milk products, processed meats, toothpaste, skin preparations, pesticides, and laxatives. Commercially available carrageenan are divided into three families depending on the position and number of sulfate groups (SO_4^{2-}): κ-(kappa), ι-(iota) and λ-(lambda) carrageenan carrying 1, 2, and 3 sulfate groups respectively (Hossain et al. 2001). λ-carrageenan does not form gels. HGs of ι-carrageenan are softer and more deformable than HGs of κ-carrageenan (Michel et al. 1997) that form the strongest gels. In fact, the compressive modulus of HG with a 40 g/L concentration of κ-carrageenan (3×10^5 g/mol MW, Fluka Chemie) is 147 kPa (Daniel-da-Silva et al. 2012), while compressive modulus of ι-carrageenan HG with concentration of 1.2 w/v% and hardened overnight with a 0.25M KCl solution is 1 kPa (Rochas et al. 1989). Tensile modulus of κ-carrageenan HG is 10 kPa (Rochas et al. 1989).

Carrageenan HGs are employed in regeneration of bone, and cartilage, and delivery of drugs and growth factors.

12.2.5 Cellulose and Its Derivatives

Cellulose is the most abundant natural polysaccharide, present in plants, animals, algae, fungi, and even minerals, and is an inexpensive, fibrous, tough substance, insoluble in water and other common solvents (Dutta et al. 2019). Cellulose provides mechanical and structural integrity to plants, and represents about 40% of their carbon content.

HGs can be prepared from native cellulose through physical crosslinking (hydrogen bonding), but, being insoluble in water, cellulose requires suitable solvents, such as N-methylmorpholine-N-oxide (NMMO), and ionic liquids (Dutta et al. 2019). Alternatively, water-soluble cellulose derivatives are synthesized in form of cellulose ethers, such as MC, CMC, and HPC, and so on, to form cellulose-based HGs through chemical and physical crosslinking (Li et al. 2020). Water-soluble cellulose derivatives offer the advantage of being generally biocompatible, and hence are added to cosmetics, food, and pharmaceutical products. Another type of cellulose-based HGs are prepared by mixing cellulose or its derivatives with biobased biodegradable polymers, such as alginate, chitin, chitosan, hyaluronan, and starch (Dutta et al. 2019).

Nanocellulose suspension (or solution) is potentially a suitable HG for bioink because it is extrudable through micro-sized nozzles, and the shape of the extruded layer is maintained over

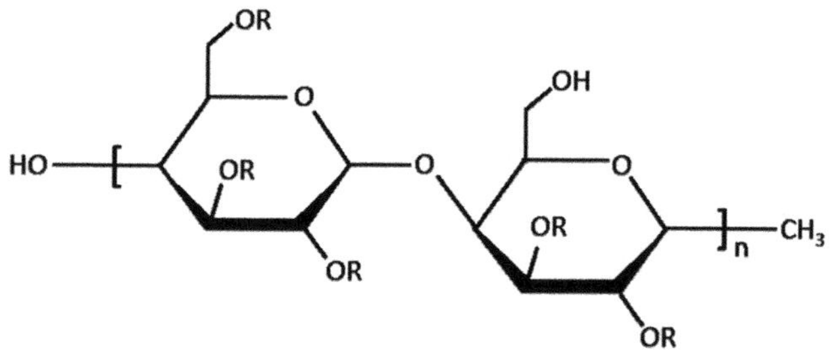

FIGURE 12.4 Chemical structure of methyl cellulose. R is H or CH_3.

time (Wang et al. 2018). As mentioned in Section 12.1, viscosity affects printability. Viscosity of a cellulose HG is largely controlled by cellulose concentration, degree of polymerization, and temperature. Nanocellulose suspensions exhibit a shear thinning behavior that, as already explained, is favorable to successful extrusion-based printing.

Cellulose HGs are utilized for delivery of drug and enzymes, and regeneration of brain, cartilage, liver, myocardial, nerve, and cell encapsulation and transplantation (Li et al. 2020).

MC (Figure 12.4) is a linear, semi-flexible, polysaccharide polymer deriving from the naturally occurring polysaccharide cellulose. It is biodegradable, and water-soluble because some hydrophobic groups in its polymeric chain are replaced with the hydroxyl group. MC gels from aqueous solution upon salt addition or heating. MC HG is usually dispensed by using pneumatic extrusion-based bioprinter at concentration around 4 w/v% (Fedorovich et al. 2008). Its unique gelation temperature at 40–50°C is close to the human body temperature, and can be utilized as a support material during bioprinting. In fact, Ahlfeld et al. (2018) developed a support material made of hydrogel ink composed of water and MC powder (M0512, Sigma-Aldrich) with 88,000 g/mol MW and 4000 cP viscosity. Solutions of MC in concentration of 6, 8, 10 w/v% were prepared. They swelled in water, and formed viscous inks that were characterized for their rheological and extrusion properties, with the most successful one having 10 w/v% concentration. MC HG can also be integrated with other hydrogels for TE applications (Piard et al. 2015). Due to its high viscosity, when utilized as a printing material (Suntornnond et al. 2017), MC demonstrates excellent shape fidelity and stackability. In fact, Li et al. (2017) prepared a robust hydrogel blend of alginate and MC powder (MW 88,000 g/mol) for bioprinting, and printed a 150-layer, 33 mm high cell-laden structure with high resolution, and good shape fidelity.

A possible strategy in regenerative medicine is developing cell culture surfaces from which to obtain intact cell sheets (CSs). Cochis et al. (2018) for the first time successfully printed CSs out of MC-based HGs. Namely, they prepared HGs by mixing MC powder (Methocel A4M, Dow) in saline solutions (Na_2SO_4 and phosphate buffered saline), and investigated their rheological behavior in order to optimize the printing process parameters. The extrusion-based AM process adopted was adequate for the preparation of CSs of different shapes for the regeneration of complex tissues.

Viable process for MC HG is EBP.

12.2.6 CHITOSAN

Chitosan (Figure 12.5) has molecular formula $(C_6H_{11}NO_4)_n$, and is a linear, semi-crystalline polysaccharide, derived by removing the acetyl group (CH_3CO) from the molecule of *chitin* (Figure 12.6), the second most abundant biopolymer after cellulose, and commonly present in invertebrates (crustacean shells and insect cuticles), mushrooms, green algae cell walls, and yeasts. Chitin has molecular formula $(C_8H_{13}O_5N)_n$, and is industrially derived mainly from crustaceans (shrimps and crabs) and fungal mycelia (Croisier and Jérôme 2013). The degree of removal of the

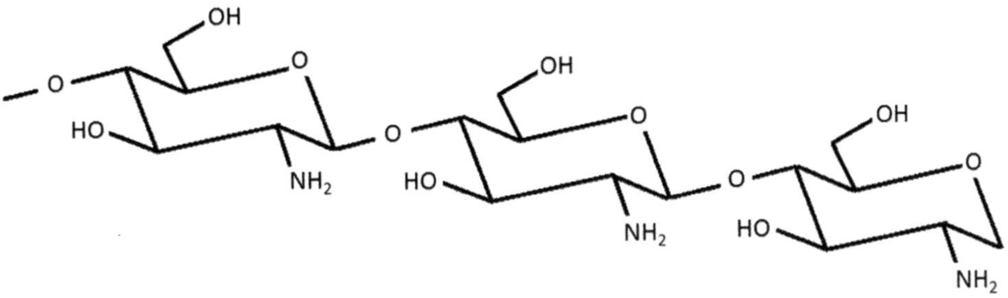

FIGURE 12.5 Chemical structure of chitosan.

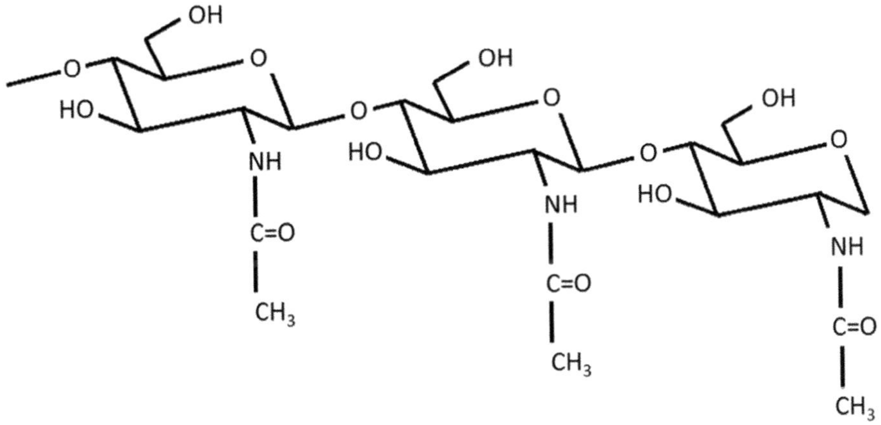

FIGURE 12.6 Chemical structure of chitin.

acetyl group influences the physical, mechanical, and biological properties of chitosan. Chitosan is a rather unique biobased polymer, because its intrinsic properties are so particular and valuable that it possesses no actual petrochemical equivalent. Most of its outstanding characteristics arise from the presence of amines along its backbone. Chitosan is the only positively charged natural polysaccharide, which is the main mechanism for chitosan to exhibit hemostatic and antibacterial activities. It is commonly insoluble in neutral conditions but easily soluble in the presence of acid. Chitosan is present in food (approved by FDA for that), cosmetics, biomedical, and pharmaceutical applications (Croisier and Jérôme 2013). Chitosan forms HGs by means of a number of ionic and chemical crosslinking methods (Li et al. 2020). Chitosan HG is antibacterial, antifungal, biodegradable (Rinaudo 2014), and nontoxic to cells.

Tensile properties of chitosan HG with 1 w/v% polymer concentration are: ultimate strength 30.6 MPa, yield strength 29 MPa, and modulus 98.9 MPa (Escobar-Sierra and Perea-Mesa 2017). The compressive modulus of chitosan featuring 600,000 g/mol MW (Jinan Haidebei Marine Bioengineering, China) is 34.9 MPa (Liu et al. 2013).

Applications of chitosan HGs encompass: regeneration of bone, cartilage, intervertebral discs, liver, myocardial, nerve, and through-cell encapsulation and transplantation; delivery of antibiotics, drugs, genes, growth factors, and proteins; therapy of liver and lung cancer; wound healing.

Wu et al. (2017) combined microextrusion and bioprinting, and formulated a chitosan HG bioink to build 10-layer microstructured scaffolds consisting in a highly flexible and organized microfiber network. Initially, 8 w/v% chitosan was dissolved in a mixed solvent (acetic acid/lactic acid/citric acid), and the resulting solution was loaded in a syringe and extruded under pressure through a 100 μm diameter nozzle at room temperature on a computer-controlled positioning platform. The chitosan ink exhibited a marked shear thinning behavior, and hence the rigidity of the chitosan filament increased by solvent evaporation, and provided sufficient structural support for building the scaffolds. The maximum tensile strength and strain at break of the HG filaments were 7.5 MPa and about 400%, respectively. After drying under vacuum for 72 h, the chitosan scaffolds were immersed in a sodium hydroxide (NaOH) solution for 2 h to neutralize the residual acids, and rinsed in water. The ionized NH_3^+ groups were converted into NH_2 groups, resulting in the disappearance of ionic repulsion and the promotion of physical crosslinking, and ultimately the physical gelation of the chitosan HG in the scaffold. The as-printed filaments displayed tensile strength and strain at failure of 3 MPa and about 180%, respectively, and the same properties in neutralized filaments were about 7.5 MPa and 400%. The authors employed the same equipment and chitosan bioink, and printed several test scaffolds: 30-layer scaffolds from 70 μm diameter filaments with area of 10 mm², total thickness of 2 mm, and square and diamond-shaped pores, and other scaffolds from 30 μm diameter filaments. All scaffolds showed high resolution, and a consistent geometry across layers. The step of acid neutralization affected the filament, causing first swelling (due to water absorption), then shrinkage that, in turn, might have caused surface wrinkling on the filament. Experiments and microscopic examination led to conclude that the printed scaffolds were biocompatible (cultivated L929 fibroblasts, i.e. biological cells synthesizing the ECM, had a survival rate of almost 100%), supported cell adhesion, proliferation, and spreading. Ultimately, it was demonstrated that it was possible to tailor bioinks in chitosan HGs to design and produce 3D tissue constructs meeting topographical, biological, and mechanical requirements.

12.2.7 COLLAGEN

Collagen, featuring molecular formula $C_{65}H_{102}N_{18}O_{21}$, is the chief protein of connective tissue in animals, and the most abundant protein in mammals. In fact it constitutes 1/3 of the total protein content in human body (Lin et al. 2019). It is a long, fibrous structural protein and, in bundles, supports most tissues and provides structure to cells from the outside. Collagen has high tensile strength, and is the main component of bone, cartilage, ligaments, teeth, and tendons. It imparts strength and elasticity to skin, and its degradation leads to wrinkles during aging. It also strengthens blood vessels, and influences tissue development (Parvizi 2010). Collagen is biocompatible, degradable, adhesive to cells, hydrophilic, flexible, weakly antigenic, and usually soluble in acidic aqueous solution. Type I collagen is predominantly found in bones and skin, whereas type II collagen is only present in cartilage. Type I collagen has limitations for bioprinting: it is liquid at low temperatures and at higher temperatures forms a fibrous structure; it has slow gelation rate, and, when deposited, it remains liquid for more than 10 min (Hospodiuk et al. 2017); gravity pulls downwards the cells inside collagen HGs when collagen is in liquid state, leading to uneven distribution of cells (Smith et al. 2004). Since collagen is the predominant component of ECM, and keeps its structural and biological integrity, collagen biomaterials can form an ECM-mimetic microenvironment and promote several cell functions: adhesion, migration, spreading, proliferation, differentiation, etc. (Li et al. 2020). Several physical and chemical methods are employed to crosslink collagen HGs, functionalize the molecular structure of collagen, and optimize its properties for a wide range of biomedical applications (Gu et al. 2019).

Lee et al. (2019) reported that for type I collagen HGs at concentrations V from 0.1 w/v% to 0.7 w/v% the tensile modulus E_t varied from 0.04 kPa to 1.6 kPa, whereas the compressive modulus E_c ranged from 1.4 kPa to 7.7 kPa, as displayed in Table 12.4, whose values indicate a quadratic dependence of E_t and E_c on V according to the equations (12.1) and (12.2):

$$Et = 3.19V^2 + 0.45V - 0.015 \qquad (12.1)$$

TABLE 12.4

Elastic Moduli of Collagen Hydrogels

Polymer Concentration	Tensile Modulus	Compressive Modulus
w/v%	kPa	kPa
0.1	0.04	1.46
0.3	0.23	2.75
0.5	0.86	4.82
0.7	1.56	7.74

Source: Data from plots in Lee, J, Song, B., Subbiah, R., et al. 2019. Effect of chain flexibility on cell adhesion: Semi-flexible model based analysis of cell adhesion to hydrogels. *Scientific Reports* 9:2463.

$$Ec = 10.19V^2 + 2.27V + 1.12 \qquad (12.2)$$

Applications of collagen HGs are: tumor models for cancer therapy; wound healing; delivery of antibiotics, drugs, growth factors, and proteins; regeneration of bone, brain, cartilage, liver, myocardial, nerve, renal, skin, tendon, and vascular constructs (Li et al. 2020).

AM technologies compatible with collagen HGs are: inkjet printing, pneumatic extrusion, matrix-assisted pulsed laser evaporation (MAPLE), and laser-based processes (Guillotin et al. 2010).

12.2.8 FIBRIN

Fibrin is an insoluble protein with molecular formula $C_5H_{11}N_3O_2$ (Figure 12.7), and is formed by the enzymatic reaction between thrombin and fibrinogen, the key proteins involved in blood clotting. It supports extensive cell growth and proliferation, plays a significant role in wound healing, and is present in skin grafts (Hospodiuk et al. 2017). In surgery, fibrin is utilized as a glue to control bleeding and bond tissues. Fibrin is a polymer whose viscoelastic properties can greatly fluctuate, depending on clot structure and biochemical properties (Weisel 2004). The fibrin network comprises filaments that are soft (differently from other filamentous biobased polymers), and forms a complex that features elastic behavior at small strains over short times, but anelastic behavior and remarkable stiffness at larger strains without breaking. In fact, the fibrin fiber is more extensible than other filamentous biopolymers, and can be stretched to more than five times its initial length without breakage, and even elongations exceeding 100% are completely recoverable upon release of stress. The tensile modulus of fibrin is equal to 4 MPa (Janmey et al. 2009). These mechanical properties, along with tunable gelation times, and large amount of fibrinogen make fibrin HGs suited for wound healing and TE in a wide range of settings (Janmey et al. 2009). Lee et al. (2019) reported that for fibrin HGs at concentrations V from 0.1-0.7 w/v% the tensile modulus E_t ranged from 0.3 kPa to 1.8 kPa, whereas the compressive modulus E_c varied from 2

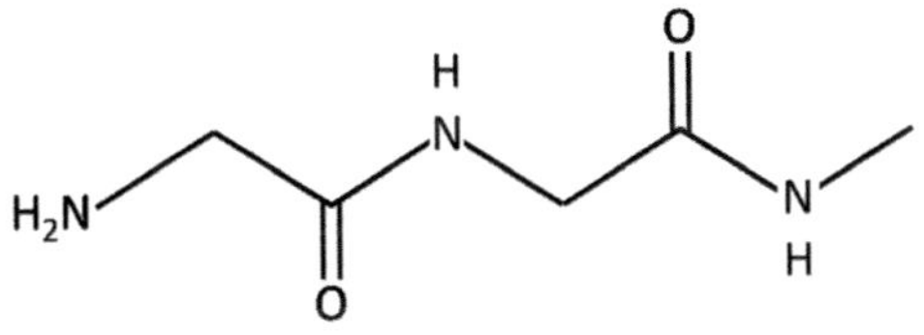

FIGURE 12.7 Chemical structure of fibrin.

TABLE 12.5
Elastic Moduli of Fibrin Hydrogels

Polymer Concentration	Tensile Modulus	Compressive Modulus
w/v%	kPa	kPa
0.1	0.30	1.98
0.3	0.53	3.64
0.5	0.85	5.77
0.7	1.79	6.38

Source: Data from plots in Lee et al. 2019. Effect of chain flexibility on cell adhesion: Semi-flexible model based analysis of cell adhesion to hydrogels. *Scientific Reports* 9:2463.

kPa to 6.4 kPa as indicated in Table 12.5, whose values indicate a quadratic dependence of E_t and E_c on V according to the equations (12.3) and (12.4):

$$Et = 4.44V^2 - 1.16V + 0.40 \tag{12.3}$$

$$Ec = 6.63V^2 + 12.98V - 0.64 \tag{12.4}$$

Fibrin HG is compatible with pneumatic extrusion.

12.2.9 GELATIN

Gelatin has molecular formula $C_{102}H_{151}O_{39}N_{31}$ (Figure 12.8), is a water-soluble, degradable, protein-based polymer derived from collagen (Gomez-Guillen et al. 2002) that promotes cell adhesion and proliferation, and features lower antigenicity compared to collagen. It is extracted from bones, skin, and connective tissues of animals. Gelatin has been used at concentrations of 2–20 w/v %. It gels below 35°C, and turns into liquid phase at 37°C, the temperature of human body. This poor mechanical stability and its relatively high enzymatic degradation rates limit use of gelatin in TE (Raucci et al. 2019). Hence, prior to printing, gelatin is usually subjected to chain modification and chemical crosslinking to enhance its mechanical properties (Zhang et al. 2005), and make its shape more stable at 37°C. Gelatin is present in drugs, food (approved by FDA), and TE. Gelatin HGs are crosslinked by physical and chemical methods (Li et al. 2020). Gelatin can form physical HGs below 27°C, but it is not stable at body temperature to function as a biomaterial due to the reversible thermal gelation, and this can be overcome by chemical modification and covalent crosslinking strategies.

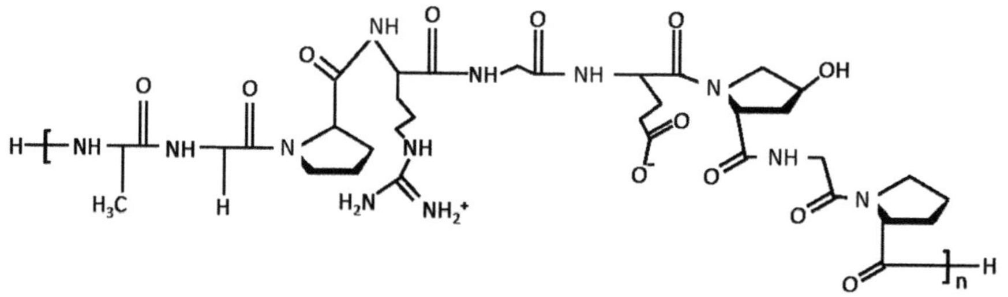

FIGURE 12.8 Chemical structure of gelatin.

TABLE 12.6

Compressive Modulus of Gelatin Methacrylate (GelMA) Hydrogels

GelMA	Methacrylate	Compressive Modulus
w/v %	% (w/v)	kPa
5	1.25	1.7
5	20	3.6
10	0.25	1.7
10	1.25	9.7
10	20	16.0
15	0.25	9.1
15	1.25	21.6
15	20	29.3

Source: Nichol, J. W., Koshy, S. T., Bae, H. et al. 2010. Cell-laden microengineered gelatin methacrylate hydrogels. *Biomaterials* 31:5536–5544.

Applications of gelatin HGs are: regeneration of bone, cartilage, corneal, myocardial, retinal, and skin; delivery of drugs, genes, growth factors, and proteins; therapy of gastric cancer and models of brain tumors (Li et al. 2020).

A common derivative of gelatin is GelMA, developed from gelatin reacting with methacrylate anhydride at 50°C. GelMA also reacts with several photoinitiators causing its chains to crosslink (Zhou et al. 2017), and provides good cell viability (Suntornnond et al. 2017). Nichol et al. (2010) investigated GelMA HG prepared with different amounts of GelMA and methacrylate, and observed that its compression modulus spanned from 1.7 kPa to 29.3 kPa, depending on the above amounts (Table 12.6). Another way to improve mechanical properties of gelatin HGs is blending. In fact, gelatin has been evaluated not only in the combinations with alginate mentioned in Section 12.2.2 but also the following: gelatin/alginate/chitosan, gelatin/alginate/fibrinogen, gelatin/chitosan, and gelatin/hyaluronan (Billiet et al. 2012). Chung et al. (2013) prepared gelatin/alginate blends that exhibited values of compression modulus similar to those of alginate HGs as-prepared and after immersion in a cell culture medium for up to 14 days at 37°C. Ramon-Azcon et al. (2013) reported a tensile modulus of 12.5 kPa and 20.9 kPa for GelMA HG unfilled and containing 0.3 mg/mL carbon nanotubes, respectively.

Gelatin and GelMA have been selected for various TE applications, from cell patterning to vascularization.

Gelatin HGs is suitable for the following AM methods: pneumatic extrusion (mostly), positive displacement extrusion, and laser-based processes.

12.2.10 HYALURONAN

Hyaluronan or hyaluronic acid (HA) has chemical formula $(C_{14}H_{21}NO_{11})_n$ (Figure 12.9), and is a linear polysaccharide found in almost all mammalian fluids and connective, epithelial, and neural tissues. It has a high capacity for lubrication, and water sorption and retention (Jeon et al. 2007). It is widely employed in synovial fluid of joints, wound healing, and treatment of osteoarthritis. Uses of HA in TE leverage its excellent biocompatibility, capacity to form flexible HGs, and tunable mechanical properties, architecture and degradation, but have to account for its slow gelation rate, and poor mechanical properties (Hospodiuk et al. 2017). HA is often chemically modified and mixed with other polymers when serving as bioink, in order to enhance its rheological and mechanical properties. F.e. HA was combined in one case with gelatin and human cardiac-derived

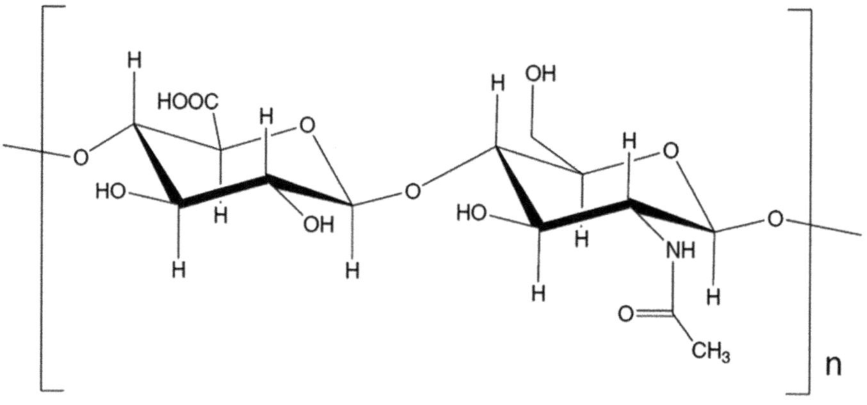

FIGURE 12.9 Chemical structure of hyaluronan.

Source: Jeon, O., Song, S. J., Lee, K.-J. 2007. Mechanical properties and degradation behaviors of hyaluronic acid hydrogels cross-linked at various cross-linking densities. *Carbohydr. Polym.* 70:251–257. Reproduced with permission from Elsevier.

cells to print a cardiac patch, and in another case with methyl cellulose to deliver stem cells (Mei and Cheng 2020).

Jeon et al. (2007) crosslinked fermentation-derived HA (MW 1.5×10^6 g/mol) with PEG diamines with two values of MW (1000 and 2000 g/mol), and measured its compressive elastic modulus that spanned from 20.7 kPa to 66.3 kPa, depending on the diamines and the MW between crosslinks.

Applications of HA HGs include: arthritis treatment; cancer therapy (thyroid cancer, tumor models); delivery of cells, drugs, genes, growth factors, and proteins; regeneration of bone, cartilage, intervertebral disc, liver, myocardial, nerve, and retina; wound healing (Li et al. 2020).

Positive displacement extrusion is an AM method compatible with HA HG.

12.2.11 PEPTIDES

Peptides are molecules consisting of 2 to 50 amino acids linked in a chain, with the carboxyl group of each acid being joined to the amino group of the next acid by a bond of the type –OC–NH–. When peptides are made up more than about 50 (the value is arbitrary) amino acids are called *proteins*. Peptides are present in animal, bacteria, plants, and animal organs: brain, cardiovascular system, kidneys, etc. Their general chemical structure is pictured in Figure 12.10.

In the human body peptides perform biological functions, acting as hormones, anti-oxidants, antimicrobial agent, and so on. Most peptides presently produced in laboratories are generated by solid-phase peptide synthesis (*synthetic peptides*) or from the enzymatic or chemical digestion of proteins (Aguilar and Purcell 2005). Applications of new peptides include formulating antibodies, drugs, enzymes, and vaccines. Moreover, peptides are extracted from beans and lentils, eggs, fish and shellfish, flaxseed, meat, milk, oats, etc., and added to health and cosmetic products for their potential anti-aging, anti-inflammatory, and muscle-building properties.

Raphael et al. (2017) studied a novel group of bioinks made up of self-assembling peptide-based HGs, and bioprinted well-defined 3D cell laden constructs with variable stiffness and enhanced structural integrity, while providing a cell-friendly extracellular matrix "like" microenvironment. Biological assays revealed that mammary epithelial cells remained viable after seven days in vitro culture, regardless of the HG's stiffness. Two peptides (PeptiGelDesign, UK) were used: Alpha1, and AlphaProB, with storage shear modulus of 10 kPa and 1 kPa, respectively after addition of culture media. Mammary epithelial cells EpH4 were cultured, and added to the HGs that were

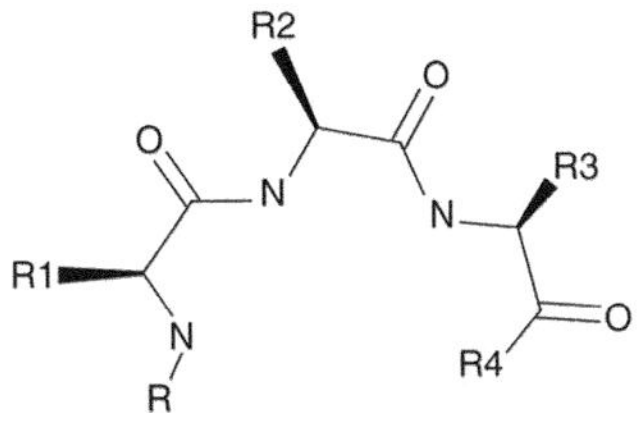

FIGURE 12.10 Chemical structure of peptide. Ri are side chain groups.

Source: Aguilar M.-I., Purcell, A. W. 2005. PEPTIDES. In *Encyclopedia of Analytical Science,* ed. A. Townshend, P. J. Worsfold, C. F. Poole, 29-36. Elsevier. Reproduced with permission from Elsevier.

FIGURE 12.11 Top view of printed cell-laded peptide-based hydrogel construct. Scale bar is 500 μm.

Source: Raphael, B., Khalil, T., Workman, V., et al. 2017. 3D cell bioprinting of self-assembling peptide-based hydrogels. *Materials Letters* 190:103–106. Reproduced with permission from Elsevier.

printed as 3D constructs at 37°C on the commercial extrusion-based AM bioprinter 3D Discovery (regenHU, Switzerland). The constructs (Figure 12.11) and were built with a 300 μm diameter filament, and consisted in a grid made of two layers of square interconnected pores of 250 μm side. Print parameters were adjusted to identify the optimal compromise among well-defined geometry, structural integrity, and cell viability. The test results revealed that the printed constructs possessed high structural integrity, and geometrical and dimensional accuracy. A large number of viable cells, irrespective of the HG's stiffness, was detected after seven days of culture, and proved that the HGs enabled cell survival and growth. AFM analysis concluded that constructs in Alpha1 were stiffer that those in AlphaProB, and surface roughness (root mean square deviation R_q) was 0.7 nm and 11 nm for soft and stiff HG, respectively. Nano-scale tensile modulus of Alpha1 and AlphaProB was 330 ± 55 MPa vs. 430 ± 240 MPa respectively, and the damping ratio (*loss tangent*) readings confirmed that Alpha 1 was softer than AlphaProB.

FURTHER READINGS

Ahmed, E. M. 2015. Hydrogel: preparation, characterization, and applications: a review. *J. Adv. Res.* 6 (2): 105–121.

Billiet, T., Vandenhaute, M., Schelfhout, J., Van Vlierberghe, S., Dubruel, P. 2012. A review of trends and limitations in hydrogel-rapid prototyping for tissue engineering. *Biomaterials* 33:6020–6041.

Hospodiuk, M., Dey, M., Sosnoski, D., Ozbolat, I. T. 2017. The bioink: a comprehensive review on bioprintable materials. *Biotechnol. Adv.* 35:217–239.

Li, H., Tan, C., Li, L. 2018a. Review of 3D printable hydrogels and constructs. *Mater. Des.* 159:20–38.

Li, H., Tan, Y. J., Liu, S., Li, L. 2018b. Three-dimensional bioprinting of oppositely charged hydrogels with super strong interface bonding. *ACS Appl. Mater. Interf.* 10:11164–11174.

Malda, J., Visser, J., Melchels, F. P. et al. 2013. 25th anniversary article: engineering hydrogels for biofabrication. *Adv. Mater.* 25:5011–5028.

Ovsianikov, A., J. Yoo, V. Mironov ed. 2018. *3D Printing and Biofabrication.* Cham: Springer.

Ozbolat, I. T. 2016. *3D Bioprinting: Fundamentals, Principles and Applications.* Academic Press.

Unagolla, J. M., Jayasuriya, A. C. 2020. Hydrogel-based 3D bioprinting: A comprehensive review on cell-laden hydrogels, bioink formulations, and future perspectives. *Appl. Mater. Today* 18:100479.

REFERENCES

Aguilar, M.-I., Purcell, A. W. 2005. Petptides. In *Encyclopedia of Analytical Science*, ed. P. Worsfold, A. Townshend and C. Poole, 29–36. Elsevier. 2nd edition.

Ahearne, M., Yang, Y., Liu, K.-K. 2008. Mechanical characterisation of hydrogels for tissue engineering applications. In *Topics in Tissue Engineering*, ed. N. Ashammakhi, R. Reis, F. Chiellini, Vol. 4, Chapter 12. Expertissues (E-book).

Ahlfeld, T., Köhler, T., Czichy, C., Lode, A., Gelinsky, M. 2018. A methylcellulose hydrogel as support for 3d plotting of complex shaped calcium phosphate scaffolds. *Gels* 4:68.

Ali, A., Ahmed, S. 2018. Recent advances in edible polymer based hydrogels as a sustainable alternative to conventional polymers. *J. Agric. Food Chem.* 66 (27): 6940–6967.

Almdal, K., Dyre, J., Hvidt, S., Kramer, O. 1993. Towards a phenomenological definition of the term 'gel'. *Polym. Gels Netw.*1 (1): 5–17.

Atabak Ghanizadeh, T., Miguel, A. H., Nicholas, R. L., Wenmiao, S. 2015. Three-dimensional bioprinting of complex cell laden alginate hydrogel structures. *Biofabrication* 7:045012.

Bertassoni, L. E., Cecconi, M., Manoharan, V. et al. 2014. Hydrogel bioprinted microchannel networks for vascularization of tissue engineering constructs. *Lab Chip.* 14:2202.

Billiet, T., Vandenhaute, M., Schelfhout, J., Van Vlierberghe, S., Dubruel, P. 2012. A review of trends and limitations in hydrogel-rapid prototyping for tissue engineering. *Biomaterials* 33:6020–6041.

Blaeser, A., Campos, D. F. D., Puster, U. et al. 2016. Controlling shear stress in 3D bioprinting is a key factor to balance printing resolution and stem cell integrity. *Adv. Healthc. Mater.* 5 (3): 326–333.

Boudou, T., Ohayon, J., Arntz, Y. et al. 2006. An extended modeling of the micropipette aspiration experiment for the characterization of the Young's modulus and Poisson's ratio of adherent thin biological samples: Numerical and experimental studies. *J. Biomech.* 39:1677–1685.

Celirix. 2018. *The Difference between Regenerative Medicine and Tissue Engineering.* https://www.celixir.com/the-difference-between-regenerative-medicine-and-tissue-engineering/ (accessed November 15, 2020).

Chung, J. H. Y., Naficy, S., Yue, Z. et al. 2013. Bioink properties and printability for extrusion printing living cells. *Biomater. Sci.* 1 (7): 763–773.

Cochis, A., Bonetti, L., Sorrentino, R. et al. 2018. 3D printing of thermo-responsive methylcellulose hydrogels for cell-sheet engineering. *Materials* 11 (4): 579.

Croisier, F., Jérôme, C. 2013. Chitosan-based biomaterials for tissue engineering. *Eur. Polym. J.* 49 (4): 780–792.

Daniel-da-Silva, A. L., Moreira, J., Neto, R. et al. 2012. Impact of magnetic nanofillers in the swelling and release properties of κ-carrageenan hydrogel nanocomposites. *Carbohydr. Polym.* 87 (1): 328–335.

Drury, J. L., Dennis, R. G., Mooney, D. J. 2004. The tensile properties of alginate hydrogels. *Biomaterials* 25:3187–3199.

Dutta, S. D., Patel, D. K., Lim, K.-T. 2019. Functional cellulose-based hydrogels as extracellular matrices for tissue engineering. *J. Biol. Eng.* 13:55.

Escobar-Sierra, D. M., Perea-Mesa, Y. P. 2017. Manufacturing and evaluation of Chitosan, PVA and Aloe Vera hydrogels for skin applications. *DYNA* 84(203). doi:10.15446/dyna.v84n203.62742.

Fedorovich, N. E., De Wijn, J. R., Verbout, A. J. et al. 2008. Three-dimensional fiber deposition of cell-laden, viable, patterned constructs for bone tissue printing. *Tissue Eng. A* 14:127.

Gasperini, L., Mano, J. F., Reis, R. L. 2014. Natural polymers for the microencapsulation of cells. *J. R. Soc. Interf.* 11:100.

Gomez-Guillen, M. C., Turnay, J., Fernandez-Diaz, M. D. et al. 2002. Structural and physical properties of gelatin extracted from different marine species: a comparative study. *Food Hydrocoll.* 16:25–34.

Gu, L., Shan, T., Ma, Y.-X., Tay, F. R., Niu, L. 2019. Novel biomedical applications of crosslinked collagen. *Trends Biotechnol.* 37 (5): 464–491.

Guillotin, B., Souquet, A., Catros S. 2010. Laser assisted bioprinting of engineered tissue with high cell density and microscale organization. *Biomaterials* 31 (28): 7250–7256.

He, Y., Yang, F., Zhao, H. et al. 2016. Research on the printability of hydrogels in 3D bioprinting. *Sci. Rep.* 6:29977.

Hospodiuk, M., Dey, M., Sosnoski, D., Ozbolat, I. T. 2017. The bioink: a comprehensive review on bioprintable materials. *Biotechnol. Adv.* 35:217–239.

Hossain, K. S., Miyanaga, K., Maeda, H., Nemoto, N. 2001. Sol-gel transition behavior of pure ε-carrageenan in both salt-free and added salt states. *Biomacromolecules* 2:442–449.

Ilinskaya, A. N., Dobrovolskaia, M. A. 2016. Understanding the immunogenicity and antigenicity of nano-materials: Past, present and future. *Toxicol Appl Pharmacol.* 299:70–77.

Janmey, P., Winer, J. P., Weisel, J. W. 2009. Fibrin gels and their clinical and bioengineering applications. *J. R. Soc. Interface* 6:1–10.

Jeon, O., Song, S. J., Lee, K.-J. 2007. Mechanical properties and degradation behaviors of hyaluronic acid hydrogels cross-linked at various cross-linking densities. *Carbohydr. Polym.* 70:251–257.

Kong, H.-J., Lee, K. Y., Mooney, D. J. 2002. Decoupling the dependence of rheological/mechanical properties of hydrogels from solids concentration. *Polymer* 43 (23): 6239–6246.

Kucukgul, C., Ozler, S. B., Inci, I. et al. 2015. 3D bioprinting of biomimetic aortic vascular constructs with self-supporting cells. *Biotechnol. Bioeng.* 112:811.

Lee, J., Song, B., Subbiah, R. et al. 2019. Effect of chain flexibility on cell adhesion: Semi-flexible model based analysis of cell adhesion to hydrogels. *Sci. Rep.* 9:2463.

Lee, K. Y., Mooney, D. J. 2012. Alginate: properties and biomedical applications. *Prog. Polym. Sci.* 37 (1): 106–126.

Li, H., Liu, S., Li, L. 2016. Rheological study on 3D printability of alginate hydrogel and effect of graphene oxide. *Int. J. Bioprinting* 2:54–66.

Li, H., Tan, C., Li, L. 2018a. Review of 3D printable hydrogels and constructs. *Mater. Des.* 159:20–38.

Li, H., Tan, Y. J., Liu, S., Li, L. 2018b. Three-dimensional bioprinting of oppositely charged hydrogels with super strong interface bonding. *ACS Appl. Mater. Interf.* 10:11164–11174.

Li, H., Tan Y. J., Leong, K. F., Li, L. 2017. 3D bioprinting of highly thixotropic alginate/methylcellulose hydrogel with strong interface bonding. *ACS Appl. Mater. Interfaces* 9:20086–20097.

Li, J., Wu, C., Chu, P. K., Gelinsky, M. 2020. 3D printing of hydrogels: rational design strategies and emerging biomedical applications. *Mate. Sci. Eng.* R 140:100543.

Lin, D. C., Yurke, B., Langrana, N. A. 2005. Use of rigid spherical inclusion in Young's moduli determination: Application to DNA-crosslinked gels. *J. Biomech. Eng.* 127:571–579.

Lin, K., Zhang, D., Macedo, M. H. et al. 2019. Advanced collagen-based biomaterials for regenerative biomedicine. *Adv. Funct. Mater.* 29 (3): 1804943.

Liu, M., Wu, C., Jiao, Y. et al. 2013. Chitosan–halloysite nanotubes nanocomposite scaffolds for tissue engineering. *J. Mater. Chem. B* 1:2078.

Malda, J., Visser, J., Melchels, F. P. et al. 2013. 25th Anniversary Article: Engineering Hydrogels for Biofabrication. *Adv. Mater.* 25:5011–5028.

Mei, X., Cheng, K. 2020. Recent Development in Therapeutic Cardiac Patches. *Frontiers in Cardiovascular Medicine* Vol. 7, Article 610364.

Michel, A.-S., Mestdagh, M., Axelos, M. 1997 Physicochemical properties of carrageenan gels in presence of various cations. *Int. J. Biol. Macromol.* 21:195–200.

Miguel, S. P., Ribeiro, M. P., Brancal, H. et al. 2014 Thermoresponsive chitosan–agarose hydrogel for skin regeneration. *Carbohydr. Polym.* 111:366.

Min, L. J., Edgar, T. Y. S., Zicheng, Z., Yee, Y. W. 2015. 3D Bioprinting and Nanotechnology. In *Tissue Engineering and Regenerative Medicine*, ed. L. G. Zhang, J. P. Fisher, K. Leong. Academic Press. 1st edition.

Mironov, V., Trusk, T., Kasyanov, V. et al. 2009. Biofabrication: a 21st century manufacturing paradigm. *Biofabrication* 1 (2): 022001.

Necas, J., Bartosikova, L. 2013. Carrageenan: a review. *Vet. Med.* 58:187–205.

Nichol, J. W., Koshy, S. T., Bae, H. et al. 2010. Cell-laden.microengineered gelatin methacrylate hydrogels. *Biomaterials* 31:5536–5544.

Normand, V., Lootens, D. L., Amici, E., Plucknett, K. P., Aymard, P. 2000. New Insight into Agarose Gel Mechanical Properties. *Biomacromolecules* 1:730–738.

Parvizi, J. ed. 2010. *High-Yield Orthopaedics*. Amsterdam: Saunders/Elsevier. 1st edition.

Paxton, N., Smolan, W., Böck, T. et al. 2017. Proposal to assess printability of bioinks for extrusion-based bioprinting and evaluation of rheological properties governing bioprintability. *Biofabrication* 9:044107.

Piard, C. M., Chen, Y., Fisher, J. P. 2015. Cell-laden 3D printed scaffolds for bone tissue engineering. *Clin. Rev. Bone Miner. Metab.* 13:245.

Ramon-Azcon, J., Ahadian, S., Estiliet, M. et al. 2013. Dielectrophoretically aligned carbon nanotubes to control electrical and mechanical properties of hydrogels to fabricate contractile muscle myofibers. *Adv. Mater.* 25:4028–4034.

Raphael, B., Khalil, T., Workman, V. et al. 2017. 3D cell bioprinting of self-assembling peptide-based hydrogels. *Mater. Lett.* 190:103–106.

Raucci, M. G., D'Amora, U., Ronca, A., Demitri, C., Ambrosio, L. 2019. Bioactivation routes of gelatin-based scaffolds to enhance at nanoscale level bone tissue regeneration. *Front. Bioeng. Biotechnol.* 7:1–11.

Rinaudo, M. 2014. Materials based on chitin and chitosan. In *Bio-Based Plastics – Materials and Applications*, ed. S. Kabasci, 63–88. Chichester: Wiley.

Rochas, C., Rinaudo, M., Landry, S. 1989. Relation between the molecular structure and mechanical properties of carrageenan gels. *Carbohydr. Polym.* 10:8–10.

Scientific American. 1999. *What is Jell-O?* https://www.scientificamerican.com/article/what-is-jell-o-how-does-i/ (accessed November 15, 2020).

Smith, C. M., Stone, A. L., Parkhill, R. L. et al. 2004. Three-dimensional bioassembly tool for generating viable tissue-engineered constructs. *Tissue Eng.* 10:1566–1576.

Suntornnond, R., An, J., Chua, C. K. 2017. Bioprinting of thermoresponsive hydrogels for next generation tissue engineering: a review. *Macromol. Mater. Eng.* 302, 1600266.

Tsakalakos, T. 1981. The bulge test: a comparison of the theory and experiment for isotropic and anisotropic films. *Thin Solid Films* 75:293–305.

Van Cauwenberghe, C. 2019. The regenerative medicine market landscape. In *Encyclopedia of Tissue Engineering and Regenerative Medicine*, Vol. 3, ed. R. El Reis. London: Academic Press.

Wang, Q., Sun, J., Yao, Q. et al. 2018. 3D printing with cellulose materials. *Cellulose* 25:4275–4301.

Webb, B., Doyle, B. J. 2017. Parameter optimization for 3D bioprinting of hydrogels. *Bioprinting* 8:8–12.

Weisel, J. W. 2004. The mechanical properties of fibrin for basic scientists and clinicians. *Biophys. Chem.* 112:267–276.

Williams, D. 2015. *Novel Material 'Celleron' Could Revolutionize 3D Bioprinting for Regenerative Medicine*. https://3dprint.com/62259/celleron-bioprinting/ (accessed November 15 , 2020).

Wu, Q., Maire, M., Lerouge, S., Therriault, D., Heuzey, M.-C. 2017. 3D printing of microstructured and stretchable chitosan hydrogel for guided cell growth. *Adv. Biosys.* 1:1700058.

Zhang, L.-M., Wu, C.-X., Huang, J.-Y. et al. 2012. Synthesis and characterization of a degradable composite agarose/HA hydrogel. *Carbohyd. Polym.* 88 (4): 1445–1452.

Zhang, Y., Ouyang, H., Lim, C. T., Ramakrishna, S., Huang, Z. M. 2005. Electrospinning of gelatin fibers and gelatin/PCL composite fibrous scaffolds. *J. Biomed. Mater. Res. B Appl. Biomater.* 72 (1): 156–165.

Zhou, M., Lee, B. H., Tan, L. P. 2017. A dual crosslinking strategy to tailor rheological properties of gelatin methacryloyl. *Int. J. Bioprinting* 3 (2): 130–137.

13 Polybutylene Succinate

To waste, to destroy our natural resources, to skin and exhaust the land instead of using it so as to increase its usefulness, will result in undermining in the days of our children the very prosperity which we ought by right to hand down to them amplified and developed.

Theodore Roosevelt

13.1 OVERVIEW OF POLYBUTYLENE SUCCINATE (PBS)

PBS is a white, highly crystalline, TP, aliphatic, polyester polymer, featuring mechanical properties similar to those of PP. Its other advantages are thermal and chemical resistance, ductility, and versatile processability comparable to common TPs (Xu and Guo, 2010a, 2010b). On the other hand, applications of PBS are hindered by high cost, softness, and low gas barrier properties. PBS is biodegradable (it naturally decomposes into water and CO_2) under industrial conditions (EN 13432), compostable, and sustainable because it can be formulated from fossil resources (maleic anhydride), and fermentation of microorganisms living on renewable feedstocks, such as food waste, glucose, hemp, microalgae, rapeseed oil, starch, sucrose, and xylose (Song and Lee 2006; Nova-Institute, n.d.). It is also approved for use with food. Its molecular formula is $(C_8H_{12}O_4)_n$ and its chemical structure is pictured in Figure 13.1. Mechanical properties of PBS depend on its MW: MW of 100,000 g/mol results in a brittle grade that can be extruded and injection molded, featuring 10% elongation at break, and notched Izod impact strength <40 J/m, whereas MW >180,000 g/mol yields a more ductile grade, processable through blowing, and exhibiting elongation at break of 270%, and notched Izod impact strength of 73 J/m (Xu and Guo 2010a). The T_m and T_g of PBS are 100–120°C and −38°C to −32°C, respectively (van Es et al. 2014; Mochane et al. 2021). The elongation at break and tensile strength of PBS are similar to the widely consumed polyolefins PP and LDPE (Candal et al. 2020). The tensile modulus of PBS ranges is 500–590 MPa according to literature data (Di Lorenzo et al. 2019). The degree of crystallinity is 35-45% (Mochane et al. 2021). A further way to tune PBS properties is through polymerization with different monomers (*copolymerization*), such as adipic acid, which results in a commercial version of PBS called poly (butylene succinate-co-butylene adipate) or PBSA, an important copolymer because its elongation and impact strength are markedly higher than those of PBS (Tserki et al. 2006).

PBS is an attractive candidate for AM, because of its processability (superior to that of PLA), ductility, and relatively low T_m. PBS is versatile thanks to its wide processing window, which is advantageous in formulating feedstocks for AM technologies requiring diverse processing conditions. Being a polyester, PBS needs to be dried before being processed at high temperature. Its maximum processing temperature is 200–230°C. Besides extrusion, injection molding, and film blowing mentioned previously, PBS can also be thermoformed and fiber spun, which translates in numerous current and potential non-AM applications, including agriculture (mulching films, etc.), composites with natural fibers, fast food packages, hygiene products such as diapers, packaging for cosmetics and food (bags, boxes, films), disposable products (plastic bags, medical articles, and tableware), fishing nets and lines, plant pots, and wood-plastic composites (Xu and Guo 2010a; Nova-Institute n.d.). PBS applications that can leverage advantages of AM processes are drug encapsulation systems and implants (bone and cartilage repair), with the latter ones being also high-value products.

A commercial source of PBS is Succinity®, a biobased succinic acid produced by Succinity GmbH, a joint venture between BASF and Corbion.

DOI: 10.1201/9781003221210-13

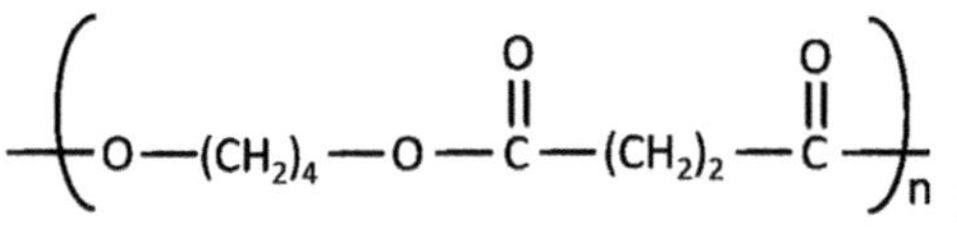

FIGURE 13.1   Chemical structure of PBS.

PBS can be combined with sustainable (PLA, PHAs, starch) and biodegradable (PBAT) polymers to improve specific properties and processability (van Es et al. 2014), as illustrated in Table 13.1. We recall that PBAT is a fossil-based copolymer. One additional feature of PBS that would benefit filaments for AM from a performance and cost standpoint is that it binds well to natural fibers without needing any bonding agent.

Table 13.2 list the properties of three grades of BioPBS™, a biobased grade of PBS produced by PTTMCC, a joint venture between Mitsubishi Chemical Corp. (Japan), and PTT Global Chemical (Thailand). BioPBS™ is polymerized from biobased succinic acid and fossil-based 1,4-butanediol, and is suited for compounding, barrier and flexible packaging, paper coating, and as a synthetic fiber, thanks to its adhesion to paper, heat resistance, impact strength, and sealing capacity. It is also approved for food contact (coffee cups, cutlery), and is compostable and home

TABLE 13.1

Improved Properties of Blends of PBS and PBSA

PBS Blends	Improved Properties
PBS/PLA	Elastic modulus, elongation at break, tensile strength, ductility
PBSA/PLA	Ductility, impact resistance, strength
PBS/PBAT	Tensile strength, toughness
PBS/carbon nanotube	Thermal and electrical conductivity, tensile modulus, storage moduli
PBS/PLA/CaSO$_4$ whiskers	Crystallization, elongation at break, impact resistance, thermal resistance
PBS/talcum	Cost, creep resistance, heat distortion temperature

Source: nova-Institute n.d., Succinity – Biobase Your Success. Available online. Reproduced with kind permission from Nova Institute.

TABLE 13.2

Properties of Commercial PBS by PTTMCC (BioPBS™)

Property	Unit	BioPBS™ FZ71	BioPBS™ FZ91	BioPBS™ FD92
Density	g/cm^3	1.26	1.26	1.24
Melt Flow Rate (190°C, 2.16 kg)	g/10 min	22	5	4
Melting Point	°C	115	115	84
Tensile Yield Strength	MPa	40	40	17
Tensile Stress at Break	MPa	30	36	24
Tensile Strength	MPa	N/A	N/A	N/A
Tensile Strain at Break	%	170	210	380
Flexural Modulus	MPa	630	650	250
Flexural Strength	MPa	40	40	18
Notched Izod Impact Strength	kJ/m^2	7	10	47
Heat Deflection Temp. (at 0.45 MPa)	°C	95	95	63
Rockwell Hardness	R Scale	107	107	56

Source: Data posted online by PTTMCC.

compostable at 30°C. Since tensile strength and modulus of BioPBS™ are not disclosed, this material cannot be directly compared to other SPs for AM. However, in comparison to the property values of commercial grades of PLA for AM listed in Table 3.9, BioPBS™ lags in tensile strength and stiffness, but leads in elongation at break and impact strength (once the impact values are converted based on 1 kJ/m^2 = 10.2 × J/m). Other suppliers of PBS and PBSA are Anqing Hexing Chemicals (China), Eastman (USA), and Zhejiang Hangzhou Xinfu Pharmaceutical (China). In 2016, the major producer of PBS Showa Denko (Japan), announced that would terminate production of its Bionolle™ by the end of December 2016.

In 2014 the price of PBS was reported at EUR 3–4/kg (van Es et al. 2014). In 2017 the distribution of the global market of PBS was the following, in decreasing order: Europe, North America, Asia Pacific, Middle East and Africa, and Latin America (Research Nester 2020). In 2020, the world production of PBS was estimated at 86,500 tonnes/year (European Bioplastics 2019). In 2018, the global PBS market was expected to grow at 10.5% CAGR from 2018 to 2023 (OMR 2019). If the demand for feedstocks containing biobased polymers will keep raising, the biobased version of PBS should also increase in sales, and possibly lead to a price reduction and consequent market growth (van Es et al. 2014). The near future of the PBS production depends on contrasting variables: on one hand, the need for heat resistant and biodegradable plastics, and blending PBS in polymers selected as feedstocks for FFF filaments to achieve this need would boost its demand (although not greatly); but, on the other hand, its complex synthesis, high cost, and large emission of greenhouse gas associated with its production hinders its diffusion. The price of fossil resources such as oil and shale gas also impacts the future of PBS: in fact, in 2019 the production of biobased succinic acid steeply dropped worldwide, because the low price of shale gas made succinic acid too expensive.

Recommended and broad reviews of PBS, its chemistry, grades, properties, copolymers, R&D, suppliers, and so on were published by Xu and Guo (2010a) and van Es et al. (2014)Xu 2010.

13.2 COMMERCIAL PBS FILAMENT FOR AM

3D Printlife sells DURA™, a sustainable filament for FFF. DURA™ has been developed by Algix3D out of PBS, PLA, and PHA, all biobased polymers, and its properties are illustrated in Table 13.3, where tensile stress and strain are reported relatively to yield and not failure. In polymers elongation at break is always higher than elongation at yield, but tensile strength may be lower or higher than tensile strength at yield. DURA™ leads the PBS grades in Table 13.2 in tensile yield strength but trails in HDT, reported by the supplier without specifying what test stress was applied (0.45 MPa or 1.81 MPa).

13.3 EXPERIMENTAL PBS FOR AM AND OTHER PROCESSES

Candal et al. (2020) investigated the thermo-rheological properties of PBS and PBSA relatively to their compatibility with FFF. Two commercial extrusion grade polymers (NaturePlast, France) in

TABLE 13.3
Properties of Algix3D DURA™ Filament for FFF

Density	MFI	Tensile Properties			HDT	T_m
		Yield Strength	Modulus	Elongation at Yield		
g/cm^3	g/10 min	MPa	MPa	%	°C	°C
1.29	12.5	29	2,754	18	120	155

Source: Data posted online by Algix3D.

pellet form were chosen: PBE003 PBS (MFI of 4–6 g/10 min at 190°C, MW of 79,250 g/mol), and PBE001 PBSA (MFI of 4–5 g/10 min at 190°C, MW of 78,600 g/mol), whose selected properties are listed in Table 13.4. The pellets were dried, and fed to a twin-screw co-rotating extruder operating at 50 rpm and 150°C nozzle temperature, and producing a 1.75 mm diameter filament. The chief difference between PBS and PBSA was that the high crystallinity of PBS led to substantial warpage, preventing its use for practical applications, whereas the less crystalline and random copolymer PBSA produced dimensionally stable and ductile printed objects. Hence, the tensile properties of printed and injected molded PBSA coupons and relative representative stress-strain curves are reported in Table 13.5 and Figure 13.2, respectively, but not those of PBS coupons. Photographs of representative test coupons at different stages of testing are also included. The superior tensile performance of molded coupons over printed coupons may be due to: (a)

TABLE 13.4

Thermal Properties and MW of PBS and PBSA

Material	MFI	T_m	T_c	T_g	Average Number MW	Average Weight MW
	g/10 min	°C	°C	°C	g/mol	g/mol
PBS	4–6	85	27	−22	19,800	79,250
PBSA	4–5	118	52	−34	12,300	78,600

Source: Candal, M. V., Calafel, I., Aranburu, N., et al. 2020. Thermo-rheological effects on successful 3D printing of biodegradable polyesters. *Additive Manufacturing* 36, 101408. Reproduced with permission from Elsevier.

TABLE 13.5

Tensile Properties of PBSA

Samples	Tensile Properties			
	Stress at Break	Stress at Yield	Modulus	Elongation at Break
	MPa	MPa	MPa	%
Printed	15.9	15.3	325	213
Injection molded	29.0	21.0	380	230

Source: Candal, M. V., Calafel, I., Aranburu, N., et al. 2020. Thermo-rheological effects on successful 3D printing of biodegradable polyesters. *Additive Manufacturing* 36, 101408. Reproduced with permission from Elsevier.

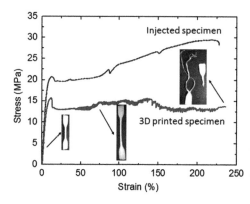

FIGURE 13.2   Tensile stress-strain curves of printed and injection molded coupons made of PBSA.

Source: Candal, M. V., Calafel, I., Aranburu, N., et al. 2020. Thermo-rheological effects on successful 3D printing of biodegradable polyesters. *Additive Manufacturing* 36, 101408. Reproduced with permission from Elsevier.

strain hardening in injection molded coupons during testing but not in printed coupons; (b) air gaps between adjacent layers of printed coupons causing porosity and stress concentration; (c) material compaction in molded coupons compensating for shrinkage; (d) cooling, shrinkage, and crystallization occurring under different conditions between printing and molding. The values of viscosity below 10^5 Pa recorded in the shear rate interval applied in printing for both polymers were consistent with the observed continuous and regular flow in the printer's nozzle. Strong interlayer adhesion between deposited layers was measured for PBS and PBSA through tearing energy: with a bed and nozzle temperatures of 25°C and 190°C, respectively, the tearing energy was 26.3–29.5 kN/m, higher than the maximum value of 25 kN/m for ABS. From pressure-volume-temperature (PVT) tests assessing the volume decrease during crystallization, the specific volume changes from 100°C to 25°C were 8.6% and 4.3% for PBS and PBSA, respectively, which translated into significant warpage for PBS preventing its use for practical uses, whereas the less crystalline PBSA produced dimensionally stable and ductile printed objects.

Ou-Yang et al. (2018) formulated PBS/PLA blends at various compositional ratios, and evaluated their suitability as a FFF filament in terms of mechanical and thermal properties, and processability (distortion of printed bars, interlayer bond strength, and appearance). They selected Ingeo™ 2002D PLA with 6 g/10 min MFI and 120,000 g/mol MW, and TH803S PBS (Xinjiang Blue Ridge Tunhe Polyester, China) featuring 8 g/10 min MFI, and 83,000 g/mol MW. The PBS/PLA blend formulations ranged from 80/20 wt% to 20/80 wt% (Table 13.6). The blends were extruded into 1.75 mm diameter filaments that in turn were printed into test coupons. Higher PLA content translated in printed bars with higher tensile strength and modulus, and lower elongation at break and impact strength, i.e. a stronger and stiffer but less ductile material. The interlayer bond strength was measured through tensile strength of dumbbell-shaped test specimens printed in the Z direction: bond strength decreased by lowering amount of PLA and raising printing temperature of the same formulation (PBS40/PLA60). No formulation exceeded all the others in all tensile

TABLE 13.6
Properties of FFF Filaments: Commercial (ABS, PC, PC/ABS) and PBS/PLA Blends

Blend wt%	T_g	T_m	Tensile Strength	Tensile Modulus	Elongation at Break	Izod Notched Impact Strength	Predicted Tensile Modulus
	°C	°C	MPa	MPa	%	kJ/m^2	MPa
PBS100	NA	113.2	41.5	554	324	7.2	N/A
PBS80/PLA20	53.7	112.0	46.8	794	356	20.9	1,067
PBS60/PLA40	53.1	112.2	45.6	1,546	297	14.7	1,567
PBS40/PLA60	55.0	112.7	51.2	2,045	159	13.9	2,044
PBS20/PLA80	54.6	111.7	55.6	2,150	93	12.9	2,464
PLA 2002D[b]	NA	N/A	43	2,700	2.4	N/A	N/A
ABS (M30™)[c]	108	N/A	28–3[a]	2,180–2,230[a]	2–7[c]	N/A	N/A
PC[c]	161	N/A	68	2,300	5	N/A	N/A
PC/ABS[c]	125	N/A	41	1,900	6	N/A	N/A

Source: Ou-Yang, Q., Guo, B., Xu, J. 2018. Preparation and Characterization of Poly(butylene succinate)/Polylactide Blends for Fused Deposition Modeling 3D Printing. *ACS Omega* 3:14309–14317.

Notes:

[a] Ranges are relative to values measured in multiple printing directions.

[b] Signori, F., Coltelli, M.-B., Bronco, B. 2009. Thermal degradation of poly(lactic acid) (PLA) and poly(butylene adipate-co-terephthalate) (PBAT) and their blends upon melt processing. *Polymer Degradation and Stability* 94 (1): 74–82.

[c] Data posted online by Stratasys.

properties and impact strength. Highest tensile strength (55.6 MPa) and modulus (2,150 MPa) were reached by PBS20/PLA80, and highest impact strength and elongation at break were displayed by PBS80/PLA20, which is consistent with the fact than PLA 2002D is stronger and stiffer than PBS TH803S, while the latter is more ductile than the former. The tensile modulus of the blends E_{blend} was predicted by applying the following Takayanagi equation (Springer 2006, 514) relative to a polymer blend in which two polymers A and B are stressed in parallel and series, and λ and φ are the volume fractions (from 0 to 1) of each polymer stressed in parallel and in series, respectively:

$$E_{blend} = (1 - \lambda)E_B + \lambda\left(\frac{\varphi}{E_A} + \frac{1 - \varphi}{E_B}\right) \qquad (13.1)$$

Assuming λ and φ were equal to one other for each material at each PBS/PLA composition, we calculated the predicted values of E_{blend}, and listed them in Table 13.6. Equation (13.1) predicted extremely accurately E_{blend} for PBS60/PLA40 and PBS40/PLA60, but overpredicted it for the remaining blends. The suitability of the blends as printing feedstock, and the optimal printing temperature were evaluated in terms of viscosity of the polymer melt at conditions simulating extruding pellets into filaments, and extruding the filament from a printer. Amount of PLA exceeding 40 wt% resulted in no distortion (caused by residual stress induced by volume shrinkage), good dimensional accuracy, and pearl-like gloss.

Badouard et al. (2019) fabricated and analyzed a composite filament for FFF containing PBS, PLLA, and flax fibers and shives (wooden refuse removed during processing of fibers), and compared it in terms of tensile properties and porosity to injection molded PBS/PLLA. The authors selected Bionolle™ 1020 MD PBS, Ingeo™ 7001D PLLA, and fibers and shives of the flax Alize variety from Normandy (France). Fibers 1 mm long featured tensile strength, modulus, and strain at break equal to 1,215 MPa, 53.8 GPa, and 2.24%, respectively. Shives were first crashed and then milled, which may have degraded the values of the above tensile properties. Fiber length showed three peaks, 20, 125 and 650 μm, whereas shive length had a 220 μm peak. PBS/PLLA was 50/50 wt%, and PBS/PLLA/flax (fiber or shives) was 50/50/10 wt%. The filament fabrication consisted in drying and compounding all constituents in a twin-screw extruder that produced composites pellets that were dried and injection molded into test coupons, and extruded into a 2.85 mm diameter filament. The tensile test results relative to PBS/PLLA/flax and PLLA/flax filaments are listed in Table 13.7. Adding PBS penalized modulus and did not improve strength, as expected since PLLA is notably stronger and stiffer than PBS. Another factor impairing tensile properties in composite filaments was porosity that was instead absent in injection molded coupons, as illustrated in Figure 10.9, because of the injection molding pressure. However, a comprehensive assessment of PBS/PLLA/flax and PLLA/flax should include performance/cost. As to PBS/PLLA/flax filaments, the one with fibers outperformed the one with shives in tensile modulus but lagged in tensile strength and elongation at break (1.3% vs. 1.8%). The authors did not measure impact strength, but elongation at break of 1020 MD (10.9%) higher than that of 7001D (2.8%) likely revealed impact strength higher in PBS/PLLA blend than in PBS/PLLA/flax.

The fact that D. Mielewski, co-author of the paper by Qahtani et al. (2019) on FFF filaments made of PBS/PLA blends, previously was a sustainability leader at Ford Motor Company indicates that sustainable polymers for AM are of interest to large companies, including carmakers. The authors fabricated and analyzed PBS/PLA filaments for FFF, and employed DoE to assess the correlation between the mechanical properties of the filament and blend compositions. The filament's printability was also determined through its flowability, and dimensional consistency. Materials evaluated were Ingeo™ 4043D PLA with MFI of 6 g/10 min (at 210°C), and BioPBS™ FZ91PM (PTT MCC) with MFI of 5 g/10 min (at 190°C). PBS/PLA blends investigated had the following compositions in wt%: 10/90, 20/80, 30/70, 40/60, and 50/50. The PBS was maximized at 50 wt% content since filaments with PBS content >50 wt% could not be successfully printed

TABLE 13.7

Tensile Properties of PLLA, PBS/PLLA, and PBS/PLLA/Flax Composites

Material	Composition	Tensile Strength	Tensile Modulus	Elongation at Break
	wt%	MPa	MPa	%
PBS Bionolle™ 1020[a]	100	32.6	589	10.9
PLLA, IM[b]	100	55.2	3,760	2.8
PBS/PLLA, IM[b]	50/50	39.1	2,070	12.5
PLLA/10F[c], filament	100/10	34.2	3,968	1.2
PBS/PLLA/10F[c], filament	50/50/10	30.3	2,786	1.8
PBS/PLLA/10S[d], filament	50/50/10	34.1	2,190	1.3

Source: Adapted from Badouard, C., Traon, F., Denoual, C., et al. 2019. Exploring mechanical properties of fully compostable flax reinforced composite filaments for 3D printing applications. *Industrial Crops & Products* 135:246–250. Reproduced with permission from Elsevier.

Notes:

[a] Phua, Y. J., Chow, W. S., Mohd Ishak, Z. A. 2011. Mechanical properties and structure development in poly(butylene succinate)/organo-montmorillonite nanocomposites under uniaxial cold rolling. *eXPRESS Polymer Letters* 5(2):93–103.

[b] IM injection molded.

[c] F fibers.

[d] S shives.

because of thermal instability (high CLTE) and excessive viscosity. The filaments were prepared through the following sequence: drying pellets of PLA and PBS, melt mixing the blends in a twin-screw extruder, collecting, running through a water bath, and drying the output strand. Since optimal content of PBS was 30 wt%, flowability was measured through the viscoelastic behavior of PBS/PLA 30/70 at varying temperatures coinciding with the range of temperatures in the printer's extruder. Generally speaking, a printing temperature too elevated may have detrimental effects: polymer decomposition, excessively fluid filament that uncontrollably leaks out of the nozzle, and filament flattening after deposition. Conversely, a printing temperature too low leads to intermittent extrusion, debonding between adjacent layers, dimensional inaccuracy, inferior mechanical properties (Khaliq et al. 2017), and longer printing time. After experimenting, 250°C was chosen as the optimal temperature because it resulted in interlayer adhesion, dimensional accuracy, and uninterrupted filament flow. The other printing settings were 65°C bed temperature, and 60 mm/s nozzle speed. The measured mechanical and thermal properties are listed in Table 13.8. Since PBS is less strong and far more ductile than PLA, adding PBS to PLA steadily reduced tensile strength (starting with 20 wt% PBS), but gradually raised impact strength. The blend PBS/PLA (10/90) outperformed PLA in tensile and impact strengths, which was attributed to crystallinity and miscibility between PBS and PLA. PLA and PBS are immiscible above 10 wt% of PBS. Although the authors recognized the influence of crystallinity on blends' tensile strength, crystallinity was excluded from the regression model they derived. The fact that two melting temperatures were recorded for each blend is explained by the fact that PBS and PLA are semi-crystalline and each features a distinct melting point. Increasing content of PBS negligibly affected T_g and T_m, by lowering them by about 2°C. In Minitab™ statistical software, a statistical analysis based on the DOE results was conducted, and, from the linear regression of test values of tensile (TS), flexural, and impact strength (IS), for each strength, an equation in three variables, V_1 (PLA wt%), V_2 (PBS wt%), and the product $V_1 \times V_2$, was derived that allowed to: (a) select the composition maximizing each strength, and (b) along with contour and surface plots, either to predict properties of specific

TABLE 13.8

Mechanical and Thermal Properties of PBS/PLA Blend for FFF

PBS/PLA Blend	Tensile Strength	Flexural Strength	Impact Strength	CTLE	T_g	$T_m{}^c$	$T_m{}^d$
wt%	MPa	MPa	J/m	µm/m°C	°C	°C	°C
0/100	66.2	113.5	32.9	78	60.8	151.1	N/A
10/90	77.0	82.3	37.5	89	60.0	148.4	113.5
20/80	64.9	85.2	41.1	98	59.9	148.7	113.2
30/70	60.2	80.1	42.5	98.5	59.3	149.5	113.5
40/60	57.1	80.4	44.4	104	59.0	149.0	113.3
50/50	47.5	63.8	54.3	115	58.5	148.9	113.2
100/0	30–42[a]	N/A	N/A	N/A	N/A	N/A	115[b]

Source: adapted from Qahtani, M., Wu, F., Misra, M. 2019. Experimental Design of Sustainable 3D-Printed Poly(Lactic Acid)/Biobased Poly(Butylene Succinate) Blends via Fused Deposition Modeling. *ACS Sustainable Chem. Eng.* 7:14460–14470.

Notes:

[a]Depending on testing direction.

[b]Brochure of BioPBS™ FZ91PM posted online.

[c]Melting of PLA.

[d]Melting of PBS.

blend formulation, or, inversely, rapidly design materials targeting a specific performance. Two of the above equations are reported below:

$$TS = 0.70V_1 - 0.34V_2 + 0.01V_1V_2 \tag{13.2}$$

$$IS = 0.34V_1 - 0.87V_2 + 0.002V_1V_2 \tag{13.3}$$

Since the properties of PLA/PBS blends are closely related to blend's morphology, and compatibility between PLA and PBS, morphology and compatibility were assessed through SEM micrographs of the fractured surface of tensile samples. PBS/PLA 10/90 did not display any clear phase separation, indicating that PLA and PBS were miscible with PBS at 10 wt%, but above that value, a structure with a PLA matrix in which PBS was dispersed as droplets was observed, which revealed immiscibility. The reader further interested in PBS/PLA blends for AM applications will find relevant references in Qahtani et al. (2019).

The Chinese patent CN10511699A, granted in 2017, disclosed a method to prepare a filament for FFF made of PBS incorporating the following ingredient (by wt%): PBS (100), reinforcing agent (10–30), compatibilizer (1–5), and a free radical crosslinking agent (0.05–0.5). The filament fabrication consists in the following steps: uniformly mixing the raw materials, wiredrawing, and three-stage cooling molding in which the temperature of the cooling media is progressively diminished. The invention claims that - compared to the state of art - its PBS filament offered improved rigidity and aging resistance, simplified wiredrawing and molding a PBS filament for FFF, ultimately facilitating its industrial-scale production (CN10511699A).

Mochane et al. (2021) compiled a review of composites made of natural fibers and PBS that can be informative for formulators of sustainable PBS composites for AM, namely fibers of bamboo, cellulose, coconut, cotton, kenaf, jute, palm, ramie, rice straw, sisal, and sugarcane bagasse. The composite preparation method, fiber modification, summary of results, and references are reported for each composite (Mochane et al. 2021). Unsurprisingly, the mechanical properties of the above composites were affected by the fiber type, fiber modification, type of fiber modifier, fiber/PBS weight ratio, and route to process the composite. Obviously, such composites cannot necessarily be adopted as feedstocks for AM without modifications to make them suitable for the processing conditions characteristic of the specific AM technology targeted. Interestingly, a combination of two fibers in a single matrix offers properties not obtainable by a single fiber (and possibly challenges), which will broaden the applications ofthe PBS/natural fiber composites. The automotive industry is one of the largest users of natural fiber and could be a target of PBS composites for AM.

Di Lorenzo et al. (2019) prepared by melt mixing and analyzed PBS/PA 11 blends with wt% compositions of 10/90, 20/80, and 40/60 with PBS the lesser ingredient. PBS was added because it is biobased, biodegradable, and compostable, and its production cost is half of that of PA 11. Rilsan® BESNO TL PA11, and Bionolle 1001MD PBS were selected. All blend formulations were immiscible (as indicated by the fact that they displayed two values of T_g), and PA 11 and PBS were arranged in a particle-matrix morphology featuring roughly spherical PBS particles suspended in the PA 11 matrix, with many small voids surrounding the PBS inclusions. The blends lagged behind unfilled PA 11 in tensile strength, strain (both yield and ultimate) and modulus, and their highest property values were associated with the 10/90 formulation, and were 910 MPa and 50 MPa for tensile modulus and strength, respectively, whereas the same properties for PA 11 were 930 MPa and 52 MPa, respectively. Some PBS/PA 11 interfacial adhesion, and interaction between amide groups in PA 11 and carbonyl groups in PBS were detected. Likely, adding a compatibilizer to increase interfacial adhesion and further promoting reaction between the functional groups of the two polymers may result in a blend with enhanced mechanical performance.

Su et al. (2019) reviewed the technical literature published from 2007 to early 2019 on PBS/PLA blends in general, and focused on the papers dealing with toughness modification through methods including simple blending, plasticization, reactive compatibilization, and copolymerization. The review also covered: the effects of PBS addition, nucleation, and processing parameters on the crystallization of PLA; the influencing factors, and degradation mechanisms of biodegradation and disintegration of PBS/PLA blends; and their recycling and application potential. The above information may be relevant to formulators of same blends for AM.

REFERENCES

Badouard, C., Traon, F., Denoual, C. et al. 2019. Exploring mechanical properties of fully compostable flax reinforced composite filaments for 3D printing applications. *Indus. Crops Prod.* 135:246–250.

Candal, M. V., Calafel, I., Aranburu, N. et al. 2020. Thermo-rheological effects on successful 3D printing of biodegradable polyesters. *Addit. Manuf.* 36:101408.

CN10511699A. *Polybutylene Succinate 3D Printing Wire and Preparation Method Thereof.* https://patents.google.com/patent/CN105111699A/en (accessed May 25, 2021).

Di Lorenzo, M. L., Longo, A., Androsch, R. 2019. Polyamide 11/poly(butylene succinate) bio-based polymer blends. *Materials* 12:2833.

EN 13432 European Standard. *Packaging – Requirements for packaging recoverable through composting and biodegradation – Test scheme and evaluation criteria for the final acceptance of packaging.* The European Committee for Normalisation (CEN). https://docs.european-bioplastics.org/publications/bp/EUBP_BP_En_13432.pdf (accessed May 25, 2021).

European Bioplastics. 2019. *Bioplastics Market Data.* https://www.european-bioplastics.org/market/ (accessed May 25, 2019).

Khaliq, M. H., Gomes, R., Fernandes, C. et al. 2017. On the use of high viscosity polymers in the fused filament fabrication process. *Rapid Prototyp. J.* 23 (4): 727–735.

Mochane, M. J., Magagula, S. I., Sefadi, J. S., Mokhena, T. C. 2021. A review of green composites based on natural fiber-reinforced polybutylene succinate (PBS). *Polymers* 13, 1200.

Nova-Institute. n.d. Succinity – Biobase Your Success. Available by contacting Nova-Institute at contact@nova-institut.de, or info@succinity.com.

OMR. 2019. *Polybutylene Succinate (PBS) Market.* https://www.omrglobal.com/industry-reports/polybutylene-succinate-market (accessed May 25, 2020).

Ou-Yang, Q., Guo, B., Xu, J. 2018. Preparation and characterization of poly(butylene succinate)/polylactide blends for fused deposition modeling 3D printing. *ACS Omega* 3:14309–14317.

Qahtani, M., Wu, F., Misra, M. 2019. Experimental design of sustainable 3d-printed poly(lactic acid)/bio-based poly(butylene succinate) blends via fused deposition modeling. *ACS Sustain. Chem. Eng.* 7:14460–14470.

Research Nester. 2020. *Polybutylene Succinate Market by End User.* https://www.researchnester.com/reports/polybutylene-succinate-market/1028. (accessed May 25, 2020).

Signori, F., Coltelli, M.-B., Bronco, B. 2009. Thermal degradation of poly(lactic acid) (PLA) and poly(butylene adipate-co-terephthalate) (PBAT) and their blends upon melt processing. *Polym. Degrad. Stab.* 94 (1): 74–82.

Song, H., Lee, S. Y. 2006. Production of succinic acid by bacterial fermentation. *Enzym. Microb. Technol.* 39:352–361.

Springer, L. H. 2006. *Introduction to Polymer Science.* Hoboken: Wiley.

Su, S., Kopitzky, R., Tolga, S., Kabasci, S. 2019. Polylactide (PLA) and its blends with poly(butylene succinate) (PBS): a brief review. *Polymers* 11:1193..

Tserki, V., Matzinos, P., Pavlidou, E., Vachliotis, D., Panayiotou, C. 2006. Biodegradable aliphatic polyesters. Part I. Properties and biodegradation of poly(butylene succinate-co-butylene adipate). *Polym. Degrad. Stab.* 91 (2): 367–376.

van Es, D. S., van der Klis, F., Knoop, R. J. I. et al. 2014. Other polyesters from biomass derived polymers. In *Bio-Based Plastics – Materials and Applications* ed. S. Kabasci, 241–274. Chichester: Wiley.

Xu, J., Guo, B.-H. 2010a. Poly(butylene succinate) and its copolymers: Research, development and industrialization. *Biotechnol. J.* 5:1149–1163.

Xu, J., Guo, B.-H. 2010b. Microbial succinic acid, its polymer poly(butylene succinate), and applications. plastics from bacteria. *Microbiol. Monogr.* 14:347–388.

14 3D Food Printing

When diet is wrong, medicine is of no use. When diet is correct, medicine is of no need.

Ayurvedic Proverb

14.1 REASONS FOR 3D FOOD PRINTING (3DFP)

Since food derives from sustainable sources (animals, plants, and trees), this chapter describes the feedstocks and processes relative to *3D food printing* (3DFP) or *food layer manufacturing* or *additive food manufacturing*. 3DFP has been heralded as "the future of food" (TNO n.d.). In brief, the reasons for extending AM to food are personalized nutrition and food customization, or, more in detail:

- Producing customized food featuring end properties that are new or superior to those of conventionally produced food, such as specific nutritional content, tailored texture, and complex geometries (Dankar et al. 2018).
- Preparing tasty and multi-ingredient food.
- Improving efficiency, consistency, and reducing cost of traditional food processing methods.
- Achieving mass customization targeting specific customers, like aging individuals with mastication and swallowing problems (Kouzani et al. 2017), athletes, children, military, and people with dietary restrictions.
- Serving the growing number of people with food allergies and intolerances.
- Improving human health.
- Preparing food for locations serving large number of meals, like catering, cruise ships, restaurants, hotels, military sites, and food in particular settings such as nursing homes and aerospace missions.

3DFP is not to be confused with robotics-based food manufacturing technologies, because the latter ones simply automate manual processes for mass production and lack features specific to AM processes. Like AM in general, 3DFP has potential to impact food supply chains (Sun et al. 2015a), and hence to have economic effects. 3DFP can also disclose new markets, such as in the case of meat "grown" and printed out of cow cells that may be consumed by vegetarians and vegans.

3DFP may include two additional steps besides the step of printing: (a) pre-printing consisting of preparing food recipes compatible with a specific printing process, and designed to obtain a food with specific characteristics; (b) post-printing to complete food preparation, such as baking, broiling, frying, freeze-drying, oven drying, slow-cooking, etc.

The first patent relative to AM applied to food materials was awarded to Yang Wu and Liu (2001), and disclosed the preparation of a complex 3D cake using an extrusion-based layering mechanism. So far food printers have limited capabilities in terms of food choice and speed, and have been run at home, in restaurants, high-end kitchens, and bakeries to prepare dishes with complicated and original shapes, perform repetitive tasks, expedite preparation, decorate desserts, etc. Current printers may require pre-made food capsules, which overall increases steps, time and cost to prepare meals. No printers capable of large throughput suited for industrial scale have been yet fabricated (Sun et al. 2018).

DOI: 10.1201/9781003221210-14

3DFP is not yet mainstream, but large food corporations (f. e. Mars and Nestlé) started filing for patents on printing food almost 20 years ago, and other large multinational firms (f.e. Barilla and Hershey's) are promoting 3DFP products. All of this may indicate that the business model for 3DFP was and is still considered profitable, or at least it is predicted to become so, which may drive further efforts to improve the relative technologies and expand its market, even if initially in niche applications (Godoi et al. 2019).

3DFP is expected to fit in specific markets that have special needs, and to trigger needs for some demographic groups. There is a growing demand for food products with personalized values of convenience, cost, nutritional values, packaging, and taste (Dankar et al. 2018).

In 2018, U.S. consumers, businesses, and government entities spent USD 1.71 trillion on food and beverages purchased in grocery stores, other retailers, and on away-from-home meals and snacks (United States Department of Agriculture 2020), therefore even a marginal share of the food market captured by a company operating in 3DFP can be significantly lucrative.

It must be clarified that 3DFP makes food products starting from food-based feedstocks not chemicals, hence 3DFP currently is seen not as a solution to food scarcity or famines. However, 3DFP may help against food scarcity, since research is being conducted to produce food from chemical compounds (Tran 2016) and, in fact, "growing" and printing a steak from cow cells has been demonstrated (Gohd 2019).

The recent overviews on 3DFP by Dankar 2018Lipton et al. (2015), Sun et al. (2015b), Dankar et al. (2018), and the book *Fundamentals of 3D Food Printing and Applications* (2019) edited by Godoi, Bhandari, Prakash, and Zhang are recommended for further reading.

14.2 FEEDSTOCK SCREENING

In screening candidate materials for printing food through an extrusion-based process, the candidate's following characteristics (or descriptors) should be assessed (Zheng et al. 2019):

- *Hardness, springiness,* and *cohesiveness*, measured through a repeated compression test on a texture analyzer such as TA-XT Plus by Stable Micro Systems (UK) that targets a specific reduction of the test coupon's initial height. *Hardness* is the peak force (N or g) reached during the first compression. *Springiness* (dimensionless value) is how much a product physically springs back after it has been compressed, and is calculated as the ratio between the height value after unloading and waiting for a specific time, and the height value before compressive loading. *Cohesiveness* (dimensionless value) expresses the material's internal adhesive force: the more cohesive the material is, the more difficult is to break it apart under some compressive or tensile stress. Cohesiveness is related to viscosity, and is measured in repeated compression testing as the ratio below (Anon. 2020):

$$\frac{(force \times displacement) \text{ during second compression cycle}}{(force \times displacement) \text{ during first compression cycle}} \qquad (14.1)$$

- *Color changes*: assessed as the value of the parameter ΔE that is a function of lightness, redness, and yellowness on a colorimeter, such as the CR-310 by Minolta (Japan)
- *Pasting properties* or *parameters*: they refer to the changes that occur in a starchy food as a result of applying heat in the presence of water while cooking it, or converting it into ground, mashed, or pureed mass, and affect digestibility, texture, and end use of the final food product (Ocheme et al. 2018). Pasting properties can be measured on a rapid visco analyzer (RVA) commercialized by Perkin Elmer, Perten (Perten n.d.), etc. The pasting parameters and RVA are explained in Balet et al. (2019) and defined below, along with their measuring units in parentheses (Figure 14.1):

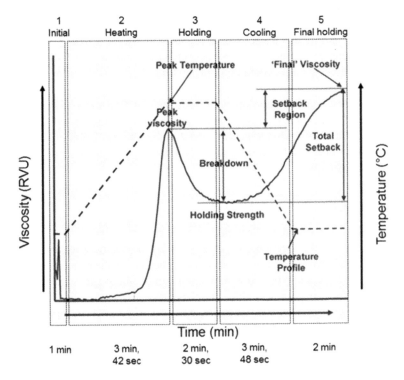

FIGURE 14.1 A typical Rapid Visco Analyser (RVA) pasting profile of a cereal, indicating the main pasting parameters measured. RVU is a viscosity unit, such that 1 RVU = 12 cP.

Source: Balet, S., Guelpa, A., Fox, G. et al. 2019. Rapid Visco Analyser (RVA) as a Tool for Measuring Starch-Related Physiochemical Properties in Cereals: a Review. *Food Anal. Methods* 12:2344–2360. Reproduced with permission from Springer Nature.

- *Breakdown viscosity* (cP or mPa s): the difference between the peak viscosity and the lowest viscosity
- *Final viscosity* (cP or mPa s): the most common parameter to define a sample's quality
- *Pasting temperature* (°C): the temperature at which starch granules begin to swell due to water uptake, or the minimum temperature required to cook a given food sample
- *Peak time* (min): time to reach peak viscosity
- *Peak viscosity* (cP or mPa s): maximum viscosity during heating. It indicates the water-holding capacity of the starch, and is often correlated with final product quality
- *Setback value* (cP or mPa s): difference between final viscosity and peak viscosity
- *Trough viscosity* (cP or mPa s): minimum apparent viscosity achieved after holding at the maximum temperature.

- *Variation in the molecular structure* (amorphous vs. ordered) and *chemical features* (f.e. formation of new substances) during cooking: detected by Fourier transform infrared (FTIR) analysis.
- *Amylose content* (AC) *in starch*: it affects the cooking and eating qualities of the grains, and the industrial properties of the starch extracted from those grains. AC is typically estimated through a colorimetric test in which starch molecularly dispersed binds with iodine: solutions of amylose and iodine turn deep blue-purple, while solutions of amylopectin and iodine take a reddish color because they show only a slight affinity towards each other (Jackson 2003). AC is a strong predictor of rice cooking and processing behavior: rice grains high in AC display high-volume expansion and degree of flakiness, cook dry, are less tender, and

become hard upon cooling, whereas rice grains with low and intermediate AC cook moist, become tender, and harden upon cooling (Panesar and Kaur 2016).

- *Texture* is a sensory attribute of food acceptability, and assessed through cohesiveness, springiness, and chewiness (Maniglia et al. 2020).

Sensorial attributes such as color, odor, off-odor, taste, and off-flavor can be measured by panelists referring to a scale of values, f.e. from 1 for most disliked to 5 for most liked. Commonly, a value of 3 is considered as the limit for marketability, and 2 as the limit for edibility (Derossi et al. 2016). Maniglia et al. (2020) illustrated examples of several pasting and other food properties and their relative measuring techniques.

14.3 FOOD VISCOSITY

Viscosity of plastics, introduced in Section 1.4, is a critical property for formulators of fluid feedstocks for 3DFP. In this section, we recall the distinction between Newtonian and non-Newtonian foods. Foods whose viscosity does not depend on the shear rate are called *Newtonian*. Typical Newtonian foods contain only low-MW compounds (like sugars), and do not contain large concentrations of dissolved polymers (like starches) and insoluble solids. Edible oils, filtered juices, honey, milk, sugar syrups, and water are examples of Newtonian fluids (Rao 2006).

In *non-Newtonian foods*, as the shear rate raises, the shear stress increases not linearly but at either increasing rate (*shear-thickening* or *dilatant fluids*) or decreasing rate (*shear-thinning fluids*). Most non-Newtonian foods feature shear thinning behavior, and include concentrated fruit juices, mayonnaise, pureed fruits and vegetables, and salad dressings (Rao 2006). A shear-thickening food is the mixture of water and cornstarch.

Examples of shear-thinning food are the homogeneous suspension of starch at 5, 10, 15, 20, 25, and 30 wt% dry concentrations whose plots of apparent viscosity (η) vs. shear rate are illustrated in Figure 14.2. All the curves plotted were fitted by the power law in equation (14.2), in which $\dot{\gamma}$ is the shear rate (1/s) and K is called *consistency index* (defined in Section 1.4):

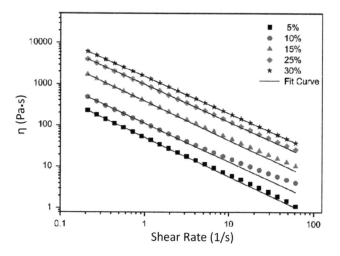

FIGURE 14.2 Viscosity versus shear rate curves of starch suspensions at different concentrations.

Source: Chen, H., Xie, F., Chen, L., Zheng, B. 2019. Effect of rheological properties of potato, rice and corn starches on their hot extrusion 3D printing behaviors. *Journal of Food Engineering* 244:150–158. Reproduced with permission from Elsevier.

$$\eta = K(\dot{\gamma})^{n-1} \tag{14.2}$$

For the specific suspensions in Figure 14.2, the values of n and K fall in the range 0.05–0.1 and 51–1,520, respectively.

14.4 AM PROCESSES FOR FOOD

14.4.1 Introduction

Food can be printed through preexisting process families: binder jetting, inkjet printing, material extrusion, and powder bed fusion (Dankar et al. 2018; Godoi et al. 2019; Rajput et al. 2019). Since Chapter 2 describes in detail all these families, in this chapter we will only touch on them.

The technology for 3DFP is very recent, since the first developed printer for 3DFP, *CandyFab*, appeared in 2005 (Octu et al. 2019). 3DFP is also pushing the technology boundaries of AM, and leveraging a new version of AM called *5D printing* (5DP), in which the AM printer builds the part in the X, Y, Z directions with translational and rotational movements (Ethereal Machines 2018). Moreover, if a type of 3DFP enables to design the composition, shape, and properties of food before printing in a way that trigger specific changes in shape and size of food after printing, that type of 3DFP can be classified as 4D printing (Section 2.11).

Table 14.1 lists for each family of 3DFP processes its feedstock types and examples, advantages, limitations, and examples of commercial printers. Since 3DFP processes work with different physical forms of feedstocks (filament, gel, liquid, paste, powder, etc.), each process is not compatible with all food types but only food featuring specific consistency and physical-mechanical properties. Table 14.2 includes some commercial food printers, their feedstocks and printing volumes.

Like for AM in general, quality and performance in 3DFP depend on three sets of variables: material properties, printing variables, and post-printing factors. The variables included in each set depend on the type of 3DFP (Liu et al. 2017). Another critical factor is interlayer binding mechanism.

Process optimization comprises fine-tuning the printing parameters and variables affecting food itself, and understanding the interaction between the two sets, in order to obtain the printed product that best meets its specific requirements. For extrusion-based 3DFP, major printing parameters are: temperature of extruder and printing platform, speed along the X, Y, Z axes, pressure and flow speed of food extruded, nozzle diameter, distance between nozzle and printing platform, layer height, and infill percent, whereas examples of material variables are: physical, chemical, mechanical (hardness, elastic and loss moduli, etc.), and rheological properties (viscosity, and flow behavior), cohesiveness, T_g, and ratio of ingredients.

Printed food is evaluated in terms of taste, texture, and nutritional value at least from a customer's perspective, whereas assessing 3DFP quality comprises evaluation of shape fidelity and mechanical properties, with the latter ones being measured differently on soft (gel) vs. hard (solid) printed material (Godoi et al. 2019).

Maniglia et al. (2020) leveraged dry heating treatment (DHT) as a simple and safe method to improve the properties of wheat starch HGs formulated as "inks" for 3DFP. Upon undergoing DHT for 2 h and 4 h at 130°C, wheat starches showed an increase in granule size but no alteration in granule's shape, surface, and the molecular functional groups. On the other hand, DHT promoted slight molecular depolymerization, and reduction of starch crystallinity. The best printability and reproducibility was displayed by the HGs based on starch dry heated for 4 h. DHT also extended the texture possibilities of printed wheat starch HGs.

14.4.2 Binder Jetting (BJ)

In BJ, a liquid bonding agent is selectively sprayed to join regions of a powder bed formed by sequentially laying down a powdered material. We recall that for BJ and all the following

TABLE 14.1
Families of 3DFP Processes

Process	Feedstock Types and Examples	Advantages	Limitations	Examples of Commercial Printer
Binder Jetting	Liquid binder and powder. Flour, protein, starch, caster and granulated sugar. Chocolate, jelly, materials that are castable to print molds (on Z Corporation printers). Powders of sugars, starch, corn flour etc. are printed as final items (on Z Corporation printers).	Complex 3D shapes, full colors, variety of flavors and textures, accurate, high resolution. Strong structural cohesion in castable feedstocks.	Few feedstocks, less nutritious products. Weak structural cohesion in powders printed as final articles.	ChefJet
Inkjet Printing	Low viscosity materials. Confectionary products. Decorations on biscuits, cookies, cupcakes. Batters, butter, cheese, chocolate, cream, dough, fat compounds, gels, jams, jellies, liquid dough, meat paste, purees, sauces, sugar icing.	Many materials, printing quality and speed, easy method, innovative decoration shapes.	Simple food design, only image decoration and surface filling on substrates, 2.5D image.	FoodJet, byFlow
Material Extrusion	Filaments (rarely), and soft materials that can be finely converted into a paste or puree. Cheese, chocolate, cookie dough, corn dough, hummus, hydrocolloids, mashed potatoes, meat puree, peanut butter, pizza.	Wide array of materials, can operate two syringes enabling many food combinations, equipment of simple design. Variable resolution depending on diameter of tip and material behavior under pressure. High structural strength for materials that harden such as chocolate. Complex geometries in low-relief objects.	Slow, seamline between layers, only simple food designs, shapes difficult to hold during post-printing. Low structural strength for materials that do not harden such as cheese, and potatoes.	Foodini, Fab@ Home, Choc Edge, CocoJet, Porimy
Powder Bed Fusion	Powdered materials. Chocolate, fat, Nesquik® powder, sugar.	Fast, complex shapes, various textures, can use recycled powder.	Limited choice of feedstocks, less nutritious products, complicated to use low-melting point materials.	CandyFab (out of business), Cornucopia

Sources: Adapted from: Liu, Z., Zhang, M., Bhandari, B., Wang, Y. 2017. 3D printing: Printing precision and application in food sector. *Trends in Food Science & Technology*, 69:83–94; Dankar, I., Haddarah, A., Omar, F. E. L., et al. 2018. 3D printing technology: The new era for food customization and elaboration. *Trends in Food Science & Technology* 75:231–242; Southerland, D., Walters, P., Huson, D. 2011. Edible 3D printing. Digital fabrication 2011 conference. NIP 27, 27th International Conference on Digital Printing Technologies pp. 819–822.

TABLE 14.2

Commercial Food Printers

Printer	Food	Printable Volume (mm)	Country	Price (2018)
Natural Machines Foodini	Pizza, spaghetti, burgers	250 × 165 × 120	Spain	USD4,000
BeeHex Deco-pod (in-store solution) BeeHex 3D Decorator BeeHex 3D Decoration Production Line BeeHex Cake Writer (in-store solution)	Substrate to be decorated: cake, cookie, cake-pop, cupcake, or icing pieces. For decoration of cookies, cakes, cookie cupcakes, donuts, pet treats. Icings available: royal icing, buttercreams, melted icings, oil-based, water-based.	N/A	USA	N/A
byFlow Focus	Chocolate, pastry, etc.	208 × 228 × 150	Netherlands	EUR3,300.
Choc Creator 2.0 Plus by Choc Edge	Chocolate	180 × 180 × 40	UK/China	EUR2,330.
Mmuse Chocolate 3D Printer	Chocolate	160 × 120 × 150	China	USD5,200
Pancakebot 2.0 Food 3D Printer	Pancakes	445 × 210 × 15	Norway	USD300
Createbot 2.0 Food Printer	Biscuit, chocolate, lotus seed paste, red and green bean pastes, etc.	150 × 150 × 100	China	USD2,349
3D Systems ChefJet & ChefJet Pro Food 3D Printer	Sugar with flavors including chocolate, vanilla, mint, sour apple, cherry, watermelon, etc.	203 × 203 × 203	USA	USD5,000–USD10,000
ZMorph VX	Chocolate, dough, cream cheese, frosting and more	250 × 235 × 165	Poland	USD3,050

Sources: https://3dsourced.com/rankings/food-3d-printer/ (accessed August 28, 2018); Liu, Z., Zhang, M., Bhandari, B., Wang, Y. 2017. 3D printing: Printing precision and application in food sector. *Trends in Food Science & Technology,* 69:83–94.

processes discussed in this chapter the term *selectively* means *in selected locations.* BJ as 3DFP features advantages such as inexpensive ingredients, fast fabrication, and ability to print meals with complex shapes, but also the downsides of expensive printers, rough surface finish, high sugar content, and possibly the need of post-processing, such as curing at temperature higher than printing temperature to strengthen the interlayer bonding. Foods compatible with BJ are caster and granulated sugar, flour, protein, and starch. Application of BJ to food is largely unpublished in technical papers, but Holland et al. (2019) authored an overview of food structures printed through BJ. A common commercial application is sugar powder bound by an ink based on water and alcohol with optional color or flavor (Holland et al. 2018). In fact, f.e., in 2013 The Sugar Lab, an American start-up, adopted the BJ-based Color Jet Printing built by 3D Systems to print on a sugar bed flavored edible binders, and prepare complex sculptural cakes for special events such as weddings (Anon. 2013)

Southerland et al. (2011) experimented with BJ-based Z-Corp printer (now part of 3D Systems), and replaced the original Z-Corp feedstock powder with a binder plus mixtures of sugar and sugar substitutes featuring a similar range of particles size, and the most successful result came from 50/50 wt% blend of icing sugar and casting sugar with 35% binder saturation.

Holland et al. (2018) utilized a small-scale powder layering device connected to an ink jet printer to test prototype powders before producing quantities typical for commercial BJ printers. Powders comprising mainly ball milled, amorphous cellulose were successful in printing 3D structures, when a polysaccharide (glucomannan or locust bean gum) was added to the ink as an ingredient of the powder component. Since cellulose powders are categorized as dietary fiber, such formulations can produce low-calorie printed food designs. The experimented inks contained (values in wt%): water (78.5–79.25), ethanol (20), emulsifier (0.5), and xanthan gum (0.2–1).

A recent review of BJ applied to 3DFP was published by Holland et al. (2019).

14.4.3 INKJET PRINTING (IP)

In IP a fluid is forced through an orifice and ejected as droplets that are selectively dispensed. Because IP printers typically process low-viscosity materials, IP is suited for fillings, graphical decorations, microencapsulation but not for complex food geometries and structures.

Typical foods deposited via IP are: cheese, chocolate, gels, jams, liquid dough, meat paste, sugar icing, etc. (Rajput et al. 2019). A short and very recent review of jetting-based 3DFP processes was authored by Vadodaria et al. (2020)

Following are some examples of applications of IP for 3DFP and their relative patents. The food giant Mars Inc. (USA), maker of M&M's® candy among other products, filed in 2003 and later patented a high-resolution IP process on edible substrates in which "fat or wax-based edible inks, which contain a colorant, a fat or wax dispersible carrier, and a fat or wax base, are used to produce high resolution images on edibles" (Shastry et al. 2009).

The multinational food company Nestlé (Switzerland) patented a process for "printing an edible ink onto a material using an IP device." The material may be an edible material, and "the ink may comprise a colorant, at least 30 [wt]% water, at least 25 [wt]% carbohydrate sweeteners and be free from both diols and triols" (Cavin et al. 2016).

Procter & Gamble (USA) leveraged 3DFP to reduce production cost. Namely, it exploited the accurate and high-resolution position control of IP printers, and in 2007 filed for a patent on printing methods providing targeted, precise, uniform, and consistent application of flavor to edible substrates, resulting in a better tasting product, and eliminating: (a) inconsistent application and waste associated with conventional methods, such as sprinkling a dry ingredient over the substrate, and (b) the costs related to emptying, cleaning, and refilling all the equipment when switching between flavors. Edible substrates can be printed with aroma, colors (two or more), flavors, images, and seasonings. The distinct flavors can be deposited in a pattern that corresponds to areas on the tongue where different taste buds are located (Wen et al. 2008).

de Grood and de Grood (2013) patented a technique, commercialized as FoodJet printing (FoodJet 2021), consisting in an array of pneumatic membrane-based nozzle jets ejecting on-demand onto a moving layer tiny drops of butters, creams, chocolate, doughs, batters, fat compounds, jams, jellies, sauces, and purees. The drops fall onto pizza bases, biscuits, and cupcakes, and permit graphical decoration, cavity deposition, and surface filling. It is an economical and innovative way for mass customization in food fabrication. Applications include graphical decoration, cavity depositing, and surface filling.

The Dutch R&D organization TNO (TNO) developed an IP-based microencapsulation process to: (a) print food with healthy amount of nutrients needed by the body without sacrificing flavor and texture; (b) modify color, flavor, and nutrients; and (c) print ingredients for oral drugs. TNO designed a new print head generating highly monodispersed droplets that turn into highly monodispersed powders after drying. The print head features 500 small nozzles, and a capacity of 100 L/h. The process consists in printing droplets of a core material through a liquid film that becomes the shell material, and forming core-shell microcapsules of controllable dimensions and shell thickness, made of waxes, fats, polymers, and aqueous dispersion, emulsions, and solutions. Examples of these microcapsules are mint syrup core in wax shell, and linseed oil in carrageenan shell. When added to food, the microcapsules modify color, flavor, and nutrients. Microencapsulation has several advantages over conventional spraying-drying processes including adjustable powder properties, high solid content, and monodisperse droplets, among others.

14.4.4 Material Extrusion (ME)

14.4.4.1 Introduction

In ME, the feedstock, typically in the form of filament, is heated and selectively dispensed through the orifice of a scanning nozzle. The extrusion mechanism is usually mechanical (piston, screw) or pneumatic. A piston directly controls the flow of viscous food through the nozzle, while screw is better suited to mix and dispense highly viscous ingredients and control the deposited volume. Pneumatic mechanism employs a simple design, and a pushing force that is only limited by the air-pressure capabilities of the mechanism, but it is also penalized by the delay of the compressed gas volume (Murphy and Atala 2014). ME is compatible with a limited number of ingredients comprising chocolate, cream cheese, and potato. Indicating D, S, and F as the nozzle diameter (mm); the moving speed of the printing head (mm/s), and the amount of food (%) delivered through the nozzle, respectively, the rate VR (mm^3/s) of volume of food deposited during printing is calculated from equation (14.3) (Maniglia et al. 2020):

$$VR = \pi \, (D/2)^2 SF \qquad (14.3)$$

Food ME can be classified as: (a) *room temperature ME*, applied to food such as dough (Yang et al. 2018a) and meat paste; (b) *melting ME*, for filament, paste-like, powder, solid pieces; and (c) *gel-forming ME*. The concentration of total solids in feedstocks range from 5 wt% (gel) to 50 wt% (paste). The optimal feedstock for ME must exhibit fast and solid-like response during extrusion, not require great force to be extruded, and keep its shape after deposition. Printing food in the form of filament from semisolid feedstocks dispensed through syringe has the advantage that post-processing is not necessary. The review of 3DFP by Dankar et al. (2018) included a well-referenced description of extrusion-based 3DFP.

14.4.4.2 Room-Temperature ME

Southerland et al. (2011) adapted a home-grade desktop printer, called RapMan 3.1, extruding plastic filaments through a heated nozzle into cold food pastes, and experimented with extrusion pressure and speed. Slower speed resulted in more precise control, and higher pressure helped clear air bubbles and small blockages. The structural strength of the extruded object depended on the feedstock: if the feedstocks (f.e. chocolate) hardened, the object could be handled, whereas materials (f.e. cheese and potato) not hardening remained fragile.

14.4.4.3 Melting ME

In melting ME, T_m of the feedstocks depends on their chemistry. F.e. fatty acids rich in carbon atoms possess high T_m. Other parameters affecting the outcome are nozzle's diameter, distance between nozzle and supporting platform, and extrusion head's moving speed. Edible filaments for ME are rare, and more frequent for pharmaceutical than nutritional uses. Herr and Cohen (2016) patented an edible TP filament for ME "that incorporates an active ingredient such as an oil extract for taste, odor or medicinal benefit [...]. The filament is made by mixing the active ingredient extraction with polyvinylpyrrolidone (PVP), starch, and super disintegrant, and spray drying the result to a pow-derized form [...] mixed with excipient ingredients including a plasticizer, colored/dyed arabic gum, a gelling agent, fillers, flour, a binding or thickening agent (which also gives the benefit of being a stabilizer), a lubricant, and a preservative, and is heated. The result is hot melt extruded into a filament" featuring 1.75 or 3 mm diameter, and "good strength, stiffness, and physical properties."

Examples of pharmaceutical applications have been reported by multiple investigators. Goyanes et al. (2016) extruded a filament of PVA loaded with drug (paracetamol [in USA known as acetaminophen] or caffeine), for printing oral capsules. Sadia et al. (2016) formulated a pharmaceutical filament based on a commercial methacrylic copolymer (Eudragit® E PO) and tribasic calcium phosphate (TCP) as a thermally stable filler, and they printed pharmaceutical tablets out of the filament.

14.4.4.4 Gel-Forming ME

Food should not gel (*gelation*) during extrusion but after that, in order to keep its shape. Gelation is promoted by different mechanisms (Godoi et al. 2019):

- *Cooling*: as it occurs with gelatin
- *Chemical crosslinking*: carried out by compounds activated by UV light, but many such compounds are harmful to people and must be removed from the food
- *Ionotropic* (promoted by ions) *crosslinking*: an example of it is printing pectin-based food simulant by formulating an ink composed of edible methoxylated pectin gel with $CaCl_2$ solution. A food simulant is a chemical with characteristics similar to the different food categories: acidic, alcoholic, fat, milk, oil, dry and watery foods. Thanks to Ca+ ions, pectin crosslinked at room temperature during printing without need of post-treatment after printing (Vancauwenberghe et al. 2018). Ionotropic crosslinking is widely applied by food industry.
- *Complex coacervate formation*: Cohen et al. (2009) selected a novel combination of hydrocolloids (xanthium gum and gelatin) and flavor agents, and independently tuned flavor and texture to print materials simulating a broad range of foods, with only a minimal number of starting materials.
- *Enzymatic crosslinking*: Schutyser et al. (2018) selected as an edible ink sodium caseinate (a nutritional protein present as an ingredient in numerous food products), and selected the enzyme transglutaminase (chosen to manufacture cheese and other dairy products) to increase the gelation temperature above room value, and make the printed construct keep its shape without refrigeration.

14.4.5 POWDER BED FUSION (PBF)

In PBF, thermal energy selectively fuses together the top regions of a powder spread on a flat bed. The advantages of PBF is that enables printing complex food items quickly without post-processing, and it is compatible with sugar and fat-based ingredients with relatively low melting points. Its downside is that it is complex to run, because it is controlled by many variables.

TNO has employed PBF, and adopted an infrared laser as a source of thermal energy for heating and melting sugar and/or fat present in the feedstock, and incorporating flavor and nutritional value to powder-based food.

In 2006, W. Oskay and L. Edman introduced a new printer called CandyFab to fabricate large 3D objects out of sugar, running a new inexpensive process called *selective hot air sintering and melting* (SHASAM) that was basically a version of PBF, and consisted in melting sugar grains by selectively directing a narrow, low-velocity beam of hot air on a layer of sugar spread on a platform that was kept warm but below the sugar's melting point, in order to minimize thermal distortion and facilitate fusion among layers. After each layer of sugar was fused, the platform lowered and was coated with a new layer of sugar. Sugar had the advantages of being inexpensive, nontoxic, and readily available, Several printers were built but their price was too high compared to that of home-grade printers entering the market at the same time, and the two inventors abandoned their project in 2009 (Oskay and Edman 2014).

14.5 FOODS FOR 3DFP

The range of food processed through 3DFP is wide and include members of all major food groups: added sugars, dairy, fruits, grains, vegetables, oils, protein food (meats, poultry, seafood, etc.), and solid fats.

In developing feedstocks for 3DFP, physical, chemical, mechanical, and rheological (viscosity and flow behavior) properties of candidate ingredients are critical and must be appropriately adjusted, because they affect applicability, printability, and post-printing (Godoi et al. 2019). Since chocolate is a popular 3DFP feedstock, the challenges in printing it are mentioned here as an example of requirements for 3DPF. Chocolate is too brittle to be wound into a hard filament, and too soft to extrude with gear and motor. Hence, melted chocolate is stored in a cartridge, melted in the printer, and extruded with a syringe. But, melted chocolate does not easily harden at room temperature, which makes it predisposed to lose its shape after being extruded; therefore, the in-printer melting temperature should be set as low as possible. Moreover, dark, white, and milk chocolate have different viscosities and properties that affect the way chocolate extrudes, sticks together, and cools (Ooi 2019).

Feedstocks for 3DFP can be divided into three groups (Rajput et al. 2019):

- *Naturally printable food*: foods that can be directly extruded from a syringe, such as butter, cheese, chocolate, icing, hummus, hydrogel, jelly, pasta dough, etc., and powder food such as starch, sugar, etc.
- *Food not naturally printable*: this must be modified to be compatible with a specific 3DFP process, f.e. by adding hydrocolloids (such as xanthan gum and gelatin) to be extruded. Examples are fruits, meat, rice, and vegetables. Traditional foods such as turkey, celery, and scallops are successfully printed after being ground and modified by additives. About meat, the company Jet-Eat can print plant-based meat (also termed *alternative meat*) on a special food printer able to match the appearance, texture, and flavor of whole muscle meat (Redefine Meat 2019). One current downside of Jet-Eat meat is its sale price of USD 33–39/kg (Gonzales 2019).
- *Alternative food*. Alternative ingredients extracted from algae, fungi, insects (Caporizzi et al. 2019), lupine seeds, and seaweed are novel sources for protein and fiber (Sun et al. 2015b). Namely, insects were studied as a protein source to replace traditional meat, and insect powders were mixed with extrudable icing and soft cheese (Walters et al. 2011) and with cereal (Caporizzi et al. 2019). Severini et al. (2018b) printed snacks from wheat flour dough enriched by ground larvae of yellow mealworms (*Tenebrio molitor*) as a novel source of proteins, and studied their main microstructural features, overall quality, and nutritional attributes as a function of formulation, time and temperature of baking. Edible insects are a good source of energy, protein, fat, vitamins, and minerals, and are consumed by more than two billion people worldwide (Caporizzi et al. 2019). More than 2,000 edible insect species have been counted. Their nutritional values widely vary depending not only on species but also other factors, such as rearing type, insect diet, etc. The ranges of values of energy and selected nutrients of edible insects are listed in Table 14.3. Additionally, residues from the existing agricultural and food processing can be converted into sustainable ingredients in form of biologically active metabolites, enzymes, and food flavor compounds.

In Section 14.1 it is mentioned that 3DFP can generate multi-ingredient products obviously obtainable by mixing ingredients prior to printing, f.e. turkey meat and celery (Lipton et al. 2010). Periard et al. (2007) printed multi-ingredient food on the open source Fab@Home Model 1 personal printer equipped with syringe/plunger deposition device. Multi-material edible 3D objects with cake frosting, chocolate, processed cheese, and peanut butter were printed by switching feedstocks inside

TABLE 14.3

Values of Energy and Specific Nutrients per 100 g of Edible Insects

Nutrients	Unit	Minimum Value	Maximum Value
Energy value	Kcal	282	762
Fat	g	7	77
Fiber	g	5	27
Protein	g	33	77

Source: Caporizzi, R., Derossi, A., Severini, C. 2019. Cereal-Based and Insect-Enriched Printable Food: From Formulation to Postprocessing Treatments. Status and Perspectives. In *Fundamentals of 3D Food Printing and Applications*, ed. F. C. Godoi, B. R. Bhandari, S. Prakash, M. Zhang, 93–116. Academic Press. Reproduced with permission from Elsevier.

TABLE 14.4

Printed Food Types. Complete References at End of This Chapter

Breakfast spreads (Hamilton et al. 2018)

Cereal snack (Severini et al. 2016; Severini et al. 2016)

Cellulose nanofiber (Lille et al. 2018)

Cheese (Lipton et al. 2010), processed cheese (Le Tohic et al. 2018), cheese puree (Southerland et al. 2011), semi-hard cheese (Kern et al. 2018)

Chocolate (Hao et al. 2010; Mantihal et al. 2017), seeded chocolate (Dankar et al. 2018)

Dough (Severini et al. 2016; Yang et al. 2018a), corn dough (Southerland et al. 2011)

Egg and rice (Anukiruthika et al. 2020)

Faba bean (Lille et al. 2018)

Fat (Shastry et al. 2009)

Fish (Kouzani et al. 2017; Wang et al. 2018)

Frosting (Rajput et al. 2019)

Fruit (Derossi et al. 2018; Severini et al. 2018a)

Gel: lemon juice gel (Yang et al. 2018b)

Hydrocolloids (Cohen et al. 2009), hydrocolloid pastes (Gholamipour-Shirazi et al. 2019)

Ice cream (Yang et al. 2017)

Icing material (Schutyser et al. 2018)

Insects (Severini et al. 2016; Caporizzi et al. 2019)

Jams (Rajput et al. 2019)

Mashed potatoes (Liu et al. 2018)

Meat (Dick et al. 2019), meat paste (Godoi et al. 2019)

Milk powder (Lille et al. 2018)

Oat (Lille et al. 2018)

Scallops (Southerland et al. 2011)

Smoothie (Severini et al. 2016)

Starch (Herr and Cohen 2016; Lille et al. 2018)

Turkey meat puree (Lipton et al. 2010)

Vegetables (Derossi et al. 2018; Severini et al. 2018a)

the syringe. It was also demonstrated that food can effectively serve as a supporting material for non-edible feedstock, by printing a block of frosting supporting a silicone bridge.

An ample list of printed food is included in Table 14.4, along with their specific literature references.

14.6 SUSTAINABILITY OF 3DFP

Since 3DFP makes food on-demand, it only consumes the needed amount of food ingredients, energy, and water for the specific "meal," and hence it may reduce the amount of food that is wasted when we buy more conventional food than we consume. Galdeano (2015) analyzed the potential environmental impacts of 3D food printers. Reducing the amount of food needed results in less water, fodder, fertilizer, pesticides, antibiotics, etc. consumed to grow food and raise animals. Octu et al. (2019) analyzed the sustainability of 3DFP through its impact on environment, society, and economy. They argued that only one economic model for the evaluation of 3DFP was currently available, and discussed the integration of the different components (environmental, social, and economic) of sustainability into the 3DFP, presenting a more integrated and holistic approach to sustainability.

Another advantage of 3DFP is that it can include as ingredients the fruits and vegetables that are fresh and good to eat but unattractive to customers because they are lumpy, misshapen, and discolored.

While some emerging countries suffer from scarcity of food, developed countries enjoy the opposite situation. In Europe the food wasted per year reaches about 77 million tons, of which 70% comes from manufacturing and households. All this waste increases the size and number of landfills, damages the environment, and has a cost to be managed. Even if a small number of households is convinced by price and performance of food printers, and adopts them, the reduction of wasted food can be significant.

If the concept of sustainability includes maintaining good health, then 3DFP is sustainable also because it can improve people's health with a diet that is balanced and/or contains a personalized amounts of nutrients. It may also be less expensive to ship food feedstocks packaged for 3DFP than food ingredients in raw form, such as fruit, meat, vegetables, etc. Preparing many meals at once with large food printers may be overall more sustainable than with conventional kitchens, if the printers are sufficiently fast.

Since 3DFP can process food typically unpopular in some cultures (such as algae and insects that are disliked in Western countries), if 3DFP drives or increases the consumption of such food in the countries where it has marginal or no market, then 3DFP will improve environmental sustainability by evening out consumption across food types geografically, and will also boost the economy of communities producing f.e. algae and insects for food, and open to them markets where they are not yet consumed.

Although 3DFP is supposed to improve health by printing food having a balanced and personalized content of nutrients, it cannot be excluded that some users will choose to print meals with unhealthy amounts of some specific ingredients such as sugar, chocolate, and salt.

The most accurate assessment of sustainability of 3DFP should be conducted comprehensively, as it is done for AM in general, by considering: (a) the impact of 3DFP on environment, economy, and society, and (b) the natural and economic resources spent in all the steps involved in 3DFP, from preparing the feedstocks to disposing of the waste. In this type of assessment, the overall effect on society should encompass the impact on number and types of jobs, health of workers and population, quality of life, education, economic development, etc. All factors weighed for 3DFP must be compared against the same factors relative to conventional food preparation. 3DFP starts with ingredients that may be more natural in composition (that is containing fewer additives) than mainstream ingredients, and hence may be more expensive than the latter ones, if the 3DFP ingredients are produced through routes that are uncommon and not optimized economically. The food for 3DFP may be sold in packages that, to be compatible with the printers, may be smaller and more expensive than packages of conventional food sold at retail scale in gallons and pounds. The type (electricity, no gas) and amount of energy and time required for printing must be compared to those spent for conventional food preparation and account for the difference in food quantities. Since 3DFP only processes food with minimal waste (f.e. meat without bones), the energy consumption in 3DFP may be more efficient than f.e. cooking a whole turkey. But, on the other hand, there is the upstream cost to prepare turkey ingredients for 3DFP without bones and skin. The 3DFP packages must be disposed of, and the amount of their waste per kilogram of the whole package should be compared to the amount of the same trash generated by conventional food. The total cost of building, running, maintaining, and servicing the printer over its lifetime, and its durability has to be taken in account.

The savings of resources with 3DFP seem evident in some instances, such as printing "artificial meat" that is developed in laboratories and not derived from farm animals (cows, hogs, chickens, etc.) with all the cost associated to raise them (water, food, space, shelter, cleaning, health, etc.), and their environmental pollution. F.e. a single 2.2-pound steak requires 5,200 gallons of water for its cow, but growing cultured meat takes up to 10 times less water and land than traditional livestock agriculture (Gohd 2019). However, reducing animal farms will cause job losses among farm workers and all the industries supporting farms, such as those providing fodder, vaccines, services, utilities, etc.

14.7 MARKET OF 3DFP

3DFP targets personalized nutrition and food customization. The market of 3DFP is minute in size, experimental and immature, and mostly composed of small firms a few of which are profitable (Rogers et al. 2019). However, the fact that some large companies are involved, and R&D is conducted by large, independent institutes indicates that currently this market offers some potential to grow, as confirmed in Section 14.8.

Firms active in the 3DFP market are 3D Systems, Barilla, BeeHex, Biozoon Food Innovations, byFlow, Choc Edge, Crafty Machines, Electrolux, Fab@Home, Hershey's, Jet-Eat/Redefine Meat, Katjes Magic Candy Factory, Modern Meadow, Natural Machines, Nestlé, NuFood, Philips, Print2Taste, Redefine Meat™, Systems & Materials Research Corporation (SMRC), and ZMorph. ESA and NASA are also interested in 3DFP.

Examples of European educational and research organizations involved in 3DFP include Eindhoven University of Technology, TNO, and Hague University of Applied Sciences (all in The Netherlands), Politecnico di Milano (Italy), Robots in Gastronomy (Spain), University College of London and The University of Nottingham (both in UK), Weihenstephan-Triesdorf University of Applied Sciences, Institute for Biomedicine of Aging, and Nuremberg University of Applied Sciences GSO (all in Germany).

Dabbene et al. (2018) developed an economic model to understand the consequences that implementing 3DFP will have on printer manufacturers, and small-batch production food firms such as restaurants, catering, and confectionery. Particularly, their model assists the former ones in speeding up and improving the adoption of the technology, and the latter ones in how to leverage 3DFP to find and boost new and current business opportunities. The model is based on the assumption that 3DFP is adopted for small batch production, and encompasses three areas: raw material suppliers, new skills, and printer producers. Through the model, the effect of one area on the others can be analyzed.

14.8 NEAR FUTURE AND CHALLENGES OF 3DFP

There is some disagreement on the amount of growth of the 3DFP market in the near future, and even uncertainty on how to forecast its growth (Rogers et al. 2019). Since 3DFP is recent, additional analysis and studies are required in order to assess the benefits of this technology, and the conditions conducive to its diffusion, but currently data lack, because of the tiny size and novelty of the 3DFP market, and reluctance of involved companies to share data. In 2018, the consultancy BIS Research predicted that the global 3DFP market would grow and reach USD 526 million by 2023 (BIS Research n.d.). In 2019, the consultancy Kenneth Research forecast that the global 3DFP market would reach USD 400 million by 2024 (MarketWatch 2019). Particularly, printed meat targets alternative (animal-free) meat, and the global market for this product is the fastest-growing segment of the food industry, and expected to surge to USD 140 billion annually by 2030 (Franck 2019).

According to Rogers et al. (2019), 3DFP will develop from a business standpoint if: (a) the product portfolio offered to consumers is considerably expanded; (b) printers permit to customize not only shape and look of food but also its composition and nutritional values; (c) customers are informed about 3DFP and its features, and are convinced to switch to it; (d) additional R&D investments are spent (Octu et al. (2019) expect a "jump" of R&D investments in the near future); (e) the product range is widened; (f) the printing process is industrialized, and scaled up. Other challenges to be tackled are speed, productivity, process optimization, quality (accuracy, resolution, etc.), cost, accessibility, and legal issues.

Increasing printing speed will attract the interest of the food industry, and businesses and organizations serving many meals at once (conference centers, cruise ships, hotels, military bases, restaurants, catering, etc.). Unfortunately, in cases where 3DFP includes pre- and post-printing steps, shrinking its overall time is more challenging.

The high cost of printed food could be in some measure offset by the printer's performance defined in terms of food variety, printing speed, ease of use, cost of supplies, ease and cost of maintenance, durability, repeatability, etc. In turn, performance varies between household-grade printers and professional-grade printers for large quantities and different types of food, and it is expected that requirements will be more numerous and demanding for the latter printers. Table 14.2 shows that current household-grade food printers sell from around USD 2,300, and feature limited range of food types. It is possible that the evolution of food printers resemble that of extrusion-based printers for plastics that have evolved in two basic families: inexpensive desktop models for personal use and prototypes, and expensive high-performance industrial models for end-use and functional parts.

Mentioning the application of 3DFP in space missions underscores that AM drives the technological progress in general, as pointed out in other chapters. In fact, recently Russian cosmonauts on the International Space Station (ISS) for the first time made synthetic meat in space, by printing cow meat from cow cells prepared by Israeli food-tech firm Aleph Farms and loaded on a spacecraft docking to ISS. The cells had been extracted from a cow through a biopsy, and then placed in a "broth" of nutrients and growth factors simulating the environment inside a cow's body, and forming the bio-ink for the printer. The bioprinter laid down layer after layer of cells that grew into a small piece of muscle tissue shaped like a steak (Gohd 2019). In 2017, it was reported that NASA had asked BeeHex, a NASA spinoff, to use its Chef 3D printer to make food in space, starting with pizza and choice of dough, sauce, and cheese (Bindi 2017). 3DFP meets the following requirements of food in space: as fresh as possible, appetizing, accurate amounts of required nutrients and possibly personalized to dietary needs, minimal energy to store and prepare, and minimal trash (Lockney n.d.).

Food consumers are also food manufacturers, and legal challenges to the diffusion of 3DFP exist, and will become more pressing as the 3DFP market grows. These challenges include: FDA regulations, food safety, labeling printed food, intellectual property, facility inspection, staff training, policy, and ethics. About FDA, as of 2018, no regulations concerning manufacturing of food 3D printed had been released (Tran 2016).

Indications are that the next few years are likely to bring newer, more mainstream applications, processes and materials (Brunner et al. 2018; Dabbene et al. 2018).

Prakash et al. (2019) expected the following near-future scenario for 3DFP. Research will continue leading to developments in food materials and printing processes. From cells, 3DFP will print meat and seafood with taste and texture similar to the natural version of them, and this way

3DFP may help against food scarcity and famine but without polluting, draining animal and natural resources, and being affected by natural problems, such as shortage of water and unfavorable weather. We argue however that preparing and packaging 3DFP feedstocks will generate waste. 3DFP feedstocks may also come from batches of food typically wasted because edible but not appealing, such as overripe fruit and meat off-cuts, as mentioned earlier.

Personalized food with customized amount of salt, fat, and sugar, fitting specific life styles without sacrificing taste will be possible. Research will be needed in the areas of pre-printing (optimizing ingredients and compositions, etc.) and post-printing (holding shape and structure upon baking, frying, etc.). Food safety and microbial stability during printing must be addressed.

Clues to the growth of 3DFP are the recent business moves of world leading companies. Printer maker 3D Systems has introduced Brill 3D Culinary Printer, a food printer that mixes ingredients together with automated precision, layer by layer, and prints edible 3D figures and embellishments in full-color and a variety of flavors, according to user's design or a design selected from a library of 3D designs (3D Systems n.d.). Italian pasta producer Barilla spun off BluRhapsody, an online seller of customized pasta printed on a Barilla-developed printer (3D Food srl 2020). International chocolate producers Hershey's has partnered with 3D Systems to develop new, innovative, and customized chocolate designs using AM.

Finally, we quote inventor/hacker P. Holman, who believes that we will eventually own food printers at home that make a meal tailored for each customer, avoids their allergens, follow their dietary restrictions, and contains their medications (Wolf 2018).

A great boost to the commercial development of 3DFP could come from large government organization, such as the Department of Defense, if they decide to adopt 3DFP even to a limited degree.

REFERENCES

3D Food srl. 2020. *YES, This Is Pasta!* https://blurhapsody.com/en/ (accessed March 6, 2020).

3D Systems. n.d. *The Brill 3D Culinary Studio Powered by 3D Systems.* https://www.3dsystems.com/culinary (accessed March 6, 2021).

Anon. 2013. *3D Systems Acquires The Sugar Lab.* https://www.3dsystems.com/sites/default/files/2016/09102013_3d_systems_acquires_the_sugar_lab.pdf (accessed March 6, 2020).

Anon. 2020. *Overview of Texture Profile Analysis (TPA).* https://texturetechnologies.com/resources/texture-profile-analysis (accessed March 6, 2020).

Anukiruthika, T., Moses, J. A., Anandharamakrishnan C. 2020. 3D printing of egg yolk and white with rice flour blends. *J. Food Eng.* 265:109691.

Balet, S., Guelpa, A., Fox, G., Manley, M. 2019. Rapid visco analyser (RVA) as a tool for measuring starch-related physiochemical properties in cereals: a review. *Food Analyt. Methods* 12:2344–2360.

Bindi, T. 2017. *NASA Astronauts May Soon be Able to 3D-Print Pizzas in Space.* https://www.zdnet.com/article/nasa-astronauts-may-soon-be-able-to-3d-print-pizzas-in-space/ (accessed March 6, 2020).

BIS Research. n.d. *Global 3D Food Printing Market.* https://bisresearch.com/industry-report/global-3d-food-printing-market-2023.html (accessed March 6, 2020).

Brunner T. A., Delley M., Denkel C. 2018. Consumers' attitudes and change of attitude toward 3D-printed food. *Food Qualit. Prefer.* 68:389–396.

Caporizzi, R., Derossi, A., Severini, C. 2019. Cereal-based and insect-enriched printable food: from formulation to postprocessing treatments. Status and perspectives. In *Fundamentals of 3D Food Printing and Applications*, ed.F. C. Godoi, B. R. Bhandari, S. Prakash, M. Zhang, 93–116. Cambridge: Academic Press.

Cavin, S., Pipe, C. J., Michel, M. 2016. *Inkjet Printing With Edible Ink.* US Patent 2016/002 1907 A1, filed March 10, 2014, and issued January 28, 2016.

Chen, H., Xie, F., Chen, L., Zheng, B. 2019. Effect of rheological properties of potato, rice and corn starches on their hot extrusion 3D printing behaviors. *J. Food Eng.* 244:150–158.

Cohen, D. L., Lipton, J. I., Cutler, M. et al. 2009. Hydrocolloid printing: a novel platform for customized food production. *Proceedings of Solid Freeform Fabrication Symposium 2009*, Austin, Texas.

Dabbene, L., Ramundo, L., Terzi, S. 2018. Economic model for the evaluation of 3D food printing. *IEEE International Conference on Engineering, Technology and Innovation* (ICE/ITMC), Stuttgart, 2018, pp. 1–7.

Dankar, I., Haddarah, A., Omar, F. E. L., et al. 2018. 3D printing technology: The new era for food customization and elaboration. *Trends Food Sci. Technol.* 75:231–242.

de Grood, J., de Grood, P. 2013. Method and device for dispensing a liquid. US Patent 8,556,392, filed November 24, 2009, and issued October 15, 2013.

Derossi, A., Caporizzi, R., Azzollini, D., Severini, C. 2018. Application of 3D printing for customized food. A case on the development of a fruit-based snack for children. *J. Food Eng.* 220:65–75.

Derossi, A., Mastrandrea, L., Amodio, M. L., de Chiara, M. L. V., Colelli, G. 2016. Application of multi-variate accelerated test for the shelf life estimation of freshcut lettuce. *J. Food Eng.* 169:122–130.

Dick, A., Bhandari, B., Prakash, S. 2019. 3D printing of meat. *Meat Sci.* 153:35–44.

Ethereal Machines 2018. *Welcome to Ethereal Machines.* https://etherealmachines.com/#/ (accessed March 6, 2020).

FoodJet 2021. *Unique in Advanced Food Depositor Solutions.* https://www.foodjet.com/ (accessed March 6, 2021).

Franck. T. 2019. Alternative meat to become $140 billion industry in a decade, Barclays predicts. https://www.cnbc.com/2019/05/23/alternative-meat-to-become-140-billion-industry-barclays-says.html (accessed August 28, 2021).

Galdeano, J. A. L. 2015. *3D Printing Food: The Sustainable Future.* Master diss. Kaunas University of Technology. https://core.ac.uk/download/pdf/41817540.pdf.

Gholamipour-Shirazi, A., Norton, I. T., Mills, T. 2019. Designing hydrocolloid based food-ink formulations for extrusion 3D printing. *Food Hydrocoll.* 95:161–167.

Godoi, F. C., Bhandari, B. R., Prakash, S., Zhang, M. 2019. An introduction to the principle of 3D food printing. In *Fundamentals of 3D Food Printing and Applications,* ed. F. C. Godoi, B. R. Bhandari, S. Prakash, M. Zhang, 1–18. Amsterdam: Elsevier.

Godoi, F. C., Prakash, S., Bhandari, B. R. 2016. 3d printing technologies applied for food design: status and prospects. *J. Food Eng.* 179:44–54.

Gohd, C. 2019. *Meat Grown in Space for the First Time Ever.* https://www.space.com/meat-grown-in-space-station-bioprinter-first.html (accessed March 6, 2020).

Gonzales, M. 2019. *Jet-Eat Wants to Redefine Meat Through 3D Printing Plants into Steaks.* https://3dprint.com/247377/jet-eat-wants-to-redefine-meat-through-3d-printing-plants-into-steaks/ (accessed March 6, 2020).

Goyanes, A., Kobayashi, M., Martínez-Pacheco, R. 2016. Fused-filament 3D printing of drug products: Microstructure analysis and drug release characteristics of PVA-based caplets. *Int. Journal of Pharmaceutics.* 514: 290–295.

Hamilton, C. A., Alici, G., in het Panhuis, M. 2018. 3D printing vegemite and marmite: redefining "breadboards". *J. Food Eng.* 220:83–88.

Hao, L., Mellor, S., Seaman, O. et al. 2010. Material characterization and process development for chocolate additive layer manufacturing. *Vir. Phy. Prototyp.* 5 (2): 57–64..

Herr, A. G., Cohen, P. E. 2016. Edible 3D printer filament. US Patent 2016/0066601 A1, filed September 6, 2015 and issued March 10, 2016.

Holland, S., Foster, T., Tuck, C. 2019. Creation of Food Structures Through Binder Jetting. In *Fundamentals of 3D Food Printing and Applications*, ed. F. C. Godoi, B. R. Bhandari, S. Prakash, M. Zhang, 257-288. Academic Press.

Holland, S., Tuck, C., Foster, T. 2018. Selective recrystallization of cellulose composite powders and microstructure creation through 3D binder jetting. *Carbohydr. Polym.* 200:229–238.

Jackson, D. S. 2003. STARCH – structure, properties, and determination, in *Encyclopedia of Food Sciences and Nutrition*, ed. B. Caballero, 5561–5567. Cambridge: Academic Press.

Kern, C., Weiss, J., Hinrichs, J. 2018. Additive layer manufacturing of semi-hard model cheese: Effect of calcium levels on thermo-rheological properties and shear behavior. *J. Food Eng.* 235:89–97.

Kouzani, A. Z., Adams, S., Whyte, D. J., et al. 2017. 3D printing of food for people with swallowing difficulties. *Proceedings of the International Conference on Design and Technology, Knowledge E,* Dubai, United Arab Emirates, pp. 23–29.

Le Tohic, C., J. O'Sullivan, J. J., Drapala, K. P., et al. 2018. Effect of 3D printing on the structure and textural properties of processed cheese. *J. Food Eng.* 220:56–64.

Lille, M., Nurmela, A., Nordlund, E., Metsä-Kortelainen, S., Sozer, N. 2018. Applicability of protein and fiber-rich food materials in extrusion-based 3D printing. *J. Food Eng.* 220:20–27.

Lipton, J., Arnold, D., Nigl, F., et al. 2010. *Multi-material food printing with complex internal structure suitable for Conventional Post-Processing*. https://www.researchgate.net/publication/266588628_ Multi-material_food_printing_with_complex_internal_structure_suitable_for_conventional_post-processing (accessed August 28, 2021). .

Lipton, J. I., Cutler, M., Nigl, F. et al. 2015. Additive manufacturing for the food industry. *Trends Food Sci. Technol.* 43:114–123.

Liu, Z., Zhang, M., Bhandari, B., Wang Y. 2017. 3D printing: Printing precision and application in food sector. *Trends Food Sci. Technol.* 69 (A): 83–94.

Liu Z., Zhang M., Bhandari B., Yang, C., 2018. Impact of rheological properties of mashed potatoes on 3D printing. *J. Food Eng.* 220:76–82.

Lockney, D. n.d. *Deep-Space Food Science Research Improves 3D-Printing Capabilities*. https://spinoff.nasa.gov/Spinoff2019/ip_2.html (accessed March 6, 2020).

Maniglia, B. C., Lima, C. D., da Matta Júnior, M. et al. 2020. Dry heating treatment: a potential tool to improve the wheat starch properties for 3D food printing application. *Food Res. Int.* 137:109731.

Mantihal, S., Prakash, S., Godoi, F. C., Bhandari, B. 2017. Optimization of chocolate 3D printing by correlating thermal and flow properties with 3D structure modeling. *Innovat. Food Sci. Emerg. Technol.* 44:21–29.

MarketWatch 2019 3D Food Printing Market Global Industry. https://www.marketwatch.com/press-release/global-3d-food-printing-market-is-expected-to-expand-at-a-cagr-of-50-during-the-period-2017–2024-and-is-expected-to-reach-usd-400-million-by-2024-2019-10-31 (accessed March 6, 2020).

Murphy, S. V., Atala, A. 2014. 3D bioprinting of tissues and organs. *Nat. Biotechnol.* 32:773–785.

Ocheme, O. B., Adedeji, O. E., Chinma, C. E., Yakubu, C. M., Ajibo, U. H. 2018. Proximate composition, functional, and pasting properties of wheat and groundnut protein concentrate flour blends. *Food Sci. Nutr.* 6:1173–1178.

Octu, G. B., Ramundo, L. Terzi, S. 2019. State of the art of sustainability in 3D food printing. *2019 IEEE International Conference on Engineering, Technology and Innovation (ICE/ITMC)*.

Ooi, T. 2019. *The Chocolate 3D Printing Guide*. https://all3dp.com/2/chocolate-3d-printer-all-you-need-to-know/ (accessed March 6, 2020).

Oskay, W., Edman, L. 2014. *The CandyFab Project*. https://candyfab.org/ (accessed March 6, 2020).

Panesar, P. S., Kaur, S. 2016. *Rice: Types and Composition*. In *Encyclopedia of Food and Health*, ed. B. Caballero, P. M. Finglas, F. Toldrá, 646–652. Cambridge: Academic Press.

Periard, D., Schaal, N., Schaal, S., Malone, E., Lipson, H. 2007. *Printing Food*. The University of Texas at Austin. https://repositories.lib.utexas.edu/handle/2152/80223 (accessed March 6, 2020).

Perten n. d. *Rapid Visco Analyser (RVA)*. https://www.perten.com/Global/Brochures/RVA/RVA%20Method %20Brochure_20151110.pdf (accessed March 6, 2020).

Prakash, S., Bhandari, B. R., Godoi, F. C., Zhang, M. 2019. Future outlook of 3D food printing. In *Fundamentals of 3D Food Printing and Applications*, ed. F. C. Godoi, B. R. Bhandari, S. Prakash, M. Zhang, 373–381. Amsterdam: Elsevier.

Rajput, H., Goswami, D., Nigam, S. G. M., Rani, R., Srivastav, P. 2019. 3D printing: advancement in food formulation. In *Recent Trends & Advances in Food Science & Post Harvest Technology*, ed.I. V. I. Chakraborty, 232–251. Delhi: Satish Serial Publishing House. https://www.researchgate.net/ (accessed March 6, 2020).

Rao, M. A. 2006. Viscosity of food: measurement and application. In *Encyclopedia of Analytical Chemistry*. New York: Wiley. doi:10.1002/9780470027318.

Rogers, H., Streich, A. 2019. 3D food printing in Europe: Business model and supply chain aspects. *Proceedings of the International Symposium on Logistics*, Würzburg, Germany.

Sadia, M., Sosnicka, A., Arafat, B. et al. 2016. Adaptation of pharmaceutical excipients to FDM 3D printing for the fabrication of patient tailored immediate release tablets. *Int. J. Pharmaceut.* 513 (1–2): 659–668.

Schutyser, M. A. I., Houlder, S., de Wit, M., Buijsse, C. A. P., Alting, A. C. 2018. Fused deposition modelling of sodium caseinate dispersions. *J. Food Eng.* 220:49–55.

Severini, C., Derossi, A., Azzollini, D. 2016. Variables affecting the printability of foods: Preliminary tests on cereal-based products. *Innovat. Food Sci. Emerg. Technol.* 38:281–291.

Severini, C., Derossi, A., Ricci, I., Caporizzi, R., Fiore, A. 2018a. Printing a blend of fruit and vegetables. New advances on critical variables and shelf life of 3D edible objects. *J. Food Eng.* 220:89–100.

Severini C., Azzollini, D., Albenzio, M., Derossi, A. 2018b. On printability, quality and nutritional properties of 3D printed cereal based snacks enriched with edible insects. *Food Res. Int.* 106:666–676.

Shastry, A. V., Collins, T. M., Suttle, J. M., et al. 2009. Edible inks for ink-jet printing on edible substrates. US Patent 7,597,752, filed June 26, 2003, and issued October 6, 2009.

Southerland, D., Walters, P., Huson, D. 2011. Edible 3D printing. *NIP & Digital Fabrication Conference, 2011 International Conference on Digital Printing Technologies*, 819–822.

Sun, J., Peng, Z., Yan, L., et al. 2015a. 3D food printing an innovative way of mass customization in food fabrication. *Int. J. Bioprint.* 1 (1): 27–38.

Sun, J., Zhou, W., Huang, D. et al. 2015b. An overview of 3D printing technologies for food fabrication. *Food Bioprocess Technol.* 8:1605–1615.

Sun, J., Zhou, W., Huang, D., Yun, L. 2018. 3D food printing: perspectives. In *Polymers for Food Applications*, ed. T. Gutierrez, 725–755. Cham: Springer.

TNO n.d. *The Future of Food*. https://www.tno.nl/media/2216/future_of_food.pdf (accessed March 6, 2020).

Tran, J. L. 2016. 3D-printed food. *Minnesota J. Law. Sci. Technol.* 17 (2): 855–880 https://scholarship.law.umn.edu/cgi/viewcontent.cgi?article=1409&context=mjlst (accessed March 6, 2020).

United States Department of Agriculture. 2020. *Food Prices and Spending*. https://www.ers.usda.gov/data-products/ag-and-food-statistics-charting-the-essentials/food-prices-and-spending/ March 17, 2020 (accessed March 6, 2020).

Vadodaria, S., Mills, T. 2020. Jetting-based 3D printing of edible materials. *Food Hydrocolloid.* 106:105857.

Vancauwenberghe, V., Verboven, P., Lammertyn, J., Nicolai, B. 2018. Development of a coaxial extrusion deposition for 3D printing of customizable pectin-based food simulant. *J. Food Eng.* 225:42–52.

Walters, P., Huson, D., Southerland, D. 2011. Edible 3D printing. In *Proceedings of 27th international conference on digital printing technologies*, October 2011, Minnesota, USA..

Wang, L., Zhang, M., Bhandari, B., Yang, C. 2018. Investigation on fish surimi gel as promising food material for 3D printing. *J. Food Eng.* 220:101–108.

Wen, L. F., Henry, W., Swaine, R. 2008. Flavor application on edible substrates. US Patent 20080075830, filed September 24, 2007, and issued March 27, 2008.

Wolf, M. 2018. *Pablos Holman Sees a Future Where We Print French Bread & Strawberries*. https://thespoon.tech/pablos-holman-sees-a-future-where-we-print-french-bread-strawberries/ (accessed March 6, 2019).

Yang, F., Zhang, M., Bhandari, B. 2017. Recent development in 3D food printing. *Crit. Rev. Food Sci. Nutr.* 57 (14): 3145–3153.

Yang, F., Zhang, M., Prakash, S., Liu, Y. 2018a. Physical properties of 3D printed baking dough as affected by different compositions. *Innovat. Food Sci. Emerg. Technol.* 49:202–210.

Yang, F., Zhang, M., Bhandari, B., Liu, Y., 2018b. Investigation on lemon juice gel as food material for 3D printing and optimization of printing parameters. *LWT Food Sci. Technol.* 87:67–76.

Yang, J., Wu, L. Liu, J. 2001. Method for rapidly making a 3D food object. US Patent 6,280,784 B1, filed February 10, 2000, and issued August 28, 2001.

Zheng, L., Yu, Y., Tong, Z., et al. 2019. The characteristics of starch gels molded by 3D printing. *J. Food Process. Preserv.* 43:13993. doi:10.1111/jfpp.13993.

15 Acrylates

The most important thing about global warming is this. Whether humans are responsible for the bulk of climate change is going to be left to the scientists, but it's all of our responsibility to leave this planet in better shape for the future generations than we found it.

Mike Huckabee

15.1 INTRODUCTION

Acrylates comprise the conjugate bases, esters, and salts of acrylic acid and compounds derived from it, and are characterized by the *acrylate ion* $CH_2=CHCOO^-$. Acrylates have two functional groups that impart them outstanding reactivity. One is the carboxylic acid group COOH that enables to derive many compounds from a wide variety of alcohols and amines. The other functional group is the double bond between two carbon atoms that enable acrylates to undergo various kinds of polymerization. Acrylates react at their double bond with themselves to form homopolymers, and with other monomers (amides, butadiene, acrylonitrile, vinyl, etc.) to form copolymers. Prevalent types of commercial acrylates are butyl acrylate, ethyl acrylate, methyl acrylate, and methyl methacrylate, whose chemical formulas are pictured in Figures 15.1, 15.2, 15.3, and 15.4, respectively. Values of selected properties of commercial versions of them are reported in Table 15.1, and range as follows: density 0.90–0.96 g/cm^3, viscosity 0.49–0.90 mPa s, MW 86–128 g/mol, and T_g −54 to 105°C. Properties such as flexibility, hardness, super-absorbency, toughness, transparency, and so on may widely vary in values among acrylates. Short-chain acrylic monomers such as methyl methacrylate produce stronger, harder, and more brittle polymers, whereas long-chain monomers such as butyl acrylate turn into tacky, softer, weaker, and more flexible polymers. Ethyl acrylate results in values of T_g and hardness that are intermediate vs. those of other acrylate monomers. By blending different acrylate monomers and tuning their ratio, formulators can meet specific end-use requirements by balancing strength and durability, hardness and ductility, and so on.

Examples of major suppliers of acrylates are the large chemical companies Arkema, BASF, Dow, Evonik, and Mitsubishi. According to a 2018 forecast by consultancy ResearchAndMarkets (PR Newswire 2018), the global market of acrylates was projected to increase at a CAGR of 3% per year in the period 2017–2023, and reach almost 5.7 million tons by the end of 2023. Another consultancy predicted that the global acrylate market would reach USD 9.9 billion by 2022, growing at a CAGR of 6.3% from 2017 to 2022 (MarketsandMarkets™ 2020).

Due to their superior reactivity and versatility, acrylates are consumed as building blocks of acrylic homopolymers and copolymers, known as *acrylics* and *polyacrylates*. The basic chemical structure of acrylics is $-[CH_2-(CR_1)-COOR_2-]_n$ with R_1 and R_2 varying in composition depending on the type of acrylate: for example R_1 and R_2 are H and CH_3 for polymethyl acrylate, respectively, and are both CH_3 for polymethyl methacrylate (Petkar 2019). Acrylics can be TP or TS, and are present in a myriad products: additives, pressure sensitive adhesives, antioxidant agents, binders, cleaning products, coatings, dental fillings, hard and soft contact lenses, intermediates, water-based paints, papers, printing inks, resins, sealants, surfactants, artificial teeth, textiles, etc. In comparison to other polymers, acrylics feature: outstanding optical clarity and good strength; chemical, weather, heat (up to 180°C), and ozone resistance; flexibility; good electrical properties; and dimension stability. Acrylics are compared to glass because of their optical clarity, and have some property advantages over window glass: acrylic can be 11 times stronger (ultimate tensile

DOI: 10.1201/9781003221210-15

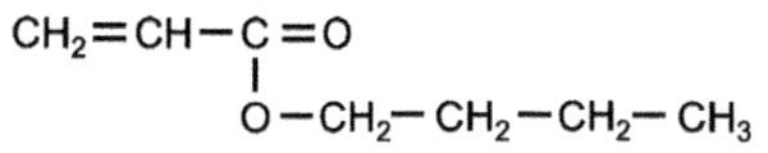

$$CH_2=CH-C=O$$
$$O-CH_2-CH_2-CH_2-CH_3$$ **FIGURE 15.1** Chemical structure of butyl acrylate.

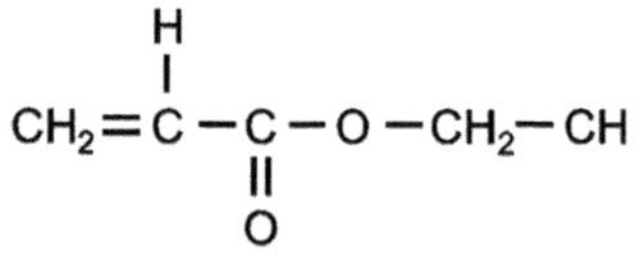

$$CH_2=C-C-O-CH_2-CH$$

FIGURE 15.2 Chemical structure of ethyl acrylate.

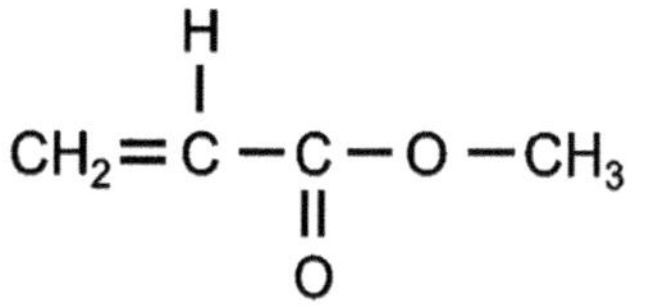

$$CH_2=C-C-O-CH_3$$

FIGURE 15.3 Chemical structure of methyl acrylate.

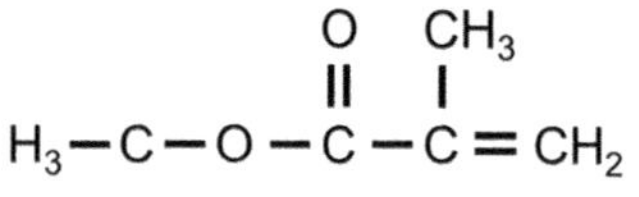

$$H_3-C-O-C-C=CH_2$$ **FIGURE 15.4** Chemical structure of methyl methacrylate.

TABLE 15.1

Properties of Commercial Acrylate Monomers

Acrylate	CAS #[a]	Chemical Formula	Density	Viscosity	MW	T_g	Freezing Point
			kg/dm^3	mPa s	g/mol	°C	°C
Butyl Acrylate	141-32-2	$C_7H_{12}O_2$	0.898	0.90	128	−54	−64
Ethyl Acrylate	140-88-5	$C_5H_6O_2$	0.922	0.56	100	−24	−72
Methyl Acrylate	96-33-3	$C_4H_6O_2$	0.956	0.49	86	10	−75
Methyl Methacrylate	80-62-6	$C_5H_8O_2$	0.939	0.53	100	105	−48

Source: Technical data sheets posted online by Dow (methyl methacrylate) and Arkema.
Note:
[a]Unique numerical identifier assigned by the Chemical Abstracts Service to each chemical substance.

strength of 11 ksi vs. 1 ksi) and will flex instead of shattering, it is a superior thermal insulator (thermal conductivity of 0.20 W/mK vs. 0.96 W/mK), and it weighs half as much. Due to the above properties and high gloss, color retention, and durability, acrylics are well suited for architectural and industrial coatings, automotive lenses and finishes, boat windshields, outdoor signs, and fluorescent street lights. Acrylic polymers encompass, among others, polybutyl acrylate, polybutyl methacrylate, polyethyl acrylate, and polymethyl methacrylate.

In the following sections, acrylate-based polymers that are cured by exposure to light (*photopolymers*) in AM processes are described. Acrylate-based photopolymers are not limited to AM, but broadly employed outside AM as UV-curable adhesives because of their strong reactivity and tunable properties (Decker 2011).

15.2 COMMERCIAL ACRYLATES FOR AM

Commercial acrylates are processed in two families of AM technologies, VP and MJ, both illustrated in Chapter 2. We recall that the former consists in curing selectively and layer by layer, by light-activated polymerization, a liquid photopolymer sitting in a vat, whereas in the latter droplets of build material are selectively deposited one layer at the time, and bonded together by chemical reaction. In both definitions, *selectively* means *in specific locations in the XY plane*. VP processes such as SL (2.5.1.1) and DLP (Sections 2.5.1.2) offer the advantages of high feature resolution (5–50 μm), fast printing, and surface finish, but require postcuring in order to convert any unreacted chemical, in order to minimize toxicity (Oskui et al. 2016), and maximize physical and mechanical properties. Some applications of VP processes comprise dental implants, and patient-specific scaffolds for tissue regeneration (Chia and Wu 2015).

Commercial acrylates react rapidly through a polymerization reaction based on successive addition of free-radical building blocks (*free-radical polymerization*). The advantage of fast reaction is offset by shrinking, warping, and curling that impair optimal dimensional fidelity between 3D digital model and printed article. Acrylates for VP have been discussed in Section 2.5.7.4, and include acrylic- and methacrylic-based resins, and other resins modified with acrylate ends.

Most common methacrylate monomers and oligomers for VP are: poly(ethylene glycol) diacrylate (Warner et al. 2016), urethane dimethacrylate (Liska et al. 2007), triethylene glycol dimethacrylate (Al Mousawi et al. 2018), bisphenol A-glycidyl methacrylate (Al Mousawi et al. 2017), trimethylolpropane triacrylate (Janusziewicz et al. 2016), and bisphenol A ethoxylate diacrylate (Credi et al. 2016). Table 15.2 lists properties and acrylate ingredients of commercial resins for VP supplied by Formlabs (USA) and 3D Systems (USA), two leading printer manufacturers. The values in Table 15.2 can serve as a target for material development, and their ranges (quite wide because those materials target different applications) are the following: tensile strength 7.7–77 MPa, tensile modulus 2.2–4.1 GPa, and elongation at break 2–85%.

Examples of acrylates present in commercial acrylic resins for MJ are trimethylolpropane formal acrylate ($C_{10}H_{16}O_4$), and urethane dimethacrylate ($C_{23}H_{38}N_2O_8$) (Section 2.7.5).

Acrylates for AM are reviewed by Bourell et al. (2017).

15.3 EXPERIMENTAL ACRYLATES IN AM

15.3.1 Introduction

Developing photocurable TS acrylic formulations for AM is mostly focused on commercial monomers and oligomers that are derived from petroleum instead of sustainable feedstocks; hence, two types of acrylates are introduced here: acrylates for AM from non-sustainable sources, serving as a benchmark for formulators of sustainable feedstocks, and sustainable acrylates for AM.

15.3.2 Non-Sustainable Acrylate-Based Feedstocks

Acrylates have been selected for AM as modifiers for precursors of aromatic polyimides (PIs) that are suited for light-based printers (*photo-printable*), and possess high values of tensile modulus and strength, and thermal stability, corrosion resistance, and low dielectric constant, similarly to poly(4,4'-oxydiphenylene pyromellitimide) (PMDA-ODA PI), a high-performance engineering PI, commercially known as Kapton™. Below a few examples of such modifiers reported in recent AM technical literature are described.

Guo et al. (2017) aimed at formulating new inks for a UV-based DLP printer, and synthetized a PI oligomer grafted with glycidyl methacrylate (GMA), later combined with vinyl pyrrolidone (NVP), lauryl methacrylate (LMA), and polyethylene glycol diacrylate (PEG400DA) as reactive diluents, and the photoinitiator Irgacure® 819. The authors tested ink formulations obtained by

TABLE 15.2

Properties and Acrylate Ingredients of Commercial Resins for VP. Post-Curing Values

Resin Commercial Name	Supplier	Acrylates in Composition	Tensile Strength	Tensile Modulus	Elongation at Break	Izod Notched Impact	HDT		Applications. Features
							At 0.45 MPa	At 1.8 MPa	
			MPa	GPa	%	J/m	°C	°C	
Accura® 60	3D Systems	Ethoxylated pentaerytritol tetraacrylate	58–68	2.7–3.1	5–13	15–25	53–55	48–50	Clear, transparent, rigid, strong parts. Investment casting patterns.
Accura® Fidelity	3D Systems	Tricyclodecane dimethanol diacrylate, dipentaerythritol pentaacrylate esters	65	2.8	5–11	25–39	63	55	High-yield investment casting patterns with ultra-low viscosity.
Accura® Phoenix	3D Systems	Tricyclodecane dimethanol diacrylate	52–77	2.6–2.9	2–7	18–24	137	103	Heat-resistant transparent prototypes: HVAC components, fans, heaters, humidifiers, small engine enclosures.
Clear	Formlabs	Urethane dimethacrylate, methacrylate monomer(s)	65	2.8	6.2	25	N/A	N/A	Rapid prototyping. Product development. Strength. Smooth finish.
Flex	Formlabs	Urethane dimethacrylate, methacrylate monomer(s)	7.7–8.5	N/A	75–85	N/A	N/A	N/A	Parts that bend and compress. Soft-touch materials. Handles, grips, overmolds. Cushioning and damping. Wearables prototyping. Packaging. Stamps
Rigid	Formlabs	Methacrylate monomer(s), isobornyl methacrylate, urethane dimethacrylate	75	4.1	5.6	18.8	88	74	Glass-filled for high stiffness. Polished finish. 50–100 μm resolution.
Tough 2000	Formlabs	Urethane dimethacrylate, methacrylate monomer(s)	46	2.2	48	40	118	108	Strong and stiff prototypes. Sturdy jigs and fixtures. ABS-like strength and stiffness.

Source: Data sheets posted online by the suppliers.

TABLE 15.3

Composition of PI-3 Acrylate-Based Ink for DLP. Values in wt%

PI-g-GMA[a]	NVP	LMA	PEG400DA	IRGACURE® 819
55	30	7.5	7.5	2

Source: Guo, Y., Ji, Z., Zhang, Y., et al. 2017 Solvent-free and photocurable polyimide inks for 3D. *J. Mater. Chem. A* 5:16307–16314.

Note:

[a]PI oligomer grafted with GMA.

varying the amount of NVP, LMA, and PEG400DA, and measured the subsequent properties across the formulations: tensile strength and strain at break were 5.8–24.9 MPa and 5.0–11.8%, respectively, hardness modulus (measured through a nano-indenter) and hardness were 1.37–3.17 GPa and 100–183 MPa, respectively. The ink formulation denoted as PI-3 with composition in Table 15.3 featured optimal mechanical properties vs. the other formulations, and high-temperature resistance, being its tensile strength 25 MPa and 20 MPa at 25°C and after 3 h at 300°C, respectively. PI-3 also met two critical requirements for being photocurable via DLP: (a) E_o, the energy to solidify PI-3, was lower than E_c, the energy applicable by the printer; (b) C_d, the cure depth and the thickness of each solidified layer (25–100 µm), was lower than D_p, the penetration depth of the UV light in PI-3. Particularly, the experimental variation of C_d as a function of D_p, E_c, and E_o followed equation (2.11) that assumed the form of equation (15.1), being in this case E_c equal to E_{max}, and D_p 19.69 mJ/cm^2 and 0.165 mm, respectively:

$$C_d = 0.165 \ln E_o - 0.492 \tag{15.1}$$

All formulations were temperature resistant; in fact, only at 336°C a 5% weight loss was observed. The decomposition temperature was 429–445°C for all tested formulations, far exceeding 359°C for a commercial ink. GMA contributed to the thermal stability of the developed inks that was superior to most commercial photosensitive resins for AM: polyurethane acrylic, epoxy acrylic, and acrylic resins.

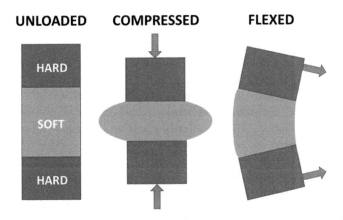

FIGURE 15.5 Article printed in one acrylate-based material featuring different properties across its volume.

Source: Based on Borrello, J., Nasser, P., Iatridis, J. C., Costa, K. D. 2018. 3D printing a mechanically-tunable acrylate resin on a commercial DLP-SLA printer. *Additive Manufacturing* 23:374–380.

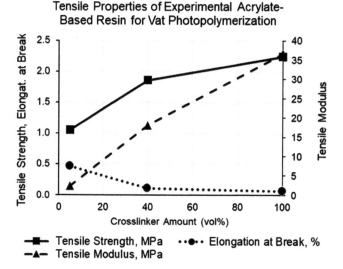

FIGURE 15.6 Variation of tensile properties of printed coupons made of ethylene glycol phenyl ether acrylate (EGPEA)-based resin with volumetric amount of 1,6-hexanediol diacrylate crosslinker.

Source: Data from Borrello, J., Nasser, P., Iatridis, J. C., Costa, K. D. 2018. 3D printing a mechanically-tunable acrylate resin on a commercial DLP-SLA printer. *Addit. Manuf.* 23:374–380.

Borrello et al. (2018) formulated a mechanically tunable acrylate resin for DLP that seemed a promising candidate for multi-material DLP, because it: (a) exhibited tunable tensile properties depending on the ratio of monomer/crosslinker in the formulation that resulted in harder and softer grades; (b) generated printed articles geometrically complicated, and made up of one material transitioning seamlessly in mechanical properties within the article's volume so that, f.e., one portion of it is flexible and soft, and the other hard and rigid (Figure 15.5). The tunable resin comprised ethylene glycol phenyl ether acrylate, plus crosslinker and photoinitiator. As illustrated in Figure 15.6 relative to printed coupons, increasing the amount of crosslinker made the resin stronger, stiffer, and less ductile, by improving tensile strength from 1.05 MPa to 2.24 MPa, and modulus from 2.23 MPa to 36.6 MPa, and lowering elongation to break from 0.46% to 0.06%. Shape fidelity, quantified as the variation in dimensions between the printed article and its 3D digital model was less than ±3%, a value comparable to existing commercial resins for DLP and SL.

T. E. Long and his team leveraged acrylates as precursors to formulate versions of Kapton™ that could be printed through SL. In one study (Hegde et al. 2017), they started with hydroxyethyl acrylate, and formulated a polymer that was successfully printed through two versions of SL: mask-projection micro-SL, and large-area mask-projection scanning SL. The printed samples exhibited good resolution, and isotropic dimensional shrinkage after post-printing, which is preferred to anisotropic shrinkage. The new feedstock is an alternative to high-performance engineering TPs characterized by tensile modulus greater than 1 GPa, T_g above 200°C, and degradation temperature surpassing 500°C, and can become a candidate for aerospace, automotive, and microelectronics components. In another study (Herzberger et al. 2018), Long and his team employed 2-(dimethylamino)ethyl methacrylate, and developed UV curable polyamic acid (PAA) salt solutions that were printed on a custom-built scanning-mask VP apparatus, and subsequently underwent thermal treatment to obtain the final all-aromatic PMDA-ODA PI. The PAA salt solutions offered the advantage that they did not require a time-consuming multistep synthesis, featured gel times below 5 s, produced gels relatively stiff and hence self-supported, and the whole method was easily carried out.

TABLE 15.4

Properties of Acrylates Used as Photopolymers for VP: Isobornyl Acrylate, 1,10-Decanediol Diacrylate, Pentaerythritol Tetraacrylate, and Multifunctional Acrylate Oligomer

Property Value	Biobased Carbon Content	Density	Viscosity	MW	Acrylate Functionality	Concentration of Double Bonds C=C
	wt%	kg/dm³	Pa s	kg/mol	N/A	mol/dm³
Minimum	10	0.98	0.01	0.21	1	2.7
Maximum	75	1.17	17	1.7	5	13

Source: Adapted from Voet, V. S. D., Strating, T., Schnelting, G. H. M., et al. 2018. Biobased Acrylate Photocurable Resin Formulation for Stereolithography 3D Printing. *ACS Omega* 3:1403–1408. https://pubs.acs.org/doi/abs/10.1021/acsomega.7b01648. Reproduced with permission from ACS. Further permissions related to the material excerpted should be directed to ACS.

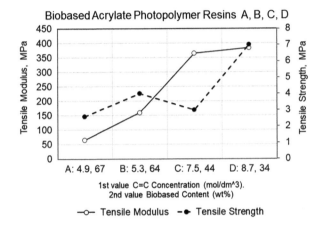

FIGURE 15.7   Tensile properties of four resins for VP from biobased acrylates: A, B, C, D. For each resin, two values are reported: first is concentration of carbon double bond, the second is biobased content.

Source: Data from Voet, V. S. D., Strating, T., Schnelting, G. H. M., et al. 2018. Biobased Acrylate Photocurable Resin Formulation for Stereolithography 3D Printing. *ACS Omega* 3:1403–1408. Reproduced with permission from ACS. Further permissions related to the material excerpted should be directed to ACS.

15.3.3   SUSTAINABLE ACRYLATE-BASED FEEDSTOCKS

A few commercial and non-commercial SPs are compatible with VP methods; therefore, the research by Voet et al. (2018) is particularly notable, because they fabricated prototypes and test coupons out of biobased acrylate photopolymer resins (BAPRs) processed on a commercial UV laser-based SL printer. The starting ingredients of the BAPR resins were four Sartomer (USA) acrylates whose ranges of properties are listed in Table 15.4: isobornyl acrylate, 1,10-decanediol diacrylate, pentaerythritol tetraacrylate, and multifunctional acrylate oligomer. These acrylates were combined together in proportions leading to four BAPR formulations (A, B, C, D) that possessed a biobased content of 34–67 wt% and viscosity adequate for processing, and were readily photopolymerized on the printer. Tensile coupons were printed, soaked, and rinsed in isopropanol, and fully cured in UV oven. Tensile strength and modulus of the BAPRs reached 2.6–7.0 MPa and 65–383 MPa, respectively (Figure 15.7). The BAPRs underperformed in comparison to Autodesk ACPR-48, a commercial non-biobased resin for VP, whose tensile strength and modulus are 19 MPa and 836 MPa, respectively. Since raising crosslink density improves tensile strength and modulus in polymers, and higher concentration of double bonds leads to a more crosslinked polymer network, not surprisingly the concentration of double bonds had a positive correlation with tensile strength and modulus, whereas biobased content was negatively correlated to those properties (Figure 15.7). Similar trends were detected through TMA: greater double bond concentration and crosslink density improved thermal stability, and raised degradation temperatures. High-viscosity resins led to high-resolution prototypes with complex and minute features (5–50 μm scale), and excellent surface finishing, comparable to commercial non-sustainable resins.

Ding et al. (2019) were the first to formulate methacrylates from biobased phenolics to widen the use of feedstock formulations for AM that were sustainable, photocured fast, and featured adequate mechanical and thermal properties. The authors started with eugenol ($C_{10}H_{12}O_2$), a liquid compound extracted from natural oils especially from basil, cinnamon, clove bud oil, and clove leaf oil, and applied a type of chemical reaction called *thiol-lene* featuring high yield and rate, and synthesized bifunctional 3,6-dioxa-1,8-octanedithiol eugenol acrylate (E) through a highly efficient and scalable process. E featured fast photocuring kinetics, and its viscosity, T_g, and thermal properties could be tailored by incorporating a sustainable reactive diluent, guaiacyl methacrylate (G), and forming a GE monomer. Vanillyl alcohol dimethacrylate (V) was added, and its aromatic nature improved tensile strength and modulus that reached 62 MPa and 3.4 GPa, respectively, well outperforming in the same properties the previously mentioned commercial ACPR-48. Two low-MW methacrylate crosslinkers were tested: renewable, bifunctional V (MW 290.3 g/mol), and trifunctional trimethylolpropane trimethacrylate (T) (MW 338.4 g/mol). T and V augmented the maximum reaction rate and shortened the gel time, with T being more effective at it, due to the higher number of methacrylates in its monomer. Four GET and four GEV formulations with different compositions were extensively and experimentally evaluated. The tensile test results revealed that all the polymers displayed stiff and brittle behavior typical of unmodified acrylic polymers. The measured properties are summarized in Table 15.5, broken up in two groups of blends, GEV and GET, according to the crosslinker V or T. The test values across all formulations were: tensile strength 33.1–61.7 MPa, tensile modulus 0.83–1.35 GPa, elongation at break 2.8–8.9%, and toughness 0.5–3.7 MJ/m³ (equal to area under the stress-strain curve up to failure). The GEV blends outperformed the GET ones in tensile strength (57% higher average), elongation

TABLE 15.5

Property Values of UV-Curable Sustainable Acrylates for SL

Acrylate	Tensile Strength	Tensile Modulus	Elongation at Break	Toughness
	MPa	GPa	%	MJ/m³
All GEV, GET	33.1–61.7	0.83–1.35	2.8–8.9	0.5–3.7
GET, range	33.1–46.3	0.83–1.35	2.8–8.2	0.5–2.2
GEV, range	44.6–61.7	1.02–1.23	6.9–8.9	1.9–3.7
GET, average	34.0	1.10	5.2	1.4
GEV, average	53.4	1.13	7.6	2.6
GEV 60/20/20 (G/E/V)	61.7	1.23	8.9	3.7

Source: Data from Ding, R., Du, Y., Goncalves, R. B., Francis, L. F., Reineke, T. M. 2019. Sustainable near UV-curable acrylates based on natural phenolics for stereolithography 3D printing. *Polym. Chem.* 10:1067–1077.

Legend: G guaiacol methacrylate, V vanillyl alcohol methacrylate, T trimethylolpropane trimethacrylate, E 3,6-dioxa-1,8-octanedithiol eugenol acrylate.

at break (46% higher average), and toughness (86% higher average), whereas the tensile modulus was not significantly different between the two groups. The superior tensile performance of GEV over GET was explained by the fact that GEV featured stronger intermolecular forces in a more homogeneous network than GET. The blend GEV 60/20/20 with 2 wt% photoinitiator ((diphenyl (2,4,6-trimethylbenzoyl) phosphine oxide) outperformed all GEV and GET blends in tensile strength (61.7 GPa), elongation at break (8.2%), and toughness (3.7 MJ/m^3), but not tensile modulus (1.23 GPa), and also displayed large renewable content, high curing rate, and low viscosity of the monomers. T resulted in higher T_g than V: the T_g for the GET and GEV formulations were 130.9°C and 107.5°C, respectively.

Sustainable plant oils, such as corn, linseed, peanut, soybean (Garrison et al. 2014), have been consumed to formulate synthetic polymers including polyethers, polyesters, polyolefins, and PUs because they possess adequate properties for engineering applications, and are economical. Plant oils are also biocompatible, and in fact they are emerging as feedstocks for formulating polymeric biomaterials for implants such as scaffolds (Miao et al. 2014).

Since studies on plant oil polymers as liquid resins for fabricating biomedical scaffolds through SL are infrequently reported, the study by Miao et al. (2016) is particularly interesting. The authors employed a novel acrylate converted from soybean oil to epoxide, and via 4DP built smart biomedical scaffolds. 4DP (Section 2.11) is any AM process printing an item whose shape, property, and functionality change as intended as a function of time from one stable state to another stable state when exposed to a specific external stimulus such as temperature and humidity f.e. (Tibbits et al. 2014). The authors utilized a novel, self-developed, table-top SL printer based on UV laser to convert the soybean oil-based acrylate into biocompatible, porous, and "smart" (temperature controlled shape-changing) scaffolds, supporting growth of human bone marrow mesenchymal (that is relative to connective and lymphatic tissue, and blood vessels) stem cells (hMSCs) that have great potential for a number of functional tissue applications. Scaffolds were shaped like a 0/90° grid with struts featuring width of about 250–440 μm, and thickness of about 100–450 μm depending on printing speed and laser frequency (from 8 kHz to 20 kHz). Particularly, increasing printing speed from 10 to 80 mm/s reduced thickness and width: at maximum speed the thickness and width were <100 μm and 250 mm, respectively, whereas at minimum speed thickness and width were 0.5 mm and 0.4 mm, respectively. The thickness and width increased with laser frequency. The compressive modulus of samples of polymerized soybean oil epoxidized acrylate printed at infill density of 70% and speed of 10 mm/s was 430–470 MPa across laser frequencies. Obviously, lower infill density translated into greater scaffold's porosity. Shape memory was experimentally demonstrated: a sample was printed flat, bent into a U shape, and locked in that shape by cooling it at −18°C; upon being heated at 37°C (human body temperature) for 60 s the sample returned flat. In terms of adhesion and proliferation of hMSCs, this novel acrylate-based epoxide outperformed non-sustainable polyethylene glycol diacrylate, a most frequently investigated resin for biomedical scaffolds, and performed as sustainable PLA and non-sustainable PCL, two polymers that are highly biocompatible, but not directly suited as feedstock resins for SL because they lack photosensitive chemical groups. The adhesion and proliferation of hMSCs may be related to contact angle, chemical groups, and so on. Particularly, in soy resin there are only hydroxyl and ester groups that are mostly non-harmful to cells.

The recent paper by Wu et al. (2020) confirmed that AM can contribute to protect the environment not only by processing biobased feedstocks, but also by reducing waste that AM upcycles as a printing feedstock. The authors formulated a biodegradable, high-performance acrylate resin for VP from waste cooking oil (WCO), a mix of oils and fats for frying and cooking food in households, restaurants, and corporations, and addressed the direct discharge of WCO into sewage lines that can be clogged due to the buildup of fats. The global production of WCO is not insignificant, and rapidly increasing, therefore this innovation can benefit the environment. China is the top producer of WCO, with 500 million tons in 2011. McDonald's, the world's largest fastfood chain, produces more than 600 ton/day of WCO. Presently, WCO is not completely wasted but recycled to produce soap and biodiesel. The authors collected WCO directly from vats in a McDonald's restaurant, converted it into acrylate resin through a simple one-step reaction (*Michael addition*), added a photoinitiator, and tested it on the commercial VP printer Solus by Reify 3D (USA). Particularly, converting WCO into acrylate resin comprised the following steps:

1. Filtering WCO using filter paper to remove insoluble ingredients
2. Reacting together acrylic acid, WCO, and boron trifluoride etherate as the catalyst
3. Heating, cooling, stirring
4. Adding hexane to dissolve the organic components, followed by washing with aqueous $NaHCO_3$ and NaCl solutions to remove unreacted acrylic acid and catalyst
5. Washing, drying, and evaporating under vacuum.

The new resin (acrylated WCO, or AWCO) enabled to print features as small as 100 μm, and was comparable in resolution to commercial resins (sold at USD 525/L) for printing high-resolution articles, and surpassed acrylated resin derived from virgin soybean oil (AESO) in storage modulus at and above room temperature, which is consistent with the higher crosslink density possessed by AWCO. The biodegradability was measured in terms of percentage weight loss after burying samples in actual soil for 11 month. The weight loss for AWCO, AESO, and the commercial acrylate resin MiiCraft BV007A (Germany) for VP were 16, 7.5, and 2.5 wt%, respectively. The authors claimed to be the first ones converting WCO into feedstock for UV-based AM, and their new resin, if commercially successful, has potential to reduce price of high-performance resins for VP.

REFERENCES

Al Mousawi, A., Dumur, F., Garra, P., et al. 2017. Scaffold based photoinitiator/photoredox catalysts: toward new high performance photoinitiating systems and application in LED projector 3D printing resins. *Macromolecules* 50:2747–2758

Al Mousawi, A., Garra, P., Schmitt, M., et al. 2018. 3-hydroxyflavone and *N*-phenylglycine in high performance photoinitiating systems for 3D printing and photocomposites synthesis. *Macromolecules* 51:4633–4641

Borrello, J., Nasser, P., Iatridis, J. C., Costa, K. D. 2018. 3D printing a mechanically-tunable acrylate resin on a commercial DLP-SLA printer. *Addit. Manufactur.* 23:374–380

Bourell, D., Kruth, J. P., Leu, M. et al. 2017. Materials for additive manufacturing. *CIRP Annal. – Manufactur. Technol.* 66:659–681

Chia, H. N., Wu, B. M. 2015. Recent advances in 3D printing of biomaterials. *J. Biol. Eng.* 9:4

Credi, C., Fiorese, A., Tironi, M., et al. 2016. 3D printing of cantilever-type microstructures by stereolithography of ferromagnetic photopolymers. *ACS Appl. Mater. Interfaces* 8:26332–26342

Decker, C. 2011. UV-radiation curing of adhesives. In *Handbook of Adhesives and Surface Preparation*, ed. Sina Ebnesajjad, 2221–2243. Norwich: William Andrew Publisher

Ding, R., Du, Y., Goncalves, R. B., Francis, L. F., Reineke, T. M. 2019. Sustainable near UV-curable acrylates based on natural phenolics for stereolithography 3D printing. *Polym. Chem.* 10:1067–1077

Garrison, T. F., Kessler, M. R., Larock, R. C. 2014. Effects of unsaturation and different ring-opening methods on the properties of vegetable oil-based polyurethane coatings. *Polymer* 55:1004–1011

Guo, Y., Ji, Z., Zhang, Y., et al. 2017. Solvent-free and photocurable polyimide inks for 3D. *J. Mater. Chem. A* 5:16307–16314

Hegde, M., Meenakshisundaram, V., Chartrain, N. 2017. 3D printing all-aromatic polyimides using mask-projection stereolithography: processing the nonprocessable *Adv. Mater.* 29:1701240

Herzberger, J., Meenakshisundaram, V., Williams, C. B., Long, T. E. 2018. 3D printing all-aromatic polyimides using stereolithographic 3D printing of polyamic acid salts. *ACS Macro Lett.* 7 (4): 493–497

Januszewicz, R., Tumbleston, J. R., Quintanilla, A. L. et al. 2016. Fabrication with continuous liquid interface production. *Proc. Natl. Acad. Sci. U.S.A.* 113:11703–11708

Liska R., Schuster M., Inführ R., et al. 2007. Photopolymers for rapid prototyping. *J. Coat. Technol. Res.* 4:505–510

MarketsandMarkets™. 2020. https://www.marketsandmarkets.com/Market-Reports/acrylate-market-172108731.html (accessed January 16, 2020)

Miao, S., Wang, P., Su, Z., Zhang, S. 2014. Vegetable-oil-based polymers as future polymeric biomaterials. *Acta Biomater.* 10:1692–1704

Miao, S., Zhou, W., Castro, J., et al. 2016. 4D printing smart biomedical scaffolds with novel soybean oil epoxidized acrylate. *Sci. Rep.* 6:27226

Oskui, S. M., Diamante, G., Liao, C., et al. 2016. Assessing and reducing the toxicity of 3D-printed parts. *Environ. Sci. Technol. Lett.* 3:1–6

Petkar, K. C. 2019. Polyacrylate nanoparticles as a promising tool for anticancer therapeutics. In *Polymeric Nanoparticles as a Promising Tool for Anti-cancer Therapeutics*, ed.P. Kesharwani, K. M. Paknikar, V. Gajbhiye, 35–56. London: Academic Press

PR Newswire. 2018. *The Global Acrylates Market is Projected to Reach Approximately 5,686 Thousand Tons by the End of 2023*. https://www.prnewswire.com/news-releases/the-global-acrylates-market-is-projected-to-reach-approximately-5–686-thousand-tons-by-the-end-of-2023--300761479.html (accessed April 3, 2020)

Tibbits, S., McKnelly, C., Olguin, C., Dikovsky, D., Hirsch, S. 2014. *4D Printing and Universal Transformation*. http://papers.cumincad.org/data/works/att/acadia14_539.content.pdf (accessed March 6, 2019)

Voet, V. S. D., Strating, T., Schnelting, G. H. M., et al. 2018. Biobased acrylate photocurable resin formulation for stereolithography 3D printing. *ACS Omega* 3:1403–1408

Warner, J., Soman, P., Zhu, W., Tom, M., Chen, S. 2016. Design and 3D printing of hydrogel scaffolds with fractal geometries. *ACS Biomater. Sci. Eng.* 2:1763–1770

Wu, B., Sufi, A., Biswas, R. G., et al. 2020. Direct conversion of McDonald's waste cooking oil into a biodegradable high-resolution 3D-printing resin. *ACS Sustainable Chem. Eng.* 8:1171–1177

Appendix A: List of Companies

This list includes major printer manufacturers, suppliers of materials, software, and services, research organizations, and universities involved in AM and mentioned in this book. The symbols ™ and ® were omitted for simplicity.

- 3D Systems, USA, https://www.3dsystems.com
- 3D4MAKERS, The Netherlands, 3d4makers.com
- 3Devo, The Netherlands, https://3devo.com/
- 3Dfactories, Germany, https://www.3dfactories.com/
- 3D-Fuel, USA, www.3dfuel.com
- 3DXTECH, USA, https://www.3dxtech.com
- 3R Recycling Inc., USA, https://3rrecyclinginc.com/
- 4 AXYZ, Canada, https://www.4axyz.com/
- ABB Robotics, Sweden, Switzerland, https://new.abb.com/products/robotics
- Advanced Laser Materials (ALM), USA, www.alm-llc.com
- Advanced Solutions, USA, https://www.advancedsolutions.com/
- Aectual, The Netherlands, https://www.aectual.com/
- Anisoprint, Luxemburg, Russia, https://anisoprint.com/
- Anqing Hexing Chemicals Co. Ltd., China
- Apis Cor, USA, https://www.apis-cor.com/
- Aprecia Pharmaceuticals, USA, www.aprecia.com/technology
- Arburg, Germany, www.arburg.com
- AREVO, USA, https://arevo.com
- Arkema, France, https://www.arkema.com
- Bambooder, The Netherlands, https://www.bambooder.nl/
- BambroTex, China, http://www.bambrotex.com/
- Barilla, 3D FOOD S.r.l., Italy, https://blurhapsody.com/en/
- BeeHex, USA, https://www.beehex.com/
- Bioinspiration, Germany, https://bioinspiration.eu/
- BioMatera, Canada, www.biomatera.com
- Biomer, Germany, http://www.biomer.de/IndexE.html
- Biozoon Food Innovations, Germany, https://biozoon.de/en/
- Black Magic 3D, USA, https://www.blackmagic3d.com/
- BLB Industries, Sweden, https://blbindustries.se/fused-granular-fabrication-fgf-2/
- BQ, Spain, https://www.bq.com/en/
- byFlow, The Netherlands, https://www.3dbyflow.com/
- CAM-LEM Inc., USA, www.camlem.com
- Carbon Inc., USA, https://www.carbon3d.com/
- Cardolite Company, USA, https://www.cardolite.com/
- CC–PRODUCTS, Germany, http://cc-products.de/
- CELLINK, Sweden, www.cellink.com
- Choc Edge, UK, http://chocedge.com/
- Cincinnati Incorporated, USA, https://www.e-ci.com/baam
- CleanGreen3D Limited, Ireland, https://cleangreen3d.com/
- colorFabb, The Netherlands, https://colorfabb.com/

- Crafty Machines, UK, http://www.craftymachines.co.uk/
- Cubic Technologies, USA, https://www.f6s.com/cubictechnologies
- Danimer Scientific, USA, https://danimerscientific.com/
- Desktop Metal, USA, https://www.desktopmetal.com/
- Diamond Plastics, Germany, http://www.diamond-plastics.de/en/products.html
- DMG Mori, Germany, Japan, https://us.dmgmori.com/products/machines/additive-manufacturing
- DSM Somos, The Netherlands, https://www.dsm.com/solutions/additive-manufacturing/en_US/products/stereolithography.html
- Eindhoven University of Technology, The Netherlands, https://www.tue.nl/en/
- Ennex, USA, http://www.ennex.com/
- EnvisionTEC, Germany, https://envisiontec.com/3d-printers/3d-bioplotter/
- EOS, Germany, https://www.eos.info/en
- ErectorBot, USA, www.ErectorBot.com
- eSUN, China, http://www.esun3d.net
- Evonik, Germany, https://3d-printing.evonik.com/en
- ExOne, USA, exone.com
- extrudr, Austria, https://www.extrudr.com/en/
- Fab@Home, https://www.fabathome.net/
- Fabrisonic, USA, https://fabrisonic.com
- Filabot, USA, https://www.filabot.com/collections/filabot-core
- Filamentive, UK, https://www.filamentive.com/
- FkuR, Germany, https://fkur.com/en/brands/bio-flex/
- Flashforge, China, https://www.flashforge.com
- Formfutura, The Netherlands, www.formfutura.com
- Formlabs, USA, formlabs.com
- Fortify, USA, https://3dfortify.com/
- Francofil, France, https://francofil.fr/
- Galactic, Belgium, www.lactic.com
- Green Cycles, Spain, www.green-cycles.com
- Green Dot Bioplastics, USA, https://www.greendotbioplastics.com/
- Hague University of Applied Sciences, The Netherlands, http://www.thehagueuniversity.com
- Hubs, The Netherlands, https://www.3dhubs.com/
- IMAKR, UK, www.imakr.com
- Impossible Objects, USA, https://www.impossible-objects.com
- Incus, Austria, https://www.incus3d.com/
- Innofil3D, The Netherlands, https://www.ultrafusefff.com/
- Institute for Biomedicine of Aging, Germany, https://www.allgemeinmedizin.uk-erlangen.de/en/research/project-partner/institute-for-biomedicine-of-aging-iba/
- Jet-Eat/Redefine Meat, Israel, https://www.redefinemeat.com/
- Kaneka, Japan, https://www.kaneka.co.jp/en/
- Kanesis, Italy, https://www.kanesis.it/
- Katjes Magic Candy Factory, UK, http://magiccandyfactory.com/
- Longer3D, China, https://www.longer3d.com/
- MakerBot Industries, USA, www.makerbot.com
- Markforged, USA, https://markforged.com
- MatterHackers, USA, https://www.matterhackers.com
- MCPP, The Netherlands, www.mccp-3dp.com
- MesoScribe Technologies Inc., USA, www.mesoscribe.com
- Modern Meadow, USA, https://www.modernmeadow.com/
- Multi3D, USA, https://www.multi3dllc.com/contact/
- Nanoscribe, Germany, www.nanoscribe.com

- Nanovia, France, https://nanovia.tech/en/
- Natural Machines, Spain, https://www.naturalmachines.com/faq
- NatureWorks, USA, https://www.natureworksllc.com
- Newlight Technologies, USA, https://www.newlight.com/
- Nova-Institute, Germany, http://nova-institute.eu/
- Novamont, Italy, https://www.novamont.com/eng
- nScrypt, USA, www.nscrypt.com
- NuFood, USA, https://www.nuvegfood.com
- Nuremberg University of Applied Sciences GSO, Germany, https://www.th-nuernberg.de/
- Optomec, USA, www.optomec.com
- Organovo, USA, https://organovo.com
- Osteopore, Singapore, https://www.osteopore.com/
- Osteopore, Australia, https://www.osteopore.com
- PHB Industrial, Brasil, https://fapesp.br/eventos/2012/07/Biopolymers/ROBERTO.pdf
- Philips Design, The Netherlands, https://www.dezeen.com/2009/09/08/food-probe-by-philips-design
- Photocentric, UK, https://photocentricgroup.us
- Politecnico di Milano, Italy, https://www.polimi.it/en
- Pollen AM, France, www.pollen.am
- Polymaker, China, https://polymaker.com/
- PopBit, China, https://www.popbit3d.com/
- Precious Plastic, The Netherlands, https://preciousplastic.com/
- Prenta Oy, Finland, https://www.prenta.fi/en
- Print2Taste, Germany, https://www.procusini.com
- Prirevo e.U., Austria, https://www.prirevo.at/
- Prodways, France, https://www.prodways.com/
- Protolabs, USA, https://www.protolabs.com/
- Protoplant, USA, https://www.proto-pasta.com/
- PTTMCC, Japan, Thailand, www.pttmcc.com/new
- Quill NanoInk Inc., USA, https://scitech.com.au/nanotechnology-surface-metrology/nanoink/
- RE PET 3D, Czech Republic, https://re-pet3d.com/
- re:3D, USA, http://www.re3d.org
- REC3D, Russia, http://rec3d.com/
- ReDeTec, Canada, https://redetec.com
- RegenHu, Switzerland, https://www.regenhu.com/
- Reify 3D, USA, http://www.reify-3d.com/products/
- RePLAy, 3R Recycling Inc. USA, https://3rrecyclinginc.com/
- Robots in Gastronomy, Spain, https://robotsingastronomy.com/
- Roboze, Italy, https://www.roboze.com/en/
- Roland DG Mid Europe Srl, Italy, https://www.rolanddg.com/en
- Sabic, Saudi Arabia, https://www.sabic.com/en
- Sartomer, France, www.sartomer.com
- Sciaky, USA, www.sciaky.com
- Sculpteo, France, https://www.sculpteo.com/en/
- SD3D, USA, https://www.sd3d.com
- Shapeways, USA, https://www.shapeways.com/
- Shenzhen Ecomann Biotechnology, China, http://ecomann.sx-gear.com/
- Sintratec, Switzerland, https://sintratec.com/
- Smart Materials 3D, Spain, https://www.smartmaterials3d.com/
- Spare Parts 3D, Singapore, https://spare-parts-3d.com/
- Stratasys, Israel, USA, https://www.stratasys.com/

- Systems & Materials Research Corporation (SMRC), USA, http://systemsandmaterials.com/
- Tethin3D, USA, https://tethon3d.com/product/porcelain-ceramic-resin/
- The University of Nottingham, UK, https://www.nottingham.ac.uk/
- Therics LLC, USA, https://www.nextsteparthropedix.com/therics
- Thermwood, USA, http://www.thermwood.com/lsam_home.htm
- TianAn Biologic Materials, China, http://www.tianan-enmat.com/
- Tianjin GreenBio Materials, China, www.tjgreenbio.com
- Titan Robotics, USA, https://titan3drobotics.com/
- TNO, The Netherlands, https://www.tno.nl/en/
- Total Corbion PLA, The Netherlands, www.total-corbion.com
- Ultimaker, The Netherlands, https://ultimaker.com/
- University College of London, UK, https://www.ucl.ac.uk/
- UPM Formi, Finland, https://www.upmformi.com/
- Veryst Engineering, USA, https://www.veryst.com/
- Victrex, UK, https://www.victrex.com/en/victrex-peek
- Voxeljet, Germany, https://www.voxeljet.com/
- Wacker, Germany, https://www.aceo3d.com/
- Weihenstephan-Triesdorf University of Applied Sciences, Germany, https://www.hswt.de/en.html
- Winsun, China, http://www.winsun3d.com/En/
- XJET, Israel, https://www.xjet3d.com/
- Yield10, USA, https://www.yield10bio.com/
- Z Corporation, USA, see 3D Systems
- Zhejiang Hangzhou Xinfu Pharmaceutical, China, http://en.yifanyy.com/zhejiang_center.html
- ZMorph, Poland, https://zmorph3d.com/
- Zortrax, Poland, https://zortrax.com

Appendix B: Standard Test Methods for Plastics Issued by ASTM and ISO

ASTM	ISO	Test Method Description
D648	75	Heat deflection temperature
D1043	458–1	Stiffness properties of plastics as a function of temperature by means of torsion test
D1044	9352	Resistance of transparent plastics to surface abrasion
D1238	1133	Melt flow rates of thermoplastics by extrusion plastometer
D1505	1183-2	Density of Plastics
D1708	6239	Tensile properties of plastics by use of microtensile specimens
D1822	8256	Tensile impact energy to break of plastics
D1894	6601	Static and kinetic coefficients of friction of plastic film and sheeting
D1922	6383–2	Resistance to tear propagation of plastic film and thin sheeting by pendulum method
D1938	6383–1	Resistance to tear propagation of plastic film and thin sheeting by single tear method
D256	180	Impact resistance of notched specimens of plastics – Izod method
D2990	899–1.2	Tensile compressive and flexural creep and creep rupture of plastics
D3418	11357–1, –2, –3	Transition temperatures and enthalpies of fusion and crystallization of polymers by differential scanning calorimetry (DSC)
D3763	6603–1.2	High-speed puncture properties of plastics
D4065, D4092	6721–1	Dynamic mechanical properties of plastics: practice and terminology
D4440	6721–10	Rheological measurement of polymer melts using dynamic mechanical procedures
D5023	6721–3, –5	Dynamic mechanical properties of plastics in flexure
D5024	N/A	Dynamic mechanical properties of plastics in compression
D5026	6721–4	Dynamic mechanical properties of plastics in tension
N/A	6721–6	Dynamic mechanical properties of plastics in shear
D5045	572	Plane-strain fracture toughness and strain energy release rate of plastic materials
D5083	3268	Tensile properties of reinforced thermosetting plastics using straight-sided specimens
D5279	6721–2, 7	Dynamic mechanical properties of plastics in torsion
D5418	6721–3, –5	Dynamic mechanical properties of plastics in flexure (dual cantilever beam)
D5729	6721	Measuring the dynamic mechanical properties of plastics in tension
D6110	179	Charpy impact resistance of notched specimens of plastics
D638	527–1, –2	Tensile properties of plastics
D695	604	Compressive properties of plastics
D732	N/A	Shear strength of plastics by punch tool

(Continued)

ASTM	ISO	Test Method Description
D785	2039–2	Rockwell hardness of plastics
D790	178	Flexural properties of unreinforced and reinforced plastics and insulating materials
D882	527–3	Tensile properties of thin plastic sheeting
E1131	11358	Compositional analysis by thermogravimetry (TGA)
E143	1827	Shear modulus
E647	15850	Fatigue crack growth rates
E794	11357-3	Melting and Crystallization Temperatures
E1356	11357–2	Measuring glass transition temperature via differential scanning calorimetry (DSC)
E1545	11359–1, –2, –3	Glass transition temperature via thermomechanical analysis (TMA)
E1640	6721–11	Glass transition temperature via dynamic mechanical analysis
E1823	12106, 12107	Terminology relative to fatigue and fracture testing

Index

Page numbers in italic and bold refer to figures and tables, respectively.